Artificial Intelligence in the Energy Industry

This comprehensive resource bridges the knowledge gap in applying modern artificial intelligence and machine learning methods to the petroleum sector. It explores how artificial intelligence (AI), machine learning (ML), and data analytics enhance efficiency, safety, and productivity in operations across various segments of the oil and gas value chain, including exploration, drilling, production, reservoir management, and renewables integration.

- Covers theoretical and practical aspects of AI/ML applications, from foundational concepts to advanced techniques.
- Features examples and A-to-Z practical workflows, empowering readers to apply what they learn directly to real-world challenges.
- Includes 500 international case studies, highlighting real-world successes, challenges, and lessons learned from a global perspective.
- Offers insights into emerging technologies like Industry 4.0, digital twins, smart fields, and Internet of Things applications, as well as more traditional areas such as drilling optimization and enhanced recovery operations.

With the growing importance of advanced data-driven approaches, this book provides value to both technical professionals looking for hands-on solutions and academics seeking a well-structured textbook for advanced studies.

Artificial Intelligence in the Energy Industry

Theory, Case Studies, and Applications

Edited by

Cenk Temizel

Salih Tutun

Celal Hakan Canbaz

Ekrem Alagoz

Emre Can Dündar

Mufrettin Murat Sari

Luigi Saputelli

Mehmet Melih Oskay

CRC Press

Taylor & Francis Group

Boca Raton London New York

CRC Press is an imprint of the Taylor & Francis Group, an **informa** business

Designed cover image: © 2026 Shutterstock

First edition published 2026
by CRC Press
2385 NW Executive Center Drive, Suite 320, Boca Raton FL 33431

and by CRC Press
4 Park Square, Milton Park, Abingdon, Oxon, OX14 4RN

CRC Press is an imprint of Taylor & Francis Group, LLC

ISBN: 978-1-041-02007-3 (hbk)
ISBN: 978-1-041-02011-0 (pbk)
ISBN: 978-1-003-61732-7 (ebk)

DOI: 10.1201/9781003617327

Typeset in Times
by SPi Technologies India Pvt Ltd (Straive)

Dedications for Family Members

To my beautiful wife, Saule, and my lovely daughters, Ada and Ela.
Cenk Temizel

To my lovely wife, Begum Basaran Tutun; and my son, Ege.
Salih Tutun

To my lovely son, Alp; my lovely wife, Ezgi; my dear brother, Serkan; and my beloved parents, Füsun and Kaya—for your unwavering love, support, and inspiration.
Celal Hakan Canbaz

To my beloved wife, Zeliha Alagoz, and our unborn child.
Ekrem Alagoz

To my beloved wife, M. Betul Dündar. Your unwavering support, patience, and wisdom fuel my every pursuit.
Emre Can Dündar

To my lovely wife, Gulhan, and kids, Ece, Emre, and Kerem.
Mufrettin Murat Sari

In memory of my beloved mother and father.
Luigi Alfonso Saputelli

To my lovely wife, Sühendan for her unflagging support.
Mehmet Melih Oskay

Contents

Foreword

In the ever-evolving landscape of technology, the integration of artificial intelligence (AI) has revolutionized various industries, including the oil and gas sector. This book, *Artificial Intelligence in the Energy Industry*, serves as a comprehensive guide to understanding and implementing AI solutions in upstream oil and gas operations.

The book begins with a solid foundation in the theory and background of AI methods, covering essential topics such as linear algebra, data science, and machine learning algorithms. Readers will gain insights into the mathematical principles that underpin these technologies and learn how to apply them using Python.

As the book progresses, it delves into the practical applications of AI in various domains within the energy industry. From reservoir characterization and geology to drilling engineering and production optimization, each chapter provides detailed workflows and real-world examples. The book highlights the transformative power of AI in enhancing decision-making, improving operational efficiency, and mitigating risks.

One of the key strengths of this book is its focus on hands-on learning. Through interactive exercises and coding instructions, readers will develop practical skills in using AI frameworks. The book also emphasizes the importance of data preparation, visualization, and ethical considerations in data science.

The applications of AI in the energy industry are vast and varied. The book explores AI-driven methodologies for reservoir simulation, well completion, enhanced oil recovery, and unconventional resources. It also addresses the role of AI in facilities management, pipeline monitoring, and renewable energy applications.

In addition to its technical content, the book provides valuable insights into the future directions of AI in the energy sector. It discusses emerging trends, challenges, and opportunities, offering a vision for the continued integration of AI in driving innovation and sustainability.

This book is an invaluable resource for petroleum engineers, geoscientists, data scientists, and software developers seeking to harness the power of AI in their work. By bridging the gap between theoretical knowledge and practical application, it equips readers with the tools and expertise needed to excel in the digital transformation of the energy industry.

As you embark on this journey through the pages of *Artificial Intelligence in the Energy Industry*, I hope you find inspiration, knowledge, and practical guidance to navigate the exciting world of AI-driven advancements in the energy sector.

Dr. Srimoyee Bhattacharya
Team Lead for Data Science and Machine Learning, Shell
Houston, TX, USA
May 25, 2025

Preface

The oil and gas industry has always been a hub of technological innovation since Drake's commercial oil well drilling in 1859. Through decades of progress in engineering, geoscience, and automation, hydrocarbons have seen fundamental changes in their exploration development and processing methods. The petroleum industry uses advanced technologies to study seismic signals and drill with automation, allowing them to operate efficiently while recovering complete resources.

The requirement for pioneering data-driven solutions has peaked in recent decades because world energy consumption keeps expanding while markets shift and environmental safety guidelines tighten and giant oil reserves decline. Artificial intelligence (AI) is the main driving force that has changed the structure of the oil and gas industry. The petroleum industry undergoes revolutionary change through AI technologies, which help organizations obtain better decision-making capabilities, optimal process production, risk management, and overall operational effectiveness. Predictive analytics and AI-driven models enable petroleum engineers and industry professionals to discover groundbreaking exploration possibilities, boost drilling performance, and achieve better production forecasting accuracy.

AI rapidly changes traditional upstream oil and gas value chain workflows, from seismic data interpretation and real-time reservoir modeling to automated drilling and predictive maintenance. Real-time analysis and processing of large amounts of geological and operational data and production data enable operators to resolve industry challenges, including declining reservoir optimization, drilling hazard mitigation, better well integrity, and reduced unplanned operational downtime. AI delivers predictive capabilities that assist engineers' in forecasting reservoir development and improving hydrocarbon recovery strategies through planning. Integrating AI with Internet of Things (IoT) sensors enables operators to conduct continuous monitoring and adaptive control, resulting in safer and more efficient upstream operations.

The exploration step for petroleum development depends on correct geological evaluation and reservoir development models to locate areas suitable for drilling activities. People who analyzed seismic data and subsurface images in the past needed significant manual help from geoscientists and petroleum engineers (Solanki, et al. 2022). AI devices use massive seismic data analysis to detect hydrocarbon indicators and accomplish this process at previously impossible speeds.

Applying machine learning and deep learning models proves beneficial in formation evaluation because they enable engineers to determine rock properties, fluid composition, and reservoir potentials. AI solutions through automation enable quicker assessments of well-log interpretations, reservoir simulations, and pore pressure forecasts, as well as cutting subsurface assessment time and labor expenditure.

The drilling operations within the oil and gas industry rank as its most advanced and financially demanding procedures. The business processes depend heavily on efficient drilling techniques that uphold operational safety standards for profit gains. Real-time drilling optimization relies heavily on AI because it uses drilling rig sensor data to recognize operational problems while predicting machine dangers and implementing automatic operational adjustments.

Drilling automation systems with reinforcement learning capabilities control drilling performance and equipment wear by changing parameters between speed rate and weight on bit and mud volume flow. The predictive maintenance systems delivered by AI help companies detect device failure before it materializes, thus reducing equipment downtime. Intelligent automation optimizes directional drilling with AI-enabled control systems through downhole analysis, resulting in real-time drilling path modifications. The procedure enables drilling operators to achieve higher efficiency with better stability and better precision in their operations.

Hydrocarbon production from wells is shifted to increase output volume and maintain operational stability. The production forecasting algorithms employing AI technology combine historical production records and current field measurements to predict the reservoir's actions and refine the operational output levels. AI systems use procedures like enhanced oil recovery (EOR) and hydraulic fracturing to identify the most effective healthy stimulation options for declining fields. These techniques allow for the most possible resource extraction from existing reservoirs. Businesses can use AI-powered monitoring systems to measure fluid flow rates, follow pressure changes, and inspect well integrity to identify problems before shutdowns happen. Companies can test tech production plans and reproduce reservoir performance using digital twins and virtual AI replicas of material assets to deploy productive strategies in their areas. The application of this system mitigates operational risks and gives better judgments.

To implement AI development in the oil and gas sector, the petroleum engineering and the AI methodological concepts must be known. This book uses comprehensive, organized information, combining theoretical knowledge with 500 real-world examples to analyze AI's effect on upstream oil and gas operations.

This book provides knowledge about core AI methods alongside machine learning, deep learning, and predictive analytics, which shows their usable forms in well-log evaluation, reservoir simulation, drilling enhancement, and production prediction. The publication demonstrates AI theory applications in industrial practice through relevant cases and programming codes, enabling experts to implement these technologies effectively in their operational processes. It also delivers interactive learning through detailed Python and AI frameworks like TensorFlow, PyTorch, and Scikit-learn coding instructions. This practical training through exercises enables readers to build functional, hands-on abilities, which include: Petrophysics, Geology, Drilling Engineering, Reservoir Engineering, Well Completion, Production, Stimulation, Enhanced Oil Recovery, Facilities, Unconventionals, Fourth Industry, and Renewable Resources.

Serving petroleum engineers and geoscientists, this resource expanding their workflow integration of AI systems, data scientists, and software developers who want to specialize in energy application fields. Using a structured method, the entire spectrum of AI concepts, from beginning professionals to advanced petroleum engineers, can learn and execute their knowledge in petroleum engineering applications.

This comprehensive guide presents engineers and industry professionals with an approachable method to implement AI solutions that drive operational performance enhancements and better decision capabilities. The book offers researchers and academics an extensive analysis of AI techniques and their industrial solutions for significant operational challenges. The energy sector now embraces a digital revolution where AI stands as its driving force. Organizations embracing AI solutions will acquire market advantages through improved resource recovery, cost reductions, and sustainability achievements in an evolving worldwide market.

Modern technology has surpassed the point where scientists view AI integration in petroleum engineering as mere futuristic speculation. The upstream oil and gas industry transforms using AI because it delivers data-driven decision support, which results in improved operational effectiveness and reduced operational risks. The text provides complete instructions to people who want to understand and implement AI solutions within the oil and gas industry. This book presents theoretical knowledge and practical sample codes to help readers develop the capabilities to understand petroleum engineering changes and effectively use AI technologies. This book provides the required knowledge, skills, and practical tools for any professional seeking AI technology to achieve innovative excellence in the energy sector.

Editors

Acknowledgments

This book is dedicated, with heartfelt gratitude, to my esteemed professors whose wisdom and mentorship have left a lasting impact.

Cenk Temizel

As an artificial intelligence (AI) researcher, I dedicate this work to those who strive to bridge the gap between AI and energy sciences. May our collective efforts pave the way for smarter, more efficient, and sustainable advancements in petroleum engineering.

Salih Tutun

This book is dedicated with profound gratitude to my esteemed professors who have enriched my life with their invaluable contributions.

Celal Hakan Canbaz

To all aspiring petroleum engineers—may your passion for energy and innovation drive the future of our industry.

Ekrem Alagoz

Dedicated to those who recognize the power of interdisciplinary collaboration. May the thoughtful integration of artificial intelligence illuminate new pathways in petroleum engineering and guide us toward sustainable solutions.

Emre Can Dündar

This book is dedicated to my valuable professors.

Mufrettin Murat Sari

To all the technicians and professionals who leave their home countries and families to work in the oil patch, your dedication and sacrifices are deeply appreciated and vital to powering our world.

Luigi Saputelli

This book is dedicated to all educators from preschool-level teachers to graduate school professors.

Mehmet Melih Oskay

About the Editors

Cenk Temizel is an energy professional with 20 years of experience. He worked at Saudi Aramco, Aera Energy LLC (a Shell-ExxonMobil Affiliate), Halliburton, and Schlumberger in the Middle East, the United States, and the United Kingdom. Before joining the industry, he was a teaching/research assistant at the University of Southern California and Stanford University. He serves as a technical reviewer for petroleum engineering journals and a member of conference committees. He has published around 150 publications in reservoir management, production optimization, enhanced recovery processes, machine learning, intelligent fields, and holds several US patents. He is the recipient of the SPE Regional Reservoir Description and Dynamics Award. He holds a BS degree (Honors) from Middle East Technical University—Ankara (2003) and an MS degree (2005) from the University of Southern California (USC), Los Angeles, CA both in petroleum engineering.

Salih Tutun is an AI researcher and faculty member at the Olin Business School, Washington University in St. Louis. He serves as a senior AI consultant on various projects, focusing on IT artifacts and products for business decision-making. He has received the Reid Teaching Awards (2021, 2022, and 2023) and teaches courses on deep learning, reinforcement learning, Python, and R. His research on Networked Pattern Recognition (NEPAR) has been widely recognized, featured by global news organizations, and appeared on the cover of *Industrial and Systems Engineering at Work* magazine. He has been a finalist in the IISE Cup competitions (2020, 2021) and won the Gold Award in 2023 for MDScan research. In 2024, he received the Olin Research Award. At Washington University, DNB Analytics, and in collaboration with companies, Salih has developed several products, including MDScan, XScanner, NEPAR, SASX, DIA, and DemansCheck. He is an external advisory board member of the SkillMeet project funded under the Horizon Europe program.

Celal Hakan Canbaz is a seasoned senior energy expert with extensive experience in both industrial and academic domains. As a researcher, instructor, and co-founder of Climpach Consulting LLC, he has developed deep expertise in SCAL analysis, reservoir engineering, and well-testing analysis. Over his career, he has contributed to over 30 projects with major oil companies, authored over 50 conference and journal papers, and written 3 books. Canbaz has delivered over

50 industrial and technical short courses covering topics such as oil and gas, energy, project management, project finance, and sustainability. His achievements include first-place awards in SPE paper contests, a US patent for a reservoir wettability characterization method, and recognition as one of the top 100 scientists in the Azerbaijan National H-index Ranking. He earned MSc and BSc degrees in Petroleum and Natural Gas Engineering from Istanbul Technical University. He is also actively involved in the SPE and EAGE organizations.

Ekrem Alagoz holds an MSc degree in Petroleum Engineering from The University of Texas at Austin. He has over eight years of experience in the oil and gas industry, focusing on drilling engineering as well as hydraulic fracturing engineering in both field operations and research and development. Currently, he works for Turkish Petroleum Corporation (TPAO), supervising drilling operations as an operator representative. Ekrem has made significant academic contributions, authoring more than 30 conference papers and journal articles, along with over 20 book chapters, covering a wide range of topics in petroleum engineering.

Emre Can Dündar is a dedicated researcher specializing in petroleum and natural gas engineering. He earned his Bachelor's degree in Mining Engineering and a Master's degree in Petroleum and Natural Gas Engineering from West Virginia University (WVU), funded by scholarships from the Turkish Ministry of National Education on behalf of the General Directorate of Mineral Research and Exploration. Emre's professional experience spans roles as a tunnel engineer in Istanbul's subway railway projects, a research assistant at WVU, and his current position as a petroleum engineer at Turkish Petroleum Corporation (TPAO), where he specializes in the design and execution of hydraulic fracturing operations. He actively contributes to academia with numerous published articles, book chapters, and peer reviews for reputable scientific journals. Alongside his technical and scholarly pursuits, Emre maintains a passion for literature and philosophy, currently pursuing a degree in Philosophy at Istanbul University and working on his first fiction novel.

Mufrettin Murat Sari is a chemistry professor at the Department of Chemistry & Biochemistry, Texas Tech University, Texas, USA. He has more than 25 years of experience in the fields of Materials Chemistry, Applied Biochemistry, Biotechnology, and Nanotechnology. He received a PhD degree from Hacettepe University in 2005. He has around 40 scientific articles, proceedings, books, book chapters, and other publications with hundreds of citations. His research interests are micro and nanomaterials, nanofluids, nanopharmaceuticals, modified surfaces, and their biotechnological, engineering, and environmental applications. Previously, as main academic appointments, he was a teaching assistant at Hacettepe University, a researcher at the Criminal Department Research Center, Ankara, Turkey, a professor at Turkish Military Academy & National Defense University, and a visiting professor at Texas A&M University, College Station. In 2015, he joined the Interfacial Phenomena in Nanotechnology and Biotechnology, IPNB, Research Group at Texas A&M University for postdoctoral study and studied the Design of 2-Dimensional Next Generation Thermal Interface Nanomaterials. Also, he has taught undergraduate and graduate level organic chemistry, biochemistry, analytical chemistry, chemistry for engineers, environmental chemistry, and general chemistry classes at Texas A&M University-Commerce, UNT-Denton, UT-Arlington, Dallas College, and the University of Nebraska until now and he still continues to teach at Texas Tech University.

Luigi Saputelli is a reservoir engineering expert advisor with 35 years of experience as reservoir, drilling, and production engineer in PDVSA, ADNOC, Hess Bapco, Halliburton, and a founding member of Frontender, an upstream energy services firm in Houston. Saputelli is a researcher, lecturer, and active volunteer in the Society of Petroleum Engineers (SPE) where he served as JPT editor (2012–2020), Production and Operations Advisory Board since 2010, founding member of Real-time Optimization Interest Group and Petroleum Data-driven Analytics technical sections since 2003. He is the recipient of the 2015 SPE International Production and Operations Award. He has published more than 100 industry papers, 6 patents, and contributed to over 5 books on digital oilfield, energy transition, reservoir management, and real-time production optimization. He holds a BSc in Electronic Engineering from the Universidad Simon Bolivar (1990), with a MSc in Petroleum Engineering from Imperial College (1996), and a PhD in chemical engineering from the University of Houston (2003).

Mehmet Melih Oskay earned his PhD degree in Petroleum Engineering from The University of Texas, Austin, TX, USA. He has taught at the University of Texas, Louisiana Tech University, Middle East Technical University, and King Fahd University of Petroleum and Minerals. His industrial experience covers Shell, Kuwait Oil Company, and Kuwait Gulf Oil Company. He has worked as Chevron's Operational Excellence Champion covering workforce safety and health, process safety, reliability and integrity, environment, efficiency, effectiveness, security, and stakeholders. He has worked with experts from Chevron and Total in line with their technical services agreements signed with KOC. He has served on the Joint Operating Committee of Kuwait Gulf Oil Company and Saudi Arabian Chevron and the Joint Operating Tender Committee which includes Saudi Arabian and Kuwait Government representatives as well. Decisions and responsibilities at that position cover A to Z all aspects of petroleum operating company activities involving KGOC as the National Operating Company and Chevron as the International Oil Companies. He has served at the SPE organizing committee for many conferences and exhibitions and held many technical session chairmanships. Currently works as a business developer and technical/safety consultant.

Contributors

Mouad Al Krmagi
Texas A&M University
College Station, Texas, USA

Ruoyu "Tagore" Cheng
Graduate Student | Alumni
Washington University
St. Louis, Missouri, USA

Caner Erden
Assistant Professor
Sakarya University
Sakarya, Turkey

Serkan Gozel
Champaign, Illinois, USA

Mehmet Gumus
Teaching Associate Professor
The University of Nevada
Reno, Nevada, USA

Elif Elcin Gunay
Clinical Assistant Professor
Purdue University
West Lafayette, Indiana, USA

Xinyi Huang
PhD student
Emory University
Georgia, USA

Muhammed Esad Kaya
Houston, Texas, USA

Sena Kır
Assistant Professor
Sakarya University
Sakarya, Turkey

Hantang Li
Washington University
St. Louis, Missouri, USA

Ferit Ozdaban
Chicago, Illinois, USA

Yasin Ozkan
Ankara, Turkey

Haowei Tang
Washington University
St. Louis, Missouri, USA

Srikanth Tangirala
Goleta, California, USA

Jie Yang
Data Analyst at Capital One
Virginia, USA

Weichao Yang
Graduate Student | Alumni
Washington University
St. Louis, Missouri, USA

Mesut Yurukcu
Odessa, Texas, USA

Tianyi Zhang
Data Strategist
Profit Insight
Kansas City, Missouri, USA

1 Theory and Background on Artificial Intelligence Methods

Mehmet Gumus, Xinyi Huang, Caner Erden, Sena Kır, Jie Yang, Tianyi Zhang, and Elif Elcin Gunay

1.1 INTRODUCTION

The energy sector experiences revolutionary changes through artificial intelligence (AI) and machine learning (ML), which enable improved data analysis and optimization while enhancing predictive maintenance in exploration, production, and renewable energy management operations. The deployment process faces numerous issues, including maintaining data quality standards, handling complex models, and promoting cross-disciplinary teamwork. Given the significant impact of AI across the energy sector, this book *Artificial Intelligence in the Energy Industry: Theory, Case Studies, and Applications* delivers a full scope analysis of technological advancements in the field. The scope of the book spans both fundamental theory and practical implementation, reflecting a dual objective: This book aims to teach readers about essential AI and ML methods pertinent to energy applications while showing how these approaches solve practical engineering and geoscience challenges. The book integrates theoretical frameworks and practical case studies to connect theoretical algorithms with real-world energy industry challenges. The book also aims to provide energy professionals, along with researchers and advanced students, the necessary understanding of leading-edge AI approaches to apply them effectively in various energy sectors, including petroleum engineering, renewable energy engineering, and power system management.

The book stands out because of its detailed examination of AI applications across multiple essential areas of the energy industry, thus enhancing expertise within each field. AI supports reservoir characterization through geological and geophysical data analysis to create accurate subsurface models; drilling engineering operations benefit from real-time analytics and predictive models which enhance drilling safety and efficiency; production optimization uses ML to boost oil and gas production with improved well management and recovery methods; facilities management gains efficiency from AI systems which conduct predictive maintenance and control processes

DOI: 10.1201/9781003617327-1

in surface infrastructure and pipelines; renewable energy sources use intelligent algorithms to handle variability in wind and solar power generation for better reliability and efficiency. The text delivers a comprehensive review of existing practices in each area while introducing case studies and innovative techniques that expand AI's capabilities. The publication advances AI development in the energy sector by sharing valuable insights about successful strategies and lessons while identifying persistent obstacles to encourage new innovation.

The book repeatedly demonstrates how AI application in the energy sector necessitates interdisciplinary collaboration. Engineering and geoscience domain knowledge combined with computational techniques must be synthesized to develop effective solutions. Solving challenges such as seismic survey interpretation, drilling optimization, and power output prediction requires advanced algorithms together with knowledge of physical principles and operational limitations. The text concentrates on adapting AI methods for domain-specific needs while demonstrating that experts from varied fields need to work together to produce valuable outcomes. Case studies demonstrate that sophisticated ML models must work alongside physics-based methodologies and expert opinions since they cannot fully replace them. Expert knowledge remains essential for directing ML algorithm implementation during complex geological and operational scenarios. The book presents a balanced methodology by integrating data-driven insights with the professional expertise of petroleum engineers, geoscientists, and energy system planners. The interdisciplinary perspective helps readers understand the significance of contextual factors in AI deployment while promoting joint efforts between computer science and engineering disciplines in the energy industry.

The book consists of ten chapters that guide the reader from basic principles to specialized energy industry applications in a stepwise manner. The first chapter establishes theoretical foundations and presents essential concepts from applied mathematics and ML that support subsequent discussions. The initial chapter introduces crucial subjects, including linear algebra and essential AI and ML algorithms, while teaching data processing to equip readers with fundamental knowledge before progressing to specialized industry case studies.

Chapters 2–6 examine different AI applications within the upstream oil and gas sector with each chapter concentrating on separate exploration and production elements. Chapter 2 investigates the role of AI applications in the fields of reservoir characterization and geomechanics while showcasing ML's role in analyzing underground data. The interpretation of subsurface data from well logs, core samples, and seismic surveys enables the creation of precise geological models, which identify hydrocarbon-bearing formations. Drilling engineering AI applications in Chapter 3 focus on real-time decision support for drilling operations, including parameter optimization and equipment failure prediction with data-driven control to prevent hazardous events and enhance penetration rates. Chapter 4 explores the role of AI in reservoir engineering management and explains how it assists reservoir modeling and production strategy through oil and gas production forecasting and optimization of reservoir management plans such as waterflooding schemes while enhancing field development decisions amidst uncertainty. Chapter 5 explores the utilization of AI applications in the domains of well completion along with production and stimulation processes. The chapter demonstrates the usage of AI and ML techniques to design and assess well completions and stimulation treatments like hydraulic fracturing while optimizing production operations through lift setting adjustments and real-time production data analysis to ensure optimal flow rates. Chapter 6 investigates AI applications in enhanced recovery methods by analyzing AI's role in advancing secondary and tertiary recovery through gas injection outcome predictions and chemical Enhanced Oil Recovery (EOR) process analysis while optimizing injection schedules with AI and tracking reservoir performance to maximize hydrocarbon extraction.

The final chapters cover upstream activities to explore facilities and emerging technologies alongside renewable energy, demonstrating the energy industry's extensive reach. Chapter 7 covers the use of AI applications within facilities operations as well as pipeline and metering systems to demonstrate how AI and ML technologies boost surface operation efficiency and safety. The text explores intelligent facilities management systems along with pipeline integrity monitoring, which includes leak detection and corrosion prediction through sensor data and smart metering for distribution optimization. Chapter 8 analyzes AI applications for extracting unconventional resources with a particular focus on shale gas and tight oil production along with other unconventional plays. This chapter explores the distinct challenges presented by unconventional resources through extensive shale operation data and complex reservoir behaviors and illustrates how AI methods including big data analytics and pattern recognition optimize key operations like well spacing, fracturing designs, and production forecasting. Chapter 9 examines AI applications within the context of Industry 4.0 and smart field operations during the digital transformation era. This section demonstrates how the combination of IoT sensors with cloud computing and AI algorithms transforms oilfields and energy facilities into fully "smart" operations. The discussion covers real-time data platforms alongside digital twins of operations and automated control systems while examining how Industry 4.0 trends affect conventional energy workflows. Chapter 10 explores AI applications in renewable resources while broadening the book's scope into the power and renewables sector.

1.2 APPLIED MATHEMATICS AND MACHINE LEARNING BASICS

1.2.1 Linear Algebra for Machine Learning

Linear algebra is an important field of mathematics that has found a wide range of applications in science and engineering. Linear algebra provides the tools to organize the data and then use that data to perform ML algorithms. It is therefore essential to have a good understanding of linear algebra for developing ML algorithms and deep learning models. In this section, the basics of linear algebra are introduced, and several useful tools are presented.

1.2.1.1 Scalars, Vectors, and Matrices

A scalar is a number. Temperature, speed, and mass are some examples of scalars. The set of all real-valued scalars is denoted by R. Formally, $x \in R$ means that x is a real-valued scalar. In this book, we shall deal with the real-valued scalars.

A vector is a list of scalar values. These values are called the entries of the vector. Indices of these entries are arranged in order. We shall denote the vectors by a lowercase letter in bold typeface such as x. For example,

$$x = \left[x_1 \, x_2 \, \vdots \, x_n \right] \in R^n \tag{1.1}$$

is a vector of n entries. x_i is the ith entry of x. Note that x_i is a scalar. We can consider vector x as being a point in a space, where x_i represents the coordinate along the ith axis. x is a zero vector if all entries are zero. The origin is identified by the zero vector. If x has at least one nonzero entry, then it is called a nonzero vector.

Example 1.1 We can identify a geometric point $(2,3,-5)$, with the vector $[23-5]$.

A matrix is a generalization of vectors from one dimensional (1D) to two dimensional (2D). We use capital letters in bold typeface to denote matrices such as A. Each entry of a matrix is denoted by two indices instead of only one. The first index identifies the row, and the second index identifies the column. For example, $A_{i,j}$ is the entry of A located in the ith row and the jth column.

$$A = \left[A_{1,1} \, A_{1,2} \ldots A_{1,n} \, A_{2,1} \, A_{2,2} \ldots A_{2,n} \vdots \vdots \ddots \vdots \, A_{m,1} \, A_{m,2} \ldots A_{m,n} \right] \in R^{m \times n} \tag{1.2}$$

is an $m \times n$ matrix with m rows and n columns. A is called a square matrix when m is equal to n. Each column of A is called a column vector. Moreover, any vector x with n entries is an $n \times 1$ matrix and thus is a single column vector.

A commonly used operation on matrices is the transpose. The transpose of a $m \times n$ matrix A is the $n \times m$ matrix, denoted by A^T, whose columns are the corresponding rows of A.

Example 1.2 Let

$$A = \left[2-134 \right], \quad B = \left[-172-3510 \right], \quad x = \left[-247 \right]$$

Then

$$A^T = \left[23-14 \right], \quad B^T = \left[-1257-310 \right], \quad x^T = \left[-247 \right]$$

The diagonal entries of an $m \times n$ matrix A are $A_{1,1}$, $A_{2,2}$, $A_{3,3}$..., which form the main diagonal of A. A diagonal matrix is a square matrix whose off-diagonal entries are zero.

Example 1.3 The matrices D and E below are 3×3 diagonal matrices

$$D = \left[-100050009 \right], \quad E = \left[1100000007 \right]$$

An identity matrix is a diagonal matrix whose diagonal entries are all one. We denote $n \times n$ identity matrix by I_n.

Example 1.4

$$I_3 = \begin{bmatrix}100010001\end{bmatrix}, \quad I_4 = \begin{bmatrix}1000010000100001\end{bmatrix}$$

1.2.1.2 Matrix and Vector Operations

Two matrices can be added if they have the same size. If A and B are $m \times n$ matrices, then the sum $A + B$ is the $m \times n$ matrix whose entries are the sum of the corresponding entries of A and B.

Example 1.5 Let $A = \begin{bmatrix}8-375-21\end{bmatrix}, \quad B = \begin{bmatrix}-172-3510\end{bmatrix}$. Then,

$$A + B = \begin{bmatrix}8+(-1)-3+77+25+(-3)-2+51+10\end{bmatrix} = \begin{bmatrix}7492311\end{bmatrix}$$

If k is a scalar and A is an $m \times n$ matrix, then kA is the $m \times n$ matrix whose entries are k times the corresponding entries of A.

Example 1.6 If A and B are the matrices in Example 1.5, then

$$2A = \begin{bmatrix}16-61410-42\end{bmatrix} \text{ and } 2A - B = \begin{bmatrix}17-131213-9-8\end{bmatrix}$$

Matrix multiplication AB can be computed when the number of columns of A is the same as the number of rows of B. Particularly, if A is an $m \times n$ matrix and B is an $n \times k$ matrix, then AB is well defined and is an $m \times k$ matrix. Note that BA is not defined when k and m are different. Furthermore, $AB \neq BA$ in general. This is an important difference between the multiplication of matrices and the multiplication of real numbers.

Before introducing the matrix product, we present the dot product (inner product) between two vectors. The dot product of two vectors of the same size, $x = \begin{bmatrix}x_1\, x_2 \vdots x_n\end{bmatrix}$ and $y = \begin{bmatrix}y_1\, y_2 \vdots y_n\end{bmatrix}$ is

$$x \cdot y = x_1 y_1 + x_2 y_2 + \ldots + x_n y_n. \tag{1.3}$$

The result of a dot product is a single number.

Example 1.7 Let $x = \begin{bmatrix}-247\end{bmatrix}$ and $y = \begin{bmatrix}351\end{bmatrix}$. Then,

$$x \cdot y = -2(3) + 4(5) + 7(1) = 21$$

If the product AB is defined, then the entry in row i and column j of AB, denoted by $(AB)_{i,j}$, is the dot product between i^{th} row of A and j^{th} column of B:

$$(AB)_{i,j} = A_{i,1}B_{1,j} + A_{i,2}B_{2,j} + \ldots + A_{i,n}B_{n,j}$$

$$\begin{aligned}&\left[A_{1,1}\, A_{1,2} \ldots A_{1,n} \vdots \vdots \ldots \vdots A_{i,1}\, A_{i,2} \ldots A_{i,n} \vdots \vdots \ldots \vdots A_{m,1}\, A_{m,2} \ldots A_{m,n}\right] \\ &\cdot\left[B_{1,1} \ldots B_{1,j} \ldots B_{1,k}\, B_{2,1} \ldots B_{2,j} \ldots B_{2,k} \vdots \vdots \vdots \ldots \vdots B_{n,1} \ldots B_{n,j} \ldots B_{n,k}\right] \\ &= \left[\vdots \quad \vdots \quad \ldots\ldots(AB)_{i,j}\ldots\ldots \quad \vdots \quad \vdots\right]\end{aligned}$$

Example 1.8 Let $A = [2 - 134]$, $B = [-1752 - 310]$. Then to find the entry in row 2 and column 3 of AB, take the dot product of row 2 of A and column 3 of B.

$$AB = [2 - 134] \cdot [-1752 - 310] = [3(5) + 4(10)] = [55]$$

Similarly, we can find the other entries of AB.

$$\begin{aligned}AB &= [2 - 134] \cdot [-1752 - 310] \\ &= [2(-1) - 1(2)2(7) - 1(-3)2(5) - 1(10)3(-1) + 4(2)3(7) + 4(-3)3(5) + 4(10)] \\ &= [-41705955]\end{aligned}$$

The dot product given in (1.3) can also be expressed as a matrix multiplication.

$$x \cdot y = x^T y = [x_1\, x_2 \ldots x_n]\,[y_1\, y_2 \vdots y_n] = x_1y_1 + x_2y_2 + \ldots + x_ny_n. \tag{1.4}$$

There are several useful properties of matrix multiplication. We list some of them below. Let A be an $m \times n$ matrix, and let B and C have sizes for which the indicated sums and products are defined.

- $A(BC) = (AB)C$ (Associative law of multiplication)
- $A(B + C) = AB + AC$ (Distributive law)
- $I_mA = A = AI_n$ (Identity for matrix multiplication)
- $(AB)^T = B^TA^T$ (Transpose of matrix multiplication)

The kth power of an $n \times n$ matrix A is computed by multiplying A by itself k times.

$$A^k = \underbrace{A \cdots A}_{k\ times}$$

Example 1.9 Let $A = [2 - 134]$. Then

$$A^2 = A \cdot A = [2 - 134]\,[2 - 134] = [1 - 61813]$$

Several linear algebra problems involve solving matrix equations analytically. This process requires manipulation of the equation in ways similar to applying ordinary algebra with real numbers. Matrix inversion serves as an essential tool for this purpose. It is defined only for square matrices. Let A be an $n \times n$ matrix. The unique matrix inverse of A, denoted by A^{-1}, is defined as the matrix such that

$$A^{-1}A = I_n = AA^{-1}.$$

Not every square matrix has an inverse. If a matrix has an inverse, then it is called an invertible (nonsingular) matrix, and a matrix that does 'not have an inverse is called a non-invertible (singular) matrix.

Example 1.10 If $A = [2714]$ and $B = [4-7-12]$, then

$$AB = [2714]\,[4-7-12] = [1001]$$

and

$$BA = [4-7-12]\,[2714] = [1001]$$

Hence,

$$B = A^{-1}.$$

If A is an $n \times n$ invertible matrix, then A^{-1} and A^T are invertible and

$$\left(A^{-1}\right)^{-1} = A \text{ and } \left(A^T\right)^{-1} = \left(A^{-1}\right)^T.$$

If B is an $n \times n$ invertible matrix, then so is AB, and

$$(AB)^{-1} = B^{-1}A^{-1}.$$

1.2.1.3 System of Linear Equations

A system of linear equations is a collection of one or more linear equations involving the same variables, $x_1, \ldots, x_n$. The general form of a system of linear equations is

$$A_{1,1}x_1 + A_{1,2}x_2 + \ldots + A_{1,n}x_n = b_1 \tag{1.5}$$

$$\vdots$$

$$A_{m,1}x_1 + A_{m,2}x_2 + \ldots + A_{m,n}x_n = b_m$$

where b_i's and $A_{i,j}$'s are coefficients, $i = 1, \ldots, m$ and $j = 1, \ldots, n$, known in advance. In (1.5), there are m equations and n variables. A solution of the system satisfies each equation in (1.5) when substituted for $x_1, \ldots, x_n$. A system of linear equations may have no solution, exactly one solution, or infinitely many solutions. The system is called consistent if there is a solution and called inconsistent if there is no solution. We can express (1.5) as a matrix–vector equation. Let

$$A = \left[A_{1,1}\, A_{1,2} \ldots A_{1,n}\, A_{2,1}\, A_{2,2} \ldots A_{2,n} \vdots \vdots \ddots \vdots A_{m,1}\, A_{m,2} \ldots A_{m,n} \right],$$
$$x = \left[x_1\, x_2 \vdots x_n \right],\ b = \left[b_1\, b_2 \vdots b_m \right] \tag{1.6}$$

Then (1.5) becomes

$$Ax = b. \tag{1.7}$$

Eq. (1.7) can be solved numerically via row reduction algorithm (Gauss–Jordan elimination). Theoretically, we can solve (1.7) for x when A is a square and invertible matrix. In this case,

$$x = A^{-1}b. \tag{1.8}$$

This process depends on finding A^{-1}, which is a useful tool in theory. However, we rarely use A^{-1} for software applications. Computing A^{-1} and $A^{-1}b$ is three times more expensive than solving $Ax = b$ by the row reduction algorithm. Besides, the row reduction algorithm leads to more accurate estimates of x as the size of the matrix A gets larger.

1.2.1.4 Linear Independence

Given vectors $v^{(1)}, v^{(2)}, \ldots, v^{(k)} \in R^n$ and given scalars $c_1, c_2, \ldots, c_k$, the vector z defined by

$$z = c_1 v^{(1)} + c_2 v^{(2)} + \ldots + c_k v^{(k)} \tag{1.9}$$

is called a linear combination of $v^{(1)}, v^{(2)}, \ldots, v^{(k)}$ with weights $c_1, c_2, \ldots, c_k$.

Example 1.11 Let $v^{(1)} = [1310]$ and $v^{(2)} = [-251]$. Then

$$z = 2v^{(1)} + 3v^{(2)} = 2[1310] + 3[-251] = [-42123]$$

is a linear combination of $v^{(1)}$ and $v^{(2)}$.

The set of all linear combinations of $v^{(1)}, v^{(2)}, \ldots, v^{(k)}$ is denoted by Span $\left\{ v^{(1)}, v^{(2)}, \ldots, v^{(k)} \right\}$ and is called a subset of R^n spanned by $v^{(1)}, v^{(2)}, \ldots, v^{(k)}$. In other words, Span $\left\{ v^{(1)}, v^{(2)}, \ldots, v^{(k)} \right\}$ is set of all vectors that can be written in the form

$$c_1 v^{(1)} + c_2 v^{(2)} + \ldots + c_k v^{(k)} \tag{1.10}$$

with $c_1, c_2, \ldots, c_k$ scalars.

We now give a geometrical description of span of a single vector. Let us take a nonzero vector $v \in R^3$. We can think of v as a point in R^3. Then Span$\{v\}$ is formed by multiplying v by any scalar. Thus, we obtain all points on a line in R^3 through v and origin.

Existence of a solution for Eq. (1.7) means that b can be written as a linear combination of the columns of A; that is, b is in the span of the columns of A. This span is named as the column space of A. Eq. (1.7) has a solution if and only if b lies in the column space of A.

The vectors $v^{(1)}, v^{(2)}, \ldots, v^{(k)}$ are said to be linearly independent if the vector equation

$$x_1 v^{(1)} + x_2 v^{(2)} + \ldots + x_k v^{(k)} = 0 \tag{1.11}$$

has only solution $x_1 = x_2 = \ldots = x_k = 0$. Otherwise, they are said to be linearly dependent. The given vectors are linearly dependent if and only if at least one of the vectors can be written as a linear combination of the other vectors.

Example 1.12 $v^{(1)} = [10]$ and $v^{(2)} = [11]$ are linearly independent vectors since the equation

$$x_1[10] + x_2[11] = [00]$$

has only solution $x_1 = x_2 = 0$.

Example 1.13 $v^{(1)} = [222]$ and $v^{(2)} = [444]$ are linearly dependent vectors since the equation

$$x_1[222] + x_2[444] = [000]$$

has a solution $x_1 = -2$ and $x_2 = 1$. On the other hand, $2[222] = [444]$. That is, $v^{(1)}$ and $v^{(2)}$ are the scalar multiple of each other. If we are given two vectors, then these vectors are linearly dependent if and only if they are the scalar multiple of each other.

We can characterize the invertible matrices via linear independence. A matrix is invertible if and only if its columns are linearly independent.

Example 1.14 Considering the vectors in Example 1.12, we define a matrix $A = [1101]$. The columns $[10]$ and $[11]$ are linearly independent. Therefore, A is an invertible matrix, and $A^{-1} = [1-101]$.

1.2.1.5 Norms

In ML, we often need to calculate the length of a vector. It is measured by a function called a norm. A norm, usually denoted by $\|\cdot\|$ is a function, which satisfies the following properties.

1. $\|x\| \geq 0$ and $\|x\| = 0$ if and only if x is a zero vector.
2. $\|cx\| = |c| \, \|cx\|$ for any scalar c.
3. $\|x + y\| \leq \|x\| + \|y\|$.

The L^p-norm is defined by

$$\|x\|_p = \left(\sum_i |x_i|^p \right)^{\frac{1}{p}}, \tag{1.12}$$

where $p \geq 1$ be a real number. Specifically, the L^2-norm is known as the Euclidean norm, which yields the distance from the origin to the point represented by the vector x.

$$\|x\|_2 = \sqrt{x_1^2 + x_2^2 + \ldots + x_n^2} = \sqrt{x^T x}. \tag{1.13}$$

A vector is called a unit vector if its length in the Euclidean norm is 1. For example, $[10], [3/54/5], [-100], [1/32/3-2/3]$ are unit vectors.

In ML, the Euclidean norm is the commonly used norm on R^n. However, it is not preferred in some applications since the Euclidean norm grows slowly in the close neighborhood of the origin. For these applications, it is significant to separate the zero vector from the vectors whose entries are close to zero. Therefore, we utilize another norm, which grows at a faster rate when the entries are close to zero. It is the L^1-norm, which is also called the Manhattan norm.

$$\|x\|_1 = \sum_i |x_i|. \tag{1.14}$$

Another norm that arises in the applications is the L^∞-norm, which returns the maximum of the absolute value of the entries.

$$\|x\|_\infty = |x_i| \tag{1.15}$$

Note that the abovementioned norms may give different results for the same vector. An appropriate norm must be chosen depending upon the problem.

Example 1.15 Let $x = [-543]$. Then

$$\|x\|_2 = \sqrt{(-5)^2 + 4^2 + 3^2} = 5\sqrt{2}$$

$$\|x\|_1 = |-5| + |4| + |3| = 12$$

$$\|x\|_\infty = |-5| = 5$$

We can use norms to describe some geometric relationships between two vectors. For x and y in R^n, the distance between x and y, denoted by $D(x,y)$, is the length of the vector $x - y$, i.e.,

$$D(x,y) = \|x - y\|_2. \tag{1.16}$$

The angle θ between the two line segments from the origin to the points represented by x and y is given by the formula:

$$x \cdot y = \|x\|_2 \|y\|_2 \cos \cos \theta \tag{1.17}$$

The nonzero vectors x and y are said to be orthogonal if the angle θ is 90°. In this case, $\cos 90° = 0$, and thus $x \cdot y = 0$. We can easily conclude from (1.17) that two nonzero vectors are orthogonal if and only if their dot product is zero.

A set of vectors $\{v^{(1)}, v^{(2)}, \ldots, v^{(k)}\}$ in R^n is called an orthogonal set if each pair of distinct vectors from the set is orthogonal, i.e., $v^{(i)} \cdot v^{(j)} = 0$ whenever $i \neq j$. If all vectors in the set are also unit vectors, then $\{v^{(1)}, v^{(2)}, \ldots, v^{(k)}\}$ is called an orthonormal set.

Example 1.16 Consider the vectors

$$v^{(1)} = [10-1], v^{(2)} = [202], v^{(3)} = [040].$$

$\{v^{(1)}, v^{(2)}, v^{(3)}\}$ is an orthogonal set since $v^{(1)} \cdot v^{(2)} = v^{(1)} \cdot v^{(3)} = v^{(2)} \cdot v^{(3)} = 0$.

Given

$$u^{(1)} = [1/\sqrt{2}0-1/\sqrt{2}], u^{(2)} = [1/\sqrt{2}01/\sqrt{2}], u^{(3)} = [010],$$

$\{u^{(1)}, u^{(2)}, u^{(3)}\}$ is an orthonormal set since $u^{(1)} \cdot u^{(2)} = u^{(1)} \cdot u^{(3)} = u^{(2)} \cdot u^{(3)} = 0$ and $u^{(1)}, u^{(2)}, u^{(3)}$ are unit vectors.

We can define a special type of matrix using orthonormal vectors. An $n \times n$ matrix A is called an orthogonal matrix if its columns form an orthonormal set of vectors. From Example 1.16,

$$A = [1/\sqrt{2}1/\sqrt{2}0001-1/\sqrt{2}1/\sqrt{2}0]$$

is an orthogonal matrix. Computationally, it is easy to find the inverse of an orthogonal matrix. The inverse of an orthogonal matrix is equal to its transpose, i.e., $A^T = A^{-1}$. Therefore,

$$AA^T = A^T A = I_n$$

There are also several norms defined for matrices. The most common one is called the Frobenius norm.

$$\|A\|_F = \sqrt{\sum_i \sum_j A_{i,j}^2} \tag{1.18}$$

Example 1.17 Let $A = \begin{bmatrix} 2 & 3 & -1 \\ 5 & -2 & 4 \end{bmatrix}$.

$$\|A\|_F = \sqrt{2^2 + 3^2 + (-1)^2 + 5^2 + (-2)^2 + 4^2} = \sqrt{59}$$

1.2.1.6 Eigenvalues and Eigenvectors

Let A be an $n \times n$ matrix. Then, a nonzero vector x is an eigenvector of A if $Ax = \lambda x$ for some scalar λ. A scalar λ is called an eigenvalue of A if there is a nonzero vector x such that $Ax = \lambda x$. In this case, x is called an eigenvector corresponding to λ.

Example 1.18 Let $A = \begin{bmatrix} 5 & -2 \\ -2 & 2 \end{bmatrix}$. Then

$$\begin{bmatrix} 5 & -2 \\ -2 & 2 \end{bmatrix} \begin{bmatrix} -2 \\ 1 \end{bmatrix} = 6 \begin{bmatrix} -2 \\ 1 \end{bmatrix}$$

and

$$\begin{bmatrix} 5 & -2 \\ -2 & 2 \end{bmatrix} \begin{bmatrix} 1 \\ 2 \end{bmatrix} = 1 \begin{bmatrix} 1 \\ 2 \end{bmatrix}.$$

Hence, $x^{(1)} = \begin{bmatrix} -2 \\ 1 \end{bmatrix}$ is an eigenvector of A corresponding to eigenvalue $\lambda_1 = 6$, and $x^{(2)} = \begin{bmatrix} 1 \\ 2 \end{bmatrix}$ is an eigenvector of A corresponding to eigenvalue $\lambda_2 = 1$.

An $n \times n$ matrix A may have repeated eigenvalues. However, A has exactly n eigenvalues, counting repeated eigenvalues. For each eigenvalue, A has infinitely many eigenvectors. For example, $\begin{bmatrix} -4 \\ 2 \end{bmatrix}$ is also an eigenvector of A corresponding to eigenvalue $\lambda_1 = 6$. In fact, any nonzero scalar multiple of $\begin{bmatrix} -2 \\ 1 \end{bmatrix}$ is an eigenvector of A.

We find real eigenvalues and eigenvectors for a real-valued matrix in Example 1.18. However, it is possible that a real-valued matrix may have complex eigenvalues and eigenvectors. For example, $B = \begin{bmatrix} 0 & 1 \\ -1 & 0 \end{bmatrix}$ has eigenvectors $x^{(1)} = \begin{bmatrix} -i \\ 1 \end{bmatrix}$ and $x^{(2)} = \begin{bmatrix} i \\ 1 \end{bmatrix}$ corresponding to $\lambda_1 = i$ and $\lambda_2 = -i$, respectively.

In several applications of linear algebra, it is important to decompose a matrix A into the matrices constructed by eigenvalues and eigenvectors of A. Specifically, we want to write A in the form:

$$A = PDP^{-1}, \tag{1.19}$$

where P is an invertible matrix whose columns are the eigenvectors of A, and D is a diagonal matrix whose diagonal entries are the eigenvalues of A. This particular decomposition is called eigendecomposition. Note that not every matrix can be written in the form of (1.19). A matrix decomposed as in (1.19) is called a diagonalizable matrix. An $n \times n$ matrix A is diagonalizable if and only if A has n linearly independent eigenvectors.

After we obtain the eigendecomposition of a diagonalizable matrix, it is convenient to find a power of the matrix.

$$A^k = \underbrace{AA\cdots A}_{k\ times} = \underbrace{PDP^{-1}PDP^{-1}\cdots PDP^{-1}}_{k\ times} = PD^kP^{-1} \tag{1.20}$$

since $P^{-1}P = I_n$.

Example 1.19 Let us consider the matrix A given in Example 1.18. The eigenvectors of A are linearly independent since $\left[-21\right]$ and $\left[12\right]$ are not scalar multiple of each other. Thus, A is diagonalizable.

$$P = \left[x^{(1)}\ x^{(2)}\right] = \left[-2112\right] \text{and } D = \left[\lambda_1 00\ \lambda_2\right] = \left[6001\right]$$

and

$$P^{-1} = \left[-2/51/51/52/5\right].$$

Hence, the eigen decomposition of A is

$$A = \left[5-2-22\right] = \left[-2112\right]. \tag{1.21}$$

Let us now compute the 5th power of A.

$$A^5 = PD^5P^{-1} = \left[-2112\right]\left[6001\right]^5\left[-\frac{2}{5}\frac{1}{5}\frac{1}{5}\frac{2}{5}\right] = \left[-2112\right]\left[6^5 001^5\right]\left[-\frac{2}{5}\frac{1}{5}\frac{1}{5}\frac{2}{5}\right]$$

$$= \left[6221-3110-31101556\right]$$

Finally, we can characterize the invertible matrices via the eigenvalues. An $n \times n$ matrix A is invertible if and only if all eigenvalues of A are nonzero. If a square matrix is diagonalizable, then from its eigendecomposition, we can determine whether it is invertible, or not. If the diagonal matrix D in eigendecomposition contains a zero diagonal entry, then the matrix is noninvertible.

1.2.1.7 Symmetric Matrices

A square matrix A is said to be symmetric if $A = A^T$; that is, A is equal to its transpose. Here, we shall consider only real symmetric matrices.

Example 1.20 Symmetric Matrices: $\left[1-2-22\right], \left[01312-13-111\right]$, $\left[4-230-25-143-1760468\right]$.

Non-symmetric Matrices: $\left[12-24\right], \left[0-1035-1034\right]$.

Symmetric matrices satisfy a number of nice properties. All eigenvalues of a symmetric matrix are real. Additionally, symmetric matrices are orthogonally diagonalizable; that is, their eigen decompositions can be formed by orthogonal matrices.

Example 1.21 $A = \left[5-2-22\right]$ from Example 1.19 is a symmetric matrix. Its eigenvalues are real. The eigen decomposition in (1.21) can be rewritten as follows.

$$\left[5-2-22\right] = \left[-2/\sqrt{5}1/\sqrt{5}1/\sqrt{5}2/\sqrt{5}\right]\left[6001\right]\left[-2/\sqrt{5}1/\sqrt{5}1/\sqrt{5}2/\sqrt{5}\right]$$
$$= QDQ^T$$

Note that Q is an orthogonal matrix and that $Q^{-1} = Q^T$. We can also conclude from this example that eigendecomposition is not unique. It depends on how we construct the matrix D and how we choose the eigenvectors corresponding to each eigenvalue of the matrix. It is conventional to construct the matrix D by writing the eigenvalues in descending order.

Next, we discuss some well-known types of symmetric matrices. An $n \times n$ symmetric matrix A is called positive semidefinite (positive definite, resp.) if $x^T Ax \geq 0$ ($x^T Ax > 0$, *resp.*) for any nonzero x in R^n. There are several ways to check positive semidefiniteness. One of them involves eigenvalues. A symmetric matrix is positive semidefinite (positive definite, resp.) if and only if all of its eigenvalues are non-negative (positive, resp.). Given the eigendecomposition of a positive semidefinite matrix $A = QDQ^T$, the matrix D contains only non-negative diagonal entries. Similarly, it contains only positive diagonal entries if A is positive definite.

Example 1.22 Let $A = \left[2111\right]$. Then for an arbitrary nonzero $x = \left[x_1\, x_2\right]$,

$$x^T Ax = \left[x_1\, x_2\right]\left[2111\right]\left[x_1\, x_2\right] = 2x_1^2 + 2x_1x_2 + x_2^2 = \left(x_1 + x_2\right)^2 + x_1^2 > 0,$$

thus A is a positive definite matrix. Alternatively, we can arrive at this conclusion from the eigenvalues of A. The eigenvalues of A are $\lambda_{1,2} = \frac{1}{2}\left(3 \pm \sqrt{5}\right) > 0$, which implies that A is positive definite.

Positive semidefinite matrices play key roles in a wide variety of applications, for example, in optimization methods and ML algorithms. Therefore, it is important to know the basics of positive semidefinite matrices to understand these applications. We now present another characterization of positive semidefinite matrices.

A symmetric matrix A is positive semidefinite if and only if $A = B^T B$ for some matrix B. Furthermore, A is positive definite if and only if B is invertible. The sufficiency part of this statement is interesting as it provides an elegant way to construct a positive semidefinite matrix. For a given matrix B, we can always conclude that $B^T B$ is a positive semidefinite matrix.

Example 1.23 Let $B = \left[1-12031\right]$. Then $B^T = \left[123-101\right]$. We obtain that

$$A = B^T B = \left[123-101\right]\left[1-12031\right] = \left[14222\right]$$

is a positive semidefinite matrix.

1.2.1.8 Singular Value Decomposition

In a previous section, we learned eigendecomposition for diagonalizable matrices. Unfortunately, this matrix decomposition is not applicable for non-square or non-diagonalizable matrices. There is another matrix decomposition called singular value decomposition (SVD), which exists for every matrix regardless of its size. SVD is a commonly used matrix factorization technique in applied linear algebra.

SVD of a matrix involves an $m \times n$ matrix in the following form.

$$\Sigma = \left[D0\ldots000\ldots0\vdots\vdots\ddots\vdots00\ldots0 \right], \tag{1.22}$$

where D is an $r \times r$ diagonal matrix and $r \leq \{m,n\}$. Diagonal entries of D are positive and called singular values of the given matrix. So, r is the number of nonzero singular values of the matrix, which also identifies the **rank** of the matrix. Singular values on the main diagonal of D written in descending order, $\sigma_1 \geq \sigma_2 \geq \ldots \geq \sigma_r > 0$. Thus, σ_1 denotes the largest singular value, and σ_r denotes the smallest singular value.

Let A be an $m \times n$ matrix. Then, there exists an $m \times n$ matrix Σ as in (1.22), and there exists an $m \times m$ orthogonal matrix U and $n \times n$ orthogonal matrix V such that

$$A = U \Sigma V^T \tag{1.23}$$

Any factorization as in (1.23), with U and V orthogonal and Σ as in (1.22), is called an SVD of A. The columns of U are called left-singular vectors of A, and the columns of V are called right-singular vectors of A.

If A is a positive semidefinite matrix, the eigendecomposition of A is $A = QDQ^T$, which is also the SVD of A from (1.23). Therefore, the singular values of A coincides with the eigenvalues of A. This is not true in general if A is not a positive semidefinite matrix.

Let us see the singular value computation for an arbitrary matrix A. As we know, $A^T A$ is a positive semidefinite matrix, and thus it is orthogonally diagonalizable. Using (1.23),

$$A^T A = V\Sigma^T U^T U \Sigma V^T = V\Sigma^T \Sigma V^T,$$

which implies that the singular values of A are the square roots of the eigenvalues of $A^T A$ since $\Sigma^T \Sigma$ is a diagonal matrix, and its entries are $\sigma_1^2, \sigma_2^2, \ldots, \sigma_r^2$. Besides, the corresponding unit eigenvectors of $A^T A$ form the columns of V. Finally, the columns of U can be computed from the following relation.

$$AV = U\Sigma$$

1.2.1.9 Applications of Singular Value Decomposition

The first application involves the sensitivity of solutions of linear systems. Let us consider a linear system $Ax = b$. We wonder how the solution of this system is affected when the entries of A are slightly changed. Computationally, it is problematic if the solution changes dramatically under a small perturbation in the entries of A because round-off errors may lead to large changes in the solution. This problem is referred to as conditioning. We quantify the sensitivity of the solution of a linear system by computing the condition number. If A is an invertible $n \times n$ matrix, then the ratio $\frac{\sigma_1}{\sigma_n}$ of largest and smallest singular values yields the condition number of A. When this number is large, the solution of the linear system $Ax = b$ is sensitive to the changes in A.

Second application is to compute the pseudoinverse of a matrix. In (1.23), it is possible that Σ contains rows or columns of zeros. In this case, we can find a more compact decomposition of A. We assume that A has r nonzero singular values. Then, we write U and V as:

$$U = \left[U_r\ U_{m-r}\right], V = \left[V_r\ V_{n-r}\right],$$

where U_r is $m \times r$ and V_r is $n \times r$. Now (1.23) becomes

$$A = \left[U_r\ U_{m-r}\right]\left[D\ 000\right]\left[V_r^T\ V_{n-r}^T\right] = U_r D V_r^T \tag{1.24}$$

since U_{m-r} and V_{n-r} are multiplied with zero rows and columns of Σ. (1.24) is called a reduced SVD of A. Note that D is an invertible matrix, so we can take the inverse of the right-hand side of (1.24). The following matrix is called the pseudoinverse of A:

$$A^+ = V_r D^{-1} U_r^T \tag{1.25}$$

We can define the pseudoinverse for non-square matrices too. The pseudoinverse is used to find a least-squares solution for an inconsistent system $Ax = b$, when A is not square. We cover least-squares solutions later in this chapter.

1.2.1.10 Trace and Determinant

Trace of a matrix A, denoted by $\operatorname{tr}(A)$, is the sum of the elements on the main diagonal.

$$\operatorname{tr}(A) = \sum_{i=1}^{n} A_{i,i} \tag{1.26}$$

Sum of the eigenvalues of a matrix is equal to the trace of the matrix. Trace is also used to write some expressions involving sum of matrix entries. For example, we can describe the Frobenius norm alternatively as

$$\|A\|_F = \sqrt{tr\left(AA^T\right)} \tag{1.27}$$

Trace has cyclic property. If A, B, C are matrices with appropriate sizes, then

$$\operatorname{tr}(ABC) = \operatorname{tr}(CAB) = \operatorname{tr}(BCA).$$

Determinant of a matrix is a function that maps the matrix to a real number. We first start with defining the determinant of a 2×2 matrix. Let $A=\left[\, A_{1,1}\ A_{1,2}\ A_{2,1}\ A_{2,2} \,\right]$. Then determinant of A, denoted by $\det(A)$, is:

$$\det(A)=A_{1,1}A_{2,2}-A_{1,2}A_{2,1} \tag{1.28}$$

Example 1.24 Let $A=\left[3426\right]$. Then

$$\det(A)=3(6)-4(2)=10 \text{ and } \operatorname{tr}(A)=3+6=9.$$

For higher-order matrices, we still need to compute the determinant of 2×2 matrices because we have a recursive formula for the determinant. Before we see this formulation, we need the following notation. A^{ij} denotes the submatrix formed by deleting i^{th} row and j^{th} column of A.

Let A be an $n\times n$ matrix. The determinant of A is:

$$\det(A)=A_{1,1}\det\left(A^{11}\right)-A_{1,2}\det\left(A^{12}\right)+\ldots+(-1)^{1+n}A_{1,n}\det\left(A^{1n}\right)$$

Example 1.25 Let $A=\left[123-134235\right]$.

$$\begin{aligned}\det(A)&=1\cdot\det\left[3435\right]-2\cdot\det\left[-1425\right]+3\cdot\det\left[-1323\right]\\&=\left(3(5)-4(3)\right)-2\left(-1(5)-4(2)\right)+3\left(-1(3)-3(2)\right)\\&=2\end{aligned}$$

Finally, the determinant of a matrix is equal to the product of all its eigenvalues. Therefore, the determinant of a matrix is zero if and only if the matrix is non-invertible.

1.2.1.11 Least-Squares Solution

In several applications, the linear system $Ax=b$ has no solution, but we are still interested in finding an x, which makes Ax as close as to b. The least-squares method aims to find an x, which minimizes $\|b-Ax\|_2$. The expression "least-squares" is motivated by minimizing the square root of a sum of squares.

Previously, we discussed that $Ax=b$ has no solution if b is not in the column space of A. We first find a $\hat{b}$ in the column space of A, which is the closest point to b, and $A\hat{x}=\hat{b}$ for some $\hat{x}$. Then the vector $\hat{x}$ is a least-squares solution of $Ax=b$. Observe that the vector $\hat{b}$ is the orthogonal projection of b onto the column space of A. Thus, $b-\hat{b}=b-A\hat{x}$ is orthogonal to the columns of A. Multiplying $b-A\hat{x}$ by A^T from left, we obtain that $A^Tb-A^TA\hat{x}$ is equal to the zero vector. This observation ultimately implies that the least-squares solution $\hat{x}$ is the solution of the linear system:

$$A^TAx=A^Tb \tag{1.29}$$

The matrix equation (1.29) is named as normal equations for $Ax = b$.

Alternatively, we can find a least-squares solution via the pseudoinverse of the matrix A defined in (1.25). Let us define

$$\hat{x} = A^{+}b = V_r D^{-1} U_r^T b \tag{1.30}$$

Then,

$$A\hat{x} = \left(U_r D V_r^T\right)\left(V_r D^{-1} U_r^T b\right) = U_r D D^{-1} U_r^T b = U_r U_r^T b$$

It is known that $U_r U_r^T b$ is the orthogonal projection $\hat{b}$ of b onto the column space of A. Hence, $\hat{x}$ in (1.30) is a least-squares solution of $Ax = b$.

Example 1.26 Let us consider the inconsistent system $Ax = b$, where

$$A = \left[100111\right],\ b = \left[236\right].$$

We will find least-squares solution using the normal equations.

$$A^T A = \left[101011\right]\left[100111\right] = \left[2112\right]$$

and

$$A^T b = \left[101011\right]\left[236\right] = \left[89\right]$$

Least-squares solution for $Ax = b$ is the solution of the system $\left[2112\right]\left[x_1\, x_2\right] = \left[89\right]$. Using the inverse of the matrix, the least-squares solution is

$$\left[x_1\, x_2\right] = \left[2112\right]^{-1}\left[89\right] = \left[2/3 - 1/3 - 1/3 2/3\right]\left[89\right] = \left[7/3 10/3\right]$$

1.2.2 Applications in Python: How to Use Concepts

We will use the NumPy package in Python for applied mathematics applications. We first import the NumPy package.

import numpy **as** np

1- Create vectors of ones and zeros of dimension 4:

   np.ones(4)

   np.zeros(4)

2- Create a matrix of size 2×3:

   A = np.matrix('1 2 3; 4 5 6')

   OR

   A = [[1 2 3], [4 5 6]]

3- Create the identity matrix of 3×3:
   np.identity(3)
4- Multiply two matrices
   A = [[2, 4],[6, 8]]
   B = [[1, 2],[3, 4]]
   np.matmul(A, B)
5- Take 4th power of a matrix
   A = [[2, 4],[6, 8]]
   np.linalg.matrix_power(A, 4)
6- Solve a system of linear equations in the form of $Ax = b$
   A = [[2, 4],[6, 8]]
   b = [[1], [2]]
   np.linalg.solve(A, b)
7- 2-norm of a vector
   b = [[1], [2]]
   np.linalg.norm(b, 2)
8- Frobenius norm of a matrix
   A = [[2, 4],[6, 8]]
   np.linalg.norm(A, 'fro')
9- Eigenvalues and eigenvectors of a matrix
   A = [[2, 4, 5],[4, 3, 7], [5, 7, -1]]
   D, P = np.linalg.eig(A)
   print(D)# Eigenvalues of A
   print(P)# Eigenvectors of A given in the order of eigenvalues in D
10- Singular value decomposition of a matrix
   A = [[2, 4, 5],[4, 3, 7], [5, 7, -1]]
   U, D, V = np.linalg.svd(A)
   print(U)# The orthonormal matrix whose columns are left-singular vectors of A
   print(D)# Singular values of A
   print(V)# The orthonormal matrix whose columns are right-singular vectors of A
11- Find least-squares solution of an inconsistent system
   A = [[1, 0], [0,1], [1,1]]
   b= [[2], [3], [6]]
   np.linalg.lstsq(A, b)[0]

BIBLIOGRAPHY

Bengio, Y. (2016). *Deep Learning*. MIT Press.

Bhatia, R. *Positive Definite Matrices*. Princeton Series in Applied Mathematics Princeton University Press, Princeton, NJ, (2007)

Lay, D. C. (2012). *Linear algebra and its applications*. Boston: Addison-Wesley.

1.3 DATA SCIENCE AND PREPARATION FOR MODELING

Data science is the art of generating information from data, and data is at the heart of data science. Today, data is produced in an incredible amount every second. Accessing data has become easier than ever before. Previously, data was a challenging stage to obtain, but now it is more important to extract meaningful information than to get it. Hence, data science is an outgoing and growing field. Data scientists work on data science and try to generate helpful information from messy, noisy, and big data. A data scientist primarily understands the data obtained in data science. Then it is analyzed with various statistical methods. If the problem is a ML problem, the problem is solved with ML methods. The outputs of data science projects are often used in decision-making, strategy development, understanding, and making sense of events. In this sense, data science studies can be used from the smallest business to an enormous scale. For example, when a company wants to segment its customers, it can use data science applications, or a meteorologist can use data science to predict future weather events. As an example of the pandemic, an epidemiology employee also carries out data science applications and shares meaningful information about the conditions and progress of the pandemic.

One of the most attractive areas of recent years, data science, has emerged in the 21st century. This profession, which did not exist until a few years ago, has become very popular due to the ease of accessing data. The data scientist was one of the highest-paid professions (Glassdoor: The 20 Best Jobs in America in 2019, 2021). Since data is in all areas, the field of study of the data scientist covers almost all fields. A data scientist can sometimes work with a doctor, sometimes with an engineer, sometimes with a legal employee. Therefore, data scientists must have at least one literacy in different fields. For example, a data scientist working on a civil engineering soil classification study needs to know what soil liquefaction means and what it can be affected by beyond preliminary knowledge. If they do not correctly understand the field information they receive from the experts, the models to be installed will not be correct. In other words, communication is one of the essential parts of data science.

The academic fields that the data scientist should know are shown as Venn diagrams by Drew Conway (The Data Science Venn Diagram, 2021). In the areas seen in Figure 1.1, the data scientist is expected to have a good degree in all fields. The next topic will share examples of tools used in these areas.

The first of the fields shown in Figure 1.1 is computer science. It is necessary to learn about the environments in which data science applications are developed in computer science. Programming knowledge in computer science will enable data scientists to build their code and create more flexible applications.

The second field is given as mathematics and statistical science. Data science can be shown as a computer application of statistics. By statistical applications, the intent is descriptive and inferential statistics. Both models that describe the data and infer from the data will be used in data science. Ceri shared the following equation for data science (Ceri, 2018).

$$\text{Data science} = \left(\text{statistics} + \text{computing} + \text{computer} + \text{communication} \right. \\ \left. + \text{sociology} + \text{management}\right) \mid \left(\text{data} + \text{media} + \text{thinking}\right)$$

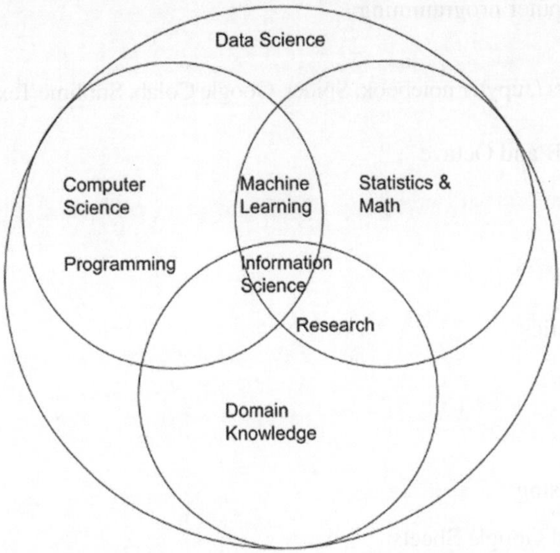

FIGURE 1.1 Data science academic fields.

In mathematics, calculus, linear algebra, and differential equations should be assimilated. For example, data scientists will need to use their mathematical knowledge to understand the mathematics behind deep learning studies.

One of the critical areas that the data scientist should know is the data preprocessing and visualization phase. The data preprocessing phase constitutes 80% of data science studies. Therefore, long periods should be spent on it.

A data scientist can be an individual researcher or a person who works as a data scientist in a business. Data scientists try to solve existing problems in enterprises by analyzing the data. However, in some cases, issues that the data scientist cannot solve may be asked to be solved by the data scientist. In this case, the tasks of the data scientist will become inoperable within the enterprise.

This subsection will discuss concepts and tasks used in data science. In this case, data scientists have a duty in all collecting, analyzing, and information extraction processes.

1.3.1 Data Scientist's Toolbox

What should be found in the data scientist's toolbox? Today, resource diversity makes it both more accessible and more challenging to work. As the number of resources increases, the level of knowledge obtained can increase to a certain extent. However, there will be no increase in the knowledge learned by increasing the resources used above a certain level. The level of understanding and using resources is more important than the level of review. New information is a burden for the person. The higher the person's load, the harder it is to move. Information is learned as it is used. There is a Chinese proverb. "I know when I see it, I know better when I hear it; I know best when I do it."

Tools for Computer programming:

- Python
    - Editors (Jupyter notebook, Spider, Google Colab, Sublime Text, Atom, etc.)
- R
- MATLAB and Octave
- Julia
- C#, C++
- Java
- Ruby

Tools for Statistics

- Scala
- Minitab
- SPSS

Data preprocessing

- Excel and Google Sheets
- Pandas, NumPy, and Scikit-Learn

Database management

- SQL
- Mongo
- Postgre

Machine learning

- Scikit-Learn (Traditional Machine Learning) (Pedregosa et al., 2011)
- Tensorflow (Martín Abadi et al., 2015; TensorFlow Developers, 2021), PyTorch (Paszke et al., 2019), Theano (Deep Learning) (Al-Rfou et al., 2016)
- MATLAB (MATLAB, 2010)
- Neurosis
- R Libraries

Channels and pages to follow

- Algebra, Education in Mathematics: https://tr.khanacademy.org/math/algebra
- Ten books listed for data mining: https://hackr.io/blog/data-science-books
- Book for Python programming language learning: Learning to program with Python 3.
- A blog page to follow: Towards Data Science https://towardsdatascience.com/

Various YouTube channel recommendations for data mining:

- 3Blue1Brownhttps://www.youtube.com/channel/UCYO_jab_esuFRV4b17AJtAw
    - Simplilearn https://www.youtube.com/user/Simplilearn
    - sentdex https://www.youtube.com/c/sentdex/playlists

- StatQuest https://youtube.com/playlist?list=PLblh5JKOoLUK0FLuzwntyYI10UQFUhsY9
- Q&A – Discussion sites:
  - Stack Overflow - https://stackoverflow.com/
  - Quora https://www.quora.com/
  - GitHub: https://github.com/

1.3.2 Understanding the Problem

As we mentioned earlier, the most crucial component of data science is the high-quality, accurate, and meaningful datasets. Most data scientists spend most of their time preparing data for analysis. It must undergo processes such as cleaning and manipulating to organize the data for analysis. It is necessary to ensure that the data is appropriate to generate meaningful information. For example, a dataset to be collected for ML requires the correct historical data to have a good training performance. Historical data is input data for the ML model. The higher quality inputs provide the better output, so the higher quality products. A data scientist should first determine the data and features they need. For example, if they are working on the car prices forecast model, an essential part of the study will be collecting data and selecting features. At this stage, statistical analysis of the dataset obtained in the problem area should be given to understand and make sense of the problem. Statistical analyses are divided into two parts. The first is descriptive statistical studies, and the it is inferential statistical studies. Data analysis studies are carried out with descriptive statistical studies. In inferential statistical studies, hypotheses about the data are produced, and conclusions are reached.

It is necessary to define the problem well before the data collection. In determining a problem, it is crucial to create good documentation by answering the what, when, where, whom, which, whose, why, and how questions. After this stage, the understanding of the data can proceed.

1.3.3 Understand Datasets with Descriptive Statistics

In descriptive statistical studies, data is known as a sample from a population. At this stage, information about the average, distribution, and range of the selection is given. The primary purpose is to summarize the data statistically. In other words, to present the dataset consisting of many lines in summary with a bit of information.

Descriptive statistics include the following stages.

- Central measures deal with the midpoint of the data. The most used of these measures is the arithmetic mean. Mode, median, geometric, and harmonic mean are also shown among the central measurements.
- Distribution measures deal with how data is distributed.
- Data types: Data can usually be divided into two types. These are categorical data, nominal or ordinal, and numerical data, which can be continuous or discrete. It is necessary to know what type it is for the data to be recognized.

1.3.3.1 Getting the Data

The studies of taking the dataset from a source and preparing it for data science cover 80–90% of the data science. Collecting data in the correct format is a critical stage. Datasets can be obtained by the researcher's means or from data collection pages. Some of these pages can be given as follows:

- UCI https://archive.ics.uci.edu/
- Google Datasets https://datasetsearch.research.google.com/
- Kaggle Datasets https://www.kaggle.com/datasets
- Amazon public datasets
  - http://aws.amazon.com/datasets#http://aws.amazon.com/datasets
- Wikipedia Datasets http://wiki.dbpedia.org/
- New York Times
  - http://developer.nytimes.com/docs#http://developer.nytimes.com/docs
- Data.gov http://data.gov/
- European Union Open Data Portal http://open-data.europa.eu/en/data/.

The sample data collection sites provided are free and open source. Some sites offer paid data. In addition, the researcher can collect this data in different formats. For example, the internet can scrape or collect data by organizing a survey. After the data is collected, preparation can be made available for analysis.

1.3.4 Data Manipulation

The data will likely have several unsuitable features. The process of seizing inappropriate properties and smoothing data is known as the manipulation process. These operations include filtering, data consolidation, and adding new features. One of the features that a data scientist should have is to enable data transformations. There are software languages developed for data manipulation operations. The most important of these languages is SQL or Structured Query Language. Data can be selected, updated, deleted, and inserted using SQL. The purpose of data manipulation operations is to reach more robust data. Therefore, data demotion or data insertion, where necessary, can be done at this stage. Pandas library is recommended for tools used for data manipulation. Pandas library is "a fast, powerful, flexible and easy to use open-source data analysis and manipulation tool, built on top of the Python programming language." (Reback et al., 2021).

1.3.5 Data Cleaning

Raw data is often dirty. In other words, raw data may contain inconducive, incomplete, or meaningless data. This data must be identified first and then processed by the specified methods. Outlier data is data located at the extreme points of the data. They are also known as "anomalous observations" in a dataset. One of the methods developed to detect this data is IQR values. When data is divided into

4, the first part is known as Q_1, the second part is Q_2, the third part is Q_3, and the fourth part is Q_4. The IQR value is the difference between Q_3 and Q_1. Moreover, it is calculated as in the following formula:

$$\text{IQR} = Q_3 - Q_1$$

After the IQR value, the outlier data is 1.5IQR smaller than Q_1 and 1.5IQR larger than Q_4. The formulas for these calculations are given below:

$$\text{Upper limit} = Q_3 + 1.5 \times \text{IQR}$$

$$\text{Lower limit} = Q_1 - 1.5 \times \text{IQR}$$

Data that does not remain within these limit values are defined as outlier data. Q_1, Q_2, Q_3, Q_4. Moreover, IQR can be shown with a boxplot. Figure 1.2 shows a boxplot sample and detecting outliers.

Likewise, when the histogram chart of the data is plotted, the status of the outlier data can be observed. If necessary, this data can be removed from the dataset, replaced with an average, or processed by different methods.

Standard deviations provide another method that can be used to detect outlier data. In this method, how much the data deviates from the average is given in standard deviations. The z-score can be calculated as in the following formula:

$$z_{score} = \frac{x - \mu}{\sigma}$$

where x is the score, μ is the mean, and σ is the standard deviation. The outlier data will be data other than z-score values −3 and +3.

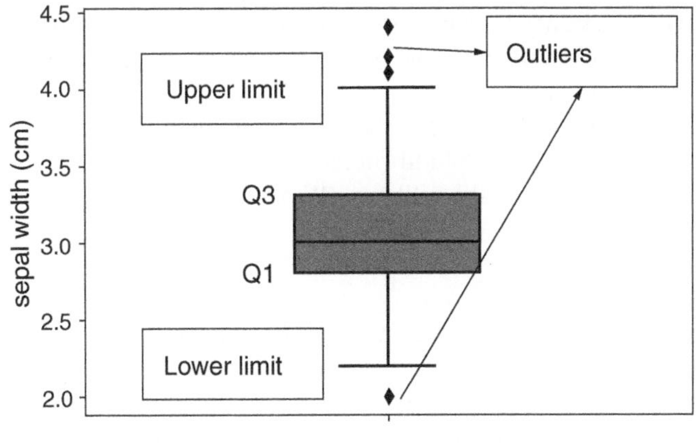

FIGURE 1.2 How to find outliers.

It is necessary to deal with the missing data and the outlier data. Many algorithms operate on the assumption that there is no missing data in the dataset. In case of missing data, the performance of learning algorithms will also decrease. Missing data typically gets "None" in the dataset. In some cases, researchers may save values such as −1, −999 as incomplete data. Therefore, before starting the analysis, attention should be paid to how the missing data is defined in the dataset.

Methods of dealing with missing data:

- Replace with centralized measurements (arithmetic mean, mode, median).
- Remove missing data.
- Fill in the missing data with unique methods.

1.3.6 Data Visualization

The use of visual tools in the disclosure of data will facilitate data analysis. The fastest and easiest process to understand the data is to chart the data. Charts have a lot to show about data. It is helpful to know the types of charts for visualizing data. Some of the basic chart types are given in Tableau, which is a visual analytics platform (What Is Data Visualization?, 2020):

- Charts
- Tables
- Graphs
- Maps
- Infographics
- Dashboards

Python has two libraries with the most common usage developed for data visualization. The first library is the matplotlib library. Dynamic or static 2D or 3D graphics can be drawn with this library. The library also offers advanced features for drawing interactive or animated graphics. Data scientists need to know Python programming languages to draw advanced graphics with matplotlib. The matplotlib library features object-managed programming. In the following section, use cases of the matplotlib library will be given (Hunter, 2007).

The second library, widely used in the Python programming language, is the seaborn library—built on the matplotlib library. Some features not in the matplotlib library are available in the Seaborn library (Waskom, 2021).

1.3.7 Data Augmentation

ML algorithms achieve better performance as the number of data increases. In cases where there is insufficient data, their learning performance is poor. Data augmentation applications are usually included in image or audio data. Visual data can be replicated if the image is in another location. A sample image processing application on data augmentation is shown in Figure 1.3. Such applications can be developed on the Scikit-Learn-image library (van der Walt et al., 2014).

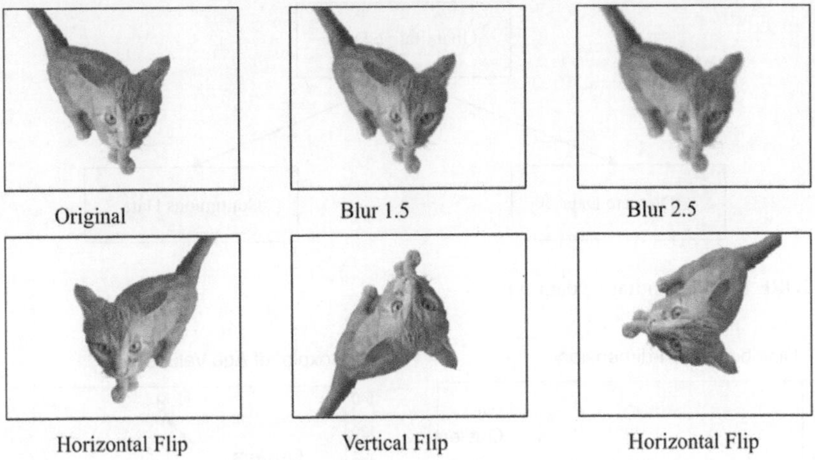

FIGURE 1.3 A sample for data augmentation.

Data Augmentation studies are carried out in natural language processing (NLP) studies like image processing applications. However, in NLP studies, a method should be applied according to the structure of the natural language. Conjunctive words can be used in these studies, which are more complex than image processing.

1.3.8 Exploratory Data Analysis and Modeling

In exploratory data analysis (EDA), questions are asked to the dataset, and the answers to these questions are given by visualizing or transforming the data. EDA studies may vary depending on the dataset and the research questions of the survey. Therefore, the discovery process can be complex and fun. Research questions need to be identified to reduce complexity. Questions are answered with information to be extracted from the dataset. As everyone knows, the correct answers are only possible with the right questions. It is necessary to see the data type of the variables to understand the dataset in data science studies. In addition, the question of whether there is a correlation between the variables can again be an EDA question. The correlation coefficient is calculated as shown in the following formula:

$$\rho_{X,Y} = \frac{E(XY) - E(X)E(Y)}{\sqrt{E(X^2) - E^2(X)}\sqrt{E(Y^2) - E^2(Y)}}$$

where $\rho_{X,Y}$ is the Pearson coefficient and E is the expectation.

As mentioned earlier, if a variable receives only a limited number of data, it is called discrete data. If the variable takes all values in a range, it is called a continuous variable. Finding those continuous or discrete variables can be given as an EDA study (see Figure 1.4).

In addition, if the distribution of the data is desired to be shown, the distribution charts and visualization of the data can be provided. In this case, charts such as

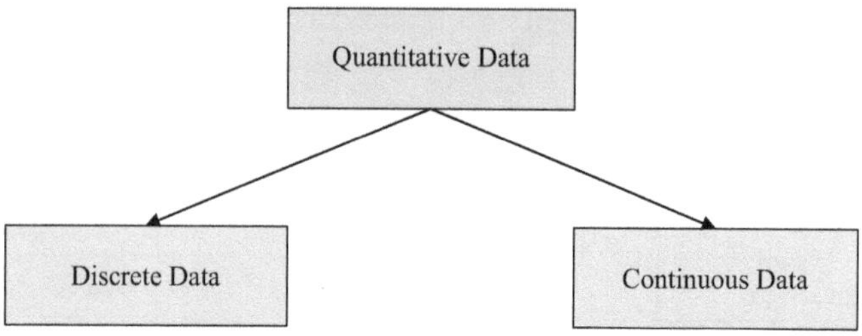

FIGURE 1.4 Quantitative data types.

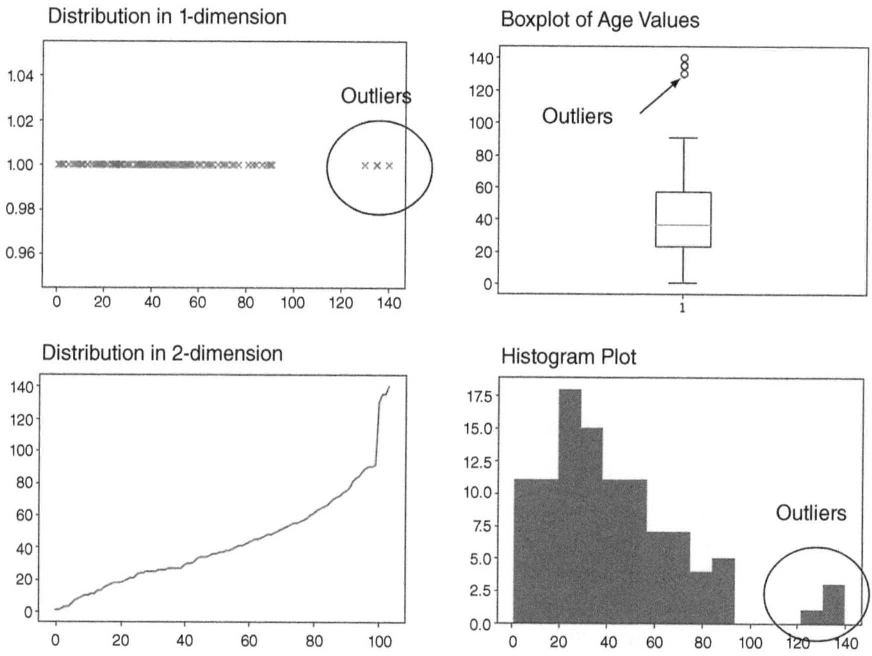

FIGURE 1.5 Outlier analysis with different visualizations.

histogram charts and scatter charts can be exported individually for each variable. Figure 1.5 provides examples of one-dimensional, two-dimensional, histograms, and box charts for age data.

```
x = [1,1,2,3,3,5,7,8,9,10,
    10,11,11,13,13,15,16,17,18,18,
    18,19,20,21,21,23,24,24,25,25,
    25,25,26,26,26,27,27,27,27,27,
    29,30,30,31,33,34,34,34,35,36,
    36,37,37,38,38,39,40,41,41,42,
    43,44,45,45,46,47,48,48,49,50,
```

51,52,53,54,55,55,56,57,58,60,
61,63,64,65,66,68,70,71,72,74,
75,77,81,83,84,87,89,90,90,91,
130, 135, 135,140]

1.3.9 Metadata and Ontologies

Metadata means data that describes the data. For example, when the content of a document containing text is evaluated as data, information such as when the document was created, who created it, how much size it takes up when it is modified are metadata of the document. The metadata of a document provides descriptive information about the record. Metadata is sometimes kept in a separate part and sometimes with data.

1.3.10 Big Data Analysis

Big data can be defined in a variety of ways. Driscoll described big data as over one terabyte. The data that businesses will use to make decisions will usually consist of big data. The discovery of hidden patterns in big data, the revealing of relationships, and the conduct of prediction studies are known as big data analytics studies. The first big data studies go back to the 1950s. Big data has 3v capability. These are volume, variety, and velocity features (Laney, 2001; McAfee et al., 2012). Hadoop application for big data transactions was developed in 2006 with an open-source license. Large data is classified as semi-structural data. Big data is non-structural data from internet records, application usage, social media content, and sensors. Therefore, the preprocessing stages of big data are essential. Big data is used in almost all sectors, such as healthcare (Miller, 2012) or financial (Corea & Cervellati, 2015).

Generally, more data provides more accurate analysis in data science. However, if the data is dirty and noisy, it will not make sense that it is too much. One of the situations to avoid in data science is overlearning. Using data science methods such as cross-verification and early stop is necessary to avoid overlearning. The generalization capability of the developed model will reveal the quality. For example, if we want to explain a bread's shape and appearance, we show the bread in a standard form. If we offer the child the square bread shape of an exceptional bread, the child will misunderstand the bread.

1.3.11 Data Science Challenges

Data science studies can be fun from an outside perspective. In fact, in a study from The Harvard Business Review (Davenport & Patil, 2012), data science is one of the most attractive professions of the 21st century due to data scientists' high incomes. However, data science studies consist of genuinely challenging and complex processes. Creating a pattern and making inferences in complex processes is not something everyone can achieve. The fact that real-world data is noisy, dirty, and irregular. That is one of the challenges faced by data scientists. Based on the previously given information about preparing the data for analysis (data preprocessing), it can be said that most of the data scientist's time is spent on these processes. Although these operations are more manageable if the data is retrieved from a single source, merging data from multiple sources compels data scientists. For example, data can be obtained by scraping a web page. The other part can be obtained from the database using SQL. In this case, merging the data can take significant time.

Another challenge for data scientists is securing the data. Nothing is 100% safe in the digital world. Therefore, taking the security importance of data is one of the data scientists' functions.

As mentioned earlier, there are many academic fields that data scientists should know. It is impossible to develop data science applications without reaching a good point in these areas. Considering that the fields mentioned earlier are now a university department at an undergraduate level, it will be understood how long each field will require. The data scientist will learn from experts in the problem field, make sense of data using statistical methods, and perform learning processes with ML methods. In other words, the data scientist is expected to be versatile and prone to interdisciplinary studies. Particular attention and accumulation are required for each of the specified fields.

The data scientist's capital is data. Businesses, brands, or societies that do not use their capital well cannot survive. Moreover, if data scientists do not use their data well, they will fail in their work. Generating information from data is a process that involves numerous attempts and failures. Sometimes the number of attempts can be too high. It still requires attention and experience to present and follow the differences between different experiments and trials in a pattern. Apart from what is mentioned here in data science, there may be personal challenges that every data scientist will experience. The most important way to deal with these challenges is to learn how to learn. Data science is a constantly updated field. To follow new developments and learn about innovations, the data science employee must have a constant state of sobriety. A data scientist who knows and understands innovations will overcome challenges.

1.3.12 Data Science Ethics

As in any business, there are ethical issues in data science. Ethical issues explain the right and wrong situations in an area. What are the rights and wrongs in data science? This section will discuss the issues to be considered in data science.

Data science is a component of AI studies. Therefore, the ethical issues in AI studies will also apply to data science. Especially in robotics, ethical issues come to the fore. The capabilities of a robot with AI need to be limited. Countries' laws, value judgments, and cultural values may affect these limits or capacities. For example, in one country, illegal, unpleasant behavior can be welcomed in another country. In this case, the right and wrong behaviors of AI applications will also change from country to country. However, some behaviors are universally inappropriate. For example, behaviors such as lying and stealing are offensive and criticized in every society.

Every society has written or unwritten rules. These rules have been developed for communities to live safely. The need to establish ethical rules for AI applications is increasing today. Because now robots can have abilities close to human abilities. In some cases, it has superior skills to man. Where should the capabilities of AI be limited? For example, should an AI-based disease diagnosis application developed in the fight against epidemic diseases be given to each patient without a doctor's supervision? Or does this diagnosis need to go through a second check-up? Or should the decisions made by AI in a legal process be implemented without the supervision of a judge or prosecutor, or is a double check necessary? The answer to this question is explained today by decision support systems. AI applications serve as a support for people to make their decisions. In other words, the decisions taken should not be taken by computer applications themselves. People must make those decisions. Apps should help people make decisions.

Unsupervised learning or reinforced learning applications, a form of data science, are exploration studies. In other words, data science studies can be carried out in cases where the data does not conclude—in this case, waiting for the machine to learn the information that is not given, and the ML processes may be involved. The main concern in AI and ethics is the capacity of machines to learn information that is not given. Although the work of machines to learn is still in its infancy, we can foresee that it will go to much more advanced levels soon.

1.3.13 Application in Python: How to Use the Concept

Python application codes for the topics mentioned in this section will be shown. Python is the most preferred programming language, especially in AI applications. One of the reasons the Python programming language is so preferred is the richness of the library. The most used libraries are NumPy, Pandas, Scikit-Learn, and matplotlib libraries.

The code in the application was run in a Jupyter Notebook environment. After the environment is installed, the necessary libraries can be installed with the following codes. Alternatively, it can be ready by setting up the Anaconda distribution on this link https://www.anaconda.com.

```
$ python3 -m pip install numpy pandas matplotlib sklearn
```

The versions of the libraries and libraries to be used in the application here are shared below.

```
import NumPy as np
import pandas as pd
import matplotlib
import sklearn as skl
np.__version__, pd.__version__, matplotlib.__version__, skl.__version__
```

```
('1.17.0', '1.2.4', '3.3.4', '0.24.2')
```

A CSV file can be downloaded from the internet to retrieve the dataset. Alternatively, the datasets available inside libraries can also be used for this application. The iris dataset will be used in this application.

```
from sklearn.datasets import load_iris
iris = load_iris()
```

Scikit-Learn datasets have their data type. The following actions can be taken to convert the data type to pandas dataframe.

```
data = iris.data # values
feature_names = iris.feature_names # column names
target = iris.target # target column values
target_names = iris.target_names # iris species
dataset = pd.DataFrame(data=data, columns=feature_names)
dataset['target'] = target
```

The head() function displays the first five rows in the dataset. At this stage, the dataset was imported. The following codes can be written for the data types in the dataset (Table 1.1).

```
dataset.head()
```

The describe() function is used to identify the dataset. The shape() function provides information about how many rows and columns the dataset has (Table 1.2).

```
dataset.describe()
```

TABLE 1.1
The Output of Head Function

	Sepal Length (cm)	Sepal Width (cm)	Petal Length (cm)	Petal Width (cm)	Target
0	5.1	3.5	1.4	0.2	0
1	4.9	3.0	1.4	0.2	0
2	4.7	3.2	1.3	0.2	0
3	4.6	3.1	1.5	0.2	0
4	5.0	3.6	1.4	0.2	0

TABLE 1.2
The Output of Describe Function

	Sepal Length (cm)	Sepal Width (cm)	Petal Length (cm)	Petal Width (cm)	Target
count	150.000000	150.000000	150.000000	150.000000	150.000000
mean	5.843333	3.057333	3.758000	1.199333	1.000000
std	0.828066	0.435866	1.765298	0.762238	0.819232
min	4.300000	2.000000	1.000000	0.100000	0.000000
25%	5.100000	2.800000	1.600000	0.300000	0.000000
50%	5.800000	3.000000	4.350000	1.300000	1.000000
75%	6.400000	3.300000	5.100000	1.800000	2.000000
max	7.900000	4.400000	6.900000	2.500000	2.000000

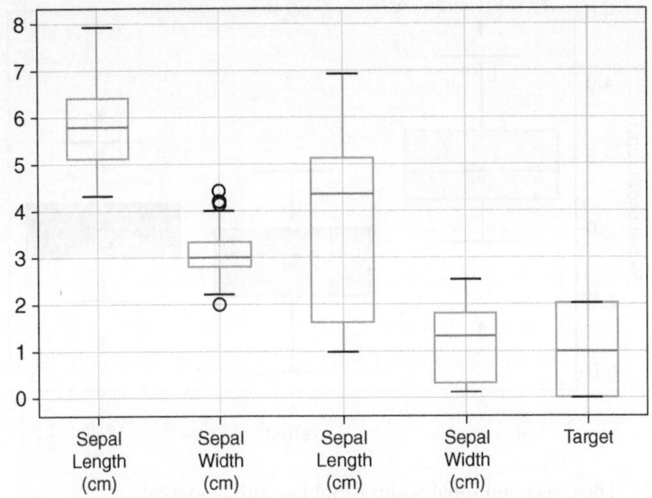

FIGURE 1.6 The results with the boxplot function.

These functions will provide information about data or missing data in the database. There is no missing data because this dataset contains 150 values for all properties. Boxplot graphs can be used for the detection of outlier data (Figure 1.6).

```
dataset.isnull().sum()
```

```
sepal length (cm)   0
sepal width (cm)    0
petal length (cm)   0
petal width (cm)    0
target              0
dtype: int64
```

```
dataset.shape
```

```
(150, 5)
```

```
dataset.boxplot()
```

There is no outlier data other than sepal width. The following codes can be used to view the sepal width outlier data (Figure 1.7).

```
import seaborn as sns
sns.boxplot(x="target",y ="sepal width (cm)",data=dataset)
```

Pair plots show changes between the two variables. The following codes draw the pair plots (Figure 1.8).

```
sns.pairplot(dataset,hue="target")
---
```

Now, we can show the outlier data in the dataset. The first method for this will be the IQR method. In the IQR method, the Q_1 and Q_3 points of the data are determined

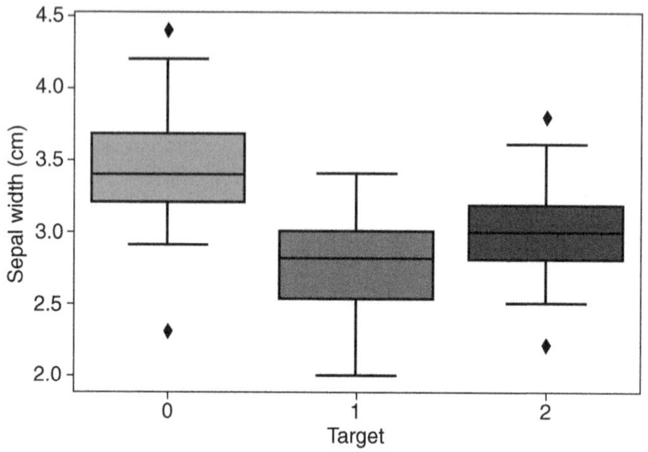

FIGURE 1.7 The target and sepal width variables with a boxplot.

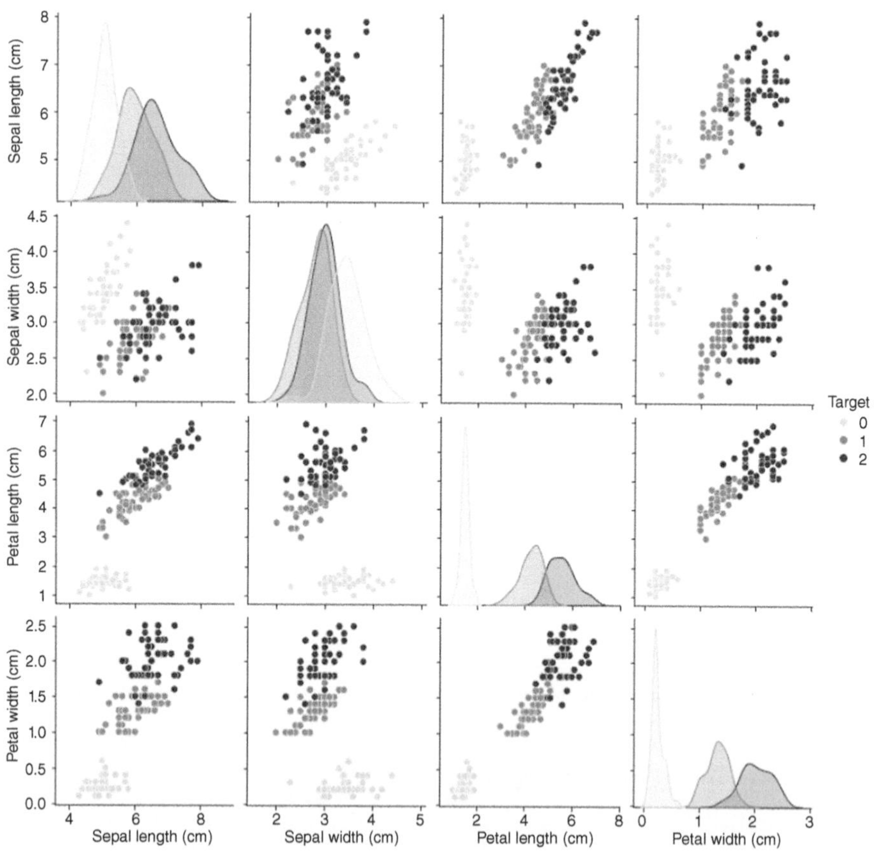

FIGURE 1.8 The results with the pairplot function.

first. The IQR value is then calculated with the formula $Q_3 - Q_1$. The values between the upper and lower limits are extracted from the dataset. After the outlier data was released, the data shape in the dataset dropped to 146 as follows.

```
print("before outlier analysis, dataset shape: ", dataset.shape)
```

```
before outlier analysis, dataset shape:(150, 5)
```

```
# Find outliers using IQRs
  Q1 = dataset.quantile(0.25)['sepal width (cm)']
  Q3 = dataset.quantile(0.75)['sepal width (cm)']
  IQR = Q3 - Q1

  upper_limit = np.where(dataset['sepal width (cm)'] >= (Q3 + 1.5*IQR))
  lower_limit = np.where(dataset['sepal width (cm)'] <= (Q1 - 1.5*IQR))
 dataset_no_outliers = dataset.drop(np.concatenate((upper_limit[0], lower_
  limit[0])))
```

```
  print(dataset_no_outliers.shape)
```

```
(146, 5)
```

At the same time, we can use the loc and iloc commands if we want to show which data is eliminated (Table 1.3).

```
dataset.iloc[lower_limit]
dataset.iloc[upper_limit]
```

We can calculate the upper and lower limits to show outlier data according to the second method, z-score. When we print these values on the screen, the upper limit is 4.36, and the lower limit is 1.74. We will reach the new dataset if we extract values other than those from the dataset (Table 1.4).

TABLE 1.3
Lower Limit of the Dataset

	Sepal Length (cm)	Sepal Width (cm)	Petal Length (cm)	Petal Width (cm)	Target
60	5.0	2.0	3.5	1.0	1

TABLE 1.4
Upper Limit of the Dataset

	Sepal Length (cm)	Sepal Width (cm)	Petal Length (cm)	Petal Width (cm)	Target
15	5.7	4.4	1.5	0.4	0
32	5.2	4.1	1.5	0.1	0
33	5.5	4.2	1.4	0.2	0

```
# Find outliers using Z-score
upper = dataset['sepal width (cm)'].mean() + 3*dataset['sepal width (cm)'].std()
lower = dataset['sepal width (cm)'].mean() - 3*dataset['sepal width (cm)'].std()
print(upper, lower)
```

```
4.3649321881434275 1.7497344785232403
```

This time, one row was removed from the dataset. This line can be shown as follows.

```
dataset_no_outliers_z_score = dataset[(dataset['sepal width (cm)']<upper) &
  (dataset['sepal width (cm)']>lower)]
dataset.loc[dataset['sepal width (cm)']>upper]
```

	Sepal Length (cm)	Sepal Width (cm)	Petal Length (cm)	Petal Width (cm)	Target
15	5.7	4.4	1.5	0.4	0

After outlier data analysis, another dataset is needed to study missing data. For this, the application will continue with another dataset.

The Python programming language can refer to two data types related to missing data. These are None and NaN values. None says there is no data type, and NaN says the data is missing. Let us go through the example to explain these.

```
# Missing Values
print("Type of None", type(None))
```

```
Type of None <class 'NoneType'>
print("Type of NaN", type(np.nan))
Type of NaN <class 'float'>
```

As you can see, the NaN data type comes as a float. None is used for non-data if the data type is none. Therefore, arithmetic operations cannot be performed with the none data type.

```
None + 1
TypeError: unsupported operand type(s) for +: 'NoneType' and 'int'
```

```
---------------------------------------------------------------------------
TypeError                    Traceback (most recent call last)
<ipython-input-95-3fd8740bf8ab> in <module>()
----> 1 None + 1

TypeError: unsupported operand type(s) for +: 'NoneType' and 'int'
```

However, arithmetic calculations can be performed with the NaN data type. The result of the operation is again returned in the NaN data type. Now, let us see the missing data operations with a new data series.

```
series_missing = pd.Series(data= np.array([None, np.nan, 1, 2, "Hello"]))
series_missing
```

```
0 None
1 NaN
2    1
3    2
4 Hello
dtype: object
```

The following functions are used to detect missing data.

- isnull(): Returns the locations where the missing data is true.
- isna(): does similar work with isnull().
- notnull(): Returns the locations where the missing data is false.

```
series_missing.isna()
0 True
1 True
2 False
3 False
4 False
dtype: bool
```

Isna() returns points that contain missing data as True. The Isnull() function functions similarly. Notnull() works in reverse with isna.

```
series_missing.notnull()
```

```
0 False
1 False
2 True
3 True
4 True
dtype: bool
```

```
# How many missing values?

series_missing.isnull().sum()
```

```
2
```

```
# Write the missing values

series_missing.loc[series_missing.isnull()]
```

```
0 None
1 NaN
dtype: object
```

```
# Write the non-missing values
    series_missing.loc[series_missing.notnull()]
```

```
2  1
3  2
4  Hello
dtype: object
```

The dropna() function is used to drop missing data.

The Fillna() function is used to fill in missing data. Some of its parameters and values are shared below.

```
    help(pd.DataFrame.fillna)
```

```
Help on function fillna in module pandas.core.frame.
    fillna(self, value=None, method=None, axis=None, inplace=False,
    limit=None, downcast=None) -> 'Optional[DataFrame]'
Fill NA/NaN values using the specified method.

Parameters
----------
value: scalar, dict, Series, or DataFrame
    Value to use to fill holes (e.g. 0), alternately a
    dict/Series/DataFrame of values specifying which value to use for
    each index (for a Series) or column (for a DataFrame).Values not
    in the dict/Series/DataFrame will not be filled. This value cannot
    be a list.

method : {'backfill', 'bfill', 'pad', 'ffill', None}, default None
    Method to use for filling holes in reindexed Series
    pad / ffill: propagate last valid observation forward to next valid
    backfill / bfill: use next valid observation to fill gap.

axis : {0 or 'index', 1 or 'columns'}
    Axis along which to fill missing values.

inplace : bool, default False
    If True, fill in place. Note: this will modify any
    other views on this object (e.g., a no-copy slice for a column in a
    DataFrame).

    # fill missing values with -1
    series_missing.fillna(value=-1)
0  -1
1  -1
2   1
3   2
4   Hello
dtype: object
```

BIBLIOGRAPHY

Abadi, Martín, Agarwal, Ashish, Barham, Paul, Brevdo, Eugene, Chen, Zhifeng, Citro, Craig, Corrado, Greg S., Davis, Andy, Dean, Jeffrey, Devin, Matthieu, Ghemawat, Sanjay, Goodfellow, Ian, Harp, Andrew, Irving, Geoffrey, Isard, Michael, Jia, Y., Jozefowicz, Rafal, Kaiser, Lukasz, Kudlur, Manjunath, … Zheng, Xiaoqiang. (2015). *TensorFlow: Large-Scale Machine Learning on Heterogeneous Systems*. https://www.tensorflow.org/

Al-Rfou, R., Alain, G., Almahairi, A., Angermueller, C., Bahdanau, D., Ballas, N., Bastien, F., Bayer, J., Belikov, A., Belopolsky, A., Bengio, Y., Bergeron, A., Bergstra, J., Bisson, V., Snyder, J. B., Bouchard, N., Boulanger-Lewandowski, N., Bouthillier, X., Brébisson, A. de, … Zhang, Y. (2016). Theano: A Python framework for fast computation of mathematical expressions. *ArXiv E-Prints, abs/1605.02688*. http://arxiv.org/abs/1605.02688

Ceri, S. (2018). On the role of statistics in the era of big data: A computer science perspective. *Statistics & Probability Letters*, *136*, 68–72. https://doi.org/10.1016/j.spl.2018.02.019

Corea, F., & Cervellati, E. M. (2015). The power of micro-blogging: How to use twitter for predicting the stock market. *Eurasian Journal of Economics and Finance*, *3*(4), 1–7.

Davenport, T. H., & Patil, D. J. (2012, October 1). Data Scientist: The Sexiest Job of the 21st Century. *Harvard Business Review*. https://hbr.org/2012/10/data-scientist-the-sexiest-job-of-the-21st-century

Glassdoor: The 20 best jobs in America in 2019. (2021). https://www.cnbc.com/2019/01/22/glassdoor-the-20-best-jobs-in-america-in-2019.html

Hunter, J. D. (2007). Matplotlib: A 2D Graphics Environment. *Computing in Science & Engineering*, *9*(3), 90–95. https://doi.org/10.1109/MCSE.2007.55

Laney, D. (2001). 3D data management: Controlling data volume, velocity and variety. *META Group Research Note*, *6*(70), 1.

MATLAB. (2010). *Version 7.10.0 (R2010a)*. The MathWorks Inc.

McAfee, A., Brynjolfsson, E., Davenport, T. H., Patil, D. J., & Barton, D. (2012). Big data: The management revolution. *Harvard Business Review*, *90*(10), 60–68.

Miller, K. (2012). Leveraging social media for biomedical research: How social media sites are rapidly doing unique research on large cohorts. *Biomedical Computation Review*.

Paszke, A., Gross, S., Massa, F., Lerer, A., Bradbury, J., Chanan, G., Killeen, T., Lin, Z., Gimelshein, N., Antiga, L., Desmaison, A., Kopf, A., Yang, E., DeVito, Z., Raison, M., Tejani, A., Chilamkurthy, S., Steiner, B., Fang, L., … Chintala, S. (2019). PyTorch: An Imperative Style, High-Performance Deep Learning Library. In H. Wallach, H. Larochelle, A. Beygelzimer, F. d'Alché-Buc, E. Fox, & R. Garnett (Eds.), *Advances in Neural Information Processing Systems 32* (pp. 8024–8035). Curran Associates, Inc. https://proceedings.neurips.cc/paper_files/paper/2019/file/bdbca288fee7f92f2bfa9f7012727740-Paper.pdf

Pedregosa, F., Varoquaux, G., Gramfort, A., Michel, V., Thirion, B., Grisel, O., Blondel, M., Prettenhofer, P., Weiss, R., Dubourg, V., Vanderplas, J., Passos, A., Cournapeau, D., Brucher, M., Perrot, M., & Duchesnay, E. (2011). Scikit-learn: Machine Learning in Python. *Journal of Machine Learning Research*, *12*, 2825–2830.

Reback, J., McKinney, W., Jbrockmendel, Van Den Bossche, J., Augspurger, T., Cloud, P., Hawkins, S., Gfyoung, Sinhrks, Roeschke, M., Klein, A., Terji Petersen, Tratner, J., She, C., Ayd, W., Naveh, S., Patrick, Garcia, M., Schendel, J., … Vetinari, H. (2021). *pandas-dev/pandas: Pandas 1.2.4* (v1.2.4) [Computer software]. *Zenodo*. https://doi.org/10.5281/ZENODO.3509134

TensorFlow Developers. (2021). *TensorFlow* (v2.8.0-rc0) [Computer software]. Zenodo. https://doi.org/10.5281/ZENODO.4724125

The Data Science Venn Diagram. (2021). Drew Conway. http://drewconway.com/zia/2013/3/26/the-data-science-venn-diagram

van der Walt, S., Schönberger, J. L., Nunez-Iglesias, J., Boulogne, F., Warner, J. D., Yager, N., Gouillart, E., & Yu, T. (2014). scikit-image: Image processing in Python. *PeerJ*, *2*, e453. https://doi.org/10.7717/peerj.453

Waskom, M. L. (2021). seaborn: Statistical data visualization. *Journal of Open Source Software*, *6*(60), 3021. https://doi.org/10.21105/joss.03021

What Is Data Visualization? Definition, Examples, And Learning Resources. (2020). Tableau. https://www.tableau.com/learn/articles/data-visualization

1.4 MACHINE LEARNING ALGORITHMS

ML is a scientific field of study in which various algorithms and techniques have been developed to enable computers to learn similarly to humans. It has been one of the fastest growing fields, enabling computers to evolve automatically through experience (Jordan and Mitchell, 2015). In other words, ML is a set of methods for creating models from data. Which type of algorithm works best depends on the type of data analyzed, the resources available, the nature of the data, and what is wanted at the end of all operations.

It is based on mathematical subfields, including optimization, statistics, probability, and ML. After the model is developed, it is anticipated that the machine will use the historical data to make future predictions. ML is different from other software engineering techniques in this regard. Because the software developer is supposed to look at the dataset and make some patterns using these old ways. These processes are impractical and excessively biased (Grigorev, 2020). Ordinary programming algorithms simply tell the computer what to do. For example, sorting algorithms transform unsorted data into data that is sorted, usually by some criteria, such as the numerical or alphabetic order of one or more fields in the data. In addition, linear regression models apply matrix transformations to minimize the least squares error between a straight line and the data. These operations are ideal statistical analyses to get the results of the operations with the data in hand.

ML types are generally divided into two as supervised and unsupervised learning. Also, reinforcement learning, which is another type of ML, will also be discussed in this section. The supervised learning algorithm learns from labeled training data, then predicts outcomes for unlabeled data. Figure 1.9 summarizes the supervised learning process.

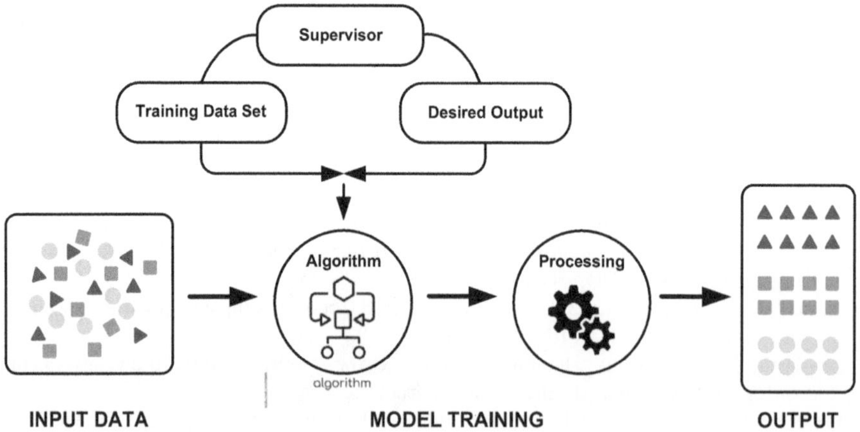

FIGURE 1.9 Supervised learning process.

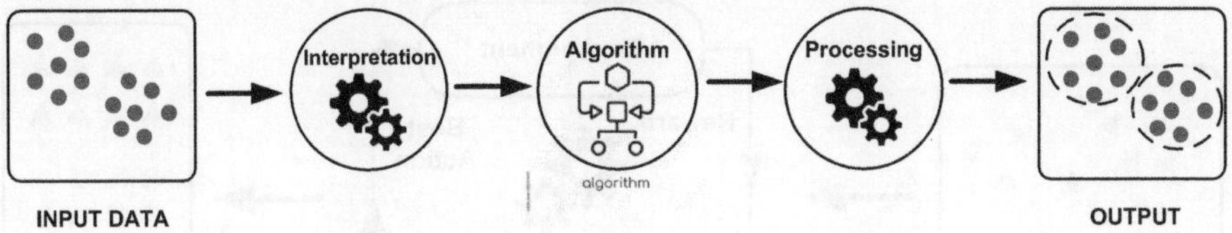

FIGURE 1.10 Unsupervised learning process.

Supervised ML refers to algorithms that need some developer supervision throughout the process. In addition to labeling the training data, the developer establishes stringent guidelines and constraints that the algorithm must adhere to. As a result, algorithms can use labeled examples to create predictions by applying previously learnt information to fresh data. The way audited algorithms operate is by specifying a collection of input data and anticipated outcomes. When the matches and predictions are accurate, the algorithm is deemed successful. Additionally, the learning algorithm may identify flaws and adjust the model by comparing its output with the right output (test dataset).

Unsupervised learning, unlike supervised learning, is learning the relationships and structures that exist in the data without labeling the data as cause–effect or input–output. Unsupervised learning refers to the ability to learn and organize information without providing an error signal to evaluate the solution (Sathya & Abraham, 2013). Unlike supervised learning, unsupervised learning does not aim to produce output in response to specific inputs but instead discovers patterns in the data. Figure 1.10 summarizes the unsupervised learning process.

Unsupervised ML algorithms are used when the information used to train is not classified or labeled. Unsupervised learning examines how systems can extract a function from unlabeled data to explain a hidden structure. When the system cannot find the correct output, it continues to search the data and infers from the datasets to explain the hidden structures from the unlabeled data.

Reinforcement learning shows how a system that perceives its environment and can make decisions on its own can learn to make the right decisions in order to reach its goal. There is an interpreter in reinforcement learning, but it does not give as much detail to the system as in supervised learning. Instead, when the learning system (agent) makes a decision, it rewards the system for the right decision and punishes the wrong one. The system improves itself with each trial based on the punishments it receives and the rewards it receives. The aim is to check whether the possible situations that the learning system has tried are the target or not and to remember all the right or wrong situations that have been tried. For example, when an agent is taught to play chess, it is sufficient to learn the rules initially. The agent then improves itself, either by playing itself or by playing with another opponent. The agent, who will make random moves at first, can then reach the level of defeating a world champion. Reinforcement learning application areas can be found in a wide range from robotics and computer games to network routing and natural language dialogue systems (Wiering and Van Otterlo, 2012). Figure 1.11 summarizes the reinforcement learning process.

One kind of ML that is investigated using a sort of trial-and-error methodology is reinforcement learning. By generating actions, seeing the outcomes, and then

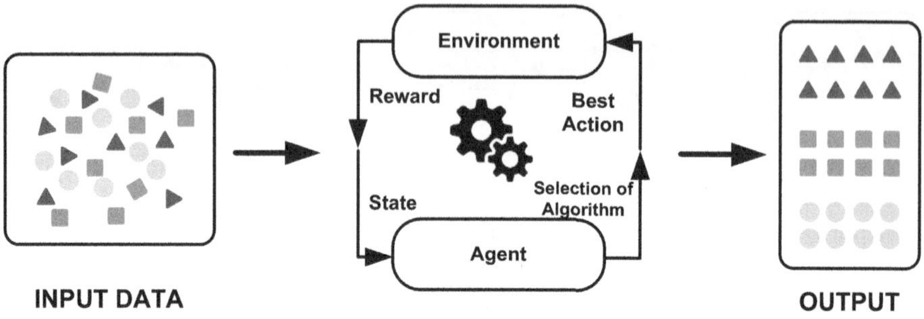

FIGURE 1.11 Reinforcement learning process.

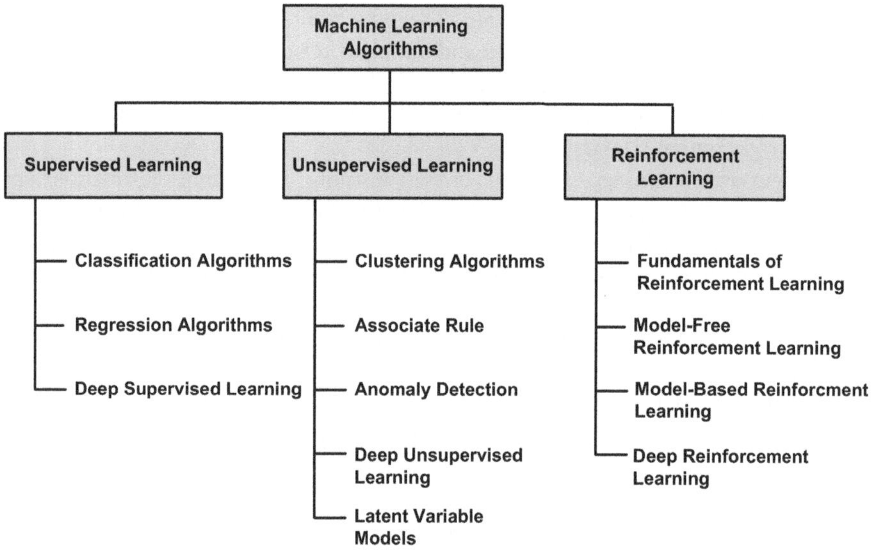

FIGURE 1.12 Content of the machine learning algorithms.

considering these implications when executing the subsequent action, the agent engages with its surroundings. This procedure keeps going until the algorithm learns and selects the best course of action. This technique enables software agents and machines to automatically identify the best course of action in a certain situation in order to optimize performance.

In this section, master ML algorithms will be discussed as in Figure 1.12.

1.4.1 Supervised Learning

Supervised learning is a branch of ML concerned with creating a function from labeled training data. It involves training the system or machine by giving the output pattern to the system to perform a function. This output (or target) pattern acts as a teacher, namely a supervisor for the machines. Typically, supervising means monitoring and directing the execution of tasks, projects, and activities.

TABLE 1.5
An example Dataset for Supervised Learning Explanation

	Input Data			Output Data	
No	Age	Height	Weight	Gender	
1	25	158	49	Female	Can be used as training dataset
2	22	176	78	Male	
3	36	183	79	Male	
4	38	165	60	Female	
5	28	169	58	Female	
6	43	192	85	Male	
7	39	175	63	Female	
8	44	172	66	Female	Test dataset
9	24	181	75	Male	
10	32	179	82	Male	

With supervised learning, real-life image classification, spam filtering, risk assessment, fraud detection, diagnostics, etc., problems can be solved. To explain the application of supervised learning with a simple example, let's try to estimate the gender of people (adults) for whom we consider height, weight, and age data. A dataset in Table 1.5 can be considered for the example. (This dataset is purely for illustration and contains synthetic data.)

After determining the data to be used, the dataset is first partitioned into training and test sets. 70% of the data of 10 people in this dataset can be used for training and 30% for testing. After learning takes place using a training pattern with a suitable learning algorithm, the performance of the predictions produced is measured with the test data. Estimation performances vary according to the convenience of the dataset and the algorithm used. The dataset in this example may not be suitable for making high-performance predictions because it consists of synthetic data and contains few data. On the other hand, the performance can be very well, but it should be noted that working with insufficient data may not always produce the desired results, as it may cause overfitting (by memorizing), which is a handicap to learning.

The aim is to correctly predict the gender of the person according to the given characteristics in the discussed example. In other words, it is the prediction of which gender class the person belongs to. Gender is a nominal (value label) data type, and this feature should be considered when choosing the algorithm to be used here.

To summarize, in supervised learning, training data consists of a set of input and output data. The input data can be a vector of features, and the output data specifies what we want the function to output. The objective may be to predict the class or value label. Depending on the type of objective, supervised learning falls roughly into two categories: Classification and regression.

1.4.1.1 Classification Algorithms

Classification means the process of classifying or categorizing data by algorithms. The output data is a label or a category such as a gender, a color, an animal, or a decision in the classification problems. In classification, the number of groups (classes)

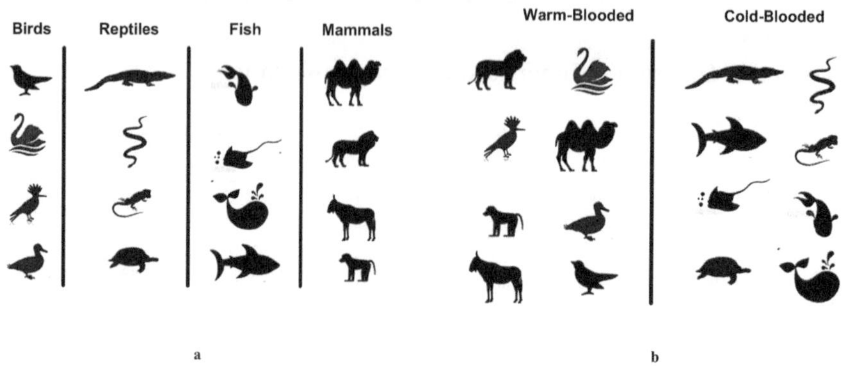

FIGURE 1.13 An example visual dataset for classification explanation

and their characteristics are certain. It is determined which class the sample belongs to according to its characteristics. Suppose we have n objects and m classes with certain characteristics. The process of placing the most similar class by looking at the characteristics of an object in question is called the classification process.

Classification algorithms first derive a pattern by considering the input and output data in the training dataset. Then, when they evaluate the input data (feature vector) on the test dataset with this pattern, they try to predict the correct output (class or label). For example, if we consider the sample dataset in Table 1.1, the classification algorithm will establish a pattern between the input data and the output data in the training dataset. Then, it will evaluate the input data in the test dataset according to this pattern and predict which gender each individual belongs to.

Imagine classifying the determined animals according to their various characteristics. Figure 1.13a shows that 16 different animals are divided into 4 different classes. In Figure 1.13b, it is seen that the same animals are divided into two different classes. It is possible to make such classifications according to similar or different characteristics. In classification, its characteristics should be introduced to the algorithm as input data and classes as output data.

The most known classification algorithms, which will be briefly described in these subsections are the K-nearest neighbor algorithm, Naïve Bayes, decision tree, and support vector machines.

1.4.1.1.1 K-Nearest Neighbor (KNN) Algorithm

The K-nearest neighbor (KNN) algorithm, which dates back to the early 1950s, is a supervised learning algorithm that predicts which class the newly given data is more likely or suitable for by looking at which class the other data are from. In other words, the data of unknown class is compared with other data in the training set, and distance measurement is made. According to the calculated distance, the closest (or best fit) class to the data that has not yet been assigned to a class is found. This algorithm, which is usually preferred for classification and can also be used for regression, is simple and easy to implement.

The main idea in the KNN algorithm is that similar points or groups of variables most likely belong to the same class. At this point, the nearness of the data of the unknown class is calculated using a determined distance criterion. The most widely used distance measure is the Euclidean distance (Hu et al., 2016).

The Euclidean distance is the length of the hypotenuse of a real triangle, calculated as in Eq. (1.31). Let $d(x_i, y_i)$ be the distance between the observation pair x_i and y_i, each containing n quantitative variables.

$$d(x_i, y_i) = \sqrt{\sum_{i=1}^{n} |x_i - y_i|^2} \tag{1.31}$$

Other distance measures used; Minkowski distance, Manhattan city block distance, Mahalanobis distance, and similarity measurement can be given as examples.

If it is desired to calculate a distance dependent on λ number of variables, the Minkowski distance measurement can be used. The Minkowski distance is calculated in Eq. (1.32) provided that $\lambda > 1$. If the value of λ is given as 2, we can see that we actually get the Euclidean distance.

$$d(x_i, y_i) = \left(\sum_{i=1}^{n} |x_i - y_i|^{\lambda} \right)^{\frac{1}{\lambda}} \tag{1.32}$$

Manhattan city block distance is the calculation of the distance between two points measured at right angles along the axes and is calculated in Eq. (1.33).

$$d(x_i, y_i) = \sum_{i=1}^{n} |x_i - y_i| \tag{1.33}$$

The Mahalonobis distance is calculated with the formula in Eq. (1.34), provided that $\vec{x}$ represents a multivariate dataset, $\vec{\mu}$ represents the mean vector of the independent variables, S the covariance matrix, and T the transpose of the matrix.

$$D(x, \mu) = \sqrt{(\vec{x} - \vec{\mu})^T s^{-1} (\vec{x} - \vec{\mu})} \tag{1.34}$$

These distance measures may be preferred for continuous variables. In the case of categorical or string variables, the Hamming distance can be used. Hamming distance is determined as the characters in the same position belonging to the expressions with the same number of characters are compared. For example, let the two expressions to be compared be "'increase' and 'decrease'." Except for the first two characters in these two expressions, the other characters are equal to each other, so the hamming distance value here is determined as 2.

The steps of the KNN algorithm are as follows:

1. The K parameter and an appropriate distance metric are determined.
2. Distances are calculated.
3. The K neighbors closest to the data to be analyzed are determined.
4. Among the K-nearest neighbors, whichever category has more is selected.

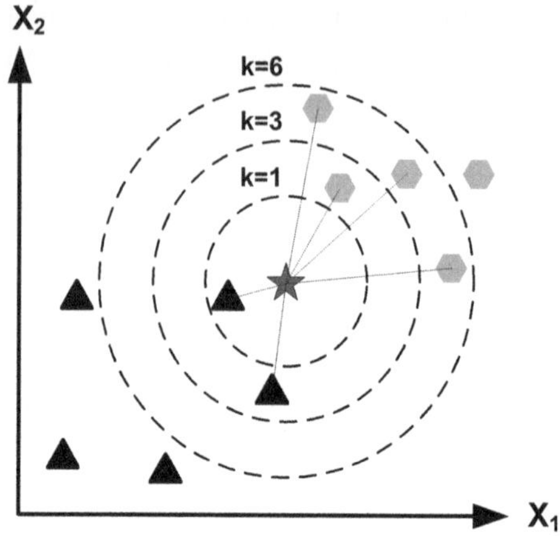

FIGURE 1.14 An illustrative example for KNN algorithm.

TABLE 1.6
An Example Dataset for the KNN Algorithm

No	Height	Weight	Class	Distance	Rank	$k = 1$	$k = 3$	$k = 6$
1	168	74	Overweight	8.49	3		O	O
2	182	96	Overweight	29.12	7			
3	177	92	Overweight	24.19	6			O
4	173	62	Normal	6.08	2		N	N
5	193	104	Overweight	40.71	9			
6	188	81	Normal	19.1	5			N
7	162	61	Normal	13.89	4			N
8	174	65	Normal	3.00	1	N	N	N
9	201	101	Normal	42.64	10			
10	189	103	Overweight	38.08	8			
11	**174**	**68**	?	-	-	**Normal**	**Normal**	**Normal**

If it is examined, these steps on the graph in Figure 1.14, two sets, triangle, and hexagon, are defined. A new shape is included in the chart, and it is desired to find which cluster it falls into. If the Euclidean distance is accepted as the distance metric, the new shape is estimated to be in the triangle class for $k = 1$, again in the triangle class for $k = 3$, but in the hexagon class for $k = 6$.

Another example in Table 1.6, 10 adults have height and weight data. In real life, an index is calculated by dividing the weight of adults by the square of their height, that is, by finding the ratio of the weight to the approximate body surface, and this is called the body mass index. According to this index, people's weight levels are categorized. In this example, it is seen that only the body information of the adults, who are divided into two different categories according to their weight and height, is given to a machine that is not given information about this index.

Accordingly, if the machine is asked to guess which class the 11th person is in, this would be as in Table 1.2, using the KNN algorithm. When the distance criterion is Euclidean and $k = 1$, it can be said that the 11th person is nearest to the 8th person, and it is in the Normal class. When $k = 3$, most of the nearest neighbors are in the Normal class (2 Normal and 1 Overweight). When $k = 6$, most of the nearest neighbors are in the Normal class, too (4 Normal and 2 Overweight). According to the Euclidean distance and $k = 1, 3, 6$ values, it can be said that the 11th person is in the Normal class.

1.4.1.1.2 Decision Trees

A decision tree is a structure that applies a series of decision rules to split a dataset with many records into smaller sets. In other words, it is a structure that uses basic decision-making procedures to divide a huge number of records into extremely small record groups. Decision trees consist of decision nodes and leaves. The decision node at the top of the tree is called the root node, and leaves represent classes (decisions). Each structure containing a decision node and its leaves is called a sub-tree.

The tree creating process is called tree induction, which consists of two phases: tree growing (splitting) and pruning. Splitting is an iterative process that involves breaking up data into smaller subsets. The first iteration considers the root node containing all the data. Subsequent iterations operate on derivative nodes (or child nodes) containing subsets of the data. Pruning refers to weeding out unwanted sub-trees or nodes after a tree has been created, making the tree more general.

Considering the example of how animals can be classified in Figure 1.5, it can be considered how these animals can be classified with decision trees according to some features. For example, considering the fact that the animals have variable/constant body temperatures, whether they have gills, and whether they lay eggs, as in Table 1.7, a small dataset can be created to express which class they are.

Based on this small and simple dataset, a decision tree can be created as in Figure 1.15. According to this tree, animal number 5 given in the dataset is a fish.

The size of the dataset given for the decision tree drawing in Table 1.3 is quite small, the number of combinations is low, and the decision tree created in Figure 1.3 is logically constructed considering this dataset. However, the datasets handled in real-life problems are too large and complex to create a decision tree with the same method. For this reason, various methods and algorithms have been proposed to

TABLE 1.7
A Sample Dataset for Classification Animals by a Decision Tree

No	Body Temperature?	Having Gills?	Laying Eggs?	Class
1	Variable	No	Yes	Reptiles
2	Variable	Yes	Yes	Fish
3	Constant	No	No	Mammals
4	Constant	No	Yes	Birds
5	Variable	Yes	No	?

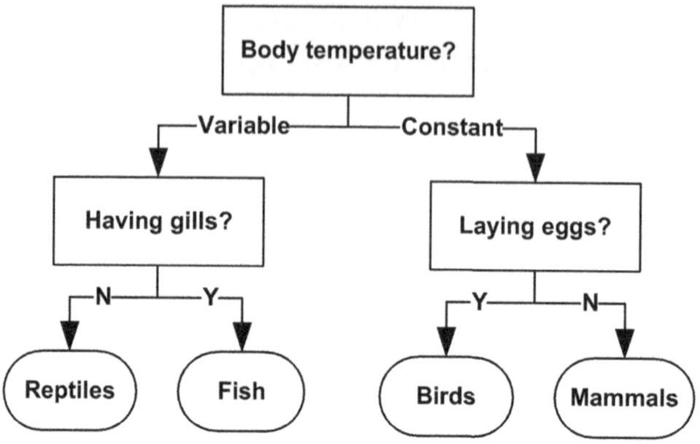

FIGURE 1.15 The decision tree for classifying the animals given in Table 1.7.

create a decision tree. The most widely used among these algorithms are the entropy-based classification algorithms (ID3, C4.5, C5.0) and classification and regression tree (CART).

1.4.1.1.2.1 ID3 Algorithm The ID3 algorithm ensures solving classification problems using entropy. Shannon's formula is commonly used for entropy calculations. Where *i* is a set of independent choices from 1 to k and pi is the probability of choice i, then entropy is calculated as in Eq. (1.35) (Ramazzotti et al., 2004).

$$E_j = -\sum_{i=1}^{k} p_i\left(p_i\right) \tag{1.35}$$

After the entropy of the system is calculated according to Eq. (1.35), the entropy of each feature is also calculated separately. Calculated entropy values are subtracted from the general entropy. It is determined which feature reduces the entropy value the most. The decrease in entropy is expressed as "gain" and is calculated by Eq. (1.36). In other words, the feature that reduces entropy the most is the attribute that gives the most gain.

$$\text{Gain}\left(S, A\right) = E_s - E_A \tag{1.36}$$

Considering the sample dataset given in Table 1.6, the data were classified categorically and transformed into the form in Table 1.8. That is, the height data between 161 and 170 are named as category 1, 171 and 180 as category 2, 181 and 190 as category 3, 191 and 200 as category 4, and 201 and above as category 5. Similarly, weight data between 61 and 70 were named as category 1, 71 and 80 as category 2, 81 and 90 as category 3, 91 and 100 as category 4, and 101 and above as category 5.

Accordingly, in order to predict with decision trees in which class the 11th data will be, it is necessary to find the root of the tree first. For this, necessary entropy calculations must be made. There are two classes, Overweight and

TABLE 1.8
A Sample Categorical Dataset Which the Adults are Classified by their Height and Weight

No	Height	Weight	Class
1	1	2	Overweight
2	3	4	Overweight
3	2	4	Overweight
4	2	1	Normal
5	4	5	Overweight
6	3	3	Normal
7	1	1	Normal
8	2	1	Normal
9	5	5	Normal
10	3	5	Overweight
11	2	3	?

Normal. In the given 10 datasets, there are 5 of each. Accordingly, the entropy of the system is calculated as follows.

$$E_S = -\frac{5}{10}\left(\frac{5}{10}\right) - \frac{5}{10}\left(\frac{5}{10}\right) = 1$$

The dataset in the height feature consists of five categories. The entropy of each category is calculated as follows.

$$E_{H1} = -\frac{1}{2}\left(\frac{1}{2}\right) - \frac{1}{2}\left(\frac{1}{2}\right) = 1$$

$$E_{H2} = -\frac{2}{3}\left(\frac{2}{3}\right) - \frac{1}{3}\left(\frac{1}{3}\right) = 0.92$$

$$E_{H3} = -\frac{1}{3}\left(\frac{1}{3}\right) - \frac{2}{3}\left(\frac{2}{3}\right) = 0.92$$

$$E_{H4} = E_{H5} = -\frac{1}{1}\left(\frac{1}{1}\right) = 0$$

$$E_H = \frac{2}{10}E_{H1} + \frac{3}{10}E_{H2} + \frac{3}{10}E_{H3} + \frac{1}{10}E_{H4} + \frac{3}{10}E_{H5} = 0.752$$

The dataset in the weight feature consists of five categories. The entropy of each category is calculated as follows.

$$E_{W1} = -\frac{3}{3}\left(\frac{3}{3}\right) = 0$$

$$E_{W2} = -\frac{1}{1}\left(\frac{1}{1}\right) = 0$$

$$E_{W3} = -\frac{1}{1}\left(\frac{1}{1}\right) = 0$$

$$E_{W4} = -\frac{2}{2}\left(\frac{2}{2}\right) = 0$$

$$E_{W5} = -\frac{3}{3}\left(\frac{2}{3}\right) - \frac{1}{3}\left(\frac{1}{3}\right) = 0.92$$

$$E_W = \frac{3}{10}E_{W1} + \frac{1}{10}E_{W2} + \frac{1}{10}E_{W3} + \frac{2}{10}E_{W4} + \frac{3}{10}E_{W5} = 0.276$$

The gain values are calculated below, and the feature with the highest gain value forms the roof of the decision tree.

$$\text{Gain}_H = E_S - E_H = 1 - 0.752 = 0.248$$

$$\text{Gain}_W = E_S - E_W = 1 - 0.276 = 0.724$$

In this example, the weight will be on the roof of the decision tree. Entropy is calculated in the same way for the five categories under this feature, and the decision tree is created.

1.4.1.1.2.2 C4.5 Algorithm The C4.5 algorithm is an extension of the ID3 algorithm and was developed to eliminate some of the deficiencies in the ID3 algorithm. One of these deficiencies is that the ID3 algorithm only works with categorical attribute data. This is a weakness for situations that need to work with numerical (continuous) data. Also, the higher the possible value variety of a categorical feature, the higher the gain for that feature will be, and this will adversely affect the accuracy of the tree. This algorithm prevents some unnecessary inferences by reducing the information gain. Therefore, a new concept called splitting value is included in the algorithm.

Where k is the number of outputs for feature A, S_j is a subset of S corresponding to the jth output, and $|S_j|$ is the number of cases in subset S_j, the splitting value is calculated by Eq. (1.37).

$$\text{Splitting value}(A) = -\sum_{j=1}^{k} \frac{|S_j|}{S}\left(\frac{|S_j|}{S}\right) \tag{1.37}$$

Gain ratio, which is dependent on gain and splitting value for feature A, is calculated by Eq. (1.38).

$$\text{Gain Ratio}(S, A) = \frac{\text{Gain}(S, A)}{\text{Splitting value}(A)} = \frac{E_S - E_A}{\text{Splitting value}(A)} \tag{1.38}$$

Apart from these calculations, another difference of the C4.5 algorithm from the ID3 is that it is possible to perform the pruning process when necessary. There are two types of pruning: pre-pruning and post-pruning. Pre-pruning can generally be thought of as not forming some branches when forming a tree. For example, some branches are not created because they would be meaningless or contain no elements. In the post-pruning, all branches of the tree are formed and then pruned according to some conditions. Here too, branches that do not contain elements can be pruned, and in some cases, statistical approaches can be used.

1.4.1.1.2.3 CART Method It was developed by Breiman et al. (1984). It has the ability to work with both numerical and categorical data. CART is based on splitting the tree into two branches at each decision node (binary tree classification). In other words, there are binary branches in such decision trees. Splitting is performed by applying a certain criterion to a node in the CART structure. For this, first, the values for which all the attributes exist are taken into account, and after all the matches, two divisions are obtained.

There are different algorithms available under the CART. The steps of the Twoing algorithm, one of these algorithms, are as follows.

Step 1a: Considering the values contained in the attributes, the training set is divided into two separate branches. These are called candidate divisions. A node t has two branches as the left and right child nodes of node t (t_L and t_R).

Step 1b: For each of the candidate divisions, the probability of the record in the training set being in the left and right of the tree (P_L and P_R) values is calculated by Eqs. (1.39) and (1.40), and the probability of a record which belongs to class j being in the left and right of the tree ($P(C_j \mid t_L)$ and $P(C_j \mid t_R)$) is calculated by Eqs. (1.41) and (1.42).

$$P_L = \frac{\text{number of records at } t_L}{\text{number of records in training set}} \tag{1.39}$$

$$P_R = \frac{\text{number of records at } t_R}{\text{number of records in training set}} \tag{1.40}$$

$$P\left(t_L\right) = \frac{\text{number of class } j \text{ records at } t_L}{\text{number of records at } t} \tag{1.41}$$

$$P\left(t_R\right) = \frac{\text{number of class } j \text{ records at } t_R}{\text{number of records at } t} \tag{1.42}$$

Step 1c: Branching occurs according to a branching criterion shown in Eq. (1.43). This value, which is symbolized by $\Phi\left(s \middle| t\right)$, is the "goodness" criterion of the candidate s split at the t node, and by maximizing it, the optimal split is achieved (Larose and Larose, 2014). After these steps are completed, the decision tree is started to create.

$$\Phi() = 2P_L P_R \sum_{j=1}^{\#\text{classes}} \left| P\left(t_L\right) - P(C_j \mid t_R) \right| \tag{1.43}$$

Step 2: By returning to step 1, the tree is continued to be created by repeating the same algorithm steps from the current branch.

TABLE 1.9
Comparison of the Features of the Most Widely Used Decision Tree Algorithms

	Algorithms		
Features	**ID3**	**C4.5**	**CART (Twoing Algorithm)**
Type of Data	Categorical	Continuous and Categorical	Data on Constant and Nominal Attributes
Speed	Low	Faster than ID3	Average
Boosting	Not supported	Not supported	Supported
Pruning	No	Pre-pruning	Post-pruning
Missing values	Can't deal with	Can't deal with	Can deal with
Formula	Gain	Splitting value and Gain ratio	Twoing Criteria
Procedure	Top-down decision tree	Top-down decision tree	Binary decision tree

The features of the three most common decision tree generation algorithms summarized in this section are compared in Table 1.9 (Shamrat et al. 2021).

1.4.1.1.3 Naïve Bayes

Naïve Bayes classification aims to determine the class of data presented to the system with a set of calculations defined according to probability principles. In Naïve Bayes classification, a certain amount of training data is presented to the system. The data given for training must have a class. With the probability operations performed on the training dataset, it is estimated which class the test data belongs to. Therefore, the size of the training dataset is important to accurately determine the class of test data. This algorithm, which calculates the probability of an item falling into a certain category using the conditional probability rule, is known as a highly effective supervised ML algorithm.

Let n sample data $X = (x_1, x_2, \ldots, x_n)$ and m classes $C = (C_1, C_2, \ldots, C_m)$ be defined in the training dataset. The posterior probability $P(C_i|X)$ given in Eq. (1.44) is maximized to find the class of any given sample. Since $P(X)$ is the same for all classes, only maximizing $P(X|C_i)P(C_i)$ is enough.

$$P(X) = \frac{P(C_i)P(C_i)}{P(X)} \tag{1.44}$$

Assuming that the feature pairs are independent and each feature has the same importance or weight, the $P(X|C_i)$ value is calculated as in Eq. (1.45).

$$P(C_i) = \prod_{k=1}^{n} P(x_k|C_i) = P(C_i)P(C_i)\ldots P(x_n|C_i) \tag{1.45}$$

For example, let's take the dataset given in Table 1.4 and find out which class the 11th sample belongs to with Naive Bayes. First, the probabilities of each class are calculated.

$$P(\text{class} = \text{overweight}) = 5/10$$
$$P(\text{class} = \text{normal}) = 5/10$$
$P(X \mid C_i)$ values for each class are calculated as follows.
$$P(\text{height} = 2 \mid \text{class} = \text{overweight}) = 1/5$$
$$P(\text{height} = 2 \mid \text{class} = \text{normal}) = 2/5$$

The absence of any data in the test dataset with weight = 3 and class = overweight does not necessarily mean that the probability $P(\text{weight} = 3 \mid \text{class} = \text{overweight}) = 0$. In such cases, Laplacian correction (or Laplace smoothing) can be used to take into account the probability $P(\text{weight} = 3 \mid \text{class} = \text{overweight})$. Laplacian correction parameter α is usually taken as 1.

$$P(\text{class} = \text{overweight}) = \frac{(\text{number of reviews weight} = 3 \text{and class} = \text{overweight}) + \alpha}{(\text{number of class} = \text{overweight}) + (\text{number of categories})\alpha}$$
$$= \frac{0+1}{5+5} = 1/10$$

$$P(\text{lass} = \text{normal}) = \frac{(\text{number of reviews weight} = 3 \text{and class} = \text{normal}) + \alpha}{(\text{number of class} = \text{normal}) + (\text{number of categories})\alpha}$$
$$= \frac{1+1}{5+5} = 2/10$$

Then the probabilities of the classes of $X = (\text{height} = 2 \text{and weight} = 3)$ are calculated as follows.

$$P(X \mid \text{class} = \text{overweight}) = (1/5)(1/10) = 1/50$$
$$P(X \mid \text{class} = \text{normal}) = (2/5)(2/10) = 4/50$$
$P(X \mid \text{class} = \text{normal}) > P(X \mid \text{class} = \text{overweight})$ therefore
$X = (\text{height} = 2 \text{ and weight} = 3)$ belongs to normal class.

1.4.1.2 Regression Algorithms

Regression is a statistical tool that tries to determine the strength of the relationship between a dependent variable and other independent variables and makes predictions according to this. Regression is used to determine whether data fit an equation. It is useful to make predictions about the future by using indicators of past behavior. Regression algorithms such as linear regression, polynomial regression, exponential regression, logistic regression, and logarithmic regression can be preferred according to the data type.

Regression analysis encompasses a wide variety of statistical methods for estimating the relationship between input variables and their associated properties. The most

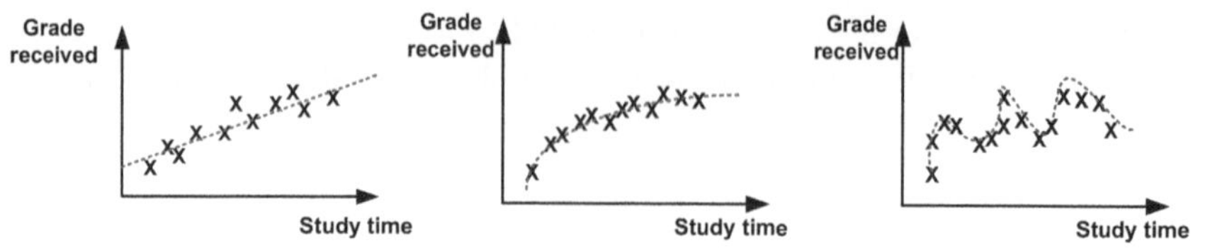

$f(x) = \beta_0 + \beta_1 x \ldots$ $f(x) = \beta_0 + \beta_1 x + \beta_2 x^2 \ldots$ $f(x) = \beta_0 + \beta_1 x + \beta_2 x^2 + \beta_3 x^3 \ldots$

FIGURE 1.16 Most common regression analysis examples.

common form is linear regression, in which a single line is drawn to best fit the given data based on a mathematical criterion such as ordinary least squares. The latter is extended by regularization (math) methods to reduce overfitting and bias, as in ridge regression. When dealing with nonlinear problems, go-to models include polynomial regression, logistic regression (often used in statistical classification), or even nonlinearity kernel ridge regression. The most common regression analysis representative graphs are seen in Figure 1.16.

1.4.1.2.1 Linear Regression Algorithm

Linear regression is essentially a linear approach to determining the relationship between two variables. One of these values is a dependent value, and the other is independent. It helps understand how a change in one variable affects the other and the relationship that causes the result to be positive or negative. It is applied where the predicted output is continuous and has a constant slope: forecasting sales, risk assessment, weather data analysis, predictive analytics, customer survey results analysis, optimizing product prices, etc.

Before attempting to fit a linear model to the observed dataset, it should be evaluated whether there is a relationship between the variables. Of course, this does not mean that one variable is the cause of the other. However, there must be a visible correlation between them. For example, the impression that a student will get higher grades the longer they study may not always be true. However, it is possible that there is a relationship between the duration of study and the grade received.

Correlation coefficients are used to calculate how strong the relationship between two variables is. It is usually denoted by "r" and has a value between −1 and 1. A positive correlation coefficient value indicates a positive relationship between variables. Similarly, a negative value indicates a negative relationship between the variables.

Linear regression comes in two varieties. Multiple linear regression and basic linear regression are these. Finding the link between a single independent variable and a matching dependent variable is the goal of the basic linear regression approach. The input is the independent variable, while the output is the matching dependent variable. Finding the association between two or more independent variables and their corresponding dependent variable is the goal of the multiple linear regression approach. Polynomial regression is another unique technique for multiple linear regression.

The linear regression algorithm performs well on linearly separable datasets and can be preferred because of its simplicity of implementation. However, the absence of a linear relationship between the variables and the state of the correlation between the variables can significantly affect the performance of the algorithm. It is best to make sure that there is a moderately strong correlation between the variables by looking at the trend.

The linear regression function is expressed as $h_\theta(x) = \beta_0 + \beta_1 x$. In this formula, $h(x)$ is the hypothesis function, that is, the output value of the function in response to the given input. Estimates made using this function are expected to be as close to the actual values as possible. We can simply calculate the difference with the cost function in Eq. (1.46).

$$J(\beta_1) = \frac{1}{2m} \sum_{i=0}^{m} \left(h\left(x^{(i)}\right) - y^{(i)} \right)^2 \quad (1.46)$$

In the function represented by this formula, m represents the number of sample pairs (x, y) used for training. The coefficient of 1/2 seen in the formula will be useful later in derivatives. An optimization-search algorithm called gradient descent is generally used to find the minimum cost. This algorithm is based on updating the parameters (β) by taking the partial derivative of the cost function. Thus, the cost function is minimized, and there are β parameters for the line that best represents the samples. The process is repeated until the smallest β value is reached in the algorithm.

1.4.1.2.2 Logistic Regression Algorithm

A logistic regression problem is a regression problem where the dependent variable is a categorical variable. Logistic regression is a statistical method used to analyze a dataset with one or more independent variables that determine an outcome. It is widely used in linear classification problems. Although it is called regression, there is a classification here. The outcome is measured with a binary variable (there are only two possible outcomes).

The logistic regression algorithm is used in binary classification problems where it results in pass or fail, true or false, etc. It is the most suitable algorithm for the need to estimate the probabilities of the dependent variable falling into one of the two categories of the response. The logistic regression algorithm can also be used in applications such as credit scoring, cancer detection, geographical image processing, handwriting recognition, image segmentation and categorization, measuring success rates of marketing campaigns, and earthquake prediction.

Binary logistic regression model is a logistic regression analysis with dependent variables containing a binary response. The odds ratio in Eq. (1.47) is used in logistic regression. The odds ratio is defined as the ratio of the probability of happening to the probability of not happening. Odds is the "P" ratio of the probability of success or occurrence, and the ratio of "$1-P$" to the probability of failure or non-occurrence. Odds value takes values between $(0, +\infty)$.

$$\text{Odds Ratio} = \frac{P}{1-P} \quad (1.47)$$

The odds ratio is a summary measure of the relationship between two variables. Variables close to odds ratio 1 do not have a significant effect on the change in y. If the coefficients of these variables are not significant, they are interpreted as "the variable is not an important risk factor." Odds ratio values greater than 1 (provided that the coefficient is significant) are interpreted as an important risk factor. So the risk increases. If the odds ratio is less than 1, it is said that the risk is decreasing. Values close to zero indicate that the factor is an important risk factor, provided that the coefficient is significant, but it is a negatively effective factor that causes y to take low values (Chen et al., 2010).

The logit function is the natural logarithm of the odds ratio. The odds value of a probability takes values between 0 and $+\infty$, while the logit value of the same probability takes values between $-\infty$ and $+\infty$. The logit function is given in Eq. (1.48), where P is the probability of the observed event being observed. Accordingly, the formula Given in Eq. (1.49) is obtained.

$$\text{Logit}(P) = \ln \ln\left(\frac{P}{1-P}\right) = y = \beta_0 + \beta_1 x \tag{1.48}$$

$$\left(\frac{P}{1-P}\right) = e^y \rightarrow P = \frac{e^y}{1+e^y} \rightarrow P(y) = \frac{1}{1+e^{-y}} = \frac{1}{1+e^{-(\beta_0+\beta_1 x)}} \tag{1.49}$$

The coefficients (β) of the logistic regression algorithm are estimated from the training data. This is done using the maximum-likelihood estimation. Maximum probability estimation is a collaborative learning algorithm used by various ML algorithms to make assumptions about the distribution of data. The best coefficients result in a model that predicts a value very close to 1 for the default class and very close to 0 for the other class.

1.4.1.3 Deep Supervised Learning

ML consists of self-training systems that make determinations by making simulations with the presented data and parameters. In deep learning, systems that discover even the parameters that need to be defined in ML in many layers with a sea of big data and make evaluations with better parameters should be considered.

Zhang et al. (2018) stated that a unified and general definition for deep learning, which was introduced by Dechter and is a subject on which many researchers have been working on, has not yet been admitted. However, a definition statement can be created based on the starting point of deep learning. Namely, deep learning is a subfield of ML. It is also known in the academic world as deep structured learning, hierarchical learning, or deep ML (Ongsulee, 2017). It is a field of study that covers artificial neural networks and similar ML algorithms with one or more hidden layers. It is a system that performs the calculations used in ML in many layers, at once, discovers even the parameters that we need to define in ML, and maybe can make evaluations with better parameters. The relations between all these terms are illustrated in Figure 1.17.

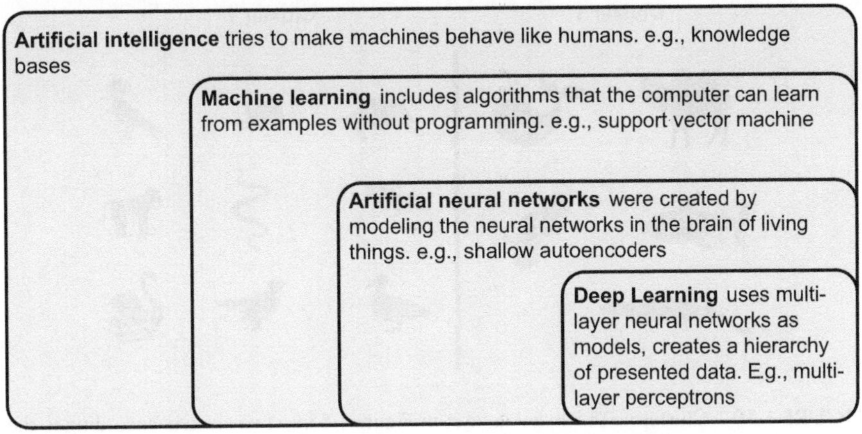

FIGURE 1.17 Relationship between artificial intelligence, machine learning, artificial neural networks, and deep learning (inspired by Goodfellow et al., 2016 and Janiesch et al., 2021).

1.4.1.4 Clustering Algorithms

Cluster analysis is a multivariate statistical method used to determine whether a dataset contains different groups and, if so, to detect these groups. It is a guiding research method for summarizing and describing data in a multidimensional space. It is a method that allows grouping observation structures in different groups that are heterogeneous or observations in similar groups that are homogeneous with appropriate methods.

Cluster analysis is a group of multivariate techniques whose main purpose is to group objects (units) based on their characteristics. Cluster analysis groups objects very similarly within a cluster, but differently between clusters. If the clustering process is successful, when a geometric drawing is made, the objects in the cluster will be very close to each other, and the clusters will be far from each other (Hair et al., 1998) as in Figure 1.18.

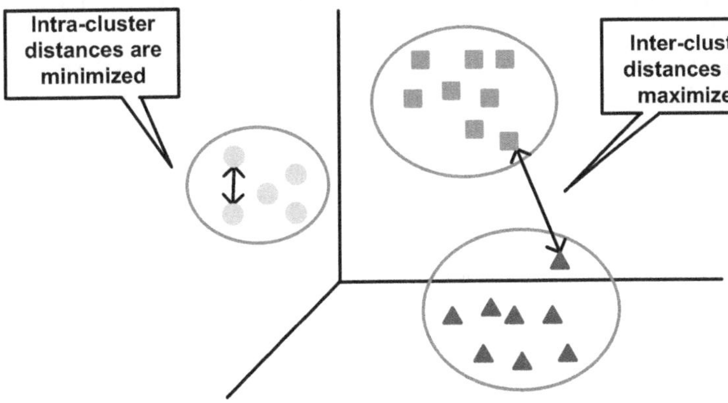

FIGURE 1.18 An example of clustering.

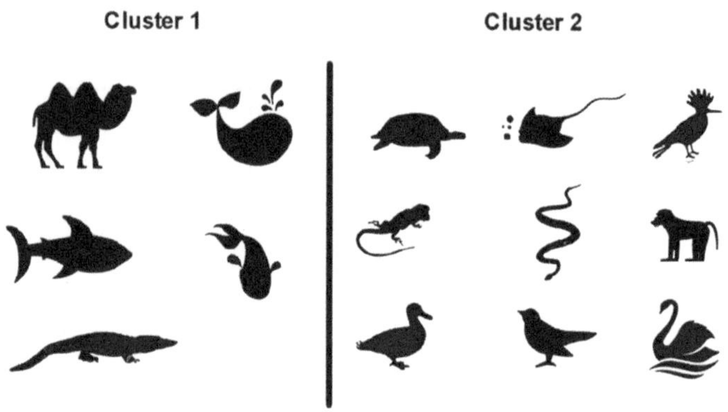

FIGURE 1.19 Clustering the visual dataset in Figure 1.5 into two clusters according to their estimated average weights.

Cluster analysis should not be confused with classification. Clustering analysis, one of the unsupervised ML techniques, does not work on a dataset with labeled data, unlike classification, which is one of the supervised ML techniques. In clustering, the number and characteristics of the groups are usually not clear. Groups are formed by looking at the characteristics of the objects or individuals to be clustered. In some cases, the number of groups can be predetermined if needed, but the characteristic is not clear. In classification, the number of groups and their characteristics are clear. It is added to the group by determining which group it belongs to according to the characteristics of the object or individual. For example, when we want to divide the animals in the visual dataset in Figure 1.13 into two clusters according to their estimated average weight, a result occurs as in Figure 1.19.

A well-known application of clustering algorithms is recommender systems. It uses clustering techniques for e-business giants recommender systems such as Netflix, Amazon, TripAdvisor, and Flipkart (Logesh et al., 2020). Clustering algorithms are also widely used in the financial industry, too. In an era of online transactions where cash usage has decreased significantly, it is necessary to determine whether transactions with cards are secure. Organizations can evaluate transactions as true or fraudulent, using historical data on customer behavior to detect fraud.

Finding parallels or differences between observed people or items is the primary goal of cluster analysis. The degree to which two items or attributes are related is referred to as similarity. Depending on the scale or type of data gathered, there are various methods for obtaining this quantitative value. Disparities, or the contrast or dissonance between two items, are measured by distance. Observations are classified based on their ability to be differentiated from one another using similarity and distance metrics.

We can express cluster analysis simply in three stages:

1. According to the data matrix, a similarity or distance matrix is obtained. The most preferred distance measurements are given as Eqs. (1.31) (Euclidean distance), (1.32) (Minkowski distance), 3 (Manhattan city block distance), and 4 (Mahalonobis distance) in Section 1.4.1.1.1.
2. Clustering is done according to the determined clustering algorithm.
3. Results are evaluated.

1.4.1.4.1 Hierarchical Clustering Algorithms

In hierarchical clustering algorithms, parameters such as the user requesting the K (number of clusters) parameter to be determined initially are not needed. The cluster hierarchy can be expressed using a standard tree structure. The root of the hierarchy represents all of the data objects to be clustered. At each level of the tree, nodes corresponding to the clusters are formed. Each level of the hierarchy corresponds to some set of clusters. The base of the hierarchy consists of the leaves of the tree, that is, single points. This cluster hierarchy is called a dendrogram. The main advantage of hierarchical clustering methods is that appropriate clusters can be obtained by ending the hierarchy at any level. This feature is the most important feature that distinguishes hierarchical clustering algorithms from divisive clustering methods. It is divided into two types as agglomerative and divisive.

1.4.1.4.1.1 Agglomerative Clustering Algorithms Agglomerative algorithms follow a bottom-up path, starting with single-point clusters and combining the most similar or closest clusters at each step.

Single linkage clustering algorithm: It is also known as the nearest neighbor technique and is based on the shortest distance principle. In this method, the closest (most similar) clusters to each other are combined considering the distance matrix. The steps start as n variables in the dataset are in n different clusters. The distance matrix and similarity matrix are calculated according to the desired distance measure of the data matrix. The highest valued units in the similarity matrix are hierarchically combined with each other, and the set ij is formed. Clusters i and j are deleted from the row related to the jth set in the similarity matrix, and the distance of the set ij remains the distance of the ith set. The operations are repeated until all clusters are combined with each other. For an illustrative example given in Figure 1.20, the clustering example is shown step by step by calculating the Euclidean distance values.

Complete linkage clustering algorithm: It is also known as the furthest neighbor technique. It is very similar to the single connection method, except that the largest distance between pairs of elements in each cluster is taken as the distance between two clusters.

Ward's Method: Ward's connection technique, also known as the minimum variance clustering method, creates clusters that will maximize homogeneity while minimizing the total within-cluster variance, instead of calculating the distances between clusters like other clustering methods. For this purpose, the error sum of squares formula is used.

Average linkage clustering algorithm: In this technique, the process is started as in single linkage and complete linkage techniques. However, the average distance between the unit in one cluster and the units in the other cluster is used as the clustering criterion. In the average linkage algorithm, the clusters are linked to each other with small variances.

Robust clustering using links (ROCK) algorithm: It is an agglomerative clustering technique based on connections between datasets. The number of links between two clusters is the number of common neighbors they have in the dataset. After calculating the links between the data, the algorithm starts with a single point for each cluster. Clusters continue to be linked until there are no linked clusters or the required number of clusters is obtained.

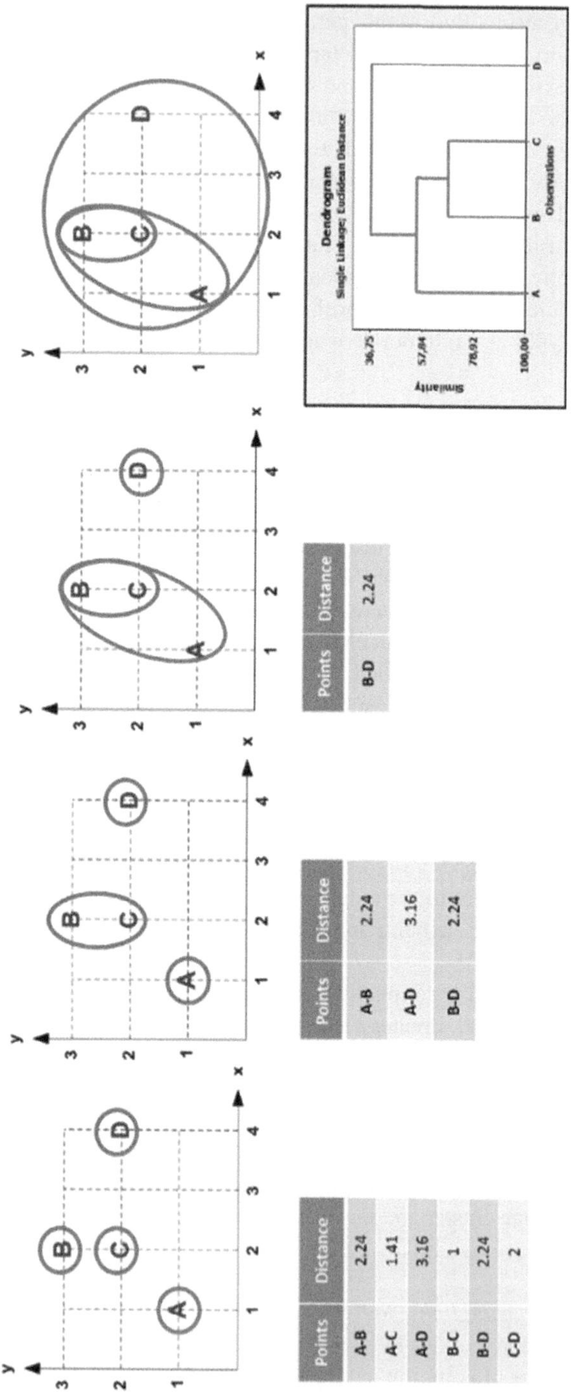

FIGURE 1.20 An example of single linkage clustering algorithm.

1.4.1.4.1.2 Divisive Clustering Algorithms Divisive algorithms follow a top-down path, starting with a set containing all points and continuing by dividing (separating) the set at each step. In this method, each step must decide which cluster to divide and how to divide it.

DIANA (divisive analysis): It was first presented by Kaufman and Rousseeuw (2009). Unlike agglomerative clustering algorithms, it has a top-down structure. So at first, all objects are considered as a single set. Then, the processes are continued until each object is a set in itself. This process continues until each object creates a cluster on its own or until a certain termination condition is met. The termination condition is to obtain the desired number of clusters or to ensure that the distance between the two closest clusters is above the given threshold value.

HCAD (hierarchical clustering based on attribute dependency) algorithm: It divises data objects based on variable dependencies. Variable dependencies are calculated with Spearman rank correlation. This measure is the strength of the relationship between the two variables. In this algorithm, the *D* matrix is created by calculating the correlations for the variable pairs in the variable list. Variables a_i and a_j with maximum dependency values in matrix D are found, and these variables are deleted from the list. The equivalence class is found. The same procedures are continued until all partitions are obtained (Anuradha and Tripathy, 2014).

1.4.1.4.2 Non-hierarchical Clustering Algorithms

In this section, non-hierarchical clustering algorithms are discussed in two parts as density-based and partitional clustering algorithms.

1.4.1.4.2.1 Density-Based Clustering Algorithms In density-based clustering techniques, a clustering process is performed according to the density of the data rather than distance-based cluster selection. Clusters are defined as areas of higher density in the database. In areas where cluster densities are sparse, there are either noisy data or data constituting the cluster boundary. It is quite suitable to use this algorithm for large databases and noisy data (Ester et al., 1996; Krigel et al., 2011).

DBSCAN (density-based spatial clustering of application with noise): DBSCAN is the best-known density-based clustering algorithm developed by directly exploiting the basic idea of such clustering algorithms. It is an algorithm that brings together points that are close to each other and marks the points found alone in low-density regions as outlier (Bhattacharjee and Mitra, 2021).

OPTICS (ordering points to identify the clustering structure): In order to solve the problem of detecting significant clusters in data of varying density, the points in the dataset are ordered. Thus, the closest points become neighbors in the ranking (Ankerst et al., 1999).

1.4.1.4.2.2 Partitional Clustering Algorithms Partitional clustering methods are a clustering technique that operates by dividing *n* objects into *k* clusters such that $k \leq n$. Each cluster must contain similar elements and have different objects from the objects outside the cluster. There must be at least one object in each of the *k* clusters thus obtained. In addition, every object in the data must be related to a cluster. The most commonly used methods for partitional clustering are *k*-means and *k*-medoids.

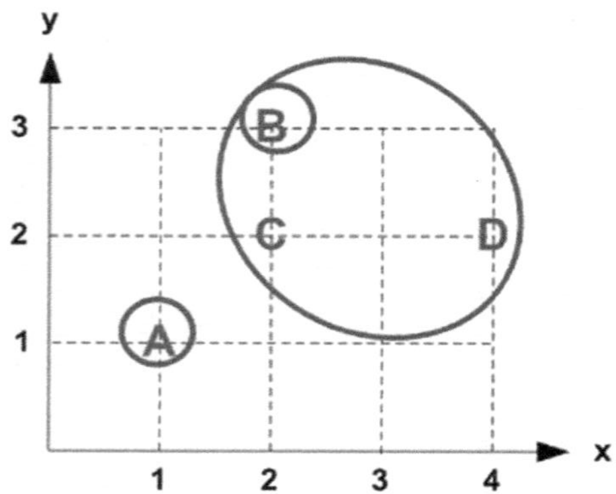

	Centroid 1	Centroid 2	Closest	New Centroid 1		New Centroid 2		Cluster 1	Cluster 2
	A	B		X	Y	X	Y		
Initial				1.00	1.00	2.00	3.00	A	B
C	1.41	1.00	Centroid 2	1.00	1.00	2.00	2.50	A	B, C
D	3.16	2.24	Centroid 2	**1.00**	**1.00**	**2.67**	**2.33**	A	B, C, D

	Centroid 1	Centroid 2	Closest	New Centroid 1		New Centroid 2		Cluster 1	Cluster 2
				X	Y	X	Y		
Initial				**1.00**	**1.00**	**2.67**	**2.33**		
A	0.00	2.13	Centroid 1	1.00	1.00	2.67	2.33	A	
B	2.24	0.94	Centroid 2	1.00	1.00	2.33	2.67	A	B
C	1.41	0.75	Centroid 2	1.00	1.00	2.22	2.44	A	B, C
D	3.16	1.37	Centroid 2	**1.00**	**1.00**	**2.67**	**2.33**	A	B ,C, D

FIGURE 1.21 An example of the *k*-means algorithm.

K-means clustering algorithm: In the algorithm proposed by MacQueen (1967), the number of clusters is initially determined, and any two points to be clustered are chosen as the centroid of the initial clusters. The distances of all the points to be clustered to the determined centroids are calculated. Each point is added to the cluster closest to the centroid. According to the points added to the cluster, the centroids to be considered in the next iteration are updated. Each iteration is compared with the previous one, and if there is no change in the clusters, the algorithm is stopped. As an example, the algorithm steps are shown for an illustrative example given in Figure 1.21. (k is determined as 2).

K-Means++ clustering algorithm: Vassilvitskii and Arthur (2006) developed this algorithm based on the idea that random assignment of the starting center points in the K-Means algorithm will worsen the result. After choosing a random starting point according to K-Means++, the distance between it and all other data is calculated. New starting points are selected by squaring this distance and making certain calculations.

K-medoids clustering algorithm: However, since the cluster centers are determined randomly at the beginning in the k-means algorithm, different clusters may occur when run again for the same input data. It is also highly affected by outliers. Therefore, the k-medoids algorithm uses the most central point in the cluster (medoid) instead of the data mean in determining the center. In this algorithm, centers are selected from among the points in the data, including the initial state (Velmurugan and Santhanam, 2010). Compared to k-means, it is a more reliable algorithm because it is less affected by outlier data, but it is not suitable for large datasets.

BIBLIOGRAPHY

Ankerst, M., Breunig, M. M., Kriegel, H. P., & Sander, J. (1999). OPTICS: Ordering points to identify the clustering structure. *ACM Sigmod record*, *28*(2), 49–60.

Anuradha, J., & Tripathy, B. K. (2014). Hierarchical clustering algorithm based on attribute dependency for attention deficit hyperactive disorder. *International Journal of Intelligent Systems and Applications*, *6*(6), 37.

Bhattacharjee, P., & Mitra, P. (2021). A survey of density based clustering algorithms. *Frontiers of Computer Science*, *15*(1), 1–27.

Breiman, L., Friedman, J., Olshen, R.A., & Stone, C.J. (1984). *Classification and Regression Trees* (1st ed.). Chapman and Hall/CRC. https://doi.org/10.1201/9781315139470

Chen, H., Cohen, P., & Chen, S. (2010). How big is a big odds ratio? Interpreting the magnitudes of odds ratios in epidemiological studies. *Communications in Statistics—simulation and Computation®*, *39*(4), 860–864.

Ester, M., Kriegel, H. P., Sander, J. Ve Xu, X. (1996). A Density-Based Algorithm for Discovering Cluster in Large Spatial Databases with Noise. 2nd International Conference on Knowledge Discovery and Data Mining, Portland, OR, 2–4 August.

Goodfellow, I., Bengio, Y., & Courville, A. (2016). *Deep learning*. MIT Press. Page: 9.

Grigorev, A. (2020). Machine Learning Bookcamp MEAP V06 (A. Books (ed.)). Copyright 2020 Manning Publications. Erişim adresi: https://www.manning.com/books/machinelearning-bookcamp

Hair, Jr. F.J., Anderson, E. R., Tatham, L. R., et al.: Multivariate Data Analysis With Readings, 5. Ed., Prentice-Hall, USA, 1998.

Hu, L. Y., Huang, M. W., Ke, S. W., & Tsai, C. F. (2016). The distance function effect on k-nearest neighbor classification for medical datasets. *SpringerPlus*, *5*(1), 1–9.

Janiesch, C., Zschech, P. & Heinrich, K. Machine learning and deep learning. *Electron Markets* 31, 685–695 (2021). https://doi.org/10.1007/s12525-021-00475-2

Jordan, M. I., & Mitchell, T. M. (2015). Machine learning: Trends, perspectives, and prospects. *Science*, 349(6245), 255–260. Erişim adresi: https://science.sciencemag.org/content/349/6245/255

Kaufman, L., & Rousseeuw, P. J. (2009). Finding groups in data: an introduction to cluster analysis. John Wiley & Sons.

Krigel, H. P., Kröger, P., Sander, J. Ve Zimek, A. (2011). Density-Based Clustering. Wire's Data Mining and Knowledge Discovery, 3, 231–240.

Larose, D. T., & Larose, C. D. (2014). *Discovering knowledge in data: an introduction to data mining* (Vol. 4). John Wiley & Sons. Page 168.

Logesh, R., Subramaniyaswamy, V., Malathi, D. et al. Enhancing recommendation stability of collaborative filtering recommender system through bio-inspired clustering ensemble method. *Neural Comput & Appl* 32, 2141–2164 (2020). https://doi.org/10.1007/s00521-018-3891-5

MacQueen, J. (1967, June). Some methods for classification and analysis of multivariate observations. In *Proceedings of the fifth Berkeley symposium on mathematical statistics and probability* (Vol. 1, No. 14, pp. 281–297).

Ongsulee, P., "Artificial intelligence, machine learning and deep learning," *2017 15th International Conference on ICT and Knowledge Engineering (ICT&KE)*, 2017, pp. 1–6, doi: 10.1109/ICTKE.2017.8259629

Ramazzotti, M., Degl Innocenti, D., Manao, G., & Ramponi, G. (2004). Entropy calculator: getting the best from your multiple protein alignments. *Italian Journal of Biochemistry*, *53*, 16–22.

Sathya, R., & Abraham, A. (2013). Comparison of supervised and unsupervised learning algorithms for pattern classification. *International Journal of Advanced Research in Artificial Intelligence*, *2*(2), 34–38.

Shamrat, F. J. M., Ranjan, R., Md, K., Hasib, A. Y., & Siddique, A. H. (2021). Performance Evaluation among ID3, C4. 5, and CART Decision Tree Algorithms. *Pervasive Computing and Social Networking: Proceedings of ICPCSN 2021*, 127. Page 137.

Vassilvitskii, S., & Arthur, D. (2006). K-means++: The advantages of careful seeding. In *Proceedings of the eighteenth annual ACM-SIAM symposium on Discrete algorithms* (pp. 1027–1035).

Velmurugan, T., & Santhanam, T. (2010). Computational complexity between K-means and K-medoids clustering algorithms for normal and uniform distributions of data points. *Journal of computer science*, *6*(3), 363.

Wiering, M. A., & Van Otterlo, M. (2012). Reinforcement learning. *Adaptation, Learning, and Optimization*, 12(3), 729.

Zhang, W. J., Yang, G., Lin, Y., Ji, C., & Gupta, M. M. (2018, June). On definition of deep learning. In *2018 World automation congress (WAC)* (pp. 1–5). IEEE.

1.5 MACHINE LEARNING RESEARCH

There are fundamental differences between ML applications and ML research. While we will talk more in detail in later sections that ML application is about engineering—on how to put ML into products and services that can better serve the companies, ML research is about the science, the backstage of the story. People in the research field

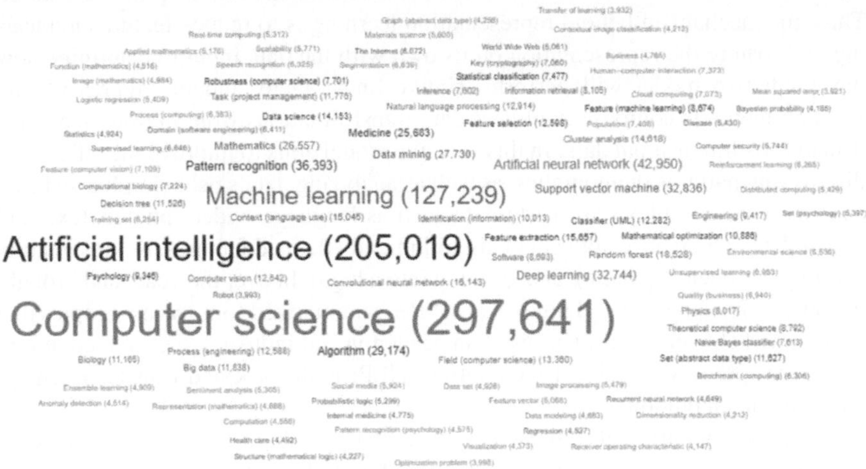

FIGURE 1.22 Insights2techinfo (2022): topics for machine learning research papers from 2010 to 2022 on the Dimensions database.

are normally called data scientists who hold a master's or a Ph.D. degree in Computer Science. They are the ones who build and, importantly, improve ML models for predictions by using algorithms and pre-existing data.

Thomas Dietterich (1997) summarized four directions for ML research that some are still applicable today. Dietterich in his paper focused on the improvement of classification accuracy, methods for scaling up supervised learning algorithms, reinforcement learning, and the learning of complex stochastic models. The topics and directions keep expanding over the past 2 decades, and ML has become more robust in different areas and industries. A. Gaurav (2022, Insights2techinfo) analyzed the topics and fields in ML research papers published between 2010 and 2022 from the Dimensions database. Among them, the most popular topics include AI, artificial neural networks, pattern recognition, feature selection, pharmaceuticals, and big data (Figure 1.22).

In the previous section, we have already discussed in-depth of algorithms for each model. Those algorithms are essential in developing and improving the accuracy and efficiency of ML methods and predictions. In the following section, we will dive into the models: Linear Factor Models (1.5.1), to explore more on the research aspect of ML.

1.5.1 Representation Learning

Representation learning, also known as feature learning, is a set of ML methods that use machines to find the representations and to learn the features or classification from raw data (Bengio et al., 2013). In order to improve the ML performance, feature engineering for data representation has been a huge effort due to its labor intensity and also limited ability to extract discriminative data from the dataset. Thus, the mechanism behind representation learning is to reduce feature engineering by learning the representation on its own with the data input to determine how this predictive model will perform. Unlike linear factor models, representation learning methods compose nonlinear data transformations from high-dimensional data to low-dimensional data. In this way, the model aims to improve the efficacy of discovering patterns or anomalies, as well as analyzing data's behavior. It also helps analyze more complex real-world data such as language, video, photos, text, and sensor data, which have not yet got algorithmically defined features. Representation learning has been expanding and growing into a big field over the years and brought into attention many areas. Bengio et al. (2013) summarized four learning topics that have remarkable empirical successes in academia and industry: speech recognition and signal processing; object recognition; NLP; multi-task and transfer learning, and domain adaptation.

As we just mentioned, representation learning is about how to improve accuracy and performance. There are three different learning models that can be categorized: supervised learning, unsupervised learning, and deep architecture methods.

Supervised learning is essentially an approach to training the system from labeled data (training data) and predicting the outcome for the unlabeled data (test data). In Section 1.4, we learned about the master ML algorithms for supervised learning, including classification, regression, and deep supervised learning.

Supervised learning has been working effectively in performing regression and classification tasks. There are many directions and industries that benefit from supervised learning, such as text categorization, spam detection, stock price indicators/predictions, face detection, weather forecasting, and predicting future prices based on the existing marketing prices (Chen et al., 2016; Crawford et al., 2015; Rodrigues et al., 2018).

However, the biggest drawback of this method is that it requires manual data labeling. It has potential issues and has room for more research and improvement. First of all, it can be too costly for a human data labeling service. According to Google Cloud AI Platform Data Labeling Service (Pricing | Data Labeling Service |, 2002), the price right now depends on the amount of data, data type, and objective. For example, the most priced service is for image classification, and it costs $870 per 1,000 units per human labeler. It would cost much more if you have more data units that needed to be labeled.

Second, data labeling is often considered a job that is error-prone, boring, and tedious, and data quality is challenging while crucial for a good model (Alonso, 2015). Alonso (2015) in his article outlined eight challenges and directions to improve label quality for supervised learning including different social aspects that would influence the quality, tasks that are designed for workers to easily understand, define and measure labeling errors as humans make errors, label complexity, quality assessments, construct the similar labeling scheme to the machine as to human, explore frameworks, and shared infrastructures among the organizations to reduce repeated errors by different groups (Figure 1.23).

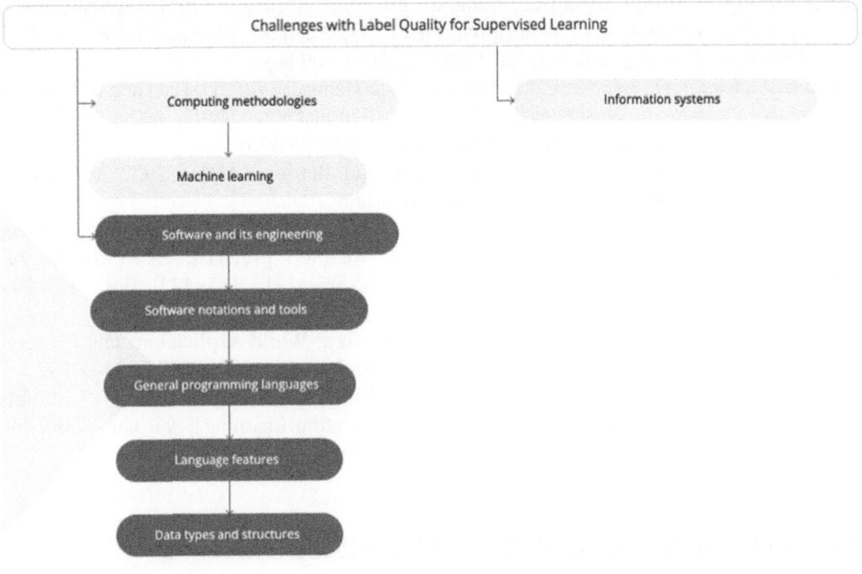

FIGURE 1.23 Challenges with label quality for supervised learning (borrowed from Alonso, 2015).

BIBLIOGRAPHY

Alonso, O. (2015). Challenges with Label Quality for Supervised Learning. *Journal of Data and Information Quality*, *6*(1), 1–3. https://doi.org/10.1145/2724721

Bengio, Y., Courville, A., & Vincent, P. (2013). Representation Learning: A Review and New Perspectives. *IEEE Transactions on Pattern Analysis and Machine Intelligence*, *35*(8), 1798–1828. https://doi.org/10.1109/tpami.2013.50

Bennett, K., and Parrado-Hernández, E. 2006. "The Interplay of Optimization and Machine Learning Research" *Journal of Machine Learning Research 7*, 1265–1281

Chen, D., Hua, G., Wen, F., Sun, J. (2016). Supervised Transformer Network for Efficient Face Detection. *Computer Vision – ECCV 2016*, 122–138. https://doi.org/10.48550/arXiv.1607.05477

Crawford, M., Khoshgoftaar, T. M., Prusa, J. D., Richter, A. N., & Al Najada, H. (2015). Survey of review spam detection using machine learning techniques. *Journal of Big Data*, *2*(1). https://doi.org/10.1186/s40537-015-0029-9

Dietterich, T.G. (1997). Machine-Learning Research. *AI Magazine*, *18*, 97–136. https://doi.org/10.1609/aimag.v18i4.1324

Fradkov, A. L. (2020). Early History of Machine Learning. *IFAC-PapersOnLine*, *53*(2), 1385–1390. https://doi.org/10.1016/j.ifacol.2020.12.1888

Gaurav. (2022, January 25). *Machine Learning Research Topics 2022*. Insights2Techinfo. Retrieved May 11, 2022, from https://insights2techinfo.com/machine-learning-research-topics-2022/

Kamiri, J., & Mariga, G. (2021). Research Methods in Machine Learning: A Content Analysis. *International Journal of Computer and Information Technology(2279–0764)*, *10*(2). https://doi.org/10.24203/ijcit.v10i2.79

Liu, X., Zhang, F., Hou, Z., Mian, L., Wang, Z., Zhang, J., & Tang, J. (2021). Self-supervised Learning: Generative or Contrastive. *IEEE Transactions on Knowledge and Data Engineering*, 1. https://doi.org/10.1109/tkde.2021.3090866

Paullada, A., Raji, I. D., Bender, E. M., Denton, E., & Hanna, A. (2021). Data and its (dis)contents: A survey of dataset development and use in machine learning research. *Patterns*, 2(11), 100336. https://doi.org/10.1016/j.patter.2021.100336

Pricing | Data Labeling Service |. (2002). Google Cloud. Retrieved May 5, 2022, from https://cloud.google.com/ai-platform/data-labeling/pricing

Rodrigues, E., Oliverira, I., Cunha, R., & Netto, M. (2018). DeepDownscale: A Deep Learning Strategy for High-Resolution Weather Forecast. *2018 IEEE 14th International Conference on e-Science (e-Science)*, 415–422, doi: 10.1109/eScience.2018.00130.

Sarker, I. H. (2021). Machine Learning: Algorithms, Real-World Applications and Research Directions. *SN Computer Science*, 2(3). https://doi.org/10.1007/s42979-021-00592-x

Turrisi da Costa, V., Fini, E., Nabi, M., Sebe, N., Ricci, E. (2022). "solo-learn: A Library of Self-supervised methods for Visual Representation Learning." *Journal of Machine Learning Research (23)* 1–6. https://doi.org/10.48550/arXiv.2108.01775

1.6 DEPLOYING MACHINE LEARNING ALGORITHMS INTO PRODUCTION

ML has gone beyond purely academic work, and with the growing interest of the public and governments, it continues to be increasingly used in standard business processes with an annual average growth of 25% (Cam et al., 2019; Davenport and Ronanki 2018; Hughes et al. 2017). Almost a decade ago, ML applications were only the focus of big technology companies, and it was almost impossible to observe any wide applications. However, thanks to technological advancements, ML solutions can be applied to many different fields. Day by day, industries and firms of various sizes apply ML to their problems. This subsection focuses on the deployment stage of the ML models into production.

The term "production" refers to making a product/service available for the use of its audience (Paleyes et al. 2022). In other words, production could be thought of as a process of placing the developed product on the shelves in the market to be delivered to the end customer. In order to clearly define all the stages in the ML workflow, we adapted Ashmore et al. (2021) definition. In their definition, ML deployment workflow includes four stages: data management, model learning, model verification, and model deployment, as illustrated in Figure 1.24.

The first stage, data management, consists of preparing data to get ready for building an ML model so that data can be input into the considered ML model. The second stage, model learning, includes the steps for model selection as well as model training. The third stage, model verification, focuses on validating whether the model achieves

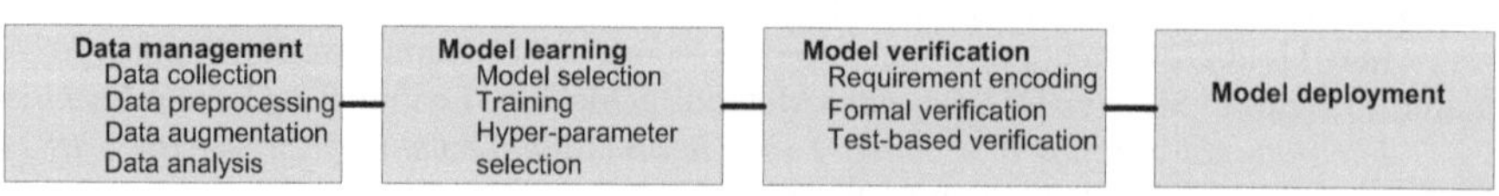

FIGURE 1.24 ML deployment workflow (adapted from Ashmore et al. (2021)).

the required performance characteristics. The last stage, model deployment, is integrating the trained ML model into any software environment that will run the model. It is important to emphasize that the apparent order of these stages is not necessarily the norm in real-life problems. Therefore, these stages can work in parallel and inform each other with feedback loops. Sections 1.6.1–1.6.4 further discuss these stages in a broader sense.

1.6.1 Data Management

The effectiveness of ML modeling depends on the quality of the train and test data, as well as the developed algorithm. Creating quality datasets is the first stage of the ML pipeline. Since data management is the first input of the overall pipeline, more successful ML models can be designed as the quality of the data increases. Data management includes data collection, preprocessing, augmentation, and analysis. Each step requires massive energy and time than expected early on. The data collection step focuses on understanding the structure of the data and organizing a clear procedure for the data acquisition. Following, the preprocessing steps include data cleaning and data dispersion to correct data and eliminate the noises that occur during data collection, such as different spellings/naming for the same entity or mislabeling the defects for a defect detection ML model. The third step, data augmentation, is needed to cover the hurdle due to the absence of labeled data.

ML model is certainly expected to perform better as the training duration increases and more trained models are built by larger training datasets. However, for real-world problems absence of data labels can be a big challenge. On the other hand, data labeling can be very tedious when a large volume of data is collected. Even though down sampling can reduce the data scale, significant time is still needed for labeling in some applications, such as network traffic analysis (Pacheco et al. 2019; Paleyes et al. 2022).

Another challenge in data labeling is reaching experts for ground truth. For instance, in the medical domain, analyzing images and labeling can only be performed by medical experts (Budd et al. 2021; Paleyes et al. 2022), or some minor defects in quality control can be detected by experienced operators (Errington and Childe 2013). In addition to difficulties due to access to experts, there is a reliability problem across experts. Usually, multiple experts are required for the same set of images to correctly label the ground truth. There might be a variation in labeling among experts, even if they are working on the same dataset. Therefore, it is hard to access a big dataset when the consistency among experts is considered (Valizadegan et al. 2013; Algan and Ulusoy 2021). Another critical issue in deploying the ML solution is the data imbalance problem, which refers to having a relatively higher number of instances for some classes than others (Xiao et al. 2012). In this case, the ML model will be biased to identify classes with a larger dataset. This situation can cause severe loss in the quality of the ML model. Taken together, data augmentation helps increase the number of datasets for the problematic, rarely observed classes. Finally, data should be analyzed to remove the biases and unexpected cases that impact data quality, such as missing values and inconsistencies across data types.

1.6.2 Model Learning

Model learning can be divided into three steps: model selection, training, and hyper-parameter selection. In the model selection step, the complexity of the model is the major factor that receives high attention. In real-world problems, practitioners tend to select simple models to avoid the complicated structure of complex models. Simple models can be easily applied to confirm the proposed ML solution. They reduce the time it takes to deploy a solution and enable developers to avoid intricate designs with reasonable quality solutions (see Mobile-Net (Howard et al. 2017)). In addition, simple models favor moderate hardware requirements (Verbraeken et al. 2020), which is one of the main factors to consider in reducing the costs of the ML solution. The ability to interpret the results obtained from the ML solution in accordance with the business area is another key point to be considered and may be more crucial than the model performance. In terms of ease of interpretation and understanding, decision trees have a wide range of applications ranging from banking to production (Paleyes et al. 2022). However, when the scale of data is increased, more complex models that depend on deep learning become prevalent.

The second step, training, can be simply described as feeding the data required for the ML model to learn. One of the most important factors to be considered is the economic cost of the training due to the powerful computing resources that the training will take place in (Liberty et al. 2020). Finally, the hyper-parameter selection is the process of deciding on the levels of the hyper-parameters. In order to accurately decide on the hyper-parameters, hyper-parameter optimization focuses on determining the optimal hyper-parameter setting. However, these algorithms add another dimensionality to the problem and require numerous training cycles for the ML model (Yang and Shami 2020).

1.6.3 Model Verification

First, the requirements of the ML model should be defined before testing the ML model. These requirements should be determined according to the needs of the target audience. The next step is the formal verification, ensuring that the ML model's functionality meets the requirements of the project under regulations. In this step, multiple tests are conducted to understand the behavior of the ML model so that requirement fulfillment is guaranteed. In the following step, test-based verification is performed to verify how well the model performs under the data that has not been previously processed. While the ideal situation would be to run these validation tests in a real-world setting, simulations are widely used for economic or safety reasons.

1.6.4 Model Deployment

While deploying ML models into production, utmost attention needs to be paid to two tasks: monitoring the system performance and maintaining its performance. For these tasks, ML and software engineering-related challenges could hurdle the deployment process. Under ML challenges, we often face data drift and concept drift problems. Data drift refers to the shift in the distribution of the input data, i.e., $P_t(x) \neq P_{t+t'}(x)$ where t' increase in time leads to a change in the distribution of input x (Duckworth et al. 2021). It occurs when the distribution of any feature changes in the testing data, which then causes differentiation in the distribution of the input for train and test data (Moreno-Torres et al. 2012). For instance, in the health sector, acute visits were reduced due to the vast increase in COVID-19 cases, and the prioritization of the operations procedures was changed (Duckworth et al. 2021). A similar shift was also observed in the shopping behaviors of the consumers toward cleaning products (Bennett et al. 2022). The other ML challenge is concept drift. Concept drift refers to the shift between the relationship between input x and output y. It could be presented as $P_{tr}(y|x) \neq P_{tst}(y|x)$ and $P_{tr}(x) = P_{tst}(x)$, where the distribution of the input x is still the same in periods tr and tst; however, the relationship (the mapping) between x and y changed (Moreno-Torres et al. 2012). For instance, assume that a house price prediction model considers the location and size of a house (m^2) as input variables to predict the price y. However, in the presence of a catastrophe such as an earthquake or inflation, the mapping between x and y can degrade.

In addition to ML challenges—data and concept drift—some software engineering problems must be considered. First, based on the nature of the prediction model, a real-time or batch mode compilation should be considered. In the case of defect detection in the factory setting, the prediction model should provide real-time, almost real-time, predictions so that the workers can take proper actions to repair the product. However, there could be cases where the prediction algorithm can be compiled in a batch mode for which an instant response is not needed. The ML model can also be compiled as a batch for those circumstances, such as demand prediction or forecasting. The second issue is to decide on the prediction service. Depending on the system's needs, cloud options run in the web browser as well as at the edge solutions, which process input directly at the source rather than depending on remote platforms. Edge solutions can also meet the latency, scalability, and data privacy issues (Merenda et al., 2020). Third, the computer resources (CPU, GPU, and memory) should be considered while deploying the ML solution. Computational resource capability limits the complexity of the developed ML solution. Fourth, latency and throughputs are other aspects of software engineering-related issues. Especially for real-time applications, both are critical performance metrics. Due to the transfer of calculations to other platforms, latency can be a problem for cloud-based solutions (Dastjerdi and Buyya 2016). Fifth, logging could facilitate troubleshooting and is useful when retraining is needed. Finally, security and privacy issues are critical points when providing ML solutions. The solution should be designed to provide an appropriate degree of security and privacy depending on the confidentiality of the data.

1.6.4.1 Deployment Patterns

The ML solution developed can be in any three forms: (i) It may be the launch of a completely new product, (ii) it may be developed to assist or completely automate an existing system, or (iii) it may be a more efficient ML solution developed against the previous ML implementation. In any of these three cases, rather than sending a vast amount of data and creating a huge traffic, two approaches can be applied:

(a) a smaller percentage of data can be passed to a developed solution, and then the percentage can slowly ramp up with monitoring, (b) rollback approach, switching the system to the former version when the algorithm struggles. Below presents three different deployment patterns considering approaches (a) and (b).

- Shadow mode deployment: In this deployment procedure, the model predictions are captured and stored for analyzing and tracking the model's performance, and they are not presented to any customer or used as an input for any other system. The objective of the shadow deployment is to observe how well the model performs compared to human judgment, and it could be thought of as a verification process of algorithm's performance (Schelter et al. 2018).
- Canary deployment: Canary deployment, seen in Figure 1.25, refers to passing a subset of the input to the developed version (V2), while the former version (V1) still serves. The traffic, the amount of data directed to the new version, can be gradually ramped up and monitored. In this deployment strategy, rather than a sudden shift from the former to a new version, a small fraction of the data is rolled out to the new version, and then this amount is increased gradually when the confidence in the performance is satisfactory (Karamitsos et al., 2020).
- Blue–green deployment: In the blue–green deployment strategy, see Figure 1.26, two different production systems, one for the older version (blue version) and the other for the new version (green version), work in parallel to avoid service unavailability. In a typical implementation, the

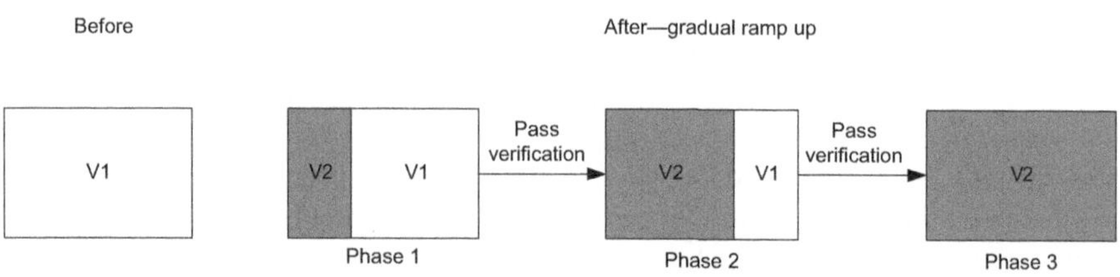

FIGURE 1.25 Canary deployment.

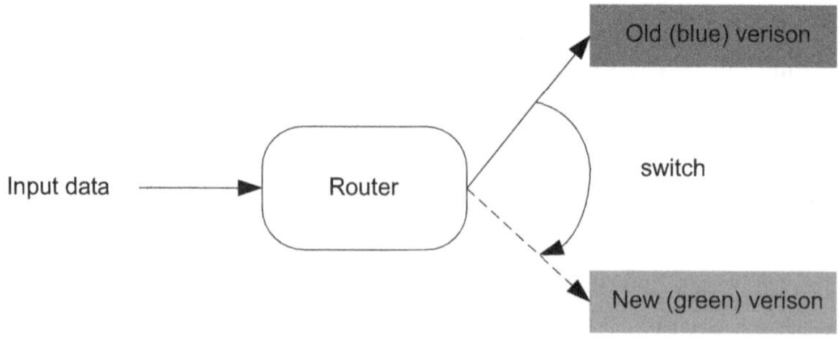

FIGURE 1.26 Blue–green deployment.

router sends all the requests to the older version, and once the new version passes the verification, the traffic is sent to the green version. In case the green version fails, the router switches the data to the blue version; thus, zero or almost zero failovers are observed. Similar to the canary deployment, a gradual ramp-up of the traffic to the green version could also be possible (Benton 2020).

1.6.5 Degree of Automation

In Figure 1.27, the stages of transition from no automation level, where the control of the system is completely dependent on humans, to the full automation level, where it is completely dependent on the ML model, are presented. It is usually suggested to increase the automation level of the system gradually. A shadow mode ML model could be used as a starting point to monitor the model's performance. After passing the shadow mode verification, ML model assistance to humans could be considered the second stage of transformation to automation. Then the system can be gradually moved to the partial automation and, finally, full automation stage (Ng, n.d.).

For example, for defect detection problems, we can think of a system in which only the labor(s) are in charge of the inspection as a starting point. Assume that an AI solution is developed to highlight the locations where defects occur to ease the inspection process for humans. Keeping the human as a decision-maker, in the next stage, an AI system can be operated in a shadow mode in the inspection process. In the following stage, the AI solution can be integrated into the decision-making process and operate in a small portion of the defect detection process. Its use percentage can be increased gradually by constantly examining the performance level. The following stage can be considered a semi-automated inspection process, where both human and AI systems collaborate in the inspection process. In the last stage, humans are not involved in any decision-making process, and the inspection process is in charge of an AI solution. Depending on the type of the problem, sometimes human-in-the-loop solutions are not feasible, such as online speech recognition or e-business. However, there are also practical applications where the best decision could be keeping both the human and the AI solution in the loop rather than full automation, such as quality control (Ng, n.d.).

1.6.6 Data Monitoring

The performance of the ML system can be tracked by using dashboards through some input and output metrics. Input metrics are needed to verify that the data is collected under the same conditions so that the data distribution can be kept the

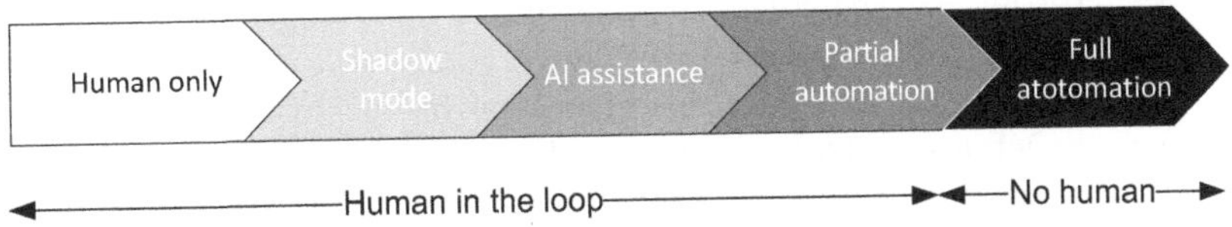

FIGURE 1.27 Degree of automation (adapted from Ng (n.d.)).

same as those used in the training stage. For a defect detection problem, the change in the illumination level might impact the performance of the developed solution. Thus, keeping track of lighting during the inspection could be beneficial. As another example, it is helpful to monitor the average input volume in speech recognition systems to examine if data is collected in the same standards that will not impact the ML solution performance for a speech recognition system. Output metrics are related to your model's performance and present your model's capability. For web search by voice, it could be beneficial to monitor how often the users proceeded with the speech recognition system first and then switched to typing for their queries.

In order to have high-performance models, both input and output metrics should be tracked. Depending on the application, there could be specific and commonly known software metrics such as memory, latency, throughput, or service load. In order to capture the specific cases that might cause a hurdle for your ML solution, the ML team should decide on some metrics to detect the problem. Once the metrics are decided, the next stage is to define the thresholds for those metrics that could trigger an alarm for maintenance.

1.6.7 Performance Evaluation

Different approaches might help define a baseline performance level for your ML algorithm, depending on the type of your data, whether structured or unstructured. Unstructured data refers to images, audio, and text data in which humans outperform ML solutions at interpreting. Therefore, it is always beneficial to have human performance for a specific set of applications that require unstructured data. However, human capability to make predictions for a given structured data, say, for example, house price prediction given the size of a house, is limited in comparison to the ML solution. In addition to human-level performance, the state-of-the-art deployed approaches can be used as a baseline solution to observe what the other studies achieved. As a third approach, a quick-and-dirty implementation of what has been published as an open-source can be an alternative approach to create a baseline (Ng, n.d.). Another option is to use a previously designed ML system as a baseline on which you can improve performance.

1.6.7.1 Error analysis

Error analysis aims to explain the gap between the current performance of the ML model and the best (perfect) performance. It is a way of finding why the algorithms do not perform well. At first, tagging your data in terms of noises or class-wise labels is helpful to find out for which cases the model needs to be improved. For instance, for defect detection algorithms, tagging information about special features of defects could be used to understand for which features there is room for performance improvement. This tag could also be related to the noises. For instance, the lighting condition or blurry images may cause a performance reduction. In order to reveal these problematic issues, the ML team brainstormed to decide what class-specific or environmental conditions, noise, might cause poor performance. Additionally, metadata could help find out the reason for the performance reduction. The information about how the data was collected may indicate any problem leading to lower performance.

TABLE 1.10
Error Analysis Example (Adapted from Ng (2011))

Example	Label	Prediction	Car Noise	People Noise	Low Bandwidth
1	"Stir fried lettuce recipe"	"Stir fry lettuce"	1		
2	"Sweetened coffee"	"Swedish coffee"		1	1
3	"Sail away song"	"Sell away some"		1	
4	"Let"s catch up!"	"Let"s ketchup"	1	1	1

TABLE 1.11
Prioritization of Errors (Adapted from Ng (n.d.))

Type	Accuracy	HLP	Gap	Percentage of Data	Performance Improvement
Clean data	93%	96%	3%	50%	1.50%
Car noise	85%	92%	7%	5%	0.35%
People noise	80%	90%	10%	35%	3.50%
Low bandwidth	83%	90%	7%	10%	0.70%

Table 1.10 exemplifies how the error analysis could be performed for a speech recognition model (Ng, n.d.). The labels in the transcript, model prediction results, and possible noise factors are tagged and audited while monitoring the model performance.

As a second step, a more in-depth analysis could be done to prioritize which error type to focus on. Therefore, (i) the percentage of errors having a particular tag, (ii) the percentage of the mislabeled data for a particular tag, and (iii) the gap between the current performance and the best possible performance could be the crucial points for increasing the model"s performance. Table 1.11 presents how the prioritization can be performed for a speech recognition case. Despite having 50% of the data clean, only 1.5% performance improvement will be achieved when the focus is on the "clean data." However, for "people noise" data, a 3.5% performance improvement in overall performance can be obtained even though its fraction is less than clean data (35%). Therefore, it is seen that the "people noise" data can be selected as the first group for performance development.

1.6.7.2 Confusion Matrix

The ML model predicts the results of a decision problem as a positive or a negative class. A positive class refers to the presence of any particular condition of interest, while a negative refers to the refusal or denial of the presence of that particular condition (Fawcett 2006). For instance, if an ML model focuses on detecting cars, the positive class is "car," while everything except a car, say a motorcycle, belongs to the negative class. Considering these two labels, the performance of the ML model can be evaluated through a confusion matrix, a.k.a.,

TABLE 1.12
Confusion Matrix

		Actual Class (Ground Truth)	
		Positive	Negative
Predicted class	Positive	True positive (TP)	False positive (FP)
	Negative	False negative (FN)	True negative (TN)

contingency table (Fawcett 2006; Davis and Goadrich 2006). A confusion matrix is very useful in evaluating the model"s performance when the data between classes is skewed in order to examine which class causes lower performance evaluation. It includes four elements: (i) True positives (TP) represent the correctly labeled positive examples, e.g., prediction of a "car" as a "car," (ii) false positives (FP) refer to incorrectly labeled examples for a particular condition, e.g., prediction of a "motorcycle" as "car," (iii) true negatives (TN) imply correctly labeled negatives, e.g., identification of a "motorcycle" as a "motorcycle," (iv) false negatives (FN) indicate incorrectly labeled positive classes as negative, e.g., identification of a "car" as "motorcycle." These four elements of a confusion matrix can be presented in Table 1.12. In Table 1.12, the actual class refers to the ground truth, in which the information about the condition is known to be true through observation or measurement.

1.6.7.3 Performance Metrics

The commonly used performance metrics to evaluate the performance of the ML model are presented below:

- Accuracy is the percentage of right predictions, and it is calculated as in Eq. (1.50).

$$\text{Accuracy} = \frac{\text{TP} + \text{TN}}{\text{TP} + \text{TN} + \text{FP} + \text{FN}} \tag{1.50}$$

- Sensitivity in Eq. (1.51) refers to the number of correctly labeled positive examples, a.k.a., true positive rate (TPR). Therefore, sensitivity shows the ability of the ML model to correctly determine positive instances.

$$\text{Sensitivity} = \frac{\text{TP}}{\text{TP} + \text{FN}} \tag{1.51}$$

- Specificity, a.k.a., true negative rate (TNR) is the proportion of true negatives as seen in Eq. (1.52). It shows the ML model's ability to correctly identify negative instances.

$$\text{Specificity} = \frac{\text{TN}}{\text{TN} + \text{FP}} \tag{1.52}$$

- Precision is the fraction of true positives to the total number of positives (true positive and false positive) as calculated in Eq. (1.53).

$$\text{Precision} = \frac{\text{TP}}{\text{FP} + \text{TP}} \tag{1.53}$$

- Recall in Eq. (1.54) is the proportion of correct positives to positive instances (TP + FN). TPR, sensitivity, and recall are the same metrics.

$$\text{Recall} = \frac{\text{TP}}{\text{TP} + \text{FN}} = \text{Sensitivity} \tag{1.54}$$

- F_1 score is a composite measure of recall and the precision value for the performance evaluation through calculating their harmonic mean, as in Eq. (1.55) (Sokolova et al., 2006).

$$F_1\,\text{score} = \frac{2 \cdot \text{precision} \cdot \text{recall}}{\text{precision} + \text{recall}} \tag{1.55}$$

- The receiver operating characteristic (ROC) curve depicts the chance of true positives against false positives, as shown in Figure 1.28 (Fawcett 2006). The performance of the ML model, the ability to distinguish between positive and negative classes, increases when the ROC curve is

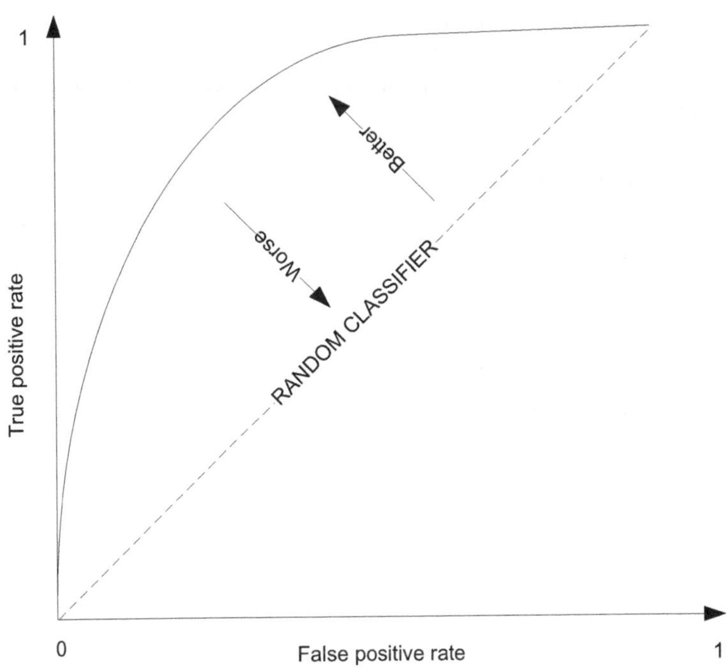

FIGURE 1.28 ROC curve.

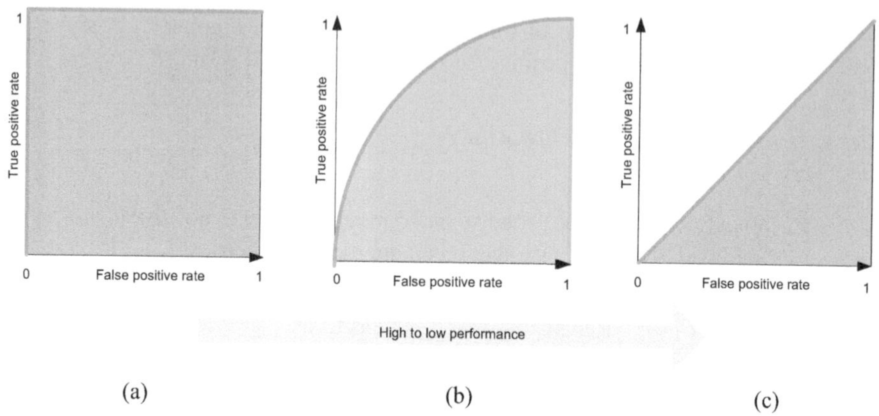

FIGURE 1.29 AUC curve.

in the upper left triangle of the random classifier, while it decreases as the ROC curve gets closer to the random classifier. The random classifier can be thought of as a flipping a coin process to decide whether the example belongs to the positive or negative class with a 50% chance.

- The area under the ROC curve (AUC) focuses on the area beneath the ROC curve to calculate the model's performance, as depicted in Figure 1.6a–c. Figure 1.29a shows the best prediction case in which the area under the curve is 1, AUC = 1. Figure 1.29b shows the case where the model has a higher chance of distinguishing positive classes against negative classes, $0.5 < AUC < 1$. Figure 1.29c shows the case where the model randomly predicts the positive and negative classes, with a 50% chance, AUC = 0.5.

In order to improve the performance of a learning algorithm, model-centric or data-centric approaches can be applied. In a model-centric approach, the data is fixed, and different models are tried to expand the model's performance. On the other hand, in the data-centric approach, the data quality is iteratively improved through error analysis or data augmentation. Both approaches should be considered when improving the ML model's performance (Ng, n.d.).

BIBLIOGRAPHY

Algan, Görkem, and Ulusoy, Ilkay. 2021. "Image Classification with Deep Learning in the Presence of Noisy Labels: A Survey." *Knowledge-Based Systems* 215: 106771. https://doi.org/10.1016/j.knosys.2021.106771

Ashmore, Rob, Calinescu, Radu, and Paterson, Colin. 2021. "Assuring the Machine Learning Lifecycle: Desiderata, Methods, and Challenges." *ACM Computing Surveys* 54 (5): 111:1–39. https://doi.org/10.1145/3453444

Bennett, Michele, Balusu, Jaya, Hayes, Karin, and Kleczyk, Ewa J. 2022. "The Silent Problem – Machine Learning Model Failure – How to Diagnose and Fix Ailing Machine Learning Models Michele Bennett, PhD, Jaya Balusu, Karin Hayes, Ewa J. Kleczyk, PhD April 22, 2022." *ArXiv Preprint ArXiv:2204.10227.*

Benton, William C. 2020. "Machine Learning Systems and Intelligent Applications." *IEEE Software* 37 (4): 43–49. https://doi.org/10.1109/MS.2020.2985224

Budd, Samuel, Robinson, Emma C., and Kainz, Bernhard. 2021. "A Survey on Active Learning and Human-in-the-Loop Deep Learning for Medical Image Analysis." *Medical Image Analysis* 71: 102062. https://doi.org/10.1016/j.media.2021.102062

Cam, Arif, Chui, Michael, and Hall, Bryce. 2019. "Global AI Survey: AI Proves Its Worth, but Few Scale Impact." *McKinsey*, no. November: 1–16.

Dastjerdi, Amir Vahid, and Buyya, Rajkumar. 2016. "Fog Computing: Helping the Internet of Things Realize Its Potential." *Computer* 49 (8): 112–116. https://doi.org/10.1109/MC.2016.245

Davenport, Thomas H., and Ronanki, Rajeev. 2018. "Artificial Intelligence for the Real World." *Harvard Business Review* 96 (1): 108–116.

Davis, Jesse, and Goadrich, Mark. 2006. "The Relationship between PR and ROC Curves." *ACM International Conference Proceeding Series*. Vol. #1551. http://portal.acm.org/citation.cfm?doid=1143844.1143874

Duckworth, Christopher, Chmiel, Francis P., Burns, Dan K., Zlatev, Zlatko D., White, Neil M., Daniels, Thomas W.V., Kiuber, Michael, and Boniface, Michael J. 2021. "Using Explainable Machine Learning to Characterise Data Drift and Detect Emergent Health Risks for Emergency Department Admissions during COVID-19." *Scientific Reports* 11 (1): 1–10. https://doi.org/10.1038/s41598-021-02481-y

Errington, Mark, and Childe, Stephen J. 2013. "A Business Process Model of Inspection in Remanufacturing." *Journal of Remanufacturing* 3 (7): 1–22. https://doi.org/10.1186/2210-4690-3-7

Fawcett, Tom. 2006. "An Introduction to ROC Analysis." *Pattern Recognition Letters* 27 (8): 861–874. https://doi.org/10.1016/j.patrec.2005.10.010

Howard, Andrew G, Zhu, Menglong, Chen, Bo, Kalenichenko, Dmitry, Wang, Weijun, Weyand, Tobias, Andreetto, Marco, and Adam, Hartwig. 2017. "MobileNets: Efficient Convolutional Neural Networks for Mobile Vision Applications." http://arxiv.org/abs/1704.04861

Hughes, G.R., Hall, Wendy, Pesenti, Jerome, and Ward, Philip. 2017. *Machine Learning: The Power and Promise of Computers That Learn by Example*. Royal Society. https://royalsociety.org/~/media/policy/projects/machine-learning/publications/machine-learning-report.pdf%0Ahttps://royalsociety.org/~/media/policy/projects/machine-learning/publications/machine-learning-report.pdf%0Ahttps://www.privacyinternational.org/

Karamitsos, Ioannis, Albarhami, Saeed, and Apostolopoulos, Charalampos. 2020. "Applying Develops Practices of Continuous Automation for Machine Learning." *Information* 11 (7): 1–15. https://doi.org/10.3390/info11070363

Liberty, Edo, Zohar Karnin, Xiang, Bing, Rouesnel, Laurence, Coskun, Baris, Nallapati, Ramesh, Delgado, Julio, et al. 2020. "Elastic Machine Learning Algorithms in Amazon SageMaker." In *SIGMOD'20*.

Merenda, Massimo, Porcaro, Carlo, and Iero, Demetrio. 2020. "Edge Machine Learning for Ai-Enabled Iot Devices: A Review." *Sensors* 20 (9): 1–34. https://doi.org/10.3390/s20092533

Moreno-Torres, Jose G., Raeder, Troy, Alaiz-Rodríguez, Rocío, Chawla, Nitesh V., and Herrera, Francisco. 2012. "A Unifying View on Dataset Shift in Classification." *Pattern Recognition* 45 (1): 521–530. https://doi.org/10.1016/j.patcog.2011.06.019

Ng, Andrew. 2011. "Advice for Applying Machine Learning." *Machine Learning*. 2011. https://see.stanford.edu/materials/aimlcs229/ml-advice.pdf

Ng, Andrew. n.d. "Introduction to Machine Learning in Production." https://www.coursera.org/learn/introduction-to-machine-learning-in-production/

Pacheco, Fannia, Exposito, Ernesto, Gineste, Mathieu, Baudoin, Cedric, and Aguilar, Jose. 2019. "Towards the Deployment of Machine Learning Solutions in Network Traffic Classification: A Systematic Survey." *IEEE Communications Surveys and Tutorials* 21 (2): 1988–2014. https://doi.org/10.1109/COMST.2018.2883147

Paleyes, Andrei, Urma, Raoul-Gabriel, and Lawrence, Neil D. 2022. "Challenges in Deploying Machine Learning: A Survey of Case Studies." *ACM Computing Surveys* 1 (1): 1–29. https://doi.org/10.1145/3533378

Schelter, Sebastian, Biessmann, Felix, Januschowski, Tim, Salinas, David, Seufert, Stephan, and Szarvas, Gyuri. 2018. "On Challenges in Machine Learning Model Management." *Bulletin of the IEEE Computer Society Technical Committee on Data Engineering*, 1–11. http://sites.computer.org/debull/A18dec/p5.pdf

Sokolova, Marina, Japkowicz, Nathalie, and Szpakowicz, Stan. 2006. "Beyond Accuracy, F-Score and ROC: A Family of Discriminant Measures for Performance Evaluation." In *AI 2006:Advances in Artificial Intelligence*, 1015–1021. https://link.springer.com/content/pdf/10.1007/11941439.pdf

Valizadegan, Hamed, Nguyen, Quang, and Hauskrecht, Milos. 2013. "Learning Classification Models from Multiple Experts." *Journal of Biomedical Informatics* 46: 1125–1135. https://doi.org/10.1016/j.jbi.2013.08.007

Verbraeken, Joost, Wolting, Matthijs, Katzy, Jonathan, Kloppenburg, Jeroen, Verbelen, Tim, and Rellermeyer, Jan S. 2020. "A Survey on Distributed Machine Learning" 53 (2): 30:1–30:33.

Xiao, Jin, Xie, Ling, He, Changzheng, and Jiang, Xiaoyi. 2012. "Dynamic Classifier Ensemble Model for Customer Classification with Imbalanced Class Distribution." *Expert Systems with Applications* 39 (3): 3668–3675. https://doi.org/10.1016/j.eswa.2011.09.059

Yang, Li, and Shami, Abdallah. 2020. "On Hyperparameter Optimization of Machine Learning Algorithms: Theory and Practice." *Neurocomputing* 415: 295–316. https://doi.org/10.1016/j.neucom.2020.07.061

2 AI Applications in Reservoir Characterization, Geology, and Geophysics

Cenk Temizel, Salih Tutun, Celal Hakan Canbaz, Ekrem Alagoz, Emre Can Dündar, Mufrettin Murat Sari, Luigi Saputelli, and Mehmet Melih Oskay

2.1 INTRODUCTION

The advent of artificial intelligence (AI) and machine learning (ML) has revolutionized the field of reservoir characterization, geology, and geophysics, offering innovative solutions to some of the most complex challenges in subsurface exploration and development. As the oil and gas industry increasingly leverages data-driven techniques, AI and ML provide powerful tools to analyze vast amounts of data, identify patterns, and make accurate predictions that were previously unattainable with traditional methods.

Reservoir characterization involves the integration of geological, geophysical, and petrophysical data to understand the subsurface and predict the behavior of hydrocarbons within a reservoir. Traditionally, this process has relied heavily on deterministic models and human interpretation, which can be time-consuming and subject to bias. AI and ML, however, bring the capability to automate and enhance these interpretations by learning from historical data and applying statistical models to predict reservoir properties, optimize well placement, and improve recovery strategies. In geology, AI and ML techniques are being used to classify rock types, predict lithology, and model complex geological structures. These technologies assist geologists in interpreting seismic data, correlating well logs, and generating geological models that are more accurate and reliable. The ability to process and analyze large datasets rapidly allows for more informed decision-making, reducing uncertainty in exploration and development. Geophysics, particularly in the area of seismic interpretation, has also seen

DOI: 10.1201/9781003617327-2

significant advancements through AI and ML. These methods enable the automatic detection of seismic features, enhance imaging techniques, and improve the resolution of subsurface models. The application of deep learning (DL) in seismic data processing is transforming the way geophysicists approach tasks such as fault detection, horizon picking, and velocity modeling. This chapter delves into the various workflows utilized in reservoir characterization, geology, and geophysics, highlighting the role of AI and ML in each domain. The subsequent sections will categorize these workflows, exploring their applications and the methodologies that underpin them. The integration of AI and ML in these fields not only enhances the accuracy and efficiency of traditional techniques but also paves the way for new approaches that continue to push the boundaries of what is possible in subsurface exploration.

2.2 CATEGORIZATION OF WORKFLOWS USED IN RESERVOIR CHARACTERIZATION, GEOLOGY, AND GEOPHYSICS

2.2.1 Petrophysics

Petrophysics, the study of the physical and chemical properties of rocks and their interactions with fluids, plays a critical role in reservoir characterization. The application of AI and ML in petrophysics has significantly improved the accuracy and efficiency of interpreting well log data, core analysis, and other subsurface measurements. Traditionally, petrophysical analysis relied on manual interpretation and deterministic models, which, while effective, often involve significant time investment and are prone to human error.

AI and ML methodologies offer a data-driven approach to petrophysical analysis, enabling the extraction of more nuanced insights from well logs, core data, and other petrophysical measurements. ML algorithms can be trained to recognize patterns in large datasets, allowing for the automatic identification of lithologies, estimation of petrophysical properties such as porosity and permeability, and prediction of fluid saturation.

One of the key applications of ML in petrophysics is the automatic classification of rock types and lithofacies. By training models on historical data, these algorithms can quickly classify new data, reducing the need for time-consuming manual interpretation. Moreover, AI can be used to integrate data from multiple sources, such as well logs and core samples, providing a more comprehensive understanding of the reservoir's properties.

In addition to classification, AI and ML are also used to predict petrophysical properties from indirect measurements. For example, models can be trained to estimate permeability and porosity from well logs, even in the absence of core data. This capability is particularly valuable in scenarios where core data is limited or unavailable. Furthermore, advanced ML techniques, such as DL, can model complex nonlinear relationships between different petrophysical properties, leading to more accurate predictions.

Another area where AI and ML are making an impact is in the interpretation of formation evaluation logs. These logs, which include data such as gamma ray (GR), resistivity, and neutron-density measurements, provide crucial information about the reservoir. AI-driven approaches can enhance the interpretation of these logs by identifying subtle features that may be overlooked by traditional methods. This leads to more accurate estimates of key reservoir parameters, such as water saturation and hydrocarbon content.

The integration of AI and ML in petrophysics also facilitates real-time data analysis, enabling more dynamic and responsive decision-making during drilling and production operations. By continuously updating models with new data, operators can make informed decisions about well placement, completion strategies, and production optimization, ultimately improving the overall efficiency and success of reservoir development.

In summary, the application of AI and ML in petrophysics is transforming the way subsurface data is analyzed and interpreted. By automating traditional workflows, enhancing the accuracy of predictions, and enabling real-time decision-making, these technologies are helping to unlock the full potential of reservoirs, driving more efficient and effective resource extraction.

2.2.2 Geology

Geology, the study of the Earth's physical structure and substance, is foundational to understanding subsurface formations and their potential for hydrocarbon reservoirs. The integration of AI and ML into geological workflows has opened new avenues for analyzing and interpreting geological data, leading to more precise and comprehensive geological models. These technologies have proven invaluable in tasks such as lithology prediction, facies classification, and the interpretation of complex geological structures, which are essential for accurate reservoir characterization.

One of the primary applications of AI and ML in geology is the automated classification of lithology. Traditionally, geologists relied on visual inspection of core samples, cuttings, and well logs to determine rock types and their distribution. This process is not only time-consuming but also subject to human bias. AI-driven models, on the other hand, can be trained on extensive datasets to recognize patterns in well log data and other geological inputs, allowing for the rapid and consistent classification of lithology across different wells and fields.

Facies classification, another critical task in geology, involves categorizing rock units based on their depositional environment and properties. ML algorithms can analyze large datasets to identify facies with greater accuracy and at a faster pace than traditional methods. By incorporating various data sources, such as well logs, seismic data, and core analysis, these models can generate detailed facies maps that are crucial for understanding reservoir heterogeneity and predicting fluid flow within the reservoir.

Geological structure interpretation, particularly in complex environments, is another area where AI and ML are making significant contributions. Seismic data, which provides a view of the subsurface, often contains noise and ambiguities that can obscure critical features like faults, folds, and stratigraphic boundaries. AI algorithms, especially those based on DL, can enhance seismic interpretation by filtering out noise and highlighting subtle geological features that might be missed by conventional methods. This leads to more accurate structural models, which are essential for identifying potential hydrocarbon traps and guiding drilling operations.

Moreover, AI and ML are increasingly being used to integrate geological data with other types of subsurface information, such as petrophysical and geophysical data. This integrated approach allows for a more holistic understanding of the subsurface, where geological models are continuously refined as new data becomes available. This dynamic modeling capability is particularly beneficial in the exploration phase, where understanding the geological context is crucial for reducing exploration risk and optimizing drilling strategies.

In addition to improving the accuracy and efficiency of traditional geological workflows, AI and ML also enable the development of new methodologies that were previously not possible. For instance, predictive modeling techniques can forecast geological features in unexplored areas by learning from known data. This capability is invaluable for frontier exploration, where data is often sparse and geological uncertainty is high.

In summary, the application of AI and ML in geology is transforming the way geoscientists analyze and interpret the Earth's subsurface. By automating routine tasks, enhancing the resolution of geological models, and enabling the integration of diverse datasets, these technologies are helping to uncover new insights into the Earth's structure and optimize the exploration and development of hydrocarbon reservoirs.

2.2.3 Seismic

Seismic interpretation is one of the most critical aspects of reservoir characterization, providing a detailed view of subsurface structures and stratigraphy. The integration of AI and ML into seismic workflows has transformed the way seismic data is processed, interpreted, and utilized, enabling more precise imaging of the subsurface and more accurate predictions of reservoir properties.

Seismic data acquisition generates vast amounts of data, typically in the form of seismic traces that represent the reflection of seismic waves off different subsurface layers. Traditional seismic interpretation involves manual picking of seismic horizons and faults, a time-consuming process that requires significant expertise. AI and ML technologies have revolutionized this process by enabling automated interpretation, reducing the time required to analyze seismic data, and increasing the consistency and accuracy of the results.

One of the primary applications of AI in seismic interpretation is horizon picking, where ML algorithms are trained to identify continuous seismic reflections that correspond to geological boundaries. These algorithms can automatically track horizons across large three-dimensional (3D) seismic datasets, even in areas where the signal is weak or noisy. This not only accelerates the interpretation process but also reduces the risk of missing subtle but important geological features.

Fault detection is another area where AI and ML have made significant strides. Faults are critical for understanding the structural integrity of reservoirs and predicting the movement of fluids within the subsurface. Traditional fault interpretation requires careful analysis of seismic data to identify discontinuities, which can be challenging in complex geological settings. AI-driven models, particularly those based on DL, can automatically detect faults by recognizing patterns in seismic attributes, such as coherence or curvature. These models are capable of identifying faults with greater accuracy and at a finer resolution than manual methods, providing more reliable inputs for reservoir modeling and simulation.

AI and ML are also being applied to enhance seismic data quality through noise reduction and signal enhancement. Seismic data often contains various types of noise, which can obscure important geological features. AI algorithms can be trained to distinguish between signal and noise, effectively filtering out unwanted data while preserving the integrity of the seismic reflections. This leads to clearer seismic images, which are essential for accurate interpretation and decision-making.

Another advanced application of AI in seismic workflows is the prediction of reservoir properties directly from seismic data. Traditionally, seismic inversion techniques have been used to derive rock properties such as porosity, density, and fluid saturation from seismic data. AI models, particularly those using DL, can enhance this process by learning complex, nonlinear relationships between seismic attributes and reservoir properties. These models can predict reservoir characteristics with a higher degree of accuracy, even in areas with sparse well control.

Seismic attribute analysis, which involves the extraction of specific features from seismic data to aid interpretation, has also benefited from AI and ML techniques. ML algorithms can analyze multiple seismic attributes simultaneously to identify patterns and correlations that may not be apparent through traditional analysis. This multi-attribute analysis can provide deeper insights into the subsurface, such as identifying subtle stratigraphic features or detecting direct hydrocarbon indicators.

In summary, the application of AI and ML in seismic interpretation has significantly enhanced the ability to visualize and understand the subsurface. By automating routine tasks, improving data quality, and enabling more accurate predictions of reservoir properties, these technologies are helping to unlock the full potential of seismic data in reservoir characterization and development. The integration of AI and ML into seismic workflows not only improves the efficiency and accuracy of traditional seismic interpretation but also opens up new possibilities for innovative approaches to subsurface exploration and modeling.

2.2.4 Lithology-Related Works

Lithology, the study and classification of rock types, is a fundamental aspect of reservoir characterization. Accurate lithological interpretation is crucial for understanding the distribution of reservoir rocks, their quality, and their potential to store and transmit hydrocarbons. AI and ML technologies have made significant contributions to lithology-related workflows by enhancing the ability to classify rock types, predict lithological variations, and integrate diverse datasets to build more accurate geological models.

One of the primary applications of AI in lithology is the automated classification of rock types based on well log data. Traditionally, lithology classification involves manual interpretation of well logs, which can be subjective and inconsistent, especially when dealing with large datasets. ML algorithms, particularly supervised learning models, can be trained on labeled datasets to recognize patterns in well log responses and automatically classify lithology with high accuracy. This automation not only speeds up the classification process but also reduces human bias, leading to more consistent and reliable results.

In addition to well log interpretation, AI and ML are used to predict lithological variations across a reservoir, particularly in areas with limited well control. By analyzing seismic attributes, core data, and other geological inputs, ML models can generate lithology maps that provide a detailed view of the spatial distribution of different rock types. These predictive models are invaluable for identifying sweet spots in a reservoir, where the combination of favorable lithology and reservoir properties can maximize hydrocarbon recovery.

AI-driven lithology prediction also plays a crucial role in unconventional reservoirs, such as shale gas and tight oil plays, where the heterogeneity of the rock can significantly impact the effectiveness of hydraulic fracturing. ML models can analyze complex datasets to predict the distribution of brittle and ductile zones, guiding the design of hydraulic fracturing operations to optimize production.

Another important application of AI and ML in lithology-related work is the integration of multiple data sources to create comprehensive lithological models. Traditional lithology interpretation often relies on a single type of data, such as well logs or core samples. However, AI techniques can integrate diverse datasets, including seismic data, petrophysical measurements, and geological information, to build more holistic and accurate lithological models. These integrated models provide a deeper understanding of the reservoir, helping to reduce uncertainty in exploration and development.

AI and ML also facilitate the identification of subtle lithological features that might be overlooked in traditional workflows. For example, unsupervised learning algorithms, such as clustering techniques, can group data points with similar characteristics, revealing patterns and trends that may indicate the presence of specific lithologies or depositional environments. These insights can be critical for understanding the distribution of reservoir rocks and predicting their behavior under different production scenarios.

Moreover, AI techniques are increasingly being used to enhance the interpretation of lithology from seismic data. By combining seismic attributes with well log data, ML models can predict lithological variations directly from seismic volumes, providing a continuous view of the subsurface that is not limited by the spacing of wells. This capability is particularly useful in frontier areas or regions with sparse well coverage, where seismic data may be the only available subsurface information.

In summary, the application of AI and ML in lithology-related works has significantly advanced the ability to classify, predict, and model lithology in reservoir characterization. By automating traditional workflows, integrating diverse datasets, and enhancing the resolution of lithological models, these technologies are helping geoscientists gain a more accurate and detailed understanding of subsurface formations. This, in turn, supports more effective exploration, development, and management of hydrocarbon resources.

2.2.5 Rock Properties

Understanding rock properties is essential for reservoir characterization, as they directly influence the reservoir's ability to store and transmit hydrocarbons. AI and ML have brought transformative changes to how rock properties are analyzed, predicted, and integrated into reservoir models. These technologies enhance the accuracy of property estimation, streamline workflows, and enable real-time decision-making in reservoir management.

Key rock properties such as porosity, permeability, and saturation are typically derived from core analysis, well logs, and seismic data. Traditionally, these properties have been estimated through deterministic methods or empirical correlations, which often require extensive calibration and can be limited by the quality of the input data. AI and ML, however, offer data-driven approaches that can model complex, nonlinear relationships between various subsurface measurements and rock properties, leading to more accurate predictions even in challenging reservoir conditions.

One of the significant applications of ML in the estimation of rock properties is the prediction of porosity and permeability from well logs. By training ML models on a dataset of well logs and corresponding core measurements, it is possible to predict these properties in wells where core data is not available. These models can account for the intricate relationships between multiple log responses and the target properties, providing a more comprehensive understanding of the reservoir's characteristics.

In addition to well log-based predictions, AI and ML are also employed to estimate rock properties from seismic data. Seismic inversion, a process that converts seismic reflection data into quantitative rock property estimates, can be enhanced using ML techniques. ML models can learn the complex mapping between seismic attributes and rock properties, resulting in more reliable inversion outcomes. This approach is particularly beneficial in regions with sparse well data, where seismic information is the primary source of subsurface knowledge.

AI and ML also facilitate the integration of various data types to estimate rock properties more accurately. For example, ML models can combine well logs, core data, and seismic attributes to generate more robust predictions of porosity, permeability, and fluid saturation. This integrated approach reduces the uncertainty in property estimation and leads to more accurate reservoir models, which are critical for effective reservoir management and production optimization.

Another innovative application of AI in rock property analysis is the prediction of mechanical properties, such as Young's modulus and Poisson's ratio, which are essential for understanding the geomechanical behavior of the reservoir. These properties influence the effectiveness of hydraulic fracturing operations and the stability of wellbores. ML models can be trained on lab-measured mechanical properties and corresponding well log data to predict these properties in other parts of the reservoir. This allows for more targeted and efficient fracturing designs, improving production rates and reducing operational risks.

Moreover, AI and ML techniques are used to detect and predict anisotropy in rock properties. Anisotropy, the variation of properties in different directions, is a critical factor in reservoir characterization, especially in unconventional plays. ML models can analyze seismic and well log data to identify and quantify anisotropy, providing insights into the directional flow of fluids and guiding the placement of wells and hydraulic fractures.

Real-time analysis of rock properties during drilling is another area where AI and ML are making a significant impact. With the help of ML algorithms, drilling data can be continuously analyzed to update rock property estimates as new information becomes available. This real-time capability enables operators to make informed decisions on-the-fly, optimizing drilling trajectories and adjusting strategies to enhance well productivity.

In summary, the application of AI and ML in the analysis and prediction of rock properties has revolutionized reservoir characterization. By providing more accurate and integrated estimates of key rock properties, these technologies support more effective exploration, development, and production of hydrocarbon reservoirs. The ability to model complex relationships, integrate diverse datasets, and provide real-time insights makes AI and ML indispensable tools in modern reservoir management.

2.2.6 Geophysics

Geophysics, the study of the physical properties and processes of the Earth, plays a crucial role in subsurface exploration and reservoir characterization. AI and ML technologies have become increasingly integral to geophysical workflows, enhancing the ability to process and interpret geophysical data, improve imaging techniques, and predict subsurface properties with greater accuracy and efficiency.

One of the most significant applications of AI and ML in geophysics is in the processing and interpretation of seismic data. Seismic surveys generate vast amounts of data that require advanced processing techniques to convert raw seismic traces into meaningful subsurface images. ML algorithms, particularly those based on DL, can be used to automate various stages of seismic data processing, including noise attenuation, multiple suppression, and velocity model building. These techniques not only improve the quality of the final seismic image but also significantly reduce the time required for data processing.

AI-driven seismic interpretation tools are capable of automatically identifying geological features such as faults, horizons, and stratigraphic boundaries from seismic data. Traditionally, these features are manually picked by geophysicists, a process that is both time-consuming and subject to interpretation biases. By training ML models on labeled datasets, these tools can learn to recognize patterns in seismic attributes and automatically highlight features of interest. This automation enables more consistent and objective interpretations, leading to more accurate reservoir models.

In addition to seismic data, AI and ML are applied to other geophysical datasets, such as gravity, magnetic, and electromagnetic (EM) data. These methods can be used to model subsurface density, magnetization, and conductivity distributions, providing valuable insights into the geological structure and composition of the subsurface. For example, ML algorithms can analyze gravity and magnetic data to detect and delineate structural features, such as faults and intrusions, that may be associated with hydrocarbon traps.

Geophysical inversion, a process used to estimate subsurface properties from geophysical measurements, is another area where AI and ML have made significant advancements. Traditional inversion techniques rely on iterative methods to fit a model to the observed data, which can be computationally expensive and sensitive to initial conditions. AI-driven inversion approaches, however, can leverage ML models trained on synthetic datasets to provide fast and accurate estimates of subsurface properties. These models can capture complex, nonlinear relationships between the observed data and the underlying properties, leading to more reliable inversion results.

Furthermore, AI and ML techniques are used to integrate different geophysical datasets into a unified interpretation. In a typical exploration scenario, multiple types of geophysical data, such as seismic, gravity, and EM, are collected to provide complementary information about the subsurface. ML algorithms can combine these datasets to produce a more comprehensive and accurate model of the subsurface, reducing uncertainty and improving the decision-making process in exploration and development.

The application of AI in geophysical monitoring is another emerging area of interest. In the context of reservoir management, geophysical monitoring techniques, such as time-lapse (4D) seismic, are used to track changes in the reservoir over time. AI models can analyze time-lapse data to detect subtle changes in the subsurface, such as fluid movement or pressure changes, which are critical for optimizing production strategies and ensuring the long-term viability of the reservoir.

Moreover, AI and ML are enhancing the resolution of geophysical models by enabling the interpretation of higher-dimensional data. For example, 3D seismic data can be analyzed using DL techniques to produce detailed models of the subsurface at a finer scale than traditional methods. This high-resolution imaging is particularly valuable in complex geological settings, where understanding small-scale features can be the key to successful exploration and production (E&P).

In summary, AI and ML are revolutionizing geophysical workflows by automating data processing, enhancing the accuracy of subsurface models, and integrating diverse datasets into a unified interpretation. These technologies are not only improving the efficiency and effectiveness of traditional geophysical methods but also opening up new possibilities for innovative approaches in subsurface exploration and reservoir management. As AI and ML continue to evolve, their impact on geophysics will likely expand, driving further advancements in the field.

2.2.7 Other Applications

Beyond the traditional domains of petrophysics, geology, seismic analysis, lithology, rock properties, and geophysics, AI and ML have found diverse applications in various other aspects of reservoir characterization and subsurface exploration. These emerging applications leverage the power of AI to tackle complex challenges, enhance predictive capabilities, and streamline decision-making processes in unconventional ways.

One such application is in the realm of well placement and optimization. AI and ML algorithms are increasingly being used to analyze vast amounts of subsurface data to identify optimal drilling locations. By integrating geological, geophysical, and petrophysical data, these models can predict the most productive zones within a reservoir, thereby maximizing hydrocarbon recovery while minimizing drilling costs. AI-driven optimization techniques can also assist in determining the best well trajectories and completion strategies, leading to improved well performance and reduced operational risks.

Another area where AI is making significant contributions is in the prediction and management of reservoir behavior over time. Reservoir simulation, a critical tool in reservoir management, involves creating complex numerical models to predict fluid flow and pressure changes within a reservoir. Traditionally, these simulations are computationally intensive and time-consuming. However, AI and ML models, particularly those using surrogate modeling techniques, can approximate the behavior of full-physics reservoir simulations with a fraction of the computational cost. This allows for faster scenario analysis, real-time forecasting, and more effective reservoir management.

AI and ML are also being applied to enhance the understanding and prediction of reservoir heterogeneity. In many reservoirs, the distribution of rock and fluid properties is highly variable and difficult to characterize using conventional methods. AI models can analyze data from multiple sources, such as core samples, well logs, and seismic attributes, to identify patterns and correlations that indicate heterogeneity. These insights can be used to build more accurate reservoir models, which are essential for optimizing production and managing risks.

In addition to improving reservoir modeling and management, AI is being utilized in the automation and optimization of drilling operations. ML algorithms can analyze real-time drilling data to predict and prevent drilling problems, such as stuck pipe, lost circulation, or wellbore instability. By identifying potential issues before they occur, AI-driven systems can help reduce non-productive time (NPT) and improve overall drilling efficiency. Moreover, AI is enabling the development of autonomous drilling systems, where AI algorithms can make real-time adjustments to drilling parameters, such as weight on bit, rotation speed, and mud flow, to optimize drilling performance.

AI and ML are also finding applications in production optimization and enhanced oil recovery (EOR). In production optimization, ML models can analyze production data to identify trends and anomalies, optimize production rates, and predict equipment failures. These models can also be used to optimize artificial lift systems, such as electric submersible pumps (ESPs) and gas lift, by predicting the best operating conditions for maximizing production. In EOR, AI-driven models can help identify the most effective recovery techniques, such as water flooding, gas injection, or chemical EOR, and optimize their implementation based on reservoir conditions.

Furthermore, AI and ML are being used to address environmental and sustainability challenges in the oil and gas industry. For example, AI models can be employed to optimize water management in hydraulic fracturing operations, reducing water usage and minimizing environmental impact. ML algorithms can also be used to monitor and predict greenhouse gas emissions from oil and gas operations, enabling more effective mitigation strategies and compliance with environmental regulations.

Finally, AI and ML are playing a crucial role in the digital transformation of the oil and gas industry. The integration of AI into digital platforms and data management systems is enabling more efficient data processing, analysis, and visualization. This digitalization allows for better data-driven decision-making, improved collaboration across teams, and the ability to respond more quickly to changing conditions in the field.

In summary, the application of AI and ML in other areas of reservoir characterization and subsurface exploration is broad and continues to expand. These technologies are not only enhancing traditional workflows but also enabling new approaches to reservoir management, drilling, production optimization, and environmental stewardship. As AI and ML continue to evolve, their impact on the oil and gas industry will likely grow, driving further innovation and efficiency improvements across the sector.

2.3 OVERVIEW OF AI METHODOLOGIES

The integration of AI, ML, and DL into reservoir characterization, geology, and geophysics represents a paradigm shift in how subsurface data is analyzed and interpreted. These methodologies provide powerful tools for handling complex, high-dimensional data, enabling more accurate predictions, automated processes, and novel insights into subsurface characteristics.

AI is a broad field encompassing various techniques and algorithms that enable machines to perform tasks that typically require human intelligence. In the context of reservoir characterization, AI can be used for decision-making, pattern recognition, and data analysis. AI techniques include rule-based systems, expert systems, and various forms of computational intelligence, such as fuzzy logic and evolutionary algorithms.

AI plays a vital role in automating routine tasks, such as data cleaning, preprocessing, and initial interpretation. For instance, AI algorithms can be used to automatically identify and correct anomalies in well log data, reducing the time and effort required for manual data processing. Additionally, AI-based decision support systems can assist engineers and geoscientists in making more informed decisions by analyzing large datasets and providing recommendations based on predefined rules or learned patterns.

ML, a subset of AI, focuses on developing algorithms that allow machines to learn from data and improve their performance over time without being explicitly programmed. ML methodologies are particularly well-suited for analyzing the vast and complex datasets typical in reservoir characterization.

2.4 APPLICATIONS OF AI

2.4.1 Petrophysics

2.4.1.1 Application-1: Well Log Interpretation Using Deep Learning Neural Networks

2.4.1.1.1 Background of the Application

This paper explores the use of DL neural networks in well log interpretation, specifically to predict the lithology of the formations encountered with the use of open-hole log data and calculate the porosity of the hydrocarbon-bearing formation using the density and neutron values. The study uses a multilayer perception (MLP) with supervised learning, specifically a forward feed network with six input neurons and two output neurons. The program was trained using a dataset that was normalized using the inverse method and divided randomly, such that 70% of the data was used for training while the rest was the test dataset. The program was able to predict the lithology of the formation and calculate the porosity with high accuracy, reducing the time of the operation during the exploration stage. If used as a real-time indicator, the program can ease the drilling operation.

2.4.1.1.2 Description of the Workflow and Machine Learning Model

In this study, a DL neural network approach is employed to interpret well log data and predict lithology and porosity. The neural network used is an MLP, which is effective in identifying patterns and making predictions from complex datasets. The MLP consists of an input layer, multiple hidden layers, and an output layer. The input layer receives normalized values from the GR, resistivity, neutron, and density logs, along with the actual values of neutron and density logs for porosity calculation. The model is trained using supervised learning, where it learns to map inputs to outputs based on labeled training data. The backpropagation algorithm is utilized to optimize the network's weights, minimizing the error between the predicted and actual outputs.

2.4.1.1.3 Advantages and Limitations of the Synchronized Algorithm

The use of an MLP for well log interpretation offers several advantages. The MLP can handle nonlinear relationships between inputs and outputs, making it well-suited for predicting lithology and porosity from well log data. It is capable of generalizing from the training data, allowing it to make accurate predictions on new data. However, the neural network requires a substantial amount of data for training to avoid overfitting, where the model memorizes the training data instead of learning patterns. Additionally, the model's performance heavily depends on the selection of hyperparameters such as learning rate, number of hidden layers, and number of neurons in each layer. Training the network can also be computationally expensive, especially with large datasets, and may require multiple iterations to achieve optimal performance. Despite these limitations, the neural network provides a robust and efficient method for well log interpretation, reducing the time and effort required compared to traditional methods.

2.4.1.1.4 Results and Discussion

The use of DL neural networks in well log interpretation can revolutionize the oil and gas industry by saving time and increasing the ease of operation. The program developed in this study was able to predict the lithology of the formation and calculate the porosity with high accuracy, reducing the time of the operation during the exploration stage. If used as a real-time indicator, the program can ease the drilling operation. The program emphasizes the importance of time and precision in the industry, and the use of machines in taking over tedious jobs can cut down human error associated with the project. The study concludes by suggesting that the program can be further improved by adding other open hole log data and basic calculations made using it.

BIBLIOGRAPHY

Gupta, Aarushi, and Utkarsh Soumya. "Well Log Interpretation Using Deep Learning Neural Networks." Paper presented at the *International Petroleum Technology Conference*, Dhahran, Kingdom of Saudi Arabia, January 2020. https://doi.org/10.2523/IPTC-19678-Abstract

2.4.1.2 Application-2: Hybrid AI and Conventional Empirical Model for Improved Prediction of Log-Derived Permeability of Heterogeneous Carbonate Reservoir

2.4.1.2.1 Background of the Application

This paper presents a systematic approach for interpreting mineralogy and porosity from petrographic thin sections of carbonate rocks, using ML algorithms. The images were acquired using optical microscopic analysis of sidewall core samples from pre-salt reservoirs of Santos Basin, Brazil. The approach involves creating image segmentation models using discrete convolutional filters, followed by artificial neural networks (ANNs) and random forest (RF) classifiers. The models were validated by chemical measurements and generate mineralogical maps or highlight porosity of petrographic thin section images. This method standardizes descriptions and reduces subjectivity and human errors during thin sections' analysis.

2.4.1.2.2 Description of the Workflow and Machine Learning Model

In this study, a hybrid approach combining ANNs and fuzzy logic, known as the adaptive neuro-fuzzy inference system (ANFIS), is used to predict the permeability of a heterogeneous carbonate reservoir. ANFIS integrates the learning capabilities of ANN with the reasoning capabilities of fuzzy logic, making it effective for modeling complex, nonlinear relationships between well log data (inputs) such as GR, resistivity (Rxo), and effective porosity (PhiE) and the rock cluster index (output). ANFIS was chosen due to its ability to handle uncertainty and approximate reasoning, thus capturing the intricate patterns within the geological data.

2.4.1.2.3 Advantages and Limitations of the Synchronized Algorithm

The ANFIS methodology provides several advantages for predicting permeability in heterogeneous reservoirs. Its ability to combine ANN's learning features with fuzzy logic's reasoning enhances the model's adaptability to varying geological conditions, improving prediction accuracy. The hybrid approach is particularly advantageous for handling complex relationships in data and minimizing computational time due to its adaptive learning algorithm.

However, the effectiveness of ANFIS heavily depends on the quality of input data and the precise tuning of its parameters. Overfitting can occur if the network is trained on an inadequate or biased dataset. Additionally, the interpretability of the ANFIS model might be limited compared to simpler empirical models, as the combination of ANN and fuzzy rules can become complex. Moreover, while ANFIS reduces computational time compared to purely neural network-based approaches, it may still require significant resources for training, especially when dealing with large datasets or numerous rock clusters. Careful management of the trade-off between complexity and interpretability is necessary for optimizing the model's performance.

2.4.1.2.4 Results and Discussion

The proposed approach using ML algorithms has been shown to be an effective method for interpreting mineralogy and porosity from petrographic thin sections of carbonate rocks. The models generated using this approach can be used to systematically interpret mineralogy and porosity from images acquired using optical microscopic analysis. The use of these models standardizes descriptions and reduces subjectivity and human errors during thin sections' analysis. This approach is potentially valuable for the oil and gas industry, as it can provide a faster and more accurate analysis of the mineralogy and porosity of reservoir rocks, which can aid in reservoir characterization and production optimization.

BIBLIOGRAPHY

Weldu, Tadesse, Shawket Ghedan, and Omar Al-Farisi. "Hybrid AI and Conventional Empirical Model for Improved Prediction of Log-Derived Permeability of Heterogeneous Carbonate Reservoir." Paper presented at the *SPE Production and Operations Conference and Exhibition*, Tunis, Tunisia, June 2010. https://doi.org/10.2118/136127-MS

2.4.1.3 Application-3: Characterization of Kerogen Content and Activation Energy of Decomposition Using Machine Learning Technologies

2.4.1.3.1 Background of the Application

This paper presents a new approach for evaluating the characteristics of organic matter in source and reservoir rocks. Traditional methods rely on well logging and experiments after core sampling to measure these characteristics. However, this study proposes a different approach that combines numerical simulations with ML technologies to estimate these characteristics.

2.4.1.3.2 Description of the Workflow and Machine Learning Model

In this study, the authors employ ANNs and support vector machines (SVMs) to characterize kerogen content and activation energy from heater temperature transient data. The ANN is adept at modeling the multidimensional and nonlinear relationships between the input temperature data and the target variables, i.e., kerogen content and activation energy. The SVM is used to classify the temperature data into different types of kerogens, leveraging its ability to find the optimal hyperplane that separates the classes in the feature space. The combination of these methods allows us to bypass time-consuming simulation runs traditionally required for inverse modeling.

2.4.1.3.3 Advantages and Limitations of the Synchronized Algorithm

The ANN provides a robust framework for estimating kerogen content and activation energy due to its capability to handle complex, nonlinear interactions within the data. Its use of multiple hidden layers allows for efficient processing of large input datasets, especially those involving time series. However, ANN requires careful tuning to prevent overfitting and can be computationally intensive during training. The SVM offers strong classification performance by maximizing the margin between classes, which is beneficial for distinguishing between different kerogen types. Nevertheless, the performance of SVM heavily relies on the selection of appropriate kernel functions and can be sensitive to the chosen parameters, sometimes requiring extensive cross-validation.

2.4.1.3.4 Summary

For the article on kerogen characterization:

- AI Techniques: ANN for regression, SVM for classification.
- Purpose: Estimate kerogen content and activation energy; classify kerogen types.
- Advantages: ANN handles complex nonlinearity, SVM excels in classification.
- Limitations: ANN requires tuning and can overfit; SVM's performance is parameter-sensitive.

2.4.1.3.5 Results and Discussion

In conclusion, the proposed method offers a simple and efficient tool for estimating the reactivity and type of kerogens in petroleum reservoirs. By reducing the number of simulation runs required in traditional inverse modeling, this approach may prove useful for source and reservoir rock evaluation. Furthermore, the successful application of ML technologies in this study suggests their potential for use in future petrophysical log data interpretation.

BIBLIOGRAPHY

Lee, Kyung Jae. Characterization of kerogen content and activation energy of decomposition using machine learning technologies in combination with numerical simulations of formation heating, *Journal of Petroleum Science and Engineering*. 188, 2020, 106860, ISSN 0920-4105, https://doi.org/10.1016/j.petrol.2019.106860

2.4.1.4 Application-4: A Comparative Analysis of Bubble Point Pressure Prediction Using Advanced Machine Learning Algorithms

2.4.1.4.1 Background of the Application

The article discusses the importance of pressure–volume–temperature (PVT) properties in the E&P lifecycle, from exploration to asset management to improved oil recovery (IOR). Accurate estimation of the bubble point pressure, which is a key PVT property, is crucial for various workflows and reservoir performance. Bubble point pressure is significant because it denotes the point at which constant composition and variable composition portions of the depletion paths merge. It also determines the existence of the incipient phase and possible gas cap when the reservoir is at saturation or near-saturated. Accurate estimation of the bubble point pressure is essential to achieve efficient gas lift design and the determination of the elements of miscibility.

2.4.1.4.2 Description of the Workflow and Machine Learning Model

In this study, advanced ML algorithms, including tree-based models like XGBoost, LightGBM, and random forest regressor, along with an MLP neural network, are employed to predict bubble point pressure (Pbp) in petroleum reservoirs. These algorithms are chosen for their ability to handle complex, nonlinear relationships between the input features and the target variable (bubble point pressure). The models utilize three datasets that capture various combinations of compositional and bulk properties, such as gas–oil ratio (GOR), gas gravity, API gravity, and reservoir temperature. To further enhance model performance, a super learner model (stacking ensemble) is employed to integrate predictions from the base models, aiming to minimize prediction error and improve robustness across different input scenarios.

2.4.1.4.3 Advantages and Limitations of the Synchronized Algorithm

- Strengths:
  - Tree-Based Models: XGBoost, LightGBM, and random forest regressor offer robustness and high accuracy, particularly effective with smaller datasets typical in PVT data.
  - MLP Neural Network: Efficiently captures nonlinear patterns in the data, providing high accuracy.
  - Super Learner: Enhances predictive capability by combining multiple models, reducing systematic error, and achieving lower mean absolute relative error (5.146% on combined data).
  - Flexibility: Models are capable of handling different data scenarios, such as only compositional or only bulk properties.
- Limitations:
  - Tree-Based Models: Can have wide error distribution for fluids with high bubble point pressure; may require parameter tuning to improve stability.
  - MLP Neural Network: Tends to under-predict at high pressure ranges; sensitive to architecture and parameter settings.
  - Super Learner: Computationally more expensive due to the need to train multiple base models and a meta-learner; relies heavily on the quality and diversity of base models.
  - Data Dependence: Model performance can be highly dependent on the quality and distribution of input data, necessitating careful preprocessing and validation.

This structure should reflect the key methodologies, advantages, and limitations discussed in the article, focusing on how these AI techniques are applied to predict bubble point pressure.

2.4.1.4.4 Results and Discussion

Bubble point pressure has been successfully predicted with high accuracy by the use of tree-based advanced ML algorithms, such as XGBoost, LightGBM, random forest regressor, and MLP (neural network) regressor. The performance of ML models is improved by the super learner model, a stacking ensemble. With limited input data, the study has shown that ML models outperform empirical correlations. When just temperature and compositional information are available, super learning models provide an accurate way to estimate bubble point pressure. In contrast to conventional empirical correlations, the study has demonstrated that ML models may be used to predict PVT parameters, particularly bubble point pressure, and can yield a more accurate result.

BIBLIOGRAPHY

Yang, Xi, Birol Dindoruk, Ligang Lu. A comparative analysis of bubble point pressure prediction using advanced machine learning algorithms and classical correlations, *Journal of Petroleum Science and Engineering*. 185, 2020, 106598, ISSN 0920-4105, https://doi.org/10.1016/j.petrol.2019.106598

2.4.1.5 Application-5: Mineralogy and Porosity Identification Using Machine Learning Algorithms

2.4.1.5.1 Background of the Application

ML offers a promising solution for extracting mineralogical and porosity data from petrographic thin section images, a critical task in oil exploration. By automating the analysis process, subjectivity and human error can be reduced, efficiency can be improved, and results can be standardized.

This study focused on developing ML models to interpret mineralogy and porosity from thin section images of carbonate rocks. Datasets from pre-salt reservoirs in the Santos Basin, Brazil were used. Models incorporating convolutional filters, ANNs, and RF classifiers were developed and achieved high accuracy in predicting mineralogy and porosity. Evaluation using 10-fold cross-validation and correlation with chemical microanalysis demonstrated the models' ability to generalize and provide reliable results. The root mean square error (RMSE) of 8.99% and the coefficient of determination of 0.82 for mineralogical predictions highlight the effectiveness of our approach.

2.4.1.5.2 Description of the Workflow and Machine Learning Model

In this study, ANNs and RF classifiers are utilized to identify mineralogy and porosity from petrographic thin section images. The ANNs are designed with an MLP architecture, featuring three hidden layers with 45 neurons each, leveraging a sigmoid activation function for nonlinear transformations. The use of ANNs enables the model to learn complex, nonlinear relationships in the image data, making them highly effective for segmentation tasks.

The RF classifier, on the other hand, operates by creating an ensemble of decision trees (DTs). A forest of 200 trees is employed, with optimization through bootstrap aggregation (bagging) to improve predictive accuracy and control overfitting. RF is particularly advantageous due to its robustness and ability to handle high-dimensional data efficiently.

Convolutional filters are applied initially to extract features from the image data, enhancing the model's capability to distinguish between different mineralogical phases and porosity. Filters such as Gaussian blur, Sobel, and Hessian are used to emphasize specific characteristics like edges and contrasts, which are crucial for accurate classification.

2.4.1.5.3 Advantages and Limitations of the Synchronized Algorithm

The application of ANNs in this study provides the advantage of modeling complex, nonlinear relationships inherent in the petrographic image data. This makes ANNs suitable for capturing the intricate patterns that distinguish mineral phases and porosity. However, they require significant computational resources and careful tuning of hyperparameters such as learning rate and network architecture, which can be computationally expensive and time-consuming.

RF classifiers offer robustness against overfitting and can efficiently handle large and high-dimensional datasets. Their ensemble approach results in improved model stability and accuracy. The main limitation of RF is its reliance on a large number of trees, which, while beneficial for accuracy, can increase computational cost and memory usage.

The convolutional filters applied for feature extraction enhance the models' ability to segment images accurately by emphasizing edges and contrasts. However, the effectiveness of these filters depends on the correct selection and tuning of parameters, such as the size of the filter and the specific attributes to be emphasized.

Overall, the combination of ANNs and RF with advanced feature extraction techniques offers a powerful approach for automated petrographic analysis, but it requires careful consideration of computational resources and parameter settings to optimize performance.

2.4.1.5.4 Results and Discussion

ML-based image segmentation models offer a quantitative approach to petrographic analysis. While the methodology can be applied to various geological settings, limitations must be considered when working with new datasets. The models' ability to classify pixels is constrained by the information available in the images. Petrographic images capture transmitted light, limiting the features that can be identified. Model accuracy depends on the training data, especially for mineralogical models. Porosity models are more adaptable and can often be used with images that employ blue dye to highlight pores, though variations in sample preparation can affect classification. Five models were developed to standardize descriptions and provide relative occurrence measures for different classes. Mineralogical models based on dataset 1, created using neural networks and RFs, achieved high accuracy in 10-fold cross-validation. Porosity models based on dataset 2 also exhibited excellent performance. A mineralogical model created using dataset 3, validated by chemical microanalysis, showed lower accuracy. This is likely due to the selection criteria for dataset 3, which included samples with similar optical properties. To evaluate generalization, model predictions were correlated with attributes obtained from different techniques. Low RMSE values indicate strong predictive capabilities. R^2 values for total mineralogical data and the "quartz" class suggest promising performance, but further data is needed to assess generalization across all classes. Attribute importance analysis revealed that Hessian-based filters are effective for extracting key features. RFs, while slightly less accurate than neural networks, offer a computational advantage. Overall, our study demonstrates the potential of ML for quantitative petrographic analysis, while highlighting the importance of data quality and model selection.

BIBLIOGRAPHY

Rubo, Rafael Andrello, Cleyton de Carvalho Mateus Fontana Carneiro, Rafael dos Santos Gioria Michelon. Digital petrography: Mineralogy and porosity identification using machine learning algorithms in petrographic thin section images. *Journal of Petroleum Science and Engineering*, 183, 2019, 106382, ISSN 0920-4105, https://doi.org/10.1016/j.petrol.2019.106382

2.4.1.6 Application-6: Machine Learning Technique for the Prediction of Shear Wave Velocity Using Petrophysical Logs

2.4.1.6.1 Background of the Application

The purpose of this work was to create a model for calculating shear wave velocity (Vs) in an Ahvaz field well using petrophysical logs. Shear wave logs are only found in a small number of wells and are costly to obtain. Finding a link between Vs and other petrophysical data has thus been the subject of numerous attempts. Tukey's approach was used in this work to preprocess the data and get rid of outliers. The Non-dominated Sorting Genetic Algorithm II (NSGA-II) was then used to estimate Vs by selecting the best features from the inputs. Vp, RT, GR, RHOB, and NPHI were the traits that were chosen. Vs was estimated by combining the least square support vector machine (LSSVM) with particle swarm optimization (PSO), genetic algorithm (GA), and cuckoo optimization algorithm (COA). Results show that the LSSVM–COA algorithm produces more accurate estimates of Vs compared to the other hybrid algorithms. The method employed in this study can be used as an efficient means to accurately estimate Vs.

2.4.1.6.2 Description of the Workflow and Machine Learning Model

In this study, the authors employ an LSSVM with a radial basis function (RBF) kernel to estimate Vs using petrophysical logs. The LSSVM is an advanced version of the traditional SVM, which simplifies the computation by solving linear equations instead of a quadratic programming problem. This method is particularly effective in capturing complex nonlinear relationships between the input logs and the target Vs.

To optimize the LSSVM parameters, three optimization algorithms—COA, PSO, and GA—are used. Each algorithm enhances the parameter tuning process, aiming to minimize the model error and improve prediction accuracy. The COA is inspired by the brood parasitism of some cuckoo species, PSO simulates social behavior among particles, and GA mimics the process of natural selection.

2.4.1.6.3 Advantages and Limitations of the Synchronized Algorithm

The LSSVM with an RBF kernel is highly effective in modeling complex, nonlinear relationships, making it suitable for geophysical applications like Vs estimation. Its ability to generalize well across different datasets is enhanced by the optimization algorithms employed. COA, PSO, and GA each offer unique advantages in terms of exploration and exploitation of the parameter space, leading to more accurate models.

However, the LSSVM requires careful selection of kernel and regularization parameters to prevent overfitting and ensure generalization. While optimization algorithms like COA, PSO, and GA improve parameter selection, they can be computationally intensive, especially when searching large parameter spaces. The performance of these algorithms is also highly dependent on their settings and the nature of the problem, requiring multiple iterations to converge to an optimal solution. Despite these challenges, the hybrid LSSVM–COA approach demonstrates superior accuracy and reliability in the prediction of Vs compared to traditional empirical and regression models.

2.4.1.6.4 Results and Discussion

The LSSVM–COA model provides more accurate and reliable results in estimating Vs compared to the empirical, regression, LSSVM–PSO, and LSSVM–GA models. The very slight difference between the error in testing and training steps suggests a higher reliability of the LSSVM–COA compared to the other hybrid algorithms. The selected features were Vp, RT, GR, RHOB, and NPHI. The results obtained from modeling reveal that the model error is considerably less than those of empirical and regression models. The method employed in this study could be used as an efficient means to accurately estimate Vs in other reservoirs.

BIBLIOGRAPHY

Anemangely, Mohammad, Ahmad Ramezanzadeh, Hamed Amiri, Seyed-Ahmad Hoseinpour. Machine learning technique for the prediction of shear wave velocity using petrophysical logs. *Journal of Petroleum Science and Engineering*, 174, 2019, 306–327, ISSN 0920-4105, https://doi.org/10.1016/j.petrol.2018.11.032

2.4.1.7 Application-7: A New Approach to Improve Neural Networks' Algorithm in Permeability Prediction of Petroleum Reservoirs Using Supervised Committee Machine Neural Network (SCMNN)

2.4.1.7.1 Background of the Application

This study uses a supervised committee machine based on a neural network (SCMNN) that combines 30 estimators to forecast permeability, a crucial parameter for assessing hydrocarbon resources. The wide range and multiple populations of the permeability data make the simple neural network's outputs inappropriate. As a result, a statistical analysis was used to separate all data into two populations: those with low and high permeability. To forecast permeability in both low and high classes, each SCMNN estimator was paired with two simple networks. One gating network, which is regarded as a classifier, categorized data into these two classes. Permeability was predicted using the SCMNN using data from a petroleum reserve in southwest Iran. R^2 (R = correlation coefficient) certified the overall fitting between the measured and anticipated permeabilities as 97.72%, which is deemed appropriate. In contrast, the best stat's R^2 for the basic network was 84.14%. Reduced bias and improved R^2 in the results show SCMNN's excellent power and efficiency.

2.4.1.7.2 Description of the Workflow and Machine Learning Model

Methodology: The paper utilizes an SCMNN to predict the permeability of petroleum reservoirs. The SCMNN is composed of multiple neural network estimators, each specialized for different permeability ranges. The model incorporates:

1. Simple Neural Networks: These networks are used to model the permeability within specific ranges (low and high permeabilities). Each neural network consists of an input layer, a hidden layer with 23 neurons, and an output layer.

2. Gating Network: A classifier that determines whether a data sample belongs to the low or high permeability category, allowing the appropriate simple neural network to make the prediction.
3. Committee Machine: Combines the outputs of multiple estimators to improve the prediction accuracy, leveraging the diversity of individual models.
4. Training Algorithm: The Levenberg–Marquardt algorithm is used to accelerate the convergence of the neural network training process, optimizing weights and biases to minimize the mean squared error (MSE).

2.4.1.7.3 Advantages and Limitations of the Synchronized Algorithm

Advantages:

- Improved Accuracy: The SCMNN approach enhances prediction accuracy by using specialized networks for different permeability ranges, achieving an R^2 value of 97.72%, significantly higher than a single neural network's R^2 of 84.14%.
- Efficiency: The committee machine architecture reduces bias and variance in the predictions, offering more reliable results than a single model.
- Fast Convergence: The Levenberg–Marquardt algorithm allows the model to converge faster compared to conventional training methods.

Limitations:

- Complexity: The SCMNN involves a complex architecture with multiple neural networks and a gating network, making it computationally intensive.
- Data Requirements: The model requires a large amount of training data to accurately distinguish between low and high permeability classes.
- Misclassification Risks: Errors in the gating network can lead to samples being incorrectly classified, resulting in predictions from the wrong network, which can reduce overall accuracy.
- Parameter Tuning: Requires careful tuning of network parameters and classifier thresholds to achieve optimal performance.

2.4.1.7.4 Results and Discussion

In conclusion, the study shows that SCMNN is a powerful estimator for predicting permeability in hydrocarbon reservoirs. The SCMNN is superior to a simple neural network, as it can produce better results with lower bias and higher R^2 values. The statistical separation of primary data into low and high permeability populations and the utilization of two supervised networks and proper classification were found to be effective in improving the accuracy of the permeability prediction. The study also suggests that more training data could further improve the accuracy of the SCMNN. Overall, the results of this study provide valuable insights for the development of more accurate and reliable permeability prediction models for hydrocarbon reservoirs.

BIBLIOGRAPHY

Karimpouli, Sadegh, Nader Fathianpour, Jaber Roohi. A new approach to improve neural networks' algorithm in permeability prediction of petroleum reservoirs using supervised committee machine neural network (SCMNN). *Journal of Petroleum Science and Engineering*, 73 (34), 2010, 227–232, ISSN 0920-4105, https://doi.org/10.1016/j.petrol.2010.07.003

2.4.1.8 Application-8: Artificial Intelligence Applied to NMR Logging for Rock and Fluid Typing in Heavy Oils

2.4.1.8.1 Background of the Application

This paper presents a case study of the successful implementation of nuclear magnetic resonance (NMR) logging and post-processing techniques based on blind source separation (BSS) using independent component analysis (ICA) for fluid typing in a heavy oil reservoir. The paper demonstrates the usefulness of NMR logging technology in the characterization and evaluation of this reservoir. Data post-processing based on BBS-ICA enables adequate differentiation between fluid components from T2 spectra. The paper discusses the input parameters for NMR logging, the BSS-ICA method employed, and the output of the work.

2.4.1.8.2 Description of the Workflow and Machine Learning Model

In this study, the authors utilize BSS with ICA to process NMR logging data for rock and fluid typing in heavy oil reservoirs. ICA is an ML technique designed to separate a multivariate signal into additive subcomponents, assuming the subcomponents are non-Gaussian and statistically independent. This approach is particularly effective in distinguishing oil from other fluids in the transverse magnetic relaxation times (T2) spectra by deconvolving the mixed signals into independent sources. The Infomax algorithm, a specific implementation of ICA, is employed to maximize the statistical independence of the estimated components. By leveraging the principles of statistical independence, this method efficiently identifies the oil component within the NMR data, facilitating accurate fluid typing.

2.4.1.8.3 Advantages and Limitations of the Synchronized Algorithm

The application of BSS–ICA, particularly the Infomax algorithm, offers several advantages in the context of NMR logging for fluid typing. It provides a robust framework for separating mixed signals without requiring prior knowledge of the mixing process, making it ideal for complex geological formations where fluid contributions overlap significantly. The method enhances the clarity of the T2 spectra, allowing for more precise differentiation between oil and water, which is critical in heavy oil reservoirs.

However, the effectiveness of ICA is contingent on the statistical properties of the input data, specifically the non-Gaussianity and independence of the source signals. While the Infomax algorithm is efficient, its performance can be sensitive to noise in the data and may require fine-tuning of parameters for optimal results. Additionally, the computational complexity of ICA can be a limitation when processing large datasets, necessitating careful consideration of computational resources. Despite these challenges, the integration of ICA into NMR logging workflows significantly improves the reliability of fluid typing in complex reservoir conditions.

2.4.1.8.4 Results and Discussion

In conclusion, this paper presents a case study of the successful implementation of NMR logging and post-processing techniques based on BSS–ICA for fluid typing in a heavy oil reservoir. The study demonstrates that NMR logging technology is a useful tool for the characterization and evaluation of heavy oil reservoirs. The BBS–ICA method for post-processing NMR data allows for a more accurate differentiation between fluid components from T2 spectra. The paper highlights the importance of selecting the appropriate input parameters for NMR logging and the significance of using advanced data processing methods such as BSS–ICA for optimal results. Overall, this study contributes to the advancement of NMR logging technology and its potential for use in the petroleum industry.

BIBLIOGRAPHY

Rojas, Pedro A. Cristea Romero, Pavlakos Alexandrina, Ergündüz Paul, Keçecioğlu Okan, Tayfun, and Server Fatih Alpay. "Artificial Intelligence Applied to NMR Logging for Rock and Fluid Typing in Heavy Oils." Paper presented at the *SPWLA 60th Annual Logging Symposium*, The Woodlands, Texas, June 2019. https://doi.org/10.30632/T60ALS-2019_LLLL

2.4.1.9 Application-9: Synthetic Well Log Generation Using Machine Learning Techniques

2.4.1.9.1 Background of the Application

Instead of running a fresh set of logs or coring and testing the entire zone of interest, this research explores the use of synthetic well logs as a more affordable and dependable way to forecast the photoelectric (PE) and unconfined compressive strength (UCS) values of an area. PE logs, which include crucial details regarding the lithology of the rocks the wellbore intersects, are typically included in contemporary well log datasets. PE logs, which are necessary to comprehend the lithology of these formations, are absent from a large number of legacy wells. Similar to this, UCS data is crucial for assessing the rock's strength; however, it is costly and time-consuming to acquire and test core samples for all zones of interest. Wireline logs from more than a hundred wells, including GR, density, neutron, resistivity, and clay volume logs, were used to create synthetic PE logs for this investigation. A 70:30 split of the data was randomly selected for the training and validation sets, respectively. The best technique based on the validation dataset's least average squared error (ASE) was chosen through model competition among a variety of ML algorithms, including linear regression, ANNs, DTs, gradient boosting (GB), and RF. UCS data, which included GR, density, porosity, neutron, clay volume, kerogen volume, compressional slowness, shear slowness, Young's modulus (static and dynamic), and Poisson's ratio, were produced utilizing wireline logs and core rebound hammer data from 14 wells. To eliminate collinearity, reduce variable redundancy, and choose the most suitable variables for the study, variable clustering was employed. The selected variables were subjected to cluster analysis in order to determine the characteristics that set data segments apart from the population. Model competition among the aforementioned suite of ML techniques was utilized to choose the champion model based on the validation dataset's least ASE after the data was randomly divided into a 70:30 training and validation split. Results show that neural networks and RFs generated the best prediction of UCS and synthetic PE logs compared to other ML algorithms used.

2.4.1.9.2 Description of the Workflow and Machine Learning Model

In this study, a variety of ML algorithms were employed to generate synthetic well logs, specifically focusing on synthetic PE and UCS logs. The models used include linear regression, DTs, GB, RFs, and ANNs. These models were chosen to handle the nonlinear relationships and interactions between the input variables and target variables effectively.

- Linear regression was used to establish a baseline for modeling the linear relationships between the inputs and the target variable. While simple, it provides interpretable results that can highlight fundamental patterns in the data.
- DTs model the data by recursively splitting the dataset based on input feature values. They are intuitive and can handle both numerical and categorical data, providing clear insight into feature importance.
- GB improves upon DTs by combining the outputs of multiple weak learners to form a strong predictive model. It iteratively minimizes a loss function, capturing complex patterns in the data.
- RFs utilize an ensemble of DTs trained on random subsets of the data, improving robustness and reducing overfitting compared to single DTs. This method is particularly effective for handling datasets with many features and interactions.
- ANNs are used for their ability to capture complex nonlinear relationships through their layered architecture, which allows them to learn hierarchical representations of the data.

2.4.1.9.3 Advantages and Limitations of the Synchronized Algorithm

The use of multiple ML algorithms provides several advantages. RFs and ANNs were found to be the most effective for predicting UCS and synthetic PE logs, respectively, due to their ability to model nonlinear relationships and interactions between features. RFs are less prone to overfitting and are efficient in handling high-dimensional data, making them suitable for the synthetic PE log prediction. ANNs, with their flexible architecture, are adept at capturing complex data patterns, which is advantageous for UCS prediction.

However, each model has its limitations. Linear regression and DTs may not perform well with highly nonlinear relationships or interactions between variables. GB and RFs require careful tuning of hyperparameters and can be computationally intensive, especially with large datasets. ANNs demand substantial computational resources and can be sensitive to the choice of architecture and training parameters.

The methodology benefits from the diversity of models applied, allowing for a comprehensive exploration of the dataset characteristics. Nevertheless, it requires significant computational power and expertise in model selection and parameter tuning to achieve optimal results.

2.4.1.9.4 *Results and Discussion*

Instead of running a fresh set of logs or coring and testing the entire zone of interest, the study demonstrates that synthetic well logs can be a dependable and less expensive option for predicting PE and UCS values. When compared to other ML algorithms, neural networks and RFs produced the greatest predictions for UCS and synthetic PE logs. Wireline logs from more than a hundred wells, including GR, density, neutron, resistivity, and clay volume logs, were used in the process of creating synthetic PE logs. Model competition among a suite of ML algorithms was used to choose the best method based on the validation dataset's least ASE. The data was randomly divided into training and validation datasets in a 70:30 split, respectively. Young's modulus (static and dynamic), Poisson's ratio, density, porosity, neutron, GR, clay volume, kerogen volume, compressional slowness, shear slowness, and core rebound hammer data from 14 wells were used in the process for creating UCS data. To eliminate collinearity, reduce variable redundancy, and choose the most suitable variables for the study, variable clustering was employed. The selected variables were subjected to cluster analysis in order to determine the characteristics that set data segments apart from the population. The data was randomly portioned into a 70:30 training and validation split, and model competition was used to select the champion model based on the least ASE of the validation dataset.

BIBLIOGRAPHY

Akinnikawe, Oyewande, Stacey Lyne, and Jon Roberts. "Synthetic Well Log Generation Using Machine Learning Techniques." Paper presented at the *SPE/AAPG/SEG Unconventional Resources Technology Conference*, Houston, Texas, July 2018. https://doi.org/10.15530/URTEC-2018-2877021

2.4.1.10 Application-10: Forecast of Prospective Oil Saturation Zones Using Machine Learning Methods

2.4.1.10.1 *Background of the Application*

The Kharyaginskoye oil field, situated in the Timan–Pechora Basin, contains Devonian age carbonate reservoirs with high heterogeneity and complex structures. Accurate evaluation of oil-saturated thicknesses is crucial due to the varying oil–water contact levels. This study integrates borehole and seismic data using ML techniques to enhance interpretation reliability. By employing multi-resolution graph-based clustering (MRGC) and a democratic neural network association (DNNA), the research effectively integrates geological and geophysical data. The approach allows for a detailed facies distribution, which is critical for reservoir modeling and reduces geological risks in well placement.

2.4.1.10.2 *Description of the Workflow and Machine Learning Model*

The study employs MRGC and DNNA to forecast oil saturation zones in Devonian carbonate deposits. MRGC is a clustering technique that leverages k-nearest neighbors (kNN) for multidimensional pattern recognition, effectively integrating geological and geophysical information. It is used to identify electrofacies with high filtration and capacitance potential. DNNA, on the other hand, is a neural network method that uses a democratic voting system to train on both well and seismic data. This approach allows for a detailed distribution of facies in the inter-well space, optimizing seismic data utilization to achieve calibrated and scaled facies distributions.

2.4.1.10.3 Advantages and Limitations of the Synchronized Algorithm

MRGC is advantageous for its ability to handle complex data structures, revealing patterns that traditional methods might overlook. It is particularly effective in identifying electrofacies, which are critical for reservoir characterization. However, its unsupervised nature may require iterative adjustments to ensure geological accuracy. DNNA excels in capturing nonlinear relationships between seismic attributes and lithofacies, improving prediction accuracy. Nevertheless, DNNA's performance relies heavily on the quality of input data, and the computational cost can be significant due to its iterative training process. While both methods enhance interpretation reliability and reduce geological risks, they demand careful calibration and validation against core data to ensure accurate predictions.

2.4.1.10.4 Results and Discussion

The application of advanced ML methods, particularly MRGC and DNNA, has significantly improved the understanding of low-permeability carbonate reservoirs. These techniques address the limitations of traditional methods in characterizing heterogeneous saturation models. The use of high-frequency seismic data combined with borehole information has increased the accuracy of lithological predictions and the identification of prospective oil zones. The integration of multi-scale geological and geophysical data has led to more reliable interpretation and minimized geological risks, aiding in the efficient placement of wells and enhancing reservoir management.

BIBLIOGRAPHY

Gusev, Sergey Igorevich, Elena Sergeyevna Kolbikova, Olga Igorevna Malinovskaya, Azat Fanisovich Garaev, and Robert Kamilevich Valiev. "Forecast of Prospective Oil Saturation Zones in the Devonian Carbonate Deposits of the Kharyaginsky Field Based on Geological and Geophysical Information Analysis by Using Machine Learning Methods." Paper presented at the *SPE Russian Petroleum Technology Conference, Virtual*, October 2021. doi:10.2118/206520-MS

2.4.2 Geology

2.4.2.1 Application-11: Geophysical Data Integration and Machine Learning for Multi-target Leakage Estimation in Geologic Carbon Sequestration

2.4.2.1.1 Background of the Application

In this study, a novel approach is presented using ML techniques to enhance seismic data and improve the detection of subsurface features. By leveraging convolutional neural networks (CNNs), the resolution of seismic images can be enhanced, while principal component analysis (PCA) and K-means clustering are employed to identify and classify seismic attributes. The methodology is validated using field data, demonstrating significant improvements in the clarity and interpretability of seismic features such as faults and channels.

2.4.2.1.2 Description of the Workflow and Machine Learning Model

In this study, CNNs are employed to enhance seismic data resolution, facilitating better detection of geological features such as faults and channels. CNNs are well-suited for image processing tasks due to their ability to capture spatial hierarchies and patterns within the data. Additionally, PCA is used for dimensionality reduction, allowing us to efficiently manage and analyze the high-dimensional seismic datasets. K-means clustering is utilized to classify seismic attributes, enabling the identification of key subsurface features. These methods work in tandem to provide a comprehensive approach to seismic data interpretation, improving accuracy and reliability.

2.4.2.1.3 Advantages and Limitations of the Synchronized Algorithm

The use of CNNs in seismic data processing provides significant advantages, including the ability to automatically extract relevant features and enhance data resolution without extensive preprocessing. CNNs are robust against noise and can learn complex patterns inherent in seismic images. However, training CNNs requires a large amount of labeled data and computational resources, which can be a limitation in some scenarios.

PCA effectively reduces the dimensionality of seismic data, making it more manageable for analysis and visualization. It simplifies the dataset by retaining only the most significant components, though it may occasionally discard subtle but important information.

K-means clustering efficiently categorizes seismic features, allowing for clear differentiation between various geological structures. While K-means is easy to implement and interpret, it requires the specification of the number of clusters, which might not be straightforward and can affect the results' accuracy.

Overall, the combination of these methods enhances the interpretability of seismic data, though considerations regarding computational cost and data requirements must be addressed to optimize performance.

2.4.2.1.4 Results and Discussion

This research successfully demonstrates the application of advanced ML techniques for seismic data enhancement and feature detection. The use of CNNs has notably enhanced seismic resolution, allowing for better visualization of subsurface structures. The integration of PCA and K-means clustering further aids in the classification and identification of significant seismic attributes. The results indicate that these methods provide a robust framework for improving seismic data interpretation, with potential applications in hydrocarbon exploration and other geophysical investigations.

BIBLIOGRAPHY

de Lima, Rafael Pires, and Youzuo Lin. "Geophysical data integration and machine learning for multi-target leakage estimation in geologic carbon sequestration." Paper presented at the *SEG International Exposition and Annual Meeting*, San Antonio, TX, September 2019. https://doi.org/10.1190/segam2019-3215405.1

2.4.2.2 Application-12: Use of Machine Learning Production Driver Cross-sections for Regional Geologic Insights in the Bakken-Three Forks Play

2.4.2.2.1 Background of the Application

Petroleum geologists have traditionally used cross-sections to understand subsurface structures and variations in log properties. This study introduces a novel application of ML to create cross-sections that visualize geologic production drivers. Using a decision-tree model trained with data from completions, geology, and spacing, the study predicts oil, gas, and water production in North Dakota's Bakken-Three Forks play. The model's predictions are explained using Shapley values, which highlight the impact of different geologic properties on production. These cross-sections reveal key production drivers for different regions within the play, such as water saturation in the Nesson area and diverse factors in the Bakken. This workflow can be adapted for other basins, providing valuable insights into subsurface variability and its impact on hydrocarbon production.

2.4.2.2.2 Description of the Workflow and Machine Learning Model

In this study, a decision-tree-based model is used to predict oil, gas, and water production in the Bakken-Three Forks play. The model incorporates data on completions, geology, and spacing to capture complex relationships between these inputs and the production outcomes. To provide interpretability, a surrogate model is trained to generate Shapley values, which quantify the contribution of each feature to the predictions. This approach helps understand the regional variations in geologic properties and their impact on production.

2.4.2.2.3 Advantages and Limitations of the Synchronized Algorithm

The decision-tree-based model excels at handling nonlinear relationships and is capable of providing transparent and interpretable results through the use of Shapley values. This makes it well-suited for understanding the influence of various geologic and completion factors on production outcomes. However, decision-tree models can be sensitive to changes in data and may require large datasets to generalize effectively. The Shapley value approach adds an additional layer of insight but can be computationally intensive, especially for large datasets, as it requires calculating contributions for each feature individually.

2.4.2.2.4 Results and Discussion

The increasing use of ML in hydrocarbon exploration requires geoscientists to interpret new types of data. This study demonstrates the use of Shapley values to explain ML model predictions, revealing the geospatial impact of factors like water saturation, clay volume, and total organic carbon (TOC) on production forecasts. These visualizations uncover regional trends and differing production drivers in the Bakken-Three Forks play. The approach can be applied to various datasets, offering a robust method for integrating statistical trends with traditional geoscientific analysis to enhance our understanding of subsurface characteristics and their effects on hydrocarbon production.

BIBLIOGRAPHY

Cross, T., K. Sathaye, and J. Chaplin. "Use of Machine Learning Production Driver Cross-Sections for Regional Geologic Insights in the Bakken-Three Forks Play." Paper presented at the *SPE/AAPG/SEG Unconventional Resources Technology Conference*, Houston, TX, July 2021. https://doi.org/10.15530/urtec-2021-5549

2.4.2.3 Application-13: Performance Evaluation of Machine Learning-Based Classification with Rock-Physics Analysis

2.4.2.3.1 Background of the Application

The study focuses on leveraging ML techniques to classify lithofacies using well log data from the Tarakan Basin in North Kalimantan, Indonesia, a known hydrocarbon-producing region. Given the diverse lithofacies and formation heterogeneities within the basin, such as differences in sedimentology and diagenetic processes, achieving reliable classification results presents a challenge. This research aims to overcome these challenges by implementing advanced supervised learning models, particularly RF and GB, to enhance the efficiency and accuracy of lithofacies prediction, thereby aiding geological analysis and decision-making processes in a more time-effective manner.

2.4.2.3.2 Description of the Workflow and Machine Learning Model

In this study, two ML techniques, RF and GB, are employed for the classification of lithofacies in well log data from the Tarakan Basin. RF, an ensemble method based on constructing multiple DTs, is utilized to enhance prediction accuracy by averaging the predictions of several trees, thus reducing overfitting and variance. GB, on the other hand, is used to sequentially optimize prediction accuracy by focusing on correcting errors made by previous models. It does this by combining the outputs of weak learners to form a strong predictive model, making it highly effective for distinguishing between lithofacies types in complex geological formations.

2.4.2.3.3 Advantages and Limitations of the Synchronized Algorithm

RF offers robustness against overfitting and high variance, making it suitable for handling the complexity of lithofacies classification. It benefits from its ensemble approach, which aggregates multiple DTs to produce more reliable predictions. However, it can become computationally expensive and may require careful tuning of parameters like the number of trees and depth to achieve optimal performance. GB excels in improving prediction accuracy by sequentially correcting errors, providing precise classification results. It is particularly advantageous in identifying subtle differences between similar lithofacies. Nevertheless, it is sensitive to overfitting and may require careful management of parameters, such as learning rate and number of iterations, to maintain a balance between bias and variance. Both methods, while powerful, require a substantial amount of computational resources and time, especially when working with large datasets typical in geological studies.

2.4.2.3.4 Results and discussion

The study successfully demonstrates the application of supervised learning methods, specifically RF and GB, in classifying lithofacies within the Tarakan Basin. These models showed high accuracy in predicting lithofacies such as shale, shaly sandstone, and sandstone, as evidenced by metrics like precision, recall, F1-score, and ROC curves. However, the classification of coal proved more challenging due to its similar physical properties to shale and the presence of thin coal beds. Despite these challenges, the ML models significantly reduced the time required for geological analysis and provided reliable results across different formations, enhancing the overall efficiency of lithofacies classification in the region.

BIBLIOGRAPHY

Antariksa, Gian, Radhi Muammar, Jihwan Lee. Performance evaluation of machine learning-based classification with rock-physics analysis of geological lithofacies in Tarakan Basin, Indonesia. *Journal of Petroleum Science and Engineering*. 208 (Part A), 2022, 109250, ISSN 0920-4105, https://doi.org/10.1016/j.petrol.2021.109250

2.4.2.4 Application-14: Machine Learning-Based Rate Optimization under Geologic Uncertainty

2.4.2.4.1 Background of the Application

Waterflooding is a widely used technique to enhance oil recovery post-primary depletion, where water is injected to maintain reservoir pressure and facilitate oil extraction. However, optimizing well rates during waterflooding is challenging due to the complex parameters, computational demands of numerical simulations, and geological uncertainties. Traditional optimization approaches, which involve multiple forward simulations, are computationally intensive and may not scale well for large fields and multiple geological realizations. This paper proposes an ML-based method for rate optimization that is fast and scalable, leveraging a reduced-order representation of permeability heterogeneity and streamlining tracing to predict the minimum time of flight (TOF) at producers. This methodology aims to maximize sweep efficiency by equalizing the arrival times of injected fluid fronts at producers, using a neural network to replace conventional simulators and significantly reduce computational time.

2.4.2.4.2 Description of the Workflow and Machine Learning Model

In this study, the authors employ an MLP neural network to predict the minimum time of flight (TOF) at the producers during waterflooding operations, considering geological uncertainties. The MLP is chosen for its ability to capture nonlinear relationships between input parameters such as well rates and geological features. The training data is generated using a reduced-order representation of the permeability heterogeneity, which is derived from a grid connectivity transformation (GCT). This transformation projects the heterogeneous field onto a lower-dimensional space using orthogonal basis functions, simplifying the complexity of the input data while preserving essential geological characteristics. The sequential least squares programming (SLSQP) optimization algorithm is used to determine the optimal well rates that minimize the variance in TOF across different geological realizations, enhancing sweep efficiency and oil recovery.

2.4.2.4.3 Advantages and Limitations of the Synchronized Algorithm

The MLP neural network is well-suited for modeling complex nonlinear interactions in the reservoir system, providing accurate TOF predictions that facilitate effective rate optimization. Its capacity to generalize from training data makes it robust for handling geological uncertainties. However, training the MLP requires a substantial amount of data and computational resources, particularly for generating the reduced-order representations. The SLSQP optimization algorithm efficiently finds optimal well rates, balancing multiple objectives and constraints. Despite its strengths, the optimization process relies heavily on the accuracy of the proxy model and the initial training data, which may limit its applicability in cases with significant changes in well configurations or geological conditions. Additionally, the need for streamline tracing during the data generation phase can be computationally intensive.

2.4.2.4.4 Results and Discussion

This study introduces an ML-driven approach to optimize waterflooding rates by predicting the time of flight (TOF) at producers, considering geological uncertainties. The proxy model, based on a neural network, provides rapid TOF predictions, enabling efficient optimization over multiple realizations. This approach significantly improves sweep efficiency and oil recovery compared to unoptimized cases or those optimized using a single geological realization. Testing on both a two-dimensional (2D) heterogeneous field and the 3D SPE Brugge benchmark demonstrates substantial enhancements in oil recovery. The proxy model's computational efficiency makes it suitable for large-scale field applications, offering a practical solution to incorporate geological uncertainty in waterflood optimization. Future work will focus on extending this methodology to more complex scenarios involving dynamic well rates and additional constraints.

BIBLIOGRAPHY

Sen, Deepthi, Hongquan Chen, Akhil Datta-Gupta, Joseph Kwon, and Srikanta Mishra. Machine learning based rate optimization under geologic uncertainty. *Journal of Petroleum Science and Engineering*, 207, 2021, 109116, ISSN 0920-4105, https://doi.org/10.1016/j.petrol.2021.109116

2.4.2.5 Application-15: Fault Diagnosis Model Based on Granular Computing and Echo State Network

2.4.2.5.1 Background of the Application

This paper presents a fault diagnosis method for a terminal guidance radar signal processing module based on granular computing and echo state network (ESN). The proposed method involves attribute reduction of the test index based on granular computing, clustering of similar distinguishability attributes, and fault diagnosis using ESN. The paper also introduces the Bienenstock–Cooper–Munro (BCM) rule and $L1/2$-norm regularization to improve the sparsification efficiency of the model. The effectiveness and superiority of the proposed method are verified through a fault diagnosis example.

2.4.2.5.2 *Description of the Workflow and Machine Learning Model*

In this method, a fault diagnosis model based on granular computing and ESN is employed to improve the efficiency and accuracy of diagnosing faults in electronic equipment. The ESN, a type of recurrent neural network (RNN), leverages a reservoir computing model where only the output weights are trained using a linear regression method. The reservoir's internal neurons are randomly sparsely connected, which simplifies the model and helps in achieving a global optimal output. This approach is particularly beneficial for handling nonlinear prediction tasks. Additionally, the model incorporates a BCM learning rule into the reservoir to adaptively adjust the weights, enhancing the reservoir's adaptability to sample data. The ESN's sparsity is further optimized using an L1/2-norm regularization term, which helps control the scale of the model and improve its generalization ability.

2.4.2.5.3 *Advantages and Limitations of the Synchronized Algorithm*

The advantages of using the ESN with the L1/2L_{1/2}L1/2-norm regularization include its ability to model complex nonlinear relationships efficiently, making it suitable for fault diagnosis in complex systems. The use of granular computing for attribute reduction simplifies the input complexity, thus enhancing the efficiency of the ESN by focusing on the most relevant features. The BCM rule improves the dynamic adaptability of the reservoir, while the L1/2L_{1/2}L1/2-norm ensures a more robust and sparse solution. However, the method's reliance on the pre-training of reservoir weights introduces an additional time requirement, although this is mitigated by performing this step before the actual training. Additionally, the method's efficiency depends on the proper configuration of the regularization parameters and the BCM rule settings, which may require iterative tuning.

This methodology highlights the balance between optimizing model accuracy and managing computational efficiency, emphasizing the need for careful tuning and pre-training to leverage the full potential of ESN and granular computing in fault diagnosis applications.

2.4.2.5.4 *Results and Discussion*

In this paper, the authors propose a fault diagnosis method for a terminal guidance radar signal processing module based on granular computing and ESN. The proposed method involves attribute reduction of the test index based on granular computing, clustering of similar distinguishability attributes, and fault diagnosis using ESN. The paper also introduces the BCM rule and $L1/2$-norm regularization to improve the sparsification efficiency of the model. The proposed method effectively diagnoses different types of faults and provides superior performance to traditional methods. The proposed method can be used to improve the fault diagnosis accuracy of other systems as well.

BIBLIOGRAPHY

Lu, Cheng, Peng Xu, and Lin-Hu Cong. Fault diagnosis model based on Granular Computing and Echo State Network. *Engineering Applications of Artificial Intelligence*, 94, 2020, 103694, ISSN 0952-1976, https://doi.org/10.1016/j.engappai.2020.103694

2.4.2.6 Application-16: Bakken Stratigraphic and Type Well Log Learning Network Exploited to Predict and Data Mine Shear Wave Acoustic Velocity

2.4.2.6.1 Background of the Application

This paper presents an ML approach to predict shear wave acoustic velocity (Vs) in the Bakken Formation of the Williston Basin, based on seven petrophysical variables obtained from well log data. The transparent open-box (TOB) learning network is used as a non-parametric, optimized data-matching supervised learning algorithm to generate high-accuracy predictions of Vs values for the Bakken formation. The best-case solution achieved root mean squared error (RMSE) = 11.1 m/s, coefficient of determination (R^2) = 0.9994. The results demonstrate that TOB is a promising tool for predicting Vs values in unconventional reservoirs where conventional correlations may not exist.

2.4.2.6.2 Description of the Workflow and Machine Learning Model

In this study, the authors utilize the TOB algorithm, a supervised learning method, to predict Vs in the Bakken formation. The TOB algorithm focuses on locally weighted data matching principles and avoids traditional correlation-based methods, providing greater insight into underlying data relationships. It uses a memetic firefly optimization approach to fine-tune prediction weights, leveraging data record matching rather than relying on hidden layers or complex network architectures. This method involves two stages of calculations: Stage 1 generates initial predictions using basic record matching, and Stage 2 applies optimization to refine these predictions, minimizing the RMSE.

2.4.2.6.3 Advantages and Limitations of the Synchronized Algorithm

The TOB algorithm offers significant advantages in its transparency and ability to provide highly accurate predictions without relying on complex correlations. This makes it particularly suitable for datasets where traditional ML methods might overfit or struggle to identify meaningful patterns. The memetic firefly optimizer further enhances the TOB by optimizing the selection of data records used for prediction, thereby improving accuracy. However, the method's reliance on data record matching means that it requires a comprehensive dataset to perform effectively. Additionally, the optimization process can be computationally intensive, especially in the presence of large and complex datasets. Despite these challenges, the TOB approach is effective for predicting shear wave acoustic velocity, as demonstrated in the study's application to the Bakken formation dataset.

This analysis captures the methodology and highlights the advantages and limitations of the approach used in the paper. Adjustments can be made to better fit the specific context of the study or to elaborate further on particular aspects of the methodology.

2.4.2.6.4 Results and Discussion

This study demonstrates the potential of the TOB learning network as a powerful tool for predicting petrophysical properties in unconventional reservoirs such as the Bakken Formation of the Williston Basin. The results of this study show that TOB can provide high-accuracy predictions of Vs values based on seven petrophysical variables obtained from well log data. The low prediction errors obtained in this study indicate that TOB can be a valuable tool in reservoir characterization and optimization. TOB has the potential to improve our understanding of complex reservoirs and enhance the accuracy of production forecasts.

BIBLIOGRAPHY

Wood, David A. Bakken stratigraphic and type well log learning network exploited to predict and data mine shear wave acoustic velocity. *Journal of Applied Geophysics*, 173, 2020, 103936, ISSN 0926-9851, https://doi.org/10.1016/j.jappgeo.2019.103936

2.4.2.7 Application-17: Thickness Prediction for High-Resolution Stratigraphic Interpretation by Fusing Seismic Attributes of Target and Neighboring Zones with an SVR Algorithm

2.4.2.7.1 Background of the Application

The thickness of fluvial sand bodies is crucial for hydrocarbon exploration and development programs. One of the most popular methods for analyzing sand thickness based on 3D seismic data is a spatially dense sampling. However, in terms of high-resolution stratigraphic interpretations, the target zone's thickness range is usually less than the seismic temporal resolution. This problem significantly affects the seismic responses in the target zone while being influenced by the neighboring zones, the upper and lower zones. Therefore, authors propose a new method of sand thickness prediction that reduces the interference of neighboring zones using a support vector regression (SVR) algorithm with ML. The sand thickness values interpreted from wells were set as supervised data, and the seismic attributes around the wells in the target and neighboring zones were input as training data. Finally, the attributes in the target and neighboring zones were inverted into the predicted sand thickness using the trained SVR model. The proposed method significantly reduces the interference of neighboring zones, and the preliminary results demonstrate its accuracy in predicting sand thickness in a real seismic dataset of Chengdao Oilfield in the Bohai Bay Basin of China.

2.4.2.7.2 Description of the Workflow and Machine Learning Model

In the paper "Thickness prediction for high-resolution stratigraphic interpretation by fusing seismic attributes of target and neighboring zones with an SVR algorithm," the authors utilize an SVR model with an RBF kernel. The SVR is applied to predict the thickness of sand bodies by fusing seismic attributes from target and neighboring zones. The SVR is selected for its ability to capture complex relationships between seismic attributes and sand thickness, providing accurate predictions despite the inherent noise in seismic data. By using the RBF kernel, the SVR model effectively handles nonlinear patterns in the data.

2.4.2.7.3 Advantages and Limitations of the Synchronized Algorithm

The SVR with an RBF kernel offers several advantages in the context of seismic data analysis for sand thickness prediction. It excels at modeling nonlinear relationships, which are common in geological data, and is robust against overfitting, particularly when the data contains noise and complex patterns. This makes SVR suitable for high-resolution stratigraphic interpretation where accurate thickness estimation is crucial for exploration and development programs.

However, the SVR methodology also presents some limitations. It requires careful tuning of hyperparameters, such as the kernel parameter and the regularization term, to achieve optimal performance. This tuning process can be computationally intensive and time-consuming, especially with large datasets. Additionally, while SVR is effective at capturing complex relationships, its performance can be sensitive to the choice of kernel and parameter settings. Therefore, achieving the best results may require iterative experimentation and validation to ensure that the model is accurately capturing the underlying geological features.

Overall, the methodology demonstrates a significant improvement in predicting sand thickness by integrating seismic attributes from both target and neighboring zones, which reduces interference and enhances prediction accuracy. However, the computational cost and the need for extensive tuning remain challenges that need to be addressed when applying this approach to large-scale seismic datasets.

2.4.2.7.4 Results and Discussion

The new proposed method of sand thickness prediction has been tested using both 2D and 3D forward models and a real seismic dataset of Chengdao Oilfield in the Bohai Bay Basin of China. The results show that the proposed method significantly reduces the interference of neighboring zones and improves sand thickness prediction. The proposed approach can be used in the future to reduce drilling risks and enhance oil recovery for hydrocarbon reservoirs related to fluvial depositional systems.

BIBLIOGRAPHY

Li, Wei, Dali Yue, Shenghe Wu, Qinglin Shu, Wenfeng Wang, Tao Long, Benhua Zhang. Thickness prediction for high-resolution stratigraphic interpretation by fusing seismic attributes of target and neighboring zones with an SVR algorithm. *Marine and Petroleum Geology*, 113, 2020, 104153, ISSN 0264-8172, https://doi.org/10.1016/j.marpetgeo.2019.104153

2.4.2.8 Application-18: Seismic Fault Detection in Real Data Using Transfer Learning from a Convolutional Neural Network

2.4.2.8.1 Background of the Application

The task of automatic seismic fault detection has been improved by DL techniques that can take advantage of large amounts of seismic data and have excellent potential for assisted fault interpretation. However, these methods require significant computational power and effort to build datasets and tune models. In this work, the authors propose using transfer learning (TL) techniques to exploit an existing classifier and apply it to other seismic data with little effort. The proposed method uses a CNN trained and tuned on synthetic seismic data as the base model. The results show that TL on a single interpreted section gives satisfying results despite the high imbalance of labeled classes.

2.4.2.8.2 Description of the Workflow and Machine Learning Model

In this method, authors use TL with CNNs to detect seismic faults in real data. CNNs are effective for image feature extraction and classification due to their ability to learn complex patterns through layers of convolutional filters and pooling operations. The base model is a CNN pre-trained on synthetic seismic data, which is adapted to real seismic data using TL techniques. This involves fine-tuning the network on a small, marked dataset from the real data, making use of pre-learned features from the synthetic data to enhance the model's performance.

Authors employ two TL strategies: full fine-tuning (FFT) of the entire network and using the CNN as a feature extractor with additional classifiers like MLP and SVM. The CNN serves as the backbone of the approach, providing a robust foundation for fault detection by leveraging its pre-trained capability to recognize seismic patterns.

2.4.2.8.3 Advantages and Limitations of the Synchronized Algorithm

The use of CNNs in conjunction with TL allows for the efficient adaptation of pre-trained models to new seismic datasets, reducing the need for large amounts of labeled real data. This methodology is particularly advantageous because it leverages the strengths of CNNs in image processing, enabling accurate fault detection while mitigating the challenges of data scarcity. Additionally, the feature extractor approach with SVM and MLP offers flexibility in fine-tuning the model for enhanced performance.

However, the approach does have limitations. Full fine-tuning of the CNN can be computationally expensive, and the effectiveness of the TL strategy heavily depends on the similarity between synthetic and real datasets. Moreover, the imbalance in fault and non-fault classes can affect model sensitivity and specificity. The feature extractor approach may require careful parameter tuning to achieve optimal results, and the performance can vary with different datasets. Despite these challenges, the methodology offers a practical solution for seismic fault detection with promising results in terms of sensitivity and specificity.

2.4.2.8.4 Results and Discussion

The proposed technique of using TL with a pre-trained CNN model for seismic fault detection has shown promising results on real data with little effort. The TL approach allows working with small datasets and adapting to different kinds of data. The proposed networks are easily tuned and trained in a few minutes on a CPU, making the technique suited for practical day-to-day use. The authors suggest that the proposed technique can be further improved by using multiple interpreted sections and by exploring more advanced TL strategies.

BIBLIOGRAPHY

Cunha, Augusto, Axelle Pochet, Hélio Lopes, Marcelo Gattass. Seismic fault detection in real data using transfer learning from a convolutional neural network pre-trained with synthetic seismic data, *Computers & Geosciences*, 135, 2020, 104344, ISSN 0098-3004, https://doi.org/10.1016/j.cageo.2019.104344

2.4.2.9 Application-19: Fault Detection and Diagnosis in Process System Using Artificial Intelligence-Based Cognitive Technique

2.4.2.9.1 Background of the Application

This paper discusses the importance of safety and reliability in chemical process industries, which require continuous monitoring and efficient detection of abnormalities. The study covers the application of both model-based and data-driven methods for fault detection and diagnosis. Model-based approaches rely heavily on the mathematical model of the system and its components, while data-driven approaches require good quality historical data from the system. Supervised learning methods in data-driven approaches require prior knowledge of the fault type, several functional forms, and fault conditions. The paper proposes a hybrid approach that combines the advantages of both methods. The proposed framework uses an incremental one-class algorithm to identify anomaly conditions and introduces a new state of fault to the current fault states if an unknown fault occurs. A dynamic shallow neural network (SNN) is used to learn and classify the fault state. The effectiveness of the proposed algorithm and model is tested using the Tennessee Eastman process and a pilot-scale system.

2.4.2.9.2 Description of the Workflow and Machine Learning Model

In this study, the authors utilize an incremental one class neural network (NN) and an SNN for fault detection and diagnosis in process systems. The incremental one class NN is designed to identify unknown fault conditions by analyzing non-faulty data and detecting anomalies without requiring labeled fault data. It uses a decision score to differentiate normal and anomalous data points, optimizing its parameters using a minimization approach. The shallow NN, on the other hand, is employed for classifying known fault states using a supervised learning approach. It dynamically updates its output nodes to accommodate newly detected fault conditions, thus enabling it to learn and classify a variety of fault types as they are identified.

2.4.2.9.3 Advantages and Limitations of the Synchronized Algorithm

The incremental one class NN offers significant advantages by being able to detect unknown faults without prior knowledge, making it highly flexible and suitable for complex systems where fault conditions are not predefined. Its unsupervised learning approach enables it to adapt to new fault conditions dynamically. However, this model may require extensive computation time when updating with new data, particularly in systems with high noise levels.

The shallow NN complements the incremental one class NN by providing robust classification of known faults through supervised learning. It can dynamically update its structure to incorporate new fault states, enhancing its adaptability. Despite its strengths, the shallow NN relies heavily on the quality of training data and may require retraining as new faults are detected. Both models collectively provide a comprehensive framework for fault detection and diagnosis but may require fine-tuning of parameters and significant computational resources to achieve optimal performance in real-time applications.

This structured description captures the methodology and the advantages/limitations of the AI approach used in the paper, reflecting the style of the example you provided.

2.4.2.9.4 Results and Discussion

The benefits of model-based and data-driven approaches to fault diagnosis and detection are combined in the suggested hybrid framework. The framework learns and classifies the fault state using a dynamic SNN and detects unknown faults using an incremental one-class approach. The Tennessee Eastman process and a pilot-scale system are used to test the efficacy of the suggested algorithm and model. The findings support the suggested framework's efficacy in identifying and categorizing known and undiscovered process operating flaws. The framework can be used for various complicated systems where it is difficult to define the mathematical models and fault situations in advance.

BIBLIOGRAPHY

Arunthavanathan, Rajeevan, Faisal Khan, Salim Ahmed, Syed Imtiaz, Risza Rusli. Fault detection and diagnosis in process system using artificial intelligence-based cognitive technique, *Computers & Chemical Engineering*, 134, 2020, 106697, ISSN 0098-1354, https://doi.org/10.1016/j.compchemeng.2019.106697

2.4.3 Seismic

2.4.3.1 Application-20: A Theory-Guided Deep Learning Formulation of Seismic Waveform Inversion

2.4.3.1.1 Background of the Application

DL techniques are becoming crucial in seismic data processing, particularly for inversion and interpretation tasks. This study introduces a theory-guided RNN designed to handle both forward and inverse seismic problems, drawing on the strengths of traditional physical models and the adaptability of DL. The authors demonstrate that under certain conditions, RNN training can approximate gradient-based full waveform inversion (FWI). Their experiments show that the adaptive moment (Adam) optimization method, with a learning rate tuned to match standard FWI updates, provides the most stable results. This approach is validated through a multidimensional 2D Marmousi model experiment, confirming its robustness and efficacy.

2.4.3.1.2 Description of the Workflow and Machine Learning Model

In this study, the authors utilize a theory-guided RNN to address the challenges in seismic waveform inversion. The RNN is designed to model dynamic processes, leveraging its ability to retain information through time, which is crucial for simulating wave propagation. The network architecture integrates the physics of wave equations with DL, allowing it to handle forward and inverse seismic problems effectively. To optimize the RNN, authors employ the adaptive moment estimation (Adam) algorithm, which adjusts the learning rates dynamically based on the gradient's first and second moments. This approach helps stabilize and accelerate the convergence of the inversion process, making it more efficient than traditional methods.

2.4.3.1.3 Advantages and Limitations of the Synchronized Algorithm

The theory-guided RNN offers significant advantages in seismic waveform inversion by incorporating physical models, which enhance its interpretability and reliability. This integration reduces the dependency on large training datasets and provides a robust framework for solving complex geophysical problems. The use of the Adam optimizer further improves the RNN's performance by providing faster convergence and better handling of non-stationary objectives. However, the methodology also has limitations. The design and training of the RNN require careful tuning of hyperparameters to achieve optimal results, and the computational cost can be high, especially for large-scale 2D or 3D models. Despite these challenges, the proposed approach successfully balances theory and data-driven insights, offering a promising solution for seismic inversion tasks.

2.4.3.1.4 Results and Discussion

This research highlights the potential of DL, specifically a theory-guided RNN, in seismic waveform inversion. The study confirms that when the RNN mimics wave propagation through a heterogeneous medium, the training process effectively becomes a form of waveform inversion. Through theoretical analysis and empirical testing, the best learning rates for various gradient-based algorithms were identified. Among the tested methods, Adam proved to be the fastest and most stable, successfully recovering detailed velocity profiles in both shallow and deep zones of the 2D Marmousi model. The findings suggest that combining data-driven and theory-guided approaches can create effective and flexible seismic inversion methods, balancing the strengths of both methodologies.

BIBLIOGRAPHY

Sun, Jian, Zhan Niu, Kristopher A. Innanen, Junxiao Li, and Daniel O. Trad. "A theory-guided deep learning formulation of seismic waveform inversion." Paper presented at the *SEG International Exposition and Annual Meeting*, San Antonio, TX, September 2019. https://doi.org/10.1190/segam2019-3215762.1

2.4.3.2 Application-21: Applications of Machine Learning to the Spatial Interpolation of Aeromagnetic Data

2.4.3.2.1 Background of the Application

This study explores the use of ML techniques, specifically random forest regression (RFR) and SVR, for interpolating aeromagnetic data. By developing relevant predictors and training these models on synthetic aeromagnetic data, the researchers were able to successfully interpolate new datasets. The approach utilized a "filter"-based method, integrating standard interpolation techniques as predictors, which enhanced the ML model's performance, producing interpolations superior to those from individual standard methods.

2.4.3.2.2 Description of the Workflow and Machine Learning Model

In this study, RFR and SVR are employed to interpolate aeromagnetic data. These ML techniques aim to predict the missing aeromagnetic values between flight lines by leveraging spatial relationships. The RF model works by averaging the predictions of numerous DTs, which helps in reducing overfitting and producing smooth interpolations. SVR, with an RBF kernel, is used for its ability to model complex, nonlinear relationships in the data. The SVR method captures intricate patterns in aeromagnetic data, while RFR provides a robust baseline due to its ensemble nature.

2.4.3.2.3 Advantages and Limitations of the Synchronized Algorithm

The application of RFR offers robustness against overfitting by averaging across multiple DTs, making it well-suited for spatial interpolation tasks. It is particularly effective in smoothing out noise and handling missing data. However, RFR can sometimes miss subtle patterns due to its averaging nature. SVR, with its RBF kernel, excels at modeling complex, nonlinear relationships and is beneficial for capturing detailed patterns in aeromagnetic data. Despite its strengths, SVR requires careful tuning of hyperparameters and can be computationally expensive, especially with large datasets. The effectiveness of both methods is enhanced by incorporating standard interpolation results as predictors, but this also increases computational complexity. Thus, while the ML approach shows promise, it requires careful parameter tuning and computational resources to achieve optimal results.

2.4.3.2.4 Results and Discussion

The research demonstrates the potential of ML methods in improving aeromagnetic data interpolation. By training RF and SVR models on synthetic datasets, the study shows that these methods can be effectively applied to new aeromagnetic datasets. While initial results using ML predictors alone were promising but not superior to standard methods, integrating results from standard interpolation methods as additional predictors led to better outcomes. Future work should involve applying these trained models to real-world data and incorporating additional related datasets, like borehole measurements and seismic data, to further refine the interpolation results.

BIBLIOGRAPHY

Naprstek, Tomas, and Richard Smith. "Applications of machine learning to the spatial interpolation of aeromagnetic data." Paper presented at the *SEG International Exposition and Annual Meeting*, San Antonio, TX, September 2019. https://doi.org/10.1190/segam2019-3215629.1

2.4.3.3 Application-22: Estimating Total Magnetization Directions Using Convolutional Neural Networks

2.4.3.3.1 Background of the Application

In this study, the authors explore the use of CNNs for predicting the magnetization directions of subsurface magnetic source bodies. Accurate interpretation of magnetic data is critical for various geological applications, and this research aims to automate the process using ML. The authors created synthetic magnetic data maps with different magnetization directions and trained two separate CNNs: one for predicting magnetization inclinations and another for predicting declinations. The optimal CNN architectures were identified, achieving high accuracy rates of 98% for declination and 100% for inclination predictions. The approach was further validated using real magnetic field data from the Black Hill Norite, Australia, demonstrating the potential of ML in geophysical data interpretation.

2.4.3.3.2 Description of the Workflow and Machine Learning Model

In this study, the authors employ CNNs to predict the magnetization directions of magnetic source bodies based on synthetic magnetic data maps. CNNs are particularly well-suited for image recognition tasks, allowing us to effectively capture spatial patterns and features in the magnetic data. Two separate CNN models are trained: one for predicting magnetization inclinations and another for predicting declinations. Each model uses a supervised learning approach, leveraging labeled datasets to learn the underlying patterns and relationships. Authors explore various CNN architectures to identify the optimal configurations for accurate predictions, with the best models achieving high test accuracies of 98% for declination and 100% for inclination predictions.

2.4.3.3.3 Advantages and Limitations of the Synchronized Algorithm

The use of CNNs in this study offers several advantages, including the ability to automatically learn complex spatial features from magnetic data, which enhances prediction accuracy without requiring extensive feature engineering. CNNs are also robust against noise in the data and can generalize well to unseen samples. However, training CNNs can be computationally intensive, especially when exploring multiple architectures and large datasets. Additionally, the model's performance is highly dependent on the quality and size of the training data. While the optimal architectures were identified for this specific application, the methodology requires substantial computational resources and may need adaptation for different datasets or geophysical scenarios. The approach demonstrated promising results, but further tuning and testing are necessary for broader application in varied geological contexts.

2.4.3.3.4 Results and Discussion

This research demonstrates the effectiveness of using CNNs to predict magnetization directions from magnetic data maps. By testing various CNN architectures, optimal models that provide high prediction accuracy for both synthetic and field data were identified. Specifically, the inclination predictor achieved 100% accuracy, while the declination predictor reached 98%. These results highlight the potential of CNNs to significantly improve the speed and accuracy of magnetic data interpretation, offering a promising tool for geophysical applications. The methodology outlined can be adapted and applied to other datasets, making it a valuable approach for future studies in the field.

BIBLIOGRAPHY

Nurindrawati, Felicia, and Jiajia Sun. "Estimating total magnetization directions using convolutional neural networks." Paper presented at the *SEG International Exposition and Annual Meeting*, San Antonio, TX, September 2019. https://doi.org/10.1190/segam2019-3216857.1

2.4.3.4 Application-23: Machine Learning-Based Seismic Data Enhancement toward Overcoming Geophysical Limitations

2.4.3.4.1 Background of the Application

Seismic data acquisition often requires high-quality, unblended, well-sampled, and broadband data, but obtaining such data is usually very costly. This study introduces an ML approach that enables the acquisition of blended, sparsely sampled, and narrowband data, which can be processed to achieve results similar to complete data. By using a deep CNN, the method suppresses blending noise, reconstructs missing traces, and extrapolates low frequencies. The performance of this approach is validated with both synthetic and real-world datasets, demonstrating its capability to enhance data quality despite the limitations of incomplete input data.

2.4.3.4.2 Description of the Workflow and Machine Learning Model

In this study, a deep CNN based on the U-Net architecture is utilized to enhance seismic data. The network consists of pairs of encoding and decoding blocks with a central bottleneck layer, designed to process blended, sparsely sampled, and narrowband seismic data. Key features of the U-Net architecture include max pooling layers for down-sampling and transposed convolutional layers for up-sampling, alongside skip connections that preserve essential information by linking encoder and decoder blocks directly. Convolutional layers, followed by rectified linear unit (ReLU) activations and batch normalization, are used to ensure efficient training and accurate predictions. The CNN effectively performs tasks such as suppressing blending noise, reconstructing missing traces, and extrapolating low frequencies from incomplete data.

2.4.3.4.3 Advantages and Limitations of the Synchronized Algorithm

The U-Net-based CNN is particularly well-suited for enhancing seismic data due to its ability to capture spatial hierarchies and detailed information through skip connections. This architecture is advantageous for dealing with incomplete data, as it can efficiently reconstruct missing information and suppress noise, leading to high-quality output data. However, the model requires a significant amount of data for training and can be computationally intensive due to its deep architecture. Additionally, the quality of the output depends on the diversity and representativeness of the training datasets. Despite these challenges, the proposed method demonstrates superior performance compared to traditional geophysical approaches, especially in handling low-frequency extrapolation and noise suppression.

2.4.3.4.4 Results and Discussion

This study applies a deep CNN, based on the U-Net architecture, to predict complete seismic data from incomplete data by simultaneously reducing blending noise, reconstructing missing traces, and extrapolating low frequencies. Validation with synthetic and field datasets confirms that the proposed method effectively derives high-quality complete data, even when initial data are significantly compressed. Notably, the method achieves comparable prediction accuracy for both extrapolated and existing frequencies, a feat not easily accomplished with traditional geophysical methods. The ML approach shows potential for significantly improving the efficiency of seismic data acquisition without compromising data quality.

BIBLIOGRAPHY

Nakayama, Shotaro, and Gerrit Blacquière. "Machine Learning Based Seismic Data Enhancement Towards Overcoming Geophysical Limitations." Paper presented at the *Abu Dhabi International Petroleum Exhibition & Conference*, Abu Dhabi, UAE, November 2020. https://doi.org/10.2118/202710-MS

2.4.3.5 Application-24: Understanding Injection-Induced Seismicity in Enhanced Geothermal Systems

2.4.3.5.1 Background of the Application

This study investigates the mechanisms of injection-induced seismicity in enhanced geothermal systems (EGS) by examining the coupled thermo-hydro-mechanical-chemical (THMC) processes that occur during EGS operations. The study proposes a methodology for predicting the occurrence of injection-induced seismicity by analyzing the pore pressure diffusion, temperature alteration, and stress-aided corrosion processes that occur during EGS. The results of this study improve our understanding of the causes of injection-induced seismicity and can help in the design of EGS projects to minimize the risk of such events.

2.4.3.5.2 Description of the Workflow and Machine Learning Model

The paper explores various theoretical and numerical models to understand injection-induced seismicity in EGS. It discusses the use of different forecasting approaches:

1. Statistical Forecasting: This approach involves using statistical models to predict seismic events based on historical data and trends. It provides a probabilistic assessment of earthquake occurrence.
2. Hybrid Forecasting: This method combines statistical models with physical simulations to improve the accuracy of predictions. By integrating multiple data sources, hybrid forecasting aims to enhance the reliability of seismic risk assessments.
3. Physically Based Forecasting: This approach relies on the physical modeling of geomechanical processes and reservoir properties. It includes the simulation of coupled THMC processes to understand fault behaviors and predict seismic events.

2.4.3.5.3 Advantages and Limitations of the Synchronized Algorithm

Advantages:

- Statistical Forecasting: Provides quick insights and probabilistic assessments, useful for preliminary analysis.

- Hybrid Forecasting: Increases prediction accuracy by leveraging diverse data sources and methodologies, offering a more comprehensive risk assessment.
- Physically Based Forecasting: Offers detailed insights into the underlying physical processes, improving understanding and long-term predictions.

Limitations:

- Statistical Forecasting: May not capture complex interactions between physical processes, potentially leading to oversimplified predictions.
- Hybrid Forecasting: While accurate, it requires significant computational resources and expertise to integrate different models effectively.
- Physically Based Forecasting: Highly dependent on the quality and availability of physical data. It can be computationally intensive and challenging to implement in real-time scenarios.

2.4.3.5.4 Results and Discussion

In this study, the authors have investigated the mechanisms of injection-induced seismicity in EGS by analyzing the THMC processes that occur during EGS operations. The results show that pore pressure diffusion, temperature alteration, and stress-aided corrosion processes are the main causes of injection-induced seismicity in EGS. They have proposed a methodology for predicting the occurrence of injection-induced seismicity by analyzing these processes. Their findings can help in the design and operation of EGS projects to minimize the risk of injection-induced seismicity. However, further research is needed to improve our understanding of these complex processes and to develop more accurate prediction models.

BIBLIOGRAPHY

Rathnaweera, Tharaka Dilanka, Wei Wu, Yinlin Ji, Ranjith Pathegama Gamage. Understanding injection-induced seismicity in enhanced geothermal systems: From the coupled thermo-hydro-mechanical-chemical process to anthropogenic earthquake prediction. *Earth-Science Reviews*, 205, 2020, 103182, ISSN 0012-8252, https://doi.org/10.1016/j.earscirev.2020.103182

2.4.3.6 Application-25: Using Machine Learning Approach for Microseismic Events Recognition in Underground Excavations

2.4.3.6.1 Background of the Application

A study on the use of ML to detect blasting and microseismic occurrences in subterranean excavations is presented in this publication. Since the gathered signals are constantly changing, conventional methods such as waveform analysis may not attain a respectable level of recognition accuracy and necessitate operational personnel with significant physical and signal processing skills. Additionally, traditional methods are ineffective and time-consuming due to the growing amount of monitoring data. The study used 10 popular ML models for microseismic/blasting event recognition, including feature selection, model training, and hyperparameter tuning, based on these factors. According to the results, the Gaussian process model performed the poorest when it came to identifying microseismic/blasting events, whereas the logistic regression (LGR) model performed the best. The study demonstrates the potential of ML in microseismic/blasting events identification and provides geological engineers with reliable prior knowledge in model selection.

2.4.3.6.2 Description of the Workflow and Machine Learning Model

Ten distinct ML models are used in this work to categorize blasting and microseismic occurrences in subterranean excavations. Backpropagation neural network (BPNN), Gaussian process classifier (GPC), DT, Naive Bayesian classifier (NBC), kNN, LGR, AdaBoost (ADA), RF, gradient boosting classifier (GBC), and SVM are among the models. A dataset comprising six features—the logarithm of seismic moment (lgM), logarithm of microseismic energy (lgE), number of triggered sensors (N), logarithm of first peak amplitude (lgA), maximum peak arrival time (T), and dominant frequency (F)—derived from microseismic monitoring data is used to train each model.

Hyperparameter tuning is conducted for each model to optimize performance, utilizing methods such as cross-validation and grid search. The study evaluates the models based on multiple performance indicators, including accuracy, precision, recall, F1 score, and AUC (area under the ROC curve), to determine the most suitable model for the task.

2.4.3.6.3 Advantages and Limitations of the Synchronized Algorithm

The diverse set of ML models employed in this study allows for a comprehensive evaluation of their capabilities in distinguishing between microseismic and blasting events. Each model has distinct advantages:

- SVM with an RBF kernel is effective at handling nonlinear relationships in data, providing robust classification performance. However, it requires careful tuning of hyperparameters such as the penalty parameter (C) and the kernel coefficient (gamma), which can be computationally intensive for large datasets.
- BPNN is capable of modeling complex patterns and interactions in the data, but it can suffer from overfitting and requires significant computational resources for training.
- LGR is simple and interpretable, offering good baseline performance with relatively low computational cost. It performs well with fewer data and linear separability.
- Ensemble methods such as ADA and GB can improve classification accuracy by combining multiple weak learners, but they are sensitive to noisy data and outliers and can be computationally expensive.
- RF provides high accuracy and robustness to overfitting by averaging multiple DTs, but it can be slower and less interpretable than single DT models.
- kNN is straightforward and effective for smaller datasets, but its performance degrades with larger datasets due to increased computational cost and sensitivity to irrelevant features.

The study highlights that while models like LGR and kNN offer simplicity and interpretability, they may not capture complex relationships as effectively as neural networks or ensemble methods. Additionally, the computational cost and requirement for hyperparameter tuning are critical factors that influence the choice of the model. Ultimately, the study recommends a balanced approach, considering both model performance and computational efficiency for practical applications in geophysical monitoring and analysis.

2.4.3.6.4 Results and Discussion

The study investigated the performance of 10 ML models for microseismic/blasting events recognition in underground excavations. The LGR model was found to be the most effective in identifying microseismic/blasting events, and the Gaussian process model was the least effective. The study demonstrates the potential of ML in microseismic/blasting events identification and provides geological engineers with reliable prior knowledge in model selection. The results of this study should encourage the use of ML for microseismic/blasting events identification and help improve the accuracy, efficiency, and reliability of the identification process in underground excavations. Future research could explore the application of ML to the prediction of rock bursts and the monitoring of tunnel deformation in underground excavations.

BIBLIOGRAPHY

Pu, Yuanyuan, Derek B. Apel, and Robert Hall, Using machine learning approach for microseismic events recognition in underground excavations: Comparison of ten frequently-used models, *Engineering Geology*, 268, 2020, 105519, ISSN 0013-7952, https://doi.org/10.1016/j.enggeo.2020.105519

2.4.3.7 Application-26: A SCiForest-Based Semi-supervised Learning Method for the Seismic Interpretation of Channel Sand-Body

2.4.3.7.1 Background of the Application

The identification and interpretation of fluvial channels in subsurface reservoirs is important for hydrocarbon E&P. Seismic attributes, including amplitude, frequency, and morphological attributes, have been employed to delineate channel features, but the results are often insufficiently calibrated. In this paper, the authors propose a semi-supervised method called SCiForest (isolation forest with split-selection criterion) algorithm by combining unsupervised isolation forest with the supervised feature selection process. The proposed method is evaluated on a fluvial channel dataset, and the results demonstrate that SCiForest is effective in improving the interpretation of channels with insufficient calibration.

2.4.3.7.2 Description of the Workflow and Machine Learning Model

In this study, a semi-supervised learning approach is used, combining an unsupervised Isolation Forest with a split-selection criterion (SCiForest) and a supervised feature selection process. The SCiForest algorithm is employed to exploit the self-contained distribution information of multiple seismic attributes. It is an improvement over the standard isolation forest, as it introduces random hyperplanes and

optimizes splits using a split-selection criterion (SDgain). This enables the model to detect both scattered and clustered anomalies, which are crucial for interpreting seismic data. The supervised feature selection process helps in identifying the most relevant attributes for seismic channel identification by optimizing their combination under the guidance of well-point samples.

2.4.3.7.3 Advantages and Limitations of the Synchronized Algorithm

The SCiForest-based method can effectively identify channels in seismic data by detecting clustered anomalies, leveraging the unsupervised algorithm's ability to utilize distribution patterns in the data. This is particularly advantageous when labeled samples are insufficient, as it uses both labeled and unlabeled data. The method outperforms traditional supervised and unsupervised algorithms by combining their strengths, allowing for a more comprehensive understanding of seismic patterns.

However, the method assumes that the distribution patterns in multiple attributes are correlated with the classification patterns of formations, which may not hold true in all cases. If the channels are too narrow or the sand-bodies are not thick enough to cause observable response differences, the method may not perform well. Additionally, while the SCiForest is robust in identifying clustered anomalies, its performance depends on the quality of input data and the initial attribute selection. This process might require substantial computational resources and expert intervention to select the most effective attributes and optimize model parameters.

2.4.3.7.4 Results and Discussion

In this paper, the authors proposed a semi-supervised method called SCiForest for the identification and interpretation of fluvial channels in subsurface reservoirs. The proposed method combines unsupervised isolation forest with supervised feature selection to improve the interpretation of channels with insufficient calibration. The results of experiments on a fluvial channel dataset demonstrate the effectiveness of the proposed method. In comparison with other state-of-the-art techniques, SCiForest provides a more accurate and efficient approach for fluvial channel identification and interpretation. The proposed method has the potential to be applied in hydrocarbon E&P, as well as other fields where subsurface reservoir characterization is required.

BIBLIOGRAPHY

Ao, Yile, Hongqi Li, Liping Zhu, Zhongguo Yang. A SCiForest based semi-supervised learning method for the seismic interpretation of channel sand-body. *Journal of Applied Geophysics*, 167, 2019, 51–62, ISSN 0926-9851, https://doi.org/10.1016/j.jappgeo.2019.04.019

2.4.3.8 Application-27: Machine Learning Assisted Velocity Auto-Picking

2.4.3.8.1 Background of the Application

The development of velocity auto-pickers has the potential to decrease the amount of time spent processing large volumes of seismic data, while increasing the number of common mid-point (CMP) gathers that are picked in a project. A new technique has been developed that utilizes seismic attributes in conjunction with an unsupervised ML clustering algorithm. The results of this method have shown that it could potentially reduce the amount of time spent on manual picking, thus driving down the cost of processing a dataset.

2.4.3.8.2 Description of the Workflow and Machine Learning Model

In this method, an unsupervised ML clustering algorithm is used to improve the velocity auto-picking process in seismic data analysis. The clustering algorithm identifies the optimal locations for velocity picks by analyzing a combination of seismic attributes, including semblance, AVO (amplitude versus offset), and continuity measures. These attributes are used to mitigate the impact of noise on the semblance-only auto-pickers. The algorithm projects the events falling into a select region of a multidimensional attribute space into the velocity–time plane, enhancing the separation between signal and noise. By using this clustering approach, the methodology effectively reduces the time required for manual velocity picking and improves the accuracy of velocity models.

2.4.3.8.3 Advantages and Limitations of the Synchronized Algorithm

The clustering-based method provides a significant advantage by reducing noise impact and allowing for automatic identification of optimal velocity picks. It enhances the accuracy of velocity models and reduces manual intervention, leading to cost savings in processing seismic data. The unsupervised nature of the clustering algorithm enables it to adapt to different datasets, making it versatile for both marine and land surveys.

However, the method has limitations. The clustering algorithm's performance depends on the choice of attributes and their filtering thresholds, requiring careful tuning for each dataset. While the method reduces the need for manual picking, the initial setup and parameter tuning can be complex. Additionally, the method may still struggle in areas with highly complex subsurface structures or extreme noise levels, where more sophisticated approaches may be required. Despite these challenges, the clustering approach represents a valuable tool for improving the efficiency and accuracy of seismic velocity analysis.

2.4.3.8.4 Results and Discussion

The new method described in this paper has been shown to be effective in reducing the time and cost of processing a seismic project. The method was tested on two different datasets, a marine dataset and a land dataset, and showed promising results in both cases. By utilizing ML, this method has the potential to greatly reduce the amount of manual picking required and thus decrease the cost of processing seismic data.

BIBLIOGRAPHY

Smith, K.J. Machine learning assisted velocity auto-picking. *SEG Technical Program Expanded Abstracts* 2017. https://doi.org/10.1190/segam2017-17684719.1

2.4.4 Lithology

2.4.4.1 Application-28: Application of Artificial Intelligence on Black Shale Lithofacies Prediction

2.4.4.1.1 Background of the Application

This paper discusses the prediction of shale lithofacies in the Marcellus Shale by AI classifiers, primarily ANN, and SVM. Seven shale lithofacies in the Marcellus Shale have been defined by mineral composition and TOC content. The relationship between lithofacies and logs is nonlinear and complex, and prediction by conventional mathematical methods is limited. The effectiveness of AI classifiers in solving complex nonlinear problems has been demonstrated. In this paper, the authors present an integrated workflow for organic-rich shale lithofacies prediction, which should be helpful for recognizing shale lithofacies in other shale-gas plays.

2.4.4.1.2 Description of the Workflow and Machine Learning Model

In this study, ANNs and SVMs are utilized for the prediction of shale lithofacies in the Marcellus Shale. ANN, inspired by the human brain's ability to process complex patterns, is employed due to its capability to handle nonlinear relationships in data. The ANN model is trained using an MLP architecture, which adjusts the weights and biases through backpropagation to minimize the error between the predicted and actual lithofacies.

SVMs with RBFRBF kernel are also implemented, which are effective in handling nonlinear separable data by transforming it into a higher-dimensional space. This allows the SVM to find an optimal hyperplane that maximizes the margin between different lithofacies classes.

Furthermore, PSO is applied to enhance the training process of ANN and SVM models. PSO optimizes the model parameters by simulating a swarm of particles that search for optimal solutions in the parameter space, improving the models' accuracy and efficiency.

2.4.4.1.3 Advantages and Limitations of the Synchronized Algorithm

The combination of ANN and SVM provides a robust framework for shale lithofacies prediction. ANN is highly flexible and can model complex, nonlinear relationships between input well logs and lithofacies, making it well-suited for this application. However, ANNs require a careful design of network architecture and are prone to overfitting, especially with small datasets.

SVM, with its RBF kernel, offers a powerful approach to classify complex data structures and is less susceptible to overfitting compared to ANNs. Nevertheless, SVMs require meticulous parameter tuning, such as selecting the penalty parameter and kernel width, which can be computationally intensive.

The integration of PSO helps to optimize these parameters, enhancing the predictive performance of both ANN and SVM models. Despite its advantages, PSO's efficiency heavily relies on its initialization and may require several iterations to converge to the best solution. As a result, while PSO reduces the computational burden compared to exhaustive search methods, it still requires significant computational resources to achieve optimal performance.

By applying these AI techniques, the study aims to improve the accuracy of shale lithofacies prediction, thereby aiding in identifying productive shale gas zones in the Marcellus Shale.

2.4.4.1.4 Results and Discussion

The authors provided an integrated methodology for predicting organic-rich shale lithofacies and thoroughly examined the advantages and disadvantages of AI classifiers in pattern recognition. The approach should assist in locating high-productive shale gas sweet spots by detecting shale lithofacies in other shale-gas areas.

BIBLIOGRAPHY

Wang, Guochang, Yiwen Ju, Timothy R. Carr, Chaofeng Li, and Guojian Cheng. "Application of Artificial Intelligence on Black Shale Lithofacies Prediction in Marcellus Shale, Appalachian Basin." Paper presented at the *SPE/AAPG/SEG Unconventional Resources Technology Conference*, Denver, Colorado, August 2014. https://doi.org/10.15530/URTEC-2014-1935021

2.4.4.2 Application-29: Evaluation of Machine Learning Methods for Formation Lithology Identification

2.4.4.2.1 Background of the Application

Identification of lithology in underground formations from well log data is a crucial task in the field of petroleum exploration and engineering. In this paper, the authors evaluate five ML algorithms, including Naïve Bayes, SVM, ANN, RF, and gradient tree boosting (GTB), to improve the accuracy of lithology identification. The authors use data from the Daniudui and Hangjinqi gas fields to train and test the models. The input parameters for each model are features selected from different well log data samples. The study employs hyperparameter optimization to obtain the best parameter set for each model and uses a validation curve to determine the parameter search range. The performance of the classifiers is evaluated using five-fold cross-validation. The results indicate that ensemble methods like GTB and RF are suitable options for supervised classification of lithology using well log data.

2.4.4.2.2 Description of the Workflow and Machine Learning Model

In this study, the authors evaluate five ML methods for formation lithology identification using well log data: Naïve Bayes (NB), SVM with an RBF kernel, ANN, RF, and GTB. These models are trained using features derived from well log data such as GR, acoustic, and density logs. The models are fine-tuned using hyperparameter optimization to improve classification accuracy. RF and GTB, as ensemble methods, leverage multiple DTs to enhance predictive performance. The SVM with an RBF kernel effectively handles nonlinear relationships in the data, while the ANN is used to capture complex patterns through its multilayered architecture.

2.4.4.2.3 Advantages and Limitations of the Synchronized Algorithm

The ensemble methods, particularly RF and GTB, are advantageous for lithology classification due to their ability to handle a small number of features with high accuracy. They are robust against overfitting and can provide reliable classification of lithology types, even in complex sandstone formations. However, these models require careful tuning of parameters and are computationally intensive, particularly GTB, which can be slower to train. The SVM with RBF kernel is effective for modeling complex nonlinear relationships but necessitates meticulous parameter tuning and can be slow with large datasets. The ANN is capable of capturing intricate data patterns but is prone to overfitting, especially with limited features, and its performance can vary with different network architectures. While NB serves as a baseline due to its simplicity, its assumption of feature independence limits its accuracy in this context.

2.4.4.2.4 Results and Discussion

The authors of this paper evaluate five ML algorithms for the formation lithology identification from well log data. The study shows that ensemble methods like GTB and RF classifiers have lower prediction errors compared to other methods. Although all the models have difficulties in distinguishing sandstone classes, GTB performs well on this task compared to other methods. The authors find that classification accuracy is similar across lithology classes for both GTB and RF models. The study suggests that ML algorithms can improve the accuracy of lithology identification from well log data and can be used in the field of petroleum exploration and engineering.

BIBLIOGRAPHY

Xie, Yunxin, Chenyang Zhu, Wen Zhou, Zhongdong Li, Xuan Liu, Mei Tu. Evaluation of machine learning methods for formation lithology identification: A comparison of tuning processes and model performances. *Journal of Petroleum Science and Engineering*, 160, 2018, 182–193, ISSN 0920-4105, https://doi.org/10.1016/j.petrol.2017.10.028

2.4.4.3 Application-30: A Machine Learning Approach to Facies Classification Using Well Logs

2.4.4.3.1 Background of the Application

This paper proposes an ML approach for facies classification using wireline logging measurements. The proposed pipeline includes a GBC and feature augmentation techniques to improve classification accuracy. The approach was tested on a dataset of more than 4000 observations from 10 different wells and achieved an f-score of 0.61, demonstrating its effectiveness in classifying facies from indirect measurements. The paper also discusses the importance of feature augmentation and the potential of adding geological constraints to further improve classification.

2.4.4.3.2 Description of the Workflow and Machine Learning Model

In this method, a GBC is used for the task of facies classification using well log data. GB is an ensemble method that combines multiple weak DT classifiers to create a strong predictive model. It is particularly effective in capturing complex patterns in the data, even with a relatively small number of features. The key innovation in this approach is feature augmentation, where additional features are generated through quadratic expansions and second-order interactions. This augmentation significantly enhances the model's ability to discriminate between different facies, improving classification accuracy.

2.4.4.3.3 Advantages and Limitations of the Synchronized Algorithm

The GBC is robust in handling a small number of features and can efficiently learn from the augmented feature set, making it suitable for facies classification. Its ensemble nature helps mitigate the overfitting typically associated with DTs. However, GBC requires careful tuning of hyperparameters, such as the number of trees and maximum features, to achieve optimal performance. The feature augmentation technique increases the complexity of the model but significantly improves the classifier's ability to separate classes. Despite these advantages, the method's performance may degrade with unbalanced datasets, and the training process can be computationally expensive due to the large number of augmented features. Additionally, post-processing steps, like median filtering, are necessary to refine predictions and correct misclassification errors.

2.4.4.3.4 Results and Discussion

In conclusion, this paper presents a successful ML pipeline for facies classification based on wireline logging measurements. The proposed approach, which includes a GBC and feature augmentation techniques, demonstrated high accuracy in classifying facies from indirect measurements. The results confirm the importance of feature augmentation and highlight the potential for incorporating geological constraints into the classification process. Future work will focus on exploring DL strategies for feature learning and classification, as well as incorporating additional geological information to further improve classification accuracy. Overall, this study provides valuable insights into the use of ML for facies classification and demonstrates its potential for improving subsurface reservoir characterization.

BIBLIOGRAPHY

Bestagini, Paolo, Vincenzo Lipari, and Stefano Tubaro. "A machine learning approach to facies classification using well logs." Paper presented at the *2017 SEG International Exposition and Annual Meeting*, Houston, TX, September 2017.

2.4.4.4 Application-31: A Comparative Study of Multilayer Perceptron Neural Network and Cluster Analysis-Based Approaches

2.4.4.4.1 Background of the Application

The objective of this study is to examine the feasibility of using multilayer perceptron neural network (MLPNN) and cluster analysis techniques to reconstruct non-cored lithofacies in the Trias Argileux Grèseux Inférieur reservoir in the Sif Fatima oil field located in the Berkine basin of Southern Algeria. The performance of the proposed models was evaluated using RMSE, mean absolute error (MAE), and correlation coefficient (R). The MLPNN model, which utilized nuclear well logging data (GR, density, potassium, and thorium) as input variables, was found to have lower RMSE and MAE values of 0.39 and 0.23, respectively, as well as stronger

R values (training = 0.87, validation = 0.78, and test = 0.92) compared to the cluster analysis model. The quantitative comparison of real and predicted electrofacies using both methods suggests that the MLPNN model is the preferred approach for rebuilding non-cored lithofacies. The MLPNN model was able to estimate lithofacies in 333 meters of non-cored lithofacies, providing significant economic benefits.

2.4.4.4.2 Description of the Workflow and Machine Learning Model

In this study, an MLPNN is utilized to predict non-cored lithofacies from nuclear well logging data, which includes GR, density, thorium, and potassium. The MLPNN is designed with an architecture of one hidden layer consisting of four neurons, and it employs a sigmoid activation function to model the complex relationships between inputs and lithofacies. The Levenberg–Marquardt algorithm is used for training the network, allowing efficient minimization of the error during the back-propagation process. Additionally, a cluster analysis approach is employed to compare its performance against the MLPNN model. This method involves partitioning the data into clusters based on similarity to predict lithofacies.

2.4.4.4.3 Advantages and Limitations of the Synchronized Algorithm

The MLPNN model is effective in capturing nonlinear relationships between well logging inputs and lithofacies, making it a powerful tool for reconstructing lithofacies in non-cored intervals. It demonstrates high performance, with low RMSE and MAE, and strong correlation coefficients in training, validation, and test phases. However, the MLPNN requires a significant amount of data for training to avoid overfitting and needs careful tuning of the network architecture and learning algorithm parameters.

Cluster analysis provides a straightforward approach to grouping similar data points, which can be useful when relationships are not easily modeled by neural networks. However, its performance is generally lower than that of MLPNN, particularly in accurately predicting the proportions of different lithofacies. The cluster analysis method is also more sensitive to the initial conditions and may not generalize well to new, unseen data. Despite these limitations, cluster analysis can serve as a useful benchmark for evaluating the effectiveness of more complex models like MLPNN.

This structured breakdown highlights the AI techniques employed in the paper, along with their benefits and drawbacks.

2.4.4.4.4 Results and Discussion

In conclusion, this study demonstrates the effectiveness of the MLPNN approach for rebuilding non-cored lithofacies in the Trias Argileux Grèseux Inférieur reservoir in the Sif Fatima oil field located in the Berkine basin of Southern Algeria. The MLPNN model, which utilized nuclear well logging data as input variables, outperformed the cluster analysis model in terms of RMSE, MAE, and correlation coefficient values. The quantitative comparison of real and predicted electrofacies using both methods suggests that the MLPNN model is the preferred approach for rebuilding non-cored lithofacies, as it was able to estimate lithofacies in 333 meters of non-cored lithofacies, providing significant economic benefits.

BIBLIOGRAPHY

Ameur-Zaimeche, Ouafi, Aziez Zeddouri, Salim Heddam, Rabah Kechiched. Lithofacies prediction in non-cored wells from the Sif Fatima oil field (Berkine basin, southern Algeria): A comparative study of multilayer perceptron neural network and cluster analysis-based approaches, *Journal of African Earth Sciences*, 166, 2020, 103826, ISSN 1464-343X, https://doi.org/10.1016/j.jafrearsci.2020.103826

2.4.4.5 Application-32: Unsupervised Machine Learning with Petrological Database

2.4.4.5.1 Background of the Application

Lunar basalts exhibit significant diversity due to variations in the lunar mantle. The current model, cumulate mantle overturn, requires specific yet undetermined parameters for effective petrological modeling. This study leverages PCA and kernel canonical correlation analysis to analyze basalt data from ApolloBasaltDB, revealing that traditional Earth-based basalt classifications are insufficient for lunar basalts. The analysis highlights significant variations in Al_2O_3 within the Procellarum KREEP Terrane, offering new insights into mineral content from major element oxide data, which are valuable for remote sensing applications.

2.4.4.5.2 Description of the Workflow and Machine Learning Model

In this study, two unsupervised ML techniques, PCA and KCA, are employed to analyze the ApolloBasaltDB dataset, which contains geochemical, mineralogical, and textural information about lunar basalts. PCA is used to reduce the dimensionality of the dataset and identify key patterns and trends in the data by transforming the original variables into orthogonal components. This technique helps reveal underlying relationships between basalt characteristics such as major element oxides and mineral modes.

KCA is utilized to group the basalt samples into distinct clusters based on their characteristics, providing insights into potential natural groupings of lunar basalts. The K-means++ algorithm, an improved version of the traditional K-means, is employed to enhance the initialization of cluster centroids, improving the convergence and stability of the clustering process. These techniques together allow for a comprehensive exploration of the complexities and heterogeneities present in the lunar basalt samples.

2.4.4.5.3 Advantages and Limitations of the Synchronized Algorithm

The combination of PCA and KCA offers a powerful methodology for analyzing complex, multidimensional geological datasets like ApolloBasaltDB. PCA excels in reducing dimensionality and uncovering significant variance patterns within the data, making it easier to interpret and visualize relationships between basalt characteristics. However, PCA assumes linearity in the data, which might not capture all nonlinear interactions between variables.

KCA is advantageous for its simplicity and effectiveness in identifying distinct groups within the dataset, providing a straightforward interpretation of clustering results. Nevertheless, K-means requires the specification of the number of clusters, which can be challenging without prior knowledge. The algorithm is also sensitive to initial centroid positions, which can impact the clustering outcome. The use of the K-means++ algorithm helps mitigate this by improving the initial placement of centroids.

Despite these advantages, both PCA and KCA have limitations. PCA may not effectively capture complex, nonlinear relationships in the data, while KCA can struggle with irregular cluster shapes or overlapping clusters. Additionally, the results of these analyses are highly dependent on the quality and completeness of the input data, which underscores the importance of careful data preparation and validation.

2.4.4.5.4 Results and Discussion

The study demonstrates that PCA and KCA can uncover intricate patterns in lunar basalts that are missed by traditional Earth-based classifications. Significant variability in Al_2O_3 content and its relationship with other elements and minerals offers fresh perspectives on lunar basalt formation. The ability to infer mineral modes from major element oxides highlights the potential for enhanced remote sensing and lunar exploration techniques. The findings advocate for updated classification methods for lunar basalts that integrate multidimensional data analysis.

BIBLIOGRAPHY

Cone, Kim A., Richard M. Palin, and Kamini Singha. Unsupervised machine learning with petrological database ApolloBasaltDB reveals complexity in lunar basalt major element oxide and mineral distribution patterns. *Icarus*, 346, 2020, 113787, ISSN 0019-1035, https://doi.org/10.1016/j.icarus.2020.113787

2.4.4.6 Application-33: Research and Application of Logging Lithology Identification for Igneous Reservoirs Based on Deep Learning

2.4.4.6.1 Background of the Application

In order to determine the lithology of igneous reservoirs, this study suggests using DL. Data logging creates the deep belief network (DBN). Network parameters including the number of restricted Boltzmann machines (RBMs), the number of neurons in each RBM's hidden layer, and the classification border are optimized using the least squares approach, while network performance is measured using the mean square error function. The trained DBN processes the logging data that has to be interpreted. The findings are further examined after being separated into four cases. The DBN with lithology correction achieves a 94.8% identification accuracy for lithology.

2.4.4.6.2 Description of the Workflow and Machine Learning Model

In this study, a DBN is used to identify lithology in igneous reservoirs from logging data. The DBN consists of several layers of RBMs that learn hierarchical representations of the input data. The model is trained using logging curves, which include resistivity, GR, density, acoustic, and neutron logs. DBN is effective in modeling the complex relationships between logging data and lithology types due to its deep architecture, allowing for automatic feature extraction and representation learning.

To optimize the performance of the DBN, the number of RBMs, the number of neurons in each hidden layer, and the classification boundary are carefully tuned.

Grey correlation analysis is employed to select the most relevant logging curves for training the network, enhancing the model's capability to distinguish between different rock types.

2.4.4.6.3 Advantages and Limitations of the Synchronized Algorithm

The DBN offers significant advantages in lithology identification by capturing complex and nonlinear relationships in logging data. Its deep architecture enables it to automatically learn features that are crucial for differentiating between various lithologies, which traditional methods struggle to achieve. The DBN's ability to process and integrate multiple logging curves results in improved accuracy in identifying igneous rock types.

However, the DBN's performance is highly dependent on the quality and quantity of training data, as well as the proper tuning of network parameters such as the number of RBMs and neurons. The model can be computationally intensive, requiring careful optimization to balance accuracy and efficiency. Additionally, the classification boundary needs to be adjusted to minimize errors in lithology identification, which may lead to misclassifications if not accurately determined. Despite these challenges, the DBN demonstrates a robust approach for logging lithology identification, offering a more automated and accurate solution compared to traditional methods.

2.4.4.6.4 Results and Discussion

In this study, DL based on the DBN is proposed to identify the lithology of igneous reservoirs. The DBN is trained by logging data, and the network parameters are optimized by the least squares method. The logging data requiring interpretation are processed by the trained DBN. The results are divided into four cases, and the corresponding lithological cross-section scheme is applied to achieve accurate lithological identification. The results show that the proposed method can effectively identify the lithology of igneous reservoirs.

BIBLIOGRAPHY

Min, Xiang, Qin Pengbo, Zhang Fengwei. Research and application of logging lithology identification for igneous reservoirs based on deep learning. *Journal of Applied Geophysics*, 173, 2020, 103929, ISSN 0926-9851, https://doi.org/10.1016/j.jappgeo.2019.103929

2.4.5 Rock Properties

2.4.5.1 Application-34: Unsupervised Learning Elastic Rock properties from Pre-stack Seismic Data

2.4.5.1.1 Background of the Application

An unsupervised ML method for estimating rock characteristics from pre-stack seismic data is presented in this research. In this method, input collects are converted to relative rock attributes using a pre-designed CNN, and a low-frequency component is then added. This system incorporates wavefield simulation and modifies neural weights and biases to minimize a loss function between data. The efficacy of this novel approach is illustrated using the Book Cliffs synthetic model, wherein subsurface structures and property values are successfully recovered. The suggested method is then used on an actual dataset for geothermal exploration in Denmark, where a comparatively considerable distance between the seismic line and well location makes inverted results less than ideal.

2.4.5.1.2 Description of the Workflow and Machine Learning Model

In this method, a CNN is employed to estimate elastic rock properties, such as acoustic impedance (AI) and the Vp/Vs ratio, directly from pre-stack seismic data. CNNs are chosen for their ability to automatically learn spatial hierarchies of features, making them suitable for complex data like seismic images. The network consists of convolutional layers that extract features and fully connected layers that transform these features into predictions of AI and Vp/Vs. A key innovation in this approach is the unsupervised learning paradigm, where no labeled data is required. Instead, a forward modeling process is integrated into the learning loop to simulate synthetic seismic data from the predicted properties, which is then compared to the true seismic data. This comparison drives the update of the CNN parameters using the Adam optimizer to minimize the data misfit, leveraging the MSE as the loss function.

2.4.5.1.3 Advantages and Limitations of the Synchronized Algorithm

The unsupervised CNN approach for seismic inversion has several advantages. It eliminates the need for labeled training data, which is often scarce or expensive to obtain in geophysical applications. The integration of forward modeling ensures that the predictions adhere to physical principles, enhancing the reliability of the estimated rock properties. Additionally, CNNs are well-suited for handling the spatial complexity inherent in seismic data, allowing them to capture intricate geological features.

However, there are also limitations to this methodology. The unsupervised learning process relies heavily on the accuracy of the forward modeling, which, if not properly implemented, can lead to suboptimal inversion results. The method also demands significant computational resources, as the forward modeling and inversion processes are computationally intensive, particularly for large datasets. Moreover, the initial random weights and hyperparameters of the CNN can affect the convergence and quality of the inversion results, requiring careful tuning and validation. Despite these challenges, the unsupervised CNN approach offers a promising alternative for seismic inversion tasks, especially in settings where labeled data is limited or unavailable.

2.4.5.1.4 Results and Discussion

In conclusion, this paper presents an unsupervised ML approach to estimate rock properties based on pre-stack seismic data, which does not require labeled examples. The proposed method involves a pre-designed CNN that transforms input gathers to relative rock properties and is capable of recovering both subsurface structures and property values very well. The approach is demonstrated to be effective using the synthetic model of Book Cliffs and a real dataset for the geothermal exploration in Denmark. Further research can be done to improve the accuracy of the method and to apply it to other geophysical inversion problems.

BIBLIOGRAPHY

Feng, Runhai, Unsupervised learning elastic rock properties from pre-stack seismic data, *Journal of Petroleum Science and Engineering*, 192, 2020, 107237, ISSN 0920-4105, https://doi.org/10.1016/j.petrol.2020.107237

2.4.5.2 Application-35: Prediction of Shear Strength Parameters of Soils Using Artificial Neural Networks and Multivariate Regression Methods

2.4.5.2.1 Background of the Application

This discussion paper provides important viewpoints on the capability of ANNs for predicting the shear strength parameters of soils. The discussers express their gratitude to the authors for their research and highlight the main controversial points of the ANN models proposed by Khanlari et al. (2012), which include reliability, robustness, and transparency.

2.4.5.2.2 Description of the Workflow and Machine Learning Model

In this study, ANNs are employed to predict the shear strength parameters of soils, specifically cohesion (C) and friction angle (φ). ANNs are chosen due to their ability to model complex, nonlinear relationships between input parameters and target variables. The input parameters include the percentages of soil passing through No. 200, No. 40, and No. 4 sieves, along with the plasticity index (PI).

Two types of ANN architectures are explored: the MLP and the RBF network. The MLP, with its layered structure and backpropagation algorithm, is used to capture intricate patterns in data, while the RBF network leverages RBFs to provide localized approximations, often resulting in faster training times.

The models are developed using various combinations of input parameters, and the optimal models are identified based on their prediction accuracy for C and φ. The MLP is highlighted as having superior performance over the RBF model in this study.

2.4.5.2.3 Advantages and Limitations of the Synchronized Algorithm

Advantages:

- Artificial Neural Networks (ANNs): ANNs are highly effective in capturing nonlinear relationships within the data, making them suitable for complex geotechnical predictions. They can handle a large number of input variables and are robust in modeling the interactions between these variables.
- Multilayer Perceptron (MLP): The MLP model, with its multiple layers and nonlinear activation functions, is particularly powerful for fitting complex datasets. It is flexible and can approximate any continuous function given enough neurons and data.
- RBF Network: RBF networks offer faster training times compared to MLPs due to their simpler structure, making them advantageous when computational resources are limited.

Limitations:

- Data Quality: The performance of ANN models is heavily dependent on the quality and quantity of the data. In this study, some models showed poor performance due to the limitations of the experimental data used.
- Model Interpretability: ANNs, while powerful, are often seen as "black boxes" because the relationships they learn are not easily interpretable. This can be a limitation in understanding the underlying physical mechanisms governing soil behavior.
- Parameter Sensitivity: The reliability of the models can be compromised if key parameters influencing soil behavior, such as initial state and consolidation history, are not included as inputs. This oversight may lead to less robust and generalizable models.
- Robustness and Generalization: The study lacks a thorough parametric and sensitivity analysis, which are crucial for understanding model robustness across varying conditions. This omission raises concerns about the models' applicability to different soil types and conditions not represented in the training data.

2.4.5.2.4 Results and Discussion

In conclusion, the authors of this discussion paper argue that the ANN models have serious limitations in terms of their reliability, robustness, and transparency. The models do not consider the key roles of the initial state and consolidation history of the soil, and they lack a physical justification of how they reflect the conceptual understanding of the problem. The authors also point out that the proposed models should have been accompanied by a physical justification of how they reflect the conceptual understanding of the problem.

BIBLIOGRAPHY

Khanlari, G.R., M. Heidari, A.A. Momeni, and Y. Abdilor. Prediction of shear strength parameters of soils using artificial neural networks and multivariate regression methods. *Engineering Geology*, 131–132, 2012, 11–18, ISSN 0013-7952, https://doi.org/10.1016/j.enggeo.2011.12.006

2.4.5.3 Application-36: Identification of Rock Pore Structures and Permeabilities Using Electron Microscopy Experiments and Deep Learning Interpretations

2.4.5.3.1 Background of the Application

By analyzing a typical water-bearing sandstone that was produced in North China, this study seeks to investigate the relationship between the macroscopic permeability of sandstone and its microscopic pore features. Images from electron microscopy investigations are used to create pixel-level annotated data for DL training. In order to investigate the connection between the macroscopic permeability parameters of the sandstone and the microscopic pore characteristics, the DL framework is utilized to examine the pore characteristics by AI-based semantic picture segmentation.

By automatically processing microscopic images, this technique overcomes the drawbacks of conventional image recognition techniques and greatly increases the precision of pore identification in rock samples.

2.4.5.3.2 Description of the Workflow and Machine Learning Model

In this study, a fully convolutional network (FCN) is utilized for semantic image segmentation to identify rock pore structures in scanning electron microscopy (SEM) images. FCN is an advanced DL architecture that performs pixel-level classification, allowing for precise segmentation of complex structures within images. The model is trained using pixel-level annotated data derived from SEM images to recognize and segment microscopic pore structures accurately.

The FCN architecture comprises convolutional, pooling, and up-sampling layers. The convolutional layers extract spatial hierarchies in the image data, while pooling layers reduce dimensionality and computational load. The up-sampling layers restore the spatial resolution to match the input image dimensions, enabling detailed pixel-level predictions.

This approach leverages the semantic image segmentation capabilities of FCNs to automatically process and analyze SEM images, significantly enhancing the accuracy and efficiency of pore identification compared to traditional image processing methods.

2.4.5.3.3 Advantages and Limitations of the Synchronized Algorithm

The FCN model excels in pixel-level semantic segmentation, making it particularly suited for identifying complex pore structures in SEM images. The use of DL allows the model to learn intricate patterns within the data, resulting in high accuracy and robust performance across various image scales.

One major advantage of FCN is its ability to handle images of varying sizes and resolutions, offering flexibility in processing diverse datasets. The network's architecture effectively captures spatial context and detail, improving segmentation outcomes over traditional methods.

However, training FCNs requires a substantial amount of labeled data and computational resources. The need for large training datasets can be a limitation, especially in specialized fields where labeled data is scarce. Additionally, while FCNs offer high accuracy, they are computationally intensive during both training and inference phases, necessitating robust hardware capabilities.

Moreover, the segmentation quality is highly dependent on the quality and diversity of the training data. Poorly annotated data or insufficient variation in training samples can impact the model's ability to generalize to new images. Despite these challenges, the integration of FCNs into the workflow for analyzing SEM images significantly enhances the potential for detailed and accurate pore structure analysis.

2.4.5.3.4 Results and Discussion

The suggested DL semantic image segmentation technique makes it possible to identify pores in SEM images and investigate the connection between the macroscopic permeability parameters and the microscopic features of sandstone pores using automated pore identification and processing in SEM images. It improves the precision of image pore identification and determines the macroscopic permeability characteristics in microscopic pictures of rocks. The technique increases picture identification accuracy while addressing the drawbacks of pore detection in SEM images using conventional image segmentation techniques. As a result, the suggested approach may be used in the future to precisely identify the permeability parameters of rock and soil.

BIBLIOGRAPHY

Yu, Qingyang, Ziwei Xiong, Chao Du, Zhenxue Dai, Mohamad Reza Soltanian, Mojtaba Soltanian, Shangxian Yin, Wei Liu, Chen Liu, Chengbin Wang, and Zeyu Song. Identification of rock pore structures and permeabilities using electron microscopy experiments and deep learning interpretations, *Fuel*, 268, 2020, 117416, ISSN 0016-2361, https://doi.org/10.1016/j.fuel.2020.117416

2.4.5.4 Application-37: Shear Wave Velocity Prediction Based on Adaptive Particle Swarm Optimization Optimized Recurrent Neural Network

2.4.5.4.1 Background of the Application

This paper proposes a novel hybrid model for predicting Vs from a series of input petrophysical log data. The model combines grey relational analysis and long short-term memory (LSTM) RNN optimized by PSO algorithm with adaptive learning strategy. The model considers the variation trend and context information with depth to improve prediction accuracy of Vs. Real petrophysical log data is used to demonstrate the effectiveness of the proposed model. Results show that the PSO algorithm with adaptive learning strategy-based LSTM network model provides higher predictive accuracy and robustness, with promising application prospects in Vs prediction studies.

2.4.5.4.2 Description of the Workflow and Machine Learning Model

In this study, a hybrid model combining LSTM networks and an adaptive PSO algorithm is employed to predict Vs using conventional petrophysical logs. The LSTM network is adept at capturing the temporal dependencies within the log data, providing robust predictions by learning long-term dependencies. To enhance the LSTM model's predictive accuracy, its hyperparameters are optimized using an adaptive PSO algorithm. This algorithm effectively searches for the optimal configuration by dynamically adjusting the exploration and exploitation of the solution space, improving the convergence speed and avoiding local optima. The combination of LSTM and adaptive PSO allows for accurate modeling of the nonlinear relationships between input logs and Vs, making it suitable for geophysical applications.

2.4.5.4.3 Advantages and Limitations of the Synchronized Algorithm

The LSTM network is highly effective in modeling temporal sequences and capturing long-term dependencies, making it ideal for predicting Vs from petrophysical logs. However, LSTM networks require careful tuning of hyperparameters to avoid issues such as overfitting and vanishing gradients. The adaptive PSO algorithm enhances the LSTM model by efficiently optimizing these hyperparameters, leading to improved model performance and prediction accuracy. Despite its strengths, the adaptive PSO algorithm's effectiveness depends on its initial settings and may require multiple iterations to converge to the best solution. This iterative process can be computationally intensive, especially when dealing with large datasets or complex models. Nevertheless, the proposed hybrid approach demonstrates significant potential in accurately predicting Vs, outperforming traditional empirical and regression models in terms of prediction accuracy and robustness.

2.4.5.4.4 Results and Discussion

In conclusion, this study proposes a novel hybrid model for predicting Vs from petrophysical log data. The proposed model combines grey relational analysis and LSTM RNN optimized by PSO algorithm with adaptive learning strategy. The model considers the variation trend and context information with depth to improve prediction accuracy of Vs. Results of processing real petrophysical log data demonstrate that the PSO algorithm with adaptive learning strategy-based LSTM network model provides higher predictive accuracy and robustness than conventional methods. The proposed model has promising application prospects in Vs prediction studies.

BIBLIOGRAPHY

Wang, Jun, Junxing Cao, and Shan Yuan. Shear wave velocity prediction based on adaptive particle swarm optimization optimized recurrent neural network, *Journal of Petroleum Science and Engineering*, 194, 2020, 107466, ISSN 0920-4105, https://doi.org/10.1016/j.petrol.2020.107466

2.4.5.5 Application-38: Estimation of Seismic Wave Velocities of Metamorphic Rocks Using Artificial Neural Network

2.4.5.5.1 Background of the Application

In this study, an ANN is used to estimate seismic wave velocities (compression and shear) of metamorphic rocks. The ANN is trained with data including chemical composition, temperature, and pressure of rocks retrieved from depths up to 10 km. The training data, collected from various literature sources, was used to develop a predictive model. This model demonstrates a strong correlation with laboratory measurements, suggesting its potential for early prediction of seismic wave parameters. This method offers a robust and cost-effective solution for applications in drilling and planetary studies.

2.4.5.5.2 Description of the Workflow and Machine Learning Model

In this study, an ANN model is employed to predict seismic wave velocities (compression and shear) of metamorphic rocks. The ANN uses an MLP architecture with three hidden layers, each containing eight neurons, optimized using the Adam optimizer and MSE as the loss function. The model is trained on data comprising the chemical composition, temperature, and pressure of rocks from various geological sites, aiming to accurately capture the complex relationships between these parameters and the resulting seismic velocities. The use of the hyperbolic tangent activation function helps in efficiently capturing nonlinear patterns within the data, providing robust predictions for seismic velocities under varying conditions.

2.4.5.5.3 Advantages and Limitations of the Synchronized Algorithm

The ANN model effectively models the complex interactions between chemical composition, temperature, and pressure, providing accurate predictions of seismic wave velocities. Its ability to learn nonlinear relationships makes it suitable for geological applications where traditional models might fall short. The choice of the hyperbolic tangent activation function and the Adam optimizer enhances the model's performance, reducing training time and improving accuracy. However, the model requires a significant amount of data for training to avoid overfitting, and the selection of network architecture, such as the number of neurons and layers, is critical and may require a trial-and-error approach. Furthermore, while the model performs well with the available data, its predictions might be limited by the absence of factors such as porosity and mineralogy, which are not considered in this study.

2.4.5.5.4 Results and Discussion

The ANN model successfully predicts the seismic wave velocities (Vp and Vs) of metamorphic rocks using inputs such as chemical composition, temperature, and pressure. The model, trained with data from different locations and depths, shows a good fit with experimental results. It was validated against known scenarios and tested with unknown samples, demonstrating its accuracy. This approach provides a preliminary but reliable estimate of wave velocities, proving useful for conditions where direct sample retrieval is challenging. Future studies will address additional factors like porosity and mineralogy to enhance model accuracy.

BIBLIOGRAPHY

Rizvi, Zarghaam Haider, Syed Jawad Akhtar, Hasan Haider, Jörg Follmann, and Frank Wuttke. Estimation of seismic wave velocities of metamorphic rocks using artificial neural network, *Materials Today: Proceedings*, 26 (Part 2), 2020, 324–330, ISSN 2214-7853, https://doi.org/10.1016/j.matpr.2019.11.320

2.4.6 Geophysics

2.4.6.1 Application-39: Machine Learning for Geophysical Characterization Of Brittleness: Tuscaloosa Marine Shale Case Study

2.4.6.1.1 Background of the Application

In the study, the authors propose an ML approach to predict the brittleness of subsurface formations, a critical property for unconventional reservoir exploration. By utilizing a diverse dataset that includes seismic attributes, rock physics, and petrophysical data, the method aims to overcome the limitations of traditional physical models. Applied to the Tuscaloosa Marine Shale (TMS), the ML model demonstrates the ability to predict brittleness distribution accurately, which can enhance the efficiency of hydraulic fracturing operations and reduce costs.

2.4.6.1.2 Description of the Workflow and Machine Learning Model

In this study, an ANN is utilized to predict brittleness in the TMS by analyzing seismic attributes. The ANN comprises three hidden layers, each with three neurons, designed to capture nonlinear relationships between input features and brittleness classes. The input features include instantaneous amplitude, instantaneous phase, and instantaneous frequency of seismic data. The ANN model is trained using one-hot encoded labels derived from the $\lambda\rho$ and $\mu\rho$ AVO attributes. The cross-entropy loss function is employed to minimize prediction errors. Data normalization and balancing are applied to ensure that the model is not biased toward any particular brittleness class, leading to more accurate predictions.

2.4.6.1.3 Advantages and Limitations of the Synchronized Algorithm

The ANN approach is advantageous for its ability to model complex, nonlinear relationships within geophysical datasets, making it well-suited for predicting brittleness zones. Its architecture allows for flexibility in handling diverse data types and integrating multiple data sources, such as seismic and petrophysical data. However, the ANN requires careful configuration and tuning to avoid overfitting and ensure generalizability. The method's performance is also sensitive to the quality and balance of input data, which necessitates preprocessing steps like data normalization and balancing. While the ANN effectively captures intricate patterns, it may demand significant computational resources and time, particularly during the training phase. Additionally, the interpretation of the model's predictions can be complex, requiring domain expertise to translate outputs into actionable geophysical insights.

2.4.6.1.4 Results and Discussion

The research highlights the importance of brittleness in evaluating unconventional resources, specifically within the TMS formation. By integrating pre-stack and post-stack seismic attributes with rock physics templates, the study shows that an ANN can effectively predict brittleness zones. This approach allows for better characterization of geomechanical properties, facilitating cost-effective drilling operations. The success of the ANN model in predicting brittleness indicates the potential of ML techniques to improve subsurface property estimation in geophysics.

BIBLIOGRAPHY

Mlella, Mark, Ming Ma, Rui Zhang, and Mehdi Mokhtariy. "Machine learning for geophysical characterization of brittleness: Tuscaloosa Marine Shale case study." Paper presented at the *SEG International Exposition and Annual Meeting, Virtual*, October 2020. https://doi.org/10.1190/segam2020-3421111.1

2.4.6.2 Application-40: Estimation of Reservoir Porosity Based on Seismic Inversion Results Using Deep Learning Methods

2.4.6.2.1 Background of the Application

The study focuses on estimating reservoir porosity across a reservoir using DL methods, specifically from seismic inversion data, despite the lower resolution compared to well-log readings. The authors propose using seismic inversion results to predict reservoir porosity by exploring the nonlinear relationship between rock properties and seismic data. A CNN is applied to this task, leveraging local information through 2D kernel filters, which improves prediction accuracy by considering the intrinsic relationship between rock properties and porosity. The approach is validated on a dataset from the Vienna Basin, showing that CNN outperforms traditional ANNs in accuracy, providing a 3D cube of reservoir porosity.

2.4.6.2.2 Description of the Workflow and Machine Learning Model

In this study, CNNs are employed to estimate reservoir porosity using seismic inversion data. The CNN model is specifically chosen for its ability to handle complex data patterns through the use of convolutional layers, which capture local information within data windows via 2D kernel filters. This allows the CNN to effectively model the intrinsic relationships between seismic data and rock properties, which are crucial for accurate porosity predictions. Additionally, ANNs are used as a comparative method to evaluate the effectiveness of the CNN approach. The CNN model uses a regression process to predict continuous porosity values and employs ReLU activation functions to introduce nonlinearity, with the sigmoid function mapping outputs to the porosity range. The model is trained using the Adam optimizer, which efficiently adjusts neural weights and biases to minimize prediction error.

2.4.6.2.3 Advantages and Limitations of the Synchronized Algorithm

The CNN methodology is advantageous for its robustness in capturing complex relationships between rock properties and porosity, outperforming traditional ANN models due to its local information processing capabilities. CNN's use of convolutional layers allows for improved noise resistance and accuracy, making it particularly suitable for geophysical data. However, CNNs require careful design and tuning of hyperparameters, such as filter sizes and the number of layers, to achieve optimal performance. Training CNNs can be computationally intensive, necessitating efficient data handling and augmentation strategies to prevent overfitting and enhance model generalization. Although ANN provides a simpler architecture, it is more susceptible to noise and lacks the local connectivity that enhances CNN performance. Despite these challenges, the CNN approach offers significant benefits for intelligent data manipulation in seismic inversion tasks, enabling accurate and rapid porosity predictions across reservoir models.

2.4.6.2.4 Results and Discussion

The research demonstrates that the limitation of well log locations can be overcome by using seismic data inversion results for predicting reservoir porosity, thereby enabling the creation of 2D sections or 3D cubes of the reservoir. The nonlinear full-waveform inversion technique employed offers high-resolution rock properties, which are effective for porosity estimation. The study compares CNN and ANN models, concluding that CNN delivers higher prediction accuracy due to its ability to utilize local data through convolutional filters. This method does not require prior knowledge or initial models, making it a data-driven approach that can be rapidly applied to different datasets once trained. The CNN's integration of local rock property information allows for improved porosity predictions, showcasing its potential for intelligent data manipulation in geophysical applications.

BIBLIOGRAPHY

Feng, Runhai. Estimation of reservoir porosity based on seismic inversion results using deep learning methods. *Journal of Natural Gas Science and Engineering*, 77, 2020, 103270, ISSN 1875-5100, https://doi.org/10.1016/j.jngse.2020.103270

2.4.6.3 Application-41: Using Deep Learning-Based Methods to Classify Salt Bodies in Seismic Images

2.4.6.3.1 Background of the Application

Salt bodies are important for the oil and gas industry as they trap hydrocarbons in the subsurface. However, identifying these salt bodies is a challenging task and requires domain experts to pick the right geological features for interpretation, which may not always provide accurate results. ML algorithms have been proposed to automate or semi-automate the process of identifying salt bodies using seismic attributes, but these methods still require domain knowledge expertise. In this paper, we propose a novel DL method based on U-Net plus Se-ResNet to identify salt bodies in seismic images. This method does not require domain knowledge expertise and is expected to extract useful features automatically from the seismic data, resulting in higher accuracy.

2.4.6.3.2 Description of the Workflow and Machine Learning Model

In this study, we employ a DL approach using a U-Net architecture enhanced with squeeze-and-excitation (SE) ResNet blocks to classify salt bodies in seismic images. The U-Net architecture is specifically designed for image segmentation tasks, with a symmetrical encoder–decoder structure that allows for the effective capture of spatial features across different scales. The addition of SE-ResNet blocks introduces adaptive feature recalibration, enhancing the network's ability to model channel-wise interdependencies and focus on the most informative features.

The training process involves a combination of binary cross entropy (BCE) and Dice loss functions for the initial phase, followed by the Lovasz-Softmax loss for optimization, improving segmentation accuracy by refining boundary predictions. A stratified training and validation split based on salt coverage ensures balanced representation in both sets, aiding in robust model evaluation.

2.4.6.3.3 Advantages and Limitations of the Synchronized Algorithm

The U-Net architecture, combined with SE-ResNet blocks, provides a powerful framework for segmenting salt bodies in seismic images due to its ability to learn complex spatial hierarchies without the need for manual feature extraction. The model's end-to-end learning capability allows for automatic discovery of relevant features directly from raw data, reducing the reliance on expert domain knowledge.

However, this DL approach requires substantial computational resources and a large volume of labeled training data to achieve optimal performance, which can be a limitation in practice. Additionally, while the SE-ResNet blocks improve feature learning, they increase the model's complexity and computational cost. The training process, particularly with Lovasz-Softmax loss, may also require fine-tuning of hyperparameters to avoid overfitting and ensure generalization across different datasets.

These sections reflect a concise summary of the methodology and its pros and cons based on the information in the paper. Adjustments and refinements can be made based on specific details and findings from the full article.

2.4.6.3.4 Results and Discussion

In this paper, a DL method is proposed for identifying salt bodies in seismic images. The proposed method is based on U-Net plus Se-ResNet and is able to extract useful features automatically from the seismic data without requiring domain knowledge expertise. After 10-fold cross-validation, a promising intersection over union (IOU) score was achieved. Our proposed method outperformed existing methods and provides a more direct and assisted process for seismic data interpretation.

BIBLIOGRAPHY

Saif ul Islam, Muhammad, Using deep learning based methods to classify salt bodies in seismic images, *Journal of Applied Geophysics*, 178, 2020, 104054, ISSN 0926-9851, https://doi.org/10.1016/j.jappgeo.2020.104054

2.4.6.4 Application-42: Unsupervised Machine Learning for Waveform Extraction in Micro Seismic Denoising

2.4.6.4.1 Background of the Application

An efficient microseismic denoising approach based on variational mode decomposition (VMD) and unsupervised ML is proposed in this study. Waveform locations are extracted from noisy raw microseismic data using the unsupervised ML approach, which provides a priori knowledge to constrain the denoising process. The VMD method is used to breakdown and reconstruct the waveforms in order to eliminate noise. The denoised waveforms are reconstructed using the decomposed intrinsic mode functions (IMFs) with high kurtosis values. The one-dimensional singular spectrum analysis (SSA) technique is used to further enhance the waveforms due to the low-frequency noise that remains in the denoised waveforms following the VMD procedure. The final result is obtained by filtering the denoised data using the window-thresholding method following the VMD and SSA processes.

2.4.6.4.2 Description of the Workflow and Machine Learning Model

In this paper, an unsupervised ML approach is applied for waveform extraction and denoising in microseismic data. The method involves using a K-means clustering algorithm to classify microseismic data into waveform and non-waveform components. This unsupervised learning step identifies key waveform components in the data, providing a priori information for further processing.

The method then applies VMD to separate the microseismic data into IMFs. The VMD process identifies and removes high-frequency noise components. A kurtosis-based thresholding approach is used to select the most significant IMFs for signal reconstruction.

Finally, SSA is employed to further suppress low-frequency residual noise. The SSA step involves decomposing the reconstructed data into a Hankel matrix, applying SVD, and selecting dominant components for noise attenuation.

2.4.6.4.3 Advantages and Limitations of the Synchronized Algorithm

The unsupervised ML method provides an automatic and adaptive way to identify key waveform components without requiring labeled training data, making it suitable for complex microseismic datasets with varying noise levels. The use of VMD allows for effective separation of signal and noise components, and the kurtosis-based thresholding ensures that only the most significant IMFs are retained, enhancing the denoising performance.

However, the choice of the number of modes in VMD and the kurtosis threshold can significantly affect the results, requiring careful parameter selection. The SSA method is effective in attenuating residual noise, but its performance depends on the appropriate selection of singular value thresholds, which can be challenging in noisy data environments.

Overall, while the methodology provides a robust framework for denoising microseismic data, it requires careful parameter tuning and may be computationally intensive, particularly for large datasets. The method's success is contingent on the effective integration of the unsupervised learning, VMD, and SSA steps to achieve optimal denoising results.

2.4.6.4.4 Results and Discussion

The proposed microseismic denoising algorithm based on the unsupervised ML and VMD method is shown to be effective in removing noise from microseismic data. Both synthetic and real microseismic datasets are used to test the algorithm and obtain appealing performance. The unsupervised ML method is able to extract key waveform components from noisy microseismic data, which is adequate for the waveform-based microseismic epicenter imaging. The VMD method is used to decompose and reconstruct the waveforms, while the SSA method is used to further improve the waveforms. The method is automatic and effective in denoising microseismic data.

BIBLIOGRAPHY

Zhou, Yatong and Guoning Wu. Unsupervised machine learning for waveform extraction in microseismic denoising. *Journal of Applied Geophysics*, 173, 2020, 103879, ISSN 0926-9851, https://doi.org/10.1016/j.jappgeo.2019.103879

2.4.6.5 Application-43: Multiple Point Geostatistical Simulation with Adaptive Filter Derived from Neural Network for Sedimentary Facies Classification

2.4.6.5.1 Background of the Application

This paper proposes a novel approach for high-precision 3D sedimentary facies classification by introducing indirect information as constraints and regularizations. The proposed method integrates seismic and logging data using an ANN and a filter-generation auto-encoder (FAE) to extract features from the data. These features are then used as input to a multiple point geostatistical simulation (MPS) algorithm to classify the sedimentary facies. The proposed method is demonstrated using a dataset from a reservoir in the western part of China, and the results show that the proposed method outperforms traditional methods in terms of accuracy and precision.

2.4.6.5.2 Description of the Workflow and Machine Learning Model

In this study, MPS is integrated with a neural network structure called the FAE to improve sedimentary facies classification. The MPS method utilizes high-order spatial statistics for simulating geological distributions, while the FAE is designed to generate adaptive filters that enhance the accuracy and applicability of MPS. The FAE consists of an encoder, a decoder, and a classifier, which are trained to generate filters that capture both seismic and sedimentary facies information. The combination of MPS and FAE allows for the incorporation of indirect information, such as seismic attributes, into the simulation process, thereby improving the classification accuracy and resolving issues related to limited direct information.

2.4.6.5.3 Advantages and Limitations of the Synchronized Algorithm

The integration of MPS with FAE provides several advantages for sedimentary facies classification. MPS is capable of handling complex spatial structures and simulating multiple sedimentary facies maps with consistent uncertainty. The use of FAE introduces adaptive filters that enhance MPS by capturing more detailed geological features, leading to improved classification accuracy. Additionally, the multidimensional distance function used in this method allows for effective incorporation of various data sources, enhancing the robustness of the simulation.

However, the methodology also has its limitations. The accuracy of the MPS depends significantly on the quality of the training images and the filters generated by the FAE. The complexity of the FAE model may require extensive computational resources and careful tuning of hyperparameters to achieve optimal performance. Furthermore, while the method shows improved results over traditional MPS, the stochastic nature of the simulations can still introduce some uncertainty and variability in the results. Balancing the weights of different distance components in the multidimensional distance function is crucial for maintaining stability in the simulations.

2.4.6.5.4 Results and Discussion

In this paper, a novel approach for high-precision 3D sedimentary facies classification is proposed by integrating seismic and logging data using an ANN, an FAE, and an MPS algorithm. The proposed method was applied to a dataset from a reservoir in the western part of China, and the results showed that the method outperformed traditional methods in terms of accuracy and precision. Our approach has the potential to improve reservoir characterization and enhance hydrocarbon recovery in the petroleum industry.

BIBLIOGRAPHY

Han, Feilong, Hongbing Zhang, Jianwen Rui, Kuiye Wei, Dailu Zhang, Wei Xiao. Multiple point geostatistical simulation with adaptive filter derived from neural network for sedimentary facies classification. *Marine and Petroleum Geology*, 118, 2020, 104406, ISSN 0264-8172, https://doi.org/10.1016/j.marpetgeo.2020.104406

2.4.6.6 Application-44: N-Hidden Layer Artificial Neural Network Architecture Computer Code

2.4.6.6.1 Background of the Application

This paper presents a MATLAB code for training ANNs with N-1 layer (*N-*hidden layer) architecture for solving geophysical problems. The ANN machine can be retrieved by the gradient optimization method, which includes Levenberg–Marquardt, steepest descent, or conjugate gradient methods. The method is applied to predict lithology and fluid identification in the oil and gas industry using Poisson's ratio as input parameters, which are calculated from compressional and shear wave velocities. The results show that the proposed method with higher numbered layer architecture outperforms the conventional two-layer architecture significantly, with a prediction error as large as 20% for the latter.

2.4.6.6.2 Description of the Workflow and Machine Learning Model

In this method, an ANN with an N-hidden layer architecture is used to solve geophysical problems such as predicting Poisson's ratio from compressional and shear wave velocities. The ANN framework is designed to overcome limitations of traditional two-layer architectures by allowing for deeper networks with more hidden layers. The network's architecture is flexible, allowing the use of various activation functions like log-sigmoid and hyperbolic tangent sigmoid functions. The network's weights and biases are optimized using gradient-based methods like the Levenberg–Marquardt algorithm, steepest descent, and conjugate gradient methods. This optimization process allows the ANN to learn complex nonlinear relationships between input patterns and target outputs effectively.

2.4.6.6.3 Advantages and Limitations of the Synchronized Algorithm

- Advantages:
    - The N-hidden layer ANN architecture provides greater flexibility and accuracy by allowing more hidden layers, which enhances the model's capacity to learn complex patterns in geophysical data.
    - The ability to choose from different optimization algorithms (Levenberg–Marquardt, steepest descent, conjugate gradient) provides versatility in training the network effectively.
    - The approach is open-source, promoting broader accessibility and adaptation across various research fields beyond geophysics.
- Limitations:
    - The ANN requires a substantial amount of computational resources, especially as the number of layers increases, which can lead to longer training times.
    - The initial choice of weights and biases is random, which may affect the speed and convergence of the learning process.
    - Careful tuning of network architecture (number of layers and neurons) and optimization parameters is crucial to avoid overfitting or underfitting the data.

This analysis provides an overview of the AI methods and their respective advantages and limitations as described in the uploaded document. The N-hidden layer ANN approach offers significant improvements over traditional two-layer networks but requires careful consideration of computational resources and parameter settings.

2.4.6.6.4 Results and Discussion

In this paper, a method is presented for predicting lithology and fluid identification in the oil and gas industry using ANNs with higher numbered layer architecture. The proposed method shows significant improvements in accuracy compared to conventional two-layer architecture with prediction errors as large as 20%. The results demonstrate the potential of the proposed method for solving geophysical problems, and the gradient optimization method can be used to retrieve the trained ANN machine. The proposed method is expected to have applications in other fields such as medical diagnosis, pattern recognition, and financial forecasting.

BIBLIOGRAPHY

Ogunbo, Jide Nosakare, Olufemi Adigun Alagbe, Michael Ilesanmi Oladapo, and Changsoo Shin. N-hidden layer artificial neural network architecture computer code: geophysical application example, *Heliyon*, 6 (6), 2020, e04108, ISSN 2405-8440, https://doi.org/10.1016/j.heliyon.2020.e04108

2.4.7 Other Applications

2.4.7.1 Application-45: Optimizing Horizontal Well Placement through Stratigraphic Performance Prediction Using Artificial Intelligence

2.4.7.1.1 Background of the Application

This paper presents a new reservoir management workflow that uses both classic and AI models to optimize horizontal well placement for a mature heavy oil field. The workflow utilizes a data-driven model based on fuzzy logic to compute a new attribute called the dynamic reservoir quality index (dRQI), as well as classic models using stratigraphic Lorenz Plots, reservoir quality index (RQI), and flow-zone indicator (FZI) to predict stratigraphic and directional heterogeneities and production performance. The workflow was successfully used to optimize horizontal well placement for 2019–2020 drilling programs.

2.4.7.1.2 Description of the Workflow and Machine Learning Model

In this methodology, a fuzzy inference system (FIS) is utilized to predict the optimal placement of horizontal wells in a heavy oil reservoir. Fuzzy logic is chosen for its ability to handle uncertainty and imprecision in geological data, providing a flexible and intuitive framework for integrating various reservoir attributes like temperature, saturation, porosity, and permeability. The FIS constructs a dRQI, which incorporates time-sensitive parameters, to evaluate the likelihood of movable oil in different regions of the reservoir.

The FIS uses fuzzy "if-then" rules to map input variables (temperature, oil saturation, and resistivity) to output variables (dRQI), allowing for complex interactions among the inputs. This approach enhances the understanding of reservoir behavior and guides the decision-making process for well placement, thereby optimizing the extraction process.

2.4.7.1.3 Advantages and Limitations of the Synchronized Algorithm

The use of FISs offers several advantages in reservoir management, particularly in dealing with the inherent uncertainties and variabilities of subsurface conditions. Fuzzy logic models are robust to imprecision and provide a comprehensible structure for modeling complex systems, which is crucial for evaluating reservoir properties dynamically.

However, the success of the FIS approach relies heavily on the quality of the input data and the design of the fuzzy rules. Developing these rules requires expert knowledge, and the system's performance is sensitive to changes in the input variables. While FIS models can capture nonlinear relationships and provide insights into the reservoir, they can also become computationally expensive, especially when scaling up for large datasets or complex reservoir models.

The methodology's reliance on dRQI means that it is crucial to have accurate and up-to-date measurements of parameters such as temperature and saturation. Furthermore, while FIS models are flexible, their interpretability can decrease as the number of rules increases, potentially leading to challenges in maintaining the system as the reservoir conditions evolve.

2.4.7.1.4 Results and Discussion

The presented workflow successfully optimized horizontal well placement for a mature heavy oil field by incorporating both classic and AI models to predict stratigraphic and directional heterogeneities and production performance. The workflow utilized a data-driven model based on fuzzy logic to compute the dRQI, as well as classic models using stratigraphic Lorenz Plots, RQI, and FZI. The workflow was validated through a lookback process on more than 400 wells, and it led to a better understanding of the reservoir intricacies as related to well placement and performance.

BIBLIOGRAPHY

Popa, Andrei and Sean Connel. "Optimizing Horizontal Well Placement Through Stratigraphic Performance Prediction Using Artificial Intelligence." Paper presented at the *SPE Annual Technical Conference and Exhibition*, Calgary, AB, Canada, September 2019. https://doi.org/10.2118/195887-MS

2.4.7.2 Application-46: A New Approach to Characterize CO_2 Flooding Utilizing Artificial Intelligence Techniques

2.4.7.2.1 Background of the Application

The use of carbon dioxide (CO_2) in miscible flooding is an effective technique for enhancing oil production. However, the flooding efficiency is highly dependent on the minimum miscibility pressure (MMP), which can be estimated through slim tube experiments or empirical equations. This paper proposes a new technique to evaluate CO_2 flooding and minimize uncertainties using AI techniques. The objective is to develop a reliable model to predict the MMP during CO_2 flooding. Actual case studies for flooding heterogeneous and anisotropic reservoirs were used to generate the MMP model. Several AI techniques were studied to estimate the CO_2-MMP for a wider range of conditions. The developed models investigate the effect of API gravity, fluid composition, and injected gas composition on the performance of CO_2 flooding operation.

2.4.7.2.2 Description of the Workflow and Machine Learning Model

In this study, various AI techniques were employed to predict the MMP during CO_2 flooding in oil reservoirs. The methods used include ANNs, RBF networks, GRNNs, and ANFIS.

- ANNs: The ANN model consists of input, hidden, and output layers and is designed to capture the nonlinear relationships between the input parameters (reservoir temperature, molecular weight of C7+, and mole fraction of C2–C6) and the output (MMP). The ANN is trained using historical data to predict MMP accurately.
- RBF Networks: The RBF model, similar in structure to the ANN, uses RBFs to approximate the function mapping between inputs and outputs. It is particularly useful for handling nonlinear data and providing smooth interpolation.
- GRNN: GRNN is a type of RBF network that performs well with smaller datasets and provides a robust prediction mechanism by leveraging the relationships captured in the training data.
- ANFIS: ANFIS combines the learning capabilities of neural networks with the fuzzy logic qualitative approach. This hybrid method aims to enhance the prediction of MMP by accommodating the uncertainties and nonlinearities in the data.

2.4.7.2.3 Advantages and Limitations of the Synchronized Algorithm

Advantages:

- ANN: Capable of modeling complex nonlinear relationships and adaptable to different types of data. It provides a strong baseline for MMP prediction with relatively high accuracy.
- RBF: Effective for nonlinear problem-solving and can provide good results with fewer parameters to tune, which makes it suitable for this type of application.
- GRNN: Requires a smaller amount of data and is less computationally expensive compared to other AI techniques. It offers fast learning and robust predictions.
- ANFIS: Combines the strengths of neural networks and fuzzy logic, allowing it to handle uncertainty and imprecision in the input data effectively.

Limitations:

- ANN: Requires extensive training data and careful tuning of hyperparameters (such as the number of layers and neurons) to avoid overfitting.
- RBF: The performance heavily relies on the selection of the RBF and the parameters involved, which can be challenging to optimize.
- GRNN: While computationally efficient, it may not perform as well with very large or very small datasets due to its reliance on the training dataset's variance.
- ANFIS: Although flexible, the model can become computationally expensive due to the fuzzy inference system, especially if the dataset is large or highly dimensional.

These AI techniques collectively aim to enhance the accuracy and reliability of MMP predictions, thus providing a significant improvement over traditional empirical and analytical methods for characterizing CO_2 flooding in complex reservoir conditions.

2.4.7.2.4 Results and Discussion

To find the best model with the highest correlation coefficient (R) value and the lowest average absolute percentage error (AAPE) for both training and testing data, various AI approaches were investigated. The error was greatly decreased following the implementation of data processing techniques. This study presents an efficient method for calculating the MMP during CO_2 flooding and improving existing numerical or analytical models to more accurately identify the local displacement efficiency. The results demonstrate that the suggested strategy works better than the existing CO_2-MMP models.

BIBLIOGRAPHY

Hassan, A., S. Elkatatny, M. Mahmoud, and A. Abdulraheem. (2018). A new approach to characterize CO2 flooding utilizing artificial intelligence techniques. In *Society of Petroleum Engineers - SPE Kingdom of Saudi Arabia Annual Technical Symposium and Exhibition 2018, SATS 2018 (Society of Petroleum Engineers - SPE Kingdom of Saudi Arabia Annual Technical Symposium and Exhibition 2018, SATS 2018)*. Society of Petroleum Engineers. https://doi.org/10.2118/192252-MS

2.4.7.3 Application-47: Automatic Detection of Pipe-Flange Reflections in GPR Data Sections Using Supervised Learning

2.4.7.3.1 Background of the Application

Ground penetrating radar (GPR) is a widely used method for investigating near-surface subsoil. However, processing and analyzing large volumes of GPR data is time-consuming and error-prone. This paper presents three methods for the automatic detection of pipe-flange signals in constant-offset GPR images, using supervised ML techniques. The methods involve SVMs with local binary patterns (LBP) and histogram of oriented gradients (HOG) descriptors, and ANNs. Due to the limited number of experimental flange patterns available, synthetic patterns were generated for training and validation. The results demonstrate the effectiveness of these methods, with ANN showing the best performance, followed by SVM with HOG, and then SVM with LBP.

2.4.7.3.2 Description of the Workflow and Machine Learning Model

In this study, SVMs and ANNs are employed for the automatic detection of pipe-flange reflections in GPR images. The SVMs are used with RBF and polynomial kernels to classify image features extracted through LBP and HOG descriptors. SVMs are chosen for their ability to handle complex, nonlinear classification problems effectively. ANNs, specifically multilayer feed-forward networks, are also utilized for their capacity to learn intricate patterns and relationships within the data, leveraging the raw pixel intensity values as inputs.

2.4.7.3.3 Advantages and Limitations of the Synchronized Algorithm

The SVMs, combined with HOG and LBP descriptors, effectively model the shape and texture features of the pipe-flange reflections, providing a robust method for automatic detection in GPR images. The RBF kernel is particularly adept at capturing nonlinear relationships, while the HOG descriptor emphasizes object shape, enhancing the SVM's accuracy. However, SVMs require careful parameter tuning, and their performance can degrade with very large datasets.

The ANN architecture, with its multiple hidden layers, can learn complex patterns from the raw data, offering a high degree of accuracy in detection. It shows superior performance compared to SVMs, especially when handling variations in image patterns not well represented in the simulated data. Nonetheless, ANNs demand significant computational resources and time for training, and they are susceptible to overfitting if not properly regularized. Additionally, the generation of synthetic samples for training these models is crucial, yet it may not always perfectly replicate real-world scenarios, leading to potential misclassifications in some cases.

2.4.7.3.4 Results and Discussion

The study demonstrates the potential of SVM and ANN methodologies for automatically detecting pipe flanges in GPR data. While the validation results were promising, the test data showed higher false negative rates, indicating that some experimental patterns were not well represented in the synthetic samples used for training. Among the methods, ANN provided the best detection performance, followed closely by SVM with HOG. Future work should focus on obtaining more experimental samples and exploring more sophisticated numerical simulations to improve detection accuracy. The study highlights the importance of refining these algorithms to handle more complex patterns and varying environmental conditions.

BIBLIOGRAPHY

Bordón, Pablo, Néstor Bonomo, and Patricia Martinelli. Automatic detection of pipe-flange reflections in GPR data sections using supervised learning. *Journal of Applied Geophysics*, 170, 2019, 103856, ISSN 0926-9851, https://doi.org/10.1016/j.jappgeo.2019.103856

2.4.7.4 Application-48: Prediction of Total Organic Carbon

2.4.7.4.1 Background of the Application

This paper presents an ML approach for the estimation of TOC from borehole logs data. Three ML models, namely RF, rotation forest, and kNN, were employed to estimate TOC from borehole logs data, including GR, resistivity, density, acoustic, and neutron. The models were compared with backpropagation ANN and SVR models. The results showed that the ML models are more efficient in estimating TOC from borehole logs data, thus reducing the need for expensive coring processes.

2.4.7.4.2 Description of the Workflow and Machine Learning Model

In this study, three ML algorithms—RF, Rotation Forest (rF), and K-Nearest Neighbors (KNN)—are utilized to estimate the TOC from conventional well logs. RF is an ensemble method that uses multiple DTs to handle regression tasks, effectively capturing the nonlinear relationships between input features and the target variable (TOC). The Rotation Forest technique enhances diversity and accuracy by applying PCA to subsets of features, thus optimizing the base classifiers. KNN is a straightforward algorithm that predicts the output by averaging the outcomes of the K nearest samples, effectively handling both regression and classification tasks due to its simplicity.

2.4.7.4.3 Advantages and Limitations of the Synchronized Algorithm

The RF algorithm is known for its robustness and ability to handle high-dimensional data efficiently, making it suitable for complex geological datasets like well logs. However, it can be computationally intensive due to the creation of numerous DTs and might require careful parameter tuning for optimal performance. rF provides an advantage by enhancing model accuracy and diversity through feature transformation using PCA, but it can become computationally expensive and may require expertise in selecting the base classifiers and rotation parameters. KNN is simple to implement and interpretable, offering flexibility in handling different types of data. Nonetheless, it is sensitive to the choice of K and can be less effective in high-dimensional spaces due to the curse of dimensionality. Overall, each model has its strengths and weaknesses, and selecting the appropriate model depends on the specific characteristics of the dataset and the problem at hand.

2.4.7.4.4 Results and Discussion

The study demonstrated that ML models, in particular, RF, rF, and KNN, can accurately estimate TOC from borehole logs data, thus minimizing the need for costly coring processes. The study also revealed that borehole logs data, when coupled with ML models, can provide an efficient and cost-effective approach for estimating TOC. This approach can significantly reduce exploration costs and improve exploration efficiency.

BIBLIOGRAPHY

Handhal, Amna M., Alaa M. Al-Abadi, Hussein E. Chafeet, Maher J. Ismail. Prediction of total organic carbon at Rumaila oil field, Southern Iraq using conventional well logs and machine learning algorithms, *Marine and Petroleum Geology*, 116, 2020, 104347, ISSN 0264-8172, https://doi.org/10.1016/j.marpetgeo.2020.104347

2.4.7.5 Application-49: TOC Content Prediction Based on a Combined Gaussian Process Regression Model

2.4.7.5.1 Background of the Application

This study proposes a combined model for accurate TOC content estimation. By classifying logging data using a Gaussian mixture model (GMM), geological conditions are represented in different categories. Independent Gaussian process regression (GPR) models are constructed based on the classified data and combined using the posterior probability given by the GMM as a weight coefficient. This combined-GPR model is compared to traditional prediction methods, such as the delta log R method and other ML algorithms, and it performs better.

2.4.7.5.2 Description of the Workflow and Machine Learning Model

In this study, authors utilize a GMM in combination with GPR to predict the TOC content in geological formations. The GMM is employed to classify the logging data into clusters that represent different geological conditions. This probabilistic model is adept at handling data with multiple distributions and provides a posterior probability that indicates the likelihood of data belonging to each classification. Subsequently, independent GPR models are constructed for each cluster. GPR is chosen for its ability to handle high-dimensional, nonlinear data with a well-defined probabilistic framework, which allows for uncertainty quantification through confidence intervals. The prediction results from these GPR models are combined using the posterior probabilities from the GMM, thus enhancing the accuracy of TOC content predictions across varying geological conditions.

2.4.7.5.3 Advantages and Limitations of the Synchronized Algorithm

The combined use of GMM and GPR in this methodology effectively captures the complex geological variations and enhances the prediction accuracy of TOC content. GMM is advantageous for its robust clustering capability, which aids in understanding different geological settings through probabilistic assignments. GPR complements this by providing a powerful regression framework that offers both predictions and confidence intervals, making it suitable for small sample sizes and nonlinear relationships. However, the methodology has limitations: GMM requires careful selection of the number of components, as misestimation can lead to suboptimal clustering. Additionally, GPR's performance is sensitive to the choice of kernel and hyperparameters, necessitating thorough tuning. The computational cost can be significant due to the iterative nature of both GMM and GPR training, especially when optimizing hyperparameters for large datasets.

This analysis mirrors the style of your example, focusing on describing the AI methods used in the paper, their advantages, and their limitations, contextualized within the study's goals and results.

2.4.7.5.4 Results and Discussion

In conclusion, this study proposes a reliable and effective solution for quantitative prediction problems in corresponding fields. By combining different models, the combined prediction model outperforms traditional single models. The proposed method can be widely used in other fields where ML algorithms are employed, and it provides an alternative way to solve quantitative prediction problems.

BIBLIOGRAPHY

Rui, Jianwen, Hongbing Zhang, Quan Ren, Lizhi Yan, Qiang Guo, Dailu Zhang. TOC content prediction based on a combined Gaussian process regression model. *Marine and Petroleum Geology*, 118, 2020, 104429, ISSN 0264-8172, https://doi.org/10.1016/j.marpetgeo.2020.104429

2.4.7.6 Application-50: A TOC Logging Evaluation Method Using Multiple Overlapping Methods Integrated with Semi-supervised Deep Learning

2.4.7.6.1 Background of the Application

This paper proposes a new method for evaluating the TOC content of shale formations using an integrated deep semi-supervised ladder network (IDLN) algorithm. The method consists of a two-step process that involves intelligent curve overlapping and automatic TOC prediction. The overlap method is employed to enhance the geological significance of the algorithm, thereby improving the accuracy of the model's predictions. The proposed method was tested on well logging data from shale formations in the Sichuan Basin, China, and the results demonstrate that it significantly improves TOC prediction accuracy, reducing the mean square error by more than 50%.

2.4.7.6.2 Description of the Workflow and Machine Learning Model

In this study, an IDLN is employed to predict TOC content from overlapping well logging curves. The IDLN combines both supervised and unsupervised learning to leverage the small amount of labeled data and a larger set of unlabeled data effectively. The DL framework is used for its ability to extract complex patterns from the input data, enhancing the prediction accuracy of TOC. Additionally, a bagging ensemble approach is integrated to further improve model robustness and accuracy by aggregating predictions from multiple networks trained on bootstrapped samples. The IDLN's noise resilience is enhanced by using Gaussian noise in the input layers, which contributes to the stability of the model.

2.4.7.6.3 Advantages and Limitations of the Synchronized Algorithm

The IDLN excels in modeling the complex, nonlinear relationships between the input features (overlapping curves) and the TOC, achieving higher accuracy and stability compared to traditional methods. Its ability to leverage both labeled and unlabeled data makes it particularly well-suited for scenarios where labeled samples are scarce. However, the IDLN requires careful tuning of hyperparameters, including the number of layers, neurons per layer, and noise levels, to optimize performance. Despite its robustness, the computational complexity of training deep neural networks, especially in the context of bagging multiple models, can be a limitation. Moreover, the semi-supervised learning approach necessitates a comprehensive pre-processing pipeline to ensure data quality, including depth matching and feature extraction.

2.4.7.6.4 Results and Discussion

In this study, a new method for evaluating the TOC content of shale formations using an IDLN algorithm is proposed. The method consists of a two-step process that involves intelligent curve overlapping and automatic TOC prediction. The results demonstrate that the proposed method significantly improves TOC prediction accuracy, reducing the mean square error by more than 50%. The new method is also more reasonable for locating non-reservoir segments. Overall, the proposed method has the potential to improve the accuracy of TOC predictions and facilitate the exploration and development of shale gas resources.

BIBLIOGRAPHY

Zhu, Linqi, Chong Zhang, Chaomo Zhang, Zhansong Zhang, Xueqing Zhou, Weinan Liu, Boyuan Zhu. A new and reliable dual model- and data-driven TOC prediction concept: A TOC logging evaluation method using multiple overlapping methods integrated with semi-supervised deep learning. *Journal of Petroleum Science and Engineering*, 188, 2020, 106944, ISSN 0920-4105, https://doi.org/10.1016/j.petrol.2020.106944

3 AI Applications in Drilling Engineering

Ekrem Alagoz, Yasin Ozkan, Haowei Tang, Hantang Li, Weichao Yang, Emre Can Dündar, Muhammed Esad Kaya, Cenk Temizel, Salih Tutun, Mehmet Melih Oskay, Celal Hakan Canbaz, Luigi Saputelli, Mufrettin Murat Sari, and M Esad Kaya

3.1 INTRODUCTION

Over the years, the oil and gas industry has witnessed a transformative shift with the integration of advanced technologies, particularly in the realm of machine learning (ML) and artificial intelligence (AI). These technologies have revolutionized drilling operations by offering unprecedented insights, predictive capabilities, and optimization strategies.

This chapter delves into the multifaceted landscape of ML and AI applications within drilling operations. From the characterization of drilling fluids to the optimization of drilling processes, these technologies play a pivotal role in enhancing operational efficiency, reducing costs, and mitigating risks.

The chapter begins with a comprehensive categorization of workflows commonly employed in drilling operations, encompassing aspects such as drilling fluids, rate of penetration (ROP), drilling optimization, and other pertinent applications. Subsequently, an overview of Artificial Intelligence (data analytics) methodologies sets the stage for a detailed exploration of various applications across these domains.

Each application is meticulously examined, providing a background context, description of the Artificial Intelligence workflow deployed, advantages and limitations of the methodology, along with notable results and recommendations. Spanning from real-time prediction of ROP to the development of autonomous drilling fluid management systems, these applications showcase the diverse array of challenges addressed and solutions provided by ML and AI technologies in the drilling sector.

By synthesizing cutting-edge research, practical implementations, and insights from industry experts, this chapter offers a comprehensive overview of the transformative impact of ML and AI in modern drilling operations. It serves as a valuable resource for industry professionals, researchers, and stakeholders seeking to harness the power of advanced technologies to optimize drilling processes, enhance productivity, and drive sustainable growth in the oil and gas sector.

DOI: 10.1201/9781003617327-3

3.2 CATEGORIZATION OF WORKFLOWS USED IN DRILLING OPERATIONS

3.2.1 Drilling Fluids

Within the realm of drilling operations, various workflows are employed to ensure the efficient and safe extraction of oil and gas resources. One crucial aspect of this process revolves around the management and optimization of drilling fluids. Drilling fluids, also known as drilling muds, serve multifaceted purposes ranging from lubricating and cleaning the drill bit to carrying rock cuttings to the surface. In this section, we delve into the categorization and application of workflows related to drilling fluids. This encompasses the characterization of fluid properties, such as viscosity and density, in real-time scenarios using advanced AI and ML techniques. Additionally, the chapter explores the development of predictive models to anticipate and mitigate challenges associated with drilling fluid performance, such as lost circulation and formation damage. By leveraging AI and ML methodologies, drilling fluid workflows aim to enhance operational efficiency, minimize downtime, and optimize overall drilling performance.

3.2.2 Rate of Penetration (ROP)

In Section 3.2.2, the focus shifts toward the crucial parameter of ROP, a key metric in drilling operations that directly impacts efficiency and cost-effectiveness. ROP refers to the speed at which the drill bit advances through the subsurface formations, and optimizing this parameter is paramount for reducing drilling time and associated expenses. This section explores various workflows and methodologies employed to analyze, predict, and optimize ROP using ML and AI techniques. By harnessing historical drilling data, real-time downhole measurements, and advanced predictive algorithms, drilling engineers can gain insights into factors influencing ROP, such as lithology, formation properties, drilling parameters, and tool conditions. Furthermore, the chapter delves into the development of ROP predictive models, enabling drillers to make informed decisions, adjust drilling parameters, and maximize penetration rates while minimizing risks such as downhole tool failure and formation damage. Through the application of AI and ML in ROP optimization, drilling operations can achieve higher efficiency, reduce operational costs, and improve overall project outcomes.

3.2.3 Drilling Optimization

In this section, the focus shifts toward drilling optimization, a critical aspect of drilling operations aimed at maximizing efficiency and minimizing costs while ensuring safety and environmental compliance. Drilling optimization encompasses a wide range of workflows and methodologies designed to enhance various aspects of the drilling process, including trajectory planning, wellbore stability analysis, and drilling parameter optimization. Leveraging AI and ML techniques, this section explores the development of advanced predictive models and optimization algorithms to optimize drilling parameters in real-time. By analyzing vast amounts of historical drilling data and incorporating real-time downhole measurements, these models can provide actionable insights into optimal drilling strategies, such as adjusting weight on bit (WOB), rotational speed, and mud properties to improve drilling performance and mitigate risks such as stuck pipe incidents, dog legs, and wellbore instability. Furthermore, drilling optimization workflows may involve the integration of advanced sensing technologies, such as in-cutter sensors and downhole vibration monitoring, to enhance real-time decision-making and improve overall drilling efficiency. Through the application of AI and ML in drilling optimization, operators can achieve significant improvements in drilling efficiency, reduced non-productive time (NPT), and enhanced wellbore quality, ultimately leading to more cost-effective and sustainable drilling operations.

3.2.4 Other Applications

In Section 3.2.4, the exploration extends beyond the traditional domains of drilling fluids, ROP, and drilling optimization to encompass a diverse array of other applications within drilling operations. These applications may include but are not limited to directional drilling, bottom hole pressure (BHP) prediction, flow rate estimation, and cement evaluation, among others. This section delves into the unique challenges and opportunities presented by these diverse applications and explores how AI and ML methodologies are being leveraged to address them. For instance, in directional drilling, AI algorithms can analyze geophysical data and drilling parameters to optimize well trajectory and improve accuracy in reaching target reservoirs. Similarly, in BHP prediction, ML models can utilize historical drilling data to forecast downhole pressures and optimize drilling parameters to prevent well control incidents. By examining these diverse applications, this section highlights the versatility and adaptability of AI and ML technologies in addressing a wide range of challenges encountered in drilling operations, ultimately contributing to improved efficiency, safety, and sustainability in the oil and gas industry.

3.3 OVERVIEW OF ARTIFICIAL INTELLIGENCE METHODOLOGIES

All the above-mentioned workflows used in the drilling technologies need analytical tools that leverage empirical data available. AI employs such methodologies that enhances the physics-based understanding of the resource by incorporating principles from big DA, data clustering, supervised and unsupervised ML. The following section illustrates a few example applications of these AI techniques in facilitating the aforementioned workflows utilized in the drilling operations by the oil and gas industry.

3.4 APPLICATIONS OF ARTIFICIAL INTELLIGENCE IN DRILLING OPERATIONS

3.4.1 Drilling Fluids

3.4.1.1 Application-1: Determination of the Rheological Properties of Invert Emulsion-Based Mud on Real Time Using Artificial Neural Network

3.4.1.1.1 Background of Application

The most basic tasks of the drilling mud are to carry the cuttings from the bottom of the well to the surface and provide well cleaning. These factors significantly affect the drilling speed, penetration rate. There are mud parameters that must be measured and kept under control during drilling to fulfill these duties. While properties such as marsh funnel viscosity, mud density, and solid percent are measured every 15–20 minutes, some parameters such as apparent viscosity, yield point, plastic viscosity, and gel strength can be measured 1–2 times a day due to time limitation on the rig site (Abdelgawad et al., 2018). Properties such as plastic viscosity and yield point were estimated using AI methods. These are the ANN (artificial neural network) method and the SaDE (Self-adaptive Differential Evolution) method proposed by Qin et al. (2009). Variables dependent on ANN were set using the SaDE method. The rheological properties of drilling mud were estimated with the combined use of SaDE–ANN.

Input parameters:

- marsh funnel time, min
- mud density, lb/cuft
- solid percent, %

Output parameters:

- plastic viscosity, cp
- yield point, lb/100 sqft
- apparent viscosity, cp
- flow index

3.4.1.1.2 Description of Artificial Intelligence Workflow

The article utilizes ANN together with SaDE to determine the drilling fluid rheological properties (yield point, plastic viscosity, flow behavior index, and fluid consistency index) based on the common measuring fluid properties: density, marsh funnel viscosity, and solid percent. To efficiently predict rheological properties, the authors use the SaDE optimizing technique to precisely set the ANN-related variables, including layer parameters, training and transferring functions, and training and testing ratios. Once these parameters are controlled and the values of the output layer are trained to be close to the experimental data, the correlations are extracted to estimate the outputs using testing data. ANN is based on a collection of connected neurons.

Neurons and their connections have weights that adjust as the training proceeds. And the weight increases or decreases the strength of the signal at a connection. To get the output, the weighted sum of inputs is taken, the bias is added, and then it is passed through an activation function (or transferring function). The equation for prediction is

$$\left[\sum_{i=1}^{N} w_{2i}\ tansig\left(\sum_{j=1}^{j} w_{1i,j}\ x_j + b_{1i}\right)\right] + b2$$

In this case, the model is trained to obtain the weights, which generated the most accurate predictions. The ANN was used to predict R_{600} and R_{300}, which are compared with the actual measured data. The equations for calculating R_{600} and R_{300} values are listed below:

$$R_{300} = YP + PV$$

$$R_{600} = PV + R_{300}$$

$$n = 3.32 * \log\left(\frac{R_{600}}{R_{300}}\right)$$

$$K = \frac{R_{600}}{1022^n}$$

3.4.1.1.3 *Advantages and Limitations of Artificial Intelligence Methodology*

The advantage of the SaDE-ANN model is that the SaDE algorithm has self-adaptive control parameters and mutation strategies based on the learning experience, which can avoid the trial-and-error procedure to reduce the time required for optimization. This technique sets precisely ANN-related variables for efficient prediction. Once the SaDE-ANN model is optimized, it can be used to predict the rheological properties in real time. Rheological properties such as density, marsh funnel viscosity, and solid percent can be measured frequently, ideally every 15–20 minutes. However, due to time constraints on drilling rigs, complete rheological measurements are typically conducted twice daily.

3.4.1.1.4 *Results and Recommendations*

The SaDe–ANN model was built using 62% of the data for training, 15% for CV, and 23% for testing as determined by the self-adaptive algorithm. The model was trained using 633 data points and compared to actual data, with a maximum average absolute percentage error (AAPE) of 5.27 for plastic viscosity and correlation coefficients above 0.9 for all properties. Testing the model with 246 data points resulted in average absolute errors less than 6% for all parameters and correlation coefficients above 0.89. The model was validated with 150 new data points, giving acceptable AAPEs of 5.39 or less and correlation coefficients higher than 0.92. The predicted mud properties can be used to calculate pressure loss, equivalent circulation density, surge and swab pressures, and detect drilling problems like pipe sticking. The mud properties can also be adjusted to improve hole cleaning performance during drilling.

The SaDe–ANN model accurately predicts spud mud rheological properties using mud density, marsh funnel viscosity, and solid percent as inputs. The model predicts the parameters with an average absolute error of less than 6% and a correlation coefficient of over 90%. The predictions can aid in real-time calculation of equivalent circulation density, surge and swab pressures, and improve hole cleaning optimization and control.

BIBLIOGRAPHY

Abdelgawad, Khaled, Elkatatny, Salaheldin, Mousa, Tamer, Mahmoud, Mohamed, and Patil, Shirish. "Real Time Determination of Rheological Properties of Spud Drilling Fluids Using a Hybrid Artificial Intelligence Technique." *Paper presented at the SPE Kingdom of Saudi Arabia Annual Technical Symposium and Exhibition*, Dammam, Saudi Arabia, April 2018. https://doi.org/10.2118/192257-MS

Qin, A.K.; Huang, V.L.; Suganthan, P.N. (2009). Differential Evolution Algorithm With Strategy Adaptation for Global Numerical Optimization, 13(2), 398–417. https://doi.org/10.1109/tevc.2008.927706

3.4.1.2 Application-2: Artificial Intelligence Strategy Minimizes Lost Circulation Non-productive Time

3.4.1.2.1 Background of Application

This study proposes a method to minimize fluid loss during well construction in the deep-water pre-salt area. A supervised learning artificial intelligence strategy is used to generalize data from 500 lost circulation events over 3 years, and a human-computer interaction interface is developed for the field engineer to make decisions based on the data. The method has been validated on data from several wells in the Santos Basin and has resulted in a 6-day time saving in well construction in two real cases.

3.4.1.2.2 Description of Artificial Intelligence Workflow

The aim of the project is to create a lost circulation material (LCM) recommendation system by classifying the operation's success status in reference to an engineered metric using a set of binary combinations of supervised machine and deep learning models through the ensemble voting method, which has been applied in the Santos Basin, Brazil.

The final output is a parameter "Success Quantitative Metric (SQM)" initially engineered by the authors, which considered the circulation loss range pre and post a specific LCM injection and is defined as follows:

- Success: Reduction to the seepage loss range (<50 bbl/h).
- Partial: Reduction to moderate loss range (50-150 bbl/h).
- Failure: Complete/severe loss range (>150 bbl/h) or no reduction.

According to the paper, the authors took the following parameters as inputs:

- Static and dynamic lost circulation pre-LCM injection (bbl/h).
- Geographic Coordinates (east and north).

- Current Well Depth.
- Utilized LCM (cross-linked pills, fluid loss squeeze, bridging agents, and cement).
- The computed SQM variable.

Missing data in loss events have been dropped rather than interpolated, which is a good endeavor if the number of missing instances is relatively small, so that it would not affect the model. They also reported that geographic coordinates introduced a high degree of bias, so the latter has been removed from the input features.

In their approach, the authors utilized a versatile set of ML families listed below, along with the respective algorithms employed:

- **Naive Bayes**: Hidden Naive Bayes (NB), weighted averaged one-dependence estimators (WAODE).
- **Instance-Based Algorithms**: Euclidean-based K-nearest neighbor (KNN), KStar instance-based classifier (KS).
- **Neural Network**: multilayer perceptron (MLP), radial basis function **neural** networks (RBFNN).

The approach is relatively simple, as they trained binary combinations of each model, and the potential combinations are summarized in Table 3.1.

The ensembles constructed have been cross-validated using a method called the K-fold cross-validation, which is a technique employed to train and test the model on multiple subsets of the data rather than a single batch, where k (number of subsets) is set to 10. Model performance assessment has been conducted using the receiver operating characteristic (ROC) metric, which tracks in a classification model the false positive rate (FPR) and true positive rate (TPR), as a scalar value; the higher the ROC, the better the model.

The authors report that the KS-MLP ensemble had the highest ROC score, thus the best performer. The models have later been deployed into an easy to use custom platform where the user can input the parameters including the LCM material that will be used and will receive the success potential of the remedial job along with other recommendations if it has been assessed as a potential failure.

TABLE 3.1
Summary Table of Binary Model Combinations Employed in the Paper

	NB	WAODE	KNN	KS	MLP	RBFNN
NB	NB	NB-WAODE	NB-KNN	NB-KS	NB-MLP	NB-RBFNN
WAODE		WOADE	WAODE-KNN	WAODE-KS	WAODE-MLP	WAODE-RBFNN
KNN			KNN	KNN-KS	KS-MLP	KS-RBFNN
KS				KS	KS-MLP	KS-RBFNN
MLP					MLP	MLP-RBFNN
RFBNN						RBFNN

3.4.1.2.3 *Advantages and Limitations of Artificial Intelligence Methodology*

Ensemble learning techniques attempt to make the performance of the predictive models better by improving their performance. Ensemble learning is a process using which multiple ML models (such as classifiers) are strategically constructed to solve a particular problem. In data science terminology, improving a model's performance means finding the right tradeoff between bias (is the model capturing the main trend?) and variance (how does the model perform on new data). Thus, tuning the model complexity sufficiently to capture the main trend while generalizing well on new observations.

Multiple ensemble methods are available in this technique toolset, among them (voting, averaging, and weighted average) and more advanced techniques such as stacking, blending, bagging, and boosting, each in a way offers an additional layer of complexity, improving performance but with a minimal risk of overfitting.

Another key observation to consider is the computational cost of running all these different combinations using a 10 K-fold cross validation, which can be quite expensive as the number of data points in the data increases. A suggestion would be to initially employ some commonly employed state-of-the-art ML or DL models with predefined hyperparameters, then run an optimization search of the latter and finally combine the best models in case where it's required.

3.4.1.2.4 *Results and Recommendations*

The work (Cristofaro et al., 2017) presents a methodology for optimizing loss control materials during drilling operations to minimize lost circulation events. A supervised learning-based artificial intelligence strategy was used to generalize data collected from 500 lost circulation events over a 3-year period. An intuitive interface was designed to enable mud engineers to interact with the data mining methodology and its related ML models. The study showed good agreement with legacy data and demonstrated the potential to save at least four rig days by directly applying the recommended cement paste. The models may be overfitted to the small amount of data, so continuous database updates and model calibration are in progress to increase reliability and robustness.

BIBLIOGRAPHY

Cristofaro, L. R. A., Longhin, G. A., Waldmann, A. A., de Sá, C. H. M., Vadinal, R. B., Gonzaga, K. A., and Martins, A. L. "Artificial Intelligence Strategy Minimizes Lost Circulation Non-Productive Time in Brazilian Deep-Water Pre-Salt." *Paper presented at the OTC Brasil*, Rio de Janeiro, Brazil, October 2017. https://doi.org/10.4043/28034-MS

3.4.1.3 Application-3: Intelligent Prediction of Stuck Pipe Remediation Using Machine Learning Algorithms

3.4.1.3.1 *Background of Application*

In order to lower downtime and expenses related to clogged pipes in oil and gas exploration activities, this research investigates the use of ML to create an expert system. In order to anticipate the solution for stuck pipes in vertical and deviated wells, ANNs and support vector machines (SVMs) were utilized to create prediction models based on field information gathered from 385 wells in Southern Iraq. Although the SVM results are more encouraging, the results demonstrate that both ANNs and SVMs can be useful. Response times and clogged pipe treatment techniques can be enhanced by these ML techniques.

3.4.1.3.2 Description of Artificial Intelligence Workflow

The main goal of the ML model in Section 3.4 is to use a wide range of parameters to decide the best action to take in a stuck pipe event. The article proposed and compared methods: ANN and SVM. There are in total 20 inputs that describe the context of the situation, and there are six possible outputs that stand for different actions. Authors have loaded the data and split it into training, testing, and validating datasetsIn the first part, a simple ANN model was established using one hidden layer with 30 nodes. For the output layer, six nodes and ReLU activation were used to suit the multiple classification problem. In the second part, a simple multiple SVM classifier was applied. Due to the lack of real data, hyperparameters were not tuned for either model.

3.4.1.3.3 Artificial Intelligence: Advantages and Limitations of Artificial Intelligence Methodology

Compared to each other, SVM provides an overall better result. And in practice, SVM also requires less computation and is faster to train. In terms of interpretability, neither of the two models is straightforward. Therefore, both models are more suitable for prediction and not for providing inference.

3.4.1.3.4 Results and Recommendations

This paper (Abbas et al., 2019) studied the use of ANNs and SVMs as expert systems for predicting stuck pipe solutions in petroleum and gas exploration operations. The study found that the accuracy of the predictive models depends on the input parameters selected and the quality and size of the dataset. The results showed that input parameters such as inclination, ROP, and flow rate have significant effects on the output. The comparison of the performance efficiency showed that the SVM method was more efficient than the ANNs method. The study also found that the proposed ML methods can predict the most effective solution for stuck pipe problems with reasonable accuracy of over 90%. The developed models can be used with real-time data and can provide more accurate solutions for problems associated with pipe sticking events.

BIBLIOGRAPHY

Abbas, Ahmed K., Flori, Ralph, Almubarak, Haidar, Dawood, Jawad, Abbas, Hayder, and Alsaedi, Ahmed. "Intelligent Prediction of Stuck Pipe Remediation Using Machine Learning Algorithms." *Paper presented at the SPE Annual Technical Conference and Exhibition*, Calgary, AB, Canada, September 2019. https://doi.org/10.2118/196229-MS

3.4.1.4 Application-4: A Machine Learning Approach to Filtrate Loss Determination and Test Automation for Drilling and Completion Fluids

3.4.1.4.1 Background of Application

Drilling fluid property characterization is an important step during drilling operations. Current methods to measure fluid properties are high-pressure and high-temperature (HPHT) filter press units to indicate filtrate loss (FL) of drilling mud. The FL can be defined as the amount of fluid filtered into the formation under certain temperature and pressure values. The paper proposes to predict FL based on plastic viscosity, yield point, and mud weight for water-based mud (WBM) and electrical stability, water content, apparent viscosity, and mud weight for oil-based mud (OBM). Both models are built using the random forest regression algorithm and utilize grid search cross-validation (CV) to tune hyperparameters. The model performance is evaluated with mean absolute error (MAE) and mean absolute percentage error (MAPE).

3.4.1.4.1.1 Prediction Model for Filtrate Loss in Oil-Based Mud (Random Forest Regression)

Inputs: Plastic viscosity, Yield point, Mud weight
Output: Filtrate loss

3.4.1.4.1.2 Prediction Model for Filtrate Loss in Water-Based Mud (Random Forest Regression)

Inputs: Apparent viscosity, Mud weight, Electrical stability, Water content
Outputs: Filtrate loss

3.4.1.4.2 Description of Artificial Intelligence Workflow

The main objective of the model is to predict FL in the drilling process. And since there's a deterministic difference between WBM and OBM, authors are using two separate random forest models to predict the FL in two types of mud. After reading the data and splitting dependent and independent variables, they split the data into the training and testing sets. Assuming they have a medium-sized dataset, and set the training size to 0.8. In the model, the x variables are plastic viscosity, yield point, and mud weight, and the output variable is the FL. There are mainly two hyperparameters that the work can tune for the random forest model: the number of trees and the depth of each tree (other hyperparameters like maximum features, maximum nodes can also be tuned in a similar way). To tune the optimal parameters for the models, 10-fold cross-validation with GridSearchCV() was used.

3.4.1.4.3 Advantages and Limitations of Artificial Intelligence Methodology

The advantage of random forest is clearly its high prediction accuracy and its relatively low requirement of a very large dataset. However, a large number of trees may cause long prediction times, which can be inconvenient for real-time applications.

3.4.1.4.4 Results and Recommendations

In this study, the researchers used ML techniques to determine the American Petroleum Institute (API) and HPHT FL of WBMs and OBMs, respectively. The aim was to see if these FL measurements can be made more efficiently and accurately. The researchers trained various ML models (random forest, XGBoost, SVM, MLP, and multi-linear regression) using fluid property readings, such as rheology, density, temperature, electrical stability, and water content. The results showed that API and HPHT FL of WBMs and HPHT FL of OBMs can be estimated with an MAE within the measurement accuracy of the mud engineer. This approach offers a way to automate FL characterization, which is often required for field reporting, and the parameters used in this approach can be obtained from relatively simple and easy to automate measurements, such as Coriolis mass-flow meters, automated pipe viscometers, water cut analyzers, and electrical stability meters.

The use of ML to determine the fluid loss (FL) in mud systems has been found to be accurate. The approach can automate FL characterization and can be used for mud maintenance guidance and field reporting. Objections to this approach include the lack of direct FL measurements and concerns over the impact of severe mud contamination. However, the current FL tests are not representative of the downhole environment, and the ML method offers a more convenient solution in cases where historical FL databases are available. Severe mud contamination is rare, and the consequences are expected to be apparent in automated mud density and rheology measurements, which can trigger corrective action.

BIBLIOGRAPHY

Sercan, Gul, Eric, van Oort, A machine learning approach to filtrate loss determination and test automation for drilling and completion fluids. *Journal of Petroleum Science and Engineering*, 186, 2020, 106727, https://doi.org/10.1016/j.petrol.2019.106727

3.4.1.5 Application-5: Estimation of Bottom Hole and Formation Temperature by Drilling Fluid Data: A Machine Learning Approach

3.4.1.5.1 Background of Application

A novel ML approach is proposed to predict bottom hole circulating temperature (BHCT) and static formation temperature (SFT) in real-time using drilling fluid data. Two different models (random forest and XGBoost) were trained using data obtained from various wells of a large independent operator, including casing and drill string design, mud reports, flow rate, bottom hole temperature readings from measurement while drilling (MWD), and temperature log data. The trained models were tested using 20% of the data with an MAPE of less than 1% for both algorithms. This approach provides high-accuracy estimates of BHCT and SFT using real-time drilling fluid data and can save time and cost in geothermal development projects.

3.4.1.5.2 Description of Artificial Intelligence Workflow

The objective of the application is to build and train ML models to predict the BHCT and the SFT based on some input features. Here is a summary of the workflow of the code:

First, the required packages are imported: pandas, numpy, skilearn, and xgboost. After loading the data, the data is split into training and testing sets using the train_test_split function from skilearn. Two functions are defined, RFR_train and XGBoost_train, to train random forest regressor and XGBoost regressor models, respectively. These functions use CV with GridSearchCV to find the best hyperparameters for the models. After obtaining the best model, predictions were made on the testing dataset using it.

3.4.1.5.3 Advantages and Limitations of Artificial Intelligence Methodology

Random forest and XGBoost are both great examples of ensemble learning. Compared to deep neural networks, the two models perform better when the data size is restricted. And the two models are more agile and easy to train compared to more complicated models.

3.4.1.5.4 Results and Recommendations

In this paper, the authors propose a ML approach to predict BHCT and SFT in real-time using drilling fluid data. The algorithm uses data from various wells obtained from a large independent operator, including casing and drill string design, daily mud reports, flow rate, and bottom hole temperature readings from MWD and temperature logs. The ML models (random forest and XGBoost) were trained with 80% of the data and tested with 20%. The results show that both BHCT and SFT can be estimated with high accuracy using this method with an MAPE of less than 1%. This approach offers a novel and more efficient way to obtain fluid temperature profiles in real time and can be used to improve and verify MWD tools and cement slurry designs, leading to cost savings in geothermal development projects.

BIBLIOGRAPHY

Gul, S., Aslanoglu, V., Tuzen, M.K., & Senturk, E. (2019). *Estimation of Bottom Hole and Formation Temperature by Drilling Fluid Data: A Machine Learning Approach.*

3.4.1.6 Application-6: Estimation of Total Dissolved Solids (TDS) Using New Hybrid Machine Learning Models

3.4.1.6.1 Background of the Application

The study aimed to predict the total dissolved solids (TDS) of aquifers using adaptive neuro-fuzzy inference system (ANFIS), SVMs, and ANN models. Various optimization algorithms were used to train the models, and the most appropriate inputs were determined through principal component analysis (PCA). The results showed that the hybrid ANFIS-MFO model was the most accurate, outperforming the ANN-MFO and SVM-MFO models by 1.4% and 3.8%, respectively. The TDS values were also analyzed for temporal variability using innovative trend analysis (ITA), with high TDS values indicating an increasing trend. The ANFIS-MFO and ANFIS-CSO models were found to have superior performance compared to the other models.

3.4.1.6.2 Description of the Workflow and Machine Learning Model

This study utilizes an ML approach to evaluate groundwater quality with TDS concentration. The primary objective of the study is to provide superior prediction accuracy in TDS estimation compared to traditional methods and standalone ML models. The study takes a hybrid ML approach combining ANFIS, SVMs, and ANNs and incorporating optimization techniques such as moth flame optimization (MFO) and cat swarm optimization (CSO) to refine the model training process. The study suggests that the model could facilitate real-time water sources monitoring and help develop sustainable water management practices and regulations with insights on patterns and factors that influence TDS levels.

Data Collection:

- The research focuses on the Yazd province in Iran, with specific geographical coordinates. The total basin area is 2,491 km^2. Data from 12 observed wells in the Yazd plain, recorded between 2002 and 2015, are utilized. The dataset consists of water quality data.

Parameter Optimization using the Taguchi Model:

- Determine the optimal values of random parameters needed for the optimization algorithms.
- Use orthogonal arrays in the Taguchi model.
- Compute the signal to noise (S/N) ratio for each parameter and level combination.
- Select the parameter combination that provides the highest S/N ratio as the optimal configuration.

Feature Reduction using PCA:

- If there are many features, employ PCA to reduce the dimensionality.
- Standardize the input data.
- Form the correlation matrix and determine the principal components.
- Select the principal components that account for the majority of the variance in the dataset (in this study, the first three PCs were chosen).

Model Building:

- Selection of various models: ANFIS, ANN, SVM, and hybrid variants.
- Configuring the soft computing models with selected optimization algorithms like MFO, cuckoo search optimization (CSO), particle swarm optimization (PSO), gravitational search algorithm (GSA), simulated annealing (SA), and grey wolf optimizer (GWO).

Model Evaluation:

- Testing the models using the testing dataset.
- Calculating statistical metrics such as root mean squared error (RMSE), MAE, percent bias (PBIAS), and Nash–Sutcliffe efficiency (NSE) for model performance evaluation.
- Using scatterplots to visually compare the performance of the different models.
- Performing uncertainty analysis using the sequential uncertainty fitting version-2 (SUFI-2) method to quantify model uncertainties.
- Evaluating trends in TDS values using ITA.

Prediction:

- Predicting TDS values using the trained models.
- Comparing the predicted values against observed data to evaluate the accuracy of predictions.
- Analyzing TDS trends for different value ranges (low, medium, and high) based on the predictions from various models.
- Drawing insights and conclusions based on the predictions and their alignment with observed data.

3.4.1.6.3 Advantages and Limitations of the Method

The study unveils several key insights into hydrological modeling. underscores the significance of random parameters in optimization, highlighting the necessity to pinpoint their optimal values. To achieve this, the research leverages the Taguchi model, a robust method that uses orthogonal arrays to identify suitable random parameters, ensuring optimum outcomes. Through the PCA model, the study reveals a significant reduction in data dimensions without major information loss, noting that the first three principal components, influenced mainly by HCO_3, Ca, and Mg, capture over 87% of input data variance. Furthermore, a comparative analysis of diverse soft computing models underscores the superior accuracy of the ANFIS–MFO model in relation to its counterparts.

3.4.1.6.4 Results and Discussion

In this study, the TDS value of aquifers in the Yazd plain was estimated using ANFIS, ANN, and SVM models. The MFO, CSO, GWO, GSA, PSO, and SA algorithms were used to train these models. The ANFIS–MFO showed the highest accuracy in the training phase, with an RMSE of 2.25, an MAE of 2.21, an NSE of 0.94, and a PBIAS of 0.12. The hybrid models (ANFIS, ANN, and SVM) performed better than the standalone models. The trend of TDS levels was determined using the ITA method; low TDS values showed a declining trend, medium TDS values showed no discernible trend, and high TDS values showed an increasing trend. Decision makers and policymakers can use these predictive models to determine whether groundwater is suitable for drinking, irrigation, and household uses. Multi-objective MFO algorithms can be used in future studies to improve the outputs' accuracy.

BIBLIOGRAPHY

Banadkooki, Fatemeh Barzegari, Ehteram, Mohammad, Panahi, Fatemeh, Saad, Sh, Sammen, Faridah Binti, and Othman, Ahmed EL-Shafie, Estimation of total dissolved solids (TDS) using new hybrid machine learning models, *Journal of Hydrology*, 587, 2020, 124989, ISSN 0022-1694, https://doi.org/10.1016/j.jhydrol.2020.124989

3.4.1.7 Application-7: Lost Circulation Prediction in South China Sea Using Machine Learning and Big Data Technology

3.4.1.7.1 Background of Application

The paper describes a study using ML algorithms and big data (BD) technology to predict lost circulation in drilling operations in the South China Sea. The study uses drilling data and considers geological characteristics, drilling fluid parameters, and drilling operation parameters. An ANN was used to conduct supervised learning and evaluate the model using accuracy, precision, F1 score, and recall metrics. The trained model was tested on 50 drilled wells over the past two years in the South China Sea, achieving an accuracy of 92%, and F1 score, recall, and precision of 89%. The study suggests that this model can be applied to other fields and can assist drilling engineers in avoiding lost circulation events by predicting lost circulation risks and choosing optimal drilling parameters.

3.4.1.7.2 Description of Artificial Intelligence Workflow

1. **Data Collection**: The first step involves gathering data from 50 drilled wells over the past two years in the South China Sea. The data includes drilling operation parameters, geological parameters, and drilling fluid properties.
2. **Data Preprocessing**:
   - **Outlier Handling**: Outliers in the dataset are identified using box plots and are replaced with average values to avoid skewing the model.
   - **Feature Encoding**: Qualitative features like lithology and lost circulation types are converted into quantitative features using one-hot encoding.
   - **Normalization**: The features are normalized to speed up the training process and improve model performance.
3. **Data Splitting**: The preprocessed data is split into training and testing sets. This allows the model to learn from one subset of the data and be validated on another.
4. **Model Building**: An ANN is constructed using TensorFlow. The ANN consists of an input layer, one hidden layer, and an output layer. The activation functions and other hyperparameters are chosen based on the problem requirements.
5. **Model Training**: The ANN model is trained using the training dataset. The model learns to predict six kinds of lost circulation risks based on the drilling mud loss rate.
6. **Model Evaluation**: After training, the model is evaluated on the testing dataset using metrics like accuracy, F1 score, recall, and precision.
7. **Prediction**: Finally, the trained model can be used to make predictions on new data to assess lost circulation risks.

3.4.1.7.3 Advantages and Limitations of Artificial Intelligence Methodology

Advantages:

1. **High Accuracy**: The model achieved an accuracy of up to 92%, making it a reliable tool for predicting lost circulation risks.
2. **Data-Driven**: Utilizes real-world data for training, making the model robust and applicable to actual drilling scenarios.
3. **Automated Decision-Making**: Can assist drilling engineers in choosing optimal drilling parameters, thereby reducing operational costs and risks.
4. **Scalability**: The model can be further refined and adapted for other drilling environments or conditions.
5. **Generalization**: The model showed good generalization abilities, indicating it could be applied to other fields or similar problems.

Limitations:

1. **Data Sensitivity**: The model's performance is highly dependent on the quality of the data. Outliers or incorrect data can significantly impact the model's predictions.
2. **Computational Cost**: Training a neural network can be computationally expensive, requiring specialized hardware for faster computation.
3. **Hyperparameter Tuning**: The performance of the ANN is sensitive to the choice of hyperparameters, which may require extensive tuning.
4. **Interpretability**: Neural networks are often considered "black boxes," making it challenging to understand how the model arrives at a specific prediction.
5. **Overfitting Risk**: Without proper validation techniques, the model could overfit the training data, making it less effective on new, unseen data.

3.4.1.7.4 Results and Recommendations

In this study, an ANN and BD technology are used to predict six types of lost circulation risks in YQ basin, a high-pressure, high-temperature offshore drilling region. The ANN model is trained with drilling operation parameters, geology parameters, and drilling fluid property parameters. The study concludes that the ANN model is effective in predicting lost circulation risks and provides guidance for the estimation of lost circulation risks prior to drilling. The study also found that a narrow safe drilling mud density window and induced fracture development are the main causes of severe lost circulation in the YQ basin. The results have established a qualitative estimation model of lost circulation in the YQ basin, which can be applied to other fields if required data are available.

BIBLIOGRAPHY

Hou, Xinxin, Yang, Jin, Yin, Qishuai, Liu, Hexing, Chen, Haodong, Zheng, Jinlong, Wang, Junxiang, Cao, Bohan, Zhao, Xin, Hao, Mingxuan, and Xun Liu. "Lost Circulation Prediction in South China Sea using Machine Learning and Big Data Technology." *Paper Presented at the Offshore Technology Conference*, Houston, TX, May 2020. https://doi.org/10.4043/30653-MS

3.4.1.8 Application-8: Data Boosted Stuck Pipe Prevention: Integrating Machine Learning Models into the Mud Plan Design Workflow

3.4.1.8.1 Background of the Application

The project aims to develop a model that predicts the likelihood of a stuck pipe event in drilling operations based on previously drilled well data. The model uses neural network models and processes over 24,000 rows of data from the Volve dataset, which includes daily drilling operation reports from 30 wells. The data is preprocessed and classified based on whether or not a stuck pipe event occurred. A natural language processing (NLP) model is used to ease the processing of the large dataset, resulting in a final dataset of about 600 rows. The dataset is then used to train the neural network model, which shows high accuracy with over 84% accuracy on the test data. The trained model is implemented in a frontend interface where the user can upload mud plan parameters and receive a stuck pipe event chance evaluation. This methodology considers historical data and geomechanical properties to reduce the chances of stuck pipe events, which are a significant cause of NPTs during drilling operations.

3.4.1.8.2 Description of the Workflow and Machine Learning Model

Data Collection: Data is collected from the Volve's dataset, which includes daily drilling operation reports of 30 wells, consisting of more than 24,000 rows of data related to geological and operational conditions.

Data Preprocessing: NLP: An NLP model is used to pre-classify the daily activity comments and identify rows related to stuck pipe events.

Feature Encoding: Categorical variables like lithology and fluid base are converted into numerical representations using one hot encoding.

Oversampling: Techniques like SVM SMOTE are used to handle the imbalance in the dataset by creating artificial points based on the statistical distribution of real points.

Dimensionality Reduction: Methods like t-SNE are applied to reduce the number of input variables into synthetic variables, easing the training process.

Model Building: A neural network model is developed, consisting of an input layer of 3 nodes, 10 hidden layers of 256 nodes each, and an output layer with a single node. The model predicts whether the inputted conditions can lead to a stuck pipe event.

Model Training: The network is trained for 10,000 epochs using the Adam optimizer and cross entropy loss function. A genetic algorithm (GA) is employed to optimize the network hyperparameters.

Model Evaluation: The model is evaluated using a testing dataset, which consists of a portion of the original dataset not used for training. The accuracy of the model is assessed based on this evaluation.

Prediction: The trained model is then used to make predictions on new data to assess the probability of a stuck pipe event occurrence.

3.4.1.8.3 Advantages and Limitations of the Synchronized Algorithm

Advantages:

High Accuracy: The model achieved an accuracy of 84.7%, making it a reliable tool for predicting stuck pipe events.

Data-Driven Insights: Utilizes extensive real-world data for training, making the model robust and applicable to actual drilling scenarios.

Automated Risk Assessment: Can assist engineers in assessing the risk of stuck pipe events, thereby enhancing operational efficiency and safety.

Optimization and Scalability: The use of GAs for hyperparameter tuning allows for optimization, and the model can be scaled and refined for different drilling environments or conditions.

Generalization: The model demonstrated good generalization abilities, indicating its applicability to various fields or similar problems.

Limitations:

Data Quality Dependency: The model's performance is highly dependent on the quality and relevance of the input data. Any inaccuracies or outliers in the data can significantly impact the model's predictions.

Computational Intensity: Training a deep neural network is computationally intensive, requiring substantial computational resources and possibly specialized hardware.

Complex Hyperparameter Tuning: The model's performance is sensitive to the choice of hyperparameters, and finding the optimal set can be time-consuming and complex, even with the aid of GAs.

Interpretability Challenges: Neural networks, being complex and multi-layered, can act as "black boxes," making it challenging to interpret and understand the model's decision-making process.

Risk of Overfitting: Without proper regularization and validation techniques, there is a risk that the model could overfit the training data, reducing its effectiveness on unseen data.

3.4.1.8.4 Results and Discussion

The project aims to develop a model that predicts the probability of a stuck pipe event in drilling operations, using statistical predictive models, particularly neural networks, based on the data of previously drilled wells. A NLP approach is used to preprocess more than 24,000 rows of data, resulting in a final dataset of 600 rows. The trained model shows an accuracy of over 84% and is implemented in a frontend for the user to upload the mud plan parameters and obtain a stuck pipe event evaluation. The NLP approach provides added value to the model and has multiple applications beyond this project. The model should be further improved to reach the production stage and be validated by the technical knowledge of the engineer. The model's accuracy can be improved by refining the dataset, modifying the network's hyperparameters, or using a different type of ML model.

BIBLIOGRAPHY

Arias, Berdugo, Andrés, Carlos, Torrado, León, Tatiana, Angie, and Carvajal, Juan Sebastián Llano. "Data Boosted Stuck Pipe Prevention: Integrating Machine Learning Models into the Mud Plan Design Workflow." *Paper Presented at the SPE Latin American and Caribbean Petroleum Engineering Conference*, Virtual, July 2020. https://doi.org/10.2118/198972-MS

3.4.1.9 Application-9: Mud Pulse Signal Demodulation Based on Support Vector Machines and Particle Swarm Optimization

3.4.1.9.1 Background of Application

A novel signal demodulation technique based on SVMs and PSO algorithm is proposed for mud pulse telemetry systems. The conventional method, using an adaptive fractionally spaced decision-feedback equalizer (FS-DFE), faces difficulties in demodulating the mud pulse signal due to high signal attenuation and low bandwidth in the mud communication channel. The proposed method overcomes these challenges by expanding the training set with correctly demodulated data and optimizing the SVM classifier's parameters with the PSO algorithm. Field experiment results show the effectiveness of the proposed method, and its computational time is much less than the signal duration.

3.4.1.9.2 Description of the Workflow and Machine Learning Model

The paper uses SVMs and PSO to demodulate mud pulse signals in measurement-while-drilling systems. The workflow involves several steps. First, the SVM classifier is trained with penalty and kernel parameters optimized by the PSO algorithm. The method also includes an innovative training set expansion scheme, which uses an error detection technique to add correctly demodulated data to the training set.

3.4.1.9.3 Advantages and Limitations of the Synchronized Algorithm

The PSO–SVM method shows superior performance in demodulating signals under challenging conditions, outperforming traditional methods like the adaptive FS-DFE. The training set expansion scheme further enhances this capability, leading to improved accuracy. However, the method also faces limitations, such as the need for real-time training to adapt to varying conditions and a dependency on the quality of the training data.

3.4.1.9.4 Results and Discussion

The paper presents a real-time and automatic signal demodulation method for mud pulse telemetry systems using SVMs and PSO algorithm. The method improves its performance by expanding the training set using the error detection method and optimizing the SVM classifier's parameters with PSO. Field experiments show that the proposed method outperforms the traditional FS-DFE method. The low computational complexity of the method enables its real-time application, and it can be extended to other modulation techniques such as binary phase shift keying and on–off keying.

BIBLIOGRAPHY

Fengzhong, Qu, Jiang, Qin, Jin, Guozheng, Wei, Yan, and Zhang, Zhujun. Mud pulse signal demodulation based on support vector machines and particle swarm optimization, *Journal of Petroleum Science and Engineering*, 193, 2020, 107432, ISSN 0920-4105, https://doi.org/10.1016/j.petrol.2020.107432

3.4.1.10 Application-10: Auto Measurement While Drilling Mud Pulse Signal Recognition Based on Deep Neural Network

3.4.1.10.1 Background of Application

In this paper, a method called stacked wavelet autoencoder (SWAE) is proposed to recognize drilling mud pulse signals with low signal to noise ratio (SNR). The method is a combination of a wavelet neural network and an autoencoder, specifically trained for mud pulse signal classification. The SWAE enhances the SNR of the output signal through signal detection and then uses the output as a signal classification attribute. The results of testing the typical dataset show that the SWAE is suitable for mud pulse signal recognition and has strong feature extraction and robustness capabilities, outperforming existing methods under the same SNR.

3.4.1.10.2 Description of Artificial Intelligence Workflow

The paper employs SWAE for recognizing mud pulse signals in drilling operations. The SWAE model integrates a wavelet neural network and autoencoder layers. Its workflow begins with the signal rectification layer, where a wavelet neural network corrects the input signal. Following this, multiple autoencoders together form the feature extraction layer, extracting key features from the rectified signal. The process concludes with a Softmax layer classifying the output signal categories.

3.4.1.10.3 Advantages and Limitations of Artificial Intelligence Methodology

The main advantage is its **enhanced recognition accuracy** in noisy environments, a crucial aspect in drilling operations where signals have low SNR. The integration of wavelet neural networks and autoencoders enables the model to be robust in feature extraction. However, the methodology also faces limitations, particularly in handling extremely low SNR conditions, as low SNR significantly reduces recognition accuracy.

3.4.1.10.4 Results and Recommendations

In this paper, a method called SWAE is proposed to address the problem of mud pulse signal recognition. The method involves removing low-frequency noises and random noises from the mud pulse signals before training the deep neural network (SWAE) for rectifying, automatically extracting features, and recognizing the signals. The experimental results show that the recognition accuracy of SWAE is better than the existing methods and has a certain degree of robustness. However, the recognition effect of SWAE is affected by the SNR of the test set, with low SNR leading to poor recognition accuracy. In future work, the authors plan to improve the denoising method to increase SNR and enhance the robustness of the recognition model, as well as to continue with theoretical research in the field of signal recognition.

BIBLIOGRAPHY

Zhang, Xingsen, Zhang, Hongxia, Guo, Jiashu, Zhu, Lianzhang. Auto measurement while drilling mud pulse signal recognition based on deep neural network, *Journal of Petroleum Science and Engineering*, 167, 2018, 37–43, ISSN 0920-4105, https://doi.org/10.1016/j.petrol.2018.04.004

3.4.1.11 Application-11: Implementation of AI and SVM Learning to Estimate the Drilling Fluid Density in High-Pressure High-Temperature Wells

3.4.1.11.1 Background of Application

This paper explores the use of AI and ML models to predict drilling fluid density in high-pressure, high-temperature wells. Four models were evaluated, and the radial basis function (RBF) algorithm was found to be the most effective. A linear equation was derived for predicting drilling fluid density with a high correlation factor, and the RBF model was found to have good accuracy and validity with experimental data. The study highlights the importance of drilling fluid rheology in wellbore management and controlling fluid loss.

3.4.1.11.2 Description of Artificial Intelligence Workflow

The workflow includes the application of the RBF algorithm, LSSVM–GA (least squares support vector machine-genetics algorithm), PSO–ANFIS, and ANFIS. Each of these models, which provided distinct equations for determining drilling fluid density, was created and evaluated in programming environments. Metrics like the R^2 and MSE were used to assess their performance, and the RBF model was ultimately chosen as the best approach because of its strong correlation factor and low error rates.

3.4.1.11.3 Advantages and Limitations of Artificial Intelligence Methodology

The key advantage is the ability to accurately predict drilling fluid density, a critical factor in wellbore management. This accuracy is demonstrated by high correlation factors and low MSE values, particularly in the RBF model. However, the limitations are inherent in the complexity of the models. Each model requires careful calibration and validation against experimental data, and their performance may vary based on the dataset and environmental conditions.

3.4.1.11.4 Results and Recommendations

This study evaluated the use of different AI and ML models to predict the drilling fluid density. ANFIS, PSO–ANFIS, LSSVM–GA, and RBF algorithms were modeled and tested. The study found that the RBF algorithm was the most effective, with a proper linear equation for the prediction of drilling fluid density of "$y = 1.0009x + 0.0034$" and a correlation factor of $R^2 = 0.9999$. The model showed appropriate accuracy and validity with experimental data. Other models also provided suitable linear equations for predicting drilling fluid density with varying levels of accuracy.

BIBLIOGRAPHY

Syah, Rahmad, Ahmadian, Naeim, Marischa Elveny, S.M., Alizadeh, Meysam Hosseini, and Khan, Afrasyab, Implementation of artificial intelligence and support vector machine learning to estimate the drilling fluid density in high-pressure high-temperature wells. *Energy Reports*, 7, 2021, 4106–4113, ISSN 2352-4847, https://doi.org/10.1016/j.egyr.2021.06.092

3.4.2 Rate of Penetration

3.4.2.1 Application-12: Performance Comparison of Algorithms for Real Time Rate-of-Penetration Optimization in Drilling Using Data-Driven Models

3.4.2.1.1 Background of the Application

This paper evaluates different optimization algorithms for real-time optimization of the ROP, a key objective function for drilling operations. A data-driven model built using ML algorithms is used to calculate the ROP based on inputs such as WOB, revolutions per minute, rock strength, and flow rate. The paper evaluates different algorithms depending on their ability to maximize the ROP and their time efficiency, including the eyeball method, random-search method, and metaheuristic algorithms. The results show that data-driven models can be used for real-time drilling by choosing the right optimization algorithm, with the simplex algorithm providing the best tradeoff between ROP improvement and computational efficiency. The ROP was improved by 30% on average with a variance of 2.5% in the test set of 14 formations.

3.4.2.1.2 Description of the Workflow and Machine Learning Model

The main objective of the article is to find optimal inputs for drilling parameters in order to maximize ROP using ML methods. Main methods used are random forest and PSO. After loading in the data, the first step of the workflow is to train a random forest model using training data to establish the relationship between input parameters (weight-on-bit, rpm, flow rate, unconfined compressive strength) and ROP. The tuning for hyperparameters such as the number of estimators and maximum depth has been outlined. The second step is using PSO to find optimal input for the model to achieve maximum ROP.

3.4.2.1.3 Advantages and Limitations of the Proposed Method

The advantage of PSO is that it's nonparametric and it works well with abstract and unknown distributions. In our case, the PSO is used to optimize the output of random forest regression, and this complicates the matter as the bias and variance of the random forest model are uncertain. If the random forest model didn't work well, the PSO would have little meaning.

3.4.2.1.4 Results and Discussion

This paper investigates different optimization strategies for real-time optimization of ROP in drilling operations. Deterministic models are commonly used for ROP optimization, but data-driven models can be more accurate. However, real-time optimization using data-driven models is challenging because of the unknown functional form, which makes optimization difficult. The paper evaluates simple algorithms, such as the eyeball method and the random-search method, as well as advanced meta-heuristic algorithms, such as the simplex method, DE, and particle-swarm method, in terms of their ROP improvement and run time. The simplex algorithm was found to have the best tradeoff, resulting in a 30% improvement in ROP with a low variance of 2.5% and a run time of 6 seconds. A Monte Carlo simulation was used to calculate the 95% confidence interval for ROP optimization. Algorithm selection will depend on rig conditions, operator preferences, and computational feasibility.

BIBLIOGRAPHY

Hegde, Chiranth, Daigle, Hugh, and Gray, K. E. "Performance Comparison of Algorithms for Real-Time Rate-of-Penetration Optimization in Drilling Using Data-Driven Models." *SPE Journal*, 23 (2018): 1706–1722. doi: https://doi.org/10.2118/191141-PA

3.4.2.2 Application-13: An ROP Predictive Model in Nile Delta Area Using Artificial Neural Networks

3.4.2.2.1 Background of the Application

The ML approach is a convenient method to automate fluid loss (FL) characterization in drilling mud systems, but there are objections raised due to the lack of direct FL measurements and potential failure in severe mud contamination cases. The current FL tests are seen as inadequate, and a more representative method for measuring FL can be used if necessary. The ML approach is still useful in cases where historical FL data is available. Severe mud contamination is rare and can be prevented, and it will result in noticeable changes in mud properties, which can be detected as outliers and prompt corrective action.

3.4.2.2.2 Description of the Workflow and Machine Learning Model

The main objective of the model is to predict ROP in the drilling process. ROP will be predicted using 24 input nodes (variables) that include bit data, drilling parameters, and formation mineral composition data.

After loading the data and separating dependent and independent variables, the dataset was split into training and testing sets with 60% allocated for training. The input variables (x) include: tricone, pdc, IADC-1, IADC-2, IADC-3, bit, new bit, rerun bit, MD, TVD, WOB, RPM, TRQ, GPM, SPP, M_Wt, p_sand, p_shale, p_sandstone, p_clay, p_siltstone, p_claystone, p_anhydrite, and p_limestone.

And the y variable is ROP. The structure of the model is predetermined without the need for tuning because the article has already given the best structure. The ANN used consists of one hidden layer with 50 nodes, employing the log sigmoid function as the activation. There is only one output node, and the activation is the hyperbolic tangent function. The function was then trained, and the model was saved for future use.

3.4.2.2.3 Advantages and Limitations of the Method

The advantage of ANN is clearly that it can capture some hidden and nonlinear relationships between dependent and independent variables. However, the disadvantage is also significant as it can be not so straightforward nor explainable. However, with a prediction accuracy high enough, it has more value in prediction than in explanation.

3.4.2.2.4 Results and Discussion

The study has concluded that an ANN technique using the back-propagation learning algorithm was successfully used to develop a model for predicting the ROP in the Nile Delta area. The model was found to provide an efficient way to determine the impact of multiple variables on the ROP while considering technological constraints. The model had a high correlation coefficient of 0.886–0.907, and its output was found to accurately predict the physical process when compared to real data. However, the model's results should only be used within the range of the data used and caution should be exercised beyond this range.

BIBLIOGRAPHY

Amer, Mustafa M., Dahab, Abdel Sattar, and El-Sayed, Abdel-Alim Hashem. "An ROP Predictive Model in Nile Delta Area Using Artificial Neural Networks." *Paper presented at the SPE Kingdom of Saudi Arabia Annual Technical Symposium and Exhibition*, Dammam, Saudi Arabia, April 2017. https://doi.org/10.2118/187969-MS

3.4.2.3 Application-14: Developing a New Rigorous Drilling Rate Prediction Model Using a Machine Learning Technique

3.4.2.3.1 Background of the Application

The study aimed to model the ROP for vertical wells in southwestern Iran using hybrid algorithms. The data collected for ROP prediction included mud logging parameters and geomechanics parameters. Tukey's approach was used to eliminate outliers, and an MLP neural network and non-dominated sorting genetic algorithm II were used to choose the most pertinent characteristics. The ROP was modeled using a combination of least-squares support-vector machines with genetic algorithms (LSSVM–GA), particle swarm optimization (LSSVM–PSO), and cuckoo optimization algorithm (LSSVM–COA). The results indicated that LSSVM–COA performed better in terms of accuracy than the other techniques. The model was determined to be the most accurate after being compared to a number of different models. After being tested on a field well, the model was discovered to be accurate and dependable, with only minor variations between the testing and training phases. The study concluded that the model can be used to predict ROP at other wells across the field, provided that the model is developed with larger sets of data across the field.

3.4.2.3.2 Description of the Workflow

In the data preprocessing phase, outliers were first detected and omitted to enhance data quality. Regarding the ROP modeling process, the paper employs LSSVM combined with optimization algorithms like the COA, PSO, and GAs. These combinations were used to model the ROP in vertical wells.

3.4.2.3.3 Advantages and Limitations

The advantages of this methodology include the use of hybrid algorithms, like LSSVM with COA, PSO, and GA, which allows for better performance compared to traditional methods. However, the limitations include the complexity of the model development process, which requires careful data preprocessing, feature selection, and the tuning of various parameters in the ML models.

3.4.2.3.4 Results and Discussion

The literature review on ROP modeling in the upstream oil industry showed a growth in publications during 2014–2019, with 60% of articles focusing on this subject published during that time period. The present research aimed to use data from vertically drilled wells in southwestern Iran for ROP modeling. The literature review showed that geomechanical parameters were largely ignored in previous ROP models. The present paper focused on a hybrid LSSVM–COA algorithm for ROP modeling based on mud logging and geomechanical parameters and compared its performance with other algorithms, including linear and nonlinear regression, Motahhari's model, MLP, and support vector regression (SVR). The Tukey method was used to remove outliers in the data to improve the accuracy of the models. The NSGA-II-MLP hybrid algorithm was used to select the dominant features for ROP modeling, and the results showed that the models improved with the increase in number of inputs, but the rate of improvement decreased as the number of inputs got larger. ROP modeling was ultimately performed using six selected parameters: UCS, FR, WOB, Depth, MD, and RPM.

The study focused on using a hybrid LSSVM–COA algorithm for modeling the ROP in an oilfield in southwestern Iran based on mud logging and geomechanical parameters. The LSSVM–COA model was compared with other models, including LSSVM–PSO, LSSVM–GA, SVR–COA, SVR–PSO, SVR–GA, MLP–COA, MLP–PSO, and MLP–GA. The results showed that the LSSVM–COA model provided the best accuracy and generalizability compared to the other models. The study also found that the COA outperformed other learning algorithms in training the SVR and MLP algorithms. The ROP modeling was performed based on six selected parameters, and the data was preprocessed using the Tukey method to remove outliers.

In conclusion, the study compared different models for predicting the ROP in a drilling field. The LSSVM–COA model, which is a hybrid ML model, was found to be the most accurate and generalizable model among all the models compared. This was concluded by comparing the LSSVM–COA model with other models such as Motahhari's, LMR, and NMLR, as well as with other hybrid ML models. The study showed that the LSSVM–COA model performed significantly better than the other models in terms of accuracy and generalizability. The proposed method has the potential to provide even more reliable and accurate predictions for ROP across the field if larger data sources are incorporated into the method, which could save costs in drilling operations.

BIBLIOGRAPHY

Mehrad, Mohammad, Bajolvand, Mahdi, Ramezanzadeh, Ahmad, Neycharan, Jalil Ghavidel. Developing a new rigorous drilling rate prediction model using a machine learning technique, *Journal of Petroleum Science and Engineering*, 192, 2020, 107338, ISSN 0920-4105, https://doi.org/10.1016/j.petrol.2020.107338

3.4.2.4 Application-15: Improving Real-Time Drilling Data Quality Using AI and ML Techniques

3.4.2.4.1 Background of the Application

The paper discusses the challenges faced by oil and gas companies in drilling operations due to increased demand for energy in remote and unconventional environments. To address these challenges, the real-time operation center (RTOC) concept is employed, where drilling data is transmitted to headquarters for live monitoring and expert advice. However, the volume of data surpasses human processing capabilities, prompting the integration of artificial intelligence and machine learning (AI) technologies to analyze real-time drilling data and enhance decision support. The paper presents a novel approach by feeding a large real-time drilling dataset into ANN, SVM, and decision tree (DT) models to improve data quality. Despite the slower speed of ANN, it outperformed DT and SVM, achieving an accuracy of 78%. The paper is one of the first to compare multiple AI techniques for enhancing real-time drilling data quality, offering a valuable guide for the industry.

3.4.2.4.2 Description of the Workflow and Machine Learning Model

Data Collection:

- Over 150,000 raw data points were collected, focusing on drilling parameters and operations.

Data Preprocessing:

- Data Cleaning: Invalid data points were removed.
- Labeling: The remaining data was categorized as valid or not valid.

Model Selection:

- Three models were considered: ANN, SVM, and DT.

Model Training:

- Each model was trained using the processed data.
- For ANN, various architectures were tested.
- SVM was adjusted with different cost-gamma combinations.
- DT was structured with over 40 terminal nodes.

Model Evaluation:

- A confusion matrix was used for assessing model performance, focusing on true positives, false positives, true negatives, and false negatives.

3.4.2.4.3 Advantages and Limitations

Advantages:

Accuracy: ANN achieved the highest accuracy (78%), indicating its potential in accurately processing real-time drilling data.

Versatility: Multiple AI techniques were compared, showcasing flexibility in approach.

Real-Time Application: The methodology allows for the processing of large datasets in real-time, which is critical for drilling operations.

Limitations:

- Data Dependency: The quality of predictions heavily depends on the quality of input data. Poor data quality can lead to inaccurate results.
- Computational Resources: Training models like ANNs can be resource-intensive, requiring significant computational power.
- Complexity in Interpretation: Models like ANN are often seen as "black boxes," making it challenging to interpret how decisions are made.
- Trade-offs between Speed and Accuracy: While ANN offers high accuracy, it is slower compared to DT and SVM. This requires a balance between speed and accuracy based on the operational needs.

3.4.2.4.4 Results and Discussion

The study employed three key AI technologies—ANN, SVM, and DT—to enhance real-time drilling data quality. Following model training, they were tested on a new dataset, with Table 3.1 detailing true positive, true negative, false positive, false negative, and accuracy. The SVM and DT models showed similar accuracies at 70% and 73%, respectively. The ANN model outperformed both with the highest accuracy at 78%. However, all models missed high/low datapoints. If these points are non-critical, ANN, despite its slower generation, could be preferred for its high accuracy, while SVM and DT offer faster alternatives, with SVM being the least accurate among the three methods.

BIBLIOGRAPHY

Al Gharbi, S. H., Al-Majed, A. A., Abdulraheem, A., Patil, S., and Elkatatny, S. M. "Improving Real-Time Drilling Data Quality Using Artificial Intelligence and Machine Learning Techniques." *Paper presented at the SPE Middle East Oil & Gas Show and Conference*, event canceled, November 2021. https://doi.org/10.2118/204658-MS

3.4.2.5 Application-16: Real-Time Prediction of Rate of Penetration While Drilling Complex Lithologies Using Artificial Intelligence Techniques

3.4.2.5.1 Background of the Application

This paper presents a new approach for predicting the ROP in real-time drilling operations. The study uses three different AI techniques (ANN, ANFIS, and SVM) to build models that can predict the ROP with high accuracy. The models were built using actual field data (1,000 data points from Well A and 972 data points from Well B). The results showed that the ANN performed better than SVM and ANFIS in predicting the ROP for unseen data, and the developed ROP-ANN model had an RMSE of less than 0.659 for both wells. The study also developed a new empirical equation for predicting the ROP with high accuracy, which can be used without the need for specialized software.

3.4.2.5.2 Description of the Workflow

The paper focuses on predicting the ROP during drilling in complex lithologies. Three AI techniques were used: ANNs, ANFIS, and SVM. There are 11 input parameters. The models were trained using 75% of data from Well A, while the remaining

25% was used for testing. The performance of each model was analyzed through a sensitivity analysis focusing on its design parameters, with the RMSE as the evaluation metric.

3.4.2.5.3 Advantages and Limitations of AI Methods

The paper highlights the high accuracy of these models in predicting ROP for complex lithologies. Notably, the ANN model was the most effective, outperforming both SVM and ANFIS in predicting ROP for unseen data. The limitations include challenges like overfitting, the need for extensive data for training, and potential issues with generalizability.

3.4.2.5.4 Results and Conclusions

This study used AI methods such as SVMs, ANFIS, and ANNs to forecast the ROP in real-time drilling operations. In order to construct the AI models, the study employed well data, such as bit design parameters, total flow area, mud density, gamma ray, and drilling parameters. The ROP for complicated lithologies could be properly predicted by all three AI algorithms, according to the data, with ANN outperforming the others with an RMSE of less than 0.66. One hidden layer, 20 neurons, fitnet as the network function, trainbr as the training function, and tansig as the transfer function were determined to be the optimal ANN settings. The developed ROP empirical correlation also predicted the ROP with high accuracy and could be used without needing specialized software.

BIBLIOGRAPHY

Elkatatny, Salaheldin. Real-time prediction of rate of penetration while drilling complex lithologies using artificial intelligence techniques. *Ain Shams Engineering Journal*, 12 (1), 2021, 917–926, ISSN 2090-4479, https://doi.org/10.1016/j.asej.2020.05.014

3.4.2.6 Application-17: The Effect of Formation Thickness on the Performance of Deterministic and Machine Learning Models for Rate of Penetration Management

3.4.2.6.1 Background of Application

In this paper, a ROP management study was conducted for a well in Southwest Iran. Several methods were compared to determine the best approach for ROP prediction, and the PSO algorithm was used for optimization of ROP. The study found that the hybrid Bingham model had the best performance in ROP prediction when a powerful mathematical tool, such as the trust-region method, was used to determine its unknown coefficients. However, the results were not consistent in thin and thick formations, so the authors suggest using the most accurate method for each formation to predict ROP. The study resulted in a 13.59% relative error in ROP prediction and a 48.30% improvement in ROP.

3.4.2.6.2 Description of Artificial Intelligence Workflow

The paper involves several methods for predicting the ROP in drilling operations. For optimizing the ROP, the PSO algorithm is employed to determine the best operational conditions. The workflow includes comparing the performance of these methods, adjusting the coefficients of the hybrid Bingham model using trust-region and multiple regression methods, and finally selecting the most accurate model as the objective function for PSO algorithm optimization

3.4.2.6.3 Advantages and Limitations of Artificial Intelligence Methodology

The traditional ROP models had limitations in accuracy and in covering all parameters affecting ROP. To address this, researchers have developed hybrid forms of traditional ROP models and improved the accuracy of ROP predictions. However, their interpretation can be complex due to the unknown functional systems. Currently, the general strategy is shifting toward using a "bag-of-algorithms" approach, where the most suitable method is selected for each specific formation

3.4.2.6.4 Results and Recommendations

In summary, this study investigated a ROP management process for a well in Southwest Iran. The results showed that the hybrid Bingham model with the trust-region method had the best accuracy in predicting ROP, but using a bag-of-algorithms strategy resulted in a better prediction with a total relative error of 13.59%. The optimization process using the PSO algorithm resulted in a total ROP improvement of 48.30%. The findings of this study can be used to improve ROP prediction accuracy and optimization in future studies, particularly in inclined and horizontal wells. The results suggest that moving from traditional ROP management methods to ML-based methods could lead to significant improvements.

BIBLIOGRAPHY

Najjarpour, Mohammad, Jalalifar, Hossein, Norouzi-Apourvari, Saeid. The effect of formation thickness on the performance of deterministic and machine learning models for rate of penetration management in inclined and horizontal wells. *Journal of Petroleum Science and Engineering*. 191, 2020, 107160, ISSN 0920-4105, https://doi.org/10.1016/j.petrol.2020.107160

3.4.2.7 Application-18: Prediction of Drilling Rate of Penetration (ROP) Using Hybrid Support Vector Regression

3.4.2.7.1 Background of the Application

The paper proposes a novel intelligent model to predict the ROP in drilling. The model is established through three stages: wavelet filtering to reduce data noise, the mutual information method to determine model inputs, and optimization of the SVR model's hyperparameters using a hybrid bat algorithm. The proposed model is validated using data from a drilling site in Central China and outperforms other well-known prediction methods.

3.4.2.7.2 Description of the Workflow and Machine Learning Model

The AI model in the paper follows a three-stage workflow. In the first stage, a wavelet filtering method is used to reduce noise in the drilling data, addressing the challenge of data quality. The second stage involves using the mutual information method to determine the model inputs, ensuring that the most relevant features are used for prediction. The final stage introduces a hybrid bat algorithm to optimize the hyperparameters of the SVR model.

3.4.2.7.3 Advantages and Limitations of the Method

The use of SVR offers better nonlinear fitting and generalization capacities, making it effective for modeling problems with small datasets. This is crucial in applications where large amounts of data are not available or feasible to collect. However, the accuracy of the SVR model is highly sensitive to the selection of its hyperparameters (c, g, and ε). To address this, the Iterative Local Search and Stochastic Inertia Weight Bat Algorithm (ILSSIWBA) is introduced for optimizing these hyperparameters, aiming to improve the robustness and reliability of the model.

3.4.2.7.4 Results and Discussion

This paper presents a new hybrid modeling method for predicting ROP in the drilling process. The method combines wavelet filtering, mutual information analysis, and SVR with an improved bat swarm algorithm. The proposed model considers the strong nonlinearity and different types of data noise in the drilling process. The method was validated using data from a drilling site in Central China and showed improved prediction accuracy compared to other well-known ROP modeling methods. However, it is noted that the obtained model is not universal and must be remodeled for use in other areas.

BIBLIOGRAPHY

Gan, Chao, Cao, Wei-Hua, Wu, Min, Chen, Xin, Hu, Yu-Le, Liu, Kang-Zhi, Wang, Fa-Wen, Zhang, Suo-Bang. Prediction of drilling rate of penetration (ROP) using hybrid support vector regression: A case study on the Shennongjia area, Central China. *Journal of Petroleum Science and Engineering*. 181, 2019, 106200, ISSN 0920-4105, https://doi.org/10.1016/j.petrol.2019.106200

3.4.3 Drilling Optimization

3.4.3.1 Application-19: Full Couple End-to-End Drilling Optimization Model Using Machine Learning

3.4.3.1.1 Background of the Application

Due to the complex nature of drilling, the drilling optimization on a single parameter like MSE or ROP doesn't include the complexity introduced by the drilling parameters that are affected by each other. Therefore, the drilling optimization should be considered one by one, and the results should be combined in a way that will result in a single outcome. The drilling parameters can be classified under three groups: control parameters, uncontrollable parameters, and response parameters. The control parameters are controlled by the engineer or drilling on the rig, and these parameters are WOB, rotation per minute (RPM), and mud flow rate. The uncontrollable parameters are those that can't be changed: strength of rock, geological parameters, and maximum mud pump power. The response parameters are those that are affected by control and uncontrollable parameters: Mechanical specific energy (MSE) and torque on bit (TOB).

The paper proposes a group of ML algorithms that are modeling each drilling parameter (ROP, TOB) and utilizing classification algorithm for torsional vibrations to ensure the predicted optimal parameters doesn't induce torsional vibrations (stick-slip). Then, these group of models are coupled to optimize controlled drilling parameters by using PSO. The model proposed in the paper utilized the meta-heuristic optimization algorithm since the model built using the random forest algorithms are unknow in functional form. The data used to train the model is collected and averaged by depth per 0.25 ft of drilled depth, and same bit is used to collect the entire dataset. Therefore, the bit type won't be an input for the model. The ratio of training and test data is 40%/60%.

The ROP model is built by using a random forest regression algorithm, and the input values are ROP, WOB, mud flow rate, and unconfined compressive strength (UCS) of the rock. The training and test dataset ratio is 50%/50%. Random search CV is used to tune hyperparameters in this model. The CV grid-search is used to optimize the final model that is created by coupling ROP, TOB, and vibration model.

The TOB model is built by using a random forest regression algorithm. The same input parameters are used to train this model to ensure that they can work jointly. Similar to the ROP model, the training and test dataset ratio is 50%/50%, and random search CV is used to tune hyperparameters in this model.

The vibration model is built by using the random forest classification algorithm. The purpose of the model is to predict specifically a type of vibrations, stick-slip. The proposed solution is prediction stick-slip severity based on a value called SSI. SSI can be measured at downhole, but in this case, ROP, TOB, RPM, WOB, and mud flow rate are used to predict SSI to avoid severe stick-slip vibrations. SSI can be calculated with the equation below. Performance of the model is measured with the F1 score.

$$\text{SSI} = (\text{Max (bit RPM)-Min (bit RPM)}) / (\text{Average (bit RPM)})$$

The final model should predict optimized drilling parameters to maximize ROP while avoiding severe stick-slip vibrations. The performance of the final model is measured by a nonparametric bootstrapping hypothesis test, and a 95% threshold value has been used to optimize the model.

3.4.3.1.1.1 ROP Model (Random Forest Regression)

Inputs

Weight on bit (WOB),
Mud flow rate,
Unconfined compressive strength (UCS)

Output

Rate of Penetration (ROP)

3.4.3.1.1.2 TOB Model (Random Forest Regression)

Inputs

Weight on bit (WOB),
Mud flow rate,
Unconfined compressive strength (UCS)

Output

Torque on bit (TOB)

3.4.3.1.1.3 Vibration Model (Random Forest Classification)

Inputs

Rate of penetration (ROP)
Torque on bit (TOB)
Rotation per minute (RPM)
Weight on bit (WOB)
Mud flow rate

Output

Constraints on drilling parameters predicted from joint model to avoid stick-slip vibration.

3.4.3.1.2 Description of the Workflow and Machine Learning Model

Different random forest models and PSO are utilized to identify the optimal drilling inputs that maximize the target function while minimizing the risk of stick-slip problems.

There are mainly two parts in the model. The first part is the random forest models that are used to predict ROP, TOB, and stick-slip using raw inputs WOB, MFR, and UCS, and the second part is the optimization model that aims to find the optimal input to maximize the target function (MSE or ROP) while avoiding stick-slip.

The workflow is straightforward. The data is initially read and split into training and testing sets. Separate models are then trained for ROP, TOB, and vibration (excluding hyperparameter tuning). The target function is defined using MSE and ROP, with inputs derived from the prediction results of the trained models. To account for stick-slip in the optimization, the Boolean output from the stick-slip prediction is multiplied with the target function. This ensures that any drilling input likely to cause stick-slip results in a target function value of zero during optimization.

The article used PSO, which is an optimization algorithm that aims to find a global optimal solution using random sample points that can interact with each other. To apply the method, the pyswarms library was utilized. The optimization algorithm requires several parameters, including c1, c2, w, n_particles, and dimensions. Due to the lack of original data, the parameter tuning process was omitted. Finally, GlobalBestPSO() was employed to identify the optimal solutions.

3.4.3.1.3 Advantages and Limitations of the ML Method

The advantage of random forest is significant since it not only can reduce bias and variance of the model significantly and solve the tree correlation problems in the normal bootstrapping method but also is relatively forgiving regarding the input data. The optimization method used serves as an effective approach to provide a non-analytical solution to a complex problem.

The limitation remains to be the lack of actual experiment and real-world testing. While using Random Forest, it remains uncertain whether lightweight Random Forest models can achieve satisfactory levels of accuracy.

3.4.3.1.4 Results and Discussion

This paper presents a new ML-based optimization model for drilling operations in the oil and gas industry. The model takes into account multiple drilling parameters, including ROP, TOB, MSE, and stick-slip vibrations while optimizing for a pre-defined key performance indicator (KPI), such as ROP or MSE. The models were built using the random forest algorithm and were coupled using an ML algorithm. An optimization algorithm was used to determine the optimal control parameters, which were then implemented ahead of the bit. The model improved ROP by 36% and reduced MSE by 49%. The optimization space was constrained to low-vibration regions to avoid excessive vibrations. The impact of ROP/MSE optimization on other drilling metrics was studied using a case study on the Charles limestone formation.

BIBLIOGRAPHY

Chiranth, Hegde, Michael, Pyrcz, Harry, Millwater, Hugh, Daigle, and Ken, Gray. Fully coupled end-to-end drilling optimization model using machine learning, *Journal of Petroleum Science and Engineering*, 186, 2020, 106681. https://doi.org/10.1016/j.petrol.2019.106681

3.4.3.2 Application-20: Advisory System for Drilling Optimization using AI and In-Cutter Sensing

3.4.3.2.1 Background of the Application

In order to improve drilling efficiency, this study suggests a novel workflow that makes use of AI and in-cutter sensing data to determine the ideal drilling parameters while taking bit wear into account. In-cutter sensing, which makes it possible to assess the forces operating on a polycrystalline diamond compact (PDC) cutter, is the fundamental capability that makes drilling optimization possible. The AI-based approach determines the ideal range of drilling parameters based on the mechanical characteristics of the rock samples by using the data from the PDC cutter and the scaled-drilling rig construction. WOB, torque, rotating speed, uniaxial compressive strength, vibrations, and the reaction force at the PDC cutter are among the input variables used to train the supervised machine-learning models. The results show that the prediction performance is enhanced when the in-cutter sensing measurement is implemented as an input variable. The reliability of the ML models is tested on samples with different mechanical properties, with the results indicating an accurate prediction of the ROP.

3.4.3.2.2 Description of the Workflow and Machine Learning Model

The main purpose of this algorithm is to predict ROP using ANN. Input variables include WOB, torque, RPM (rotational speed), psi (uniaxial compressive strength of rock), vibration, and Fs (force perpendicular to the cutter's face).

After categorizing input variables and target variables, the dataset is split into training, validation, and test sets, with 70%, 15%, and 15% of the dataset, respectively. The article does not mention other features besides the six above; therefore, all analyses are performed assuming there are only six input features.

The model used in the article is ANN, but specific architecture is not discussed, including the number of layers and the number of neurons in each layer. The activation functions and other hyperparameters are also not mentioned, but should be nonlinear. The metrics used to evaluate the model are RMSE, AAPE, and r^2.

3.4.3.2.3 Advantages and Limitations of the RF Method

The advantage of ANN is its performance, as it works well in capturing nonlinear correlation between features in this problem. However, as the architecture of the model is not clear, parameters selected are primarily subjective, which may lead to varying performance when different parameters are applied.

3.4.3.2.4 Results and Discussion

This paper demonstrates that the integrated approach of in-cutter sensing and AI-based decision-making can facilitate the drilling optimization process. The in-cutter sensing technology enables the drilling industry to monitor the drilling process in real-time, providing valuable information to improve drilling efficiency. The AI-based solution provides a reliable and rapid decision to identify the optimum drilling parameters with respect to physics constraints. In the future, this approach could be integrated with the drilling control system, allowing the drilling process to be automated and optimized.

BIBLIOGRAPHY

Koulidis, Alexis, and Ahmed, Shehab. "Advisory System for Drilling Optimization Using Artificial Intelligence and In-Cutter Sensing Data." *Paper Presented at the SPE Symposium Leveraging Artificial Intelligence to Shape the Future of the Energy Industry*, Al Khobar, Saudi Arabia, January 2023. https://doi.org/10.2118/214468-MS

3.4.3.3 Application-21: Real-Time ROP Optimization Using Machine Learning and Downhole Vibration Monitoring

3.4.3.3.1 Background of application

The paper presents a case study involving a real-time field test of an ML ROP prediction and optimization algorithm, along with a vendor-neutral vibration monitoring system, conducted on ten wells in Northeastern British Columbia, Canada. The ML model features a unique auto-calibration mechanism to address prediction bias in real-time. The study aimed to accelerate the adoption of AI by rig personnel and involved drilling ten wells in three formations using one rig. Initial training sessions with the rig team were conducted during the first two laterals, while subsequent tests were carried out in real-time advisory mode. The objectives included developing trust in the ML model, validating the vibration monitoring tool with real-time downhole data, and gathering feedback for AI uptake. The ML ROP model initially met success criteria in most formations but required an automated calibration procedure to adjust prediction bias for improved accuracy. This enhancement broadened the model's application across different formations and basins. The vibration monitoring system provided real-time data to both the rig and office, offering valuable insights for situational awareness and decision-making. Rig personnel feedback influenced the implementation of new functionalities, demonstrating successful collaboration between operators and service companies in advancing AI technology for drilling operations.

3.4.3.3.2 Description of ML Workflow

Objective: Testing the ML ROP prediction/optimization system and validating real-time vibration monitoring tools.

Process:

1. A 10-well campaign across three target formations was drilled using one rig.
2. Initial laterals were drilled with office engineers present to train and gain buy-in from the rig team.
3. The ML ROP model initially passed the success criteria in two of three formations.
4. Development of an automated calibration procedure to adjust the ML model's bias in real-time.

3.4.3.3.3 Advantages and Limitations

Advantages:

1. Improved Accuracy: The automated calibration factor (ACF) significantly improved the model's prediction accuracy.
2. Real-Time Adaptability: The model could adjust predictions in real-time, enhancing its practical applicability.
3. Operational Insights: The vibration monitoring tool provided crucial real-time data, aiding in better decision-making and situational awareness.

Limitations:

1. Data Sensitivity: The model's accuracy is highly dependent on the quality and relevance of the training data.
2. Complexity in Real-Time Calibration: The need for real-time adjustment of the model adds complexity to its deployment.
3. Limited Scope: The study was confined to specific geographical locations and formations, which may limit the generalizability of the findings.

3.4.3.3.4 Results and Recommendations

The positive outcomes of field tests and the introduction of an ACF have instilled confidence in the drilling advisory system. The ML model for ROP prediction is actively utilized on the rig, showing close alignment with driller performance in factory drilling pads. Future efforts involve developing an ML model for ROP prediction in rotary steerable systems (RSS) without a positive displacement motor (PDM) and creating models that meet acceptance criteria in vertical and intermediate hole sections. Ongoing work also includes integrating the drilling advisory system with hardware for direct transmission of recommendations to auto-driller parameter set points. This integration aims to enhance drilling operations' efficiency and safety by closing the loop and reducing human intervention on the rig.

BIBLIOGRAPHY

Robertson, Ryan, Deans, Aidan, Singh, Kriti, Braga, Daniel, Kamyab, Mohammedreza, Cheatham, Curtis, and Borges, Tatiana. "Field Test Results for Real-time ROP Optimization Using Machine Learning and Downhole Vibration Monitoring - A Case Study." *Paper presented at the SPE/IADC International Drilling Conference and Exhibition*, Stavanger, Norway, March 2023. https://doi.org/10.2118/212568-MS

3.4.3.4 Application-22: Combining Drilling Big Data and Machine Learning Method to Improve the Timeliness of Drilling

3.4.3.4.1 Background of the Application

The paper proposes a method to determine the working conditions and potential accidents in oil drilling wells by using logging BD and ML. The method starts with preprocessing the logging data to ensure accuracy, then the data is input into the AI method to train and form a database. The trained database is then used to judge the working conditions of the well being drilled and whether there is an accident. The method was tested using logging data from Bohai Oilfield, and the results showed that the relative error of the four working conditions was less than 5%, and the average calculation time was faster than traditional methods. This new method saves a lot of manual processing time and helps find common accidents during drilling in a timely manner.

3.4.3.4.2 Description of the Workflow and Machine Learning Model

The paper integrates ML with BD to enhance drilling efficiency in oil wells. The approach involves preprocessing vast spans of logging data to ensure accuracy, followed by inputting this data into an AI method for training through iterations. Finally, the preprocessed logging data is input into this training set, enabling the determination of current working conditions and accident occurrences based on the AI's output.

3.4.3.4.3 Advantages and Limitations of the RF Method

Advantages: The ML method, specifically the SVM, offers a substantial advantage in speed and accuracy over traditional methods. It can process and analyze data quickly, reducing the time and manual effort required for drilling condition assessments. However, one of the limitations is the data accuracy and quality required for effective training of the ML model. Additionally, the system's efficacy is contingent on the comprehensiveness and diversity of the training set.

3.4.3.4.4 Results and Discussion

This paper is about using ML, specifically the SVM method, in the evaluation of drilling conditions in the oil drilling industry. It is a new approach that does not rely on human judgment and is based on pure data. The method was tested using actual data and was found to be significant for improving drilling efficiency, especially in high-risk and high-cost offshore drilling.

The paper presents a new method that combines ML with data analysis for the assessment of drilling conditions in the oil drilling industry. The proposed method allows for real-time monitoring and quicker response to abnormal data, compared to traditional methods. This method can help improve the efficiency, safety, and economy of offshore drilling. In future research and development, the method can be further applied to real-time risk monitoring, risk warnings, and other areas to continuously improve the safety and economy of offshore drilling technology.

BIBLIOGRAPHY

Zhao, Ying, Sun, Ting, Yang, Jin, Yin, Qishuai, Wei, Hongshu, Liu, Zhengli, Li, Zhong, and Huang, Yi. "Combining Drilling Big Data and Machine Learning Method to Improve the Timeliness of Drilling." *Paper presented at the SPE/IADC International Drilling Conference and Exhibition*, The Hague, The Netherlands, March 2019. https://doi.org/10.2118/194111-MS

3.4.3.5 Application-23: Autonomous Drilling Fluid Management System—Development of Fluid Advisory System and First Lab Trial

3.4.3.5.1 Background of Application

The creation of a fluid advisory system that uses a variety of algorithms and techniques to optimize the drilling fluid treatment process is described in the study. Based on manual and real-time observations, the system determines the ideal amount of chemical additions to maintain target drilling fluid qualities while taking inventory levels into account. Using ML, predictive analytics for model selection, and incorporating predictive models into a prescriptive model, the process entails building a fluid property model (digital twin). This prescriptive model solves an optimization problem and suggests additive concentrations to get particular fluid properties.

Small-scale laboratory tests were used to validate the autonomous fluid management algorithms, and then the Equinor Sandsli Automatic Drilling Fluid Laboratory conducted a yard-scale testing. It was shown that the fluid advisory system and remote control of drilling fluid treatment equipment could sustain desired characteristics such weight, rheology, API fluid loss, pH, hardness, and chlorides. In order to provide autonomous drilling fluid management, the system seeks to optimize real-time drilling fluid treatment for use in Integrated Operations Level 3 and above. This approach reduces personnel requirements, ensures consistent fluid performance, improves rig supply chain efficiency, lowers NPT, enhances safety, reduces operational costs, and lowers the carbon footprint. The study marks the first development and experimental validation of a fluid advisory system for maintaining multiple properties of drilling fluid in a remote-controlled mixing facility.

3.4.3.5.2 Description of the Workflow and Machine Learning Model

1. Objective: Develop and validate an autonomous system for drilling fluid management.

2. Data Handling: Historical data was integrated into a master table, and lab tests were conducted for data collection.
3. Model Development:
   a. A digital twin of the fluid property model was created using ML.
   b. Various ML methods, including linear regression, SVR, random forest, XGBoost, and feedforward neural networks, were evaluated.
   c. A feedforward neural network was selected, and its hyperparameters were tuned.
4. Model Application:
   a. The ML model was incorporated into a prescriptive model that suggests modifications in additive concentrations.
   b. An optimization problem was set and solved to determine the quantity of additives required.
   c. The system was validated through lab experiments.

3.4.3.5.3 *Advantages and Limitations of the AI algorithm*

Advantages:

1. Data-Driven Accuracy: The ML model provides precise calculations for additive quantities.
2. Automation: Reduces manual intervention and enhances consistency in fluid management.
3. Real-Time Adaptability: Capable of adjusting to real-time changes in fluid properties.

Limitations:

1. Complex System Integration: The integration of ML models into real-time systems is complex.
2. Data Quality Dependency: The system's accuracy is highly reliant on the quality of input data.
3. Generalization: The findings are specific to the tested fluids and might not generalize across different fluid types or drilling conditions.

3.4.3.5.4 *Results and Recommendations*

The fluid advisory system employs mathematical formulas and machine-based algorithms to alert about the required type and quantity of chemicals for maintaining multiple drilling fluid properties within specifications. This includes the development and testing of digital tools such as remotely controlled hardware, sensors, and algorithm-based notifications. These tools are crucial for ensuring service delivery quality during the transition to remote operations.

The fluid advisory system, compared to traditional methods based on pilot testing and experience, offers higher consistency and a focus on multiple properties. Initial lab trials highlight the impact of uncertainties in measurement instruments, mixing processes, and flow loop conditions on final fluid properties.

The study acknowledges the need for further testing and adjustments to optimize the algorithm and address limitations observed in small-scale lab testing.

In conclusion, the research marks a significant step toward a revolutionary change in drilling operations, showcasing the potential of an autonomous drilling fluid management system. The proposed fluid advisory system demonstrates substantial advantages, including cost reduction, safety improvement, and operational efficiency enhancement, paving the way for a new era in drilling operations.

BIBLIOGRAPHY

Ettehadi, Reza, Onegova, Elizaveta, Fevang, Finn Øivind, Knizhnik, Alexey, Postovalov, Sergey, Brevik, Jan Ove, Thompson Jr, Charles, Egorenkova, Tatyana, and Kaageson-Loe, Nils. "Autonomous Drilling Fluid Management System - Development of Fluid Advisory System and First Lab Trial." *Paper Presented at the SPE Annual Technical Conference and Exhibition*, San Antonio, TX, October 2023. https://doi.org/10.2118/215047-MS

3.4.3.6 Application-24: Real-Time Solution for Down Hole Torque Estimation and Drilling Optimization in High Deviated Wells Using Artificial Intelligence

3.4.3.6.1 Background of Application

This project presents a new approach to improve the accuracy of downhole torque measurements by combining AI technology with the challenges faced in drilling deep and highly deviated wells. AI techniques such as ANNs, SVMs, and ANFISs will be used to estimate the wellbore torque based on surface measurements, which have a major impact on the downhole torque values. A user-friendly software will be developed to control the surface drilling variables, such as WOB, ROP, and flow rate, to maintain the downhole torque within a safe zone and prevent stuck pipe situations. The model will be validated through a real drilling scenario in northern Africa, where data on WOB, surface torque, stand-pipe pressure, and ROP were collected from a well that experienced a costly stuck pipe situation. The proposed model is expected to save the well and avoid NPT.

3.4.3.6.2 Description of Artificial Intelligence Workflow

The paper describes an AI model for estimating downhole torque in high deviated wells, using techniques like ANNs and ANFIS. The model's inputs include surface drilling parameters like WOB (Klbs), ROP (ft/hr), standpipe pressure (psi), and surface torque (Kft.lbs). The model is developed through appropriate AI training and testing, and applied through Matlab programming

3.4.3.6.3 Artificial Intelligence Methodology Advantages and Limitations

The AI model's advantages include its ability to predict real-time downhole torque, which is crucial for optimizing the drilling process and preventing stuck pipe situations. This approach helps to avoid expensive stuck situations, prevent downhole equipment loss, and control downhole torque from the surface. The model, integrated with a user-friendly software interface and the driller's cyber control system, offers a solution for automated drilling systems.

3.4.3.6.4 Results and Recommendations

The research developed two AI downhole torque models using ANN and ANFIS technology with areal drilling data from North Africa, which showed excellent results in predicting downhole torque in high deviated sections. The developed AI ANN model had a high correlation coefficient and low AAPE. An empirical correlation was generated and validated with high accuracy. The downhole torque model can be used to optimize surface drilling variables in real-time and enhance drilling efficiency.

BIBLIOGRAPHY

Elzenary, Mahmoud Nader. "Real-Time Solution for Down Hole Torque Estimation and Drilling Optimization in High Deviated Wells Using Artificial Intelligence." *Paper Presented at the SPE Symposium: Artificial Intelligence - Towards a Resilient and Efficient Energy Industry*, Virtual, October 2021. https://doi.org/10.2118/208646-MS

3.4.4 Other Applications

3.4.4.1 Application-25: Machine Learning for Improved Directional Drilling

3.4.4.1.1 Background of Application

The development of an AI system was described for improving the efficiency of directional drilling in terms of optimized ROP, less tortuous borehole, less personnel on board, and consistency across operations. Historical data from directional drilling operations was gathered and filtered to train ANNs and refine them using reinforcement learning methods. The resulting neural network showed competence in replicating the decisions of expert directional drillers with an error of less than 3%. The algorithm will be further tested in simulation and potentially integrated into drilling operations for fully automated directional drilling.

3.4.4.1.2 Description of Artificial Intelligence Workflow

The main goal of the ML model is to use a reinforcement learning framework to direct the drilling process. The inputs of the model are: rotary torque, differential pressure, hook load, and ROP. And the outputs of the model are the drilling parameters: rotary speed, flow rate, WOB, and top driver center position drilling process.

The training process is simple and in two steps. In the first step, a traditional neural network is used to learn from historical data with the objective of imitating expert directional drillers. In the second step, the model is connected to a drilling simulator and further trained using reinforcement learning to optimize tool face stability (minimizing standard deviation), reduce deviation from the planned wellbore trajectory, minimize tortuosity, and maximize the rate of penetration (ROP). The actual workflow is as follows.

After loading the historical data, the dataset is divided into training, validation, and testing subsets. To maintain simplicity and interpretability, the tuning of hyperparameters and regularization techniques is omitted. A simple LSTM model with four output nodes is employed for the time-series data, and the trained model is saved for future use. Although specific implementation details of the reinforcement learning approach are not fully provided in the article, a basic demonstration is presented. A custom object, *learningAgent*, is defined to store previous states, perform actions, receive feedback, and iteratively improve the model. And to actually apply the reinforcement learning, one will need to write another application that makes the *learningAgent* interact with the environment.

3.4.4.1.3 Advantages and Limitations of Artificial Intelligence Methodology

Using reinforcement learning is a very practical way to improve a deep learning model that aims to direct an action. And it can suit very complicated real patterns. The disadvantage of the model is also related to its complexity. With more parameters that need to be trained, it can be computationally intensive, and it might not be suitable for simple problems. What's more, it can require a lot of maintenance in the future.

3.4.4.1.4 Results and Recommendations

An ML system was developed for controlling the tool face angle and guiding the drill bit sliding in directional drilling. Historical data from 14 horizontal wells in Appalachia and the Permian Basin was gathered, filtered, and used to train ANNs. A generative adversarial network and a feed-forward MLP network were used to predict future differential pressure and rotary torque based on current tool face, WOB, and other factors. A computational model for drill string physics was used within a reinforcement learning framework to optimize the drilling process. The ML approach can lead to partial or complete automation of directional drilling.

The paper (Pollock et al., 2018) presents the potential for using ML in directional drilling, with the goal of creating an advisory system for drillers and ultimately an automated drilling rig. The researchers worked with experts in directional drilling to compile and clean data from multiple drilling operations. The data was used to train an ANN that predicts the choices of expert drillers with an error of 3% or less. The system was then incorporated into a reinforcement learning framework based on drilling simulation to learn the best course of action. Future efforts will focus on integrating the system with geosteering information, hook position, and real-world testing, with the ultimate goal of improving the economics of directional drilling operations. The full integration of the automated drilling system will involve linking it to the control of drilling operations equipment.

BIBLIOGRAPHY

Pollock, Jacob, Stoecker-Sylvia, Zachary, Veedu, Vinod, Panchal, Neil, and Hani Elshahawi. "Machine Learning for Improved Directional Drilling." *Paper Presented at the Offshore Technology Conference*, Houston, TX, April 2018. https://doi.org/10.4043/28633-MS

3.4.4.2 Application-26: Application of Machine Learning to Accidents Detection at Directional Drilling

3.4.4.2.1 Background of the Application

The directional drilling is simply drilling non-vertical wellbores. The directional drilling is done by controlling the direction and deviation of the bit to reach a predetermined target location. To achieve that, the directional drilling tools (MWD)

transfer telemetry data collected by the magnetometer at downhole to the surface. The telemetry data is continuous and numeric.

The paper proposes a classification algorithm to identify drilling accidents while direction drilling from MWD data by using Gradient boosting of DTs. The algorithm is divided into two sections: analogues search model and leave one out CV on the standard quality metrics for the binary classification problem. The supervised learning technique is utilized with gradient boosting of DTs. The MWD data used to train and test the model includes depth of drill bit, torque on the motor, weight on the hook, input pressure, rotation speed, volume of input flow, depth of the bottom hole, gas content, and WOB. In addition, the database included the type of accident that occurred (pipe stuck, wash-out, drill pipe break, mud loss, shale collars, gas, and water show) and the type of drilling operation conducted during the accident (tripping in, tripping out, drilling, and reaming).

The model is trained by using statistical values (mean, variance, slope angle, absolute deviation, and relative coefficients of MWD time-series) of different window sizes (1 hour, 2 hours, or anything you want!). As a first step, the analogue search model identified the similarities between the window sizes for the same accident types and drilling operations. If these coincide in terms of accident types and drilling operations, these targets are referred to as ground truth. As a second step, CV is utilized to validate predictions. After training the model based on lessons used as an input in the analogue search model, the model predicts the accident type based on similarity among the entries if the threshold value is higher than 0.7 (this is what is used in the paper)

The model is evaluated based on two common metrics: the area under both the ROC curve (area under curve (AUC)) and the precision–recall curve (PR AUC). The ROC AUC curve is created based on TPR and FPR values. PR AUC is created based on precision and recall values. In the final model, ROC AUC is used to quantify the performance of the model.

$$\mathrm{TPR} = \frac{\mathrm{TP}}{\mathrm{TP}+\mathrm{FN}} \qquad \mathrm{FPR} = \frac{\mathrm{FP}}{\mathrm{FP}+\mathrm{TN}}$$

$$\mathrm{Precision} = \frac{\mathrm{TP}}{\mathrm{TP}+\mathrm{FP}} \qquad \mathrm{Recall} = \mathrm{TPR}$$

Input parameters:

- **Main numerical data (this is the data type used to calculate statistics of windows)**
    - Depth of drill bit,
    - Torque on the motor,
    - Weight on the hook,
    - Input pressure,
    - Rotation speed,
    - Volume of input flow,

  - Depth of the bottom hole,
  - Gas content,
  - Weight on bit.
- **Classification data (This is the data used to identify accidents for each type of drilling operation)**
  - **Accident types**
    - Pipe stuck,
    - Wash-out,
    - Drill pipe break,
    - Mud loss,
    - Shale collars, gas,
    - Water show.
  - **Type of drilling operation**
    - Tripping in,
    - Tripping out,
    - Drilling,
    - Reaming.

Output parameters:

- Accident type.

3.4.4.2.2 Description of the Workflow and Machine Learning Model

The main goal of the ML model is to use a tree-based method with an ensemble learning method (boosting) to predict accidents in the drilling process. Normally, such a problem would require a classification tree model, but in this article, the author used regression trees instead. To transform a continuous output into a binary classification, the optimal threshold must be determined using cross-validation (CV).

Due to the absence of the original dataset used in the article, some assumptions are made. In the absence of information regarding the actual sampling frequency of the sensors, it is assumed that each row in the dataset represents statistical properties of sensor data from a 2-hour time window at a given well. Preprocessed data, rather than raw data, is utilized for prediction. It is further assumed that all relevant parameters—such as drill depth, motor torque, weight on hook, input pressure, rotation speed, input volume, bottom hole depth, gas content, weight on bit (WOB), and drilling phase—are available.

The workflow is straightforward: all data is read and the dependent and independent variables are separated. The dataset is then split into training (80%), testing (10%), and validation (10%) sets. A GradientBoostingRegressor() is initialized with a learning rate of 0.1, n_estimators set to 20, and a maximum depth of 3. Hyperparameter tuning is omitted due to the lack of access to real data.

After training the boosting trees model, it can be used for predictions. However, since regression trees are employed, a custom prediction function is required that takes both the model's output and a threshold as inputs. The article's author used leave-one-out cross-validation (LOOCV) to find the optimal threshold, and the same approach is followed here.

Due to the custom prediction function, a custom loss function must be defined. Then, make_scorer() is used to create a customized scoring function compatible with GridSearchCV(), enabling effective hyperparameter tuning despite the non-standard prediction setup.

3.4.4.2.3 Advantages and Limitations of the Synchronized Algorithm

The advantage of boosting tree model is its good ML property. It provides both low bias and low variance. Tree-based methods are also less time-consuming, less demanding, and easier to interpret. However, in this case, using regression trees might be inefficient since classification trees could be used directly and more appropriately for the task. Ensemble learning can also make models lose their interpretability and become less straightforward.

3.4.4.2.4 Results and Discussion

A real-time analogue search model has been developed to detect anomalies and find analogues in a historical data database. The model uses a gradient boosting classification model to compare and rank the real-time MWD data with data from the historical database. The analysis showed that the model has a high quality with significant ROC AUC and PR AUC values, outperforming a random guess classifier. The use of aggregated statistics as input for the model improved the number of analogues found. The model's robustness was also tested, showing that it can still accurately identify analogues even with reasonable distortion of the MWD signal.

BIBLIOGRAPHY

Ekaterina, Gurina, Nikita, Klyuchnikov, Alexey, Zaytsev, Evgenya, Romanenkova, Ksenia, Antipova, Simon, Igor, Victor, Makarov, and Koroteev, Dmitry. "Application of machine learning to accidents detection at directional drilling." *Journal of Petroleum Science and Engineering*, 184, 2020, 106519, https://doi.org/10.1016/j.petrol.2019.106519

3.4.4.3 Application-27: Predictive Model for Bottom Hole Pressure Based on Machine Learning

3.4.4.3.1 Background of the Application

The goal of this work is to use ML to create a predictive model for multiphase wellbore flows. A dataset created from numerical simulations of full-scale transient wellbore flows was used to train the ANN model. One of the important wellbore flow characteristics, bottomhole pressure (BHP), is then predicted by the neural network. The use of neural networks to investigate extremely transitory processes in wellbores, where the majority of characteristics of interest can be represented as interdependent time series of variables, is what makes this work novel. For complicated wellbore geometries and flows, the neural network with two hidden layers showed that it could forecast the BHP with a 5% normalized RMSE. The model is tested on data affected by noise and shows a slightly increased prediction error but still captures the essential features of the transient process. The description of the model, analysis of various test use cases, and possible future research directions are discussed.

3.4.4.3.2 Description of the Workflow and Machine Learning Model

The main objective of the model is to predict BHP using ANN. The inputs are WHP (wellhead pressure), volumetric flow rate at standard conditions (of surface water, surface oil, and surface gas), and well geometry (md, tvd, d)

After reading in the data and splitting dependent and independent variables, the dataset was split into training and testing sets. Assuming a medium-sized dataset, the author set the training size to 0.8. In our model, the x variables are whp, qo_t, qw_t, qg_t, md_i, tvd_i, d, and the combinations, as well as the transformations of these variables. According to the article, the total number of predictive parameters is 76, but most of them are not explained in the article.

The main methods used in the article to tune the model are modifying the structure of the model by adding a different number of nodes in the hidden layers. The 100–50 structure recommended in the article was adopted. The metrics used to evaluate the models are mean squared error (MSE), RMSE, MAE, MAPE, sum of absolute error (SAE), symmetric absolute percentage error (SAPE), and normalized RMSE.

3.4.4.3.3 Advantages and Limitations of the Synchronized Algorithm

The main obstacle to us was the absence of a clear indication of what variables were used to make the prediction. But the main structure of the model is clear. The advantage of an ANN is its ability to discover nonlinear relationships among variables, and it provides exceptional predictive accuracy on real data.

3.4.4.3.4 Results and Discussion

An ML method for predicting the BHP in multiphase wellbore flows is presented in this work. Data from a physics-based numerical simulator that captures the intricate physical events influencing the BHP's behavior is used to train the deep neural network. Using input parameters like WHP, surface flow rates, and wellbore shape, the trained network forecasts the BHP. It is shown that the model can anticipate how variables will change over time and can detect physical correlations inherent to the events under consideration. The prediction is almost instantaneous, making it suitable for real-time applications in field operations. The potential of the approach to be trained on actual field data is left for future work, but the results so far show that it can serve as an alternative to traditional solutions for the inverse problem of determining the BHP from surface measurements.

This work is about developing an ML algorithm to predict the BHP in multiphase wellbore flows. The algorithm uses an ANN trained on a dataset generated by a physics-based simulator. The model has two hidden layers with specific parameters and the rectified linear unit (ReLU) activation function, batch normalization, and AdaDelta optimization. The model can accurately predict the BHP for different well geometries and complex flow regimes, with a mean normalized RMSE below 5%. The model was also tested on noisy data and was still able to capture the basic features of the noisy BHP behavior.

BIBLIOGRAPHY

Spesivtsev, Pavel, Sinkov, Konstantin, Sofronov, Ivan, Zimina, Anna, Umnov, Alexey, Yarullin, Ramil, and Vetrov, Dmitry. Predictive model for bottomhole pressure based on machine learning, *Journal of Petroleum Science and Engineering*, 166, 2018, 825–841, ISSN 0920-4105, https://doi.org/10.1016/j.petrol.2018.03.046

3.4.4.4 Application-28: Flow Rate Estimation Using Dynamic Artificial Neural Networks with Ultrasonic Level Measurements

3.4.4.4.1 Background of the Application

The estimation of the flow rate of non-Newtonian drilling fluids in a Venturi rig using a dynamic artificial neural network (DANN) is covered in this research. Three distinct learning algorithms—back propagation through time (BPTT), real-time recurrent learning (RTRL), and extended Kalman filter (EKF)—were used in a simulation study. Out of the three algorithms, RTRL was the most accurate and least complicated, according to the data. Because it estimates the current output using past inputs and outputs, the DANN with RTRL learning method was determined to be the most accurate model when compared to static ANN and SVR models in the experimental investigation.

3.4.4.4.2 Description of the Workflow and Machine Learning Model

The main objective of the model is to predict flow rate using different measurements. In the article, the authors concluded that the best approach to predict flow rate is using a recurrent neural network with RTRL. After reading the data and separating dependent and independent variables, the dataset was split into training and testing sets. Assuming a medium-sized dataset, the training set size was set to 0.8. The model's input variables (x) include upstream level measurement, level measurement at the throat, and downstream level measurement. The output of the model is the mass flow rate. The structure of the neural network is relatively simple. For each timestep, there are 3 input nodes, 1 output node, and 25 nodes in the hidden layer. The maximum number of iterations (epochs) was also set to 500 to align with the article.

3.4.4.4.3 Advantages and Limitations of the Synchronized Algorithm

RTRL, compared with BPTT, is much more efficient in terms of storage usage. And RNN is a much more suitable model than regular ANN in analyzing time series data. However, the exploding/disappearing gradient problem remains unresolved. The longer the time series is, the bigger the problem can be. Therefore, models such as LSTM can be considered for use in future studies.

3.4.4.4.4 Results and Discussion

The paper discusses the use of DANN to estimate the flow rate of non-Newtonian drilling fluids in an open channel Venturi rig for safe and efficient drilling operations. The simulation of three different learning algorithms—BPTT, RTRL, and EKF—showed that RTRL was the most accurate, less complex, and fastest algorithm. Hence, RTRL was selected for the practical implementation of DANN. In the experiment, the flow rate was estimated using non-intrusive ultrasonic level measurements in the Venturi rig, and the results showed that the DANN model was more accurate than static ANN and SVR models. The findings suggest that ultrasonic level measurements in an open Venturi channel could be an alternative to expensive flow meters to measure drilling fluid flow.

BIBLIOGRAPHY

Chhantyal, Khim, Hoang, Minh, Viumdal, Håkon, Mylvaganam, Saba. Flow Rate Estimation using Dynamic Artificial Neural Networks with Ultrasonic Level Measurements. *Proceedings of The 9th EUROSIM Congress on Modelling and Simulation, EUROSIM 2016, The 57th SIMS Conference on Simulation and Modelling SIMS* 2016. http://doi.org/10.3384/ecp17142561

3.4.4.5 Application-29: Non-Newtonian Fluid Flow Dynamics in Rotating Annular Media: Physics-Based and Data-Driven Modeling

3.4.4.5.1 Background of the Application

In this study, the authors report on the modeling of non-Newtonian fluid flow dynamics in a rotating annular medium. The focus is on the pressure loss ratio (PLR) in the presence of rotation compared to when the system is stationary. An ML-based predictive model using an LSSVM optimized with a coupled simulated annealing (CSA) algorithm is developed and tested against 730 experimental data. Computational fluid dynamics (CFD) simulations are also performed to validate the results. The study finds that both the LSSVM and CFD models are more accurate than a widely used empirical correlation for PLR. The LSSVM model is the most accurate, while the CFD approach captures the nonlinear behavior of PLR. The effects of fluid viscosity, axial velocity, and rotational velocity on PLR dynamics are also studied, and a dimensionless parameter, similar to the Strouhal number, is used to identify specific regimes where PLR shows distinct behavior. The study sheds light on the role of shear-thinning and its interplay with inertial forces on the pressure loss in rotating yield-power law fluids.

3.4.4.5.2 Description of the Workflow and Machine Learning Model

The objective of the model is to predict PLR using fluid density, yield stress, fluid behavior index, consistency index, axial velocity, diameter ratio, eccentricity, and inner cylinder rotational velocity. The model chosen was LSSVM, which is a regression method using support vectors. The LSSVM regressor written by Danny Vanpouche was obtained through GitHub. After reading the data and separating dependent and independent variables, the dataset was split into training and testing sets. The size of the training set was set to 0.7, and the rest of the data was allocated to test and validation group. There are many hyperparameters that can be tuned: gamma, sigma, c, and d. However, tuning was beyond the scope of the original article, so default settings were used for training. After training, the intercept and coefficients for the regression model were obtained.

3.4.4.5.3 Advantages and Limitations of Synchronized Algorithm

The LSSVM model is highly efficient and elegant. It is an exceptional replacement for ordinary regression methods. As shown by the experimental results, this method outperformed CFD, a conventional numerical model solving a similar problem. The limitation of the implementation is the lack of exact information on the hyperparameter tuning process.

3.4.4.5.4 Results and Discussion

In this study, the researchers analyzed the behavior of PLR in annular media with a rotating inner cylinder. They developed a machine-learning model (CSA-LSSVM) based on 730 experimental data points and compared its performance to CFD simulations and a widely used empirical correlation. The results showed that the LSSVM model was the most accurate tool for estimating PLR, while the CFD simulations captured the underlying physical behavior of PLR more reliably. The researchers also studied the effects of fluid viscosity, axial velocity, and rotational velocity on PLR and found that the PLR showed different behaviors for different values of the Strouhal number. The results were confirmed through a global sensitivity analysis.

BIBLIOGRAPHY

Ershadnia, Reza, Amooie, Mohammad Amin, Shams, Reza, Hajirezaie, Sassan, Liu, Yuanhong, Jamshidi, Saeid, and Soltanian, Mohamad Reza. Non-Newtonian fluid flow dynamics in rotating annular media: Physics-based and data-driven modeling. *Journal of Petroleum Science and Engineering*, 185, 2020, 106641, ISSN 0920-4105, https://doi.org/10.1016/j.petrol.2019.106641

3.4.4.6 Application-30: A Mathematical Casing Cutting Model and Operation Parameters Optimization of a Large-Diameter Deep-Water Hydraulic Cutter

3.4.4.6.1 Background of the Application

The application of AI-based sound wave analysis to evaluate drilling equipment condition and forecast equipment failure is covered in the article. A microphone is used to record sounds, the short-time Fourier transform is used to convert the time domain sound wave to the frequency domain, and the time and frequency domain signals are combined to create a two-dimensional spectrogram. This spectrogram is used to train an AI model that can differentiate between typical and unusual equipment sounds. To avoid equipment failure, an alert is sent to the operational crew if unusual noises are heard. According to the findings, the technique worked well for equipment that made noise in the ultrasonic range but less well for those that made noise in the audible sound range.

3.4.4.6.2 Description of the Workflow and Machine Learning Model

The main concept of this application is to use an Auto Encoder for anomaly detection. An Auto Encoder can reconstruct input data, and when a larger-than-expected reconstruction error occurs, it may indicate an anomaly. Using this mechanism, anomalies can be detected. Assuming preprocessed data is available, the first step is to train the Auto Encoder on normal signals. After achieving a sufficiently low reconstruction error, a new dataset containing mixed signals (normal + abnormal) is used to find the optimal error threshold for classification. With this threshold established, the model can be tested and deployed in practical applications.

3.4.4.6.3 *Advantages and Limitations of the Synchronized Algorithm*

As demonstrated in the article, Auto Encoder can deliver good results in classification tasks. However, it requires a lot of data to train, and it's relatively harder to explain the mechanism of classification.

3.4.4.6.4 *Results and Discussion*

The article discusses the results of using AI-based sound wave analysis to assess the health condition of drilling equipment and predict equipment failure by detecting anomalous sound patterns in both time and frequency domains. The method involves capturing sounds via a microphone, converting the time domain sound wave to its frequency domain using short-time Fourier transform, and using a two-dimensional spectrogram to train an AI model to differentiate between normal and abnormal equipment sounds. The results showed that the proposed method achieved better performance in separating normal and anomalous sounds in some machine types, such as the train in the ultrasound region, while underperforming in others such as the conveyor in the ultrasound region. The results also showed that the audible region performed better in detecting anomalous sounds in the toy conveyor, while the ultrasound region performed better in detecting anomalous sounds in the toy train.

The ability to detect anomalous sounds in machines using AI can be improved if the machine produces sounds in the ultrasound region. The study showed that a toy train with an ultrasound signature yielded better results in detecting sound anomalies due to a clear pattern above the 16 kHz audible threshold, while a toy conveyor had degraded performance due to a lack of sounds above the 16 kHz threshold. A toy car showed no significant improvement in detection due to minimally produced sounds in the ultrasound region. However, the spectrogram of an oil well drilling machine had a strong signature in the ultrasound region, indicating that AI-aided anomalous sound detection could help detect early signs of drilling failure.

BIBLIOGRAPHY

Alsaif, Saif Abdulmohsen, and Alothman, Tameem Saud. "Predict Drilling Equipment Failure Using AI-Based Sound Waive Analysis Methodology." *Paper Presented at the Offshore Technology Conference*, Houston, TX, May 2022. https://doi.org/10.4043/31828-MS

3.4.4.7 Application-31: Probabilistic Neural Network with Bayesian-Based, Spectral Torque Imaging and Deep Convolutional Autoencoder for PDC Bit Wear Monitoring

3.4.4.7.1 *Background of the Application*

An ML system is proposed to monitor drill bit wear in real-time operations using a combination of probabilistic neural network (PNN) and Bayes theorem, along with feature extraction by deep convolutional autoencoder (DCAE) and power spectral density (PSD) of surface torque image. The study was performed using wells from Brazilian ultradeep water pre-salt fields, with the results indicating that DCAE feature extraction is more efficient in discriminating bit wear state compared to PSD torque image and torque raw data. The numerical experiments revealed that PNN was the best ML classification model, with an accuracy of 87%, a precision of 83%, a recall of 91%, and an F1-score of 86%. The system was tested in real-time simulation with five wells and was able to capture the optimal moment for replacing a worn-out drill bit. The results show that the proposed system can aid drilling engineers in monitoring PDC drill bit wear in real-time operations.

3.4.4.7.2 *Description of the Workflow and Machine Learning Model*

The main objective of the model is to use PNN and DCAE to monitor the bit wear during the drilling process. The model builds a DCAE for feature extraction, trains it on a dataset that contains 28 × 28 images (derived from drilling data) with binary labels, and uses the trained DCAE to extract features from the dataset. The encoder would produce a compressed image of size 4 × 4 × 16. And the output tensor will be flattened to facilitate later operations.

The model then trains a PNN on the extracted features and their corresponding labels. The PNN is built with trainable posterior weight distributions and a prior weight distribution that is a normal distribution of mean = 0 and standard deviation = 1 (an assumption different from the original article to simplify the example). The loss function was set to be binary cross-entropy, and the model was trained for 5 epochs using a batch size of 10 and an RMSprop optimizer with a learning rate of 0.001.

3.4.4.7.3 *Advantages and Limitations of the Synchronized Algorithm*

According to the article, the combination of DCAE and PNN works well compared to the baseline model and other models. One of the greatest advantages of using PNN as the classification model is that the training speed is much faster than that of the back propagation-based models.

3.4.4.7.4 *Results and Discussion*

An ML model is proposed to monitor the status of PDC drill bits in real-time drilling operations. The surface torque signal is transformed into PSD images, and then a DCAE is used for feature extraction. The PNN classifier is used with the Bayes theorem to check the bit status every 30 minutes and predict the best time for replacement of a worn-out drill bit. The results from field tests in five wells showed the effectiveness of the system, with 87% accuracy, 83% precision, 91% recall, and 86% F1-score. However, the application is limited to specific oilfields with the same drilling conditions, and the DCAE PSD image training process may take a long time.

BIBLIOGRAPHY

Agostini, Cristiano Eduardo and Sampaio, Marcio Augusto. Probabilistic Neural Network with Bayesian-based, spectral torque imaging and Deep Convolutional Autoencoder for PDC bit wear monitoring. *Journal of Petroleum Science and Engineering*, 193, 2020, 107434, ISSN 0920-4105, https://doi.org/10.1016/j.petrol.2020.107434

3.4.4.8 Application-32: Application of Artificial Neural Networks to Optimum Bit Selection

3.4.4.8.1 Background of the Application

In drilling engineering, the optimum bit selection is crucial and is determined by the lowest cost per foot, considering the bit cost and performance, as well as penetration rate. Conventional methods use computer programs based on mathematical models and data from previously drilled wells in the area. This study uses ANNs and fractal geostatistics to improve the optimum bit selection process. A back-propagation ANN model was developed using real rock bit data from several wells in a carbonate field and trained using gamma ray and sonic log data. The model was tested with various drilling scenarios in different lithologic units and showed satisfactory results.

3.4.4.8.2 Description of the Workflow and Machine Learning Model

The objective of the application is to build and train machine-learning models to predict rock bit types and help optimize bit selection. The structure of the ANN is straightforward. There's only one hidden layer with 210 nodes. The learning rate was set to 0.2, and the momentum factor was set to 0.085 according to the article. All the activation functions were set to the sigmoid function. The original article didn't mention the output format in detail. Three output nodes were used in the interpretation, each representing a different aspect of the IADC code.

3.4.4.8.3 Advantages and Limitations of the Synchronized Algorithm

ANNs are highly useful for improving forecasting accuracy. However, it can be difficult to train on smaller datasets, and its black-box nature makes it harder to interpret.

3.4.4.8.4 Results and Discussion

The paper proposes a new approach for optimum rock bit selection in drilling engineering, combining ANNs and fractal geostatistics. The accuracy and efficiency of the developed ANN model were tested on real and synthetic cases and were found to provide satisfactory results. The model requires input from sonic and neutron log data and is found to estimate rock bit properties correctly for different drilling scenarios, including wells drilled both within and outside known boundaries of a field. It was concluded that the developed program is suitable as a front-end system for rock bit selection, which could help engineers in making informed decisions.

BIBLIOGRAPHY

Yılmaz, Serkan, Demircioglu, Cem, and Akin, Serhat. Application of artificial neural networks to optimum bit selection. *Computers & Geosciences*, 28 (2), 2002, 261–269, ISSN 0098-3004, https://doi.org/10.1016/S0098-3004(01)00071-1

3.4.4.9 Application-33: Automatic Interpretation of Cement Evaluation Logs from Cased Boreholes Using Supervised Deep Neural Networks

3.4.4.9.1 Background of the Application

The proposed system is aimed at assisting human interpreters in evaluating cement integrity in cased boreholes using well logging data. It is based on deep convolutional neural networks trained on a dataset of 60 km of interpreted well log data, where the task is to classify bond quality (six ordinal classes) and hydraulic isolation (two classes) in 1 m depth segments of each well based on the surrounding 13 m of log data. The performance of the networks is compared to reference interpretations and human reinterpretations, showing that the networks match the reference 51.6% of the time for bond quality and 86.7% for hydraulic isolation, compared to a random-guess baseline of 16.7% and 50%, respectively. The networks' performance is somewhat better than human reinterpretations, but the subjectivity inherent in the interpretation process is the main reason for not achieving an even better match.

3.4.4.9.2 Description of the Workflow and Machine Learning Model

The main purpose of this application is to use a deep neural network to classify the cement evaluation logs. There are in total three inputs going through three different sets of layers. A synthetic dataset was generated using random numbers from a standard normal distribution to simulate the shape of the original data used in the article. After reading and splitting dependent and independent variables, the data was divided into training and testing sets with a training size set to 0.2 based on the dataset size. The model is built with three input layers, each corresponding to one of the three data types. Each input layer is followed by a set of convolutional and pooling layers, which extract features from the input data. The resulting feature maps are flattened and concatenated into a single feature vector, which is then passed through a set of dense layers, which compute the final output value.

The model was trained using the RMSProp optimizer, with a batch size of 16 and for 5 epochs. The parameters are from the original article.

3.4.4.9.3 Advantages and Limitations of the Synchronized Algorithm

The advantage of the neural network with a more complicated structure is that it can process different types of data using different layers. This would enhance the performance of the model when facing more complicated input. However, there are also disadvantages related to its complexity. More parameters mean it takes more time to train the model, and the variance of the model can be high when facing test data.

3.4.4.9.4 Results and Discussion

The authors propose a deep neural network-based system for automatic well log interpretation. The system is trained using a dataset of manually interpreted well log data to draw the connection between the data and interpretations. The performance of the system is quantified by comparing its interpretation with the reference interpretation. The system performs well with a match rate of 51.6% for bond quality and 86.7% for hydraulic isolation, outperforming the random-guess baseline. However, the system is limited in its ability to interpret data that is not similar to the data seen during training and may not be able to handle new types of data without additional training.

The authors propose a deep learning-based system for automatically interpreting cement evaluation logs. The system uses deep convolutional neural networks trained on a dataset of around 60 km of interpreted well log data. The networks are tasked with classifying the bond quality and hydraulic isolation based on the well log data in each 1 m depth segment. The results show that the network's interpretation performs well, with an exact match with reference interpretation of 51.6% for bond quality and 86.7% for hydraulic isolation. The networks' interpretation was compared with manual reinterpretation and found to match the reference interpretations better, although further testing with multiple reinterpretations is needed to conclude this. However, interpretation subjectivity remains a major challenge, and the dataset used should be quality-controlled to ensure internal consistency.

BIBLIOGRAPHY

Viggen, Erlend Magnus, Merciu, Ioan Alexandru, Løvstakken, Lasse, and Måsøy, Svein-Erik. Automatic interpretation of cement evaluation logs from cased boreholes using supervised deep neural networks. *Journal of Petroleum Science and Engineering*, 195, 2020, 107539, ISSN 0920-4105, https://doi.org/10.1016/j.petrol.2020.107539

3.4.4.10 Application-34: Research on Remote Intelligent Control Technology of Throttling and Back Pressure in Managed Pressure Drilling

3.4.4.10.1 Background of the Application

In this work, an improved particle swarm optimization PID neural network (IPSO-PIDNN) model was proposed to control the pressure in managed pressure drilling (MPD). The wellhead back pressure was taken as the controlled parameter, and the mathematical model of the throttle valve was calculated. The IPSO-PIDNN model was compared with traditional PID neural network (PIDNN) and particle swarm optimization PID neural network (PSO-PIDNN) using remote communication and VR technology. The results showed that the IPSO-PIDNN model has good self-learning, high optimization quality, high control accuracy, no overshoot, fast response, and short regulation time, providing a technical guarantee for well control safety in MPD.

3.4.4.10.2 Description of the Workflow and Machine Learning Model

Data Collection:

- The initial step involves gathering data from various drilling operations. This data encompasses drilling parameters, geological parameters, and other relevant metrics.

Data Preprocessing:

- Outlier Handling: Outliers in the dataset are identified using statistical methods and are managed to ensure they don't distort the model's performance.
- Feature Encoding: Categorical features, such as lithology types or bit designs, are transformed into numerical representations using techniques like one-hot encoding.
- Normalization: Features are scaled to ensure they have a similar scale, which aids in the convergence of the model during training.
- Data Splitting: The processed data is divided into training and testing sets to train the model and validate its performance, respectively.

Model Building: Three primary models are considered:

- ANN: Constructed with multiple layers, including input, hidden, and output layers. The architecture and hyperparameters are optimized for the best performance.
- ANFIS: Built based on subtractive clustering techniques. The model's parameters, like cluster radius, are fine-tuned for optimal results.
- SVM: A model that uses hyperplanes to classify or predict data. It's optimized using parameters like the regularization parameter, kernel function, and others.

Model Training: Each model is trained using the training dataset, learning to predict the ROP based on the given features.

Model Evaluation: Post-training, models are evaluated on the testing dataset using metrics like RMSE to gauge their accuracy and reliability.

Prediction: The trained models can predict the ROP for new drilling operations, aiding in real-time decision-making.

3.4.4.10.3 Advantages and Limitations of the Synchronized Algorithm

Advantages:

- Precision: Models, especially the ANN, have shown high accuracy in predicting the ROP, making them valuable tools for drilling operations.
- Versatility: Multiple models (ANN, ANFIS, SVM) provide a range of tools that can be chosen based on the specific requirements of the drilling operation.
- Real-time Decision-making: Predictions from these models can guide drilling engineers in real-time, optimizing operations, and minimizing risks.
- Adaptability: The models can be retrained and adapted to different drilling environments or conditions, ensuring their continued relevance.
- Data-Driven Insights: Leveraging real-world data ensures that the models are grounded in actual drilling scenarios, enhancing their reliability.

Limitations:

- Data Dependency: The efficacy of the models is closely tied to the quality and quantity of the data. Inaccurate or insufficient data can lead to suboptimal model performance.
- Computational Intensity: Training models, especially ANFIS, can be computationally demanding, necessitating robust hardware.
- Hyperparameter Sensitivity: Models, especially SVM, require careful tuning of hyperparameters, which can be a time-consuming process.
- Black Box Dilemma: Neural networks, in particular, can be challenging to interpret, making it hard to discern how they derive specific predictions.
- Potential for Overfitting: Without rigorous validation and regularization, there's a risk that the models might overfit the training data, reducing their generalization capability on new data.

3.4.4.10.4 Results and Discussion

In this paper, a new control model called IPSO-PIDNN was proposed to improve BHP control in MPD. The authors analyzed the principles and factors that affect BHP control, selected the wellhead back pressure as the controlled parameter, and calculated the mathematical model of the choke valve. The IPSO-PIDNN model was designed to improve the optimization quality of the traditional PSO algorithm, providing faster response time and adjustment time for pressure control. The model was tested using actual production data from an oil field and showed promising results with high control accuracy and no overshoot. The authors concluded that the IPSO-PIDNN model can provide technical support for the well control safety in MPD.

BIBLIOGRAPHY

Zhang, He, Qiu, Yongzhi, Liang, Haibo, Li, Yunan, and Yuan, Xiru. Research on remote intelligent control technology of throttling and back pressure in managed pressure drilling. *Petroleum*, 7 (2), 2021, 222–229, ISSN 2405-6561, https://doi.org/10.1016/j.petlm.2020.04.003

3.4.4.11 Application-35: Hybrid Machine Learning for Pullback Force Forecasting During Horizontal Directional Drilling

3.4.4.11.1 Background of the Application

This paper proposes a hybrid ML model using NSGA-II and SVM to predict pullback force in horizontal directional drilling (HDD) construction. The model is tested on two practical HDD projects in China, and the results show that it has better accuracy and stability than benchmark models with an MAPE of less than 7%. The study also concludes that ML models can predict pullback force more accurately than empirical models and are feasible for practical application. The paper also discusses the practical application of ML models in HDD construction and future development directions.

3.4.4.11.2 Description of the Workflow and Machine Learning Model

The proposed model starts with preprocessing data sourced from a CSV file, which includes various HDD-related parameters. This data is then divided into training and testing subsets, followed by normalization of the feature variables. This step is essential because SVM is sensitive to the scale of the data. Following preprocessing, a hybrid machine-learning approach is initiated. Here, the nondominated sorting genetic algorithm II (NSGA-II) operates combined with SVR. The genetic algorithm serves as an optimizer for the hyperparameters of the SVR, specifically the regularization (C) and epsilon parameters. The SVR, once optimized, is employed in a regression context, with the aim of predicting the pullback force in HDD construction. Finally, the model's predictive accuracy is evaluated on the testing set using MAPE, achieving a result of less than 7%.

3.4.4.11.3 Advantages and Limitations of the Synchronized Algorithm

Advantages:

- **Hybrid Model**: The combination of NSGA-II with SVM provides a two-fold advantage. NSGA-II efficiently searches for the optimal hyperparameters, ensuring that the SVM model performs at its best.
- **Nonlinear Relationships**: SVMs can capture nonlinear relationships in the data by using different kernel functions. This is essential for complex tasks like predicting pullback force in HDD construction.
- **Regularization**: With the hyperparameter C, the SVM inherently has a regularization mechanism that helps prevent overfitting to the training data.
- **Robustness**: SVMs are relatively less sensitive to outliers, which can be beneficial if the dataset contains unexpected values or noise.
- **Global Optimization**: The genetic algorithm (NSGA-II) aims to find global optimal values for hyperparameters rather than getting stuck in local optimal values.

Limitations:

- **Computational Complexity**: SVMs, especially with large datasets and complex kernel functions, can be computationally expensive in terms of both memory and time.
- **Difficulty of Interpretability**: SVM models, particularly with nonlinear kernels, lack the interpretability of some other models. It might be hard to understand the rationale behind predictions.
- **Parameter Sensitivity**: Although NSGA-II helps optimize hyperparameters, SVMs can still be quite sensitive to the chosen parameters. If NSGA-II does not converge well, the performance of the SVM might suffer.
- **Genetic Algorithm Convergence**: While genetic algorithms are powerful, there's no guarantee that they'll always converge to the global optimum. Depending on the initialization and genetic operators (crossover, mutation), the algorithm may converge to suboptimal solutions.
- **Cost of Hybrid Approach**: Introducing a genetic algorithm for hyperparameter optimization adds a layer of complexity and computational overhead to the modeling process. Simple grid or random search approaches might sometimes be more efficient for certain problems.
- **Data Scaling Necessity**: SVMs require feature scaling for optimal performance. This can introduce an additional preprocessing step and potential issues if not done correctly.

3.4.4.11.4 Results and Discussion

This work proposes an ML model based on a multi-objective optimizer to predict the pullback force in HDD construction. The proposed model's accuracy and stability are better than those of the two benchmark models, and the MAPE is less than

7%. The model is data-driven and can be used in the construction phase, but not in the design phase, without sufficient data. Future research directions include transfer learning to explore commonness among projects and the application of parallel computing to improve the model's calculation speed in real-time monitoring and early warning.

BIBLIOGRAPHY

Lu, Hongfang, Iseley, Tom, Matthews, John, and Liao, Wei. Hybrid machine learning for pullback force forecasting during horizontal directional drilling. *Automation in Construction*. 129, 2021, 103810, ISSN 0926-5805, https://doi.org/10.1016/j.autcon.2021.103810

3.4.4.12 Application-36: Supervised Data-Driven Approach to Early Kick Detection During Drilling Operation

3.4.4.12.1 Background of the Application

In summary, this work focuses on the importance of monitoring downhole parameters for early kick detection in the offshore drilling industry. The aim is to ensure safety and prevent incidents like the 2010 Macondo blowout. The study employs an ANN and a binary classifier to build a data-driven kick detection model using experimental data generated from monitoring downhole parameters during a kick event. The parameters such as mud flow-out rate, conductivity, density, and downhole pressure were combined to detect the kick events. Abnormal pressure and flow regimes were found to be the most significant parameters, but relying on them alone could lead to false alarms. The model was tested on data from two different laboratory-scale drilling simulators.

3.4.4.12.2 Description of the Workflow and Machine Learning Model

In this work, ANN, a binary classifier, and downhole monitoring of drilling flow parameters are combined to build data-driven kick detection models. There are mainly three parts in the model. The first part is to simulate the dataset based on the workflow, which consists of six datasets with 1,510 data points. And this dataset is randomly split into six datasets (data1, data2, data3, data4, data5, and data6). Then the rules follow as:

1. A training dataset (say data1) yields a trained model.
2. There are two remaining unique datasets (data2 and data3), and each is separately used for validation.
3. For 1st validation done (data2), there are four test datasets (data3, data4, data5, and data6). This yields four test instances.
4. For 2nd validation done (data3), there are four test datasets (data2, data4, data5, and data6). This yields another four test instances.
5. The bs are second runs and repeat processes to monitor for repeatability and variability in the process.
6. Therefore, a trained model was validated with two different datasets and tested with four remaining datasets each. This yields eight test instances for a single trained model.
7. For three trained models (three different data), it becomes a total of 24 test instances.

The second part is to run all 24 instances and record their performance. The third part is the result of a specific example of 24 instances, the result also shows the predicted value and label corresponding to its original dataset. The workflow is straightforward. First, 24 instances are simulated and the data normalized. Then, a simple feedforward ANN is constructed with five input layer nodes, no hidden layer, and one output neuron, using tanh as the activation function. For the second part of the code, it will run to the final result and present it. For the third part of the code, I used an example of instance 12, and it will show the final predicted value and label.

3.4.4.12.3 Advantages and Limitations of the Synchronized Algorithm

The advantage of ANNs is significant since ANNs are capable of modeling complex, nonlinear relationships between inputs and outputs, which other simpler models might not capture effectively. The use of activation functions like the "tanh" function in our model introduces nonlinearity, which can help model complex patterns in the data. Once trained, ANNs can make predictions in real time, which is useful in many applications such as kick detection. Also, ANNs have a high degree of parallelism as they perform numerous operations simultaneously. This is particularly beneficial when dealing with large-scale, high-dimensional data.

The limitation remains to be the lack of actual experiment and real-world testing. The example provided is the simplest and may not capture more complex relationships within the data.

3.4.4.12.4 Results and Discussion

In order to detect kick events, the authors of this work have modeled the surface parameters of WOB, RPM, torque, pump pressure, and ROP using ANNs. In order to enable early kick detection, they have also employed ANNs to model downhole measured quantities, which is a novel technique. The authors assessed the model's performance using a confusion matrix and discovered that it had a 100% success rate in identifying kick events, with 0% false alarms, 0% missed detections, and early detection within 5 seconds of the influX. Finding the most important parameter for recognizing a kick event is made easier by the internal structure of the ANN model, which displays the weight of each parameter. The consistently positive-valued weights of downhole pressure and mud flow-out rate showed that they are more significant in identifying a kick event. The authors have also noted the limitations of using single-input parameter-based models, which are susceptible to noise, uncertainties, disturbances, and faults, and could be dangerous in the event of sensor failures. The authors recommend considering other parameters in future studies.

The study described two experimental drilling systems and presented a methodology for using supervised learning in early kick detection in drilling operations. The study found that a simple feedforward ANN with four input nodes and a tanh activation function can effectively detect kicks. The study also highlighted the importance of factors such as missed detection, false detection, and detection time in building a reliable kick detection model and identified the most significant parameters in achieving a learning goal. However, a more complex neural network architecture may be required to differentiate between different anomalous drilling events. The study is a preliminary success based on laboratory observations, and further validation with field data is necessary.

BIBLIOGRAPHY

Muojeke, Somadina, Venkatesan, Ramachandran, Khan, Faisal. Supervised data-driven approach to early kick detection during drilling operation. *Journal of Petroleum Science and Engineering*, 192, 2020, 107324, ISSN 0920-4105, https://doi.org/10.1016/j.petrol.2020.107324

3.4.4.13 Application-37: Cutter Layout Optimization for Reduction of Lateral Force on PDC Bit using Kriging and Particle Swarm Optimization Methods

3.4.4.13.1 Background of the Application

In this study, the authors propose a model to optimize the cutter layout of PDC bits to reduce the lateral force on the bits. The lateral force reduction plays an important role in reducing drilling vibration and improving the service life of PDC bits. The authors build models of wear and force for PDC cutters and use a zero-point traversal algorithm to calculate the unknown parameters in the force and wear models. The results show that the proposed optimal approach is efficient and can effectively reduce the lateral force on PDC bits, leading to improved stability and increased service life.

3.4.4.13.2 Description of the Workflow and Machine Learning Model

The workflow integrates a force and wear model for individual cutters, employing a Kriging surrogate model and PSO method. The process starts with defining the optimization model, considering the WOB, TOB, and lateral force. The optimization model aims to minimize the mean lateral force across the bit's lifetime by optimizing cutter layout parameters. After modeling, the Kriging model is created from initial design points and their functional values, which serve as the basis of the optimization process. The PSO method then searches for the optimal solution that results in the smallest lateral force.

3.4.4.13.3 Advantages and Limitations of the Synchronized Algorithm

One significant advantage is its high computational efficiency and ability to perform multi-parameter linkage adjustments. This approach allows for a significant reduction in lateral force, enhancing the bit's stability. However, limitations include the complexity of the process, requiring iterative calculations and optimization, which could be computationally intensive. Additionally, the accuracy of the model depends on the initial design points.

3.4.4.13.4 Results and Discussion

This paper presents a new approach to optimizing the layout of PDC bits in order to reduce the lateral force during drilling. The process takes into account the wear condition of each cutter and aims to retain a small lateral force even as the bit wears. A wear model and force model of PDC cutters were obtained, and the zero point traversal (ZPT) algorithm was proposed to calculate the unknown parameters. The optimization approach used the ZPT method, Kriging surrogate model, and PSO to find the assembly of cutter layout parameters that resulted in the smallest lateral force. The optimization case for a PDC bit with 16 cutters showed a significant reduction of lateral force and improvement in the stability of drilling and the service life of the PDC bit. The paper focuses on abrasive wear during drilling in homogeneous formations, but the force analysis and layout optimization of PDC bits under impact fracture and thermal wear conditions are also important areas for future study.

BIBLIOGRAPHY

Ma, Yachao, Huang, Zhiqiang, Li, Qin, Zhou, Yi, Peng, Shijin. Cutter layout optimization for reduction of lateral force on PDC bit using Kriging and particle swarm optimization methods. *Journal of Petroleum Science and Engineering*, 163, 2018, 359–370, ISSN 0920-4105. https://doi.org/10.1016/j.petrol.2018.01.001

3.4.4.14 Application-38: Automated Drilling Data Quality Control Using Application of AI Technologies

3.4.4.14.1 Background of the Application

This work presents an automated data quality control workflow for drilling sensor data using AI algorithms. The workflow aims to resolve complex quality issues and harmonize sensor drilling data to prepare it for advanced analytics. The approach uses ML algorithms, statistics, signal processing, and rule-based engines to detect and classify data quality issues, including error values, outliers, bias, drifts, noise, and missing values. The workflow scores and treats data quality issues on a context-specific basis, resulting in a data quality and preparation engine that organizes drilling data for further analysis. The workflow was able to recover more than 90% of a drilling dataset made up of 18 offshore wells that would otherwise not be usable for analytics. The workflow automates the end-to-end process for data quality control of drilling sensor data and allows for the detection and recovery of wrong or missing data, preserving the consistency of the dataset across all channels.

3.4.4.14.2 Description of the Workflow and Machine Learning Model

The workflow employs context-specific algorithms and ML to address issues like error values, outliers, bias, drifts, noise, and missing values in sensor data. It standardizes the data and recovers the maximum volume of data while avoiding information loss. The process involves data standardization (ensuring uniform units, channel labels, and timestamps). The model identifies and treats quality issues, resampling time series with gaps and different sampling rates. This approach helps organize drilling data for advanced analytics and measure the quality of the dataset.

3.4.4.14.3 Advantages and Limitations of the Synchronized Algorithm

The methodology's strengths lie in its ability to recover over 90% of a drilling dataset, which is achieved by resolving specific issues like resampling time series with gaps and different sampling rates, and smart imputation of wrong/missing data. In this way, it can preserve the dataset's consistency across all channels. On the other hand, the ML-based data imputation approach can improve data recovery but faces challenges related to modeling complexity, especially when dealing with long sequences or rare events.

3.4.4.14.4 Results and Discussion

The present work involves an automated data quality control workflow for drilling sensor data using AI algorithms. The workflow has been applied to 18 anonymized wells provided by ADNOC, and the results show a recovery of 90% of the data. The data retrieval routine was run with a resampling rate of 5 seconds, the null ratio (quality threshold for discarding data) was set to 80%, and the max time gap (to avoid excessive data interpolation) was set to 60 seconds. The results show that the approach was effective in recovering more than 10% of values that would have been lost in other channels, and a similar improvement was observed at the well level. It is noted that increasing the max time gap will increase the data recovery but also introduce bias through data interpolation. Further investigation of ML models for data retrieval is ongoing.

In this work, an automated workflow has been developed to improve the quality of operational drilling data for use in ML and analytics. The workflow addresses common data quality issues such as irregular sampling, inconsistent names and units, missing values, and depth compensation. The procedure uses ML algorithms, statistics, signal processing, and rule-based engines to detect and classify data quality issues, and score and treat them on a context-specific basis to recover maximum data while avoiding information loss. The methodology was applied to a real dataset of 18 wells and showed that it allowed the recovery of more than 90% of measurements. The next steps include extending the procedure to real-time data acquisition and using ML-based data imputation to further improve the results.

BIBLIOGRAPHY

Battocchio, Francesco, Sreekantan, Jaijith, Arnaout, Arghad, Benaichouche, Abed, Al Shamsi, Juma Sulaiman, Awad, Mohamad Abdul Salam, Alnuaimi, Ahmed, Mohamed, and Peraza, Luis Ramon Baptista. "Automated Drilling Data Quality Control Using Application of AI Technologies." *Paper Presented at the Abu Dhabi International Petroleum Exhibition & Conference*, Abu Dhabi, UAE, November 2021. https://doi.org/10.2118/207598-MS

3.4.4.15 Application-39: Evolving Neural Conditional Random Fields for Drilling Report Classification

3.4.4.15.1 Background of the Application

This work focuses on using Recurrent Neural Networks and evolutionary computing to extract important information from daily drilling reports in oil and gas prospecting. The goal is to improve decision-making in drilling operations and avoid accidents and environmental damages. The proposed approach outperforms other techniques, and experiments show an improvement in Macro-F1 scores by over 47%. The use of an ensembling technique results in an additional improvement of 3% in the best scenario. These results hold promise for future research in this field.

3.4.4.15.2 Description of the Workflow and Machine Learning Model

The workflow of the AI model described in the paper involves a contextual-driven approach based on recurrent neural networks. The workflow can be outlined in two main steps: first, computing a feature vector for each sample in the drilling report, and second, learning the relevant patterns corresponding to each class. The model incorporates contextual information during learning and prediction, which is achieved through contextualizing word embeddings using RNNs.

3.4.4.15.3 Advantages and Limitations of the Synchronized Algorithm

The paper highlights the evolutionary computing to combine partially trained models using cyclical learning rates. This method allows the model to find several local optima, enhancing its overall effectiveness. Additionally, the model includes a conditional random field layer to consider labels assigned to adjacent sentences, thereby increasing its robustness in predicting sequences.

3.4.4.15.4 Results and Discussion

This work focuses on classifying sentences in drilling reports into a set of operations. The problem requires learning the sequence of events to provide reliable results, making it suitable for contextual-based classifiers. The results show that contextual models are sensitive to the length of the sequence and that contextual information and CRF-based models are essential for the classification task. The experimental results show that the proposed approaches outperform the baseline and that both GA and GP with ensemble learning outperform their single counterpart versions. Sampling from minima using cyclical learning rates was found to be effective.

In this work, an approach for drilling event classification was presented, which considers contextual information through an attention mechanism, features extracted from neighboring sentences, and conditional random fields. The proposed model was compared to a fastText baseline and showed improvement in accuracy and Macro-F1 scores by 10.06% and 49.98%, respectively. The evolutionary ensemble approach provided a gain of up to 3.31% in Macro-F1 scores, highlighting its relevance for the classification task. Future work will focus on exploring other optimization techniques, decision fusion strategies, and deep learning approaches for feature extraction.

BIBLIOGRAPHY

Ribeiro, Luiz C.F., Afonso, Luis C.S., Colombo, Danilo, Guilherme, Ivan R., Papa, João P. Evolving Neural Conditional Random Fields for drilling report classification. *Journal of Petroleum Science and Engineering*, 187, 2020, 106846, ISSN 0920-4105, https://doi.org/10.1016/j.petrol.2019.106846

3.4.4.16 Application-40: Drilling Stuck Pipe Classification and Mitigation in the Gulf of Suez Oil Field

3.4.4.16.1 Background of the Application

The drilling pipe stuck issue is a major problem in the oil and gas industry, causing significant financial losses. Developing a reliable classification model for drilling pipe stuck is crucial for decision-makers in the petroleum drilling rig. In this study, 12 ML techniques were used for drilling pipe stuck classification and mitigation in the Gulf of Suez oil fields. The key contribution of the study is to automate pipe stuck classification using ML algorithms and mitigate the pipe stuck cases using genetic algorithm optimization. Out of the 12 AI techniques used, the most reliable algorithm was found to be extremely randomized trees, with 100% classification accuracy based on a testing dataset.

3.4.4.16.2 Description of the Workflow and Machine Learning Model

The main purpose of this application is to predict drilling pipe stuck, which is a classification problem using multiple ML models, from single models to ensemble ones. Seven input variables are used in this case, including mud weight, bit diameter, rotary speed, flow rate, motor bending angle, WOB, and ROP. The output value is a multi-class variable indicating the extent of drilling pipe stuck ranging from 0 to 1, where 0–0.4 means non-stuck, 0.4–0.7 means partially stuck, and 0.7–1 means stuck.

According to the article, all input data are normalized using min–max scaling, which is also performed in the work. Given that no details about models are disclosed in the article, all models are using default or popular parameter values. A total of 12 ML models are built, trained, and evaluated individually, including single models (SVM, DT, logistic regression, KNN, Gaussian NB, and ANN) and ensemble models (Random forest, Xgboost, Ada boost, bagging, stochastic gradient boosting, and extra trees).

3.4.4.16.3 Advantages and Limitations of the Synchronized Algorithm

The dataset only contains seven input features, and the total data size is not clear, which may have a potential lack of data problem. Model details are not clearly explained in the article, especially about crucial parameters in some models. Extra trees model has an impossibly perfect 1.00 accuracy, which may involve an overfitting problem or the dataset is not generalized enough.

3.4.4.16.4 Results and Discussion

In conclusion, the study has shown that ML techniques can be effectively used for drilling pipe stuck classification and mitigation in the Gulf of Suez oil fields. The results of the study suggest that extremely randomized trees are the most reliable algorithm for this purpose. The study also presents a public open dataset for the drilled wells at the Gulf of Suez, which can be used for future experiments, algorithms' validation, and analysis. The proposed model can provide a proactive prediction tool that can early predict the stuck occurrence based on the key drilling stuck predictors, and can avoid the stuck cases and optimize the drilling parameters.

BIBLIOGRAPHY

Elmousalami, H.H. and Elaskary, M. (2020). Drilling stuck pipe classification and mitigation in the Gulf of Suez oil fields using artificial intelligence. *Journal of Petroleum Exploration and Production Technology*, 10, 2055–2068.

3.4.4.17 Application-41: Forecasting the Abnormal Events at Well Drilling with Machine Learning

3.4.4.17.1 Background of the Application

This study introduces a machine-learning method that predicts the likelihood of six different kinds of drilling accidents in real time using data from drilling telemetry. The algorithm is trained on 125 historical drilling mishaps from 100 distinct Russian oil and gas wells, and it is based on a bag-of-features representation of the time series. According to validation, the model has a 40% FPR and can predict 70% of drilling mishaps.

3.4.4.17.2 Description of the Workflow and Machine Learning Model

The model's workflow involves two key stages: feature generation and classification. Initially, the model uses a bag-of-features approach to transform time-series data into a histogram of labels predicted for the time series intervals. This transformation is achieved through wavelet transform and clustering algorithms. The second stage is the classification problem, where the model predicts the probability of various types of drilling accidents using gradient boosting DTs.

3.4.4.17.3 Advantages and Limitations of Synchronized Algorithm

The methodology demonstrates several advantages, such as the ability to predict 70% of drilling accidents with a relatively high FPR of 40%. Its real-time operation and integration with existing telemetry systems make it suitable for real-world oil gas applications. However, the model also has limitations. The high FPR indicates a need for further refinement to reduce false alarms. Additionally, the complexity of the model might limit its applicability in scenarios with limited data availability.

3.4.4.17.4 Results and Discussion

In conclusion, a data-driven and physics-informed algorithm for drilling accident forecasting has been presented, addressing the partial prevention of drilling accidents during well construction. The bag-of-features representation of the time series and machine-learning model have shown promising results in forecasting drilling accidents in real-time. Recommendations for improving the model's quality and interpretability have also been provided. Overall, this paper contributes to the development of AI solutions in the oil and gas industry and highlights the potential of ML for improving safety and efficiency in drilling operations.

BIBLIOGRAPHY

Gurina, E., Klyuchnikov, N., Antipova, K., & Koroteev, D.A. (2022). Forecasting the abnormal events at well drilling with machine learning. *Applied Intelligence*, 52, 9980–9995.

3.4.4.18 Application-42: Machine Learning Applied on Fishing Occurrence Prediction

3.4.4.18.1 Background of the Application

In this paper, the authors present a method for predicting the demand for fishing services in the oil and gas industry using data mining and ML. The oil and gas industry can benefit from these tools to modernize its processes and increase efficiency. The method involves preprocessing the input data, selecting the most suitable ML model, and analyzing the results. The results can be used for several purposes, such as planning and scheduling, reducing NPT, and forecasting rig schedules. The authors also mention that the methods used in this work can be applied to other industries.

3.4.4.18.2 Description of the Workflow and Machine Learning Model

The paper focuses on predicting fishing occurrences in the oil industry. The objective of the model is to predict the demand for fishing services in oil wells based on historical data. The model relies on historical data, which includes various aspects of well operations. Then, various ML techniques, including neural networks, DTs, linear and polynomial regression, are tested to determine the most suitable method for the data at hand. These models are trained using a portion of the data and then tested for accuracy using another set. Finally, the trained model is used to predict future occurrences of fishing events in oil wells

3.4.4.18.3 Advantages and Limitations of the Synchronized Algorithm

Advantages:

- Efficiency: The AI approach significantly enhances the efficiency of predicting fishing occurrences.
- Adaptability and Improvement: The model is capable of continuous improvement and can be adapted to new scenarios, making it a dynamic tool.

Limitations:

- A major challenge is the availability of historical data. If the data is not comprehensive enough, it results in an inexperienced model incapable of achieving satisfactory results. In this study, only 2 years of historical data were available, which was not sufficient for implementing sophisticated models, especially due to the high standard deviation in the number of fishing occurrences.

3.4.4.18.4 Results and Discussion

This paper describes a simple solution for forecasting the demand for fishing services in the oil industry using ML. The main challenge in developing this solution is obtaining a sufficient and compatible historical database for training the model. In this case, due to a lack of rich data, the model is based on a single variable, the type of activity performed in the well during each fishing event. The results obtained from the simple model were reasonable, and the use of a Python script makes it easy to quickly run a new forecasting for each new schedule.

The paper discusses the use of ML to predict fishing activity in well operations. The solution presented is simple and easy to understand, but the challenge lies in the availability and size of historical data. The current data is not rich enough, but the idea of sharing data between smaller companies could help consolidate a broader database. The lack of rich data demanded a solution dependent on a single variable, the kind of activity performed during each fishing operation. Despite this, the results obtained with the single input model seem reasonable. In the future, with the enrichment of history data, a multi-variable solution could be used to predict the fishing occurrence rate, considering not only the well's activity but also its features. The solution could be useful for operations forecasting and supporting a program of fishing occurrence prevention. Another intended derivative work is the analysis of the correlation between fishing occurrence and project complexity. The work serves as a guide for those who want to use ML for upstream applications and will be a starting point for further work.

BIBLIOGRAPHY

Filho, Peixoto, Tito, Flavio, and Filardo, Juarez Guaraci. "Machine Learning Applied on Fishing Occurrence Prediction." *Paper Presented at the Offshore Technology Conference Brasil*, Rio de Janeiro, Brazil, October 2019. https://doi.org/10.4043/29700-MS

3.4.4.19 Application-43: Real-Time Drill String Vibration Characterization Using Machine Learning

3.4.4.19.1 Background of the Application

This paper presents a new method for detecting and characterizing downhole shocks and vibrations in real-time during drilling operations, using only surface measurements and ML. The technique uses historical and simulated well data to build a classifying model that can detect and characterize downhole dynamics like stick-slip and lateral shocks. The new technique has been validated using recorded data, showing its capability to accurately detect and characterize downhole dynamics from surface measurements, reducing the need for expensive downhole sensors.

3.4.4.19.2 Description of the Workflow and Machine Learning Model

The paper presents an innovative ML approach for real-time characterization of drillstring vibrations in drilling operations. The core of this methodology is building a classifying model using historical and simulated well data. The system first gathers high-frequency surface measurements, then analyzes the frequency spectrum of the data and extracts relevant features. The next step is to use algorithms (such as SVM or bagged DTs) on the extracted features to build a classifying model. Finally, applying the model to real-time data can help detect vibrations.

3.4.4.19.3 Advantages and Limitations of the Synchronized Algorithm

Advantages:

Cost-Effective: Eliminates the need for expensive downhole sensors, relying instead on surface measurements.

Real-Time Monitoring: Enables immediate detection and characterization of downhole conditions, allowing for timely interventions.

Limitations:

Data Dependency: The accuracy of predictions depends on the quality and relevance of the training data.

Complexity in Interpretation: High-frequency data analysis requires sophisticated algorithms and can be complex to interpret.

3.4.4.19.4 Results and Discussion

This paper introduces a new technique for detecting and characterizing downhole shocks and vibrations during oil drilling operations. The method uses surface measurements and ML, eliminating the need for costly downhole sensors. The technique is trained on historical or simulated data to build a classification model that can recognize stick-slip and lateral shocks in real-time. The system can be used to classify the severity levels of these phenomena, estimate equivalent drilling time, and trigger an automated advisory system, resulting in improved drilling efficiency and tool life duration.

BIBLIOGRAPHY

Millan, Ekaterina, Ringer, Maurice, Boualleg, Riadh, and Li, Denis. "Real-Time Drillstring Vibration Characterization Using Machine Learning." *Paper Presented at the SPE/IADC International Drilling Conference and Exhibition*, The Hague, The Netherlands, March 2019. https://doi.org/10.2118/194061-MS

3.4.4.20 Application-44: Machine Learning Lessons Learnt in Stick-Slip Prediction

3.4.4.20.1 Background of the Application

This paper explores the use of ML algorithms to predict downhole vibrations in drilling operations. The authors analyzed drilling mechanics data from thousands of wells over a period of more than 10 years. The data was prepared by cleaning and correcting for time-dependent nature, imputing missing values, handling outliers and anomalies, and creating variables for better capturing data patterns and time-series dependencies. The authors tested several ML algorithms, including gradient boosting, random forest, and deep learning models, and determined the best algorithm based on the results. The performance of the algorithms was evaluated using a moving window to capture the time-series aspect of the data. The paper provides a discussion on the evaluation of the benefits and drawbacks of each algorithm and ways to improve the accuracy of downhole vibration prediction, with a summary of each finding and recommendations for the industry.

3.4.4.20.2 Description of the Workflow and Machine Learning Model

The data, consisting of time-indexed sensor readings from rig sites, is converted into a fixed length of sequences. Features are then calculated from a rolling window to capture the behavior of changes in the system. Random forest, gradient boosting, and deep learning models are trained with this data. The real-time sensor data is prepared in a similar format, and relevant features are extracted. This processed data is then passed through the trained models for inference, which predicts the severity level of stick-slip events.

3.4.4.20.3 Advantages and Limitations of the Synchronized Algorithm

The use of ensemble models like random forest and gradient boosting is beneficial due to their ability to minimize errors. However, they struggle with missing values in sensor data, leading to reduced accuracy. Deep learning models, particularly

CNN-LSTM hybrids, show promise in predicting severe stick-slip classes, yet they still confuse low, medium, and high classes.

3.4.4.20.4 Results and Discussion

This paper focuses on using ensemble models and deep learning models to predict stick-slip in drilling operations. The data used is multi-variate time-series sensor data, and the stick-slip is divided into four classes. The results show that for a small dataset, the gradient boosting algorithm works well but does not consider the time-series nature of the data and thus has some performance loss. If a larger dataset is available, the CNN-LSTM model is a better candidate for stick-slip classification as it identifies time-dependent relationships in the data. Other methods such as improved feature engineering, use of time series statistical models, and better data preprocessing may also improve the model performance.

BIBLIOGRAPHY

Gupta, Soumya, Chatar, Crispin, and Celaya, José R. "Machine Learning Lessons Learnt in Stick-Slip Prediction." *Paper Presented at the Abu Dhabi International Petroleum Exhibition & Conference*, Abu Dhabi, UAE, November 2019. https://doi.org/10.2118/197584-MS

3.4.4.21 Application-45: Drilling Performance and Data Quality Control with Live Digital Twin

3.4.4.21.1 Background of the Application

The paper describes the use of a dynamic digital well model to optimize drilling parameters and determine actual downhole loads. The implementation of the digital platform improved compliance with desired regimes, ensured safety, and enabled faster decision-making. The digital twin allows for real-time analysis of activities and optimal drilling parameter selection based on specific mechanical energy. Quality control of sensors is important for evaluating loads and identifying downhole friction factor values. Trend analysis of friction factors and drilling parameters can prevent overloads and downhole hazards, reduce NPT, and evaluate well circulation and conditioning activities.

The introduction of a digital ecosystem and dynamic digital twin improved well construction management by accelerating decision-making and minimizing the negative effects of the human factor. A single digital platform allows field and office personnel to work together and know the true downhole loads and optimal operating parameters. A method for assessing the quality of measuring device readings and determining the true WOB optimizes drilling parameters in real-time. The specific mechanical energy is calculated based on effective downhole loads and abnormal increases notify the driller to adjust parameters. Safe corridors and the operational roadmap are dynamically updated according to the current state of the wellbore.

3.4.4.21.2 Description of the Workflow and Machine Learning Model

The paper presents a dynamic digital twin and a digital ecosystem for drilling operations. The digital twin simulates the drilling environment to evaluate operational parameters effectively. The system dynamically adjusts drilling parameters based on real-time analysis, minimizing risks and enhancing efficiency.

3.4.4.21.3 Advantages and Limitations of the Synchronized Algorithm

Advantages:

By continuously monitoring operational parameters, the system minimizes risks associated with drilling and optimizes drilling efficiency. Also, the system significantly accelerates operational responses and decision-making processes.

Limits:

The deployment of such a comprehensive digital system requires extensive training for personnel. While the system offers substantial improvements, it may not fully account for all unexpected downhole conditions and anomalies, potentially leading to operational challenges.

3.4.4.21.4 Results and Discussion

The implementation of a digital platform and dynamic digital twin improved the well construction management process by accelerating operational response and decision-making. The method of assessing the quality of sensor readings allowed optimization of technological parameters during drilling. The ecosystem reduced accidents and complications by 25% and improved efficiency in detecting deviations in sensor measurements. The digital platform enabled real-time drilling control, eliminated the need for constant manual monitoring, and allowed for objective evaluation of work inefficiencies. On-bottom efficiency increased by 15%, and the optimization of off-bottom operations reached 30%. On-line automatic drilling analysis and visualization of statistics on deviations allowed for minimizing the negative influence of the human factor.

BIBLIOGRAPHY

Karpov, Roman Borisovich, Zubkov, Denis Yurjevich, Murlaev, Aleksandr Vitalyevich, and Valiullin, Khaydar Bulatovich. "Drilling Performance and Data Quality Control with Live Digital Twin." *Paper Presented at the SPE Russian Petroleum Technology Conference*, Virtual, October 2021. https://doi.org/10.2118/206527-MS

3.4.4.22 Application-46: Application of Machine Learning for Lithology-on-Bit Prediction Using Drilling Data in Real-Time

3.4.4.22.1 Background of the Application

The paper introduces the concept of hybrid modeling, which integrates analytical correlations with ML algorithms to obtain stable and consistent solutions from different datasets. This approach addresses the limitations of data quality that often affect drilling analysis and leads to unstable and unpredictable results. The authors demonstrate the effectiveness of this approach in improving predictivity in drilling analysis.

3.4.4.22.2 *Description of the Workflow and Machine Learning Model*

The paper presents the development of a hybrid modeling method for lithology-on-bit prediction using real-time drilling data. This approach combines analytical correlations and ML algorithms to derive stable solutions. ML algorithms, specifically random forest and gradient boosting, are chosen for their ability to predict lithologies with minimal delay. A significant focus is also placed on the interpretability and stability of the model, moving away from black-box models toward more understandable and controllable solutions.

3.4.4.22.3 *Advantages and Limitations of the Synchronized Algorithm*

The methodology's advantages include improved prediction accuracy and reduced blind spot distance within reservoirs. The use of hybrid modeling enables a more adaptable and interpretable approach, enhancing the algorithm's applicability in different scenarios. However, there are notable limitations. The method shows high sensitivity to input data quality, and the accuracy significantly drops when the number of training wells is insufficient. The approach also relies heavily on the correct identification of lithology from logging data and the appropriate processing of this data. In short, careful management of data quality is necessary.

3.4.4.22.4 *Results and Discussion*

The paper presents the development of a stable and interpretable ML algorithm for identifying lithology while drilling. The proposed algorithm achieved up to 85% accuracy and can be integrated into geosteering and drilling monitoring software. The paper also discusses the importance of feature selection in model tuning and proposes a hybrid modeling method that integrates stationary analytical variables with ML algorithms to improve model stability and predictability. Finally, the paper suggests further improving model accuracy by stacking with other algorithms and implementing neural networks.

BIBLIOGRAPHY

Zhekenov, Temirlan, Nechaev, Artem, Chettykbayeva, Kamilla, Zinovyev, Alexey, Sardarov, German, Tatur, Olga, Petrakov, Yuriy, and Sobolev, Alexey. "Application of Machine Learning for Lithology-on-Bit Prediction using Drilling Data in Real-Time." *Paper Presented at the SPE Russian Petroleum Technology Conference*, Virtual, October 2021. https://doi.org/10.2118/206622-MS

3.4.4.23 Application-47: Sensitivity Analysis of Factors Controlling the Cement Hot Spot Temperature and the Corresponding Well Depth

3.4.4.23.1 *Background of the Application*

This study describes a new systematic approach to optimize cement formulation for high-pressure, high-temperature wells, which uses an ML and CFD simulation technique to determine the effects of five design variables on the hot spot temperature and its corresponding depth. The results indicate that casing size and cement circulation rate have the greatest impact, and increasing well depth, geothermal gradient, and casing size all increase the hot spot temperature and depth, while increasing flow rate and wellbore size cause a reduction. The study provides practical guidelines for improved cement job design in the field.

3.4.4.23.2 Description of the Workflow and Machine Learning Model

The paper uses a combination of ML and CFD simulation techniques. Specifically, the study investigates the effects of five design variables—wellbore depth, geothermal gradient, casing size, hole size, and flow rate—on the hot spot temperature and its corresponding depth in cement design. The data from CFD simulations were used to train multiple linear regression (MLR) models. These models were then employed to assess how changes in these variables influence the hot spot temperature and depth.

3.4.4.23.3 Advantages and Limitations of the Synchronized Algorithm

The methodology's advantages include the effective use of ML to handle large datasets and make intelligent decisions, which is particularly useful in complex fields like petroleum engineering. And MLR is particularly advantageous for understanding the relative influence of multiple variables and identifying outliers. However, the limitations include a limited accuracy of the regression model predictions due to the limited number of datasets (30 in this study) and computational constraints.

3.4.4.23.4 Results and Discussion

The paper proposes a new approach to improve cement job design by coupling ML and CFD simulation to study the effects of operational variables on the hot spot temperature and depth. Five operational variables (depth, geothermal gradient, casing size, hole size, and flow rate) were studied, and multiple linear regression models were established using the ML method. The results of sensitivity analyses show that the hot spot temperature and depth increase with increasing wellbore depth and geothermal gradient, while they decrease with increasing hole size and flow rate. Increasing casing size has a significant impact on the hot spot temperature and depth. The effects of all five operational variables on the hot spot temperature and depth are synchronized.

BIBLIOGRAPHY

Wang, Rui, Kuru, Ergun, Yan, Yifei, Yang, Xinxiang, and Yan, Xiangzhen. Sensitivity analysis of factors controlling the cement hot spot temperature and the corresponding well depth using a combined CFD simulation and machine learning approach. *Journal of Petroleum Science and Engineering*, 208, Part D, 2022, 109617, ISSN 0920-4105, https://doi.org/10.1016/j.petrol.2021.109617

3.4.4.24 Application-48: Review and Analysis of Supervised Machine Learning Algorithms for Hazardous Events in Drilling Operations

3.4.4.24.1 Background of the Application

The application of supervised ML to investigate dangerous drilling incidents is reviewed and bibliometrically analyzed in this research. Researchers discovered that ANNs were the most often used algorithm, with SVMs, random forests, and deep learning gaining traction. The use of algorithms, their advantages and disadvantages, and input parameters for estimation or detection are all covered in the review. The study also emphasizes the necessity of a sizable, publicly available database including information from many oilfields in order to create ML algorithms that will improve drilling operations.

3.4.4.24.2 Description of the Workflow and Machine Learning Model

The use of supervised ML to investigate dangerous drilling incidents is thoroughly reviewed in this research. According to the analysis, the most widely used algorithm among researchers is ANNs. Furthermore, SVMs, random forests, and deep learning have all become more popular recently. The review looks at how these algorithms were applied, their advantages and disadvantages, and the input parameters that helped the ML algorithms identify. The study highlights the necessity of a sizable, publicly available database containing information from many oil fields in order to create ML algorithms.

3.4.4.24.3 Advantages and Limitations of the Synchronized Algorithm

The advantage of deep learning is that it removes the requirement for manually created engineering features by capturing hierarchical representations of the information in the training data. It is especially well-suited for tasks where human expert analysis is constrained, such as those involving very large datasets and quickly shifting system behaviors. Deep learning's high computing resource requirements, such as training time, memory for data storage, and related power and cost requirements, are a major drawback. The study also points out that platforms like Amazon Web platforms, Google Cloud, Azure, Alibaba Cloud, and Baidu Cloud are accessible to facilitate the use of deep learning in spite of these obstacles.

3.4.4.24.4 Results and Discussion

The paper presents the results of a bibliometric analysis and review of the literature on the use of supervised ML for hazardous events during drilling. The analysis indicates that ANN is the most popular algorithm, and deep learning, random forest, and SVM are gaining momentum. Deep learning is recommended for large amounts of data, while random forest and SVM are recommended for non-large-scale classification problems. The study highlights the need for publicly accessible large databases to aid in the development of ML algorithms and the mitigation of gaps in their usage. The increasing availability of different computing platforms is seen as having aided research in the use of ML.

The detailed review on hazard-related events during drilling concludes that ANN is the most popular algorithm and can be improved by applying fuzzy logic. Proper denoising and exploiting data trend/dynamic nature is recommended to aid ML. Input parameters used in predicting/detecting various hazardous events have been presented, and the probability of their occurrence is influenced by the drilling state/phase. SVM performed better than ANN in some cases, but deep learning is expected to perform better for BD.

The conclusions based on the review on deep learning of drilling parameters are: (i) the availability of computing platforms and the development of deep learning algorithms are expected to lead to better results with BD compared to traditional ML; (ii) deep learning is being used for various drilling parameters, including lithology identification, drilling rig state determination, and generating logging parameters; and (iii) the most commonly used deep learning algorithms are CNN and RNN, including variants such as LSTM-RNN.

The study identified several gaps in the use of ML in drilling data, including the lack of a publicly accessible global database of drilling data, the small size of the datasets used, and the fact that most datasets are not publicly available. In addition, most of the deep learning algorithms used were not primarily customized for drilling activity. The study recommends the creation of a publicly accessible database with data from different oilfields to help researchers and industrial users obtain more generalizable results, perform transfer learning to compensate for a new field with limited data, and encourage the development of ML algorithms in line with the unique needs of drilling data/tasks.

BIBLIOGRAPHY

Osarogiagbon, Augustine Uhunoma, Khan, Faisal, Venkatesan, Ramachandran, and Gillard, Paul. Review and analysis of supervised machine learning algorithms for hazardous events in drilling operations. *Process Safety and Environmental Protection*, 147, 2021, 367–384, ISSN 0957-5820, https://doi.org/10.1016/j.psep.2020.09.038

3.4.4.25 Application-49: Use of Machine Learning and Data Analytics to Detect Downhole Abnormalities While Drilling Horizontal Wells, with Real Case Study

3.4.4.25.1 Background of the Application

The study aimed to develop an intelligent system to predict surface drilling torque and detect downhole abnormalities ahead of time. Three models were trained and tested using actual field data, and the best model was coupled with a classification tool to measure how close an observation is to the predictive trend. The system successfully detected abnormalities in Well-1 and Well-2 before the crew observed any abnormalities. This system could help the drilling crew to take corrective action promptly.

3.4.4.25.2 Description of the Workflow and Machine Learning Model

The approach involved training and testing three models: random forest, ANN, and functional network, using an 80/20 training-to-testing data ratio. Their prediction capability was assessed based on the *R* coefficient and average AAPE. The best model was then used to predict the surface drilling torque on the last day leading up to a stuck pipe incident in Well-1, representing a normal and healthy trend. This model was combined with the Mahalanobis distance as a classification tool to label observations as "NORMAL" or "ANOMAL." The system successfully detected abnormalities in Well-1 and Well-2 several hours before they were observed by the drilling crew.

3.4.4.25.3 Advantages and Limitations of the Synchronized Algorithm

It offered a proactive approach to detecting downhole abnormalities, allowing for early intervention and preventing incidents. The use of ML models, particularly random forest, performs well in predicting surface drilling torque and identifying anomalies. However, the methodology's effectiveness depended on the quality and range of input data, as well as the chosen thresholds for anomaly detection. The Mahalanobis distance used for classification had to be carefully calibrated. This approach required a careful balance between sensitivity to anomalies and the avoidance of false positives.

3.4.4.25.4 Results and Discussion

This work presents an intelligent system that uses ML to predict the surface drilling torque and detect downhole abnormalities in real-time. The system uses actual field data from two wells to train and test a random forest model, which achieves high accuracy in predicting the surface torque. The intelligent system is able to detect anomalies ahead of time, with a continuous alarm being generated in Well-1 9 hours before any abnormality was observed, and in Well-2 7 hours before.

BIBLIOGRAPHY

Alsaihati, A., Elkatatny, S., Mahmoud, A. A., and Abdulraheem, A. (August 28, 2020). "Use of Machine Learning and Data Analytics to Detect Downhole Abnormalities While Drilling Horizontal Wells, With Real Case Study." *ASME Journal of Energy Resources Technology*. April 2021; 143(4): 043201. https://doi.org/10.1115/1.4048070

3.4.4.26 Application-50: Application of Machine Learning Models for Real-Time Prediction of the Formation Lithology and Tops from the Drilling Parameters

3.4.4.26.1 Background of the Application

This study explores the use of ML models (ANN, ANFIS, and FNN) to predict lithology changes and formation tops in real-time while drilling. The study found that the optimized ANN model outperformed ANFIS and FNN in predicting lithology distribution and formation tops with over 98.1% accuracy for both training and testing data. While ANFIS and FNN accurately predicted the top of the carbonate/shale formation in the training data, they did not predict the top of all formations accurately in the testing data.

3.4.4.26.2 Description of the Workflow and Machine Learning Model

The paper focuses on using three ML models: ANNs, ANFIS, and functional neural networks (FNN). These models were trained on datasets consisting of six drilling parameters, and the goal is to predict lithology changes and formation tops in real-time during drilling operations. The study used data from two different wells, with the models first being trained on data from one well and then tested on another. The training involved optimizing the models based on the mean squared deviation and correlation coefficient between the predicted and actual formations, with a focus on real-time prediction.

3.4.4.26.3 Advantages and Limitations of the Synchronized Algorithm

The study highlights the high accuracy of the optimized ANN model in predicting lithology distribution and formation tops: the ANN model achieves over 98.1% accuracy in both training and testing data. This model outperformed both ANFIS and feedforward neural network (FNN) in these tasks, which is due to the ANN's ability to predict formation characteristics in real-time using readily available drilling parameters. However, the limitations are evident in the ANFIS and FNN models, which, despite their accuracy in training data, were less effective in predicting all formation tops accurately in the testing data. This indicates weak potential generalizability and model robustness, especially in varying drilling conditions.

3.4.4.26.4 Results and Discussion

This study evaluated the use of ML models (ANN, ANFIS, and FNN) to predict lithology changes and formation tops while drilling through four different formations. The models were trained on data from Well-A and tested on data from another well (Well-B). The optimized ANN model predicted lithology and formation tops with over 98.1% accuracy in both training and testing data, outperforming the ANFIS and FNN models. Although ANFIS and FNN accurately predicted the carbonate/shale formation top for training data, they were not as accurate in predicting all formations for testing data compared to ANN.

BIBLIOGRAPHY

Mahmoud, Ahmed Abdulhamid, Elkatatny, Salaheldin, and Al-AbdulJabbar, Ahmad. Application of machine learning models for real-time prediction of the formation lithology and tops from the drilling parameters. *Journal of Petroleum Science and Engineering*, 203, 2021, 108574, ISSN 0920-4105, https://doi.org/10.1016/j.petrol.2021.108574

4 AI Applications in Reservoir Engineering and Management

Cenk Temizel, Salih Tutun, Celal Hakan Canbaz, Ekrem Alagoz, Emre Can Dündar, Mufrettin Murat Sari, Luigi Saputelli, and Mehmet Melih Oskay

4.1 INTRODUCTION

The advent of artificial intelligence (AI), machine learning (ML), and deep learning (DL) in the field of reservoir engineering management marks a significant evolution in how oil and gas reservoirs are developed and managed. These technologies have revolutionized traditional workflows by providing advanced tools for data analysis, predictive modeling, and optimization, leading to more efficient and effective reservoir management.

Reservoir engineering, a critical discipline in the oil and gas industry, focuses on maximizing hydrocarbon recovery while ensuring the efficient and sustainable use of resources. It involves the application of engineering principles to the evaluation, development, and production of oil and gas reservoirs. Traditionally, reservoir engineers have relied on physics-based models, empirical correlations, and manual interpretation of data to make decisions about reservoir development. However, the complexity of subsurface environments, coupled with the vast amounts of data generated from reservoirs, often makes these traditional approaches time-consuming and limited in accuracy.

AI, ML, and DL offer powerful solutions to these challenges by enabling the analysis of large, complex datasets, and identifying patterns and relationships that may not be evident through conventional methods. These technologies can be applied across various aspects of reservoir engineering, from reservoir characterization and simulation to well testing, production forecasting, and EOR strategies. By automating routine tasks and providing more accurate predictions, AI and ML help reservoir engineers to make more informed decisions, optimize production, and reduce operational risks.

The application of AI and ML in reservoir engineering management also facilitates real-time decision-making, which is increasingly important in modern reservoir operations. With the ability to continuously analyze incoming data, AI-driven systems can provide real-time insights into reservoir performance, allowing for timely

DOI: 10.1201/9781003617327-4

adjustments to production strategies and improving overall reservoir management. Additionally, these technologies enable the integration of multidisciplinary data, combining geological, geophysical, and engineering information to create more comprehensive and accurate reservoir models.

Various workflows in reservoir engineering management where AI, ML, and DL have significantly contributed will be explored in this chapter. These workflows will be categorized into key areas, including reservoir engineering applications, reservoir simulation works, well test applications, and PVT (Pressure–Volume–Temperature) analysis. Each section will discuss how methodologies have enhanced these workflows, providing examples of their application and the benefits they bring to reservoir management.

Furthermore, an overview of AI, ML, and DL methodologies commonly used in reservoir engineering management will be provided, highlighting their strengths, limitations, and potential future developments. As the oil and gas industry continues to embrace digital transformation, the role of AI, ML, and DL in reservoir engineering management will undoubtedly expand, driving further innovation and efficiency in the development and management of hydrocarbon resources.

4.2 CATEGORIZATION OF WORKFLOWS USED IN RESERVOIR ENGINEERING MANAGEMENT

Reservoir engineering management encompasses a range of workflows aimed at optimizing the development, production, and sustainability of hydrocarbon reservoirs. AI, ML, and DL technologies have introduced new methodologies and tools that enhance these workflows, making them more efficient and effective. This section categorizes the primary workflows in reservoir engineering management where these technologies have been applied, including reservoir engineering applications, reservoir simulation works, well test applications, and PVT applications.

4.2.1 Reservoir Engineering Applications

Reservoir engineering applications involve the analysis and management of reservoirs to maximize hydrocarbon recovery. AI and ML have been transformative in this area by improving the accuracy of reservoir models, optimizing production strategies, and enabling real-time monitoring.

Reservoir Modeling: AI and ML techniques are used to create more accurate and detailed reservoir models by integrating various types of data, such as geological, geophysical, and petrophysical information. ML algorithms can process large datasets to identify patterns and correlations, leading to improved reservoir characterization and modeling.

Production Optimization: AI models analyze production data to identify trends, optimize production rates, and enhance recovery techniques. ML algorithms can predict the impact of different operational strategies on reservoir performance, allowing engineers to make data-driven decisions for optimizing production.

Enhanced Oil Recovery (EOR): AI and ML are applied to design and optimize EOR techniques, such as water flooding, gas injection, and chemical flooding.

By analyzing historical data and simulations, these technologies can identify the most effective EOR strategies and predict their impact on recovery efficiency.

Real-Time Monitoring: AI-driven systems enable continuous monitoring of reservoir performance by analyzing real-time data from sensors and field instruments. This capability allows for timely adjustments to operational strategies, improving overall reservoir management and reducing risks.

4.2.2 Reservoir Simulation Works

Reservoir simulation involves creating computational models to predict the behavior of reservoirs under various scenarios. AI and ML methodologies enhance reservoir simulation by improving accuracy, reducing computational time, and enabling advanced analysis.

Model Calibration: ML algorithms can be used to calibrate reservoir simulation models by comparing simulation results with observed data. This process helps in adjusting model parameters to achieve more accurate predictions.

Surrogate Modeling: AI-driven surrogate models approximate the behavior of complex reservoir simulations with reduced computational cost. These models use ML techniques to predict reservoir responses based on simplified representations, enabling faster scenario analysis and decision-making.

History Matching: AI and ML are employed to improve history matching, the process of adjusting simulation models to match historical production data. ML algorithms can identify the best-fitting models more efficiently than traditional methods, leading to more accurate reservoir predictions.

Scenario Analysis: AI techniques allow for the exploration of multiple scenarios and their impacts on reservoir performance. This capability enables engineers to evaluate different development strategies and operational decisions quickly.

4.2.3 Well Test Applications

Well testing involves evaluating well performance to make informed decisions about well operations and reservoir management. AI and ML technologies enhance well testing by improving data analysis, interpretation, and prediction.

Pressure Transient Analysis: ML algorithms can analyze pressure transient data to determine reservoir properties, such as permeability and storativity. These models can interpret complex pressure data more accurately and efficiently than traditional methods.

Production Testing: AI-driven systems analyze production test data to evaluate well performance and predict future production rates. These systems can identify trends and anomalies in production data, helping engineers optimize well operations.

Well Performance Forecasting: ML models can predict future well performance based on historical test data and operational conditions. These forecasts assist in planning production strategies and managing reservoir resources.

Anomaly Detection: AI techniques are used to detect anomalies in well test data, such as equipment malfunctions or unexpected changes in reservoir behavior. Early detection of these anomalies helps in addressing issues before they impact well performance.

4.2.4 PVT Applications

PVT analysis involves studying the Pressure–Volume–Temperature relationships of reservoir fluids to understand their behavior and optimize production. AI and ML technologies improve PVT analysis by enhancing data interpretation and fluid property prediction.

Fluid Property Prediction: ML algorithms can predict fluid properties, such as viscosity and density, based on the PVT data and other reservoir parameters. These predictions are crucial for designing production processes and optimizing reservoir management.

PVT Data Interpretation: AI techniques are employed to interpret complex PVT data, including phase behavior and fluid composition. These methods can extract valuable insights from the PVT data, aiding in the development of accurate fluid models.

Reservoir Simulation Integration: AI-driven PVT models are integrated with reservoir simulation models to provide a comprehensive understanding of fluid behavior. This integration enhances the accuracy of reservoir simulations and improves production forecasting.

Real-Time PVT Analysis: AI systems enable the real-time analysis of PVT data, allowing for immediate adjustments to production strategies and reservoir management based on current fluid conditions.

In summary, AI, ML, and DL technologies are increasingly being integrated into various workflows in reservoir engineering management. By enhancing reservoir modeling, simulation, well testing, and PVT analysis, these technologies contribute to more efficient and effective reservoir management, ultimately leading to improved hydrocarbon recovery and resource optimization.

4.3 OVERVIEW OF AI METHODOLOGIES

Artificial intelligence (AI), machine learning (ML), and deep learning (DL) have revolutionized the way reservoir engineers analyze, model, and optimize subsurface systems. This section presents a concise overview of these methodologies, outlining their fundamental principles and highlighting their relevance to reservoir engineering management.

AI encompasses the broader field of simulating intelligent behavior in machines, aiming to replicate human-like reasoning and decision-making. Within this domain, machine learning represents a subset focused on algorithms that enable systems to learn from data and improve over time without being explicitly programmed. Supervised, unsupervised, and reinforcement learning are the primary paradigms within ML. These techniques allow for prediction, classification, clustering, and anomaly detection tasks across reservoir datasets. Deep learning, a specialized branch of ML, uses multilayered neural networks to capture complex nonlinear relationships in large volumes of structured or unstructured data. Architectures such as convolutional neural networks (CNNs), recurrent neural networks (RNNs), and their variants have shown remarkable performance in areas like seismic interpretation, permeability prediction, and production forecasting.

The integration of AI in reservoir engineering workflows has enabled a paradigm shift from conventional physics-based models to hybrid and data-driven approaches. Traditional simulation and well test analysis workflows are increasingly enhanced by intelligent algorithms that can reduce computational costs, accelerate history matching, and support real-time decision-making. The versatility of these methods allows their application across diverse domains including reservoir characterization, dynamic modeling, PVT behavior estimation, and enhanced oil recovery optimization. Section 4.4 will illustrate these applications in detail through a range of case studies and state-of-the-art implementations across different reservoir engineering domains.

4.4 APPLICATIONS OF AI

The integration of AI, ML, and DL into reservoir engineering management brings transformative benefits but also faces several challenges. One major challenge is the quality and availability of data. Reservoir engineering relies on diverse data sources, such as well logs, seismic surveys, and production data, which can be inconsistent or incomplete. To address this, efforts must focus on improving data acquisition, enhancing data cleaning processes, and utilizing data augmentation techniques to fill gaps and ensure the reliability of AI-driven models.

Another challenge is the integration of AI models with traditional reservoir engineering workflows. Many AI-driven systems operate independently, which can complicate their incorporation into existing practices. To overcome this, hybrid approaches that blend AI with traditional methods should be developed, alongside creating tools that facilitate seamless data exchange between AI systems and conventional techniques. Additionally, the interpretability of AI models remains a critical issue. Ensuring that these models provide transparent and understandable insights is essential for gaining trust and making informed decisions.

Looking ahead, future research and development should focus on enhancing the efficiency and adaptability of AI technologies. This includes advancing computational techniques to reduce resource requirements and developing adaptive models that can continuously learn from new data. Addressing ethical and regulatory considerations, investing in skill development, and exploring new research areas will also be crucial. By tackling these challenges and embracing innovative solutions, the field of reservoir engineering management can fully leverage the potential of AI, ML, and DL technologies to drive further advancements and efficiencies.

4.4.1 Reservoir Engineering Applications

4.4.1.1 Application-1: Predicting Gas Production Using Machine Learning Methods

4.4.1.1.1 Background of the Application

Predicting well performance and production in the oil and gas industry has historically been challenging, but advancements in big data and ML offer new solutions. This study explores the application of supervised regression methods to predict gas production in undrilled wells. Using data from Jonah Energy's gas field in Sublette County, Wyoming, the authors aimed to forecast well performance and annual gas production during the first production year. They incorporated various geological, drilling, and production data to develop their models. Through extensive data cleansing and conditioning, they tested several ML algorithms, including linear regression, neural network regression, boosted decision tree, and binned decision tree. The binned decision tree model performed best, achieving a coefficient of determination of 0.63 and a squared error of 0.36. This study demonstrates the feasibility and accuracy of ML models in predicting gas production, highlighting the importance of data-driven methodologies.

4.4.1.1.2 Description of the Workflow and Machine Learning Model

This paper introduces an integrated approach combining ML and geostatistical methods to generate stochastic models for reservoir property estimation. Kriging is used for spatial correlation, while Random Forest (RF) constrains simulation with secondary data sources. The approach focuses on capturing uncertainty quantification and feature importance, using ensemble ML algorithms like bagging to enhance conditional distribution estimation.

4.4.1.1.3 Advantages and Limitations of the Synchronized Algorithm

The combined use of ML and geostatistics leverages the strengths of both, allowing for nonlinear trend modeling without extensive prior data transformation. While ensemble methods improve performance by integrating diverse expert opinions, they require careful implementation to avoid overfitting and ensure accurate simulation results. The approach demonstrates improved predictive accuracy and efficiency, although it relies heavily on the quality and comprehensiveness of input data.

4.4.1.1.4 Results and Discussion

The analysis concluded that the Decision Tree model was the most effective for predicting initial gas production, reducing prediction error significantly from 0.95 to 0.3. Further refinement through feature engineering and ensemble methods could potentially enhance accuracy. While neural network regression also provided competitive results, the simplicity and speed of Decision Trees made them preferable. Ensemble methods such as decision forest were used to ensure robustness and avoid overfitting, confirming the decision tree's suitability. This model, leveraging its straightforward implementation and scalability, stands out as the best choice for future deployment in predicting gas production from new wells. The study emphasizes the practicality and reliability of Decision Tree models, especially in datasets with physical constraints.

BIBLIOGRAPHY

Tania Mukherjee, Thomas Burgett, Telha Ghanchi, Colleen Donegan, and Taylor Ward. n.d. Predicting gas production using machine learning methods. *SEG International Exposition and 88th Annual Meeting*. https://doi.org/10.1190/segam2019-3215692.1

4.4.1.2 Application-2: Machine-Learning-Based Methods for Estimation and Stochastic Simulation

4.4.1.2.1 Background of the Application

This paper introduces an innovative approach that integrates ML with traditional geostatistics for the stochastic modeling of reservoir properties. Unlike conventional methods such as sequential simulation, which may be limited in utilizing multiple secondary data sources and generating complex reservoir properties, the proposed approach combines Kriging for spatial correlation with RF for integrating secondary data. This hybrid method not only enhances uncertainty quantification in simulations but also evaluates the importance of various secondary variables.

4.4.1.2.2 Description of the Workflow and Machine Learning Model

In this research, ML techniques including multiple linear regression and neural networks were applied to optimize well completions and designs in the Duvernay shale. The workflow involved analyzing various well completion variables and predicting future performance outcomes. By evaluating data from 262 wells, the study identified key drivers of well performance and areas for cost savings.

4.4.1.2.3 Advantages and Limitations of the Synchronized Algorithm

The integration of multiple linear regression and neural networks allows for the comprehensive analysis of complex datasets, providing insights into well performance drivers. This methodology eliminates biases introduced by excluding wells and achieves high predictive accuracy. However, the approach requires careful management of overfitting risks and may demand substantial computational resources for large datasets. Despite these challenges, the methodology offers significant potential for cost savings and performance improvements.

4.4.1.2.4 Results and Discussion

The study presents a novel method that effectively combines ML and geo-statistics to estimate and simulate reservoir properties. By leveraging ensemble ML and bagging techniques, the approach accommodates non-linear relationships and multiple secondary features without requiring prior data transformation. This method demonstrates promising results in generating stochastic models that adhere to spatial correlation requirements, offering potential applications in realistic reservoir characterization scenarios. Future work involves testing the method on datasets with nonlinear relationships and further refining sensitivity analyses with variogram parameters.

BIBLIOGRAPHY

Arnab Dhara, Whitney Trainor-Guitton, and Ali Tura. n.d. Machine-learning-based methods for estimation and stochastic simulation. *SEG International Exposition and 88th Annual Meeting* https://doi.org/10.1190/segam2018-2991178.1

4.4.1.3 Application-3: Machine Learning Applied to Multi-Well Test Analysis and Flow-Rate Reconstruction

4.4.1.3.1 Background of the Application

Permanent downhole gauges (PDGs) offer a continuous record of flow rate and pressure, valuable for reservoir analysis despite challenges such as variable flow rates, noise, and large data volumes. This study extends ML methods previously applied to single-well pressure analysis to multi-well systems and flow-rate reconstruction. The developed ML models, tested on both real and synthetic datasets, demonstrated flexibility and effectiveness in capturing well interference and reconstructing flow rates.

4.4.1.3.2 Description of the Workflow and Machine Learning Model

This study investigates ML techniques, including K-Nearest Neighbors (KNN), Support Vector Regression, Kernel ridge regression, RF, Adaptive Boosting, and Collaborative Filtering, to predict reservoir fluid properties. Using PVT data from the Niger Delta, models were trained to estimate saturation pressure, formation volume factor, oil viscosity, and condensate gas ratio, demonstrating potential improvements over standard industry correlations.

4.4.1.3.3 Advantages and Limitations of the Synchronized Algorithm

ML techniques offer the ability to learn complex relationships in large datasets without explicit programming. The study found that these methods provided better predictions for certain fluid properties compared to standard correlations. However, the effectiveness of ML depends on the quantity and quality of the data, and techniques like Adaptive Boosting require careful tuning to achieve optimal results.

4.4.1.3.4 Results and Discussion

The research extends the ML framework for single-well pressure analysis to multi-well systems, successfully capturing well interactions and predicting pressure contributions. A new ML model for flow-rate reconstruction showed promising results on synthetic and real data, offering an effective alternative to traditional methods. The study illustrates the adaptability of ML in different applications, confirming its potential as a robust tool for PDG-data analysis.

BIBLIOGRAPHY

Chuan Tian and Roland N. Horne. "Machine Learning Applied to Multiwell Test Analysis and Flow Rate Reconstruction." *Paper Presented at the SPE Annual Technical Conference and Exhibition*, Houston, TX, September 2015. https://doi.org/10.2118/175059-MS

4.4.1.4 Application-4: Reservoir Engineer and Artificial Intelligence Techniques for Data Analysis

4.4.1.4.1 Background of the Application

The paper explores the use of AI techniques to address data transformation challenges in the petroleum industry. It covers data analysis steps such as problem exploration, space analysis, and data preparation, leading to the creation of data mining models. The study includes two applications: identifying potential hydrocarbon-producing zones in the Colombian PETROLEA field using Fuzzy Learning Vector Quantization (FLVQ) and screening EOR methods using a Fuzzy Rule Diagnosis System.

4.4.1.4.2 Description of the Workflow and Machine Learning Model

The paper presents an ML approach using Support Vector Machines (SVMs) with Radial Basis Function (RBF) kernels for recognizing micro-seismic events in underground excavations. SVM is chosen for its ability to handle complex patterns in seismic data, while feature selection and data preprocessing are crucial steps in enhancing model performance.

4.4.1.4.3 Advantages and Limitations of the Synchronized Algorithm

SVM with RBF kernels offers robustness against overfitting and can model complex relationships in seismic data. However, the method requires careful parameter tuning and can be computationally intensive with large datasets. Feature selection is critical to reduce dimensionality and improve accuracy, though it may limit the model's adaptability to new data.

4.4.1.4.4 Results and Discussion

The paper showcases intelligent data analysis techniques for solving reservoir engineering problems using Top-down and Bottom-up strategies. The Bottom-up approach identifies hydrocarbon zones in the PETROLEA field with a 95% accuracy using the FLVQ algorithm. The Top-down approach applies a fuzzy rule diagnosis system for EOR screening. These techniques improve process optimization, cost reduction, and result accuracy, emphasizing the importance of high-quality data and expert knowledge.

BIBLIOGRAPHY

O. Tapias, C. P. Soto, J. Sandoval, H. H. Pérez, and A. Bejarano. "Reservoir Engineer and Artificial Intelligence Techniques for Data Analysis." *Paper Presented at the SPE Asia Pacific Oil and Gas Conference and Exhibition*, Jakarta, Indonesia, April 2001. https://doi.org/10.2118/68743-MS

4.4.1.5 Application-5: Predicting Carbonate Formation Permeability Using Machine Learning

4.4.1.5.1 Background of the Application

Characterizing the permeability of carbonate formations is crucial for simulating subsurface fluid flow, especially since these formations hold a significant portion of the world's hydrocarbon reserves. Traditional methods often require large core samples, which are difficult to obtain. Instead, researchers have used mercury injection capillary pressure (MICP) measurements on smaller samples, like drill cuttings, to infer permeability. However, converting capillary pressure data directly to permeability is challenging due to the need for further interpretations and assumptions about pore structures. This study leverages ML, which does not impose spatial distribution restrictions on pore sizes, to predict permeability from MICP data. Using a dataset of 193 carbonate samples, the authors propose two new predictive models with high accuracy, achieving an R^2 score greater than 0.96 on an average. This data-driven approach offers a robust tool for permeability prediction in carbonate formations.

4.4.1.5.2 Description of the Workflow and Machine Learning Model

This paper examines ML techniques, including neural networks and decision trees, for production forecasting. These methods are used to predict production rates based on historical data and reservoir parameters. The study highlights the importance of feature engineering and model selection in achieving accurate forecasts.

4.4.1.5.3 Advantages and Limitations of the Synchronized Algorithm

ML techniques offer significant potential for improving production forecasting accuracy. They can capture complex relationships in data but require careful feature selection and parameter tuning. The computational requirements of these methods can be substantial, necessitating efficient data handling and processing.

4.4.1.5.4 Results and Discussion

This study demonstrates the efficacy of ML in predicting carbonate formation permeability from MICP measurements. By incorporating both supervised and unsupervised learning algorithms, the researchers developed physics-driven models that outperformed purely mathematical ones. The physics-driven models captured a broad range of permeability values accurately, with an average R^2 score of 0.96, though their accuracy decreased for samples with permeability higher than 100 millidarcies due to limited input data. Despite these limitations, the proposed models provide a quick and reliable method for characterizing formation permeability, underscoring the potential of ML in enhancing the understanding of subsurface properties.

BIBLIOGRAPHY

Huy Tran, Ahmed Kasha, Ibnelwaleed Hussein, A. Sakhaee-Pour. Predicting carbonate formation permeability using machine learning. *Journal of Petroleum Science and Engineering*, 195, 2020, 107581, ISSN 0920-4105, https://doi.org/10.1016/j.petrol.2020.107581

4.4.1.6 Application-6: Reservoir Uncertainty Analysis: The Trends from Probability to Algorithms and Machine Learning

4.4.1.6.1 Background of the Application

Reservoir development has faced the challenge of accurately characterizing reservoir parameters due to subsurface geological complexities. This paper reviews the advances and current practices in reservoir uncertainty modeling, examining statistical, probabilistic, and stochastic methods. The recent incorporation of AI algorithms, such as genetic algorithms (GAs), artificial neural networks (ANNs), and Bayesian networks, is highlighted. These AI methods improve the confidence in reservoir models and parameters, optimizing uncertainty modeling by leveraging ML's pattern recognition capabilities. The paper provides a comprehensive review and analysis of various uncertainty analysis methods, showcasing the benefits of AI in enhancing reservoir uncertainty modeling.

4.4.1.6.2 Description of the Workflow and Machine Learning Model

This paper examines innovative AI applications in reservoir engineering, including DL and reinforcement learning. These techniques are used to optimize reservoir operations and enhance decision-making processes. The study highlights the potential of AI to transform traditional reservoir engineering practices.

4.4.1.6.3 Advantages and Limitations of the Synchronized Algorithm

Innovative AI applications offer significant potential for transforming reservoir engineering, providing improved optimization and decision-making capabilities. However, these methods require substantial computational resources and expertise in model development. The quality of input data is crucial for ensuring reliable and accurate predictions.

4.4.1.6.4 Results and Discussion

The paper concludes that there is no universally correct methodology for reservoir uncertainty analysis, as different methods yield varying results depending on the case. Deterministic methods rely heavily on available data, showing limitations when data is insufficient. Probabilistic approaches consider a broader set of models and scenarios, though they can be time-consuming. AI algorithms significantly reduce modeling and analysis run time while integrating geological and dynamic uncertainties. The AI approach enables the hybridization of deterministic and non-deterministic methods for better reservoir uncertainty modeling, offering a promising future for integrating these advanced techniques into the industry.

BIBLIOGRAPHY

Maureen Ani, Gbenga Oluyemi, Andrei Petrovski, and Sina Rezaei-Gomari. "Reservoir Uncertainty Analysis: The Trends from Probability to Algorithms and Machine Learning." *Paper Presented at the SPE Intelligent Energy International Conference and Exhibition*, Aberdeen, Scotland, UK, September 2016. https://doi.org/10.2118/181049-MS

4.4.1.7 Application-7: The Use of Capacitance–Resistance Models for Rapid Estimation of Waterflood Performance and Optimization

4.4.1.7.1 Background of the Application

The paper presents the capacitance–resistance model (CRM) as an efficient method for evaluating waterflood performance in reservoirs. This semi-analytical approach, grounded in signal processing, utilizes variations in injection rates to predict production responses at producers. The CRM, which uses production and injection rate data along with bottomhole pressure, simplifies and accelerates performance predictions compared to traditional numerical simulations. The model is applied at different control volumes—entire field, individual producer drainage areas, and injector-producer pairs—offering flexibility and speed. Once calibrated with historical data, CRM can be used to optimize oil production by reallocating water injection rates. The study verifies the CRM's accuracy against numerical simulations and demonstrates its ability to maximize oil output in case studies.

4.4.1.7.2 Description of the Workflow and Machine Learning Model

The study explores AI and ML applications in reservoir analysis, focusing on SVMs and neural networks. These methods are used to analyze complex datasets and predict reservoir behavior. The approach emphasizes data integration and feature selection to enhance model performance.

4.4.1.7.3 Advantages and Limitations of the Synchronized Algorithm

AI and ML techniques offer robust tools for reservoir analysis, providing improved accuracy and insights into reservoir behavior. However, these methods require careful data preprocessing and model calibration. Computational demands can be challenging, necessitating efficient data management and processing strategies.

4.4.1.7.4 Results and Discussion

The study introduces three formulations of CRM, providing analytical solutions and physical interpretations of the model parameters. Case studies demonstrate that the calibrated CRMs yield results comparable to those from 3D numerical simulations, rapidly matching historical data and predicting future performance. The models effectively optimize water injection to enhance oil recovery, even accounting for errors in measurement data. The CRM's ability to quickly assess waterflood performance and inform injection strategies offers significant advantages for reservoir management. The study recommends starting with a field-wide CRM and progressively using more detailed models for individual wells and injector–producer pairs.

BIBLIOGRAPHY

M. Sayarpour, E. Zuluaga, C.S. Kabir, Larry W. Lake. The use of capacitance–resistance models for rapid estimation of waterflood performance and optimization. *Journal of Petroleum Science and Engineering*, 69 (3–4), 2009, 227–238, ISSN 0920-4105, https://doi.org/10.1016/j.petrol.2009.09.006

4.4.1.8 Application-8: A Deep-Learning-Based Surrogate Model for Data Assimilation in Dynamic Subsurface Flow Problems

4.4.1.8.1 Background of the Application

This paper introduces a DL-based surrogate model designed to predict dynamic subsurface flow in channelized geological formations. Utilizing deep convolutional and RNN architectures, specifically a residual U-Net and a convolutional long short-term memory (LSTM) network, the surrogate model was trained on global pressure and saturation maps from simulated oil-water flow in numerous (1500) 2D-channelized system realizations. The resulting recurrent R-U-Net model accurately predicts dynamic pressure and saturation maps and well rates for new geological realizations. The model's high accuracy was validated against both individual and ensemble geological realizations. The surrogate model was subsequently employed for data assimilation (history matching) in channelized systems, using the randomized maximum likelihood method and CNN-PCA parameterization for permeability fields. This approach demonstrated a significant reduction in prediction uncertainty, with high-fidelity numerical simulations aligning closely with the surrogate model predictions. The study's results suggest the recurrent R-U-Net model can support more formal posterior sampling methods in realistic scenarios.

4.4.1.8.2 Description of the Workflow and Machine Learning Model

This paper presents an algorithm for detecting multiple partial blockages in liquid pipelines using inverse transient analysis. The method simulates transient flow and analyzes pressure waves to identify blockage locations and severity, integrating finite-volume methods (FVMs) and optimization techniques.

4.4.1.8.3 Advantages and Limitations of the Synchronized Algorithm

The algorithm provides an accurate, nonintrusive method for detecting blockages, but it requires significant computational resources and is sensitive to measurement noise, which can affect accuracy.

4.4.1.8.4 Results and Discussion

The study confirms the effectiveness of the recurrent R-U-Net surrogate model for history matching in subsurface flow problems. Compared to traditional simulation-based methods, the surrogate model dramatically reduces computational cost while maintaining high accuracy in predicting dynamic pressure, saturation fields, and well rates. The randomized maximum likelihood method, combined with the surrogate model, successfully generated posterior geomodels with significantly reduced prediction uncertainty. These posterior models closely matched the high-fidelity simulation results, affirming the surrogate model's reliability. The surrogate-based history matching procedure demonstrated that complex geological features and well connectivity could be accurately captured and predicted, thus facilitating more efficient reservoir management and decision-making. The study highlights the potential of DL-based surrogate models to enable advanced data assimilation techniques in practical subsurface applications.

BIBLIOGRAPHY

Meng Tang, Yimin Liu, Louis J. Durlofsky. A deep-learning-based surrogate model for data assimilation in dynamic subsurface flow problems, *Journal of Computational Physics*, 413, 2020, 109456, ISSN 0021-9991, https://doi.org/10.1016/j.jcp.2020.109456

4.4.1.9 Application-9: A Stochastic Learning-from-Data Approach to the History-Matching Problem

4.4.1.9.1 Background of the Application

History matching is the process of adjusting uncertain reservoir model attributes to align with actual production data. This process is inherently complex and admits multiple solutions, playing a crucial role in reservoir management by improving the reliability of reservoir performance forecasts. The study introduces a stochastic learning-from-data approach to address this problem. This data-driven algorithm integrates components that manage both petrophysical and global uncertain attributes. By utilizing patterns from carefully selected solutions, the algorithm generates new models that aim to match historical data. The approach is validated using the UNISIM-I-H benchmark based on the Namorado Field in Brazil, showing its capability to produce multiple solutions that not only fit historical data but also provide reliable performance forecasts. Compared to existing methods, the proposed approach achieves competitive results with fewer simulations required.

4.4.1.9.2 Description of the Workflow and Machine Learning Model

This paper discusses advanced numerical modeling techniques for simulating multiphase flow and transport in porous media. It explores a range of approaches including the Volume-of-Fluid (VOF) method, the level set (LS) method, and phase field lattice Boltzmann method (LBM). These methods are designed to capture the complex interactions within multiphase systems accurately.

4.4.1.9.3 Advantages and Limitations of the Synchronized Algorithm

The advanced numerical modeling methods presented in this paper provide significant improvements in simulating complex multiphase flow, allowing for more accurate predictions of flow dynamics. However, these methods can be computationally expensive and may require extensive calibration to specific reservoir conditions.

4.4.1.9.4 Results and Discussion

The study presents a novel stochastic learning-from-data method to tackle the history-matching challenge, demonstrating its effectiveness through experiments on a complex synthetic benchmark. The approach consistently generates high-quality solutions that offer improved history matches and reliable future performance predictions. The methodology's strength lies in its ability to handle both petrophysical and global attributes, producing multiple viable solutions efficiently. Future work may explore applying this method to other challenging cases, incorporating additional components to handle categorical attributes, and updating current components with advanced ML techniques such as spectral clustering or LSTM networks. The results affirm the approach's potential in enhancing reservoir management practices through better history matching and forecasting accuracy.

BIBLIOGRAPHY

Cristina C.B. Cavalcante, Célio Maschio, Denis Schiozer, Anderson Rocha. A stochastic learning-from-data approach to the history-matching problem. *Engineering Applications of Artificial Intelligence*. 94, 2020, 103767, ISSN 0952-1976, https://doi.org/10.1016/j.engappai.2020.103767

4.4.1.10 Application-10: Compressibility Predictions Using Digital Thin-Section Images of Rocks

4.4.1.10.1 Background of the Application

In this study, numerical simulations are employed on petrographically characterized thin-section images to predict the three-dimensional (3D) elastic compressibility of sandstones. The research focuses on predicting two significant statistics of compressibility curves under uniaxial boundary conditions in laboratory settings: the minimum compressibility at 1500 psi (10 MPa) depletion stress and the maximum compressibility. A novel Digital Rock workflow is introduced, which uses a segmented two-dimensional (2D) thin-section image to simulate stress fields and predict compressibility. The study also establishes linear and non-linear relationships between log base 10 compressibility and in-situ porosity, allowing for compressibility predictions in the absence of laboratory measurements or 2D images. The application of these methods to samples from different fields with laboratory measurements showed that the Digital Rock workflow is most effective for predicting minimum compressibility, while the proposed non-linear relationship with in-situ porosity is best for predicting maximum compressibility. These predictions can assist in making more informed economic decisions during field development planning.

4.4.1.10.2 Description of the Workflow and Machine Learning Model

This study presents a Digital Rock workflow to predict compressibility using 2D thin-section images of rocks. The workflow involves numerical simulations to estimate stresses and predict key compressibility metrics, such as minimum and maximum compressibility.

4.4.1.10.3 Advantages and Limitations of the Synchronized Algorithm

The Digital Rock workflow offers a novel approach for predicting rock compressibility, which is critical for reservoir management. While it shows good agreement with laboratory measurements, the approach is limited by the quality of the input images and the complexity of the simulations required.

4.4.1.10.4 Results and Discussion

This work developed methods for predicting the compressibility of sandstones at a depletion stress of 1500 psi (10 MPa) and the maximum compressibility of compressibility curves obtained from laboratory measurements. Empirical relationships between log base 10 compressibility and porosity were established, with the non-linear relationship showing better correlation than the linear one. The non-linear relationship includes geological information, such as rock sorting, impurities cementation factor, and can be calibrated with laboratory measurements specific to a field or basin. Additionally, a Digital Rock workflow was developed to predict compressibility using petrographically characterized thin-section images. The application of these methods to various samples indicated that the Digital Rock workflow provided the best predictions for minimum compressibility, while the non-linear relationship with porosity was most effective for maximum compressibility. Numerical simulations on thin-section images also provided valuable insights into the rock's behavior under stress, highlighting areas with high stress concentrations that are not easily inferred from the images alone.

BIBLIOGRAPHY

Vishal Das, Nishank Saxena, Ronny Hofmann. Compressibility predictions using digital thin-section images of rocks. *Computers & Geosciences*, 139, 2020, 104482, ISSN 0098-3004, https://doi.org/10.1016/j.cageo.2020.104482

4.4.1.11 Application-11: Applications of Smart Proxies for Subsurface Modeling

4.4.1.11.1 Background of the Application

In this study, authors developed and validated smart proxy models using AI and ML techniques for reservoir simulation tasks such as history matching, sensitivity analysis, and uncertainty assessment. Proxy models that significantly reduce computational time while maintaining high accuracy were created by employing ANNs. The models were tested on two case studies: one for calibrating a reservoir simulation model to predict well production and another for the fast-track modeling of CO_2 EOR to predict reservoir pressure and phase saturation distribution. The results demonstrate that the smart proxy models can achieve accurate predictions while saving approximately 98.9% of computational time compared to traditional numerical simulation models, highlighting the effectiveness of the proposed workflow for various subsurface modeling applications.

4.4.1.11.2 Description of the Workflow and Machine Learning Model

In this paper, ML techniques are applied to predict the compressibility of porous media. The study uses various regression models, including SVMs and neural networks, to analyze the relationship between porosity and compressibility.

4.4.1.11.3 Advantages and Limitations of the Synchronized Algorithm

ML models provide an efficient method for predicting compressibility in porous media. However, the accuracy of these models depends heavily on the quality and quantity of training data. Additionally, the models may require significant computational resources for training and prediction.

4.4.1.11.4 Results and Discussion

This study demonstrates that ML-based proxy models can substantially improve the efficiency of reservoir simulation models, offering a fast and accurate alternative for various tasks such as sensitivity analysis, production optimization, and uncertainty assessment. By following a structured approach to data preprocessing, feature selection, and model development, proxy models can accurately replicate the results of complex reservoir simulations with significantly reduced computational costs. The two case studies presented underscore the versatility and robustness of proxy models for subsurface applications, though the specific steps may vary depending on the problem at hand. The findings confirm the potential of these models to enhance the performance of reservoir management workflows, ultimately supporting more effective and efficient decision-making processes in the oil and gas industry.

BIBLIOGRAPHY

Alireza Shahkarami and S. Mohaghegh. Applications of smart proxies for subsurface modeling, *Petroleum Exploration and Development*, 47 (2), 2020, 400–412, ISSN 1876-3804, https://doi.org/10.1016/S1876-3804(20)60057-X

4.4.1.12 Application-12: Predicting Porosity, Permeability and Water Saturation Applying an Optimized Nearest-Neighbour, Machine-Learning and Data-Mining Network of Well-Log Data

4.4.1.12.1 Background of the Application

This paper presents a novel methodology for predicting petrophysical properties such as permeability (k), water saturation (Sw), and effective porosity (EP) from well log data using a transparent open box (TOB) learning network. The approach leverages a standardized well log representation to facilitate accurate predictions and detailed data mining. The TOB algorithm, a sophisticated data-matching technique, is optimized to handle complex, non-linear datasets typical of well logs. By combining screening-level sampling with detailed zoomed-interval analysis, the method achieves high prediction accuracy. This comprehensive system not only provides precise petrophysical predictions but also allows for the transparent inspection of the data, enhancing error analysis and revealing influential data records and intervals.

4.4.1.12.2 Description of the Workflow and Machine Learning Model

The study focuses on estimating the elastic properties of rocks using digital rock physics methods. It involves the use of micro-computed tomography (micro-CT) imaging combined with numerical simulations to predict properties such as bulk modulus and shear modulus.

4.4.1.12.3 Advantages and Limitations of the Synchronized Algorithm

Digital rock physics provides a powerful tool for estimating the elastic properties of rocks, offering high accuracy and detailed insights. However, the method is limited by the resolution of the micro-CT images and the computational resources required for simulations.

4.4.1.12.4 Results and Discussion

This research introduces an advanced standardized well log network, optimized with the TOB learning network, to accurately predict petrophysical metrics. The method demonstrates its effectiveness through a case study on the Hassi R'Mel gas field, achieving significant prediction accuracy improvements. The TOB network, configured for both broad screening and detailed interval analysis, provides reliable and transparent predictions of permeability, water saturation, and porosity. The system surpasses traditional ML tools by offering comprehensive data mining capabilities and clear insight into prediction contributions. The findings suggest that the TOB methodology could serve as a benchmark for evaluating well logs and enhancing petrophysical evaluations in various geological settings.

BIBLIOGRAPHY

David A. Wood. Predicting porosity, permeability and water saturation applying an optimized nearest-neighbor, machine-learning and data-mining network of well log data. *Journal of Petroleum Science and Engineering*, 184, 2020, 106587, ISSN 0920-4105, https://doi.org/10.1016/j.petrol.2019.106587

4.4.1.13 Application-13: Robust Optimization of the Locations and Types of Multiple Wells Using CNN Based Proxy Models

4.4.1.13.1 Background of the Application

The paper focuses on optimizing the placement and types of multiple wells under geological uncertainty using CNN-based proxy models. The primary challenge in well placement optimization is the high computational cost associated with full reservoir simulations. To address this, the authors incorporate physical information, specifically streamline time of flight (TOF) maps, into the training data for CNN, significantly enhancing the model's predictive accuracy. The developed proxy models demonstrate a strong correlation with full reservoir simulation results in both 2D and 3D models, achieving determination coefficients of 0.934 and 0.932, respectively. Integrating these proxy models with a particle swarm optimization (PSO) algorithm allows for robust well placement optimization while drastically reducing computation time to 19.3% and 7.7% of the original in the 2D and 3D fields, respectively.

4.4.1.13.2 Description of the Workflow and Machine Learning Model

This paper presents a DL approach for predicting reservoir properties. CNNs and recurrent neural networks (RNNs) are employed to analyze seismic and well log data, offering enhanced accuracy in property predictions.

4.4.1.13.3 Advantages and Limitations of the Synchronized Algorithm

DL algorithms provide powerful tools for capturing complex patterns in reservoir data, leading to more accurate predictions. However, these models require substantial computational resources and large datasets for effective training, and are prone to overfitting if not carefully managed.

4.4.1.13.4 Results and Discussion

The study introduces a novel proxy model leveraging streamline TOF maps and CNNs to accurately predict Net Present Value (NPV) for various well placement scenarios across multiple reservoir models. The inclusion of TOF maps enhances the model's ability to capture fluid dynamics efficiently, making it a valuable tool for complex fields. The well placement optimization, conducted using PSO and the proposed proxy model, demonstrates promising results in both 2D and 3D fields, maintaining high accuracy and significantly reducing computational costs. Specifically, the optimization process is streamlined to 19.3% of the computational time in 2D fields and 7.7% in 3D fields compared to full reservoir simulations. This computational efficiency, particularly evident in the 3D field, underscores the proxy model's potential for practical applications in large-scale, heterogeneous reservoir optimization.

BIBLIOGRAPHY

Joonyi Kim, Hyungjun Yang, Jonggeun Choe. Robust optimization of the locations and types of multiple wells using CNN based proxy models. *Journal of Petroleum Science and Engineering*, 193, 2020, 107424, ISSN 0920-4105, https://doi.org/10.1016/j.petrol.2020.107424

4.4.1.14 Application-14: A Comparative Analysis of Bubble Point Pressure Prediction Using Advanced Machine Learning

4.4.1.14.1 Background of the Application

Accurate estimation of bubble point pressure is crucial in petroleum reservoir management due to its impact on reservoir performance and various engineering applications. This study evaluates the performance of advanced ML models, including Extreme Gradient Boosting (XGBoost), LightGBM, Random Forest Regressor, and Multi-Layer Perceptron (MLP) neural network, in predicting bubble point pressure. Additionally, a novel super learner model, which combines multiple ML algorithms to enhance prediction accuracy, is employed. The study utilizes three datasets with different predictors: compositional data, bulk properties (Gas–Oil Ratio, gas gravity, American Petroleum Institute (API) gravity, and reservoir temperature), and a combination of both. Results indicate that the ML models, particularly the super learner, outperform traditional empirical correlations, offering accurate predictions even with limited input data.

4.4.1.14.2 Description of the Workflow and Machine Learning Model

This paper explores the application of AI techniques, such as DL and ensemble methods, in the numerical simulation of complex reservoir systems. The study focuses on improving the accuracy and efficiency of simulations by integrating AI-driven models with traditional reservoir simulation approaches.

4.4.1.14.3 Advantages and Limitations of the Synchronized Algorithm

AI techniques offer significant advantages in enhancing the accuracy of reservoir simulations, particularly in complex environments. However, these methods require substantial computational resources and careful integration with existing simulation models to avoid inconsistencies.

4.4.1.14.4 Results and Discussion

In this research, three datasets containing compositional data, bulk properties, and a combination of both were prepared to predict bubble point pressure using advanced ML algorithms. Tree-based models (XGBoost, LightGBM, Random Forest Regressor) and MLP neural networks were applied, with the super learner model significantly enhancing robustness and accuracy. The study found that the super learner consistently provided the most accurate predictions across all datasets, especially when both compositional data and bulk properties were used. While the tuned Dindoruk and Christman correlation performed well, it did not match the precision of the ML models. The findings suggest that using super learners with a combination of tree-based and MLP algorithms is highly effective for global bubble point pressure prediction, providing a reliable and accurate alternative to traditional empirical methods.

BIBLIOGRAPHY

Xi Yang, Birol Dindoruk, Ligang Lu. A comparative analysis of bubble point pressure prediction using advanced machine learning algorithms and classical correlations. *Journal of Petroleum Science and Engineering*, 185, 2020, 106598, ISSN 0920-4105, https://doi.org/10.1016/j.petrol.2019.106598

4.4.1.15 Application-15: A Data-Driven Approach to Predict Compressional and Shear Wave Velocities in Reservoir Rocks

4.4.1.15.1 Background of the Application

Compressional (Vp) and shear (Vs) wave velocities are crucial for understanding reservoir characteristics such as lithology, porosity, fluid content, and mechanical properties of rocks. Obtaining these velocities can be challenging and expensive when relying solely on direct measurements from well logs. This study investigates the potential of intelligent systems to predict Vp and Vs in reservoir rocks using data from conventional well logs. The authors utilized three advanced AI models—ANN, Adaptive Neuro-Fuzzy Inference System (ANFIS), and Least Square Support Vector Machine (LSSVM)—to develop predictive models. The dataset for this research was sourced from the Norwegian North Sea Basin. The performance of the models was assessed using Pearson's correlation coefficient (R^2) and root mean square error (RMSE). Among the models, LSSVM demonstrated the highest accuracy, outperforming traditional methods and providing precise predictions for both Vp and Vs. The findings highlight the effectiveness of data-driven approaches for reservoir characterization and operational decision-making in the petroleum industry.

4.4.1.15.2 Description of the Workflow and Machine Learning Model

This research applies ML models to estimate porosity and permeability in reservoir rocks. Techniques such as neural networks and gradient boosting machines are used to analyze geological data and predict these critical reservoir properties.

4.4.1.15.3 Advantages and Limitations of the Synchronized Algorithm

ML models provide effective tools for estimating porosity and permeability, offering high accuracy and the ability to handle large datasets. However, the models require careful validation and may be sensitive to the quality of the input data, which can impact their reliability.

4.4.1.15.4 Results and Discussion

This research explored the application of three distinct data-driven models—ANN, ANFIS, and LSSVM—to predict compressional and shear wave velocities using well log data. The dataset, sourced from the Volve field in the Norwegian North Sea, confirmed that Vp and Vs can be accurately predicted using conventional well log parameters. A Comparative analysis revealed that the LSSVM model surpassed ANN and ANFIS in predictive accuracy, achieving the highest R^2 and the lowest RMSE values. The intelligent models developed in this study significantly outperformed existing empirical correlations and regression equations. The results underscore the superiority of data-driven approaches for predicting geomechanical properties across various data ranges, independent of geological settings and depositional environments. Implementing such advanced models is crucial for enhancing the precision of reservoir characterization, ultimately aiding geoscientists, reservoir engineers, and drilling operations.

BIBLIOGRAPHY

Teslim Olayiwola, Oluseun A. Sanuade. A data-driven approach to predict compressional and shear wave velocities in reservoir rocks. *Petroleum*, 7 (2), 2021, 199–208, ISSN 2405-6561, https://doi.org/10.1016/j.petlm.2020.07.008

4.4.1.16 Application-16: Artificial Neural Networks in the Domain of Reservoir Characterization: A Review from Shallow to Deep Models

4.4.1.16.1 Background of the Application

In recent times, ML methods have gained popularity across various engineering domains, including reservoir characterization. Among these methods, ANNs stand out for their success in many fields, including the estimation of reservoir properties using geological and geophysical data. By analyzing the correlation between different data sources, ANN models can predict reservoir properties by understanding complex nonlinear relationships within the input data. The evolution of ANN models from shallow to deep architectures has enhanced their capability, leading to the development of hybrid models that combine ANN with other soft computing methodologies. DL models, in particular, are gaining traction due to their ability to automatically extract features, handle high-dimensional data, and emulate human brain-like problem-solving by learning through multiple levels of abstraction. This review explores the application and evolution of ANN in reservoir characterization, highlighting the significant impact and future potential of these models in intelligent interpretation and oil reservoir analysis.

4.4.1.16.2 Description of the Workflow and Machine Learning Model

The study explores AI-enhanced strategies for reservoir management, focusing on the integration of ML algorithms to optimize production and improve decision-making processes. Techniques such as reinforcement learning and BO are applied to develop more effective management plans.

4.4.1.16.3 Advantages and Limitations of the Synchronized Algorithm

AI-enhanced strategies offer significant potential for improving reservoir management, particularly in terms of optimizing production and reducing operational risks. However, the success of these strategies depends on the availability of high-quality data and the robustness of the AI models used.

4.4.1.16.4 Results and Discussion

The comprehensive survey underscores the growing importance of ANN approaches in reservoir characterization, reflecting their evolution over the past 25 years. Various ANN models, from shallow networks to advanced DL architectures, have been applied to predict reservoir properties with considerable success. The survey reveals that while conventional ANN models are effective, they require careful tuning of hyperparameters and feature selection to handle complex reservoir data efficiently. Hybrid models, combining ANN with techniques like fuzzy logic and GAs, have shown promise in addressing the limitations of traditional ANN, such as overfitting and local minima issues. The emergence of DL models marks a significant advancement, offering enhanced capability in feature extraction and handling high-dimensional data. Despite the challenges, the future of ANN in reservoir characterization looks promising, with continuous improvements expected in modeling efficiency and accuracy, driven by the integration of more sophisticated ML techniques and better data preprocessing methods.

BIBLIOGRAPHY

Pallabi Saikia, Rashmi Dutta Baruah, Sanjay Kumar Singh, Pradip Kumar Chaudhuri. Artificial Neural Networks in the domain of reservoir characterization: A review from shallow to deep models, *Computers & Geosciences*, 135, 2020, 104357, ISSN 0098-3004, https://doi.org/10.1016/j.cageo.2019.104357

4.4.1.17 Application-17: Lessons for Machine Learning from the Analysis of Porosity–Permeability Transforms for Carbonate Reservoirs

4.4.1.17.1 Background of the Application

For reservoir characterization, it is essential to predict permeability from porosity measurements in heterogeneous carbonate reservoir facies. Jerry Lucia's 1995 model is important for establishing porosity–permeability transforms for carbonate lithologies and connecting them to rock fabric. ML models have recently been put out as a substitute method for permeability prediction and rock fabric classification. In addition to enhanced versions of traditional algorithms like ordinary least squares (OLS), ML makes use of methods like gradient boosting regressors, RFs, support vector machines, and neural networks. Large datasets can be processed by these algorithms, which enable quick pattern recognition, predictive inference, and estimation of the relative significance of variables influencing certain results. Despite their accuracy, ML models are often viewed as "Black Boxes," making their predictions difficult to interpret and understand by end-users.

4.4.1.17.2 Description of the Workflow and Machine Learning Model

This paper discusses AI-driven techniques for fracture characterization in reservoirs, employing ML algorithms such as support vector machines and neural networks to analyze seismic and well log data. The goal is to improve the accuracy of fracture detection and characterization.

Advantages and Limitations of the Synchronized Algorithm

AI-driven techniques provide powerful tools for fracture characterization, offering improved accuracy and efficiency compared to traditional methods. However, these techniques require extensive data preprocessing and may be sensitive to noise and data quality, which can impact their performance.

4.4.1.17.3 Results and Discussion

This work created methods to improve the transparency and resilience of ML investigations of reservoir facies and related petrophysical characteristics, especially permeability. Characterizing the dataset to comprehend its statistical distribution is the initial stage, and porosity–permeability are examined using histogram plots. Transformations like the Box-Cox transform are advised because many ML algorithms require data to have roughly normal distributions. Following data processing, porosity–permeability regularized linear regression was carried out and verified using test-training splits and cross-validation. The work brought attention to the problem of heteroscedasticity, which can negatively affect ML algorithms and is indicated by variations in the distribution of regression residuals. The analysis concluded that lithofacies interpretations do not significantly improve permeability predictions over simple porosity-only relations and emphasized the importance of understanding the data's nature to ensure accurate ML model predictions.

BIBLIOGRAPHY

F. Jerry Lucia. "Rock-fabric/petrophysical classification of carbonate pore space for reservoir characterization." *AAPG Bulletin* 79(9) (1995): 1275–1300.

Frank Male, Ian J. Duncan. Lessons for machine learning from the analysis of porosity-permeability transforms for carbonate reservoirs. *Journal of Petroleum Science and Engineering*, 187, 2020, 106825, ISSN 0920-4105, https://doi.org/10.1016/j.petrol.2019.106825

4.4.1.18 Application-18: An Integrated Closed-Loop Solution to Assisted History Matching and Field Optimization with Machine Learning Techniques

4.4.1.18.1 Background of the Application

This study addresses the significant computational demands associated with traditional petroleum reservoir modeling workflows, including reservoir characterization, history matching, and field development optimization. The approach integrates various optimization techniques, ML algorithms, and model-order reduction methods to create an efficient closed-loop workflow. By employing BO for history matching and comparing several optimization algorithms such as PSO, genetic algorithm (GA), and their hybrid (GSO), authors aim to reduce computational costs and improve efficiency. Proxy models are developed using multivariate regression, ANNs, and extreme gradient boosting to replace full-physics simulations, thereby significantly decreasing the CPU time. Additionally, the parameter space is reduced using the Karhunen–Loève expansion, making the problem more manageable. This integrated workflow is demonstrated using data from the South Cowden reservoir, where waterflooding has been implemented.

4.4.1.18.2 Description of the Workflow and Machine Learning Model

The paper presents AI-assisted techniques for downhole flow measurement, focusing on the integration of ML algorithms to enhance the accuracy of flow rate predictions. These techniques are particularly useful in challenging environments where traditional methods fall short.

Advantages and Limitations of the Synchronized Algorithm

AI-assisted techniques improve the accuracy of downhole flow measurements, but they require significant computational power and expertise in ML to implement effectively.

4.4.1.18.3 Results and Discussion

This research presents a comprehensive integrated closed-loop solution for reservoir characterization, history matching, and production optimization using ML techniques. Key accomplishments include:

1. Efficient History Matching: Implementation of BO for history matching, achieving satisfactory results within computational limitations.

2. Model Order Reduction: Utilization of Karhunen–Loève expansion to reduce problem complexity and facilitate efficient workflow execution.
3. Proxy Flow Model Development: Construction of highly accurate proxy models using ANN and XGB, with ANN showing superior accuracy with larger datasets, while XGB performs better with limited data.
4. Optimization of Well Controls: Application of PSO, GA, and GSO algorithms to optimize well controls, demonstrating varying levels of effectiveness.

The integrated approach merges static and dynamic parameters within a closed-loop system, improving reservoir management processes. Future studies will extend these methods to full 3D models and explore more advanced model order reduction techniques to further decrease the computational time. Additionally, proxy models could be incorporated into history matching processes to enhance efficiency further. Potential challenges such as the applicability of Kernelized Least Squares Estimation (KLE) to large reservoirs and the cost of generating training datasets for proxy models need to be addressed in future studies.

BIBLIOGRAPHY

Zhi Chai, Azor Nwachukwu, Yevgeniy Zagayevskiy, Shohreh Amini, Srinath Madasu. An integrated closed-loop solution to assisted history matching and field optimization with machine learning techniques, *Journal of Petroleum Science and Engineering*, 198, 2021, 108204, ISSN 0920-4105, https://doi.org/10.1016/j.petrol.2020.108204

4.4.1.19 Application-19: Application of Machine Learning Techniques for Selecting the Most Suitable Enhanced Oil Recovery Method

4.4.1.19.1 Background of the Application

The paper addresses the necessity of a reliable method for selecting the most suitable EOR method in reservoir engineering. ML techniques are highlighted as effective tools for initial EOR screening. The study develops various ML models using a comprehensive database of over 1000 global EOR projects. These models include shallow and deep ANNs, Naïve Bayes (NB), Decision Tree (DT), Random Forest (RF), and Principal Component Analysis (PCA). The RF and deep ANN models achieved the highest accuracy at 90%, outperforming other methods. The study also discusses the limitations and challenges of ML in EOR screening, recommending that ML results be cross-checked with other methods or expert knowledge.

4.4.1.19.2 Description of the Workflow and Machine Learning Model

This research investigates the application of ML models, such as SVMs and deep neural networks, to predict multiphase flow in oil and gas reservoirs. The study highlights the advantages of these models in handling complex fluid dynamics.

4.4.1.19.3 Advantages and Limitations of the Synchronized Algorithm

ML applications provide robust tools for predicting multiphase flow, but their success depends on the quality of the training data and the complexity of the model. These methods also require extensive computational resources.

4.4.1.19.4 *Results and Discussion*

The study provides a thorough investigation into the application of ML for selecting appropriate EOR methods and discusses the associated challenges and limitations. Various ML models were developed and tested on a large dataset of EOR projects. The RF and deep ANN models showed the best performance, achieving an average accuracy of 90%. While ML offers a fast and unbiased screening tool, it should not be the sole method for prediction due to inherent limitations such as data imbalance and lack of comprehensive input features. Future research should address these limitations to enhance the reliability of ML-based EOR screening.

BIBLIOGRAPHY

Yasaman Cheraghi, Shahin Kord, Vahid Mashayekhizadeh. Application of machine learning techniques for selecting the most suitable enhanced oil recovery method; challenges and opportunities. *Journal of Petroleum Science and Engineering*, 205, 2021, 108761, ISSN 0920-4105, https://doi.org/10.1016/j.petrol.2021.108761

4.4.1.20 Application-20: An Efficient Bi-Objective Optimization Workflow Using the Distributed Quasi-Newton Method and Its Application to Well-Location Optimization

4.4.1.20.1 *Background of the Application*

In the field of petroleum engineering, solving multi-objective optimization problems, such as those involving complex reservoir simulations, can be computationally intensive. Traditional optimization methods often struggle due to the high cost of evaluating objective functions without benefiting from gradient information. This paper introduces a novel workflow utilizing the distributed quasi-Newton (DQN) method, which is both parallelized and derivative-free, to efficiently address bi-objective optimization problems. By sharing simulation results across different optimization tasks, the DQN method optimizes a weighted average of two objectives concurrently. This approach is validated on synthetic examples and tested on well location optimization problems, showing significant improvements in efficiency and robustness compared to other methods.

4.4.1.20.2 *Description of the Workflow and Machine Learning Model*

This study evaluates AI-based methods for optimizing production in oil and gas reservoirs, focusing on the use of reinforcement learning and GAs. The research aims to enhance production efficiency by optimizing well placement and production strategies.

4.4.1.20.3 *Advantages and Limitations of the Synchronized Algorithm*

AI-based methods offer significant potential for production optimization, but they require careful tuning and validation to avoid suboptimal solutions. Additionally, the computational demands of these methods can be high.

4.4.1.20.4 *Results and Discussion*

This paper presents an efficient bi-objective optimization workflow using the DQN method for field development optimization. The DQN method demonstrates the ability to construct Pareto fronts accurately and efficiently for problems with analytical objectives. Key advantages include shared information among optimization threads and the reuse of simulation results to optimize multiple objectives concurrently. The performance of the DQN method is comparable to, and sometimes surpasses, the best existing methods like Bi-level Model Adaptive Sampling Design (BiMADS++). Future improvements could include adaptive weighting factors for threads and employing support-vector regression to estimate sensitivity matrices, potentially further reducing the number of simulations required for convergence.

BIBLIOGRAPHY

Yixuan Wang, Faruk Alpak, Guohua Gao, Chaohui Chen, Jeroen Vink, Terence Wells, and Fredrik Saaf. "An Efficient Bi-Objective Optimization Workflow Using the Distributed Quasi-Newton Method and Its Application to Well-Location Optimization." *SPE J* 27 (2022): 364–380. https://doi.org/10.2118/203971-PA

4.4.1.21 Application-21: Numerical Comparison between ES-MDA and Gradient-Based Optimization for the History Matching of Reduced Reservoir Models

4.4.1.21.1 Background of the Application

This study explores and compares two methods for history matching in reservoir models: Ensemble Smoother with Multiple Data Assimilation (ES-MDA) and a gradient-based optimization method. These methods were applied to reduced models based on the GPSNet-type framework for waterflooding scenarios. Utilizing the MATLAB Reservoir Simulation Using the MRST toolbox, both techniques were implemented and their performance evaluated in terms of history-matching accuracy and computational efficiency. The findings highlight the strengths of each approach, with ES-MDA being effective when precise gradient information is unavailable, and gradient-based optimization proving highly efficient when exact gradients are accessible.

4.4.1.21.2 Description of the Workflow and Machine Learning Model

The paper explores the optimization of reservoir management using AI techniques, including ML and DL algorithms. The focus is on improving decision-making processes and optimizing production strategies.

4.4.1.21.3 Advantages and Limitations of the Synchronized Algorithm

AI techniques provide powerful tools for optimizing reservoir management, but their success depends on the availability of high-quality data and the robustness of the algorithms used. Additionally, these methods require significant computational resources.

4.4.1.21.4 Results and Discussion

The study compared the performance of Ensemble Smoothing Model-Data Assimilation (ES-MDA) and gradient-based optimization methods for history matching in reduced reservoir models. Gradient-based optimization, using exact gradients, demonstrated superior accuracy and computational efficiency. However, when exact gradients were not available, ES-MDA provided competitive results, offering a viable alternative with reasonable computational demands. ES-MDA performance was found to be sensitive to ensemble size and number of iterations, enabling tuning according to available computational resources. Overall, the results suggest that both methods have distinct advantages and can be effectively used depending on the specific requirements and constraints of the reservoir modeling task.

BIBLIOGRAPHY

M. Reverón A. Borregales, H. H. Holm, O. Møyner, S. Krogstad, and Lie, K.-A. "Numerical Comparison Between ES-MDA and Gradient-Based Optimization for History Matching of Reduced Reservoir Models." *Paper Presented at the SPE Reservoir Simulation Conference, On-Demand*, October 2021. https://doi.org/10.2118/203975-MS

4.4.1.22 Application-22: History Matching Complex 3D Systems Using Deep Learning-Based Surrogate Flow Modeling and CNN-PCA Geological Parameterization

4.4.1.22.1 Background of the Application

This study explores the use of DL techniques for geological parameterization and fast surrogate flow modeling, which can significantly improve the feasibility of rigorous history matching algorithms. Principal Component Analysis (PCA) combined with CNN-PCA for geological parameterization and a recurrent residual U-Net for surrogate flow modeling are incorporated into three history matching procedures: Rejection Sampling (RS), Randomized Maximum Likelihood with Mesh Adaptive Direct Search Optimization (MADS-RML), and ensemble smoother with multiple data assimilation (ES-MDA). The results demonstrate that both MADS-RML and ES-MDA can produce history matching results comparable to RS, with MADS-RML showing greater accuracy but higher computational demand. This framework can be used to enhance a variety of history matching methods.

4.4.1.22.2 Description of the Workflow and Machine Learning Model

This research discusses AI-driven approaches to enhance oil recovery, focusing on the application of neural networks and reinforcement learning to optimize recovery processes. The study emphasizes the potential of AI in improving recovery rates and reducing operational costs.

4.4.1.22.3 Advantages and Limitations of the Synchronized Algorithm

AI-driven approaches offer promising solutions for enhancing oil recovery, but they require careful model tuning and validation. Additionally, these methods can be computationally intensive and may require large datasets.

4.4.1.22.4 Results and Discussion

3D CNN-PCA was applied for geological parameterization and a recurrent R-U-Net for flow surrogate modeling to perform data assimilation on a complex 3D reservoir model. This approach enabled efficient and accurate history matching by reducing the computational burden associated with traditional high-fidelity simulations. Three history matching algorithms were evaluated: RS, MADS-RML, and ES-MDA. The findings indicate that MADS-RML provides more accurate posterior uncertainty quantification than ES-MDA, though it requires more computational resources. This study highlights the potential of integrating DL-based surrogate models with various history matching techniques to improve reservoir modeling and prediction. Future work should focus on further accelerating the RML method and exploring its application to more complex systems, including those involving coupled flow and geomechanics.

BIBLIOGRAPHY

Meng Tang, Yimin Liu, and Louis J. Durlofsky. "History Matching Complex 3D Systems Using Deep-Learning-Based Surrogate Flow Modeling and CNN-PCA Geological Parameterization." *Paper Presented at the SPE Reservoir Simulation Conference, On-Demand*, October 2021. https://doi.org/10.2118/203924-MS

4.4.1.23 Application-23: Bayesian Optimization for Field Scale Geological Carbon Sequestration

4.4.1.23.1 Background of the Application

This study introduces a framework for applying BO to manage well operations in geological carbon sequestration. The integrated parallel accurate reservoir simulator (IPARS) simulator, which models the physical processes involved in CO_2 sequestration, is coupled with IBM's BO (IBO) for optimizing CO_2 injection strategies. BO uses Gaussian process regression to create a probabilistic model of the objective function and an acquisition function to determine sampling points based on uncertainty. The IBO method supports parallel executions, handles high-dimensional problems better, and is more robust compared to traditional BO. The approach is demonstrated on the Cranfield site using field data, showing that IBO achieves competitive results with significantly fewer model evaluations compared to GAs and covariance matrix adaptation evolution strategies. Additionally, BO provides uncertainty quantification and extends naturally to optimization under uncertainty.

4.4.1.23.2 Description of the Workflow and Machine Learning Model

The paper presents innovative AI techniques for fracture characterization in reservoirs, using ML algorithms to analyze seismic and well log data. The goal is to improve the accuracy of fracture detection and characterization.

4.4.1.23.3 Advantages and Limitations of the Synchronized Algorithm

Innovative AI techniques provide enhanced accuracy in fracture characterization, but they require extensive data preprocessing and are sensitive to noise. These methods also demand significant computational resources.

4.4.1.23.4 Results and Discussion

The presented framework is a pioneering effort to integrate high-fidelity physical models with ML techniques for field-scale geological carbon sequestration. By using Gaussian process regression and an acquisition function, the framework efficiently explores the solution space while maintaining accurate physical modeling.

The study's results demonstrate that BO can achieve competitive objective values with fewer function evaluations compared to GAs and evolution strategies. The Bayesian approach also provides a natural method for uncertainty quantification. The successful benchmarking against traditional optimization methods highlights the potential for further applications of BO in carbon sequestration and similar fields.

BIBLIOGRAPHY

Xueying Lu, Kirk E. Jordan, Wheeler, Mary F., Pyzer-Knapp, Edward O., and Matthew Benatan. "Bayesian Optimization for Field Scale Geological Carbon Sequestration." *Paper Presented at the SPE Reservoir Simulation Conference, On-Demand*, October 2021. https://doi.org/10.2118/203950-MS

4.4.1.24 Application-24: Evaluation of Machine Learning Methods for the Prediction of Multiphase Production Rates

4.4.1.24.1 Background of the Application

The paper evaluates the use of ML methods for predicting multiphase production rates in oil and gas operations. Traditional multiphase meters, while accurate, are costly and require maintenance. Virtual flow metering (VFM) offers a cost-effective alternative by leveraging data from existing sensors and production history to estimate flow rates through mathematical models. This study explores the effectiveness of data-driven algorithms, including ANNs, Gated Recurrent Units (GRU), and XGBoost, in predicting oil, gas, and water production rates. The results show that ML algorithms can accurately estimate these rates, demonstrating the potential of VFM for continuous production monitoring. Among the tested methods, DL techniques provided the most precise results with reasonable training times.

4.4.1.24.2 Description of the Workflow and Machine Learning Model

This study explores AI-assisted optimization of reservoir production, focusing on the use of ML algorithms to predict production rates. The research highlights the potential of AI in improving production efficiency.

4.4.1.24.3 Advantages and Limitations of the Synchronized Algorithm

AI-assisted optimization offers significant benefits for reservoir production, but the accuracy of these methods depends on the quality of the training data. Additionally, these approaches require substantial computational power and expertise.

4.4.1.24.4 Results and Discussion

The study investigates various data-driven methods for analyzing and forecasting production time-series data, demonstrating that VFM is a practical and reliable tool for predicting production rates directly from data. Following are the key findings of the study:

1. Emphasizing the importance of flow modeling in intelligent production monitoring.

2. Evaluating the performance of different ML algorithms in practical forecasting applications.
3. Highlighting that advanced methods, while computationally intensive, offer significant accuracy improvements.
4. Recommending that simple neural networks or XGBoost can provide baseline estimates for most production data scenarios.

Future work should focus on real-life applications, particularly addressing the challenges of noisy data, which was filtered out in this study. Special attention is needed for measurements that fall outside the training range to enhance the robustness of VFM methods.

BIBLIOGRAPHY

Anton Gryzlov, Liliya Mironova, Sergey Safonov, and Muhammad Arsalan. "Evaluation of Machine Learning Methods for Prediction of Multiphase Production Rates." *Paper Presented at the SPE Symposium: Artificial Intelligence - Towards a Resilient and Efficient Energy Industry, Virtual*, October 2021. https://doi.org/10.2118/208648-MS

4.4.1.25 Application-25: Deep Learning Assisted Doppler Sensing for Hydrocarbon Downhole Flow Velocity Estimation

4.4.1.25.1 Background of the Application

Accurate measurement of downhole fluid flow velocities is crucial for optimizing oil and gas production. Traditional mechanical spinner-based sensors face challenges in complex flow conditions. This paper introduces an advanced DL framework to estimate spinner phase velocities from ultrasonic Doppler sensor data, which are robust and suitable for real-time applications in various logging conditions. The framework demonstrates strong performance on benchmark datasets, providing reliable and automatic flow velocity estimations for both wireline and logging-while-drilling operations.

4.4.1.25.2 Description of the Workflow and Machine Learning Model

This paper presents a DL framework for estimating hydrocarbon downhole flow velocities using Doppler sensing data. The framework utilizes a feedforward backpropagation network to predict spinner fluid velocities, improving the accuracy of downhole flow measurements.

4.4.1.25.3 Advantages and Limitations of the Synchronized Algorithm

DL provides a powerful tool for estimating downhole flow velocities, offering improved accuracy and robustness compared to traditional methods. However, these models require significant computational resources and are sensitive to the quality of input data.

4.4.1.25.4 Results and Discussion

The innovative DL framework developed in this study successfully estimates spinner velocities from the Doppler-based sensor data, proving both methods to be comparable and effective. The framework's strong predictive accuracy highlights its potential for deployment in downhole environments, particularly in harsh conditions with significant solids that can impair mechanical sensors. This advancement offers a reliable alternative for velocity measurements in production logging, particularly under challenging drilling scenarios.

BIBLIOGRAPHY

Klemens Katterbauer, Alberto Marsala, Virginie Schoepf, and Linda Abbassi. "Deep Learning Assisted Doppler Sensing for Hydrocarbon Downhole Flow Velocity Estimation." *Paper Presented at the SPE Europec featured at 82nd EAGE Conference and Exhibition*, Amsterdam, The Netherlands, October 2021. https://doi.org/10.2118/205183-MS

4.4.2 Reservoir Simulation Works

4.4.2.1 Application-26: Algorithm for Detecting Multiple Partial Blockages in Liquid Pipelines by Using Inverse Transient Analysis

4.4.2.1.1 Background of the Application

This study addresses the issue of partial blockages in liquid pipelines caused by asphaltene precipitation and scale deposition, which can adversely affect pipeline efficiency, economy, and safety. Traditional detection methods often involve expensive and technically challenging installation of devices along the pipeline. This research focuses on a cost-effective, non-intrusive remote measurement technique using the pressure-wave technology. The developed method involves solving both direct and inverse problems related to transient flow in pipelines with blockages. The direct problem is solved using a second-order Godunov-type FVM to simulate transient flow, while the inverse problem is tackled through an analytical and optimization approach to determine blockage distribution based on pressure responses at measuring points. The proposed method efficiently and accurately identifies multiple irregular partial blockages, demonstrating its effectiveness through numerical results.

4.4.2.1.2 Description of the Workflow and Machine Learning Model

This paper presents an algorithm for detecting multiple partial blockages in liquid pipelines using inverse transient analysis. The method involves simulating transient flow in pipelines with blockages, analyzing pressure waves, and solving the inverse problem to identify blockage locations and severity. The algorithm integrates the Godunov-type FVM for the direct problem and an optimization technique for the inverse problem.

4.4.2.1.3 Advantages and Limitations of the Synchronized Algorithm

The proposed algorithm provides a nonintrusive and accurate method for detecting blockages in pipelines. It is effective in both simple and complex blockage scenarios, making it versatile. However, the method requires high computational resources and is sensitive to measurement noise, which can impact the accuracy of blockage detection.

4.4.2.1.4 *Results and Discussion*

This paper presents a comprehensive study on detecting partial blockages in liquid pipelines using inverse transient analysis with the pressure-wave technology. The method is applicable to both uniform and irregular partial blockages. A second-order Godunov scheme was implemented to simulate transient flows and establish the relationship between pressure-wave profiles and blockage parameters. The core challenge was solving the inverse transient problem to determine the blockage profile using valve pressure responses. The proposed backward integration method, combined with Genetic Algorithm optimization, effectively calibrates the profile of multiple uniform and irregular blockages. The approach has shown promising accuracy and potential for future enhancements, including ML applications, large-scale parallel implementation, and experimental validation.

The detailed methodologies, numerical simulations, and optimization strategies outlined in this study provide a robust framework for the accurate detection and characterization of partial blockages, which is crucial for maintaining pipeline efficiency and safety.

BIBLIOGRAPHY

C. Zhang, J. J. Zhang, C. B. Ma, and G. E. Korobkov. "Algorithm for Detecting Multiple Partial Blockages in Liquid Pipelines by Using Inverse Transient Analysis." *SPE J* 26 (2021): 3011–3039. https://doi.org/10.2118/205519-PA

4.4.2.2 Application-27: Automatic Well Test Interpretation Based on Convolutional Neural Network for a Radial Composite Reservoir

4.4.2.2.1 *Background of the Application*

This study introduces an automated well test interpretation method for radial composite reservoirs using CNN. The proposed method converts pressure variation data and its derivatives into log-log plots, which are then input into a CNN. The network, optimized by minimizing the mean square error and using dropout to prevent overfitting, is trained to automatically interpret well test data, yielding reservoir parameters such as mobility ratio, storativity ratio, dimensionless composite radius, and a dimensionless group for well storage and skin effects. The effectiveness and accuracy of this method are validated using actual field data from the Daqing Oilfield, demonstrating superior performance over traditional analytical and least square methods.

4.4.2.2.2 *Description of the Workflow and Machine Learning Model*

This paper proposes an automatic well test interpretation method based on CNN for radial composite reservoirs. The approach transforms pressure and derivative data into a format suitable for CNN input, enabling the automatic extraction of reservoir parameters with high accuracy.

4.4.2.2.3 *Advantages and Limitations of the Synchronized Algorithm*

The CNN-based interpretation method offers significant advantages in terms of accuracy and efficiency, particularly in handling complex well test data. However, it requires extensive training data and computational resources, and may be less effective in cases with insufficient data quality.

4.4.2.2.4 Results and Discussion

The research presents a CNN-based method for automatically interpreting well test data in radial composite reservoirs. By inputting pressure change and derivative data, the trained CNN provides accurate reservoir parameters, significantly improving interpretation efficiency and enabling non-experts to perform well test analyses. Field data from the Daqing Oilfield confirm the method's robustness and precision, outperforming traditional analytical and least square approaches. However, the method requires adaptations for different reservoir types due to variations in well test curve characteristics and parameter needs, indicating future research directions.

BIBLIOGRAPHY

Daolun Li, Xuliang Liu, Wenshu Zha, Jinghai Yang, and Detang Lu. Automatic well test interpretation based on convolutional neural network for a radial composite reservoir. *Petroleum Exploration and Development*, 47 (3), 2020, 623–631, ISSN 1876-3804, https://doi.org/10.1016/S1876-3804(20)60079-9

4.4.2.3 Application-28: Machine Learning Applied to Optimize Duvernay Well Performance

4.4.2.3.1 Background of the Application

This study explores the application of ML techniques, specifically multiple linear regression and neural networks, to optimize completions and well designs in the Duvernay shale formation. By analyzing various well completion variables, the developed workflow predicted future well performance and identified optimal solutions, potentially saving over a million dollars per well and improving well performance by over 50%. This methodology, which was validated using data from 262 Duvernay wells, is not limited to this formation and can be applied to other basins. The approach effectively inferred relationships between numerous predictor variables and well performance, demonstrating high predictive power while avoiding overfitting.

4.4.2.3.2 Description of the Workflow and Machine Learning Model

This research applies transfer learning using CNNs to detect seismic faults in real data. By leveraging pre-trained CNN models, the study reduces the need for extensive labeled datasets, focusing on feature extraction and fault classification in seismic images.

4.4.2.3.3 Advantages and Limitations of the Synchronized Algorithm

Transfer learning with CNNs allows for efficient feature extraction and fault detection in seismic data, benefiting from pre-trained models. This approach reduces the dependency on large labeled datasets and accelerates model training. However, transfer learning requires careful adaptation of pre-trained models to specific datasets and may not capture all domain-specific features.

4.4.2.3.4 Results and Discussion

The study concluded that the neural network model successfully explained 78% of the variance in well performance for the test wells, while the multiple linear regression model explained 67%. It was found that significant improvements in well performance, ranging from 19% to 97%, could be achieved by optimizing controllable parameters using the neural network model. The analysis revealed that expensive fracturing procedures, such as using ceramic or resin-coated proppant, and orienting wells parallel to the minimum horizontal stress provided no significant benefit. The total fracture tonnage was confirmed as a key driver of well performance, and the fracture pump rate was suggested for further investigation. The workflow is adaptable and can be applied to different fields with various input variables to reduce uncertainty and maximize well performance.

BIBLIOGRAPHY

Braden Bowie "Machine Learning Applied to Optimize Duvernay Well Performance." *Paper Presented at the SPE Canada Unconventional Resources Conference*, Calgary, AB, Canada, March 2018. https://doi.org/10.2118/189823-MS

4.4.2.4 Application-29: Automatic Well Test Interpretation Based on Convolutional Neural Network for Infinite Reservoir

4.4.2.4.1 Background of the Application

Automatic well test interpretation is a crucial technique for estimating well and reservoir properties like permeability and skin factor. Historically, researchers have struggled to develop reliable automatic interpretation tools. This paper introduces a novel approach using CNNs to automate well test interpretation for infinite acting reservoirs. The proposed CNN model takes pressure change and pressure derivative data from log-log plots as inputs and outputs a dimensionless group. The CNN model's effectiveness and accuracy were tested using data from the Daqing oil field in China. The results demonstrated that CNN is a promising method for automatic well test interpretation, showing potential for application in more complex scenarios.

4.4.2.4.2 Description of the Workflow and Machine Learning Model

The study utilizes CNNs to estimate total magnetization directions from geophysical data. CNNs are employed for their ability to handle spatial dependencies and extract meaningful features from complex datasets, facilitating the accurate estimation of magnetization directions.

4.4.2.4.3 Advantages and Limitations of the Synchronized Algorithm

CNNs provide advantages in capturing spatial features and dependencies, offering high accuracy in estimating magnetization directions. The method leverages the hierarchical structure of CNNs to learn relevant features, but it requires substantial computational resources and careful tuning of network parameters to prevent overfitting.

4.4.2.4.4 Results and Discussion

This study demonstrates the feasibility and effectiveness of using CNNs for automatic well test interpretation in infinite acting wells. The CNN model outputs the dimensionless group 2log*f*0 (SDCe)2\log(S_{DC}e)2log(SDCe), which, when combined with input pressure change and derivative curves from log-log plots, allows for the automatic derivation of well characteristics without manual intervention.

Field case studies and parameter analysis validated the method's efficiency and accuracy. While this initial application focused on infinite reservoirs, the method shows promise for extending deep neural networks to more complex reservoir conditions in future research. This advancement significantly enhances the efficiency of transient pressure data interpretation.

BIBLIOGRAPHY

Xuliang Liu, Daolun Li, Jinghai Yang, Wenshu Zha, Ziqi Zhou, Liping Gao, Jiahang Han. Automatic well test interpretation based on convolutional neural network for infinite reservoir. *Journal of Petroleum Science and Engineering*, 195, 2020, 107618, ISSN 0920-4105, https://doi.org/10.1016/j.petrol.2020.107618

4.4.2.5 Application-30: An Improved Deconvolution Algorithm Using B-splines for Well-Test Data Analysis

4.4.2.5.1 Background of the Application

In the field of petroleum engineering, the analysis of well-test data is crucial for understanding reservoir properties and performance. Traditional deconvolution algorithms, such as those based on B-splines and the Laplace transformation, face challenges due to the discontinuities in production rate functions. These challenges can lead to significant computational time and potential failure in numerical Laplace inversion. To address these issues, the authors present an improved deconvolution algorithm that operates in the real-time integral space rather than the Laplace space, thus avoiding the requirement for continuity in production rate functions. This new approach employs piecewise analytical integration to compute the sensitivity matrix elements and combines it with a binary search method to enhance computational speed. By incorporating regularization techniques, the algorithm effectively handles data errors, ensuring the accurate reconstruction of wellbore pressure and its derivative. Numerical tests demonstrate the algorithm's superior accuracy, speed, and stability compared to previous methods, making it a robust tool for well-test data analysis.

4.4.2.5.2 Description of the Workflow and Machine Learning Model

This research employs DL techniques for lithology identification in igneous reservoirs using logging data. The approach focuses on developing models capable of distinguishing between different lithologies by analyzing complex patterns in the logging measurements.

4.4.2.5.3 Advantages and Limitations of the Synchronized Algorithm

DL models excel in handling complex, high-dimensional data, making them suitable for lithology identification. They can learn intricate patterns from logging data, improving classification accuracy. However, these models require large amounts of training data.

4.4.2.5.4 Results and Discussion

The improved deconvolution algorithm developed by the authors addresses the inherent issues of traditional methods by maintaining the computation in the real-time integral space, thus eliminating the need for Laplace transformation and its associated problems. By using piecewise analytical integration, the algorithm significantly enhances computational speed and ensures successful deconvolution even in the presence of large data discontinuities. The inclusion of a regularization method further mitigates the impact of data errors, resulting in accurate wellbore pressure and derivative reconstructions. Field application and numerical experiments confirm the algorithm's high accuracy, computational efficiency, and error tolerance. Overall, this improved algorithm is better suited for handling large datasets and provides a reliable tool for reservoir engineers to perform well-test data analysis.

BIBLIOGRAPHY

Wenchao Liu, Yuewu Liu, Guofeng Han, Jianye Zhang, Yizhao Wan. An improved deconvolution algorithm using B-splines for well-test data analysis in petroleum engineering. *Journal of Petroleum Science and Engineering*, 149, 2017, 306–314, ISSN 0920-4105, https://doi.org/10.1016/j.petrol.2016.10.064

4.4.2.6 Application-31: Artificial Neural Network (ANN) to Estimate Reservoir Parameters in Naturally Fractured Reservoirs Using Well Test Data

4.4.2.6.1 Background of the Application

This study leverages ANNs to estimate reservoir parameters using well test data in Naturally Fractured Reservoirs (NFRs). The ANN is trained using theoretical pressure derivative curves generated through the Stehfest algorithm. Instead of using normalized pressure derivative data, Chebyshev polynomial coefficients are employed as input to the ANN. Among various training algorithms tested, the Levenberg–Marquardt (LM) algorithm provided the lowest Mean Relative Error (MRE). The results demonstrated that using Chebyshev polynomial coefficients improved the accuracy of the ANN compared to traditional methods. Field data further validated the ANN's performance, showing consistency with results obtained from well test software.

4.4.2.6.2 Description of the Workflow and Machine Learning Model

This paper explores the evolution of reservoir uncertainty analysis, focusing on the transition from probabilistic and stochastic methods to the integration of AI algorithms. AI methods, including GAs, ANNs, and Bayesian networks, are increasingly used to improve reservoir modeling. These methods enhance uncertainty quantification by capturing complex patterns in large datasets, thus providing more reliable reservoir models.

4.4.2.6.3 Advantages and Limitations of the Synchronized Algorithm

AI approaches offer significant advantages in reservoir modeling, such as improved accuracy and reduced computational time compared to traditional methods. However, AI models depend heavily on the quality of input data and require careful tuning to avoid overfitting. Despite these challenges, AI methods provide a promising alternative for capturing uncertainties in reservoir properties and enhancing model predictions.

4.4.2.6.4 *Results and Discussion*

The study concludes that using Chebyshev polynomial coefficients as input significantly enhances the learning process of ANNs, leading to a more accurate estimation of reservoir parameters. This method allows for the inclusion of more data points without increasing the complexity of ANNs, thus addressing issues related to data normalization and transformation. While conventional polynomial coefficients increased the MRE, Chebyshev polynomials improved performance. Sensitivity analysis indicated that the LM algorithm, despite its computational intensity, yielded the lowest MRE. The ANN trained with Chebyshev polynomials was validated using both test data and a field case, showing superior accuracy compared to other methods. The study underscores the efficacy of this novel approach in well test analysis for NFRs.

BIBLIOGRAPHY

M. Adibifard, S.A.R. Tabatabaei-Nejad, E. Khodapanah. Artificial Neural Network (ANN) to estimate reservoir parameters in Naturally Fractured Reservoirs using well test data. *Journal of Petroleum Science and Engineering*, 12, 2014, 585–594, ISSN 0920-4105, https://doi.org/10.1016/j.petrol.2014.08.007

4.4.2.7 Application-32: Artificial Neural Network-Based Time-Domain Interwell Tracer Testing for Ultralow-Permeability Fractured Reservoirs

4.4.2.7.1 *Background of the Application*

This study focuses on improving the efficiency and cost-effectiveness of interwell tracer testing in ultralow-permeability fractured reservoirs, particularly during the middle and late stages of development. Traditional interwell tracer testing can be costly and sometimes ineffective due to the reduced number of tracers used. By leveraging the advancements in AI, specifically ANNs, the authors developed a method to predict future interwell parameters using historical tracer data and production information. The research was conducted in the eastern Ordos Basin, where ANN-based approach demonstrated a significant reduction in error rates, achieving a relative error of just 1.94%. This innovative method offers a reliable, low-cost solution for monitoring and predicting interwell dynamics in challenging reservoir conditions, potentially revolutionizing the monitoring processes in the petroleum industry.

4.4.2.7.2 *Description of the Workflow and Machine Learning Model*

This study examines the application of ML techniques in petroleum engineering, particularly in reservoir characterization and production optimization. Techniques such as SVMs, ANNs, and decision trees are employed to analyze complex datasets and predict the reservoir behavior. These methods facilitate better decision-making by identifying key parameters influencing reservoir performance.

4.4.2.7.3 *Advantages and Limitations of the Synchronized Algorithm*

ML techniques provide powerful tools for handling large, complex datasets in petroleum engineering, offering improved accuracy in predictions. However, these methods require careful data preprocessing and parameter tuning to ensure robustness and reliability. The computational demand of ML algorithms can also be a limitation, necessitating efficient data management and processing strategies.

4.4.2.7.4 Results and Discussion

The paper concludes that the proposed ANN-based method for time-domain interwell tracer testing is both effective and economical for monitoring interwell dynamics in the later stages of reservoir development. By incorporating historical tracer test results and production data, the ANN was able to accurately predict key interwell parameters, demonstrating minimal errors. The study highlights the significant potential of this approach for the continuous and systematic monitoring of ultralow-permeability fractured reservoirs. The findings suggest that expanding the learning samples and adapting the neural network structure to other reservoir types could further enhance the applicability and accuracy of this method, providing a robust tool for the oil and gas industry to optimize reservoir management and improve recovery rates.

BIBLIOGRAPHY

Cheng Jing, Xiaowei Dong, Wenhao Cui, Zhenzhen Dong, Long Ren, Xiaoyu Gu, Yanlong He, Chunsheng Pu, Xiaoyu Liu. Artificial neural network–based time-domain interwell tracer testing for ultralow-permeability fractured reservoirs. *Journal of Petroleum Science and Engineering*, 195, 2020, 107558, ISSN 0920-4105, https://doi.org/10.1016/j.petrol.2020.107558

4.4.2.8 Application-33: Integrated Nonlinear Regression Analysis of Tracer and Well Test Data

4.4.2.8.1 Background of the Application

This study focuses on improving the conventional analysis of pressure transient tests by integrating it with tracer test data through nonlinear least squares regression. Traditional pressure transient tests often match models designed for homogeneous systems, failing to capture reservoir heterogeneities. Tracer tests, however, can be more sensitive to these variations due to the nature of convective flow. By correlating permeability and dispersivity, the researchers developed a combined model that integrates multi-fracture tracer test results with dual-porosity pressure models. Experimental data from a fractured geothermal reservoir were used to test this approach. The findings indicate that combining tracer and pressure transient test analyses provides a more accurate description of reservoir characteristics, particularly in capturing the effects of changing injection-production well depths on flow paths.

4.4.2.8.2 Description of the Workflow and Machine Learning Model

This paper discusses AI-driven approaches for reservoir simulation and optimization, focusing on GAs and ANNs. These techniques are applied to optimize well placement and production strategies, using large datasets to model reservoir behavior under different scenarios. The integration of AI methods allows for more flexible and efficient simulation processes.

4.4.2.8.3 Advantages and Limitations of the Synchronized Algorithm

AI-driven methods enhance the ability to model complex reservoir systems and optimize production strategies. While these techniques offer flexibility and improved performance, they also require significant computational resources and expertise in model calibration. The accuracy of AI models depends on the quality of input data and the appropriateness of the selected algorithms.

4.4.2.8.4 Results and Discussion

The integration of pressure transient and multi-fracture tracer test analyses using nonlinear least squares regression offers significant improvements over conventional methods. The combined approach effectively couples permeability and dispersivity, providing better matches for experimental data from a fractured geothermal reservoir model. The study demonstrates that fractal models do not significantly enhance analysis beyond what is achieved by conventional models. The combined method improves the physical interpretation of experiments, showing minimal variability in obtained solutions. This integrated approach addresses the ambiguity in interpreting pressure transient data and offers a more comprehensive understanding of reservoir heterogeneities, leading to more accurate reservoir characterizations.

BIBLIOGRAPHY

Serhat Akin. Integrated nonlinear regression analysis of tracer and well test data, *Journal of Petroleum Science and Engineering*, 39 (1–2), 2003, 29–44, ISSN 0920-4105, https://doi.org/10.1016/S0920-4105(03)00038-X

4.4.2.9 Application-34: Modelling Viscosity of Liquid Dropout Near a Wellbore Region in Gas Condensate Reservoirs Using Modern Numerical Approaches

4.4.2.9.1 Background of the Application

Liquid dropout occurs in gas condensate reservoirs below the dew point pressure near the wellbore region due to production depletion. Accurate determination of liquid dropout viscosity (μc) is crucial for forecasting production and optimizing future recoveries in gas condensate reservoirs. Traditional correlation estimation methods fall short in accuracy due to the complex behavior of gas condensate reservoirs near the wellbore. This study employs modern numerical approaches, including LSSVM and ANNs, to model condensate phase viscosity near the wellbore. Using over 300 viscosity datasets from literature and experimental studies, new models were developed. These models outperform traditional correlations, providing better accuracy in predicting liquid dropout viscosity and can be integrated into commercial reservoir simulation packages.

4.4.2.9.2 Description of the Workflow and Machine Learning Model

This research explores the use of advanced ML techniques in subsurface modeling, particularly in the context of reservoir simulation and uncertainty quantification. Techniques such as DL and ensemble methods are used to improve the accuracy and efficiency of subsurface models.

4.4.2.9.3 Advantages and Limitations of the Synchronized Algorithm

The application of advanced ML techniques in subsurface modeling offers significant improvements in accuracy and efficiency. However, these techniques require large datasets and extensive computational resources, which can be a limitation in certain applications.

4.4.2.9.4 Results and Discussion

PVT calculations, production projections, simulation studies, and future recoveries all depend on accurate condensate viscosity modeling in gas condensate reservoirs. The prediction performance of current literature methodologies is subpar in the vicinity of the wellbore region. This study used ANNs and LSSVM to construct numerical AI-based models. Condensate viscosity changes with changes in API gravity, pressure, temperature, solution gas to oil ratio (Rs), and gas specific gravity were faithfully replicated by both models. The findings show that the suggested AI models outperform conventional correlations in terms of robustness, efficiency, and dependability. To further minimize error, future research could investigate improving these models using Coupled Simulated Annealing (CSA) or GAs. The simplicity and flexibility of the developed models make them suitable for implementation in PVT calculations for gas condensate reservoirs, enhancing accuracy and reliability.

BIBLIOGRAPHY

Foad Faraji, Johnson Obunwa Ugwu, Perk Lin Chong, Farhad Nabhani. Modelling viscosity of liquid dropout near wellbore region in gas condensate reservoirs using modern numerical approaches, *Journal of Petroleum Science and Engineering*, 185, 2020, 106604, ISSN 0920-4105, https://doi.org/10.1016/j.petrol.2019.106604

4.4.3 Well Test Applications

4.4.3.1 Application-35: Forecasting Multiple-Well Flow Rates Using a Novel Space-Time Modeling Approach

4.4.3.1.1 Background of the Application

Predicting the flow rate of oilfields and producing wells is crucial for petroleum engineers, as it assists in regulating production responsibilities in daily and annual production programs and aids economic authorities in planning annual budgets and long-term programs. Various methods have been developed for analyzing oil production data, with reservoir simulation techniques standing out due to their accuracy and robustness. These techniques incorporate the areal and vertical distributions of non-uniform rock properties, coupled rock-fluid properties, and resulting rates, pressures, and saturations. Although reservoir simulation models require extensive data and are time-consuming, they are essential for projects requiring high accuracy, especially for unconventional and naturally fractured reservoirs where many unknowns remain. Simpler methods may also be employed for production data analysis when comprehensive data is not available.

4.4.3.1.2 Description of the Workflow and Machine Learning Model

This study explores the application of ML techniques for reservoir characterization, focusing on the integration of AI algorithms such as SVMs and neural networks. The aim is to improve the accuracy of reservoir property predictions by analyzing geological and petrophysical data.

4.4.3.1.3 Advantages and Limitations of the Synchronized Algorithm

ML techniques offer significant advantages in reservoir characterization, including improved prediction accuracy and the ability to handle large datasets. However, these methods require the careful tuning of parameters and can be computationally demanding, necessitating efficient data management.

4.4.3.1.4 Results and Discussion

The study highlights the importance of accurate forecasting of multiple-well flow rates and demonstrates a novel space–time modeling approach. By employing hierarchical and spatial clustering techniques, the study categorizes wells based on their production flow rate patterns and geographical proximity. The Space–Time Autoregressive Moving Average (STARMA) model, combined with clustering, provides reliable forecasts of production flow rates with reasonable accuracy. The proposed methodology's effectiveness is validated through cross-validation techniques, showing its potential for practical application in oilfield production management. Future research may focus on enhancing the model by incorporating additional factors such as well maintenance operations and new well additions to further improve forecasting accuracy and adaptability to diverse conditions.

BIBLIOGRAPHY

Rouhollah Ahmadi, Jamal Shahrabi, Babak Aminshahidy. Forecasting multiple-well flow rates using a novel space-time modeling approach. *Journal of Petroleum Science and Engineering*, 191, 2020, 107027, ISSN 0920-4105, https://doi.org/10.1016/j.petrol.2020.107027

4.4.3.2 Application-36: Multiscale Modeling of Compartmentalized Reservoirs Using a Hybrid Clustering-Based Non-local Approach

4.4.3.2.1 Background of the Application

A rapid and precise way to assess oil and gas reservoirs is to portray the reservoir as a network of distinct compartments with neighbor and non-neighbor links. In this work, a physics-based multiscale modeling approach is combined with a hybrid framework tailored to reservoir analysis for autonomously identifying clusters in space utilizing spatial and temporal field data. The reservoir is represented by an adjacency matrix that characterizes the connectivities of these compartments in a novel and rigorous non-local formulation for flow in porous media. The framework can handle a variety of data formats and noisy or missing values by utilizing unsupervised clustering techniques and suitable distance measurements. Finally, the clusters are delineated using the SVM approach. The proposed workflow is demonstrated on a major field with a long production history, showing that this approach can effectively divide the reservoir into distinct compartments and provide probabilistic forecasts for different field development scenarios.

4.4.3.2.2 Description of the Workflow and Machine Learning Model

This paper presents numerical simulations of polymer flooding in porous media, using a combination of finite volume and finite element methods. The study aims to optimize the flooding process by accurately modeling the flow dynamics of polymers through porous reservoirs.

4.4.3.2.3 Advantages and Limitations of the Synchronized Algorithm

The numerical simulation approach provides valuable insights into the polymer flooding process, enabling the optimization of EOR techniques. However, the complexity of the simulations and the need for accurate input parameters are significant challenges in applying this method broadly.

4.4.3.2.4 Results and Discussion

This study presents a novel non-local formulation for flow in porous media, where the reservoir is segmented into compartments with distinct static and dynamic properties. The hybrid framework developed in this research successfully integrates spatial and temporal clustering with a physics-based multiscale modeling approach. The approach was applied to a major field with a long production history, demonstrating its capability to automatically detect clusters and perform efficient history matching, significantly reducing the uncertainty of global reservoir parameters. The history matching process achieved a close agreement with the field's actual pressure data, reducing the pressure mismatch by nearly an order of magnitude. The ensemble of history-matched models provided accurate probabilistic forecasts in a blind test study. This framework can be extended to various field cases, including fractured reservoirs and 3D network models, making it a robust tool for reservoir management and field development applications.

BIBLIOGRAPHY

Soheil Esmaeilzadeh, Amir Salehi, Gill Hetz, Feyisayo Olalotiti-Lawal, Hamed Darabi, David Castineira. Multiscale modeling of compartmentalized reservoirs using a hybrid clustering-based non-local approach. *Journal of Petroleum Science and Engineering*, 184, 2020, 106485, ISSN 0920-4105, https://doi.org/10.1016/j.petrol.2019.106485

4.4.3.3 Application-37: Surrogate Permeability Modelling of Low-Permeable Rocks Using Convolutional Neural Networks

4.4.3.3.1 Background of the Application

Low-permeable rocks' permeability is an essential mechanical characteristic that is important for forecasting a number of engineering problems, such as gas production and CO_2 geo-sequestration. These rocks are extremely heterogeneous and have very low permeability, which makes it difficult to determine permeability accurately both experimentally and numerically. The purpose of this work is to use 3D CNNs to quickly assess the permeability of low-permeable porous material. The study creates low-permeable porous media samples by utilizing an upgraded quartet structure generation set (QSGS) that includes pore size distribution (PSD) and anisotropy. A dataset of 4,500 samples was produced for CNN modeling using the LBM 3D, 19 Velocity Model (D3Q19) model. A parametric study assessed the impact of CNN architecture and training epochs on performance, revealing that the CNN architecture significantly influences predictive accuracy, achieving a high correlation coefficient and low errors.

4.4.3.3.2 Description of the Workflow and Machine Learning Model

This paper introduces a DL framework for estimating hydrocarbon downhole flow velocities using Doppler sensing data. The study employs a feedforward backpropagation network to predict spinner fluid velocities, offering enhanced accuracy in flow measurements.

4.4.3.3.3 Advantages and Limitations of the Synchronized Algorithm

DL offers robust performance in estimating downhole flow velocities, but the models are sensitive to the quality of input data and require significant computational power. Overfitting can be a concern if the models are not properly tuned.

4.4.3.3.4 Results and Discussion

This study employs CNN modeling to explore the relationship between permeability and the microstructure of porous media. Utilizing an improved Quasi-Solid Gas Simulation (QSGS) algorithm considering Particle Size Distribution (PSD) and anisotropy, the research generated porous structures, and the 3D, 19 Velocity Model (D3Q19) model of LBM was used to calculate their permeability. A parametric study highlighted the influence of the CNN architecture and training epochs on model performance. Key conclusions include:

- CNN architecture significantly affects performance, with architecture 3 demonstrating the lowest average validation loss.
- Training epochs beyond 20 can lead to over-fitting, indicating that 20 epochs are optimal for training the CNN.
- The CNN model showed excellent predictive performance, with high R-value and low MAE and RMSE on the testing set, though it tended to slightly overestimate permeability.
- Lower permeability groups exhibited higher testing errors compared to higher permeability groups, attributed to the rock's intrinsic microstructure and CNN settings during model training.

Future work should extend this methodology to real images of coal, shale, and sandstones to enhance CNN's industrial application reliability and investigate further ways to improve CNN performance.

BIBLIOGRAPHY

Jianwei Tian, Chongchong Qi, Yingfeng Sun, Zaher Mundher Yaseen. Surrogate permeability modelling of low-permeable rocks using convolutional neural networks. *Computer Methods in Applied Mechanics and Engineering*, 366, 2020, 113103, ISSN 0045-7825, https://doi.org/10.1016/j.cma.2020.113103

4.4.3.4 Application-38: Advanced Numerical Modeling and Algorithms for Multiphase Flow and Transport

4.4.3.4.1 Background of the Application

The study delves into the complex realm of multiphase flow and transport phenomena, crucial in both natural and engineering contexts. Recognizing the significant advancements in this field over recent decades, the paper emphasizes the persistent challenges in accurately modeling and efficiently simulating these processes. The special issue aims to underscore the latest progress and the ongoing challenges in mathematical and numerical modeling, algorithms, and computational techniques for multiphase flow and transport. It features contributions from prestigious global institutions, covering various pivotal topics, indicating ongoing research directions and future trends.

4.4.3.4.2 Description of the Workflow and Machine Learning Model

This paper discusses advanced numerical modeling techniques for simulating multiphase flow and transport in porous media. It explores a range of approaches including the VOF method, the LS method, and phase field LBM. These methods are designed to capture the complex interactions within multiphase systems accurately.

4.4.3.4.3 Advantages and Limitations of the Synchronized Algorithm

The advanced numerical modeling methods presented in this paper provide significant improvements in simulating complex multiphase flow, allowing for more accurate predictions of flow dynamics. However, these methods can be computationally expensive and may require extensive calibration to specific reservoir conditions.

4.4.3.4.4 Results and Discussion

The special issue on advanced numerical modeling and algorithms for multiphase flow and transport showcases significant strides in the field, presenting novel models, algorithms, and methods. The findings contribute to a deeper understanding of complex multiphase flow and transport processes, aiming to tackle related scientific and engineering challenges. The articles included reflect not only the current research landscape but also the promising future trends.

BIBLIOGRAPHY

Elsevier Special Issue: Advanced numerical modeling and algorithms for multiphase flow and transport. https://doi.org/10.1016/j.jcp.2019.109222

4.4.3.5 Application-39: Efficient Deep Learning Techniques for Multiphase Flow Simulation in Heterogeneous Porous Media

4.4.3.5.1 Background of the Application

This study explores advanced deep learning techniques for simulating multiphase flow in heterogeneous porous media. By leveraging the inherent sparsity of the underlying discrete systems, the methods serve as efficient alternatives to traditional full-order system solvers. Specifically, for the flow problem, a neural network architecture incorporating convolutional and locally connected layers is developed to enable model reduction. Additionally, local mass conservation is preserved through a custom loss function that ensures the physical properties of the velocity solutions being learned. For the saturation problem, a residual neural network is proposed to account for the sparse interactions between inputs and outputs, allowing for accurate long-term predictions. The trained networks, particularly for two-phase flow, demonstrate a significant potential in both accuracy and computational efficiency, validated through various numerical experiments.

4.4.3.5.2 *Description of the Workflow and Machine Learning Model*

This special issue focuses on advanced numerical modeling and algorithms for multiphase flow and transport. It includes topics such as diffuse/sharp interface approaches, equilibrium and interfacial dynamics, and DL applications in multiphase simulations.

4.4.3.5.3 *Advantages and Limitations of the Synchronized Algorithm*

Advanced numerical models provide more accurate simulations of multiphase flow and transport processes, but they are computationally intensive and require extensive calibration to achieve the best results. DL approaches can speed up simulations but may suffer from overfitting if not properly managed.

4.4.3.5.4 *Results and Discussion*

In order to approximate flow and transport equations in both single-phase and two-phase flow settings, this work presents effective deep neural networks as surrogate models. When compared to fully connected networks, these networks significantly reduce the trainable parameters by taking use of the sparsity of the discrete systems. The DL models use convolutional and locally linked layers together with a proprietary loss function that ensures local mass conservation to estimate the mapping from source terms or relative permeability fields to velocity solutions for the flow problem. To mimic saturation dynamics over time, residual networks with sparsely customized layers are created for addressing the transport problem. With this method, long-term solutions may be predicted iteratively with impressive efficiency and precision. Future research will focus on integrating these methods with multiscale model reduction techniques and developing more sophisticated DL tools for larger, coupled flow and transport systems.

BIBLIOGRAPHY

Yating Wang, Guang Lin. Efficient deep learning techniques for multiphase flow simulation in heterogeneous porous media, *Journal of Computational Physics*, 401, 2020, 108968, ISSN 0021-9991, https://doi.org/10.1016/j.jcp.2019.108968

4.4.3.6 Application-40: Fast and Robust Algorithm for Compositional Modeling: Part I—Stability Analysis Testing

4.4.3.6.1 *Background of the Application*

The paper presents a robust algorithm for stability analysis, which is essential in determining the phase state of a given fluid system under specified conditions of pressure, temperature, and composition. The focus is on leveraging the tangent-plane-distance

(TPD) approach within a reduced parameter space, leading to significant improvements in robustness and efficiency of calculations. The transformation utilized in this method results in a smoother TPD surface with a single minimum, enhancing the accuracy and speed of stability determinations, particularly for compositional reservoir simulations.

4.4.3.6.2 Description of the Workflow and Machine Learning Model

This study investigates ML methods for predicting reservoir performance, focusing on neural networks and decision trees. These techniques are applied to forecast production rates and identify key factors influencing reservoir behavior. The approach emphasizes data integration and feature selection to enhance predictive accuracy.

4.4.3.6.3 Advantages and Limitations of the Synchronized Algorithm

ML methods provide powerful tools for predicting reservoir performance, offering improved accuracy and insights into key influencing factors. However, these techniques require careful data preprocessing and model calibration. The computational requirements can be significant, necessitating efficient data handling and processing.

4.4.3.6.4 Results and Discussion

The proposed algorithm for stability analysis offers notable flexibility in terms of speed and accuracy, outperforming conventional methods, especially in handling interaction coefficients. The robustness of the method is highlighted by its ability to effectively manage complex multicomponent mixtures, reducing the computational difficulties typically associated with the ragged Gibbs free energy surface. The method's ability to include all necessary interaction coefficients and its superior performance in reducing the TPD surface complexity underscore its advantages for efficient phase behavior computations in compositional modeling.

BIBLIOGRAPHY

Abbas Firoozabadi and Huanquan Pan. "Fast and Robust Algorithm for Compositional Modeling: Part I—Stability Analysis Testing." *SPE J*, 7 (2002): 78–89. https://doi.org/10.2118/77299-PA

4.4.3.7 Application-41: Deep-Learning-Based Surrogate Model for Reservoir Simulation with Time-Varying Well Controls

4.4.3.7.1 Background of the Application

The paper introduces a new DL-based reduced-order modeling (ROM) framework for subsurface flow simulation, based on an embed-to-control (E2C) approach. This framework utilizes an auto-encoder to reduce the system to a low-dimensional subspace and a linear transition model to approximate the system's evolution. The authors enhance the E2C framework with a physics-based loss function to ensure consistency with governing flow equations and accuracy in key well quantities like fluid production rates. Applied to oil-water flow in a heterogeneous reservoir with nine wells, the E2C ROM shows significant computational efficiency, achieving speedups over 1000 times compared to full-order simulations. Training involves 300 high-fidelity simulations and takes 10–12 minutes on a Tesla V100 GPU. The ROM delivers accurate predictions for global saturation and pressure fields and well responses over time, with higher accuracy for a larger number of training simulations.

4.4.3.7.2 Description of the Workflow and Machine Learning Model

The study explores ML applications in reservoir characterization, utilizing techniques such as SVMs and neural networks. These methods analyze geological and petrophysical data to identify reservoir properties and improve characterization accuracy. The approach emphasizes data integration and feature selection to enhance model performance.

4.4.3.7.3 Advantages and Limitations of the Synchronized Algorithm

ML offers robust tools for integrating diverse datasets and improving reservoir characterization accuracy. However, the success of these methods depends on the quality and representativeness of input data. Model complexity and computational demands can be challenging, requiring the careful management of data preprocessing and model training.

4.4.3.7.4 Results and Discussion

The E2C ROM developed in this study successfully provides accurate and efficient predictions for subsurface flow simulations under varying well controls. The model demonstrates significant computational speedup while maintaining accuracy, particularly in predicting well production rates and global pressure and saturation fields. The physics-based loss function plays a crucial role in improving the model's reliability. Despite its effectiveness, the current E2C ROM is limited to 2D systems and requires a larger number of training simulations compared to existing Proper Orthogonal Decomposition (POD)-based ROMs. Future work will focus on extending the model to 3D systems and exploring methods to reduce the number of necessary training simulations while maintaining accuracy. The promising results indicate the potential of the E2C ROM framework for practical applications in reservoir management and optimization.

BIBLIOGRAPHY

Zhaoyang Larry Jin, Yimin Liu, Louis J. Durlofsky. Deep-learning-based surrogate model for reservoir simulation with time-varying well controls. *Journal of Petroleum Science and Engineering*, 192, 2020, 107273, ISSN 0920-4105, https://doi.org/10.1016/j.petrol.2020.107273

4.4.3.8 Application-42: Machine Learning Methods to Speed Up Compositional Reservoir Simulation

4.4.3.8.1 Background of the Application

Compositional reservoir simulation is a critical tool for reservoir engineers, heavily relying on accurate phase behavior computations, which can be computationally intensive. This paper proposes using ML techniques to solve phase equilibrium problems, reducing computational costs. Classifiers determine phase stability, while regression models predict equilibrium coefficients based on fluid composition, pressure, and temperature. The approach, demonstrated for two-phase equilibria, significantly reduces the CPU time compared to traditional methods.

4.4.3.8.2 Description of the Workflow and Machine Learning Model

This paper explores AI techniques, including neural networks and fuzzy logic, for data analysis in reservoir engineering. A FLVQ model is designed for pattern recognition, enabling the identification of potential hydrocarbon-producing zones. The approach involves data preprocessing, model training, and application of fuzzy clustering methods to enhance prediction accuracy.

4.4.3.8.3 Advantages and Limitations of the Synchronized Algorithm

The use of AI techniques such as FLVQ offers advantages in handling uncertainty and identifying patterns within complex datasets. These methods provide robust predictive models that can adapt to varying reservoir conditions. However, the success of such models depends on the quality of input data and the expert's ability to fine-tune parameters. The approach's computational complexity can also be a limitation, necessitating efficient data preprocessing and model optimization.

4.4.3.8.4 Results and Discussion

The study presents a rapid, non-iterative method for phase equilibrium calculations, ideal for compositional reservoir simulations. ML models provide explicit, accurate predictions of phase stability and equilibrium coefficients, avoiding time-consuming iterations. This method, applicable to any Equation-of-State (EoS) model, greatly accelerates simulations, particularly beneficial for applications like gas condensate reservoir depletion and gas miscible flooding, where phase behavior calculations dominate simulation time.

BIBLIOGRAPHY

Vassilis Gaganis and Nikos Varotsis. "Machine Learning Methods to Speed up Compositional Reservoir Simulation." *Paper Presented at the SPE Europec/EAGE Annual Conference*, Copenhagen, Denmark, June 2012. https://doi.org/10.2118/154505-MS

4.4.4 PVT Applications

4.4.4.1 Application-43: Practical Application of Machine Learning on Fast Phase Equilibrium Calculations in Compositional Reservoir Simulations

4.4.4.1.1 Background of the Application

In reservoir simulations, accurately describing fluid phase behavior is crucial, typically achieved using EoS models. However, these models involve phase equilibrium calculations that are computationally expensive. This study enhances an ANN model to predict phase stability accurately while significantly reducing computational costs. The improved ANN model predicts saturation pressures, allowing for stability determination by comparing given and predicted pressures. Another ANN model is developed to provide better initial guesses for phase splitting calculations, reducing nonlinear iterations. The models, which feature multi-layer neural networks for more efficient training and accurate predictions, also incorporate an adaptive data generation process for optimal training data quality. The effectiveness of these models is validated through tests with a three-component hydrocarbon mixture and applied successfully in compositional reservoir simulations, demonstrating their practical value in speeding up phase equilibrium calculations.

4.4.4.1.2 Description of the Workflow and Machine Learning Model

This paper discusses advanced modeling techniques for simulating reservoir behavior. The focus is on the application of finite element and finite difference methods to solve complex geological formations and predict fluid flow behavior.

4.4.4.1.3 Advantages and Limitations of the Synchronized Algorithm

These advanced modeling techniques offer precise simulation results, which are crucial for optimizing reservoir performance. However, they are computationally expensive and require significant expertise to set up and interpret accurately.

4.4.4.1.4 Results and Discussion

This study presents improved ANN models for accelerating phase equilibrium calculations in compositional reservoir simulations. By focusing on predicting saturation pressures instead of treating stability as a classification problem, the proposed models require significantly less training data, leading to faster data generation and model training. The enhanced ANN models, validated with a three-component hydrocarbon mixture and applied to two different compositional simulation cases, show that they can maintain simulation accuracy while drastically reducing computational costs for stability tests and phase splitting calculations. The results demonstrate that using ANN models in compositional simulations can achieve the same accuracy with significantly lower computational costs, making them a valuable tool for reservoir simulations.

BIBLIOGRAPHY

Kun Wang, Jia Luo, Yizheng Wei, Keliu Wu, Jing Li, Zhangxin Chen. Practical application of machine learning on fast phase equilibrium calculations in compositional reservoir simulations, *Journal of Computational Physics*, 401, 2020, 109013, ISSN 0021-9991, https://doi.org/10.1016/j.jcp.2019.109013

4.4.4.2 Application-44: Prediction of Dead Oil Viscosity: Machine Learning vs. Classical Correlations

4.4.4.2.1 Background of the Application

The paper addresses the challenge of predicting dead oil viscosity, a critical parameter in reservoir engineering that is notoriously difficult to estimate accurately using traditional black oil correlations. Experimental determination of dead oil viscosity is both costly and time-consuming, prompting the need for an accurate and rapid predictive model. This study introduces six ML models—RF, LightGBM, XGBoost,

Multilayer Perceptron (MLP) Neural Network, Stochastic Real-Valued (SRV), and SuperLearner—to predict dead oil viscosity. Utilizing over 2000 PVT data points, the models cover a wide range of viscosities from light to heavy oils. The study finds that a functional form based on the API gravity and temperature yields the best performance, suggesting that additional parameters might be unnecessary. Among the tested models, the SuperLearner algorithm outperforms both other ML models and traditional empirical correlations, demonstrating its potential to replace empirical models for a wide range of viscosities.

4.4.4.2.2 Description of the Workflow and Machine Learning Model

This study explores the application of data-driven methods, including ML and neural networks, to optimize oil production. The aim is to predict production rates and identify key factors influencing reservoir performance.

4.4.4.2.3 Advantages and Limitations of the Synchronized Algorithm

Data-driven methods provide valuable insights into production optimization, but their effectiveness heavily depends on the quality and volume of available data. Additionally, these methods require substantial computational resources.

4.4.4.2.4 Results and Discussion

This study explores the application of multiple ML algorithms, including RF, LightGBM, XGBoost, MLP Neural Network, SRV, and SuperLearner, to predict dead oil viscosity based on various functional forms. The results indicate that the SuperLearner and XGBoost models show the most promise for accurately predicting dead oil viscosity, offering a cost-effective alternative to laboratory measurements. The SuperLearner model, in particular, enhances prediction accuracy by integrating the strengths of various base learners. The study highlights that with adequate training on previous data patterns, these ML models can effectively predict unknown instances, confirming the superior performance of the SuperLearner model. The simplest functional form, relying solely on the API gravity and temperature, is sufficient for accurate predictions, negating the need for additional variables. The SuperLearner model is shown to be robust, reliable, and applicable across a wide range of viscosities, making it a valuable tool for reservoir simulations and EOR processes.

BIBLIOGRAPHY

F. Hadavimoghaddam, M. Ostadhassan, E. Heidaryan, M.A. Sadri, I. Chapanova, E. Popov, A. Cheremisin, and S. Rafieepour. Prediction of Dead Oil Viscosity: Machine Learning vs. Classical Correlations. *Energies* 2021, 14, 930. https://doi.org/10.3390/en14040930

4.4.4.3 Application-45: Implementation of Multilayer Perceptron (MLP) and Radial Basis Function (RBF) Neural Networks to Predict Solution Gas–Oil Ratio

4.4.4.3.1 Background of the Application

In this study, the accurate estimation of the PVT properties of reservoir oils is essential for reservoir calculations, performance prediction, and designing optimal production conditions. This research aims to create reliable and intelligent models using

MLP and RBF neural networks to estimate the gas–oil ratio based on bubble point pressure, reservoir temperature, oil gravity (API), and gas specific gravity. Utilizing 710 experimental data sets from various global crude oil samples, the performance of these models was compared against well-known empirical correlations through statistical and graphical error analyses. Results showed that the proposed models significantly outperformed traditional empirical correlations, with the RBF model demonstrating superior accuracy and efficiency over the MLP model.

4.4.4.3.2 Description of the Workflow and Machine Learning Model

The study investigates the use of ML algorithms, including RFs and SVMs, to optimize EOR processes. By analyzing historical production data and reservoir characteristics, these models aim to improve recovery rates and reduce operational costs.

4.4.4.3.3 Advantages and Limitations of the Synchronized Algorithm

ML-based approaches provide significant improvements in optimizing EOR processes. However, the accuracy of these models depends heavily on the quality of input data and the robustness of the training process, requiring careful validation and calibration.

4.4.4.3.4 Results and Discussion

This study successfully developed MLP and RBF neural network models to predict the solution gas–oil ratio of oil reservoirs. Using a comprehensive dataset of 710 experimental data points from diverse geographical locations, the models incorporated inputs such as bubble point pressure, reservoir temperature, oil gravity, and gas specific gravity. The results indicated that both MLP and RBF models are highly reliable for predicting the solution gas–oil ratio within specified ranges of these parameters. Notably, the RBF model showed greater efficiency and accuracy compared to the MLP model. Statistical analysis confirmed the validity of the models, with less than 3% of the data points identified as outliers. The study concludes that the developed models provide accurate and reliable predictions and can be effectively used when experimental methods are unavailable.

BIBLIOGRAPHY

Aref Hashemi Fath, Farshid Madanifar, Masood Abbasi, Implementation of multilayer perceptron (MLP) and radial basis function (RBF) neural networks to predict solution gas–oil ratio of crude oil systems, *Petroleum*, 6 (1), 2020, 80–91, ISSN 2405-6561, https://doi.org/10.1016/j.petlm.2018.12.002

4.4.4.4 Application-46: Modeling Viscosity of CO_2 at High Temperature and Pressure Conditions

4.4.4.4.1 Background of the Application

The study applies various ML techniques to predict the viscosity of CO_2 under diverse thermodynamic conditions. Utilizing a comprehensive dataset of 1124 experimental points covering temperatures from 220 K to 673 K and pressures from 0.1 to 7960 MPa, the research models CO_2 viscosity as a function of temperature and density. The techniques employed include MLP, gene expression programming (GEP), and the group method of data handling (GMDH). Among the MLP models trained with different backpropagation algorithms, the LM algorithm demonstrated superior performance with an RMSE of 0.0012 mPa·s and a coefficient of determination (R^2) of 0.9999. The MLP-LM model outperformed existing ML models for CO_2 viscosity prediction, and the GEP-derived correlation surpassed previous explicit correlations.

4.4.4.4.2 Description of the Workflow and Machine Learning Model

This research focuses on optimizing reservoir simulation models using AI techniques. The study employs reinforcement learning and GAs to refine simulation parameters and improve model accuracy, particularly in complex reservoir environments.

4.4.4.4.3 Advantages and Limitations of the Synchronized Algorithm

AI-driven optimization techniques enhance the accuracy and efficiency of reservoir simulations. However, they require extensive computational power and careful tuning to avoid convergence issues, which can be a limitation in certain applications.

4.4.4.4.4 Results and Discussion

This research successfully developed robust and user-friendly ML models to accurately predict CO_2 viscosity over a wide range of temperatures and pressures, using temperature and density as inputs. A dataset of 1128 experimental points was used for model calibration and validation. The MLP model optimized with the LM algorithm (MLP-LM) emerged as the most accurate, exhibiting the lowest RMSE values and highest R^2 scores across training, testing, and overall datasets. The MLP-LM model outperformed previous best models, and the new Genetic Programming (GEP)-based explicit correlation showed better accuracy than existing correlations. Trend analysis indicated that both temperature and density positively correlate with CO_2 viscosity, aligning with experimental observations. The study provides a reliable tool for predicting CO_2 viscosity, enhancing the design and optimization of processes involving CO_2.

BIBLIOGRAPHY

Menad Nait Amar, Mohammed Abdelfetah Ghriga, Hocine Ouaer, Mohamed El Amine Ben Seghier, Binh Thai Pham, Pål Østebø Andersen. Modeling viscosity of CO2 at high temperature and pressure conditions, *Journal of Natural Gas Science and Engineering*, 77,2020,103271, ISSN 1875-5100, https://doi.org/10.1016/j.jngse.2020.103271

4.4.4.5 Application-47: Interpreting Pressure and Flow Rate Data from Permanent Downhole Gauges with Convolution-Kernel-Based Data Mining Approaches

4.4.4.5.1 Background of the Application

The paper presents a novel method for analyzing data from PDGs, which are often plagued by noise, gaps, and outliers. This method utilizes a data mining approach based on convolution kernels to uncover the true underlying data signal. The technique is noteworthy for its ability to perform deconvolution, thereby computing the reservoir model even when the data is noisy. Originally designed for natural language processing, the convolution kernel method decomposes data into simpler parts and evaluates them using a kernel function.

The data mining process is split into two stages: learning and prediction. During the learning stage, PDG data is used to train the convolution-kernel-based algorithm until it converges, producing a reservoir model in the form of polynomials in a high-dimensional space. In the prediction stage, this model is used to predict pressure responses to various flow rate histories, which helps reveal the underlying reservoir model. The method has been tested with both synthetic and real field data, demonstrating robustness in handling problematic data situations such as significant noise, outliers, incomplete production histories, and unknown initial pressures.

4.4.4.5.2 Description of the Workflow and Machine Learning Model

This study discusses advances in ML for reservoir management, focusing on neural networks and ensemble methods. These techniques are applied to optimize reservoir operations and enhance decision-making. The integration of ML approaches allows for more dynamic and responsive reservoir management strategies.

4.4.4.5.3 Advantages and Limitations of the Synchronized Algorithm

ML advances offer significant benefits for reservoir management, including improved optimization and decision-making capabilities. However, these methods require substantial computational resources and expertise in model calibration. The quality of input data is crucial for ensuring the reliability and accuracy of predictions.

4.4.4.5.4 Results and Discussion

The study concludes that the convolution kernel method is effective in analyzing PDG data even in the presence of various data quality issues. The method successfully handles outliers and aberrant segments without requiring preprocessing, tolerates moderate levels of noise, and performs well with incomplete production histories by using effective rate corrections. Additionally, the method can determine the initial reservoir pressure through optimization, even when the initial pressure is unknown or inappropriate.

Several key findings were highlighted:

1. The method can tolerate moderate outliers and aberrant segments, maintaining prediction accuracy without prior data cleaning.
2. Effective rate readjustment helps improve prediction accuracy when dealing with incomplete production histories.
3. An outer optimization loop can effectively estimate the initial reservoir pressure, enhancing the model's prediction capability.

Overall, the convolution kernel method proves to be a robust and reliable tool for interpreting PDG data, offering significant advantages over traditional deconvolution algorithms, especially in challenging data scenarios.

BIBLIOGRAPHY

Yang Liu and Roland N. Horne. "Interpreting Pressure and Flow Rate Data from Permanent Downhole Gauges with Convolution-Kernel-Based Data Mining Approaches." *Paper Presented at the SPE Annual Technical Conference and Exhibition*, New Orleans, LA, September 2013. https://doi.org/10.2118/166440-MS

4.4.4.6 Application-48: Applying Machine Learning Techniques to Interpret Flow Rate, Pressure and Temperature Data from Permanent Downhole Gauges

4.4.4.6.1 Background of the Application

In this study, various ML techniques were applied to interpret flow-rate, pressure, and temperature data collected from PDGs. These gauges continuously record data, providing valuable insights into reservoir conditions. The research focused on developing efficient and accurate models for deconvoluting and interpreting these data streams. By formulating the problem as a linear regression on the parameters connecting nonlinear flow-rate features with pressure targets, the study achieved significant computational efficiency without sacrificing accuracy. Additionally, kernel methods and Kernel ridge regression were employed to address complex reservoir behaviors and overfitting issues, leading to the successful recovery of detailed reservoir characteristics such as wellbore storage and boundary effects. The study also explored the potential of using temperature data for pressure history reconstruction and co-interpretation when flow-rate data are unavailable.

4.4.4.6.2 Description of the Workflow and Machine Learning Model

The paper explores AI techniques for EOR, including SVMs and GAs. These methods are used to model reservoir behavior and optimize recovery strategies. The study emphasizes the importance of integrating AI with traditional reservoir engineering approaches to improve recovery efficiency.

4.4.4.6.3 Advantages and Limitations of the Synchronized Algorithm

AI techniques offer promising tools for enhancing oil recovery, providing improved modeling and optimization capabilities. However, the success of these methods depends on the quality and representativeness of input data. Computational demands and model complexity can be challenging, requiring efficient data management and processing.

4.4.4.6.4 Results and Discussion

This research demonstrated that ML techniques are highly effective for interpreting PDG data. Key conclusions include the ability to formulate the problem as a linear regression, which provides a closed-form solution and rapid computation. The linear approach performed well on simple datasets with low pressure variations, effectively denoising and deconvolving pressure signals without breakpoint detection. For more complex datasets with higher data variation, Kernel ridge regression successfully captured detailed reservoir behaviors, balancing feature expansion with regularization to control prediction errors. Additionally, ML models were shown to effectively utilize temperature data for reconstructing pressure histories and co-interpreting temperature and pressure data in the absence of flow-rate data, highlighting the versatility and robustness of these techniques in reservoir analysis.

BIBLIOGRAPHY

Chuan Tian and Roland N. Horne. "Applying Machine Learning Techniques to Interpret Flow Rate, Pressure and Temperature Data from Permanent Downhole Gauges." *Paper Presented at the SPE Western Regional Meeting*, Garden Grove, CA, April 2015. https://doi.org/10.2118/174034-MS

4.4.4.7 Application-49: Application of Machine Learning Ideas to Reservoir Fluid Properties Estimation

4.4.4.7.1 Background of the Application

This paper explores the application of various ML techniques to predict reservoir fluid properties. The study utilizes a dataset comprising PVT data from 296 oil and 72 gas reservoirs in the Niger Delta. The ML methods examined include KNN, Support Vector Regression, Kernel ridge regression, RF, Adaptive Boosting (Adaboost), and Collaborative Filtering. These techniques were employed to predict properties such as saturation pressure, formation volume factor, oil viscosity, and condensate gas ratio. The results indicate that these ML approaches often outperform traditional industry correlations, suggesting significant potential for enhancing fluid property predictions in the oil and gas industry.

4.4.4.7.2 Description of the Workflow and Machine Learning Model

This study utilizes ML techniques such as ANNs for generating synthetic well logs. The approach involves training ANNs on existing well log data to predict missing or incomplete logs, enhancing subsurface characterization in regions with limited data availability.

4.4.4.7.3 Advantages and Limitations of the Synchronized Algorithm

ANNs offer powerful capabilities for capturing nonlinear relationships in well log data, providing accurate predictions of missing logs. However, the success of ANNs depends on the quality and quantity of training data, and the models can be prone to overfitting. The approach requires significant computational resources, particularly for training complex networks with multiple layers.

4.4.4.7.4 Results and Discussion

The study evaluates six ML techniques—Collaborative Filtering, Random Forest, Adaptive Boosting, Kernel ridge regression, Support Vector Regression, and KNN—on a dataset of oil and gas PVT data from the Niger Delta. The results show promising performance in predicting oil PVT parameters such as bubble point pressure, oil formation volume factor, and saturated oil viscosity. Ensemble methods like Adaboost and RF were particularly effective, particularly in predicting saturated oil viscosity without needing dead oil viscosity data. However, the methods were less effective with the limited gas PVT data. Overall, the study suggests that ML techniques, including those adapted from other fields like consumer product recommendation systems, hold potential for improving reservoir fluid property estimations. Further research with larger datasets and more advanced techniques like DL is recommended to validate and extend these findings.

BIBLIOGRAPHY

Chukwuma Onwuchekwa "Application of Machine Learning Ideas to Reservoir Fluid Properties Estimation." *Paper Presented at the SPE Nigeria Annual International Conference and Exhibition*, Lagos, Nigeria, August 2018. https://doi.org/10.2118/193461-MS

4.4.4.8 Application-50: Reliable Models for Determining the Pressure–Volume–Temperature (PVT) Properties Using Artificial Intelligence

4.4.4.8.1 Background of the Application

In the petroleum industry, accurately determining PVT properties is essential for predicting the performance of oil and gas reservoirs. Traditional methods to measure these properties are expensive and time-consuming, prompting the development of various predictive correlations. However, these often show significant deviations from actual values. This paper introduces reliable models utilizing artificial intelligence to estimate PVT properties, specifically bubble point pressure, formation volume factor, and solution gas–oil ratio. Using ANNs and a dataset of 250 samples, the models were evaluated for reliability through Average Absolute Percentage Error (AAPE) and the coefficient of determination (R-value). The results indicated that these models, with an average error of about 7.7%, outperform traditional correlations like those by Standing and Al-Marhoun, which exhibit errors around 11.7%. The AI models enhance production management by providing accurate PVT property estimates, thereby reducing the time and cost associated with laboratory measurements.

4.4.4.8.2 Description of the Workflow and Machine Learning Model

This paper presents models to determine the PVT properties using AI techniques. The study employs ANNs to estimate bubble point pressure, formation volume factor, and solution gas–oil ratio based on reservoir pressure, temperature, and hydrocarbon densities.

4.4.4.8.3 Advantages and Limitations of the Synchronized Algorithm

The AI-based models offer improved accuracy and reduce estimation errors for PVT properties. However, they require a significant amount of data and computational resources. The performance of the models is sensitive to the quality of input data, which can affect prediction reliability.

4.4.4.8.4 Results and Discussion

This study presents robust models for predicting the PVT properties of Sudanese crude oils using ANNs. The models were fine-tuned for optimal performance, validated using a dataset of 250 samples, and evaluated for accuracy. The developed ANN models predict bubble point pressure with an average error of 11.39%, significantly improving the accuracy of the PVT property estimation compared to traditional methods. The errors in estimating the solution gas–oil ratio and formation volume factor were reduced from about 20% to 10% and from 6.6% to 2.5%, respectively. These AI-driven models can be integrated into commercial software to enhance the reliability of simulation results, thus improving the overall efficiency of production management.

BIBLIOGRAPHY

Amjed Hassan, Abdulazeez Abdulraheem, and Abdelmjeed Mohamed. "Reliable Models for Determining the Pressure-Volume-Temperature PVT Properties Using Artificial Intelligence Technique." *Paper Presented at the International Petroleum Technology Conference*, Dhahran, Kingdom of Saudi Arabia, January 2020. https://doi.org/10.2523/IPTC-19884-Abstract

5 AI Applications in Well Completion, Production, and Stimulation

Ekrem Alagoz, Haowei Tang, Ferit Ozdaban, Yasin Ozkan, Emre Can Dündar, Mehmet Melih Oskay, Celal Hakan Canbaz, Luigi Saputelli, Mufrettin Murat Sari, Salih Tutun, and Cenk Temizel

5.1 INTRODUCTION

In the realm of oil and gas exploration and production, the integration of cutting-edge technologies has become paramount in driving efficiency, productivity, and profitability. Among these technologies, machine learning (ML) and artificial intelligence (AI) stand out as transformative tools capable of revolutionizing traditional workflows. This chapter delves into the myriad applications of ML and AI in the critical domains of well completion, production, and stimulation within the petroleum industry.

The introduction of ML and ai into these domains represents a paradigm shift, offering unprecedented insights and predictive capabilities to optimize operations and maximize resource recovery. By leveraging vast amounts of data generated throughout the lifecycle of a well, from exploration to abandonment, operators can make informed decisions in real-time, leading to enhanced reservoir characterization, improved well performance, and increased profitability.

This chapter serves as a comprehensive guide to the utilization of ML and AI methodologies across various workflows, ranging from well completion optimization to artificial lift systems. Through a systematic exploration of these applications, readers will gain a deeper understanding of how advanced analytics can be harnessed to address challenges such as reservoir heterogeneity, production decline, and operational inefficiencies.

By illuminating the synergies between ML and data analytics (DA) techniques and traditional petroleum engineering practices, this chapter underscores the transformative potential of digital technologies in reshaping the future of well operations. As the industry continues to evolve in response to technological advancements and market dynamics, the insights offered in this chapter will be invaluable in guiding stakeholders toward a more sustainable and competitive future.

DOI: 10.1201/9781003617327-5

5.2 CATEGORIZATION OF WORKFLOWS USED IN THIS CHAPTER

5.2.1 Well Completion and Optimization

Well completion and optimization represent pivotal stages in the lifecycle of an oil or gas well, with significant implications for overall productivity and profitability. In this section, we delve into the application of ML and AI methodologies to streamline and enhance the completion process. By leveraging advanced DA and predictive modeling, operators can optimize wellbore configurations, select optimal completion techniques, and mitigate risks associated with reservoir uncertainties. Through the integration of real-time monitoring systems and intelligent algorithms, operators can also optimize production rates, minimize downtime, and extend the economic lifespan of wells. This section explores the various techniques and strategies employed in well completion and optimization, highlighting the transformative impact of AI and ML in maximizing reservoir recovery and operational efficiency.

5.2.2 Well Stimulation

Well stimulation techniques play a crucial role in enhancing the productivity of oil and gas reservoirs, particularly in unconventional formations where hydraulic fracturing (HF) (fracking) is prevalent. In this section, the application of ML and AI in optimizing well stimulation processes is investigated. By analyzing historical production data, geomechanics properties, and fluid dynamics, AI algorithms can predict the most effective fracturing designs and injection parameters tailored to specific reservoir conditions. Moreover, ML models can identify correlations between stimulation treatments and production outcomes, enabling operators to fine-tune their strategies for optimal performance. Through the integration of real-time monitoring and feedback loops, AI -driven stimulation systems can dynamically adjust fracturing operations in response to reservoir changes, ultimately maximizing recovery rates while minimizing environmental impact and operational costs. This section elucidates the role of AI and ML in revolutionizing well stimulation practices, ushering in a new era of efficiency and sustainability in reservoir development.

5.2.3 Fracture Identification

Fracture identification is a critical aspect of reservoir characterization, essential for optimizing well performance and maximizing hydrocarbon recovery. The application of ML and AI techniques in fracture identification processes is explored in this section. By analyzing seismic data, well logs, and other geophysical measurements, AI algorithms can detect and characterize subsurface fractures with unprecedented accuracy and efficiency. ML models trained on vast datasets can recognize complex patterns and anomalies indicative of fracture networks, aiding in the identification of sweet spots and optimal well placement. Furthermore, AI -driven fracture identification systems can facilitate real-time monitoring and interpretation of fracturing operations, enabling operators to adjust their strategies on the fly for enhanced reservoir connectivity and production rates. Through the integration of advanced imaging technologies and predictive analytics, AI and ML are revolutionizing fracture identification methodologies, empowering operators to unlock the full potential of their reservoirs with greater precision and confidence.

5.2.4 Production Aspects

Production optimization lies at the heart of efficient oil and gas operations, necessitating constant innovation and adaptation to extract maximum value from reservoirs. In this section, the integration of ML and AI techniques is explored to address various production challenges. Leveraging real-time data streams from sensors, downhole gauges, and surface facilities, AI algorithms can analyze production trends, detect abnormalities, and forecast future performance with unprecedented accuracy. ML models trained on historical production data can identify optimal operating conditions, recommend production strategies, and optimize reservoir management practices. Additionally, AI -driven predictive analytics can facilitate proactive maintenance scheduling, reducing unplanned downtime and operational costs. By harnessing the power of AI and ML, operators can achieve higher production rates, lower lifting costs, and improved asset integrity, ultimately enhancing the economic viability and sustainability of oil and gas production operations.

5.2.5 Artificial Lift Systems

Artificial lift systems play a crucial role in maintaining production rates and extending the economic lifespan of oil and gas wells, particularly in mature fields with declining reservoir pressure. In this section, the application of ML and AI methodologies is examined for optimizing artificial lift systems. By analyzing historical production data, downhole pressure measurements, and equipment performance records, AI algorithms can identify optimal lift methods, predict equipment failures, and recommend maintenance schedules to maximize uptime and efficiency. ML models trained on operational data can also optimize lift system parameters in real-time, adjusting pump speeds and fluid injection rates to match changing reservoir conditions and production demands. Furthermore, AI -driven predictive maintenance algorithms can anticipate equipment degradation and recommend preemptive interventions, reducing costly downtime and mitigating production losses. Through the integration of AI and ML technologies, artificial lift systems can be optimized to enhance production rates, minimize operating costs, and prolong the economic viability of oil and gas assets.

5.3 OVERVIEW OF AI METHODOLOGIES

A comprehensive overview is provided of the methodologies underpinning AI, ML, and DA in the context of oil and gas applications. Fundamental principles and techniques used in AI —such as neural networks, natural language processing, and expert systems—are explored, with emphasis on their relevance to reservoir engineering, production optimization, and asset management. The utility of various ML algorithms, including regression analysis, decision trees, and clustering methods, is elucidated in tasks like pattern recognition, predictive modeling, and anomaly detection. Additionally, the role of DA in extracting actionable insights from large volumes of structured and unstructured data is discussed, supporting informed decision-making and operational optimization. Through this comprehensive overview, readers will gain a deeper understanding of the AI methodologies driving innovation and efficiency in the oil and gas industry.

5.4 APPLICATIONS OF ARTIFICIAL INTELLIGENCE (AI)

5.4.1 Well Completion and Optimization

5.4.1.1 Application-1: Preventing Slugging by Tuning Choke through Machine Learning

5.4.1.1.1 Background of Application

This paper discusses an ML approach to mitigate or avoid slugging in gas-lift wells, a phenomenon when multi-phase fluid production suffers from sudden bursts of gas, leading to spikes in pressure and problems at the separator. The production choke and gas injection choke are two points at which the operator may influence the slug. However, the operator must know that a slug is going to occur in advance so that avoidance actions can be implemented. The paper proposes a long short-term memory (LSTM) network to forecast the well state, which includes the wellhead pressure and other variables, 5 hours in advance, given typical field instrumentation of the well. This paper presents a successful forecast of slugs in three different wells in the Brage oilfield.

5.4.1.1.2 Description of AI Workflow

In this tutorial, slugging is forecasted by developing a Vanilla LSTM model for multivariate time series prediction. The LSTM, a type of artificial neural network (ANN), is utilized to capture temporal dependencies and patterns within the data. Unlike standard feedforward neural networks, LSTM has feedback connections. Such a recurrent neural network can process single data points and entire data sequences.

First, random samples were generated and basic data preprocessing was conducted. Subsequently, the Vanilla LSTM structure was explored in det AI l and the model was constructed incrementally. Finally, a prediction of wellhead pressure was performed for a specific time step $Pt + \tau$.

5.4.1.1.3 Advantages and Limitations of AI Methodology

In time-series modeling challenges involving long-term memory, LSTM offers unquestionable advantages. In the lengthy sequence training procedure, it can resolve the gradient explosion and gradient disappearance problems. An LSTM's heavy reliance on stepwise forecasts is one of its primary drawbacks when it comes to time series forecasting. As demonstrated in the example, wellhead pressure at time *t* was predicted using the preceding four variables. As a result, long-range forecasting using an LSTM might not work.

5.4.1.1.4 *Results and Recommendations*

The LSTM network helps forecast slugs successfully 5 hours in advance, which is a significant improvement in avoiding or mitigating slugging in gas-lift wells. The next step is to use an optimization algorithm to answer the question of how to proactively change the production choke and gas injection choke to reduce the upcoming slug. Although this final step remains to be done due to operational restrictions, work is ongoing to collect the required data.

BIBLIOGRAPHY

Bangert, Patrick. "Preventing Slugging by Tuning Choke through Machine Learning." *Paper Presented at the International Petroleum Technology Conference*, Dhahran, Kingdom of Saudi Arabia, January 2020. https://doi.org/10.2523/IPTC-19931-Abstract

5.4.1.2 Application-2: Real-Time Drill String Vibration Characterization Using Machine Learning

5.4.1.2.1 *Background of Application*

The paper proposes a new technique for detecting and characterizing drill string vibrations in real-time from solely surface measurements using an ML approach. This technique uses the frequency spectrum of the surface data and extracts several features from previously drilled wells or simulated data to build a classifying model. The classifying model is trained to create a link between a specific set of frequency spectrum features extracted from the surface signal and a given vibration mode and its severity level. Validation of the new technique on recorded data demonstrates its capability to detect and characterize downhole dynamics such as stick-slip and lateral shocks from surface measurements.

5.4.1.2.2 *Description of AI Workflow*

Based on the literature reviews and the nature of the data, fast Fourier transform (FFT) was applied to the torque signal. The top four peak frequencies and their corresponding power values were then extracted and used as features. It's important for us to transform raw data into useful features that help us to understand the model better and increase its predictive power. It's clear that all the input variables are time series data, and the target variable: 'vibration_type' is made of four types of vibrations, including no/low vibrations, medium vibrations, high vibrations, and severe vibrations. As mentioned in the paper, the system builds a specific classifying model for each type of vibration mode, so it will build four different models for each type of vibration.

After completing feature engineering, both bagged decision tree (DT) and support vector machine (SVM) models were constructed to predict vibration types. The bagged DT model yielded the highest confidence score among the models evaluated. However, it should be noted that all evaluations were conducted using a simulated dataset.

5.4.1.2.3 *Advantages and Limitations of AI Methodology*

Bagged DT model is trained with the bagging algorithm and based on DT models. Bagging is a ML ensemble meta-algorithm designed to improve the stability and accuracy of ML algorithms used in statistical classification and regression. It also reduces variance and helps to avoid overfitting.

However, tree-based models are more effective on tabular data, since the raw data is a time-series data, readers can think of modifying or improving the feature engineering part or using deep learning (DL) models in time-series prediction or audio processing such as LSTM for prediction.

5.4.1.2.4 Results and Recommendations

The proposed real-time workflow can detect and characterize drillstring stick-slip and lateral shocks from high-frequency surface measurements using an ML approach. The technique uses the frequency spectrum of the surface data and extracts several features from previously drilled wells or simulated data to build a classifying model. The model is trained to create a link between a specific set of frequency spectrum features extracted from the surface signal and a given vibration mode and its severity level. Validation of the new technique on recorded data demonstrates its capability to detect and characterize downhole dynamics such as stick-slip and lateral shocks from surface measurements. The proposed technique has the potential to significantly reduce nonproductive time, improve the rate of penetration (ROP), and minimize the risk of downhole tool failure, thus reducing the cost of drilling operations.

BIBLIOGRAPHY

Millan, Ekaterina, Ringer, Maurice, Boualleg, Riadh, and Li, Denis. "Real-Time Drillstring Vibration Characterization Using Machine Learning." *Paper Presented at the SPE/IADC International Drilling Conference and Exhibition*, The Hague, The Netherlands, March 2019. https://doi.org/10.2118/194061-MS

5.4.1.3 Application-3: Identification of Coal Structures Using Geophysical Logging Data: Investigation by Kernel Fisher Discriminant Analysis

5.4.1.3.1 Background of Application

Because it is intimately linked to the porosity and permeability of the reservoirs, coal structure is crucial to the enrichment and effective development of coalbed methane (CBM) reservoirs. Since it affects HF and the overall effectiveness of CBM extraction, accurate coal structure identification is a crucial problem in CBM exploration and development. Conventional techniques for determining coal structure from mining seam observations or drill cores are costly and have limitations. As a result, geophysical logging has emerged as the most practical and economical method. Nevertheless, a number of linear correlations that were developed to explain the connections between well logs and coal structure are unable to adequately capture nonlinear linkages between them. To address this issue, a new quantitative coal structure identification model based on the kernel Fisher discriminant analysis (KFD) method was developed using geophysical logging data. The model was trained, tested, and optimized using 178 logging data sets from 15 CBM wells in Qinshui Basin, China. The KFD-based identification model has high prediction accuracy and exhibits strong capability during the modeling and generalization in the determination of nonlinear relationships, providing an efficient way for the coal structure prediction in basic research of coal reservoirs.

5.4.1.3.2 Description of AI Workflow

Data Collection: Geophysical logging data from 15 CBM wells were collected, focusing on characteristics relevant to coal structures.

- Normalization: Data was normalized to mitigate scale discrepancies among features.
- Feature Selection: Relevant features were selected based on their significance to coal structure classification.
- Model Building: The KFD model was chosen for its capability in handling nonlinear classification tasks.
- Model Training: The model was trained with the preprocessed dataset, learning the complex patterns associated with different coal structures.
- Model Evaluation: The KFD model's performance was validated against a held-out test set, demonstrating high accuracy.
- Prediction: The trained model can now be applied to new logging data for coal structure classification, facilitating targeted exploration efforts.

5.4.1.3.3 Advantages and Limitations of AI Methodology

Advantages:

- High Accuracy: Demonstrated effective in identifying coal structures, crucial for optimizing exploration strategies.
- Handling Nonlinear Relationships: KFD's nonlinear approach captures complex patterns not possible with linear models.
- Application Potential: The methodology has broad implications for geoscience, offering a robust tool for various classification tasks.

Limitations:

- Data Quality Dependence: The model's success heavily relies on the quality and comprehensiveness of the logging data.
- Complexity: Requires a deep understanding of ML and geophysical principles to implement and interpret the results effectively.

5.4.1.3.4 Results and Recommendations

This study proposed a method for identifying coal structure that utilizes the KFD analysis ML approach with geophysical logging data. The KFD method showed excellent performance in accurately identifying coal structure, with a high identification accuracy of 93.8% for the total data. The validation result from the blind well test further demonstrated the strong generalization ability of the built KFD model. Overall, the KFD method is recommended as a reliable and efficient tool for identifying coal structure. It can extract key classification features of coal structure from logging data and handle the highly complex nonlinear relationship between coal structure and geophysical logging data.

BIBLIOGRAPHY

Shi, Jinxiong, Zeng, Lianbo, Dong, Shaoqun, Wang, Jipeng, Zhang, Yunzhao. Identification of coal structures using geophysical logging data in Qinshui Basin, China: Investigation by kernel Fisher discriminant analysis, *International Journal of Coal Geology*, 217, 2020, 103314, ISSN 0166-5162, https://doi.org/10.1016/j.coal.2019.103314

5.4.1.4 Application-4: An Accelerated Gradient Algorithm for Well Control Optimization

5.4.1.4.1 Background of Application

This paper discusses the optimization of injection and production control settings in hydrocarbon reservoirs through simulation-assisted methods. Although gradient-based algorithms are typically preferred over derivative-free algorithms, the high number of dimensions in well control optimization problems makes gradient calculation difficult. Recent studies have focused on improving gradient approximation methods, but less attention has been paid to the framework in which these approximations are utilized. This paper proposes an improved framework using an adaptive moment estimation technique and compares it to the steepest descent framework in two case studies of increasing complexity. The proposed framework showed significant improvements in convergence speeds and optimal value compared to the steepest descent framework. The effect of gradient approximation accuracy and bound constraint handling techniques was also investigated.

5.4.1.4.2 Description of AI Workflow

The paper combines the SPSA gradient approximation with the Adam framework, aiming to be more efficient than the traditional steepest descent framework. The experiments indicated that increasing the number of perturbations used for calculating the average SPSA gradient did not affect the initial iteration of Adam but significantly impacted the steepest descent. To address this limitation, the first iteration was modified to a steepest descent (ascent) iteration. This ensures that the first proposed solution is guided by the SPSA, and subsequent iterations are based on the Adam update rule.

5.4.1.4.3 Advantages and Limitations of AI Methodology

There are some advantages listed in the paper: The algorithm benefits from the efficiency of SPSA gradient approximation and the flexibility of the Adam framework, making it more effective in dealing with complex optimization problems. However, its performance may be influenced by the quality of gradient approximation, as it depends on the number of perturbations and the precision of gradient calculations.

5.4.1.4.4 Results and Recommendations

The proposed adaptive moment estimation technique for gradient-based optimization of injection and production control settings in hydrocarbon reservoirs showed significant improvements in convergence speeds and optimal value compared to the steepest descent framework. The technique was less sensitive to gradient approximation accuracy, and the projection method outperformed the logarithmic transformation method for bound constraint handling. The proposed method provides a promising framework for future research into well control optimization problems.

BIBLIOGRAPHY

Arouri, Yazan, Sayyafzadeh, Mohammad. An accelerated gradient algorithm for well control optimization. *Journal of Petroleum Science and Engineering*, 190, 2020, 106872, ISSN 0920-4105, https://doi.org/10.1016/j.petrol.2019.106872

5.4.1.5 Application-5: Application of Machine Learning Methods to Well Completion Optimization: Problems with Groups of Interactive Inputs

5.4.1.5.1 Background of Application

In order to account for the impact of geographically diverse geological conditions on hydrocarbon production, this paper suggests a statistical model that examines the relationship between shale oil production and completion parameters, such as completion lateral length, total proppant, and number of HF stages. In order to pursue geological homogeneity, the authors create a non-parametric regression technique that combines a generalized additive model with a fused LASSO regularization. They also provide an alternating augmented Lagrangian method for estimating model parameters. The suggested approach may account for and examine the relationships between the completion parameters, as well as control or eliminate the heterogeneous non-completion effects. The authors apply their method to the analysis of a real case from a Permian Basin US onshore field and show how their model can account for the interaction between the completion parameters. Their results provide key findings on how completion parameters affect oil production and can lead to optimal well completion designs.

5.4.1.5.2 Description of AI Workflow

In order to investigate the relationships between shale oil output and well completion parameters, the study presents a statistical model. A non-parametric regression technique is created for the search of geological homogeneity by fusing fused LASSO regularization with a generalized additive model. Parameter estimations are done using an alternating augmented Lagrangian technique. This approach works especially well for examining how different completion parameters interact.

5.4.1.5.3 Advantages and Limitations of AI Methodology

The methodology showcases several advantages: The method investigates the relationships between production and key completion parameters, using a GAM. It accounts for geological confounding effects on production, which is a significant advancement over previous studies.

One limitation is that the model's effectiveness depends on accurately capturing the heterogeneous geological effects, which can be complex.

5.4.1.5.4 Results and Recommendations

In this paper, the authors propose a novel method for developing and applying a statistical non-parametric regression model that comprises a generalized additive model to investigate the functional associations between production and key completion parameters and a random effect to account for geological confounding effects on production. The method and its results provide a useful tool and guidance to completion engineers for well completion designs that maximize well productivity and save completion cost. Their approach demonstrates the potential of ML and statistical data mining techniques to solve challenging problems in the oil industry that require extensive field data analysis.

BIBLIOGRAPHY

Zhou, Peng, Lu, Ligang, Sang, Huiyan, and Dindoruk, Birol. "Application of Machine Learning Methods to Well Completion Optimization: Problems with Groups of Interactive Inputs." *Paper Presented at the SPE Annual Technical Conference and Exhibition*, Dubai, UAE, September 2021. https://doi.org/10.2118/206174-MS

5.4.1.6 Application-6: Estimation of Total Dissolved Solids (TDS) Using New Hybrid Machine Learning Models

5.4.1.6.1 Background of the Application

This paper investigates how completion parameters, such as lateral length, frac type, number of stages, proppant, and fluid volume, affect the production performance of unconventional wells. Unconventional reservoirs, like shale oil and natural gas, are complex in nature and have ultra-tight producing zones, which require effective well completion and stimulation treatments for successful and economical production. The goal of this study is to create ML models that use completion and areal parameters as input to predict production. The models employ tree ensemble extreme boosting machine (XGBOOST), ANN, and support vector machine (SVM). The study proposes an ML workflow, from feature engineering to interpretation of the models, to quantify the effects of completion parameters on the production rate of unconventional wells. The results show that it is possible to quantify the impact of completion parameters on production performance, even when subsurface parameters such as reservoir and geological data are limited. This study provides a framework for completion optimization in unconventional reservoirs.

5.4.1.6.2 Description of the Workflow and Machine Learning Model

The workflow employs various predictive models such as neural networks, SVMs, and DT-based ensemble models. These models correlate completion parameters like lateral length, frac type, number of stages, proppant, and fluid volume with the production rate of each well. The process involves feature engineering and interpretation of these parameters, leading to the identification of key performance indicators and optimum values for the best-performing completions.

5.4.1.6.3 Advantages and Limitations of the Method

The methodology showcases several advantages: The study utilizes AI for optimizing parameters in HF, such as stage numbers and proppant amount. Also, fuzzy logic and backward propagation are used for optimizing horizontal well placement through stratigraphic performance prediction.

Limitation includes data availability: The study acknowledges the complexity of running full-field simulations for unconventional resources, especially due to the diverse characteristics of shale formations. Additionally, the study indicates a requirement for region-specific analysis, suggesting the need for subdividing large fields into representative regions based on geomechanical properties.

5.4.1.6.4 Results and Discussion

This paper demonstrates the application of a data-driven ML workflow for completion optimization in unconventional reservoirs using limited formation variability attributes. The workflow shows that areal and geographical information can be used as input for predictive models, serving as proxies when reservoir characteristics are limited. True vertical depth (TVD) and county names are identified as strong proxies for the target sweet spot location for the well trajectory in terms of vertical depth and spatial location. The proposed workflow enables the quantification of the impact of completion parameters on production performance, especially when subsurface parameters are limited. The feature importance values obtained from the models can be used to optimize completion parameters in field development business plans. The workflow is successfully applied to the Eagle Ford basin, and the same approach can be applied to other basins.

BIBLIOGRAPHY

Baki, Sohrat, Temizel, Cenk, and Dursun, Serkan. "Well Completion Optimization in Unconventional Reservoirs Using Machine Learning Methods." *Paper Presented at the SPE Annual Technical Conference and Exhibition*, Dubai, UAE, September 2021. https://doi.org/10.2118/206241-MS

5.4.1.7 Application-7: Machine Learning and Artificial Intelligence Provides Wolfcamp Completion Design Insight

5.4.1.7.1 Background of Application

This paper presents an ML approach to evaluate the impact of completion and frac design on multi-fractured horizontal Wolfcamp B well production and economic performance. The authors assembled a database consisting of reservoir, completion, frac, and production information from 301 wells in a 10-county area in the Permian Basin. Two independent self-organizing data mappings were performed to determine the characteristics of wells assigned to each mapping bin. These two mappings were then combined into a reservoir type vs frac stage type matrix to remove systematic errors in measurement, bias, and inconsistencies in the database. A feed-forward neural network model was developed from the mapped data to estimate Wolfcamp B production and economics for completion and frac designs.

5.4.1.7.2 Description of AI Workflow

The methodology included two key steps: First, self-organizing data mappings were conducted on completion and frac stage parameters, as well as on reservoir and geology characteristics. This helped in organizing wells into different bins based on their characteristics. Second, these mappings were combined into a reservoir type vs frac stage type matrix. Finally, an ANN model was developed from this mapped data, which was used to estimate production and economics for different completion and frac designs in Wolfcamp B.

5.4.1.7.3 Advantages and Limitations of AI Methodology

The methodology showcases several advantages: 1. The data modeling indicates that Wolfcamp B well performance can be increased by 30–50% compared to standard designs. 2. Data preprocessing using self-organizing mapping helps suppress systematic errors and improve stability. 3. The ANN model can quickly forecast well production and optimize well completion designs. 4. This method can be applied to any formation, basin, or well type, given sufficient data. Notable limitations are not specifically mentioned.

5.4.1.7.4 Results and Recommendations

The ML method used by the authors to assess how completion and frac design affect the production and economic performance of multi-fractured horizontal Wolfcamp B wells offers valuable insights. It was discovered that reservoir data was crucial to comprehending the effects of frac design and completion. The best combinations of proppant intensity, fluid intensity, treatment rate, frac stage spacing, and perforation clustering are found in the most successful completion types. It is a unique and innovative approach that is quick and inexpensive to evaluate reservoir and completion parameters independently before integrating them into a production matrix. Any well type and formation can use the authors' methodology and process as long as there is enough data available.

BIBLIOGRAPHY

Shelley, Robert, Oduba, Oladapo, and Melcher, Howard. "Machine Learning and Artificial Intelligence Provides Wolfcamp Completion Design Insight." *Paper Presented at the SPE Hydraulic Fracturing Technology Conference and Exhibition*, Virtual, May 2021. https://doi.org/10.2118/204199-MS

5.4.1.8 Application-8: Machine Learning Approaches for Formation Matrix Volume Prediction from Well Logs: Insights and Lessons Learned

5.4.1.8.1 Background of the Application

Accurate resource assessment requires precise determination of formation rock type and petrophysical properties. Well-log data provides valuable insights into formation characteristics such as porosity, shale content, and saturation, essential for quantifying hydrocarbon reserves. While traditional physics-based interpretation methods can be time-consuming and challenging, particularly in heterogeneous formations, ML offers a more efficient and automated approach.

In this study, five supervised ML algorithms were employed to analyze well-log data from four wells in the Volve field. The goal was to accurately predict sandstone, carbonate, and shale mineral volumes in real-time. By leveraging multiple well-log parameters, ML models identified patterns and behaviors within the dataset.

Random forest (RF), gradient boosting, support vector regression (SVR), K-nearest neighbor, and ANNs were trained and evaluated using a robust framework that included random and grid search, cross-validation, and data splitting. All models demonstrated effective mineral volume prediction, with K-nearest neighbor, RF, and gradient boosting consistently outperforming the others. RF, in particular, excelled in minimizing overfitting due to its ensemble nature.

While ML models can significantly enhance reservoir characterization, they are not a complete replacement for traditional methods. Domain expertise remains crucial for guiding the application of ML algorithms, especially in scenarios with varying geological conditions. Nonetheless, the results highlight the potential of ML as a valuable tool for petrophysical evaluations, offering time-saving benefits and improving accuracy.

5.4.1.8.2 Description of the Workflow and Machine Learning Model

This paper presents an algorithm for detecting multiple partial blockages in liquid pipelines using inverse transient analysis. The method involves simulating transient flow in pipelines, analyzing pressure waves, and solving the inverse problem to identify blockage locations and severity. The algorithm integrates finite-volume methods and optimization techniques.

5.4.1.8.3 Advantages and Limitations of the Synchronized Algorithm

The algorithm provides a nonintrusive and accurate method for detecting blockages in pipelines. It is effective in both simple and complex blockage scenarios, making it versatile. However, the method requires high computational resources and is sensitive to measurement noise, which can impact the accuracy of blockage detection.

5.4.1.8.4 Results and Discussion

Accurate reservoir characterization and reserve estimation hinge on the effective interpretation of open-hole well logs. This research presents an ML-based approach to automate well-log processing and mineral volume calculation. By employing five supervised ML techniques and addressing dataset imbalances, a robust model capable of predicting mineral content with high accuracy was developed.

Key findings include:

- ML's Potential: Ensemble methods like RF and gradient boosting demonstrate superior performance in well-log regression analysis and mineral volume calculation.
- Alternative Models: While ANNs can be used, other models may offer better efficiency and accuracy, potentially saving computational time.
- Model Limitations: Distinguishing between sandstone and carbonate minerals remains a challenge for ML models, but they excel at predicting shale content. Geological variations and borehole conditions can impact well-log data and model performance.
- Data Considerations: Drilling conditions and local geology influence input log accuracy and ML model effectiveness. Models trained on one well may require adaptation for other reservoirs.

As ML technologies advance, their application in formation evaluation is expected to grow. While ML can be a valuable tool, it is not a substitute for traditional interpretation methods. By leveraging ML in conjunction with domain expertise, the E&P industry can optimize decision-making and operations.

BIBLIOGRAPHY

Kannaiah, Pamidi Venkata Durga, Maurya, Neetish Kumar. Machine learning approaches for formation matrix volume prediction from well logs: Insights and lessons learned. *Geoenergy Science and Engineering*, 229, 2023, 212086, ISSN 2949-8910, https://doi.org/10.1016/j.geoen.2023.212086

5.4.1.9 Application-9: Machine Learning Advanced Algorithm to Enhance Production Optimization: An ANN Proxy Modelling Approach

5.4.1.9.1 Background of the Application

This paper describes an innovative methodology for generating fast and highly accurate simulation models using advanced analytics and DL techniques. The methodology is used to create proxy models that replicate two types of simulators: Fluid-dynamics for the well/gathering network and Thermo-dynamics for process treatment. The paper discusses the fundamental prerequisites for training the neural network, including gathering enough training samples that cover uniformly the multidimensional input space. The paper also describes the random walk learning algorithm that has been developed to explore the possible input domain. The efficacy of the methodology is demonstrated by an overall 10^3–10^4 time reduction and a mean error smaller than 1%, resulting in significant time savings and unlocking the full potential of production optimization analysis.

5.4.1.9.2 Description of the Workflow and Machine Learning Model

The study utilized an ANN approach to create proxy models for production optimization. ANNs are a type of ML algorithm inspired by the human brain. They are capable of learning complex patterns and relationships within data. In this case, the ANN was trained to replicate the behavior of a physical simulator, allowing for faster and more efficient production optimization calculations. The backpropagation algorithm was used to adjust the ANN's weights and biases during training, ensuring that the model could accurately predict simulator outputs.

5.4.1.9.3 Advantages and Limitations of the Synchronized Algorithm

Advantages:

- Speed: ANNs can significantly reduce simulation time, enabling faster decision-making and optimization.
- Accuracy: The study demonstrated that ANN-based proxy models can achieve high accuracy, with a mean error of less than 1%.
- Flexibility: ANNs can be adapted to model various types of physical simulators and analyze different scenarios.

Limitations:

- Data Quality: The accuracy and generalizability of ANN models depend on the quality and quantity of training data.
- Complexity: ANNs can be complex to train and interpret, requiring specialized expertise.
- Overfitting: Careful hyperparameter tuning and validation are necessary to prevent overfitting, where the model becomes too specialized to the training data and performs poorly on new data.

5.4.1.9.4 Results and Discussion

This paper has presented an innovative methodology for generating fast and highly accurate simulation models using advanced analytics and DL techniques. The methodology has been successfully applied to create proxy models that replicate two types of simulators: Fluid-dynamics for well/gathering network and Thermo-dynamics for process treatment. The proxy models have been used to unplug physical simulators from the production optimization pipeline, resulting in an overall 10^3–10^4 time reduction and a mean error smaller than 1% while still maintaining high accuracy. This has unlocked the full potential of what-if and production optimization analysis through the evolutionary tool used, giving the possibility to optimize the production system online and real time, facing every different operating scenario and maximizing the value of the asset.

BIBLIOGRAPHY

Cadei, Luca, Rossi, Gianmarco, Montini, Marco, Fier, Piero, Milana, Diletta, Corneo, Andrea, Lancia, Lorenzo, Loffreno, Danilo, Purlalli, Elisabetta, Carducci, Francesco, Nizzolo, Riccardo, and Sophia, Gustavo. "Machine Learning Advanced Algorithm to Enhance Production Optimization: An ANN Proxy Modelling Approach." *Paper Presented at the International Petroleum Technology Conference*, Dhahran, Kingdom of Saudi Arabia, January 2020. https://doi.org/10.2523/IPTC-19992-MS

5.4.1.10 Application-10: Prediction of the Productivity Ratio of Perforated Wells Using Least Squares Support Vector Machine with Particle Swarm Optimization

5.4.1.10.1 Background of the Application

The productivity ratio is a critical metric for evaluating perforated completion efficiency. This study introduces a novel approach that combines three-dimensional finite element simulation and ML to accurately and rapidly predict this ratio.

A dataset was initially generated by simulating various perforation parameters using finite element analysis. Subsequently, a least squares support vector machine (LSSVM) model was developed to predict productivity ratios. Particle swarm optimization (PSO) was employed to optimize the LSSVM parameters.

The PSO–LSSVM model outperformed other combinations of optimization methods and ML algorithms, demonstrating rapid convergence, high accuracy, and strong generalization. This research offers valuable insights into optimizing perforation design and predicting the productivity of complex completions.

5.4.1.10.2 Description of the Workflow and Machine Learning Model

This study focuses on the optimization of HF fluid using ML techniques. Data from six horizontal Marcellus Shale wells were analyzed to optimize the concentration of friction reducer (FR) used in the HF process. The project employed XGBoost as the primary ML model to predict surface treating pressure and optimize FR concentration, ensuring effective and cost-efficient operations.

5.4.1.10.3 Advantages and Limitations of the Synchronized Algorithm

The XGBoost model demonstrated high accuracy in predicting surface treating pressure, making it effective for optimizing the HF fluid. However, when applied to adjacent wells, the model showed reduced performance, necessitating retraining for different wells. The method is computationally intensive, and the results are sensitive to the quality of input data, making real-time adjustments necessary.

5.4.1.10.4 Results and Discussion

This study introduces a novel method to predict the productivity ratio of perforated wells using three-dimensional finite element simulation and ML. Leveraging numerical simulation data, an LSSVM-based predictive model was developed.

PSO significantly outperformed the traditional grid search algorithm in optimizing LSSVM parameters, reducing training time by over half. Additionally, the PSO–LSSVM model demonstrated superior predictive accuracy and generalization compared to models based on ANN and RF.

This research validates the effectiveness and feasibility of combining PSO and LSSVM for predicting productivity ratios. It offers valuable insights into perforation optimization design and provides a new approach for predicting the productivity of complex completions and geothermal wells.

BIBLIOGRAPHY

Wang, H.; Zhang, C.; Zhou, B.; Xue, S.; Wang, F. Prediction of the Productivity Ratio of Perforated Wells Using Least Squares Support Vector Machine with Particle Swarm Optimization. *Applied Sciences* 2023, 13, 12978. https://doi.org/10.3390/app132412978

5.4.1.11 Application-11: Hydraulic Fracturing Fluid Optimization Using Machine Learning Approach

5.4.1.11.1 Background of the Application

FRs play a crucial role in reducing tubular drag during high-flow-rate HF operations. While low-viscosity slick water fracturing fluids use low FR concentrations, shale gas reservoirs often require higher concentrations, typically ranging from 0.5 to 2.0 gallons per thousand gallons (gpt). However, determining the optimal FR concentration for each stage remains a challenge. Overuse of FRs can lead to significant economic losses.

This study aimed to optimize FR usage in six horizontal Marcellus Shale wells by analyzing data from over 340 HF stages. A model was developed using various ML techniques to predict surface treating pressure and optimize FR concentration.

In the first phase, XGBoost emerged as the best-performing model for predicting surface treating pressure in stage 1 of Boggess 1H. While this model showed promise for other stages of Boggess 1H, it proved less effective for adjacent wells. This highlights the need for tailored models for individual wells.

5.4.1.11.2 Description of the Workflow and Machine Learning Model

This study focuses on the optimization of HF fluid using ML techniques. Data from six horizontal Marcellus Shale wells were analyzed to optimize the concentration of FR used in the HF process. The project employed XGBoost as the primary ML model to predict surface treating pressure and optimize FR concentration, ensuring effective and cost-efficient operations.

5.4.1.11.3 Advantages and Limitations of the Synchronized Algorithm

The XGBoost model demonstrated high accuracy in predicting surface treating pressure, making it effective for optimizing the HF fluid. However, when applied to adjacent wells, the model showed reduced performance, necessitating retraining for different wells. The method is computationally intensive, and the results are sensitive to the quality of input data, making real-time adjustments necessary.

5.4.1.11.4 Results and Discussion

The study utilized ML techniques to analyze data from the MSEEL project and optimize HF operations. In three distinct phases, models were developed and evaluated to predict surface treating pressure and optimize FR concentration. In phase 1, XGBoost emerged as the most effective model for predicting surface treating pressure in stage 1 of Boggess 1H. In phase 2, the model successfully generalized to other stages of Boggess 1H. However, in phase 3, the model failed to generalize to adjacent wells, highlighting the need for well-specific models. To further optimize operations, The inverse slope method was employed to adjust FR concentration. The research demonstrates the importance of real-time monitoring and intervention for enhancing HF efficiency and cost-effectiveness.

BIBLIOGRAPHY

Johar, A., Fathi, E., Adenan, M. F., and Belyadi, F. "Hydraulic Fracturing Fluid Optimization Using Machine Learning Approach: A Case Study of Marcellus Shale." *Paper Presented at the SPE Eastern Regional Meeting, Wheeling*, West Virginia, October 2023. https://doi.org/10.2118/215920-MS

5.4.2 AI/ML Application in Stimulation Operations

5.4.2.1 Application-12: Evaluating the Effects of Acid Fracture Etching Patterns on Conductivity Estimation Using Machine Learning Techniques

5.4.2.1.1 Background of the Application

This paper evaluates the impact of acid fracture etching patterns on conductivity estimation using ML techniques. The authors used a dataset of 97 experiments and 563 data points collected from an extensive literature review of acid fracture experiments. The data was cleaned, missing values were filled in, outliers and failed experiments were excluded, and the dataset was standardized. ML models were developed to predict acid fracture conductivity, including classification techniques to determine the most likely etching patterns, and a linear regression-based model was developed as a baseline for comparison with other models generated through ridge regression.

5.4.2.1.2 Description of the Workflow and Machine learning model

According to the literature reviews and the data types, standardization was selected to scale the raw data. The input variables contain both categorical and numerical variables. As mentioned in the paper, an etching classification model was first built using Random Forest (RF), followed by the development of a regression model for conductivity prediction using ridge regression. The model result is not ideal because all the variables are simulated with the same distribution.

5.4.2.1.3 Advantages and Limitations of the Synchronized Algorithm

The RF model is based on DT models and is trained using the bagging technique. An ensemble learning technique for classification, regression, and other problems, RFs, also known as random decision forests, works by building a large number of DTs during training. The class chosen by the majority of trees is the RF's output for classification problems. The tendency of DTs to overfit to their training set is compensated for by random decision forests. Nevertheless, compared to the boosting model generally, the prediction accuracy is lower. In situations when the independent variables are highly correlated, ridge regression is a technique for estimating the coefficients of multiple-regression models. By giving undesirable variables a coefficient of 0, it can effectively eliminate them. However, prediction accuracy is less than tree-based models in general.

5.4.2.1.4 Results and Discussion

Detailed empirical conductivity correlations and models were developed in this work for three carbonate rock types. The authors note that the development of a single empirical correlation for all rock types often results in significant errors in conductivity estimation, as different rock types result in different etching patterns that cannot be explained by a single correlation. The authors conclude that ML techniques can be used effectively to predict acid fracture conductivity and that considering the impact of etching patterns on conductivity can improve the accuracy of predictions.

BIBLIOGRAPHY

Desouky, Mahmoud, Aljawad, Murtada Saleh, Alhoori, Hamed, and Al-Shehri, Dhafer. "Evaluating the Effects of Acid Fracture Etching Patterns on Conductivity Estimation Using Machine Learning Techniques." *Paper Presented at the SPE Europec*, Virtual, December 2020. https://doi.org/10.2118/200527-MS

5.4.2.2 Application-13: Prediction of Optimum Injection Rate for Carbonate Acidizing Using Machine Learning

5.4.2.2.1 Background of the Application

The purpose of this paper is to use ML to estimate the ideal injection rate for carbonate acidizing. The ideal injection rate produces a prominent wormhole dissolution pattern in linear core flood studies. However, in order to determine the ideal injection rate, models need reaction kinetic parameters. This work uses a variety of data points from linear core floods to create a new, user-friendly ML model that predicts the ideal injection rate based on three factors: acid concentration, rock, and injection rate. 170 experimental data points gathered from Lavoux, Indiana, and Desert Rose limestone make up the data used in this investigation. The results show that the newly developed model is capable of predicting the pore-volume-to-breakthrough (PVBT) along with the optimal injection rate with 90% accuracy. This model will reduce the number of laboratory experiments needed to obtain the PVBT curve and narrow the search for the optimal injection rate. Furthermore, the results from this model can be upscaled to the field scale.

5.4.2.2.2 Description of the Workflow and Machine Learning Model

The paper built and evaluated the performance of three different ML methods, using eight features to predict the PVBT. A successful model can be very helpful in reducing the number of lab experiments needed to obtain the PVBT curve much more efficiently.

The dataset contains 170 experimental data of HCI strength, length, diameter, porosity, permeability, temperature, pressure, injection rate, and PVBT from the Indiana, Desert Rose, and Lavoux limestone. The three methods are feed-forward neural network using the Levenberg–Marquardt algorithm, fuzzy inference system, and SVM. Models for each of the methods are compared using the correlation coefficient, absolute average percentage error, or root mean squared error (RMSE) to select the best-performing one and its corresponding parameters.

5.4.2.2.3 Advantages and Limitations of the Method

Due to the unavailability of the original dataset used in the paper, the dataset was generated based on the reported statistics using a normal distribution function. This approach impacted the results, as certain statistical characteristics and inter-feature correlations were lost in the process. However, since the true distribution of the data and the inter-feature correlations are unknown, no further corrective actions could be taken to address this limitation.

The author used MATLAB to realize the three methods. The functions are trainlm, GENFIS2, and SVM. Due to the unavailability of GENFIS2's source code and the abstract nature of the underlying fuzzy inference system, it was not feasible to implement the exact functionality in Python. As an alternative to the *trainlm* function, we utilized a TensorFlow-based module implementing the Levenberg–Marquardt algorithm for training the models. For the third method, support vector regression (SVR) was applied to perform the regression task. The Levenberg–Marquardt neural network is faster to converge than either the GN or gradient descent on its own. It can handle models with multiple free parameters, and it can find an optimal solution even if the initial guess is far. However, it can perform slowly for very large sets.

5.4.2.2.4 Results and Discussion

In conclusion, this paper provides a powerful tool that efficiently predicts the optimum injection rate for carbonate acidizing using ML. This tool will help production engineers design successful acid jobs by reducing the number of laboratory experiments needed to obtain the PVBT curve and by providing the optimal injection rate for a dominant wormhole dissolution pattern. The model can be upscaled to the field scale to predict the optimal injection rate for carbonate formations.

BIBLIOGRAPHY

Sidaoui, Ziad, Abdulraheem, Abdulazeez, and Abbad, Mustapha. "Prediction of Optimum Injection Rate for Carbonate Acidizing Using Machine Learning." *Paper Presented at the SPE Kingdom of Saudi Arabia Annual Technical Symposium and Exhibition*, Dammam, Saudi Arabia, April 2018. https://doi.org/10.2118/192344-MS

5.4.2.3 Application-14: Intelligent Prediction of Acid-fracturing Performance in Carbonates Reservoirs

5.4.2.3.1 Background of the Application

A common method for improving fracture conductivity in carbonate reservoir stimulation is acidizing joint HF. However, a thorough understanding of acid–rock interactions, rock strength, stress applied to the rock, and their impact on fracture conductivity is necessary to forecast acid fracturing performance for optimal fracturing project design. Numerous prediction limitations exist for acidized fracture conductivity with closure stress in the models that are currently available. AI is proposed to get a more accurate forecast for acid-fracture conductivity. In this study, intelligent models that can provide an accurate design for acid-fracture conductivity are developed using ANNs and adaptive neuro-fuzzy inference systems (ANFIS). Published experimental data from the literature is used to tr AI n the models. The results showed that both ANN and ANFIS models outperformed the currently available models.

5.4.2.3.2 Description of the Workflow

The paper describes the use of AI techniques such as ANN and ANFIS to predict acid fracturing performance in carbonate reservoirs. The methodology involves splitting the data into a training set (70%) and a testing set (30%). The AI models, ANN and ANFIS, are then trained with the experimental data. The ANN model employs a structured layer of neurons and processes the input parameters, while ANFIS combines fuzzy logic with neural networks to model complex systems. The models' performance is evaluated based on their accuracy in predicting acid fracture conductivity, with the results showing that both ANN and ANFIS models outperform existing models.

5.4.2.3.3 *Advantages and Limitations*

Advantages include more accurate predictions compared to traditional methods, as evidenced by their lower error rates and higher R^2 values. They are capable of handling complex relationships between various parameters. However, there are also limitations to these methodologies. The AI models may lack transparency in how they derive their results, as the physical relationships in the data are not always evident. Also, the accuracy of these models is highly dependent on the comprehensiveness of the training data; any gaps in the data can lead to skewed results.

5.4.2.3.4 *Results and Discussion*

The performance of acid fracturing in carbonate has been precisely captured by AI techniques. A variety of methods, including SVMs, fuzzy logic, and ANNs, were investigated in order to determine which model will perform better than the correlations that are now available. With an AAPE of 1.36, an SSE of 0.86, an R^2 of 0.998, and an RMSE of 0.18, ANFIS subtractive clustering outperformed both ANN and ANFIS grid partitioning at the majority of cluster radii, offering the best match for the data. However, as the physical relationship is frequently invisible to the user, AI techniques should be developed and used carefully. The developed models usually cover a limited range of data, and using the models beyond these limits introduces a tremendous amount of uncertainty to the estimation process.

BIBLIOGRAPHY

Eliebid, Mohammed, Hassan, Amjed M., Mahmoud, Mohamed, Abdulraheem, Abdulazeez, and Elkatatny, Salaheldin. "Intelligent Prediction of Acid-fracturing Performance in Carbonates Reservoirs." *Paper Presented at the SPE Kingdom of Saudi Arabia Annual Technical Symposium and Exhibition*, Dammam, Saudi Arabia, April 2018. https://doi.org/10.2118/192356-MS

5.4.2.4 Application-15: Hydraulic Fracturing Candidate-Well Selection by Interval Type-2 Fuzzy Set and System

5.4.2.4.1 *Background of the Application*

This study is focused on selecting a target formation(s) among a vast number of zones/sub-layers within a reservoir for HF. The candidate-well selection (CWS) process for HF is complex, nonlinear, non-equilibrium, and adherent with uncertainty. In this paper, the authors propose a novel methodology to address the problem of selecting an appropriate well or formation for HF treatment. The methodology uses Interval Type-2 Fuzzy Logic and Systems (IT2-FLSs), which are very effective for dealing with uncertainties. The procedure of applying this novel study in the area of HF CWS is illustrated through a case study in a carbonate reservoir.

5.4.2.4.2 *Description of the Workflow and Machine Learning Model*

The paper introduces a novel workflow for selecting candidate wells for HF using IT2-FLSs. The workflow involves defining the decision objective (selecting an optimal HF candidate), identifying fuzzy variables, developing membership functions for these variables, establishing domain expert rules, and finally, executing an evaluation procedure using IT2 fuzzy inference. By leveraging the fuzzy logic system to manage the inherent uncertainties in the decision-making process, the model provides an adaptable approach than traditional Type-1 Fuzzy Logic Systems (T1-FLSs).

5.4.2.4.3 Advantages and Limitations

The use of IT2-FLS leads to smoother response surfaces and more reliable decision-making in well selection, as demonstrated by better performance metrics like prediction efficiency error, mean squared error (MSE), and R-square values. However, one limitation is the complexity involved in implementing this system. It requires domain knowledge for setting up appropriate fuzzy variables and rules.

5.4.2.4.4 Results and Discussion

The results of this study demonstrate that the proposed methodology of using IT2-FLS for HF CWS is an effective approach. The IT2-FLS method is better at handling linguistic and numerical uncertainties than the classical T1-FLS method. The developed methodology is supported by simulation results, qualitative observations, and quantitative measures of errors. The authors have also shown that the proposed methodology provides a reliable model of HF CWS. The investigation presented in this paper is the first research that applied such a cutting-edge approach and tried to fill the gap between recent developments in uncertainty management through the utilization of IT2-FLSs in HF CWS.

BIBLIOGRAPHY

Zoveidavianpoor, Mansoor, Samsuri, Ariffin, and Shadizadeh, Seyed Reza. "Hydraulic Fracturing Candidate-Well Selection by Interval Type-2 Fuzzy Set and System." *Paper Presented at the International Petroleum Technology Conference*, Beijing, China, March 2013. https://doi.org/10.2523/IPTC-16615-MS

5.4.2.5 Application-16: Deep Recurrent Neural Network DRNN Model for Real-Time Multistage Pumping Data

5.4.2.5.1 Background Information

A new deep recurrent neural network (DRNN) model was developed to predict response variables like surface pressure in real-time during HF. The process is complex, and the measured data is often uncertain and noisy, so ML-based techniques are the best approach to resolve nonlinear variations. The proposed method uses an LSTM network-based approach using surface pumping data, which consists of real-time data captured within each stage, such as surface treating pressure, fluid pumping rate, and proppant rate. The method can predict diverter pressure response and provides high-resolution and accurate results, which are useful in oil and gas applications like HF and matrix acidizing.

5.4.2.5.2 Description of the Workflow

This paper employs DRNN to predict surface pressure responses during HF operations. The model incorporates all known surface variables like surface treatment pressure, fluid pumping rate, and proppant rate. Moreover, it employs LSTM networks to predict response variables like surface pressure responses. This work represents the first attempt to use memory-retentive recurrent neural networks to predict surface pressure responses in real-time during the pumping stages.

5.4.2.5.3 Advantages and Limitations of AI Methods

The DL neural network model described in the paper shows capability in predicting surface pressure, with prediction errors within 10% of actual values. This approach allows real-time simulation of response variables, providing engineers with an accurate representation of downhole and reservoir conditions. However, this DL approach has several challenges. Multi-step forecasting involves computational issues, such as an increase in the size of RNNs with the exponential expansion of the search space. Although LSTM can solve problems of exploding gradients, internal temporal correlations within data must be considered for predictions.

5.4.2.5.4 Results and Conclusions

During HF, the suggested DRNN model can accurately forecast reaction variables like surface pressure in real time. Throughout the pumping stages, a dependable, stable, and precise numerical solution is produced by the LSTM-based method employing surface pumping data. Complex physics and uncertainty surrounding the precision of the measured data can be handled by the model. The suggested approach gives engineers a precise picture of the wellbore and reservoir conditions by simulating response variables in real-time. This work is the first attempt to use a memory-preserving RNN variation to forecast a response variable during a pumping stage. The LSTM model is capable of modeling the surface pressure in an HF process well, and the surface pressure predictions obtained were within 10% of the actual values.

BIBLIOGRAPHY

Madasu, S. and Rangarajan, K.P. (2018). Deep Recurrent Neural Network DRNN Model for Real-Time Multistage Pumping Data. Day 3 Wed, November 07, 2018.

5.4.2.6 Application-17: Designing a Robust Proppant Detection and Classification Workflow

5.4.2.6.1 Background of Application

The distribution of proppant particles in HF treatments is important for understanding the effectiveness of the treatment, as well as identifying unstimulated and under-stimulated zones within the reservoirs of interest. Proppant particles can be found in drilling fluid return or in cores sampled from the subsurface. In order to identify and categorize proppant particles and other particles of interest, this study uses an ANN-based process that is restricted to naturally occurring calcite or other minerals from underlying rocks. Controlled test cases and real subsurface samples are used to identify and validate a variety of features of interest that facilitate the categorization process. The ANN classifier is compared to various classification techniques such as SVMs, k-nearest neighbors, and naïve-Bayes classifiers. For upcoming test programs, including real-time applications, a methodology is suggested to analyze subsurface samples and measure proppant distribution. Proppant distribution from a Permian Basin case study is shared based on this workflow, and proppant distribution using this proposed method is compared with results from an independent workflow on a similar dataset that does not utilize ML.

5.4.2.6.2 Description of AI Workflow

Data Collection:

Subsurface rock samples post-HF were collected and prepared for analysis, emphasizing the importance of high-quality, relevant data for accurate classification.

Data Preprocessing:

Samples underwent cleaning, sieving, and high-resolution imaging. The preprocessing stage ensured that the data was in a suitable form for feature extraction and further analysis.

Model Selection:

The workflow explored several ML models, including an ANN, k-nearest neighbors, Naïve Bayes, and SVMs, highlighting a comparative approach to select the best classifier.

Model Training:

The ANN model was trained with features derived from the image processing of rock samples, emphasizing the role of specific attributes like hue, roundness, and translucence in the classification process.

Model Evaluation:

Performance was assessed through comparative tests, with the ANN classifier showing notable efficiency in classifying proppant particles, underscoring the effectiveness of ML in this application.

5.4.2.6.3 Advantages and Limitations of AI Methodology

Advantages:

- High Efficiency: The ANN model demonstrated superior capability in accurately classifying proppant particles, crucial for optimizing HF.
- Data-Driven Insights: The approach provides a robust framework for understanding subsurface conditions, facilitating better decision-making in drilling operations.

Limitations:

- Data Quality Dependency: The success of the methodology hinges on the quality of the collected data and preprocessing steps.
- Computational Resources: The training and evaluation of complex models like ANN demand significant computational power and time.

5.4.2.6.4 Results and Recommendations

The proposed workflow provides a robust and effective method for identifying and classifying proppant particles and various particles of interest, limited to naturally occurring calcite or other minerals from subsurface rocks. The ANN classifier, in particular, demonstrates superior performance compared to other classification methods, including k-nearest neighbors, naïve-Bayes classifiers, and SVMs. This workflow can be used for processing subsurface samples and quantifying proppant distribution for future test programs, including real-time applications. The Permian Basin case study demonstrates the effectiveness of this proposed method for proppant distribution analysis.

BIBLIOGRAPHY

Maity, Debotyam, Ciezobka, Jordan. Designing a robust proppant detection and classification workflow using machine learning for subsurface fractured rock samples post hydraulic fracturing operations. *Journal of Petroleum Science and Engineering*, 172, 2019, 588–606, ISSN 0920-4105, https://doi.org/10.1016/j.petrol.2018.09.062

5.4.2.7 Application-18: Quantifying the Impact of Fracturing Chemicals on Production Performance Using Machine Learning

5.4.2.7.1 Background of the Application

Because of its substantial influence on production performance in unconventional reservoirs, the use of chemicals in HF completion design has emerged as a crucial study topic. In order to solve this multivariable problem, this research uses ML to investigate the actual effects of these substances on production performance. Using a data-driven methodology, the study offers a framework to assist completion engineers in choosing the best chemicals and supporting the fracture treatment design process. The authors have attempted to dissect the many chemical groups that are employed in HF in order to independently assess their effects. The study offers suggestions for process optimization in finishing design and explains the causal relationship between chemical composition and design and the influence on production.

5.4.2.7.2 Description of the Workflow and Machine Learning Model

In this paper, the authors adopted an ML approach to quantify the impact of HF chemicals on production performance. The workflow began with collecting various data related to HF. Subsequently, different feature selection methods like F_Regression, decision tree regression, mutual info regression, correlation-based feature selection, and recursive feature elimination were applied to identify the factors most impacting production performance. This process included the analysis of short-term (6 months), mid-term (12 months), and long-term (24 months) production data.

5.4.2.7.3 Advantages and Limitations of the Method

This method is particularly suited for analyzing the effect of chemicals on production performance, as it can handle large datasets and identify key influencing factors.

5.4.2.7.4 Results and Discussion

The study results show that even with a variety of input variables like petrophysical parameters and completion and well design metrics, chemicals make it to the top five significant variables in 11 out of the 18 scenarios. Therefore, the significance of chemicals in predicting production performance is established. The analysis also shows that certain chemicals are significant at the beginning of production and lose their importance with time. Using ML, the study offers a comprehensive understanding of the effects of the most widely employed chemical types in shale completions, offering a genuinely innovative perspective on each product's unique contribution. By establishing a cause-and-effect link between chemical composition and design and the influence on production, the study advances the grasp of this idea and offers suggestions for process optimization in finishing design.

BIBLIOGRAPHY

Rastogi, Ayush and Sharma, Akash. "Quantifying the Impact of Fracturing Chemicals on Production Performance Using Machine Learning." *Paper Presented at the SPE Liquids-Rich Basins Conference - North America*, Odessa, TX, November 2019. https://doi.org/10.2118/197095-MS

5.4.2.8 Application-19: Estimating Ultimate Recovery for Multistage Hydraulically Fractured Horizontal Wells

5.4.2.8.1 Background of the Application

Predicting long-term production and estimated ultimate recovery (EUR) in unconventional wells is difficult, and estimating reserves is a crucial component of reservoir development. In the oil and gas sector, a production prediction that is both precise and dependable is crucial for decision-making. There are a number of ways to estimate EUR, but each has drawbacks. A popular technique for calculating EUR in conventional reservoirs is traditional decline curve analysis (DCA). However, it is not appropriate for forecasting production trends in low-permeability plays in unconventional reservoirs. Numerous methods have been devised to establish trustworthy estimations of reserves in order to get around this restriction. A methodology is presented to predict EUR for multistage hydraulically fractured horizontal wells by incorporating completion data, demonstrating superior performance compared to many existing approaches.

5.4.2.8.2 Description of the Workflow and Machine Learning Model

The workflow utilizes ANN predictive models to EUR for multistage fractured horizontal wells. This method integrates production history data and DCA with ANN models. Sixteen models were developed, each varying in input parameters, structure, and required production history data period, ranging from 3 months to 2 years. These models demonstrated high accuracy in predicting EUR, combining forecasts from DCA models with well completion data to improve the predictions.

5.4.2.8.3 Advantages and Limitations of the ML Method

The methodology showcases several advantages: The study's methodology improves the accuracy of estimating reserves. It addresses the shortcomings of DCA, especially in the context of unconventional wells where transient flow dominates for much of the well's life. Limitation includes data availability: The methodology requires a certain amount of production history data (ranging from 3 months to 2 years) to accurately forecast future production.

5.4.2.8.4 Results and Discussion

The developed workflow in this study provides a more reliable and accurate approach to predicting EUR for multistage hydraulically fractured horizontal wells. The workflow combines production history matching and forecasting using DCA models with ANN predictive models, which incorporate completion data to improve EUR predictions. The developed models showed high accuracy in predicting EUR given only 3 months to 2 years of production data. This study highlights the importance of considering completion data along with the traditional DCA method for reserve estimation in unconventional reservoirs. The end product of this work is a comprehensive workflow to predict EUR that can be implemented in different formations.

BIBLIOGRAPHY

Alarifi, Sulaiman A., and Miskimins, Jennifer. "A New Approach To Estimating Ultimate Recovery for Multistage Hydraulically Fractured Horizontal Wells by Utilizing Completion Parameters Using Machine Learning." *SPE Prod & Oper* 36 (2021): 468–483. https://doi.org/10.2118/204470-PA

5.4.2.9 Application-20: Machine-Learning-Based Well Production Prediction

5.4.2.9.1 Background of the Application

This paper proposes a novel method for efficient and robust post-history production forecasting in shale gas reservoirs. The method, called latent-space learning-based direct forecast approach (LS-LDFA), uses a convolutional autoencoder for dimensionality reduction and low-order representation of time-series data. The ML proxy is trained offline, and the online post-history production forecast can be efficiently achieved by inputting history data. The proposed method is compared with model-based history matching using two examples of increasing complexity, and the results show that LS-LDFA achieves high robustness and computational efficiency simultaneously.

5.4.2.9.2 Description of the Workflow and Machine Learning Model

The paper focuses on an ML model for shale gas production prediction. The proposed model, LS-LDFA, incorporates a convolutional autoencoder to process multi-well time-series data. This approach regularizes the data into a low-order representation within a latent space. Once the ML proxy is trained, it enables efficient online post-history production forecasting by inputting historical data. This method is tested on a multi-fractured horizontal well and a naturally fractured reservoir model with multi-well-pad production.

5.4.2.9.3 Advantages and Limitations of the RF Method

The methodology showcases several advantages: The LS-LDFA model is more robust and computationally efficient compared to traditional methods. Also, the ML model can handle complex scenarios like multi-fractured wells and naturally fractured reservoir models effectively. Limitation includes various uncertain parameters like fracture number, permeability, and porosity that need to be accurately estimated, adding complexity to the model.

5.4.2.9.4 Results and Discussion

In conclusion, the proposed LS-LDFA method provides an efficient and robust approach for post-history production forecasting in shale gas reservoirs. The method does not require the computationally expensive history matching step, and instead employs an ML proxy that can efficiently generate reliable forecasts using input history data. The comparative studies show that LS-LDFA achieves high accuracy and computational efficiency simultaneously, which makes it a promising option for practical applications in the oil and gas industry.

BIBLIOGRAPHY

Xiao, Cong, Wang, Guangdong, Zhang, Yayun, Deng, Ya. Machine-learning-based well production prediction under geological and hydraulic fracture parameters uncertainty for unconventional shale gas reservoirs. *Journal of Natural Gas Science and Engineering*, 106, 2022, 104762, ISSN 1875-5100, https://doi.org/10.1016/j.jngse.2022.104762

5.4.2.10 Application-21: Insights to Fracture Stimulation Design in Unconventional Reservoirs Based on Machine Learning Modeling

5.4.2.10.1 Background of application

In this work, the authors present a study on the application of ML techniques to predict oil production in shale reservoirs. A dataset comprising 121 horizontal wells in the Montney formation, Canada, was utilized, encompassing well location, stimulation design, and production data. Four supervised learning algorithms—RF, adaptive boosting, SVM, and neural network—were applied to predict 12-month BOE production. Additionally, feature selection techniques were employed to identify the most influential parameters on production outcomes. The results indicate that ML techniques can accurately predict the production rate of shale reservoirs. The most significant factors affecting production were found to be the well latitude, longitude, well TVD, proppant pumped per well, well lateral length, and fluid injected per well.

5.4.2.10.2 Description of ML Workflow

The workflow involves a data mining process using well production data from the Montney Formation. Key steps include feature selection using recursive feature elimination with cross-validation, data preprocessing, and the application of various

ML algorithms like RF, adaptive boosting, SVM, and neural networks. The models predict the first-year oil production by analyzing crucial factors like well location, depth, proppant pumped per well, and fluid injected per well.

5.4.2.10.3 Advantages and Limitations of AI Methodology

The methodology showcases several advantages: It provides a detailed insight into the impact of various fracturing parameters on oil production. Additionally, RFECV effectively identifies the most influential factors for accurate predictions.

At the same time, there are some limitations: The model's accuracy is heavily reliant on the quality and comprehensiveness of the dataset. The model's effectiveness is specific to the dataset used (Montney Formation), which may limit its applicability to other reservoirs.

5.4.2.10.4 Results and Recommendations

The efficacy of ML techniques in predicting oil production in shale reservoirs was demonstrated using a dataset of 121 horizontal wells in the Montney formation, Canada. Four supervised learning algorithms were applied to predict 12-month BOE production. The results showed that the RF algorithm outperformed other algorithms with an R-squared value of 0.83. Feature selection techniques identified six significant parameters affecting production. The findings suggest that ML techniques can be valuable tools for predicting oil production in shale reservoirs, which can help optimize well completion and HF design.

BIBLIOGRAPHY

Wang, Shuhua, Chen, Shengnan. Insights to fracture stimulation design in unconventional reservoirs based on machine learning modeling. *Journal of Petroleum Science and Engineering*, 174, 2019, 682–695, ISSN 0920-4105. https://doi.org/10.1016/j.petrol.2018.11.076

5.4.3 AI Application of Fracture Identification

5.4.3.1 Application-22: Conductive Fracture Identification Using Neural Networks

5.4.3.1.1 Background of the Application

This paper presents a study on the impact of various factors on the performance of ML models. The study includes an analysis of different algorithms, feature sets, and data preprocessing techniques. The results show that the choice of algorithm has the most significant impact on model performance, followed by feature selection and data preprocessing. The study also evaluates the impact of imbalanced datasets on model performance and proposes methods to mitigate this issue.

5.4.3.1.2 Description of the Workflow and Machine Learning Model

Based on literature reviews and data characteristics, raw features were transformed into categorical variables. For instance, the alteration degree was categorized using cutoff values at 0, 10, 50, and 100. Therefore, the input variables contain only categorical variables. As mentioned in the paper, a backpropagation neural network will be built for conductive fracture identification. The model result is not ideal because all the variables are simulated with the same distribution.

5.4.3.1.3 Advantages and Limitations of the RF Method

The back propagation neural network was the first and simplest type of ANN devised. In this network, the information moves in only one direction—forward—from the input nodes, through the hidden nodes (if any) and to the output nodes. It can obtain strong nonlinearity by building a deep network with activation functions; however, it is more suitable for image or text data. For classic tabular data, tree-based models seem to be a better solution.

5.4.3.1.4 Results and Discussion

In conclusion, this study highlights the importance of carefully selecting the appropriate algorithm, feature set, and data preprocessing techniques for ML tasks. The results show that the choice of algorithm can have a significant impact on model performance and that the use of imbalanced datasets can negatively affect model accuracy. It is recommended that future studies explore additional methods for addressing class imbalance and evaluate the performance of other algorithms and feature selection techniques. Overall, the findings of this study can aid researchers and practitioners in making informed decisions when designing ML models.

BIBLIOGRAPHY

Thomas, L.A. and La Pointe, P.R. (1195). Conductive fracture identification using neural networks. ARMA-95-0627. *The 35th U.S. Symposium on Rock Mechanics (USRMS).*

5.4.3.2 Application-23: Using Machine Learning to Identify the Highest Wet Gas

5.4.3.2.1 Background of Application

This study presents the petroleum analytics learning machine (PALM), which was used to automatically classify hydraulic fractures using ML techniques. The objective was to identify correlations between fracture classes, technological improvements, and gas and condensate production from 2009 to 2013. PALM classified time-series fracture data recorded during 1,842 hydraulic fracture stimulation stages in 156 horizontal wells in the wet gas Marcellus shale. The results showed that wells with more than one-quarter Frac-Class 4 fractures outperformed wells dominated by more than three-quarters Frac-Class 5 fractures, particularly in the higher liquids area. The study also found that improvements in completion technologies, such as deeper landing zones and shorter stage lengths, contributed to the observed production improvement.

5.4.3.2.2 Description of the Workflow and Machine Learning Model

The workflow for this project involved data simulation, dataset splitting, and model training. The dataset was simulated using five continuous features and one categorical target variable. The continuous features included SurfaceTreatingPressure, JobCleanRate, JobSlurryRate, JobDensity(ppa)*10, and DownHoleJobDensity(ppa)*10, while the categorical target variable was class. After simulating the dataset, it was split into training and testing sets using train_test_split function from scikit-learn. The training dataset was then used to train an SVM model. The SVM model is a popular ML model commonly used for classification tasks. The SVM model works by finding a hyperplane that best separates data points in different classes. The SVM model tries to maximize the margin between the hyperplane and the data points closest to it. The SVM model was trained using the scikit-learn library. Scikit-learn is a widely used ML library in Python. It provides a range of ML algorithms, including SVM, that are easy to use and efficient.

5.4.3.2.3 Advantages and Limitations of the AI Algorithm

The SVM model has several advantages that make it an attractive choice for classification tasks. One of the main advantages of the SVM model is its ability to handle high-dimensional data. This makes it useful for tasks that involve a large number of features. Additionally, the SVM model is effective at separating data points in nonlinearly separable datasets. This means that it can be used for datasets where the classes overlap or are not well separated.

Another advantage of the SVM model is its ability to handle large datasets. This is because the SVM model only uses a subset of the training data to find the hyperplane, rather than using the entire dataset. This makes it computationally efficient for large datasets. However, there are also some limitations to using an SVM model. One limitation is that the performance of the SVM model is sensitive to the choice of kernel function. Different kernel functions can lead to different hyperplanes, which can affect the accuracy of the model. Additionally, SVM models can be computationally expensive to train on large datasets. This is because the model needs to find the hyperplane that maximizes the margin, which can be a complex optimization problem.

Lastly, SVM models may not perform well on datasets with noisy or overlapping classes. This is because the model tries to find a hyperplane that separates the classes as cleanly as possible. If the classes overlap or are not well separated, the model may not be able to find a hyperplane that accurately separates the classes. In summary, while the SVM model has several advantages that make it an attractive choice for classification tasks, it is important to carefully consider its limitations and make sure that it is an appropriate choice for the specific dataset and task at hand.

5.4.3.2.4 Results and Recommendations

This study highlights the potential of ML techniques in optimizing HF in the wet gas Marcellus shale. The PALM was used to automatically classify hydraulic fractures and identify correlations between fracture classes, technological improvements, and gas and condensate production. The results showed that wells with more than one-quarter Frac-Class 4 fractures outperformed wells dominated by more than three-quarters Frac-Class 5 fractures, particularly in the higher liquids area. This study also identified improvements in completion technologies, such as deeper landing zones and shorter stage lengths, that contributed to the observed production improvement. The findings from this study could help gas companies optimize their HF operations and improve production in the wet gas Marcellus shale.

BIBLIOGRAPHY

Anderson, R.N., Xie, B., Wu, L., Kressner, A.A., Frantz, J.H., Ockree, M.A., McLane, M.A. (2016) Using machine learning to identify the highest wet gas producing mix of hydraulic fracture classes and technology improvements in the Marcellus shale. *Unconventional Resources Technology Conference*, pp. 1–13.

5.4.3.3 Application-24: Application of Machine Learning for Closure Pressure Determination

5.4.3.3.1 Background of Application

In HF, determining the closure pressure of the rock is essential for optimal treatment design. Analytical methods for determining closure pressure, however, can be subjective and require an experienced analyst to interpret the data. In this paper, the authors introduce a method using ANNs, a branch of ML, to predict closure pressure with objective and accurate results. The ANN is designed using critical parameters on which closure pressure depends, and the model then learns from the patterns in the data and predicts the required output. The results show that the ANN was able to predict the closure pressure with reasonable accuracy.

5.4.3.3.2 Description of AI Workflow

According to the literature reviews and the data types, an ANN model was built with a list of bottom hole pressures as input, a single hidden layer, and one predicted closure pressure as output. Due to the lack of input data, bottom hole pressures were simulated for each well by generating a sequence of random data. As illustrated in the paper, out of the 5 wells, 3 were used for training and 2 for testing

As mentioned in the paper, the target is to find the optimal lambda for the model. The model by default uses back propagation to optimize the weights. A model with L2 regularization was built, using MSE as the metric to minimize the error. The L2 parameter was set using different lambda values, and the test MSEs were calculated.

5.4.3.3.3 Advantages and Limitations of AI Methodology

Since the dataset is simple, with only bottom hole pressures as predictors and one closure pressure, it is good to use a simple architecture as an ANN to estimate the closure pressure. However, it is not explicitly explained why the author transformed the regression problem into a classification problem. And with the limited data and no further evaluation, the performance of the model cannot be accurately estimated.

5.4.3.3.4 Results and Recommendations

In conclusion, the authors have introduced a new approach to determine the closure pressure of rock using ANNs. The traditional analytical methods for determining closure pressure can be prone to subjectivity, and human error can lead to erroneous interpretations. The ANN method presented in this paper provides an objective and accurate way of predicting closure pressure, and the results show that it is a promising method for optimizing HF treatment design.

BIBLIOGRAPHY

Nande, Soumitra "Application of Machine Learning for Closure Pressure Determination." *Paper Presented at the SPE Annual Technical Conference and Exhibition*, Dallas, TX, September 2018. https://doi.org/10.2118/194042-STU

5.4.3.4 Application-25: Machine Learning for Improved Directional Drilling

5.4.3.4.1 Background of Application

This paper introduces a novel method for predicting fracture pressure using ANN models. Accurately predicting fracture pressure is crucial for maintaining the safety and efficiency of drilling operations. Traditional models for predicting fracture pressure have limitations, including applicability only to certain types of formations or pressure mechanisms. By contrast, ANN models can consider a wide range of parameters and can accurately predict fracture pressure from real-time surface drilling parameters alone. In this paper, the authors demonstrate the effectiveness of their ANN approach in predicting fracture pressure, achieving high precision and outperforming traditional models.

5.4.3.4.2 Description of AI Workflow

The workflow involves using an **ANN** to predict fracture pressure from real-time surface drilling parameters. The ANN model is structured with input, hidden, and output layers. The input parameters include weight on bit (WOB), rotary speed (RPM), ROP, torque, mud weight (MW), and pore pressure. The model employed the optimal selection of layers, neurons, training functions, and network functions. The study utilized a dataset of 3925 data points across various formations, with 80% used for training the ANN model and 20% for testing its predictions.

5.4.3.4.3 Advantages and Limitations of AI Methodology

The ANN model demonstrates significant advantages in predicting fracture pressure. Its high accuracy is evidenced by an excellent R value and a low AAPE, indicating its superior performance over models like the Matthews and Kelly model. However, the model's limitations include overfitting with too many neurons and increased computational time with more hidden layers.

5.4.3.4.4 Results and recommendations

In conclusion, this paper presents a useful and effective method for predicting fracture pressure using ANN models. The authors were able to demonstrate the high precision and accuracy of their approach, which outperformed traditional models in predicting fracture pressure. The results of this study suggest that ANN models can be a valuable tool for improving the safety and efficiency of drilling operations.

BIBLIOGRAPHY

Ahmed, S., Abdulmalek, Elkatatny, Salaheldin, Ali, Abdulwahab, Z., Abdulraheem, Abdulazeez, and Mahmoud, Mohamed. "Artificial Neural Network ANN Approach to Predict Fracture Pressure." *Paper Presented at the SPE Middle East Oil and Gas Show and Conference*, Manama, Bahrain, March 2019. https://doi.org/10.2118/194852-MS

5.4.3.5 Application-26: Benchmarking Operator Performance in the Williston Basin Using a Predictive Machine Learning Model

5.4.3.5.1 Background of the Application

In recent years, the application of imaging logging data has become more and more widespread, providing borehole image information with high reference value for discerning the degree of fracture development. However, whether using conventional logging data or imaging logging data, the current interpretation methods still require manual selection or solution of multiple parameters, and generally suffer from low efficiency, empirical, and subjective influences that can easily lead to misjudgment. Therefore, it is necessary to find a method to discriminate the fracture development grade that is less affected by parameter uncertainty, more accurate in fracture evaluation, and more efficient in identification.

5.4.3.5.2 Description of the Workflow and Machine Learning Model

The convolutional neural network algorithm is an efficient DL method for image discrimination and speech research with efficient DL methods. The algorithm does not need to build a logging interpretation model or calculate various parameters reflecting fracture characteristics. The network will automatically find the implicit relationship between different levels of fractures and logging response data, which is difficult to express mathematically. It is difficult to express the relationship in mathematical formulas. Convolutional neural networks can extract the features of the input information more efficiently and comprehensively because of local connectivity and weight sharing. This paper introduces convolutional neural networks to avoid the defects of overfitting and local optimal solutions. Therefore, this paper introduces the convolutional neural network technique to identify the cracks with different degrees of development.

The convolutional neural network model for fracture development recognition consists of an input layer, a convolutional layer, a fully connected layer, and an output layer. For the geological and logging profiles of the selected study area, logging curves and fracture discrimination data with high correlation with fractures are selected. After preprocessing, fractures of different development levels and their corresponding logging data are selected to form the learning and testing samples.

5.4.3.5.3 Advantages and Limitations of the Synchronized Algorithm

Determining the grade of a fractured reservoir requires manual synthesis to analyze multiple logs and calculate multiple fracture-related parameters, which is a large workload and low efficiency. The results can be biased or uncertain if the parameters or models are not selected properly. In contrast, convolutional neural network technology only requires multiple logging curves to be input, without many intermediate loops and human intervention. This approach enables the simultaneous use of multiple logging methods, allowing them to complement each other and provide a more comprehensive understanding of the characteristics of each fracture grade. The fracture grading is accurate.

5.4.3.5.4 *Results and Discussion*

(1) The multi-layer network of a convolutional neural network with multiple convolutional kernels, weight sharing, and non-full connectivity makes the network run faster and less likely to fall into local optima. The convolutional neural network algorithm can make full use of the information about fracture characteristics in the logging curve when evaluating fractured reservoirs in a hierarchical manner, and the extraction process is objective and free from human interference, reducing the judgment error and calculation error caused by humans.

(2) The convolutional neural network algorithm has good consistency in judging the degree of fracture development compared with the results obtained by other methods, but it is more efficient and convenient to use, and it can be used as an effective supplement to the existing methods to improve the efficiency of logging interpretation, which has certain promotion value.

(3) The selection of learning samples has a large impact on the grading results of fractures, and it is necessary to select known wells with strong correlation with test wells and extract training data from them.

BIBLIOGRAPHY

Cross, T., Sathaye, K., Darnell, K., Niederhut, D., and Crifasi, K. "Benchmarking Operator Performance in the Williston Basin using a Predictive Machine Learning Model." *Paper Presented at the SPE/AAPG/SEG Unconventional Resources Technology Conference*, Virtual, July 2020. https://doi.org/10.15530/urtec-2020-2750

5.4.3.6 Application-27: Using Artificial Intelligent Technique to Identify Fractures

5.4.3.6.1 *Background of the Application*

The identification and prediction of natural fractures is difficult to solve without other additional data. AI techniques including a whole set of data processing and sensing such as neural networks and fuzzy logic that have emerged in recent years play an important role in geology.

In this paper, the ANN technique and fuzzy logic are used to evaluate the fracture distribution around the wellbore and to identify the presence of fracture zones from logs. The fracture identification method is a combination of fracture response characteristics to different logs, so this technique has great potential in the field of fracture identification.

5.4.3.6.2 *Description of the Workflow and Machine Learning Model*

The network model consists of seven input parameters and one output parameter. The number of nodes in the hidden layer is determined by the BP algorithm and the action function using the test data, which is essentially the backward transmission of error and successive corrections to complete the training until the error between the expected value and the actual value is minimized. Before training with the actual fracture data, data with the same (similar) fracture distribution density values and random distribution (i.e., no correspondence with depth) were used for training in order to find the structure of the control network that better fits the statistical pattern of the original data. The selection of control points was carried out using data from 20 points in each well, and the entire network was trained 11 000 times to minimize the error. It can be seen that the expected values of the network structure using random data points are in good agreement with the predicted values of the actual cross-check, and the predicted fracture densities are in good agreement with the distribution of fracture densities reflected by the actual imaging logging interpretation.

The ranking results show that in this study area, first, the natural gamma value and the highest score of uranium content in the energy spectrum indicate that the sensitivity to fracture is the most obvious. This is mainly due to the fact that on the one hand, the natural gamma in this area reflects the different lithological characteristics (e.g. dolomite and quartzite have lower natural gamma values on the one hand and high gamma values on the other hand).

On the other hand, due to the active groundwater, the uranium elements accumulate on the fracture wall and cause a sharp increase in uranium values. The other parameters are not considered in the modeling because they are weakly reflected in the fractures. The other parameters are not considered in the modeling.

5.4.3.6.3 Advantages and Limitations of the Synchronized Algorithm

ANN technique is a traditional and mature neural network technique. ANN can be generalized, and after learning from the initialized inputs and their relationships, it can also infer unknown relationships from unknown data, thus allowing the model to generalize and predict unknown data.

The limitation is that the learning process between observations cannot be observed, and the output is difficult to interpret, which can affect the credibility and acceptability of the results.

5.4.3.6.4 Results and Discussion

The developed fuzzy logic and neural network expert system was used to identify the fracture development in the longitudinal direction (around the wellbore) by imaging log data and core data, and the application and effectiveness of the method were illustrated by examples. In fractured reservoirs, the lithological, physical, and electrical properties are complex, and the use of neural networks to establish nonlinear relationships between natural fractures and multiple parameters of logging response avoids many artificial factors and has advantages in fracture identification. It has some advantages in fracture identification. The fuzzy logic can better solve the multi-parameter selection problem, and the most sensitive parameter to the fracture response can be selected to speed up the training of the network and improve the recognition accuracy.

BIBLIOGRAPHY

Chang, Zhang Ji, Yuzhong, Xing, Li Hui, Zheng. Using Artificial Intelligent Technique to Identify Fractures. *Well Logging Technology*, 2005, 29 (1).

5.4.3.7 Application-28: A Graphics Processing Unit-Based, Industrial Grade Compositional Reservoir Simulator

5.4.3.7.1 Background of the Application

In recent years, submerged oil and gas reservoirs have become a key exploration and development area for oil and gas resources. This paper focuses on fracture determination prediction by coherent body and multiple attributes such as curvature, ant body, arc length, instantaneous frequency, root mean square amplitude, and reflection intensity and establishes an effective post-stack seismic identification model suitable for fractures in subduction mountain metamorphic rocks in DMT area.

At present, the three main methods of fracture prediction are geological, logging, and seismic. Since seismic data has the advantage of continuous observation in space, the use of seismic data for fracture prediction becomes an extremely critical means and technique. The model is suitable for post-stack seismic identification of subduction mountain metamorphic fractures.

5.4.3.7.2 Description of the Workflow and Machine Learning Model

Due to the multi-solution nature and uncertainty in seismic exploration, reservoir prediction using a single seismic attribute cannot characterize the reservoir well, and a combination of multiple types of seismic attributes is needed for oil and gas prediction. A more typical BP neural network algorithm is used to predict the fracture development zone in the study area using the seven types of fracture-sensitive seismic attributes extracted above.

ANN is a mathematical structure model that introduces biological principles into mathematical algorithms and consists of three parts: input, neuron, and output. In the training phase, the data is input from the input; the neurons are assigned certain weights to each input, and the destination data is output under the influence of the excitation function assigned by the neurons; the residuals between the actual data and the simulation results are calculated, and the initial assigned weights are adjusted by the neural network algorithm until a limited error range is reached. In the prediction phase, the input data are mapped to the output data through the established network.

5.4.3.7.3 Advantages and Limitations of the Synchronized Algorithm

The application of BP neural network is not only a simple operation and processing but also critical for the selection of variables whether to apply a professional perspective. Innovative application and bold experimentation with geological data are the key to the successful use of ML tools.

The limitation is that it is difficult to make professional interpretative judgments by the quality of the model output.

5.4.3.7.4 Results and Discussion

1. The submerged mountain reservoir in the study area mainly develops metamorphic lithology, which is influenced by multi-period tectonic movements, and its fractures are also multi-period, extensive, and multi-directional, and the degree of development is uneven; coupled with post-formation and weathering, leaching and alteration, the secondary pore and microfracture dual-media reservoirs developed along the fractures are formed.

2. Seismic attributes such as coherence, ant body, and reflection intensity are more sensitive to fracture. Although various single seismic attributes can express different physical characteristics and geological information of seismic signals, they can often only express a certain aspect of the geological body, and the prediction results often have multiple solutions. There is no correspondence between them and fracture development. The extent of fracture distribution and development can be predicted more accurately by using the geological information provided by post-stack seismic attributes.
3. The BP nerve network algorithm is used to quantitatively or semi-quantitatively predict the distribution and development of fractures using seven common seismic attributes. . The average error of the validation wells is 8%, which is a small error. The results are valid and reliable. The fracture development zones are the Dongshengbao and Fahabu tectonic zones, Jinganbao; the secondary development zones are the Biantai, Pinganbao, Jing.
4. This provides a basis for the later comprehensive reservoir prediction.
5. Constrain the well points and use the BP nerve network with multiple seismic attributes as input information. The BP nerve network prediction method with multiple seismic attributes as input information can be used to the above seismic attributes into quantitative fracture density, which improves the accuracy and precision of fracture.

BIBLIOGRAPHY

Esler, K., Gandham, R., Patacchini, L., Garipov, T., Samardzic, A., Panfili, P., Caresani, F., Pizzolato, A., and Cominelli, A. "A Graphics Processing Unit-Based, Industrial Grade Compositional Reservoir Simulator." *SPE J.* 27 (2022): 597–612. https://doi.org/10.2118/203929-PA

5.4.3.8 Application-29: Non-Newtonian Fluid Flow Dynamics in Rotating Annular Media: Physics-Based and Data-Driven Modeling

5.4.3.8.1 Background of the Application

The correct identification of fractures and seamhole fillings is the key to the exploration and development of carbonate reservoirs and plays a pivotal role in the comprehensive evaluation of the reservoirs. It is important to find an accurate and fast method to identify fractures and fracture fillings for the understanding and research of fracture-type carbonate reservoirs.

Based on the previous research results, this paper presents the first method of identifying fractures and seam fillings in carbonate reservoirs (PSO-LSSVM) based on PSO and LSSVM.

In order to demonstrate the superiority of this method, the logs of two wells in the eastern margin of the Bin Caspian Basin were selected as sample data, and two methods, PSO-LSSVM and BP neural network, were used to identify fractures and seam fills.

5.4.3.8.2 Description of the Workflow and Machine Learning Model

In order to solve the problem that the learning and recognition ability of ANN and SVM are not high enough, this paper applies LSSVM to fracture and seam hole filling for the first time. The PSO algorithm is used to perform parameter. The global optimal parameters can be obtained, which improves the generalization ability of LSSVM for small sample data. The generalization ability of LSSVM for small sample data is improved. After normalizing the learning and test samples, the PSO algorithm is used to optimize the LSSVM parameters. The optimal parameters are then used to construct the LSSVM recognition model using the optimal parameters; finally, the accuracy of the recognition model is verified using the test samples. Finally, the accuracy of the recognition model is verified using test samples.

5.4.3.8.3 Advantages and Limitations of the Method

LSSVM is an SVM derivative method proposed by Suykens and Vandewalle (1999), which successfully introduces least-squares estimation into SVM based on the statistical learning theory of SVM and the principle of minimizing structural risk, improves the generalization ability to small sample data, and better solves the problems of long learning time, randomness of learning results and insufficient generalization ability of neural networks, which are widely used in the modeling of complex nonlinear systems. Compared with the standard SVM inequality constraint and solving quadratic programming problems, LSSVM selects the error squared term as the optimization objective, takes the equation constraint as the constraint condition, and solves the linear equation problem, which simplifies the operation process and improves the computational speed and accuracy.

5.4.3.8.4 Results and Discussion

1. The key to the problem of fracture and seamhole filling identification lies in the sample selection, and good learning samples can improve the accuracy and precision of the identification model. By analyzing the logging response characteristics of fracture and seam filling in carbonate reservoirs on the eastern edge of the Bin Caspian Basin, five sensitive parameters, namely resistivity difference ratio (RTC), secondary porosity (RP), density ratio (C1), acoustic time difference ratio (C2), and compensation neutron ratio (C3), are extracted based on conventional logging curves, and learning samples and test samples are selected from them.
2. The PSO–LSSVM method utilizes the PSO algorithm for global parameter search, which improves the generalization ability of LSSVM to sample data. In addition, choosing the appropriate number of populations and the number of population evolution can save time of LSSVM parameter search.
3. The two methods, PSO–LSSVM and BP neural network, are used to establish fracture and crevice filling recognition models, respectively, and test the accuracy of the two recognition models by using test samples. The application examples show that the PSO–LSSVM method has better learning ability and recognition effect than the BP neural network, and the recognition results obtained by PSO–LSSVM method are in good agreement with the

results obtained from FMI electrical imaging logging images and core data. It is feasible to use this method for fracture and seam filling identification in carbonate reservoirs, and it is important in the case of limited imaging logging data and core samples.

BIBLIOGRAPHY

Suykens, J., Vandewalle, J. Least Squares Support Vector Machine Classifiers. *Neural Processing Letters* 9, 293–300 (1999). https://doi.org/10.1023/A:1018628609742

Tian, J., Qi, C., Sun, Y. et al. Permeability prediction of porous media using a combination of computational fluid dynamics and hybrid machine learning methods. *Engineering with Computers* 37, 3455–3471 (2021). https://doi.org/10.1007/s00366-020-01012-z

5.4.4 AI Applications in Production Aspects

5.4.4.1 Application-30: Advanced Machine Learning Methods for Production Data Pattern Recognition

5.4.4.1.1 Background of the Application

This paper proposes an automated quality assessment module for rate measurements in test separators using advanced analytics and real-time data. The module employs a suite of advanced ML algorithms, such as semi-supervised learning and active learning, to tackle the challenge of acquiring labeled data for training models created through conventional pattern recognition algorithms. The proposed workflow involves data preprocessing, similarity matrix creation, interactive initial labeling, semi-supervised learning, active learning, and iteration between semi-supervised learning and active learning. The output of this work is a quality assessment module that can analyze high-frequency data from test separators and provide real-time feedback on the quality of rate measurements. The underlying assumption is that the extracted features capture the important characteristics of a test and allow us to distinguish between good and bad tests. The proposed workflow reduces the amount of expert time required to label data, making the analysis procedure more efficient and productive.

5.4.4.1.2 Description of the Workflow and Machine Learning Model

The workflow presented in the paper involves a suite of advanced ML algorithms, such as semi-supervised learning and active learning, to tackle the challenge of acquiring labeled data for training models created through conventional pattern recognition algorithms. The proposed workflow consists of the following steps:

1. Data Preprocessing: Clean the data, remove any irrelevant features, and perform feature engineering to extract the salient features of each test.
2. Similarity Matrix Creation: Use a similarity matrix to visualize the structure of the data and identify clusters of similar data.
3. Interactive Initial Labeling: Use the similarity matrix to interactively select the first few data points for labeling by an expert.
4. Semi-Supervised Learning: Use the labeled data and the structure contained in all the unlabeled data together to propagate the label information and obtain class probabilities for the unlabeled samples.

5. Active Learning: Use the predicted probability distributions for class membership to identify the data points with the highest uncertainty and present them to the user for additional label input.
6. Iterate between Semi-supervised Learning and Active Learning: Continue iterating between steps 4 and 5 until the change in predicted labels is below a small threshold.
7. Train Classifier: Train a classifier using the labeled data and the class probabilities for the unlabeled samples.

The ML model used in the workflow is a semi-supervised learning algorithm called label propagation. It is an iterative algorithm that propagates labels across the graph of data points based on their similarity. The algorithm uses a kernel function to compute the similarity between data points, and the propagation of labels is done using a strategy similar to PageRank. The label propagation algorithm is known for its ability to handle large datasets and can be used for classification and clustering tasks.

5.4.4.1.3 Advantages and Limitations of the Method

Advantages:

- The proposed workflow reduces the amount of expert time required to label data, making the analysis procedure more efficient and productive.
- The combination of semi-supervised learning and active learning allows for the most efficient use of expert time in labeling data.
- The workflow can handle complex datasets that do not display a clear structure.
- The use of similarity matrices makes it easy to visualize the structure in the data and select a good initial set for labeling.

Limitations:

- The success of the workflow depends on the quality of the initial labeled data and the similarity matrix.
- The workflow relies on the assumption that the extracted features capture the important characteristics of a test and allow us to distinguish between good and bad tests.
- The workflow may require a considerable amount of computational resources, especially for large datasets.
- The workflow may require domain expertise to interpret the results of the analysis.

5.4.4.1.4 Results and Discussion

This work explored the use of advanced ML algorithms, such as semi-supervised learning and active learning, to develop an automated quality assessment module for rate measurements in test separators. The proposed workflow involves data preprocessing, similarity matrix creation, interactive initial labeling, semi-supervised learning, active learning, and iteration between semi-supervised learning and active learning. The results demonstrate that the proposed workflow can handle complex datasets that do not display a clear structure and reduce the amount of expert time required to label data, making the analysis procedure more efficient and productive. The underlying assumption is that the extracted features capture the important characteristics of a test and allow us to distinguish between good and bad tests. Further work is needed to validate the proposed workflow on a larger dataset and to investigate the generalizability of the approach to other domains. The proposed workflow represents a promising approach for developing advanced analytics models that can make use of large amounts of data and reduce the amount of expert time required for data labeling.

BIBLIOGRAPHY

Subrahmanya, Niranjan, Xu, Peng, El-Bakry, Amr, and Reynolds, Carmon. "Advanced Machine Learning Methods for Production Data Pattern Recognition." *Paper Presented at the SPE Intelligent Energy Conference & Exhibition*, Utrecht, The Netherlands, April 2014. https://doi.org/10.2118/167839-MS

5.4.4.2 Application-31: Prediction of Toxicity Effects of Oil Field Chemicals Using Adaptive Genetic Neuro-Fuzzy Inference Systems

5.4.4.2.1 Background of the Application

This paper discusses the use of oil field chemicals in various stages of oil production, such as drilling, production, stimulation, and enhanced oil recovery, and the toxic effects they have on the environment. The study aims to develop a screening protocol to assess and manage the inherent risks of these chemicals, and to develop models and tools for screening, predicting, and monitoring their application. An adaptive genetic neuro-fuzzy inference system (AGNFIS) is developed to assess the toxicity of oilfield chemicals. The system has proven to have better classification and regression capability and is able to handle high-dimensional features.

5.4.4.2.2 Description of the Workflow and Machine Learning Model

The paper introduces an AGNFIS aimed at assessing the toxicity of chemicals used in the oil field. The system integrates the strengths of neural networks and fuzzy logic, enhanced with a genetic algorithm to optimize the neuro-fuzzy parameters, thereby improving prediction accuracy and generalizability. This system, trained based on fuzzy rules, classifies new data or predicts its toxicity using a Gaussian membership function.

5.4.4.2.3 Advantages and Limitations of the Synchronized Algorithm

There are some advantages listed in the paper: The study demonstrates that the genetic algorithm provides optimal parameters for the neuro-fuzzy model. This GA-Neuro-Fuzzy method proves less cumbersome than conventional mathematical modeling for material classification

5.4.4.2.4 Results and Discussion

This study demonstrated that an AGNFIS is a useful tool for determining an oilfield chemical's toxicity. The system can handle high-dimensional features and has improved regression and classification capabilities. The fuzzy rule-based model was trained on the available dataset, and the Gaussian membership function selected for this application is used to classify the new or unknown data into the proper class or predict its toxicity. The method aims to eliminate the current, extremely time-consuming, and laborious animal testing, and it is less complicated than the traditional computational modeling typically used for chemical classification and characterization.

BIBLIOGRAPHY

Ibrahim, H. D., and Odedele, T. O. "Prediction of Toxicity Effects of Oil Field Chemicals Using Adaptive Genetic Neuro-Fuzzy Inference Systems." *Paper Presented at the SPE Western Regional Meeting*, Garden Grove, CA, April 2018. https://doi.org/10.2118/190039-MS

5.4.4.3 Application-32: Using Machine Learning for Building Multivariate IPR Models from High Frequency Streaming Data

5.4.4.3.1 Background of the Application

In this research, a novel method for creating asset productivity models using high-frequency data obtained from SCADA systems is presented. The strategy entailed creating an algorithm that could create multivariable, IPR-like models. Data retrieval, filtering, model fitting, and results display are all handled by the algorithm. It can accurately forecast short- and medium-term changes in productivity and takes depletion-driven changes into consideration. Other variables, such as environmental conditions, depletion rates, and sensor recalibration, can be added if needed, thanks to the multivariable method. A gas condensate field limited to a daily nominal rate of about 200 MMscfd of dry gas was used to evaluate the method. The resultant code has excellent prediction capabilities and effectively constructed IPR curves. This approach gives a tool to automate the creation of mathematical models for data-rich assets. It shows how DA techniques allow building versatile programs that can be rapidly deployed in the field.

5.4.4.3.2 Description of the Workflow and Machine Learning Model

The Python code shown outlines an effectively designed AI pipeline that is ready to predict gas rates using high-frequency data from producing wells. At the center of this process is the use of multivariable regression, enhanced by the computed addition of polynomial features, which are designed to effectively navigate the intricate nonlinear interactions between variables like relative time and wellhead pressure in relation to gas production rates. This advanced process is contained in an ML pipeline that combines regression modeling with Ridge and data standardization with StandardScaler to provide an efficient modeling approach.

5.4.4.3.3 Advantages and Limitations of the Synchronized Algorithm

The many advantages of this technology include its ability to accurately forecast outcomes by modeling complex relationships in a detailed way using polynomial transformations. Using Ridge Regression adds an essential regularization layer, which minimizes the overfitting risk—a common issue when working with high-dimensional datasets. Or to simplify the problem, when polynomial features lead to an increase in model complexity. The application of data transformations and modeling is ensured by the pipeline approach, which improves the reliability and integrity of the model. However, there are certain drawbacks to this strategy. Because combining polynomial characteristics increases model complexity, careful parameter tweaking is necessary. Furthermore, the effectiveness of the model is closely related to the variables given, suggesting that if more variables are included, but the output would be more accurate. The delicate trade-offs involved in this AI endeavor are highlighted by navigating the balance between over fitting and parameter tuning and leveraging model complexity for increased accuracy.

5.4.4.3.4 Results and Discussion

In conclusion, this paper demonstrates the potential of using ML techniques to create multiple models that can provide insightful recommendations to operators and engineers. The presented algorithm was able to build IPR curves with great prediction capabilities and was successfully applied in both high and low demand wells. The approach described in this paper addresses the challenge of modeling and regularly updating the production capabilities of an asset using high-frequency data. It provides a tool to automate the creation of mathematical models for data-rich assets and shows how DA techniques can allow building versatile programs that can be rapidly deployed in the field.

BIBLIOGRAPHY

Fiorillo, del Pino, Antonio, Mario. "Using Machine Learning for Building Multivariate IPR Models from High Frequency Streaming Data." *Paper Presented at the SPE Latin American and Caribbean Petroleum Engineering Conference*, Virtual, July 2020. https://doi.org/10.2118/199132-MS

5.4.4.4 Application-33: Data Conditioning and Forecasting Methodology Using Machine Learning on Production Data for a Well Pad

5.4.4.4.1 Background of the Application

This paper presents a novel approach to conditioning and predicting production data for a well pad using an ensemble of ML and statistical techniques. The methodology involves outlier detection, missing data imputation, and smoothing. The study explores the complex nonlinearity of features and estimates wellbore oil production with ML and deep-learning models. The paper also presents a comparison of conventional SVR and MLP methods with modified SVR and LSTM algorithms.

5.4.4.4.2 Description of the Workflow and Machine Learning Model

The research primarily involves data conditioning and time series forecasting. Initially, the data conditioning process includes outlier detection, handling of missing data, data imputation, and smoothing. By employing ML like multilayer perceptrons, principal component analysis, and SVR to process the data, it becomes suitable for further analysis. Following this, the time series forecasting is based on the processed datasets, utilizing a modified SVR structure and LSTM algorithms for comparative analysis.

5.4.4.4.3 Advantages and Limitations of the Synchronized Algorithm

There are some advantages listed in the paper: the models can generate effective processing and prediction of complex, high-dimensional datasets. This is especially evident in data preprocessing and forecasting, where methods like MLP and SVR efficiently handle missing and outlier values.

5.4.4.4.4 Results and Discussion

In conclusion, the study demonstrates the efficacy of the proposed methodology in conditioning and predicting production data for a well pad. The use of advanced ML methods such as MLP and SVR with radial basis function kernel resulted in the detection and imputation of more than 60% of missing data, including average ΔP tubing, average downhole pressure and temperature, average annulus pressure, and water injection. The modified SVR with time-series data structure and LSTM algorithms achieved an R2 of 98%, with LSTM outperforming SVR in terms of RMSE. The paper highlights the importance of data conditioning in accurate predictions and the potential of ensemble ML techniques in the field of production data analysis.

BIBLIOGRAPHY

Bagheri, Maryam, Zhao, Haoran, Sun, Manyang, Huang, Li, Madasu, Srinath, Lindner, Peggy, and Toti, Giulia. "Data Conditioning and Forecasting Methodology Using Machine Learning on Production Data for a Well Pad." *Paper Presented at the Offshore Technology Conference*, Houston, TX, May 2020. https://doi.org/10.4043/30854-MS

5.4.4.5 Application-34: Machine-Learning Predictions of the Shale Wells' Performance

5.4.4.5.1 Background of the Application

The ultra-low permeability nature of shale reservoirs poses significant challenges in forecasting the EUR due to the complexity of flow regimes and the indeterminate onset of the boundary-dominated flow. To overcome these challenges, the authors propose a novel unsupervised ML methodology for reliable EUR estimation based on field-wide analyses. The approach involves automatic identification of the optimal number of features present in the data based on non-negative matrix/tensor factorization coupled with *k*-means clustering, incorporating regularization and physics constraints. The input data to the ML algorithm is the available (public) production history from existing unconventional reservoirs. The proposed approach provides fast and accurate estimations of well performance without presumptions about the state of the well or the flow regime.

5.4.4.5.2 Description of the Workflow and Machine Learning Model

The paper presents an innovative ML methodology for predicting the performance of shale wells. This approach uses a non-negative matrix factorization coupled with k-means clustering to identify and learn the decline signatures from field-wide production history. The workflow begins by classifying wells into groups based on their production profiles. Then, it decouples these profiles to identify constituent signals, which helps forecast the performance of new wells.

5.4.4.5.3 Advantages and Limitations of the Synchronized Algorithm

The advantages include fast and accurate estimations of well performance. This approach is particularly beneficial in shale reservoirs, where traditional methods often fall short due to complex fracture networks and flow regimes. The method also allows for the automatic identification of the optimal number of features in the data. However, there are limitations as well. The accuracy of predictions depends on the availability of mature wells. Without sufficient data, the model's predictions may be less reliable. Additionally, the approach cannot accommodate changes in operational conditions and their impact on well performance.

5.4.4.5.4 Results and Discussion

The authors demonstrate the effectiveness of their proposed ML approach in reliably estimating EUR based on field-wide analyses using historical production data. The approach can identify poorly performing wells, which could benefit from early refracturing. The authors conclude that their approach provides a powerful tool for fast and accurate estimations of well performance without presumptions about the state of the well or the flow regime and thus can help optimize production and reduce operational costs.

BIBLIOGRAPHY

Mehana, Mohamed, Guiltinan, Eric, Vesselinov, Velimir, Middleton, Richard, Hyman, Jeffrey D., Kang, Qinjun, Viswanathan, Hari. Machine-learning predictions of the shale wells' performance. *Journal of Natural Gas Science and Engineering*, 88. 2021. 103819, ISSN 1875-5100, https://doi.org/10.1016/j.jngse.2021.103819

5.4.5 AI Applications in Production Aspects

5.4.5.1 Application-35: Use of AI to Reduce Operational Disruptions of ESP

5.4.5.1.1 Background of the Application

This paper explores the use of AI technology and the Internet of Things (IoT) to develop a predictive maintenance model for electrical submersible pumps (ESPs). ESPs are a highly effective method for boosting oil production from onshore and offshore fields, but their failure can result in prolonged downtime and costly production losses. The traditional diagnostic methods can only offer retrospective views of any degraded process conditions, and an early warning system is needed to take corrective action before an ESP's failure. Data-driven analytics powered by massive data sets can be used efficiently thanks to recent developments in AI technology. With the use of data collected from the system, AI technology, specifically DL and neural networks, can be very successful in identifying anomalous behavior in complicated physical systems, like ESPs, and provide decision support for fixing or managing the underlying problems.

5.4.5.1.2 Description of the Workflow and Machine Learning Model

Here, a back propagation neural network model is used to predict the operation of ESP pumps, using the seven parameters generated during the operation of ESP pumps and the working conditions of ESP pumps in the corresponding cases as training data, and establishing a back-propagation model with relu as the activation function to detect abnormalities in the data set (sent by ESP pumps) before the failure of ESP pumps. This model can reduce the economic losses caused by well stoppage and repair of ESP pumps through corresponding measures after detecting the abnormalities of ESP pumps.

5.4.5.1.3 Advantages and Limitations of the Synchronized Algorithm

Forward propagation and back propagation are the most mainstream and basic models in DL. In the process of using the training dataset for training, by continuously improving the weights and adjusting the bias, a model is finally determined that can predict whether the ESP pump will fail or not by the seven parameters generated during the operation of the ESP pump.

5.4.5.1.4 Results and Discussion

After several months of testing the predictive maintenance solution in the pilot installation, the conclusion was drawn that the system performs exceptionally well and can detect multiple kinds of anomalies. Although these new types of complex anomalies were not known and therefore difficult to interpret as to what the root cause of the anomaly would be, they further substantiated the use of ML-based neural network technology as an analytical tool for implementing a predictive maintenance solution for ESPs. The outcome of this research provides an innovative approach to reduce the operational disruptions of ESPs and to increase their availability to prevent production disruptions. The future of AI technology and IoT provides a platform for managing large data sets and can be an effective means of analyzing the relational behavior of complex systems.

BIBLIOGRAPHY

Jansen van Rensburg, Nico "Usage of Artificial Intelligence to Reduce Operational Disruptions of ESPs by Implementing Predictive Maintenance." *Paper Presented at the Abu Dhabi International Petroleum Exhibition & Conference*, Abu Dhabi, UAE, November 2018. https://doi.org/10.2118/192610-MS

5.4.5.2 Application-36: Artificial Lift Selection Using Machine Learning

5.4.5.2.1 Background of the Application

This paper proposes the use of supervised ML in artificial lift selection to minimize the life-cycle cost of artificial lift wells. By incorporating past performances and lessons learned from installations, this approach can improve the selection process and reduce the impact of human biases. In order to identify the "Good" and "Bad" wells based on their life-cycle cost, the data is pre-processed and organized into a structured dataset. The classification models are then trained and validated using this data. The most straightforward and precise model is used to evaluate the performance of existing wells and choose future artificial lifts. The productivity of the fields will be enhanced by the pattern for selecting artificial lift that has been found. Furthermore, new data may be incorporated into the existing dataset, and the model can be further integrated to respond to the dynamic changes in the fields' circumstances, thanks to ML's limitless learning capacity.

5.4.5.2.2 Description of the Workflow and Machine Learning Model

According to the literature reviews and the data types, eight significant variables were selected as predictors, including "depth," "flow_rate," "density," "pressure," "power_grid," "service_life," "availability," and "noise_level." It's important for us to transform raw data into useful features that help us to understand the model better and increase its predictive power. It's clear that all the variables are ordinal categorical data types. For example, the target variable, "methods" is made of four types of artificial lift methods including beam pump (BP), gas lift (GL), ESP, and progressive cavity pump (PCP). Categorical variables were encoded into numerical variables, with methods transformed into 1, 2, 3, and 4, corresponding to four different artificial lift methods. After completing feature engineering, both DT and RF models were built to predict artificial lift methods. The dataset was split, and the *DecisionTreeClassifier* model was instantiated using the criterion *gini_index*. The highest confidence score was obtained from the RF model among those evaluated. The output of this model (*Y_pred*) was selected for further prediction.

5.4.5.2.3 Advantages and Limitations of the Synchronized Algorithm

An ensemble learning technique for classification, regression, and other tasks is called a RF or random decision forest. It works by building a large number of DTs (n_estimators=100) during training and producing the class that is the mean prediction (for regression) or the mode of the classes (for classification) of the individual trees. Regarding Table 5.1, It is known that various predictors impact the selection of artificial lift methods. Readers should build models based on real-world data.

RF reduces overfitting problem in DTs and reduces the variance and therefore improves the accuracy. It can be used to solve both classifications and regression problems. One of the most significant merits of RF is that it works well with both categorical and continuous variables. However, RF is more complex as it requires much more computational resources and long training periods.

TABLE 5.1
Artificial Lift Selection Parameters

Well Parameters	Production Conditions	Fluid Properties	Reservoir Parameters
Depth	Flow Rates	Density	Pressure
Tortuosity	Decline Rates	Viscosity	Temperature
Tubulars	GOR	Solid Production	DAQ requirement
	Water Cut	Scales	Productivity
Surface Facilities	**Profitability Factors**	**Supplier Factors**	**HSE Considerations**
Power grid	Service Life	Availability	Noise Level
Gas Availability	CAPEX	Ease of Maintenance	CO_2 Emission
Pipeline	OPEX	Access to Spare Parts	Well Integrity
Capacity constraints	Production Deferment	Staff Experience	Staff Competency

5.4.5.2.4 Results and Discussion

In conclusion, this paper shows that the supervised ML process is more comprehensive than the selection made by conventional approaches. The ML algorithm provides a more reliable and accurate selection method for artificial lift wells by minimizing life-cycle cost. The AL selection model uses the DT algorithm, which offers intuitive comprehension in addition to good prediction accuracy and training speed. The decision-making process used by this algorithm is comparable to that of humans. The engineer will therefore be able to explain the model with ease. Compared to man's curated selection process, the ML process offers a more thorough approach and is a promising instrument that can assist in resolving difficult human issues. In the end, it gives the oil and gas sector a long-lasting competitive edge.

BIBLIOGRAPHY

Ounsakul, Thanawit, Sirirattanachatchawan, Thum, Pattarachupong, Wiwat, Yokrat, Yaovanart, and Ekkawong, Peerapong. "Artificial Lift Selection Using Machine Learning." *Paper Presented at the International Petroleum Technology Conference*, Beijing, China, March 2019. https://doi.org/10.2523/IPTC-19423-MS

5.4.5.3 Application-37: A Method of Intelligent Scheme Optimization for Non-stop System Intermittent Production Units

5.4.5.3.1 Background of the Application

In order to avoid problems brought on by void pumping or prolonged system stops, the non-stop-system intermittent pumping (NSSIP) technology has been provisionally used in a few low-production wells in the Daqing oilfield in China. Nevertheless, NSSIP production schemes are primarily based on experience decisions, the pump's feed and fetch are out of balance, and it's questionable if NSSIP's low energy usage feature actually works. To enhance the pump's feed and fetch balance and optimize the overall benefit, an intelligent scheme optimization method for NSSIP is presented in this work. The process consists of the following: the creation of a physical model of bottom hole flowing pressure buildup; the calibration of pump efficiency vs dynamometer working conditions; the training of the CNN + SNN + HDG model for dynamometer recognition; the development of a purposeful non-stop pumping scheme and data acquisition scheme for typical experimental wells; the measurement of load, displacement, and production rate data of pumping wells by NSSIP in real time; and the creation of data models for correlation analysis of different parameters. The key parameters that affect the pump balance of feed and fetch and the economic benefit of a single well are screened out, and the results and models are integrated into the ensemble DT model to optimize the most reasonable pumping time and frequency for NSSIP schemes. The method is expected to be of great importance to the realization of cost reduction and efficiency increase for China's petroleum industry.

5.4.5.3.2 Description of the Workflow and Machine Learning Model

The chapter presents a method of intelligent scheme optimization for non-stop system intermittent production units. The method involves three models of which the third is an ML model. The goal of the model is to predict the production rate based on the input parameters such as stroke, stroke frequency, max load, min load, and pump efficiency. The model is trained on a dataset of 1000 rows and eight columns using regression techniques. The dataset is loaded into a panda's data frame and is preprocessed before training the model.

5.4.5.3.3 Advantages and Limitations of the Synchronized Algorithm

The main advantage of the method is that it is a ML model, which can learn from the dataset and make predictions without explicit programming. It can handle large amounts of data and can produce accurate predictions. However, the method has some limitations. The model is only as good as the dataset it is trained on, so if the dataset is biased or incomplete, the model will not perform well. Also, the model may not perform well if the input parameters are not well defined or if the data is noisy. Finally, the method may require a lot of computational resources and time to train the model.

5.4.5.3.4 Results and Discussion

This research proposes an intelligent scheme optimization approach for NSSIP that may effectively reduce pumping system failures due to long-term system stopping or void pumping. Additionally, it can save more energy and significantly improve pumping and system efficiency. According to the findings, NSSIP might improve system efficiency by an average of 7.8% over regular pumping and 4.4% over regular intermittent pumping. Additionally, compared to intermittent, NSSIP might save an average of 22.14 kWh of energy each day. It is hypothesized that the intermittent scheme is mainly impacted by fluid supply capacity. NSSIP could maintain pump fullness, improve system efficiency, and save energy simultaneously for low-permeability and low-production wells.

BIBLIOGRAPHY

Wang, Cai, Xiong, Chunming, Shi, Junfeng, Han, Qiqing, Zhang, Jianjun, Zhao, Ruidong, Lei, Gang, Zhang, Xishun, and Sun, Yizhen. "A Method of Intelligent Scheme Optimization for Non-stop System Intermittent Production Units." *Paper Presented at the Abu Dhabi International Petroleum Exhibition & Conference*, Abu Dhabi, UAE, November 2019. https://doi.org/10.2118/197661-MS

5.4.5.4 Application-38: Prediction of Maximum Possible Liquid Rates Produced from Plunger Lift by Use of a Rigorous Modeling Approach

5.4.5.4.1 Background of the Application

The paper presents a reliable predictive model to estimate the maximum possible liquid production rates using plunger lift. The model is designed to use tubing size and well depth as affecting parameters for accurate prediction. The study uses a simple approach to detect outlier data points, and on this basis, it is found that the experimental data is reliable. The model is efficient in estimating the maximum possible rate by use of plunger lift as a function of tubing size and well depth.

5.4.5.4.2 Description of the Workflow and Machine Learning Model

The workflow of the model involves using the LSSVM approach together with the coupled simulated annealing (CSA) optimization technique. The model takes tubing size and well depth as input parameters and predicts the maximum liquid production rate. Data for developing the model is gathered from the literature and then divided into a training set and a test set. The LSSVM mathematical scheme is utilized to handle the linear equations, making it a faster and more adequate option compared to traditional SVM tools.

5.4.5.4.3 Advantages and Limitations of the Synchronized Algorithm

The paper uses a novel feature selection mechanism based on CSA optimization, which enhances the model's accuracy. The LSSVM model is particularly adept at characterizing nonlinear behavior, which is crucial in this context. However, the model does not consider the effects of fluid properties due to their minimal impact on the maximum possible liquid production rate using a plunger-lift system, which could be a limitation.

5.4.5.4.4 Results and Discussion

The study has introduced a new predictive model that accurately predicts the maximum liquid rates that may be produced from plunger lift by using tubing size and well depth as influencing parameters. A precise and dependable mathematical procedure was used to create the model. The results show that the model is effective and dependable in determining the highest rate that can be achieved using plunger lift. Only a small percentage of the experimental data appear to be beyond the applicability zone of the recently established model for predicting the highest feasible rate, according to the straightforward method employed to identify outlier data points.

BIBLIOGRAPHY

Kamari, A., Bahadori, A., and Mohammadi, A. H. "Prediction of Maximum Possible Liquid Rates Produced From Plunger Lift by Use of a Rigorous Modeling Approach." *SPE Prod & Oper* 32 (2017): 7–11. https://doi.org/10.2118/180912-PA

5.4.5.5 Application-39: In-Silico Model for Predicting the Corrosion Inhibition Efficiency of Steel Inhibitors

5.4.5.5.1 Background of the Application

The computational framework for examining the characteristics of organic corrosion inhibitors—which prevent steel from corroding in acidic solutions—is presented in this paper. Modules including data preprocessing, descriptor selection, and model construction make up the framework. Using cutting-edge ML techniques like SVMs, RFs, and gradient boosting machines (GBMs), a reliable predictive model for several classes of corrosion inhibitors was created. A new integrated ensemble method was used to choose the descriptors. The corrosion inhibition efficiency of inhibitors may be predicted with a much greater degree of accuracy by the model based on the GBM algorithm.

5.4.5.5.2 Description of the Workflow and Machine Learning Model

The workflow comprises modules like data preprocessing, descriptor selection, and model building. The study uses GBM, RF, and SVMs to develop a robust predictive model for multiple classes of corrosion inhibitors. A novel integrated ensemble technique is used for selecting the most relevant descriptors. The model, particularly the one based on the GBM algorithm, showed significantly higher accuracy in predicting the corrosion inhibition efficiency.

5.4.5.5.3 Advantages and Limitations of the Synchronized Algorithm

The model achieves high accuracy across multiple classes of molecules, overcoming the limitations of previous studies that focused on single molecule classes. The use of diverse ML algorithms and the integrated ensemble approach for feature selection enhances the model's predictive power. However, there are limitations, such as the model's potential to fail for molecules whose efficiencies drop after a certain concentration. Moreover, the model may not be as effective in conditions other than those tested.

5.4.5.5.4 Results and Discussion

The study successfully developed a computational framework for predicting the corrosion inhibition efficiency of inhibitors for steel in acidic medium. The results show that the model based on the GBM algorithm performs significantly better than other ML algorithms. The study provides a promising approach for designing and discovering novel corrosion inhibitors, which can help combat corrosion in industries and reduce economic losses due to corrosion.

BIBLIOGRAPHY

Agarwal, Abhishek, Rathore, Pradeep, Jain, Vinay, and Beena Rai. "In-Silico Model for Predicting the Corrosion Inhibition Efficiency of Steel Inhibitors." *Paper Presented at the CORROSION 2019*, Nashville, TN, March 2019.

5.4.5.6 Application-40: A Predictive Model for Steady-State Multiphase Pipe Flow: Machine Learning on Lab Data

5.4.5.6.1 Background of the Application

The paper proposes a new method for calculating the pressure drop in a pipeline based on ML algorithms. The existing methods, which rely on empirical correlations or first-principles mechanistic models, have limited applicability and accuracy. The proposed method involves segmenting the pipe and using three surrogate models based on ML algorithms to calculate the pressure gradient in each segment. The models are trained on a representative lab data set from the open literature and predict the liquid holdup, flow pattern, and pressure gradient. The proposed method extends the applicability and accuracy of the existing methods.

5.4.5.6.2 Description of the Workflow and Machine Learning Model

In the paper, the workflow of the AI model is centered around the segmentation of pipelines and the calculation of pressure gradients in each segment. The first model predicts the liquid holdup in the segment, the second identifies the flow pattern, and the third estimates the pressure gradient. Various ML algorithms, including RF, gradient boosting DTs, SVM, and ANN, are employed and compared. This approach achieved a high coefficient of determination ($R^2 = 0.95$) using the gradient boosting algorithm.

5.4.5.6.3 Advantages and Limitations of the Synchronized Algorithm

The use of larger databases for training allows these algorithms to cover a broad range of input parameters, offering better applicability compared to standard techniques. The model surpasses traditional methods like Mukherjee and Brill correlation ($R^2 = 0.92$) and a combination of Ansari and Xiao mechanistic models ($R^2 = 0.91$). However, the paper also implies the limitations of these methods, such as the dependency on extensive data, and the variability between algorithms.

5.4.5.6.4 Results and Discussion

Compared to the conventional accessible methods, the suggested approach for calculating pressure gradients based on ML algorithms trained on lab data achieves excellent accuracy and can be applied to a larger variety of input parameters. The suggested methodology can transition from lab to field circumstances without the need for further retraining, thanks to the usage of ML algorithms and the larger database used for their training. When compared to correlations and mechanistic models, the suggested approach produces scores on field data that are on par with or even higher.

BIBLIOGRAPHY

Kanin, E.A., Osiptsov, A.A., Vainshtein, A.L., Burnaev, E.V. A predictive model for steady-state multiphase pipe flow: Machine learning on lab data. *Journal of Petroleum Science and Engineering*, 180, 2019, 727–746, ISSN 0920-4105, https://doi.org/10.1016/j.petrol.2019.05.055

5.4.5.7 Application-41: Investigation on Sulfur Solubility in Sour Gas at Elevated Temperatures and Pressures with an Artificial Neural Network Algorithm

5.4.5.7.1 Background of the Application

The T-S fuzzy neural network (T-S FNN), an effective model for examining the solubility of elemental sulfur in sour gas mixtures, is presented in this research. The model takes into account a number of variables, such as temperature, reservoir pressure, and the mole fraction of carbon dioxide, hydrogen sulfide, and methane as input parameters and sulfur solubility as target parameters. The model is trained and tested using several experimental sulfur solubility data sets. With an overall average absolute relative deviation of 5.35%, the result shows that the predicted solubility data and the experimental sulfur solubility are in excellent agreement. This model is therefore workable and efficient.

5.4.5.7.2 Description of the Workflow and Machine Learning Model

The paper describes an AI model known as the T-S FNN. The workflow of this model involves several steps: The model considers reservoir pressure, temperature, and the mole fractions of CH_4, H_2S, and CO_2 as input parameters. The data sets are divided into training, testing, and validation sets, normalized to a range of −1 to 1.

The T-S FNN consists of an antecedent network with four layers (input, membership function, product inference rule, and weighted average defuzzifier) and a consequent network. It uses a gradient descent learning algorithm to adjust parameters during the training process.

5.4.5.7.3 Advantages and Limitations of the Synchronized Algorithm

There are some advantages listed in the paper: The T-S FNN model achieves an overall AARD of 5.35%, indicating strong prediction accuracy. By considering multiple influential factors like temperature, pressure, and gas composition, the model can achieve accurate predictions. However, limitation includes limited H_2S content range in datasets: The model is trained with H_2S content ranging only from 1 to 26.62%, which may limit its prediction accuracy for higher H_2S concentrations.

5.4.5.7.4 Results and Discussion

In summary, an efficient intelligent algorithm model, T-S FNN, is presented in this paper to predict the solubility of elemental sulfur in sour gas. Compared with empirical models that tend to produce large errors, the T-S FNN model is more accurate, with an overall average absolute relative deviation of 5.35%. Additionally, it provides a new idea for the prediction of sulfur solubility in sour gas and helps operators to develop sour gas reservoirs better.

BIBLIOGRAPHY

Fu, Liang, Hu, Jinghong, Zhang, Yuan, Li, Qian. Investigation on sulfur solubility in sour gas at elevated temperatures and pressures with an artificial neural network algorithm. *Fuel*, 262, 2020, 116541, ISSN 0016-2361, https://doi.org/10.1016/j.fuel.2019.116541

5.4.5.8 Application-42: Fault Identification Using a Chain of Decision Trees in an Electrical Submersible Pump Operating in a Liquid-Gas Flow

5.4.5.8.1 Background of the Application

In this study, a specific assembly of Classification and Regression Trees (CART) is proposed as a strategy for identifying and categorizing early problems in a pumping system. Just the pressure, flow, torque, and temperature are monitored in the "Chain of Decision Trees" methodology, which is the suggested approach for observing system activity. There are two steps in the algorithm. Whether the system is in a fault state is determined by the first, and if it is, the sort of fault is determined by the second. The goal of the suggested method is to increase the balance between classification and failure cause interpretation.

5.4.5.8.2 Description of the Workflow and Machine Learning Model

Data Collection: Operational data from ESP systems are gathered, focusing on various parameters essential for fault analysis.
Data Preprocessing:

Normalization: Data is normalized to ensure consistency in scale across different measurements.

Feature Selection: Relevant features are selected based on their predictive power for fault detection.

Model Building: A Chain of Decision Trees is developed, designed to first detect the presence of a fault and then classify its type.

Model Training: The model is trained on a dataset simulating various fault conditions, learning to distinguish between normal operation and different types of faults.

Model Evaluation: The CDT model's performance is assessed using standard metrics, demonstrating its effectiveness in fault detection and classification.

Prediction: The trained model can be applied to new data to predict and classify faults in real-time, enhancing ESP system reliability.

5.4.5.8.3 Advantages and Limitations of the Synchronized Algorithm

Advantages:

- Enhanced Accuracy: The CDT approach shows high precision in fault detection and classification.
- Interpretability: DTs provide understandable decision paths, aiding in fault diagnosis.
- Real-time Application: Enables prompt fault identification and classification, crucial for operational efficiency and safety.

Limitations:

- Data Quality: Model performance heavily depends on the quality and diversity of the training data.
- Model Complexity: Managing a chain of models can be complex, requiring careful tuning and validation.
- Adaptability: While effective, the model may need retraining for different ESP systems or operational conditions.

5.4.5.8.4 Results and Discussion

In conclusion, the proposed methodology using the Chain of Decision Trees approach improves the overall accuracy and reduces individual misclassification compared to a single DT. The proposed methodology takes advantage of historical data for future predictions and has shown promising results in detecting and classifying incipient faults in a pumping system. The proposed methodology can be used in various industrial applications, such as petroleum, chemical, nuclear, and geothermal industries, where multistage centrifugal pumps are used for gas-liquid two-phase flow.

BIBLIOGRAPHY

Castellanos, Mauricio Barrios, Serpa, Alberto Luiz, Biazussi, Jorge Luiz, Verde, William Monte, Natache do Socorro Dias Arrifano Sassim. Fault identification using a chain of decision trees in an electrical submersible pump operating in a liquid-gas flow. *Journal of Petroleum Science and Engineering*, 184, 2020, 106490, ISSN 0920-4105, https://doi.org/10.1016/j.petrol.2019.106490

5.4.5.9 Application-43: Regression-Based Finite Element Machines for Reliability Modeling of Downhole Safety Valves

5.4.5.9.1 Background of the Application

In offshore wells, the downhole safety valve (DHSV) is a crucial component that stops blowouts, which are uncontrolled releases of gas and oil into the environment. Determining the dependability of these valves can aid in decision-making, workover demand assessment, and blowout event prediction. FEMaR, a parameterless finite element machine for regression problems that requires no training step, is provided. In order to assess the dependability of DHSVs for comparison with conventional statistical techniques, A number of ML models are also assessed. The experimental evaluation, which was carried out using a dataset gathered from a Brazilian oil and gas corporation, demonstrated that ML techniques can beat the statistical methodologies taken into consideration and can produce promising results even when information is filtered. Such findings, also investigated using uncertainty analysis, provide evidence that economic resources can be saved and safety increased in offshore well operations.

5.4.5.9.2 Description of the Workflow and Machine Learning Model

Data Collection:

Gathers operational data from DHSVs, including pressure readings, valve open/close cycles, and environmental conditions.

Data Preprocessing:

Outlier Handling: Identifies and corrects anomalies in the data to prevent model distortion.
Feature Encoding: Transforms categorical variables into numerical formats for processing.
Normalization: Scales features to a uniform range to aid model convergence.
Data Splitting: Segregates the dataset into training and testing sets for model training and validation.

Model Building:

Regression Models: Utilizes ML regression models to predict DHSV reliability metrics.
Evaluation Metrics: Employs metrics like MSE to assess model accuracy.

Model Training and Evaluation:

Trains the models on historical data, evaluating their performance on unseen data to ensure reliability in predicting DHSV operational outcomes.

Prediction:

Enables predictive maintenance by forecasting DHSV reliability, facilitating timely interventions.

5.4.5.9.3 Advantages and Limitations of the Synchronized Algorithm

Advantages:

Enhances DHSV maintenance strategies through accurate reliability predictions.
Allows for the optimization of DHSV designs and operational parameters based on predictive insights.

Limitations:

Dependent on the quality and completeness of historical operational data.
Requires computational resources for processing and model training.

5.4.5.9.4 Results and Discussion

In order to measure the dependability of DHSVs for comparison with conventional statistical approaches, this work assessed a number of ML models and proposed FEMaR, a parameterless finite element machine for regression issues. The study showed that ML methods can outperform the statistical methodologies taken into consideration in this work and can produce encouraging findings even when information is suppressed. The findings imply that these methods can improve the safety of offshore well operations while also conserving financial resources.

BIBLIOGRAPHY

Colombo, Danilo, Lima, Gilson Brito Alves, Pereira, Danillo Roberto, Papa, João P. Regression-based finite element machines for reliability modeling of downhole safety valves. *Reliability Engineering & System Safety*, 198, 2020, 106894, ISSN 0951-8320, https://doi.org/10.1016/j.ress.2020.106894

5.4.5.10 Application-44: Machine Learning Models to Automatically Validate Petroleum Production Tests

5.4.5.10.1 Background of the Application

Because of the dynamic and intricate nature of the process, this research suggests an autonomous tool based on ML models to swiftly validate production tests, which are required to assess well behavior in oil fields. Using six classification algorithms and 13 wells in an offshore oil field in Brazil, the program automatically classified fresh production tests as either valid or invalid based on the well's historical data. Training and validation sets of the dataset were separated, and the outcomes were assessed using the validation set. The study identified that certain variables, such as IGLR (injection gas/liquid ratio), oil flow rate, and the pressure loss between the wellhead and platform, were representative for most of the wells and had a significant influence on production well test validation.

5.4.5.10.2 Description of the Workflow and Machine Learning Model

The workflow contains four main steps: data collection, data preprocessing, model training, and model evaluation. Initially, datasets are created for each well inspection based on valid and invalid production tests. The data preprocessing phase aims to remove outliers. Next, six classification algorithms (logistic regression, naive bayes, k-nearest neighbors, DT, RF, and SVM) are employed to automatically label new production tests as valid or invalid. Lastly, the results are assessed using a test set, comparing the model's output with actual test labels.

5.4.5.10.3 Advantages and Limitations of the Synchronized Algorithm

There are some advantages listed in the paper: The model can rapidly classify and identify issues, providing more confidence in each production test. And these data-driven approaches are more efficient with reduced computational demands. However, there are limitations, such as a small dataset potentially leading to large variations in AUC metrics during cross-validation. Additionally, the differing behaviors of each well may require individualized adjustments in the feature selection stage.

5.4.5.10.4 Results and Discussion

From the results of the case study, it is evident that the proposed tool is effective in automating the validation process of production tests, reducing subjectivity and increasing confidence in the results. The SVM and logistic regression models were found to perform the best, and accurate results were achieved, with at least five out of six tests correctly classified in 70% of the wells analyzed. The study also highlights the potential of data-driven models, such as ML, in solving complex problems in the oil industry.

BIBLIOGRAPHY

de Almeida, Machado Duque, Maria Clara, Souza Chaves, Gabriela, de Oliveira Monteiro, Danielle, Velasco Medani, Luciana, and Filho, Virgílio José Martins Ferreira. "Machine Learning Models to Automatically Validate Petroleum Production Tests." *Paper Presented at the SPE Latin American and Caribbean Petroleum Engineering Conference*, Virtual, July 2020. https://doi.org/10.2118/199112-MS

5.4.5.11 Application-45: Well Performance Prediction Using Machine Learning Approaches

5.4.5.11.1 Background of the Application

A study on the creation of a model utilizing big DA, such as data mining and ML, to forecast the average gas production rate for unconventional gas reservoirs during their first year of operation is presented in this paper. The study employed a database of well completion design parameters and well characteristics for 603 horizontal wells and concentrated on the Montney formation in the Canadian province of British Columbia. The study identified a subset of the most important parameters after evaluating the association between a number of parameters and the rate of gas production. In order to determine the optimal strategy for estimating the average gas production rate, the study lastly performed a comparison analysis of ANN models with various training algorithms.

5.4.5.11.2 Description of the Workflow and Machine Learning Model

Initially, the researchers constructed a database containing oil well completion design parameters. Then, they assessed the relationship between various parameters and natural gas production, identifying a set of the most significant parameters. Finally, they compared different ANN models to determine the best predictive algorithm. The benchmarks for comparison of these models included RMSE, MAE, MRE, R^2, and Execution Time.

5.4.5.11.3 Advantages and Limitations of the Synchronized Algorithm

There are some advantages listed in the paper: A major advantage is its ability to handle large volumes of data and discover meaningful patterns within them. For instance, in this study, the analysis of extensive well data enabled researchers to identify key factors affecting natural gas production.

5.4.5.11.4 Results and Discussion

According to the study, the best forecasts for the average gas production rate during the first year of operation were made using a one-hidden-layer ANN model with 14 neurons in the hidden layer and the Bayesian regularization technique. Additionally, the study demonstrated that eight factors, such as well geometry, completion interval, and completion design parameters, had a significant impact on the average gas production rate. An interesting direction for future research is the study's use of big DA, such as data mining and ML, to create a model for forecasting the rate of gas production in unconventional reservoirs.

BIBLIOGRAPHY

Rahmanifard, Hamid, Alimohamadi, Hamzeh, and Gates, Ian. "Well Performance Prediction in Montney Formation Using Machine Learning Approaches." *Paper Presented at the SPE/AAPG/SEG Unconventional Resources Technology Conference*, Virtual, July 2020. https://doi.org/10.15530/urtec-2020-2465

5.4.5.12 Application-46: Well Performance Prediction in Montney Formation Using Machine Learning Approaches

5.4.5.12.1 Background of the Application

The oil and gas industry employ various pumps in different stages of the oilfield life cycle. However, there is no established method to monitor the fault states of pumps as they operate based on sensor feedback. This often results in premature or reactive maintenance, leading to wasteful downtime and expenses. In this paper, a ML-based neural network model is used to identify different fault states in a triplex pump from measured pressure sensor data. The model is trained on 1500 runs of suction and discharge pressure data recorded for eight different fault states and nominal condition. The results show that the model has over 90% accuracy in predicting the correct fault class solely from the suction and discharge pressure data. This work demonstrates the accuracy of ML techniques in predicting different faults in pumps in operation and highlights the importance of using engineering domain knowledge for feature extraction and manipulation in creating insightful ML models.

5.4.5.12.2 Description of the Workflow and Machine Learning Model

The paper described a model for monitoring and identifying failure states of pumps used in oil fields. Initially, they set up a pump in a lab environment, introducing eight distinct failure states such as damages to the discharge valve, stuffing box, and suction valve. Using this data, the researchers trained a neural network model to predict the pump's failure category based on suction and discharge pressure data. This model, after being tested through multiple cross-validations, achieved an accuracy rate of over 90%.

5.4.5.12.3 Advantages and Limitations of the Synchronized Algorithm

There are some advantages listed in the paper: the use of ML techniques enables accurate predictions of different failures in operating pumps, requiring only monitoring of suction and discharge pressure data. The model demonstrated a 95% accuracy rate in predicting pump failure states on unseen pressure sensor data.

5.4.5.12.4 Results and Discussion

This paper presented a ML-based approach for condition-based monitoring of triplex pumps. The results show that the model accurately predicts different faults in pumps in operation using suction and discharge pressure data. The work highlights the importance of using engineering domain knowledge for feature extraction and manipulation in creating insightful ML models. Overall, the model has potential for deployment in the oil and gas industry to optimize maintenance and reduce downtime and expenses.

BIBLIOGRAPHY

Chakraborty, Ishita, and Kluk, Daniel J. "Condition Based Monitoring of Pumps Using Machine Learning." *Paper Presented at the Offshore Technology Conference*, Houston, TX, May 2020. https://doi.org/10.4043/30648-MS

5.4.5.13 Application-47: A Machine Learning-Based System for Self-Diagnosis Multiphase Flow Meters

5.4.5.13.1 Background of the Application

This study suggests an unsupervised ML algorithm-based anomaly detection method for multiphase flow meters (MPFM). MPFMs are small, multi-sensor devices that measure a well's water, gas, and oil flow rates continuously without requiring the phases to be physically separated. However, in order to guarantee precise flow rate measurements, quality assessment methods are vital. The complexity and unpredictability of MPFM data can be efficiently managed by the suggested approach, which can also identify the malfunctioning sensor and identify outliers. The suggested method is made to be implemented embeddedly.

5.4.5.13.2 Description of the Workflow and Machine Learning Model

In this paper, an ML-based MPFM diagnostic system is described. The workflow primarily involves data preprocessing, feature extraction, application of anomaly detection algorithms, and root cause analysis. Through the feature extraction process, useful information is extracted from the raw data. Anomaly detection algorithms are used to identify anomalies in the data, while the root cause analysis module determines the faulty sensors. This process enables effective identification and analysis of anomalies.

5.4.5.13.3 Advantages and Limitations of the Synchronized Algorithm

There are some advantages listed in the paper: the model can handle the complexity and variability of MPFM data, effectively pinpointing faulty sensors. This method does not require training on historical data, making it suitable for implementation. However, some limitations are also highlighted, such as the need to balance between detection capability and time complexity of anomaly detection algorithms.

5.4.5.13.4 Results and Discussion

The proposed ML-based system for auto-diagnosis is an effective tool to increase the overall confidence in MPFMs. The system evaluates the quality of the sensor measurements and detects the module that is malfunctioning. The anomaly detection system for MPFM presented in this paper is based on unsupervised learning algorithms and is able to identify sneaky anomalous events that could pass unnoticed even to domain experts. The root cause analysis algorithm identifies the faulty module that causes the anomaly. The proposed approach has the potential to significantly improve the reliability and accuracy of MPFMs in the oil and gas industry.

BIBLIOGRAPHY

Barbariol, Tommaso, Feltresi, Enrico, and Susto, Gian Antonio. "A Machine Learning-Based System for Self-Diagnosis Multiphase Flow Meters." *Paper Presented at the International Petroleum Technology Conference*, Dhahran, Kingdom of Saudi Arabia, January 2020. https://doi.org/10.2523/IPTC-19865-MS

5.4.5.14 Application-48: Machine Learning Application for Oil Rate Prediction in Gas Lift Wells

5.4.5.14.1 Background of the Application

When assessing system performance, reservoir and production engineers consider well production rates. On the other hand, production rate monitoring is typically not done in real time. Although they are acknowledged to have inherent errors, flowmeters are utilized in some situations. As a result, the industry uses correlations to assign production to wells. It has been recognized throughout time that a number of technological and financial problems make the widely utilized correlations insufficiently effective. The development of a correlation that reliably forecasts oil rate in artificial GL wells using ML techniques is the main topic of this article. The goal is to offer a straightforward, broadly applicable answer. Various intelligent algorithms are employed, including the ANFIS, SVMs, and ANN. The newly developed model using ANN shows that it can predict the flow rate with 99% accuracy, an interesting outcome as such accuracy has not been reported in the literature usually.

5.4.5.14.2 Description of the Workflow and Machine Learning Model

The paper builds AI algorithms for predicting the oil rate of artificial GL wells. After the data preparation process, a dataset of 1,500 separator test points is obtained. Several ML models, including ANNs, ANFIS, and SVM, were then trained using this data. After determining the optimal parameters, these models were used to predict oil well production rates, showing that the ANN model achieved a 99% accuracy rate, outperforming other models.

5.4.5.14.3 Advantages and Limitations of the Synchronized Algorithm

The methodology showcases several advantages: this approach offers a highly accurate predictive tool, especially in predicting the oil rate of artificial GL wells. The ANN model particularly stands out for its superiority, providing predictions with 2.56% MAPE and a high correlation coefficient ($R^2 = 0.99$).

5.4.5.14.4 Results and Discussion

When compared to earlier ai models, the correlation created with ANN performs several times better and surpasses all existing empirical correlations. By offering a functional equation that can be applied to any field data, the study eliminates any doubts or misunderstandings surrounding the idea of AI knowledge and software. In addition to validating the well tests, this endeavor offers an industrial perspective on the function of data-driven computational models for the production reconnaissance scheme as a useful instrument to lessen production provision concerns.

BIBLIOGRAPHY

Khan, Mohammad Rasheed, Alnuaim, Sami, Tariq, Zeeshan, and Abdulraheem, Abdulazeez. "Machine Learning Application for Oil Rate Prediction in Artificial Gas Lift Wells." *Paper Presented at the SPE Middle East Oil and Gas Show and Conference*, Manama, Bahrain, March 2019. https://doi.org/10.2118/194713-MS

5.4.5.15 Application-49: Machine Learning for Progressive Cavity Pump Performance Analysis: A Coal Seam Gas Case Study

5.4.5.15.1 Background of the Application

Computer Science Technology has been instrumental in simulating gas and petroleum networks. Wellhead chokes are crucial for maintaining stable downstream pressure and balancing gas well productivity. Traditional choke flow models have limitations in accurately predicting well performance parameters. Recent advancements in ML have led to the development of new models, such as the 2010 Choke Gas Equation, which offer improved accuracy compared to traditional methods. However, these models may still struggle in deviating conditions. This study explores the application of random forest regression (RFR) and Gaussian process regression (GPR) for gas production estimation. By analyzing multiple variables such as choke size, pressures, CGR, and temperatures, these ML algorithms demonstrate superior predictive accuracy compared to traditional methods. The results highlight the potential of ML to enhance gas production estimation and address the limitations of conventional models.

5.4.5.15.2 Description of the Workflow and Machine Learning Model

This paper introduces a new approach to choke flow modeling using ML algorithms. Specifically, RFR and GPR are applied to predict gas production rates based on various parameters such as choke size, flowing pressure, and gas-liquid ratios. The ML models are compared to traditional regression models, demonstrating improved accuracy in predicting well performance.

5.4.5.15.3 Advantages and Limitations of the Synchronized Algorithm

RFR and GPR models provide more accurate predictions compared to traditional regression methods, especially in handling high variability in data. However, these models are sensitive to noise in the data and require extensive computational resources for training and validation. The complexity of the models also makes them more challenging to implement in real-time applications.

5.4.5.15.4 Results and Discussion

While predictive performance has improved with advancements in ML, analytical and theoretically based approaches offer valuable insights into the underlying process. A hybrid approach, combining analytical models and ML, can provide the best results.

Simple models can be implemented using MS-Excel or more sophisticated tools like MATLAB/Python. Gaussian Process Regression is an example of a robust analytical model that can reveal underlying relationships. However, ML models excel at modeling the data generation process and often provide superior gas forecasting results.

The Gilbert and operator company Equation 1 models demonstrate this trade-off. Their reliance on a limited number of parameters can lead to inaccuracies in low gas rates and high CGR/BSW scenarios. A hybrid approach that incorporates both analytical and ML models can help address these limitations and improve overall gas forecasting accuracy.

BIBLIOGRAPHY

Jauregui, Leal, Alonso, Jairo, Lopez, Arevalo, Jose, Alfredo, Atwi, Mohammed, and Leal, Daniel Alejandro. "A New Approach to Choke Flow Models Using Machine Learning Algorithms." *Paper Presented at the International Petroleum Technology Conference*, Riyadh, Saudi Arabia, February 2022. https://doi.org/10.2523/IPTC-22168-MS

5.4.5.16 Application-50: Application of Automated Machine Learning for Multi-variate Prediction of Well Production

5.4.5.16.1 Background of the Application

This paper presents a study that applies various ML methods to a sanitized unconventional oil and gas production dataset. The objective of the study is to predict an optimal set of variables that maximize production. The study employs five ML models to identify the set of variables that maximize the predicted production of barrels of oil equivalent (BOE). The study also introduces the initiative to establish concepts and best practices for predictive and prescriptive analytics in domains of reservoir simulation, description, and asset management.

5.4.5.16.2 Description of the Workflow and Machine Learning Model

The paper discusses a workflow for the multivariate prediction of well production. The workflow begins with data preprocessing, which includes anomaly detection, consistency checks, and missing value handling. This is followed by feature selection and ranking. Subsequently, multiple ML models such as generalized linear models

(GLMs), DL, DTs, RFs, and GBMs are built and trained. Finally, response optimization is conducted on the best-performing models to enhance prediction accuracy.

5.4.5.16.3 Advantages and Limitations of the Synchronized Algorithm

The methodology showcases several advantages: The workflow is capable of handling large volumes of complex data, providing highly accurate predictions. The integration of different types of ML models effectively captures complex patterns in the data.

Limitation includes data availability: The handling of missing data and identification of outliers could impact the accuracy of the models. Moreover, the training and optimization of models require substantial computational resources.

5.4.5.16.4 Results and Discussion

This study applied various ML methods, including GLM, DL, DT, RF, and gradient boosted machine (GBM), to a dataset of 470 horizontal wells with 15 predictor variables and one response variable, produced barrels of oil equivalent (BOE). The study found that the ensemble-based DT methods, RF and GBM, demonstrated superior performance compared to other models. The study also demonstrated the importance of mapping variables with severely skewed and asymmetric distributions to a symmetrical normal score domain. The findings of this study can be applied to real field case studies with significantly larger and more complex data sets in the domain of reservoir simulation, description, and asset management.

BIBLIOGRAPHY

Maucec, Marko, and Garni, Saad. "Application of Automated Machine Learning for Multivariate Prediction of Well Production." *Paper Presented at the SPE Middle East Oil and Gas Show and Conference*, Manama, Bahrain, March 2019. https://doi.org/10.2118/195022-MS

6 AI Applications in Enhanced Recovery

Emre Can Dündar, Ekrem Alagoz, Ferit Ozdaban, Mehmet Melih Oskay, Celal Hakan Canbaz, Luigi Saputelli, Mufrettin Murat Sari, Salih Tutun, and Cenk Temizel

6.1 INTRODUCTION

Artificial intelligence (AI), machine learning (ML), and deep learning (DL) have emerged as powerful tools with the potential to revolutionize various industries. The oil and gas sector, particularly enhanced oil recovery (EOR) operations, is increasingly exploring the applications of these technologies to improve efficiency, reduce costs, and optimize production. This chapter delves into the integration of AI methodologies within EOR workflows. It provides a comprehensive overview of the different categories of EOR applications, including EOR screening, carbon dioxide (CO_2) projects, water flooding, and unconventional and other applications. Furthermore, the chapter explores the fundamental concepts of AI and highlights their specific use cases within the context of enhanced recovery. By understanding the potential of AI in EOR, industry professionals can harness these technologies to unlock new opportunities, enhance decision-making processes, and drive sustainable production.

6.2 CATEGORIZATION OF WORKFLOWS USED IN ENHANCED RECOVERY

6.2.1 EOR Screening

EOR screening plays a pivotal role in assessing the feasibility and benefits of implementing advanced recovery techniques in a reservoir. Traditional methods often rely on manual analysis of geological, reservoir, and production data, a process that is both time-intensive and susceptible to human error. The integration of AI, ML, and DL technologies offers a transformative approach to this process. These technologies enable the automation of data analysis, allowing algorithms to process vast datasets efficiently while uncovering patterns and trends that might elude human analysts. Additionally, ML models can be trained on historical data to predict the potential success of various EOR methods based on specific reservoir characteristics, while AI tools can refine screening criteria by identifying the most influential factors affecting EOR performance. Furthermore, these algorithms can automatically pinpoint

DOI: 10.1201/9781003617327-6

reservoirs that meet predefined criteria, making them suitable candidates for EOR. By incorporating AI into EOR screening, operators can optimize resource allocation and focus on the most promising projects, ensuring more informed and effective decision-making.

6.2.2 CO_2 Projects

CO_2 injection is a prominent EOR method that enhances oil recovery by injecting carbon dioxide into reservoirs. Despite its effectiveness, optimizing CO_2 injection processes poses significant challenges due to factors like reservoir heterogeneity, fluid properties, and the strategic placement of injection wells. Integrating AI, ML, and DL technologies can provide critical insights and support for CO_2 injection projects. These advanced tools can predict CO_2 plume movement with high accuracy by training models on reservoir simulation data, enabling operators to better understand the dynamics within the reservoir. They also aid in optimizing CO_2 injection rates, striking a balance between maximizing oil recovery and minimizing CO_2 leakage. Furthermore, AI applications can monitor CO_2 storage efficiency and identify potential leakage risks, ensuring environmental safety. Additionally, these technologies assist in determining optimal well placement for both injection and production, maximizing oil recovery. By leveraging AI, operators can significantly enhance the efficiency and environmental sustainability of CO2 injection projects, leading to increased oil production and reduced ecological impact.

6.2.3 Water Flooding

Water flooding is a widely adopted EOR method that enhances oil production by injecting water into reservoirs to displace oil. However, optimizing water flooding operations is a complex task influenced by factors such as reservoir heterogeneity, fluid properties, and strategic well placement. The integration of AI, ML, and DL technologies offers significant advancements in managing these challenges. AI models, trained on reservoir simulation data, can accurately predict water flooding performance, enabling operators to forecast outcomes more reliably. These technologies also aid in optimizing water injection rates to maximize oil recovery while minimizing issues like early water breakthrough. Additionally, AI tools enable continuous monitoring of water breakthrough and help identify potential operational challenges. Moreover, AI can guide the placement of injection and production wells, ensuring maximum oil recovery from the reservoir. By leveraging AI, operators can improve the efficiency of water flooding projects, resulting in increased oil production and reduced water usage, while addressing key operational complexities effectively.

6.2.4 Unconventional and Other Applications

Unconventional reservoirs, such as shale gas and tight oil, along with other EOR applications, present unique challenges due to their complex geological structures, low permeability, and high fluid viscosities. The integration of AI, ML, and DL technologies offers transformative opportunities to address these challenges and maximize resource extraction. These technologies enable advanced data analysis and modeling, helping to analyze extensive datasets, improve the understanding of reservoir properties, and develop accurate geological models. AI also facilitate production prediction by utilizing historical production data to forecast future outputs and optimize production strategies. Moreover, they support data-driven decision-making for well placement, completion design, and production optimization, while enabling real-time monitoring of production performance through sensor data to promptly identify and mitigate potential issues.

In specific applications, such as shale gas, AI aid in optimizing well placement, refining fracking designs, and predicting production trends. For tight oil, these technologies enhance water injection strategies, prevent production decline, and improve reservoir characterization. Similarly, in geothermal energy, they contribute to reservoir modeling, production optimization, and risk assessment. By incorporating AI, operators can make informed decisions, reduce risks, and improve efficiency across unconventional and EOR applications, unlocking the potential of challenging reservoirs and advancing energy production technologies.

6.3 OVERVIEW OF AI METHODOLOGIES

AI, ML, and DL are rapidly evolving technologies with the potential to transform various industries, including the oil and gas sector. AI refers to the development of intelligent systems capable of reasoning, learning, and problem-solving. ML is a subset of AI that focuses on training algorithms on data to make predictions or decisions. DL, a further specialization of ML, involves training artificial neural networks (ANNs) with multiple layers to learn complex patterns from data.

AI encompasses a wide range of techniques, including supervised and unsupervised learning, reinforcement learning (RL), convolutional neural networks (CNNs), recurrent neural networks (RNNs), and generative adversarial networks (GANs). These methodologies offer powerful tools for analyzing large datasets, identifying patterns and trends, and making data-driven decisions.

6.4 APPLICATIONS OF AI

6.4.1 EOR Screening

6.4.1.1 Application-1: Selection of EOR/IOR Opportunities Based on ML

6.4.1.1.1 Background of the Application

This document presents an analysis of hydrocarbon recovery methods for various reservoirs. The analysis includes the use of cluster analysis to identify patterns and trends in recovery methods, as well as screening criteria for new drainage strategies. The paper also discusses the importance of understanding input parameters such as porosity, temperature, pressure, permeability, gravity, and viscosity in the context of hydrocarbon recovery.

6.4.1.1.2 *Description of the Workflow and Machine Learning Model*

In this code, reservoir data is processed by first loading it from a CSV file, followed by displaying basic statistics for parameters including porosity, temperature, pressure, permeability, gravity, and viscosity. The data is then visualized using histograms to show the distribution and frequency of each parameter. This method provides a clear initial understanding of the dataset and highlights any significant patterns or anomalies.

6.4.1.1.3 *Advantages and Limitations of the Synchronized Algorithm*

This workflow efficiently summarizes and visualizes key data characteristics, making it easier to identify trends and outliers. The use of descriptive statistics and histograms allows for a straightforward interpretation of the data. However, this method is limited by its reliance on basic statistical measures and simple visualizations, which may not be able to show complex interactions between variables or deeper information about the data.

6.4.1.1.4 *Results and Discussion*

In conclusion, this paper provides valuable insights into hydrocarbon recovery methods and the use of cluster analysis to identify patterns and trends. The screening criteria presented in this paper can be used to evaluate new drainage strategies and guide decision-making in the field of reservoir engineering. However, further research is needed to explore the potential of AI methods in this field and to better understand the complex relationships between input parameters and hydrocarbon recovery.

BIBLIOGRAPHY

Alvarado, Vladimir, Ranson, Aaron, Hernández, Karen, Manrique, Eduardo, Matheus, Justo, Liscano, Tamara, and Prosperi, Natasha. "Selection of EOR/IOR Opportunities Based on Machine Learning." *Paper Presented at the European Petroleum Conference*, Aberdeen, United Kingdom, October 2002. https://doi.org/10.2118/78332-MS

6.4.1.2 Application-2: A Novel EOR Screening Methods to Integrate Theoretical Screening Criteria and Real Field EOR Practices Using ML Algorithms

6.4.1.2.1 *Background of the application*

This paper proposes an automated EOR screening algorithm based on ML and real EOR data sets. The algorithm utilizes a nine-dimensional feature space, including oil gravity, viscosity, composition, saturation, formation type, net thickness, permeability, reservoir depth, and temperature. The ML methods used include Bayesian optimal classifier, k-nearest neighbor (kNN) classifier, minimum mean distance classifier, discriminant analysis, and ANNs. The algorithm's performance is evaluated using real EOR data sets, and the results show that the Bayesian optimal classifier with Gaussian parametric PDF estimator has the best performance compared to the other classifiers. The proposed algorithm provides a reliable and efficient approach to EOR screening, enabling the selection of the most suitable EOR method for a specific reservoir condition.

6.4.1.2.2 *Description of the Workflow and Machine Learning Model*

In this method, several types of classifiers such as Bayesian classifiers, kNN classifiers, Discriminant Analysis, and ANNs are used. These algorithms assist in deciding the best EOR method based on different reservoir properties. By employing these advanced ML techniques, the most suitable EOR method can be correctly suggested for given conditions. Each classifier looks at the data in a different way, which helps in making more reliable predictions and better decisions for managing oil reservoirs.

6.4.1.2.3 *Advantages and Limitations of the Synchronized Algorithm*

Using a mix of Bayesian classifiers, kNN, discriminant analysis, and ANN gives an advantageous approach to EOR screening. Bayesian classifiers provide a probability for each EOR method, helping to understand the chances of success. kNN is simple and good at finding patterns in the data. Discriminant analysis works well when data can be separated linearly, and ANN can handle complex data relationships. However, each method has its downsides. Bayesian classifiers need good initial knowledge, kNN can be slow with large datasets, discriminant analysis assumes data is linear, and ANN can be prone to overfitting and needs a lot of computational power. Despite these issues, using all these methods together makes for an effective EOR screening process.

6.4.1.2.4 *Results and Discussion*

This paper presents a comprehensive study of an automated EOR screening algorithm based on ML and real EOR data sets. The algorithm utilizes a nine-dimensional feature space and several ML methods to classify reservoirs into one of the EOR methods. The Bayesian optimal classifier with a Gaussian parametric PDF estimator has the best performance compared to other classifiers, providing a reliable and efficient approach to EOR screening. The proposed algorithm eliminates arbitrary decision-making, enables the integration of different types of data, and provides accuracy and fast computation. Future work will focus on feature selection and conditioning to improve classification performance.

BIBLIOGRAPHY

Tarrahi, Mohammadali; Afra, Sardar, and Surovets, Irina. "A Novel Automated and Probabilistic EOR Screening Method to Integrate Theoretical Screening Criteria and Real Field EOR Practices Using Machine Learning Algorithms." *Paper Presented at the SPE Russian Petroleum Technology Conference*, Moscow, Russia, October 2015. https://doi.org/10.2118/176725-MS

6.4.1.3 Application-3: Application of Using Fuzzy Logic as an AI Technique in the Screening Criteria of the EOR Techniques

6.4.1.3.1 *Background of the Application*

This paper presents a new screening tool for EOR methods based on fuzzy logic. The screening tool was developed using statistical analysis of a database comprising rock and fluid properties for 347 EOR projects worldwide. The database was collected from the 2004, 2010, 2012, and 2014 Worldwide EOR Survey of the Oil

& Gas Journal. The screening tool includes various EOR methods and associated fuzzy logic parameters. The tool was validated by comparing its results with those of commercial EOR software. The results show that the screening tool is reliable and consistent. Case studies were conducted to investigate the applicability of EOR in some Egyptian oil reservoirs. The results of the screening for the reservoirs under study are presented and analyzed.

6.4.1.3.2 Description of the Workflow and Machine Learning Model

In this method, an EOR screening tool is used, and fuzzy logic is employed to evaluate the suitability of various EOR methods based on reservoir rock and fluid properties. Membership functions are defined for each input, and a set of fuzzy rules is established to assess the compatibility of fourteen different EOR techniques. This approach leverages fuzzy inference to handle uncertainties and provide an informed assessment of EOR method applicability, enhancing decision-making in reservoir management.

6.4.1.3.3 Advantages and Limitations of the Synchronized Algorithm

The fuzzy logic-based methodology succeeds in handling the inherent uncertainties and imprecision in reservoir data, offering a flexible and intuitive framework for EOR screening. It automates the evaluation process, providing quick and reliable assessments based on defined criteria. However, the effectiveness of this method heavily depends on the accuracy and coverage of the input data and membership functions. Inadequate or biased data can lead to suboptimal recommendations, emphasizing the need for thorough data collection and validation to ensure robust and accurate EOR screening results.

6.4.1.3.4 Results and Discussion

In conclusion, this paper presents a new screening tool for EOR methods based on fuzzy logic. The tool was developed using statistical analysis of a database comprising rock and fluid properties for 347 EOR projects worldwide. The tool was validated by comparing its results with those of commercial EOR software. The results show that the screening tool is reliable and consistent. Case studies were conducted to investigate the applicability of EOR in some Egyptian oil reservoirs, and the results indicate that immiscible gas injection is a good option for these reservoirs. However, miscible methods are not feasible in most of these oil reservoirs. The developed screening tool can help in selecting the most suitable EOR method for a given reservoir, which can lead to significant improvements in oil recovery.

BIBLIOGRAPHY

Nageh, Mohamed, Ela, El, Abu, Mahmoud, Tayeb, El, Sayed, El, and Sayyouh, Helmy. "Application of Using Fuzzy Logic as an Artificial Intelligence Technique in the Screening Criteria of the EOR Technologies." *Paper Presented at the SPE North Africa Technical Conference and Exhibition*, Cairo, Egypt, September 2015. https://doi.org/10.2118/175883-MS

6.4.1.4 Application-4: Prediction of the Best EOR Method by Artificial Intelligence

6.4.1.4.1 Background of the Application

The paper presents a model for predicting the best EOR method based on seven input parameters, including primary oil in place, porosity, permeability, temperature, oil density, viscosity, and initial oil saturation. The model uses the adaptive network-based fuzzy inference system (ANFIS), clustering, and normalizing techniques to design a proficient model for comparing the behavior of different EOR methods in terms of cumulative oil production. The model was tested on more than 200 oil fields from around the world where at least one of the EOR methods was applied. The results of the model were discussed based on a case study of an Iranian oil field.

6.4.1.4.2 Description of the Workflow and Machine Learning Model

In this method, ANFIS is used alongside clustering techniques to identify the optimal EOR method. ANFIS integrates fuzzy logic with neural networks (NNs), adeptly managing complex data relationships. Clustering improves prediction accuracy by grouping similar reservoir properties. By analyzing key parameters such as porosity, permeability, and oil density, this method reliably predicts the most suitable EOR method for specific field conditions, facilitating more informed decision-making in oil reservoir management.

6.4.1.4.3 Advantages and Limitations of Synchronized Algorithm

The combination of ANFIS and clustering is a powerful solution for predicting EOR methods. ANFIS effectively models nonlinear relationships and handles data uncertainty, while clustering improves the accuracy by grouping similar data points. However, ANFIS requires significant computational resources and can be complex to actually use. Clustering's effectiveness depends on selecting an appropriate number of clusters. Despite these limitations, the integrated use of ANFIS and clustering offers a balanced method, leveraging the strengths of both techniques to optimize EOR strategies with a blend of precision and practicality.

6.4.1.4.4 Results and Discussion

The paper presents a model for predicting the best EOR method based on seven input parameters, using ANFIS and clustering and normalizing techniques. The model aims to provide a cost-effective and efficient way to predict the cumulative oil production rate for a field in five different EOR methods and predict the best performance for field application. The results of the model were discussed based on a case study of an Iranian oil field. The study found that hydrocarbon injection is the best EOR method for the field under study, as it leads to the highest cumulative oil production. The designed model is a valuable tool for reservoir engineers to predict and compare the effectiveness of different EOR methods without the need for costly and time-consuming laboratory work.

BIBLIOGRAPHY

Esmaeilnezhad, E., Ranjbar, M., Nezamabadi-Pour, H., & Shoaei Fard Khamseh, F. (2013). Prediction of the Best EOR Method by Artificial Intelligence. *Petroleum Science and Technology*, 31(16), 1647–1654. https://doi.org/10.1080/10916466.2010.551235

6.4.1.5 Application-5: Screening and Optimization of Polymer Flooding Projects Using Artificial-Neural-Network (ANN) Based Proxies

6.4.1.5.1 Background of the Application

This paper discusses the use of various input parameters and AI methods in reservoir engineering. Input parameters include well performance, relative permeability, rheology, pressure, and saturation, while AI methods such as ANNs and particle swarm optimization (PSO) are used for reservoir engineering applications. The output of the work discussed in the paper varies depending on the specific research question and methodology, with some studies optimizing CO2 storage and oil recovery and others predicting polymer flooding performance.

6.4.1.5.2 Description of the Workflow & Machine Learning Model

This paper develops ANN-based proxies for screening and optimizing polymer flooding projects. The proxies utilize synthetic production histories generated from a high-fidelity numerical simulation model. The ANN models are designed for both forward-looking predictions and inverse design problems. The forward-looking ANN acts as a forecasting tool, predicting time-based project responses, while the inverse-looking ANN helps design projects that achieve desired oil recoveries. The proxies are trained on a dataset that includes reservoir properties, fluid characteristics, and project design parameters. A PSO algorithm is integrated with the ANN to maximize the net present value (NPV) of polymer injection projects.

6.4.1.5.3 Advantages and Limitations of the Synchronized Algorithm

The ANN-based proxies provide an efficient and cost-effective method for screening and optimizing polymer flooding projects. The forward-looking ANN can quickly predict project outcomes, while the inverse-looking ANN offers valuable insights for project design. However, the accuracy of the models depends on the quality of the training data, and non-unique solutions may arise in inverse design problems. The integration of PSO further improves optimization, but it can be computationally expensive and sensitive to parameter settings.

6.4.1.5.4 Results and Discussion

In conclusion, this paper highlights the importance of utilizing advanced methods such as AI and ML in reservoir engineering. By incorporating a wide range of input parameters and applying AI methods such as ANNs and PSO, reservoir engineers can optimize field development planning, predict well performance, and study processes such as steam-assisted gravity drainage (SAGD) and cyclic steam stimulation. The use of these advanced methods can lead to more accurate and efficient reservoir engineering practices, ultimately resulting in increased oil recovery and lower production costs.

BIBLIOGRAPHY

Sun, Qian, Ertekin, Turgay. Screening and optimization of polymer flooding projects using artificial-neural-network (ANN) based proxies. *Journal of Petroleum Science and Engineering*, 185, 2020, 106617, ISSN 0920-4105, https://doi.org/10.1016/j.petrol.2019.106617

6.4.1.6 Application-6: EOR Screening and Early Production Forecasting in Heavy Oil Fields: A Machine Learning Approach

6.4.1.6.1 Background of the Application

This paper proposes an ML approach to EOR screening and financial evaluation. The approach uses a compound NN that incorporates regression and classification capabilities to predict the cash flow during the first years of the project and provide estimations that can support further decision-making. The input parameters for the model are area, permeability, depth, residual oil saturation, the lithology of producing formation, project evaluation time, producer and injector well count. The paper describes the AI methods employed, including genetic algorithms (GAs) and Auto Keras library, and shows the output and units used.

6.4.1.6.2 Description of the Workflow and Machine Learning Model

This paper introduces an ML approach for EOR screening and early production forecasting in heavy oil fields. The study uses a combination of ANNs and gradient boosting algorithms to select the optimal EOR method and predict oil production during the first 5 years of a project. The model integrates GAs and auto-ML techniques to enhance the selection of parameters and improve forecasting accuracy. Data from over 900 fields were used to train the model, which then provides recommendations based on a dimensionality reduction process and NN classification.

6.4.1.6.3 Advantages and Limitations of the Synchronized Algorithm

The ML model offers a powerful tool for selecting the optimal EOR method and forecasting production in heavy oil fields. The use of NNs and gradient boosting provides high accuracy in both classification and regression tasks. However, the model's effectiveness depends on the quality of the training data, and the dimensionality reduction process may overlook some complex interactions between variables. Additionally, while the model shows high accuracy in predictions, its application to new fields requires careful calibration to ensure reliable results.

6.4.1.6.4 Results and Discussion

In conclusion, the proposed approach provides a valuable tool for EOR project evaluation, as it considers variables that are not considered in traditional screening techniques and can include several projects into the analysis, in order to have a better and data-supported result. The model's capability of predicting the cash flow during the first years of the project allows the evaluator to forecast technical and financial results with greater reliability. The selection of the correct way forward plan during a field development will be critical in order to reach the highest potential of the E&P asset.

While this methodology considers the use of homogeneous properties across the reservoir, it could be possible to incorporate heterogeneity map as an input for the model. The main limitation of this approach would be acquiring a consistent dataset with the required parameters. Further research can address this limitation.

BIBLIOGRAPHY

Vélez, Muñoz, Andrés, Eduardo, Consuegra, Romero, Felipe, and Arias, Carlos Andrés Berdugo. "EOR Screening and Early Production Forecasting in Heavy Oil Fields: A Machine Learning Approach." *Paper Presented at the SPE Latin American and Caribbean Petroleum Engineering Conference*, Virtual, July 2020. https://doi.org/10.2118/199047-MS

6.4.1.7 Application-7: Rule-Based Systematic EOR Screening Methodology for ADNOC Oil Reservoirs

6.4.1.7.1 Background of the Application

A novel screening process for EOR techniques in Abu Dhabi carbonate reservoirs is presented in this research. A wider variety of possible injectants for miscible gas injection are considered throughout the screening process, such as CO_2, nitrogen, sour gas streams, and sweet hydrocarbon gas. The procedure uses equation-of-state (EOS)-based minimum miscibility pressure (MMP) computations to determine miscibility conditions for hundreds of combinations of injection gas compositions and reservoir fluids. The reservoirs' high salinity and temperature provide a problem for conventional chemical EOR techniques that use surfactants and polymers. Recent advancements in research and development have created new prospects for foam and polymer systems. The screening process yields a list of EOR methods applicable to certain areas of each major oil reservoir.

6.4.1.7.2 Description of the Workflow and Machine Learning Model

This paper introduces a rule-based systematic EOR screening methodology specifically developed for ADNOC's oil reservoirs. The methodology integrates ML algorithms, data analysis, and response surface models to enhance the screening process for EOR techniques across a portfolio of reservoirs. The approach involves evaluating a wide range of EOR methods, including miscible gas injection (CO_2, nitrogen, and sour gas), polymer flooding, and foam-based techniques, with a focus on high-temperature, high-salinity carbonate reservoirs. The methodology also leverages EOS models for MMP calculations to assess gas injection feasibility across different reservoirs.

6.4.1.7.3 Advantages and Limitations of the Synchronized Algorithm

The rule-based screening methodology provides a structured approach for evaluating EOR potential in a diverse set of reservoirs. The integration of ML models and EOS-based MMP calculations enhances the accuracy of screening outcomes and helps identify the most suitable EOR techniques for different reservoir conditions. However, the methodology relies heavily on the availability of high-quality data, and the performance of ML models can be affected by data limitations. Additionally, the complexity of carbonate formations in ADNOC's portfolio requires continuous refinement of the screening criteria to account for new technological developments and changing reservoir conditions.

6.4.1.7.4 Results and Discussion

In conclusion, the new EOR screening procedure presented in this paper provides a systematic approach to matching EOR methods with Abu Dhabi carbonate reservoirs. The process considers a broader range of potential injectants for miscible gas injection and estimates miscibility conditions using EOS-based MMP calculations. The reservoirs' high temperature and high salinity pose a challenge for traditional chemical EOR methods, but recent R&D progress has opened up opportunities for polymer and foam systems. The screening process yields a list of EOR methods applicable to certain areas of each major oil reservoir. The successful implementation of an appropriate EOR technique relies on the right well placement, with the right monitoring in place and the right level of injection profile control.

BIBLIOGRAPHY

Mogensen, Kristian, and Al-Keebali, Aaesha. "Rule-Based Systematic EOR Screening Methodology for ADNOC Oil Reservoirs." *Paper Presented at the Abu Dhabi International Petroleum Exhibition & Conference*, Abu Dhabi, UAE, November 2019. https://doi.org/10.2118/197821-MS

6.4.1.8 Application-8: A New Machine Learning Ensemble Model for Class Imbalance Problem of Screening Enhanced Oil Recovery Methods

6.4.1.8.1 Background of the Application

This paper presents a comprehensive review of ML techniques that have been used for EOR. The authors discuss various approaches that have been proposed for addressing the challenges associated with EOR, such as imbalanced data sets, noisy data, and limited data availability. The paper provides an overview of the different ML methods that have been applied to EOR, including decision trees (DTs), support vector machines (SVMs), NNs, and ensemble learning. The authors also highlight the advantages and limitations of each method and provide recommendations for future research in this field.

6.4.1.8.2 Description of the Workflow and Machine Learning Model

This paper addresses the class imbalance problem in the screening of EOR methods using a novel ML ensemble model. The proposed model, called B2S, integrates three ensemble learning techniques: bagging, boosting, and stacking. The goal is to improve the classification accuracy in an imbalanced dataset, which consists of various EOR methods. The B2S model aims to balance variance and bias, thereby enhancing predictive accuracy for all EOR methods, including those with low representation in the dataset. Mutual information (MI) is used for nonlinear feature selection, and hyperparameters are optimized using random search.

6.4.1.8.3 Advantages and Limitations of the Synchronized Algorithm

The B2S ensemble model offers several advantages, such as the ability to handle imbalanced datasets more effectively compared to traditional models. By integrating bagging, boosting, and stacking, the model benefits from both reduced variance and bias, leading to high classification accuracy. However, the complexity of the model increases the risk of overfitting, especially when dealing with noisy or outlier data. The random initialization of hyperparameters helps mitigate this risk, but it still requires careful tuning to avoid potential pitfalls in model performance.

6.4.1.8.4 Results and Discussion

This paper reviews the current state of ML techniques for EOR. Various approaches proposed to address challenges such as imbalanced datasets, noisy data, and limited data availability are discussed. An overview of different ML methods applied to EOR, including decision trees, support vector machines, neural networks, and ensemble learning, is provided. The advantages and limitations of each method have been highlighted, and recommendations for future research in this field have been provided. Overall, the review indicates that ML techniques have the potential to significantly improve the efficiency and effectiveness of EOR processes.

BIBLIOGRAPHY

Pirizadeh, Mohsen, Alemohammad, Nafiseh, Manthouri, Mohammad, Pirizadeh, Meysam. A new machine learning ensemble model for class imbalance problem of screening enhanced oil recovery methods. *Journal of Petroleum Science and Engineering*, 198, 2021, 108214, ISSN 0920-4105, https://doi.org/10.1016/j.petrol.2020.108214

6.4.1.9 Application-9: Application of Machine Learning Techniques for Selecting the Most Suitable EOR

6.4.1.9.1 Background of the Application

This paper explores the limitations of using artificial intelligence and machine learning (AI) methods for EOR screening. While AI methods provide valuable insight into the EOR selection process, they should not be relied upon as the sole method for prediction. The authors discuss the unrealistic assumption of EOR screening as a classification problem by ML methods, the lack of sufficient data for generalizable learning, unbalanced data, and the absence of representation of economic issues in existing EOR datasets. This paper also mentions various AI methods used for EOR screening, including fuzzy DTs and NNs.

6.4.1.9.2 Description of the Workflow and Machine Learning Model

This paper discusses the application of various ML techniques to screen for the most suitable EOR methods. A large database of over 1,000 EOR projects from around the world was compiled, and several ML models, including Shallow and Deep ANNs, Naïve Bayes (NB), DT, Random Forest (RF), and principal component analysis (PCA), were developed to predict the best EOR method for candidate reservoirs. Among the models, RF and Deep ANN showed the highest accuracy, with an average of 90%. The ML models were trained on the gathered data and then tested on non-observed data to validate their predictive performance.

6.4.1.9.3 Advantages and Limitations of the Synchronized Algorithm

The ML approach provides a fast and effective tool for EOR screening, offering a high level of accuracy in predicting the most suitable recovery method. The RF and Deep ANN models, in particular, demonstrated strong performance in classifying different categories of EOR methods. However, the paper also highlights several limitations, such as the reliance on available data, which may not always represent all possible scenarios, and the challenges posed by unbalanced datasets. Moreover, the study emphasizes that while ML offers valuable insights, it should not be the sole method relied upon for EOR screening, and the results should be cross-checked with expert knowledge and other screening approaches.

6.4.1.9.4 Results and Discussion

This work concludes by pointing out the drawbacks of employing AI techniques for EOR screening and suggesting that in order to improve the dependability of the findings, future studies concentrate on offering more significant and pertinent input features. Additionally, the authors recommend that AI techniques be cross-checked with other screening methods or with expert knowledge rather than being used as the exclusive way for prediction. All things considered, this work sheds light on the opportunities and difficulties of applying AI techniques for EOR screening and promotes more research on the subject.

BIBLIOGRAPHY

Cheraghi, Yasaman, Kord, Shahin, Mashayekhizadeh, Vahid. Application of machine learning techniques for selecting the most suitable enhanced oil recovery method. challenges and opportunities. *Journal of Petroleum Science and Engineering*, 205, 2021, 108761, ISSN 0920-4105, https://doi.org/10.1016/j.petrol.2021.108761

6.4.1.10 Application-10: EOR Screening and Early Production Forecasting in Heavy Oil Fields: A Machine Learning Approach

6.4.1.10.1 Background of the Application

This paper presents an application of AI and ML techniques for screening EOR methods. By using a large dataset of heavy and extra-heavy oil reservoirs, the authors demonstrate how AI and ML can be used to identify the optimal EOR method based on input parameters such as area, permeability, depth, residual oil saturation, lithology of producing formation, project evaluation time, and producer and injector well count. The authors also show how GAs and NNs can be used to improve the accuracy of the results.

6.4.1.10.2 Description of the Workflow and Machine Learning Model

The workflow begins with the compilation and preprocessing of a comprehensive dataset comprising numerous heavy and extra-heavy oil field development projects. Key reservoir and operational parameters—such as formation area, average permeability, reservoir depth, residual oil saturation, lithology type, evaluation time frame, and well count—are used as inputs. The machine learning model integrates genetic algorithms (GA) with artificial neural networks (ANN) to establish correlations between these parameters and successful EOR methods. The GA component is responsible for optimizing the selection of input features and the structure of the neural network, improving model convergence and reducing overfitting. The ANN is then trained to learn nonlinear relationships and forecast both the optimal EOR method and the expected early production profile. This hybrid model allows for rapid and efficient screening of multiple EOR scenarios under varying conditions, enabling practical application in early-phase field evaluations.

6.4.1.10.3 Advantages and Limitations of the Synchronized Algorithm

One of the key advantages of the synchronized GA-ANN algorithm is its ability to process large and diverse datasets while capturing complex, nonlinear relationships between reservoir characteristics and EOR outcomes. By integrating genetic optimization, the model effectively navigates the high-dimensional input space to identify the most relevant features, which enhances the accuracy and robustness of the neural network's predictions. Moreover, the model provides reliable early production forecasts, offering insights into potential cash flows and supporting economic feasibility studies in the initial phases of field development. However, the algorithm also has limitations. Its performance is highly dependent on the quality and representativeness of the input dataset. Additionally, the interpretability of neural networks remains a challenge, as the "black-box" nature of the model can hinder full understanding of the decision-making process. Computational demands for training and optimization may also be significant, especially for large-scale implementations.

6.4.1.10.4 Results and Discussion

In conclusion, the use of AI and ML techniques for EOR methods screening provides a valuable tool for decision-making in the oil and gas industry. The ability to statistically analyze variables that are not considered in traditional screening techniques, and the inclusion of several projects into the analysis, results in better accuracy and convergence of the model. Furthermore, the capability of the model to predict cash flow during the first years of the project allows evaluators to forecast technical and financial results with greater reliability. This methodology provides a preliminary analysis for EOR project evaluation, providing valuable estimations that can support further decision-making.

BIBLIOGRAPHY

Vélez, Muñoz, Andrés, Eduardo, Consuegra, Romero, Felipe, and Arias, Carlos Andrés Berdugo. "EOR Screening and Early Production Forecasting in Heavy Oil Fields: A Machine Learning Approach." *Paper Presented at the SPE Latin American and Caribbean Petroleum Engineering Conference*, Virtual, July 2020. https://doi.org/10.2118/199047-MS

6.4.2 CO_2 Projects

6.4.2.1 Application-11: Applying Hybrid Support Vector Regression and Genetic Algorithm to Water Alternating CO_2 Gas EOR

6.4.2.1.1 Background of the Application

This paper presents a time-dependent multiple support vector regression (SVR) proxy model coupled with GA to optimize a water alternating gas (WAG) CO_2 process in real time. The proxy model was trained utilizing a database of an insignificant number of runs and was validated through blind runs to check its reliability with unseen datapoints. The hybridization of GA with the multiple SVRs proxy allowed the investigation of optimum WAG design parameters with high accuracy and reasonable CPU time. The paper reports the optimization results of GA in its best run, demonstrating that a high WAG ratio and small slug sizes are favorable in this process. The main advantage of the developed proxy is the noticeable reduction in the simulation time compared to the numerical simulator.

6.4.2.1.2 Description of the Workflow and Machine Learning Model

In this method, the WAG CO2 injection process is modeled using Support Vector Regression (SVR) combined with a Genetic Algorithm (GA). The dataset, featuring injection time, water and gas injection rates, and half-cycle time, is loaded and standardized for better model performance. The data is split into training and testing sets. An SVR model is trained using a grid search with cross-validation to find the best hyperparameters, evaluated by mean squared error (MSE). Finally, a GA identifies the optimal WAG CO_2 process parameters to maximize field oil production total (FOPT).

6.4.2.1.3 Advantages and Limitations of the Synchronized Algorithm

The hybrid SVR-GA method has several advantages, like handling complex nonlinear relationships for accurate predictions and improving training efficiency through standardization and cross-validation. The GA explores various solutions to find the optimal WAG CO_2 parameters. However, this method has limitations, including its computational expense, especially with large datasets, and dependency on the data and hyperparameter quality. The stochastic nature of GA may require multiple runs for linear and consistent results.

6.4.2.1.4 Results and Discussion

This study shows how to effectively use GAs and multiple SVR to optimize a WAG CO_2 process in real time. The hybridization with GA enables the exploration of optimal WAG design parameters with manageable CPU time, and the proposed multiple SVRs proxy model can accurately predict the numerical simulator performance. A high WAG ratio and tiny slug sizes are the ideal parameters for the WAG CO_2 process that were developed in this investigation. Establishing a proxy has the primary benefit of drastically cutting down on simulation time. By maintaining the necessary precision over a fair amount of time, the established proxy produced the intended results.

BIBLIOGRAPHY

Amar, N.M., Zeraibi, N., Ghahfarokhi, A. Applying hybrid support vector regression and genetic algorithm to water alternating CO2 gas EOR. *Greenhouse Gas Sci Technol.* 10:613–630 (2020); https://dx.doi.org/10.1002/ghg.1982

6.4.2.2 Application-12: Prediction of Supercritical CO_2/Brine Relative Permeability in Sedimentary Basins During Carbon Dioxide Sequestration

6.4.2.2.1 Background of the Application

This paper presents a genetic algorithm-radial basis function (GA-RBF) model for predicting CO_2/brine relative permeability. The input parameters include porosity, permeability, interfacial tension, salinity, pressure, threshold capillary pressure, temperature, and rock type. The model was tested using experimental data and achieved satisfactory results with a relative deviation of less than 5%. The proposed GA-RBF model provides a promising alternative for predicting CO_2/brine relative permeability in subsurface reservoirs.

6.4.2.2.2 Description of the Workflow and Machine Learning Model

In this method, a GA-RBF model is used to predict CO2/brine relative permeability. The GA sets the RBF network's hyperparameters, making sure for optimal performance. Input parameters, such as porosity, permeability, interfacial tension, salinity, pressure, threshold capillary pressure, temperature, and rock type, are normalized. The GA searches for the best length scale for the RBF kernel, improving the model's ability to capture complex data patterns. Once optimized, the RBF network is trained on the normalized data to predict CO_2/brine relative permeability accurately.

6.4.2.2.3 Advantages and Limitations of the Synchronized Algorithm

The GA-RBF model provides robust predictions by setting hyperparameters for optimal accuracy. This approach effectively handles nonlinear relationships and interactions in geophysical data. However, the GA can be computationally expensive due to multiple iterations required for parameter optimization. Additionally, handling the normalization of input and data scaling is important to maintain a high prediction accuracy. Despite these difficulties, the GA-RBF model's precise permeability predictions make it a valuable tool in CO_2 sequestration studies.

6.4.2.2.4 Results and Discussion

In this paper, a GA-RBF model is proposed for predicting CO2/brine relative permeability in subsurface reservoirs. The model was tested using experimental data, and the results showed that the model achieved satisfactory accuracy with a relative deviation of less than 5%. The proposed model provides a promising alternative to existing methods for predicting CO_2/brine relative permeability. Future work may focus on further testing and optimization of the GA-RBF model to improve its accuracy and applicability in different geological settings.

BIBLIOGRAPHY

Prediction of supercritical CO2/brine relative permeability in sedimentary basins during carbon dioxide sequestration. Greenhouse Gas *Sci Technol.* 5:756–771 (2015); https://doi.org/10.1002/ghg.1524

6.4.2.3 Application-13: Application of Artificial Intelligence in CO_2-EOR Process Optimization

6.4.2.3.1 Background of the Application

This paper explores the use of AI in optimizing the carbon dioxide-enhanced oil recovery (CO_2-EOR) process. CO_2-EOR is a technique that can improve oil production while also promoting geological sequestration. The paper proposes an optimization method based on AI tools such as GAs and PSO, which can be used to support decision-making about well location, well exposure, and well control. The optimization algorithm used in this study involves determining the optimal well placement and control strategy to maximize the project's economic value. The paper concludes that the use of AI tools can be effective in optimizing the CO_2-EOR process, resulting in increased profitability for oil production from matured fields.

6.4.2.3.2 Description of the Workflow and Machine Learning Model

In this approach, a GA and PSO are used to optimize the CO2-EOR policy by determining the best well placement and control strategy. Input parameters include pressure, saturation, and a control vector (well location, perforation intervals, production, and injection rates). GA generates and improves potential solutions through selection and mutation, while PSO adjusts particle positions to find the optimal configuration. The workflow involves loading data from a CSV file, defining the objective function, and running both algorithms to produce a CO_2-EOR policy aimed at maximizing economic value.

6.4.2.3.3 Advantages and Limitations of the Synchronized Algorithm

GAs and PSO handle complex, high-dimensional search spaces well, making them ideal for CO_2-EOR optimization. GA's evolutionary method and PSO's swarm intelligence ensure thorough exploration and robust solutions without relying on derivatives. However, these methods are computationally expensive, requiring many iterations and careful parameter tuning. The success of the optimization heavily depends on the data quality, with poor data leading to negative results and a risk of overfitting, particularly with GA.

6.4.2.3.4 Results and Discussion

This study has shown that the optimal management of hydrocarbon reserves, specifically in the case of CO_2-EOR, can be effectively approached through the use of AI technologies. The optimization models suggested in this work can be applied to new and challenging scenarios, and AI features like learning capacity and abstraction can be utilized as an extension of manual search techniques. The suggested optimization algorithm makes it possible to boost the profitability of oil production from established fields, which is crucial when oil prices are low and the expenses of applying

EOR are relatively high. The results of this study suggest that the combination of reservoir modeling and AI tools can provide an improved approach to optimizing the CO_2-EOR process.

BIBLIOGRAPHY

Janiga, D., Stopa, J., Wojnarowski, P. Application of Artificial Intelligence in CO2-EOR Process Optimization. *16th International Multidisciplinary Science Geoconference SGEM 206*. https://www.researchgate.net/publication/316664691

6.4.2.4 Application-14: Pressure of Different Gases Injected into Large-Scale Coal Matrix: Analysis of Time–Space Dependence and Prediction Using Light Gradient Boosting Machine

6.4.2.4.1 Background of the Application

This paper presents a gas injection pressure prediction model based on the light gradient boosting machine (LGBM) algorithm. The model aims to predict the gas injection pressure at unknown times and arbitrary locations based on measured gas injection pressure data. The key factors affecting the gas injection pressure are the spatial position, gas injection volume, and gas type. The LGBM model is based on a DT and effectively overcomes the limitations of the gradient boosting DT algorithm in processing large amounts of data. The proposed model efficiently predicts the gas injection pressure, accurately captures the spatiotemporal differences in the pressure distribution during gas injection, and identifies the changes caused by gas differences.

6.4.2.4.2 Description of the Workflow and Machine Learning Model

In this method, an LGBM model is used to predict gas injection pressure in a coal matrix. The workflow involves loading data from a CSV file, preprocessing it, and training the LGBM model. This model is chosen for its efficiency and ability to handle complex data, enabling accurate predictions based on historical data. The trained model provides insights for optimizing gas injection strategies. Accurate predictions are generated, which can guide decision-making in gas injection processes. This method leverages LGBM's strengths, offering a scalable solution for predicting gas injection pressure.

6.4.2.4.3 Advantages and Limitations of the Synchronized Algorithm

The LGBM model is very good in predicting gas injection pressure due to its ability to handle large datasets and model complex relationships. Its gradient boosting framework iteratively enhances prediction accuracy by focusing on previous errors, ensuring that subtle data patterns are captured. Additionally, LGBM's speed and efficiency make it ideal for large-scale applications. However, LGBM requires careful hyperparameter tuning, which can be time-consuming and computationally expensive. There is a risk of overfitting, particularly with noisy or insufficient data. The quality of input data is crucial, as poor data can lead to inaccurate predictions.

6.4.2.4.4 *Results and Discussion*

This study presents a gas injection pressure prediction model based on the LGBM algorithm that efficiently predicts the gas injection pressure, accurately captures the spatiotemporal differences in the pressure distribution during gas injection, and identifies the changes caused by gas differences. The model effectively overcomes the limitations of the gradient boosting DT algorithm in processing large amounts of data, resulting in significantly improved accuracy and prediction efficiency. The temporal dimension is the most important factor affecting the pressure in the coal matrix and reflects the effect of the specific gas injection technology on the pressure. The proposed model can be further improved by incorporating additional factors such as in situ stress, coal wettability, and temperature.

BIBLIOGRAPHY

Zhou, Bin, Xu, Jiang, Han, Feng, Yan, Fazhi, Peng, Shoujian, Li, Qixian, Jiao, Feng. Pressure of different gases injected into large-scale coal matrix: Analysis of time–space dependence and prediction using light gradient boosting machine, *Fuel*, 279, 2020, 118448, ISSN 0016-2361, https://doi.org/10.1016/j.fuel.2020.118448

6.4.2.5 Application-15: Predicting Permeability Changes with Injecting CO_2 in Coal Seams During CO_2 Geological Sequestration: A Comparative Study among Six SVM-Based Hybrid Models

6.4.2.5.1 *Background of the Application*

This paper discusses a variety of research studies related to coal and energy, which utilize optimization algorithms and ML techniques to make predictions and simulations. The studies cover topics such as coal permeability changes, electricity prices, and CO_2 injection in coal seams. The paper does not provide a comprehensive list of input parameters or units for all parameters discussed.

6.4.2.5.2 *Description of the Workflow and Machine Learning Model*

This paper presents a comparative study of six SVM-based hybrid models to predict permeability changes in coal seams during CO_2 geological sequestration. The models integrate SVM with various intelligent optimization algorithms, including artificial bee colony (ABC), cuckoo search (CS), PSO, differential evolution (DE), gray wolf optimizer (GWO), and a combination of DE and GWO (DE–GWO). These models are trained on a dataset containing permeability measurements influenced by factors such as CO_2 injection pressure, effective stress, temperature, buried depth, and coal rank.

6.4.2.5.3 *Advantages and Limitations of the Synchronized Algorithm*

The SVM-based hybrid models offer robust predictions for permeability changes in coal seams, with the DEGWO–SVM model providing the highest accuracy. However, the performance of these models is sensitive to the choice of optimization algorithm and the quality of the training data. Each model has strengths and weaknesses depending on the specific characteristics of the coal seam and the CO_2

injection process. The DEGWO-SVM model, in particular, balances accuracy and computational efficiency, making it a strong candidate for real-time applications.

6.4.2.5.4 *Results and Discussion*

In conclusion, this paper highlights the importance of utilizing advanced optimization algorithms and ML techniques in the field of coal and energy research. These tools can help predict and simulate various outcomes, such as coal permeability changes, electricity prices, and CO_2 injection in coal seams. While the paper does not provide a comprehensive list of input parameters or units for all parameters discussed, it demonstrates the potential for these methods to improve the understanding of the complex processes involved in coal and energy production.

BIBLIOGRAPHY

Yan, Hao, Zhang, Jixiong, Rahman, Sheik S., Zhou, Nan, Suo, Yu. Predicting permeability changes with injecting CO2 in coal seams during CO2 geological sequestration: A comparative study among six SVM-based hybrid models. *Science of The Total Environment*, 705, 2020, 135941, ISSN 0048-9697, https://doi.org/10.1016/j.scitotenv.2019.135941

6.4.2.6 Application-16: Petrophysical Characterization of Deep Saline Aquifers for CO_2 Storage Using Ensemble Smoother and Deep Convolutional Autoencoder

6.4.2.6.1 *Background of the Application*

This paper discusses the use of various data-driven and AI-based methods for modeling and simulating geological CO_2 storage. The paper covers several studies that utilize different input parameters and AI methods, including ensemble-based kernel learning, correlation-based adaptive localization, and DL electromagnetic inversion with CNNs. The outputs of these studies include simulations of CO_2 storage and plume migration, estimates of CO_2 storage potential, and uncertainty quantification. The paper emphasizes the importance of accurately modeling and simulating CO_2 plume migration to ensure safe and effective CO_2 storage.

6.4.2.6.2 *Description of the Workflow and Machine Learning Model*

This paper focuses on the petrophysical characterization of deep saline aquifers for CO_2 storage. The study utilizes a combination of ensemble smoother (ES) and deep convolutional autoencoder (DCAE) methods to predict porosity and permeability. The DCAE reduces the dimensionality of seismic data, enabling the ES to efficiently handle large geophysical datasets. By integrating borehole observations, seismic data, and geostatistical realizations, the method updates the petrophysical models every time new data are available, improving the accuracy of CO_2 plume predictions during storage.

6.4.2.6.3 *Advantages and Limitations of the Synchronized Algorithm*

The integration of the ES with DCAE provides a powerful approach to managing large geophysical datasets, allowing for more accurate and timely predictions of CO_2 plume behavior. The reduction of data dimensionality by the DCAE ensures that the ES remains computationally feasible. However, the accuracy of the predictions depends on the quality of the training data, and the complexity of the method may require significant computational resources. Additionally, the approach assumes that the aquifer properties remain constant over time, which may not always be the case in dynamic reservoir conditions.

6.4.2.6.4 *Results and Discussion*

In conclusion, this paper presents a comprehensive overview of various data-driven and AI-based methods for modeling and simulating geological CO_2 storage. The studies discussed in the paper demonstrate the potential of these methods to improve the understanding of CO_2 storage and plume migration and to provide more accurate estimates of CO_2 storage potential. However, the paper also highlights the need for further research to address uncertainties in model parameters and to improve the accuracy of CO_2 plume migration simulations. Overall, this paper provides valuable insights into the current state of research on geological CO_2 storage and the potential of data-driven and AI-based methods to advance this field.

BIBLIOGRAPHY

Liu, Mingliang, Grana, Dario. Petrophysical characterization of deep saline aquifers for CO2 storage using ensemble smoother and deep convolutional autoencoder. *Advances in Water Resources*, 142, 2020, 103634, ISSN 0309-1708, https://doi.org/10.1016/j.advwatres.2020.103634

6.4.2.7 Application-17: A Fuzzy Logic Approach for Estimating Recovery Factors of Miscible CO_2-EOR Projects in the United States

6.4.2.7.1 *Background of the Application*

A thorough resource database for evaluating recoverable hydrocarbons from CO_2 EOR is presented in this research. With an emphasis on CO_2 EOR, the study seeks to offer a thorough resource database for evaluating the recoverable hydrocarbons generated by CO_2 EOR. Data about oil reservoirs, such as reservoir depth, porosity, permeability, and oil saturation, are probably included in the input parameters.

6.4.2.7.2 *Description of the Workflow and Machine Learning Model*

This paper introduces a fuzzy logic approach to estimate recovery factors (RFs) for miscible CO_2-EOR projects in the United States. The study utilizes a Mamdani-type fuzzy inference system, which incorporates reservoir properties, fluid characteristics, and operational parameters as inputs. The fuzzy logic model is designed to handle the uncertainties and complexities inherent in predicting RFs, offering a more flexible and intuitive alternative to traditional reservoir simulation and laboratory experiments. The model uses nine input variables and is trained on data from 24 field applications to predict RFs with reasonable accuracy.

6.4.2.7.3 *Advantages and Limitations of the Synchronized Algorithm*

The fuzzy logic model provides a quick and cost-effective method for estimating RFs, making it an attractive alternative to more data-intensive and computationally expensive approaches. The flexibility of the fuzzy logic system allows it to accommodate incomplete or uncertain data, which is common in reservoir studies. However, the model's accuracy is limited by the quality and quantity of the input data, and the predictive power may decrease when applied to reservoirs with conditions significantly different from those used in the training set. Additionally, the fuzzy logic approach may require expert knowledge to properly define the membership functions and rule sets.

6.4.2.7.4 Results and Discussion

In conclusion, this paper provides a valuable resource for those interested in the assessment of recoverable hydrocarbons produced by CO_2-EOR. The comprehensive resource database can be used to evaluate the potential of CO_2-EOR in various oil reservoirs. However, the paper does not explore the use of AI methods for this purpose, leaving room for further research in this area.

BIBLIOGRAPHY

Karacan, C. Özgen. A fuzzy logic approach for estimating recovery factors of miscible CO2-EOR projects in the United States. *Journal of Petroleum Science and Engineering*, 184, 2020, 106533, ISSN 0920-4105, https://doi.org/10.1016/j.petrol.2019.106533

6.4.2.8 Application-18: Application of an Artificial Neural Network in Predicting the Effectiveness of Trapping Mechanisms on CO_2 Sequestration in Saline Aquifers

6.4.2.8.1 Background of the Application

This paper discusses the evaluation of carbon dioxide (CO_2) storage integrity with rate allocation in a multi-layered aquifer. The study focuses on the characteristics of the aquifer, injected CO_2, and formation brine. The authors evaluate the CO_2 storage integrity using a compositional simulator and analyze the impact of different parameters on the storage integrity. The study concludes that the CO_2 storage integrity can be maintained with proper rate allocation and injection strategies.

6.4.2.8.2 Description of the Workflow and Machine Learning Model

This paper discusses the application of an ANN model to predict the effectiveness of trapping mechanisms during CO_2 sequestration in saline aquifers. The ANN-GCS (Geological CO_2 Sequestration) model is developed and trained using Python with Keras, utilizing data from synthetic aquifers. The model aims to predict residual trapping and solubility trapping indices by optimizing the relationships between geologic formation characteristics and CO_2 storage efficiency. The ANN architecture consists of eight hidden layers, with each layer comprising 64 nodes. The model demonstrates high accuracy, as reflected in its coefficient of determination (R^2) and root mean squared error (RMSE) values during validation on field data from the Pohang Basin, Korea.

6.4.2.8.3 Advantages and Limitations of the Synchronized Algorithm

The ANN-GCS model offers a fast and accurate approach to predicting CO_2 trapping mechanisms in saline aquifers, making it an efficient tool compared to conventional numerical simulations. The model's ability to handle complex geologic data and provide real-time predictions enhances its practical utility for large-scale CO_2 sequestration projects. However, the accuracy of the model is dependent on the quality and quantity of training data. Additionally, while the ANN approach reduces computational costs, the black-box nature of NNs can limit interpretability, and the model may require significant computational resources for large datasets.

6.4.2.8.4 Results and Discussion

In conclusion, the study evaluates the feasibility of CO_2 storage in a multi-layered aquifer by analyzing the impact of different parameters on the storage integrity. The authors demonstrate that the integrity can be maintained with proper rate allocation and injection strategies. The study highlights the importance of understanding the characteristics of the aquifer, injected CO_2, and formation brine for successful CO_2 storage. The findings of the study can be used to guide the design and implementation of CO_2 storage projects.

BIBLIOGRAPHY

Song, Youngsoo, Sung, Wonmo, Jang, Youngho, Jung, Woodong. Application of an artificial neural network in predicting the effectiveness of trapping mechanisms on CO2 sequestration in saline aquifers. *International Journal of Greenhouse Gas Control*, 98, 2020, 103042, ISSN 1750-5836, https://doi.org/10.1016/j.ijggc.2020.103042

6.4.2.9 Application-19: Robust Optimization of CO_2 Sequestration through a Water Alternating Gas Process under Geological Uncertainties

6.4.2.9.1 Background of the Application

This paper explores the potential for carbon capture and storage (CCS) through the injection of CO_2 into geological formations. The study covers a range of input parameters related to CO_2 storage, including rock and fluid properties, as well as injection parameters. The paper provides insights into the mechanisms by which CO_2 is stored in the subsurface and discusses the importance of understanding its behavior in geological formations. The study also discusses several possible outcomes resulting from CCS, including EOR and large-scale deployment of these technologies.

6.4.2.9.2 Description of the Workflow and Machine Learning Model

This paper presents a robust optimization workflow for CO_2 sequestration through a WAG process in a heterogeneous sandstone reservoir located in the Cuu Long Basin, Vietnam. The study integrates AI techniques with geological modeling and reservoir simulation to optimize CO_2 trapping under geological uncertainties. A big-loop modeling framework is utilized, which involves generating multiple geological realizations and applying an AI optimizer to determine optimal WAG parameters, such as cycle lengths of water and CO_2 injection. The optimization framework also compares robust optimization, which accounts for geological uncertainties, with traditional nominal optimization workflows.

6.4.2.9.3 *Advantages and Limitations of the Synchronized Algorithm*

The robust optimization workflow offers several advantages, including improved CO_2 trapping efficiency and reduced risks associated with geological uncertainties. The AI-driven optimization ensures that the WAG process is fine-tuned to maximize CO_2 storage performance while accounting for the variability in reservoir characteristics. However, the approach requires extensive computational resources due to the large number of simulations and geological realizations needed to achieve robust results. Additionally, the complexity of the workflow may make it difficult to implement in real-time operational settings without advanced computing infrastructure.

6.4.2.9.4 *Results and Discussion*

The study shows that CCS through the injection of CO_2 into geological formations is a promising technology for mitigating climate change and reducing greenhouse gas emissions. The research provides insights into the mechanisms by which CO_2 is stored in the subsurface and the importance of understanding its behavior in geological formations. The study also highlights the potential economic benefits of CCS technologies, including EOR and the potential for large-scale deployment of these technologies. Overall, the study suggests that CCS is a critical technology for achieving global climate goals and reducing the impact of greenhouse gas emissions.

BIBLIOGRAPHY

Thanh, Hung Vo, Sugai, Yuichi, Nguele, Ronald, Sasaki, Kyuro. Robust optimization of CO2 sequestration through a water alternating gas process under geological uncertainties in Cuu Long Basin, Vietnam, *Journal of Natural Gas Science and Engineering*, 76, 2020, 103208, ISSN 1875-5100, https://doi.org/10.1016/j.jngse.2020.103208

6.4.2.10 Application-20: Assessment of Enhanced Oil Recovery and CO_2 Storage Capacity Using Machine Learning and Optimization Framework

6.4.2.10.1 *Background of the Application*

This study presents an approach for improving numerical simulations of CO_2 production and storage performance in the Pennsylvanian Upper Morrow sandstone reservoir, which is situated in Ochiltree County, Texas's Farnsworth Unit (FWU). The suggested framework enhances field performance and CO_2 storage at the same time by combining optimization techniques with ML algorithms for reduced order modeling.

In order to successfully match the observed performance, the model is calibrated using historical data from an active CO_2 flood. The framework integrates geological, geophysical, and engineering data. In the calibrated model, uncertain factors like relative permeability exponents and reservoir rock characteristics are changed to take into consideration possible variations in wettability. A proxy model is created by combining multi-layer and RBF NN techniques in order to optimize the objective function, which considers variables like oil recovery, CO_2 storage, and NPV. Iterative training and calibration processes are used to the proxy model until it satisfies the predetermined validation requirements. After obtaining a suitable proxy model, the best solution for the predetermined objective function is found using a hybrid strategy that combines evolutionary and ML optimization algorithms. Gas oil ratio, WAG cycle, production rates, bottom hole pressure of producers and injectors, amount of CO_2 purchased, volume of recycled gas, and the location of new infill wells are among the uncertain and control variables taken into account in the optimization research.

The sensitivity analysis of the results reveals the influence of the control variables on the optimal outcomes. The predictive study indicates the feasibility of developing a reliable and robust ML optimization algorithm for maximizing both oil production and CO_2 storage in the FWU reservoir, considering aqueous-gaseous-mineral phases.

6.4.2.10.2 Description of the Workflow and Machine Learning Model

This paper presents an optimization methodology for CO_2 storage and EOR performance using ML and numerical simulation frameworks. The study combines hybridized multi-layer and RBF NN methods to develop a proxy model for the FWU in Texas. The proxy model predicts oil recovery, CO_2 storage, and NPV under various operational scenarios. The hybrid ML framework is integrated with optimization algorithms, including evolutionary and PSO, to co-optimize field performance and CO_2 storage. The model is calibrated using historical CO_2 flood data and geological, geophysical, and engineering data.

6.4.2.10.3 Advantages and Limitations of the Synchronized Algorithm

The hybrid ML and optimization framework provides a powerful tool for maximizing both oil recovery and CO_2 storage in mature reservoirs. The proxy model significantly reduces computational time compared to full-scale reservoir simulations, allowing for rapid scenario testing and optimization. However, the accuracy of the model is dependent on the quality of the training data and the proper tuning of hyperparameters. The use of evolutionary algorithms such as PSO can be computationally expensive, especially when optimizing complex objective functions. Additionally, the model's performance may vary with different reservoir conditions, necessitating continuous calibration and validation.

6.4.2.10.4 Results and Discussion

In the Pennsylvanian Upper Morrow sandstone reservoir in the FWU, Texas, an optimization framework for co-optimizing CO_2 storage and CO_2-EOR performance, as well as project NPV is presented in this research. To determine the best combinations of control variables chosen from a sensitivity analysis, 160 optimization trials using the PSO evolutionary algorithm were conducted. According to the prediction analysis, a strong ML optimization method that can be relied upon to optimize a developmental plan to maximize oil production, project NPV, and CO_2 storage in aqueous-gaseous phases inside the FWU may be developed.

BIBLIOGRAPHY

You, Junyu, Ampomah, William, Kutsienyo, Eusebius Junior, Sun, Qian, Balch, Robert Scott, Aggrey, Wilberforce Nkrumah, and Cather, Martha. "Assessment of Enhanced

Oil Recovery and CO2 Storage Capacity Using Machine Learning and Optimization Framework." *Paper Presented at the SPE Europec featured at 81st EAGE Conference and Exhibition*, London, England, UK, June 2019. https://doi.org/10.2118/195490-MS

6.4.2.11 Application-21: Optimization of Water-Alternating-CO_2 Injection Field Operations Using a Machine-Learning-Assisted Workflow

6.4.2.11.1 Background of the Application

This paper presents a Gaussian-SVR-based workflow for handling the multi-objective optimization (MOO) problem in a water-alternating-CO_2 injection project. The workflow is capable of handling a large number of control parameters and is computationally efficient in simultaneously optimizing hydrocarbon recovery and carbon storage volume using a Pareto-front-based method. The Gaussian-SVR model is used to learn the complicated structure between input parameters and simulation outcomes, and the proposed self-adjusting protocol improves proxy quality when the size of the training database is limited. The workflow is proven to be capable of constructing a Pareto front of considered objectives, and the generated Pareto repository is validated by comparing the results with simulation outcomes, revisiting the same parameters obtained by the workflow.

6.4.2.11.2 Description of the Workflow and Machine Learning Model

This paper introduces a machine-learning-assisted workflow for optimizing WAG-CO_2 injection field operations. The workflow leverages Gaussian-SVR as a proxy model to predict field performance metrics such as oil recovery, CO_2 storage, and NPV. To enhance computational efficiency, the proxy model is coupled with an iterative self-adjusting protocol that minimizes the size of the training dataset while maintaining high accuracy. Bayesian optimization is used to fine-tune the hyperparameters of the proxy models, and multi-objective particle swarm optimization (MOPSO) is applied to generate Pareto fronts for decision-making. The approach is tested on the FWU field, a mature oil reservoir undergoing CO_2-WAG.

6.4.2.11.3 Advantages and Limitations of the Synchronized Algorithm

The machine-learning-assisted workflow significantly reduces computational overhead compared to traditional numerical simulation methods. The use of Gaussian-SVR provides a robust proxy model that can handle the nonlinear relationships between control parameters and field performance. The iterative self-adjusting protocol ensures that the size of the training dataset remains manageable, improving computational efficiency without sacrificing accuracy. However, the methodology is dependent on the quality of the initial training data and the careful selection of hyperparameters. Additionally, while the workflow demonstrates strong performance in handling MOO, it may require extensive computational resources when applied to large-scale reservoir models with a high number of control variables.

6.4.2.11.4 Results and Discussion

In this paper, it is demonstrated that the Gaussian-SVR-based workflow is an efficient method for handling the multi-objective optimization (MOO) problem in a water-alternating-CO2 injection project. The results show that the Gaussian-SVR model is capable of learning the complicated structure between input parameters and simulation outcomes, improving the prediction performance of the proxy model when there are many variables to control. Additionally, the proposed self-adjusting protocol improves proxy quality when the size of the training database is limited. The proposed workflow is proven to be capable of constructing a Pareto front of considered objectives, and the generated Pareto repository is validated by comparing the results with simulation outcomes, revisiting the same parameters obtained by the workflow.

Overall, the workflow is computationally efficient and can handle a large number of control parameters, making it a valuable tool for optimizing hydrocarbon recovery and carbon storage volume in a water-alternating-CO_2 injection project.

BIBLIOGRAPHY

You, Junyu, Ampomah, William, Tu, Jiawei, Morgan, Anthony, Sun, Qian, Wei, Bing, and Wang, Dianlin. "Optimization of Water-Alternating-CO2 Injection Field Operations Using a Machine-Learning-Assisted Workflow." *SPE Res Eval & Eng* 25 (2022): 214–231. https://doi.org/10.2118/203913-PA

6.4.3 Water Flooding

6.4.3.1 Application-22: Application of Foam Floods for Enhancing Heavy Oil Recovery Through Stability Analysis and Core Flood Experiments

6.4.3.1.1 Background of the Application

This paper investigates the stability of foams and the ultimate oil RF in surfactant alternating gas (SAG) injection compared to WAG injection. The effects of surfactant type, salt/alkaline concentration/type, and SAG ratio on foam stability and oil RF were studied. The results show that N_2 foams are more stable than CO_2 foams for all types of surfactants, with anionic surfactant SDS exhibiting the best foamability. Salt/alkaline concentration/type significantly affects foam stability, and the ionic nature of surfactant also plays a crucial role. The least squares support vector machine (LS-SVM) model was employed for data analysis, and the reproducibility of the experiments was confirmed. Core floods demonstrate that SAG injection results in higher ultimate oil RFs than WAG injection, indicating its applicability for EOR in sandstone reservoirs. However, for surfactants SDS and CTAB, the ultimate RF decreases with increasing SAG ratio. The authors recommend the analysis of foam stability for the successful design of foam flooding in EOR processes.

6.4.3.1.2 Description of the Workflow and Machine Learning Model

In this method, an LS-SVM, a supervised learning algorithm for regression and classification, is used. It minimizes a cost function that balances training error and model complexity, using kernel functions to transform data into a higher-dimensional space. This allows it to capture complex, nonlinear relationships. In the application, LS-SVM predicts foam stability and oil RF in EOR processes, learning from historical data to optimize foam flooding strategies and improve operational efficiency.

6.4.3.1.3 Advantages and Limitations of the Synchronized Algorithm

LS-SVM has notable advantages, including its ability to handle nonlinear data and deliver accurate predictions with fewer samples, ideal for experimental settings. Its use of kernel functions effectively models complex interactions. However, LS-SVM can be computationally expensive for large datasets and requires careful optimization of kernel parameters. Despite these difficulties, its robustness and predictive accuracy make LS-SVM a valuable tool for optimizing foam flooding in EOR operations.

6.4.3.1.4 Results and Discussion

In conclusion, this study provides valuable insights into the optimization of SAG injection for EOR. The results show that N_2 foams are more stable than CO_2 foams for all types of surfactants, with anionic surfactant SDS exhibiting the best foamability. Salt/alkaline concentration/type significantly affects foam stability, and the ionic nature of surfactant also plays a crucial role. The LS-SVM model was employed for data analysis, and the reproducibility of the experiments was confirmed. Core floods demonstrate that SAG injection results in higher ultimate oil RFs than WAG injection, indicating its applicability for EOR in sandstone reservoirs. However, for surfactants SDS and CTAB, the ultimate RF decreases with increasing SAG ratio. The authors recommend the analysis of foam stability for the successful design of foam flooding in EOR processes.

BIBLIOGRAPHY

Shokrollahi, A., Ghazanfari, M.H., Badakhshan, A. Application of Foam Floods for Enhancing Heavy Oil Recovery Through Stability Analysis and Core Flood Experiments. *Can. J. Chem. Eng*. 92:1975–1987, 2014. 10.1002/cjce.22044

6.4.3.2 Application-23: Predicting Field Production Rates for Waterflooding Using a Machine Learning-Based Proxy Model

6.4.3.2.1 Background of the Application

The paper discusses the use of a learning-based data-driven forecast approach for predicting future reservoir performance. The approach uses ML algorithms, but the specific methods employed are not explicitly discussed. The paper proposes that this approach can improve upon traditional reservoir modeling techniques, which can be computationally expensive and time-consuming. The authors demonstrate the effectiveness of the approach by applying it to a synthetic reservoir model and comparing the results to those obtained using traditional modeling techniques.

6.4.3.2.2 Description of the Workflow and Machine Learning Model

This paper presents an ML-based proxy model for predicting field production rates in waterflooding projects. The study utilizes a conditional deep convolutional generative adversarial network (cDC-GAN) to model the dynamic fluid distribution in heterogeneous reservoirs under waterflooding. The proxy model uses permeability distribution and forecast time as inputs, while water saturation is the desired output. The cDC-GAN learns the relationship between the inputs and outputs through a generative-discriminative framework, where the generative model attempts to produce realistic fluid distribution, and the discriminative model differentiates between generated and real data. This model allows for rapid prediction of water and oil saturation distributions, as well as production rates, with less computational effort than traditional reservoir simulations.

6.4.3.2.3 Advantages and Limitations of the Synchronized Algorithm

The cDC-GAN proxy model offers several advantages, including reduced computational time and the ability to handle high-dimensional data without the need for feature engineering. It is particularly useful for waterflooding optimization and uncertainty analysis. However, the performance of the model relies on the quality of the training data and can be sensitive to the configuration of the generative and discriminative networks. Additionally, while the model demonstrates strong interpolation capabilities, its extrapolation ability may be limited in scenarios not covered by the training data.

6.4.3.2.4 Results and Discussion

The paper presents a novel approach for predicting future reservoir performance using a learning-based data-driven forecast approach. The approach is shown to be effective when applied to a synthetic reservoir model, and the authors suggest that it could be applied to real-world reservoirs as well. However, further research is needed to determine the effectiveness of the approach on a larger scale and in more complex reservoirs. Overall, the approach shows promise for improving upon traditional reservoir modeling techniques and providing more accurate predictions of future reservoir performance.

BIBLIOGRAPHY

Zhong, Zhi, Sun, Alexander Y., Wang, Yanyong, Ren, Bo. Predicting field production rates for waterflooding using a machine learning-based proxy model. *Journal of Petroleum Science and Engineering*, 194, 2020, 107574, ISSN 0920-4105, https://doi.org/10.1016/j.petrol.2020.107574

6.4.3.3 Application-24: Artificial Neural Network-Based Production Forecasting for a Hydrocarbon Reservoir under Water Injection

6.4.3.3.1 Background of the application

This paper presents a novel approach to predicting oil, gas, and water production rates in a water flooding reservoir using ANN models trained by Bayesian regularization. Thirteen features based on knowledge of reservoir flow physics were extracted and used as input to the models to improve their prediction performance. The models' architecture and training algorithm were optimized to minimize MSE and maximize the coefficient of determination. The models were validated using a blind test set, and their performance was evaluated using MSE and the coefficient of determination. The results demonstrate that the physics-based feature extraction significantly improves the prediction performance of the NN models.

6.4.3.3.2 *Description of the Workflow and Machine Learning Model*

This paper introduces an ANN-based production forecasting model designed for hydrocarbon reservoirs under water injection. The study leverages a Bayesian regularization algorithm to train the NN, which is optimized for predicting oil, gas, and water production rates. The model uses production history data, pressure measurements, and injection rates as inputs. The ANN architecture is developed with a focus on capturing complex nonlinear relationships between input parameters and production outcomes. Through feature extraction and expansion based on fluid flow physics, the model enhances prediction accuracy and generalization capabilities.

6.4.3.3.3 *Advantages and Limitations of the Synchronized Algorithm*

The ANN-based model offers a powerful tool for production forecasting, particularly for waterflooding scenarios where traditional methods struggle to manage the inherent uncertainties. The Bayesian regularization algorithm improves the model's ability to generalize from noisy data, reducing the risk of overfitting. However, the model's performance depends on the quality of the training data, and it may require significant computational resources due to the complexity of the ANN architecture. The approach is particularly beneficial in cases where data availability is limited, but its success hinges on the careful selection of input features and training strategies.

6.4.3.3.4 *Results and Discussion*

This paper presents a novel approach for predicting oil, gas, and water production rates in a water flooding reservoir using ANN models trained with Bayesian regularization. The models were optimized to minimize MSE and maximize the coefficient of determination, and their performance was evaluated using a blind test set. The results demonstrate that the physics-based feature extraction significantly improves the prediction performance of the models. The feature that improved the models' prediction performance the most out of all of them was the delay in the time step term. Since many production prediction techniques, such as numerical reservoir modeling, do not consider the historical injection rate as a factor impacting the current production rate, this discovery is extremely important.

BIBLIOGRAPHY

Negash, Berihun Mamo, Yaw, Atta Dennis. Artificial neural network-based production forecasting for a hydrocarbon reservoir under water injection. *Petroleum Exploration and Development*, 47 (2), 2020, 383–392, ISSN 1876-3804, https://doi.org/10.1016/S1876-3804(20)60055-6

6.4.3.4 Application-25: Data-Driven Optimization for Fine Water Injection in a Mature Oil Field

6.4.3.4.1 *Background of the Application*

This paper presents a data-driven method for optimizing water injection plans in mature oil fields using separated-layer water injection technology. The method involves history matching using performance data, ML clustering to judge how to adjust water injection, and intelligent optimization to obtain the final optimal plan for injection adjustment. Compared with conventional methods, this methodology and workflow involve automated, intelligent, data-driven, and ML algorithms.

6.4.3.4.2 Description of the Workflow and Machine Learning Model

This paper introduces a data-driven approach to optimize fine water injection in mature oil fields. The method combines traditional numerical simulation and optimization algorithms with data-driven techniques. Real-time monitoring data of layered injection and production are integrated into the workflow using ML algorithms. The approach utilizes automatic identification methods for layered injection-production flow relationships, multi-layer, and multi-direction production splitting techniques, as well as a PSO algorithm to optimize detailed water injection plans. The system continuously updates the geological model parameters with the aid of data assimilation technologies, achieving a dynamic match between the model and observed well performance data.

6.4.3.4.3 Advantages and Limitations of the Synchronized Algorithm

The data-driven optimization method provides a systematic approach for fine-tuning water injection in mature oil fields. It leverages real-time data and ML to optimize injection rates and improve oil recovery. The use of PSO allows for efficient exploration of multiple injection scenarios, reducing the risk of suboptimal solutions. However, the approach requires significant computational resources, particularly when dealing with large datasets from complex faulted reservoirs. Additionally, the accuracy of the model depends heavily on the quality of the real-time monitoring data and the proper calibration of the ML algorithms.

6.4.3.4.4 Results and Discussion

This study developed a data-driven approach using separated-layer water injection technology to optimize water injection plans in mature oil fields. The approach involves history matching, ML clustering, intelligent optimization, and simulation of injection allocation. The results show that the approach increases cumulative oil production and reduces water cut. The methodology and workflow involve automated, intelligent, data-driven, and ML algorithms. The approach provides a practical and effective solution for optimizing water injection plans in mature oil fields.

BIBLIOGRAPHY

Jia, Deli, Liu, He, Zhang, Jiqun, Gong, Bin, Pei, Xiaohan, Wang, Quanbin, Yang, Qinghai. Data-driven optimization for fine water injection in a mature oil field. *Petroleum Exploration and Development*, 47 (3), 2020, 674–682, ISSN 1876-3804, https://doi.org/10.1016/S1876-3804(20)60084-2

6.4.3.5 Application-26: Waterflooding Optimization under Geological Uncertainties by Using Deep Reinforcement Learning Algorithms

6.4.3.5.1 Background of the Application

This paper presents the use of deep RL algorithms to optimize waterflooding in a three-dimensional (3D) three-phase reservoir under geological uncertainties. Four algorithms, including deep Q-network (DQN), double DQN (DDQN), dueling DDQN, and deep deterministic policy gradient (DDPG), are employed to determine the water injection rate that maximizes the NPV of the waterflooding. The study also includes a comparison of the expected cumulative oil production, oil production rates, water cuts, and average reservoir pressures in 50 reservoir realizations under various control scenarios.

6.4.3.5.2 Description of the Workflow and Machine Learning Model

This paper explores the use of deep RL algorithms to optimize waterflooding operations under geological uncertainties. The study applies four state-of-the-art deep RL algorithms: DQN, DDQN, dueling DDQN, and DDPG. These algorithms adjust water injection rates to maximize the NPV of a waterflooding project. The reservoir simulation data, including oil, water, and gas production rates, as well as reservoir pressures, are used to train the RL agents. The study incorporates geological uncertainties by generating multiple reservoir realizations using a geostatistical technique.

6.4.3.5.3 Advantages and Limitations of the Synchronized Algorithm

The deep RL algorithms demonstrate strong potential in optimizing waterflooding under uncertain geological conditions. The DDPG algorithm, in particular, converges faster than traditional optimization methods like PSO while still providing competitive results. However, the performance of the RL algorithms is sensitive to the exploration-exploitation trade-off, and finding the global optimum can be challenging. Additionally, tuning the hyperparameters of the RL models can be computationally expensive, and the algorithms require significant computational resources to handle large-scale reservoir simulations.

6.4.3.5.4 Results and Discussion

The study demonstrates that deep RL algorithms are useful for optimizing waterflooding in a 3D three-phase reservoir under geological uncertainties. The algorithms enable the determination of the optimum water injection rate that decelerates the decline rate of oil production, supplies reservoir energy quickly, and achieves the maximum expected NPV. The results also show that the geological uncertainties should be considered to manage risks and make better decisions. In future studies, the algorithms should be tested for more practical field-scale applications in the oil and gas industry and improved with the development of AI technology.

BIBLIOGRAPHY

Ma, Hongze, Yu, Gaoming, She, Yuehui, and Gu, Yongan "Waterflooding Optimization under Geological Uncertainties by Using Deep Reinforcement Learning Algorithms." *Paper Presented at the SPE Annual Technical Conference and Exhibition*, Calgary, AB, Canada, September 2019. https://doi.org/10.2118/196190-MS

6.4.3.6 Application-27: Deriving Alkali Polymer Parameter Distributions from Core Flooding by Applying ML in a Bayesian Framework to Simulate IOR

6.4.3.6.1 Background of the Application

This paper proposes a methodology for conditioning numerical models of alkali polymer (AP) core flooding simulation to various observed data using a Bayesian framework. The proposed workflow involves defining objective functions with error tolerances for the different observed data sets, conditioning the numerical model to each of the objective functions, and rejecting all parameter combinations that do not conform to the intersection area of all objective functions. The workflow also includes using ML to derive statistical models that allow for sampling of millions of parameter combinations. The methodology leads to conditioning the prior parameter distributions to various observed data, ensuring that the posterior parameter distribution complies with a Bayesian framework.

6.4.3.6.2 *Description of the Workflow and Machine Learning Model*

This paper applies ML within a Bayesian framework to derive posterior parameter distributions for AP flooding core flood simulations. The study uses RF algorithms to generate statistical models that predict the outcomes of core flood experiments based on a set of uncertain physico-chemical parameters. These models are trained using data from 2,000 simulation runs, generated through Latin hypercube sampling (LHS). The approach focuses on conditioning prior parameter ranges to observed data, including pressure, oil production, and chemical concentration measurements. The ultimate goal is to identify parameter combinations that conform to multiple objective functions reflecting different observed responses.

6.4.3.6.3 *Advantages and Limitations of the Synchronized Algorithm*

The use of RF algorithms within a Bayesian framework allows for efficient conditioning of uncertain parameters to multiple observed data sets. This approach significantly narrows down the range of acceptable parameter combinations, ensuring a more accurate and consistent simulation of core flood experiments. However, the complexity of the workflow and the need for multiple iterations of resampling and rejection steps make the methodology computationally intensive. Additionally, while the approach effectively handles uncertainties in the input parameters, its success is contingent on the quality of the observed data and the appropriateness of the selected objective functions.

6.4.3.6.4 *Results and Discussion*

The proposed methodology for AP core flooding simulation involves conditioning numerical models to various observed data using a Bayesian framework. The workflow includes defining objective functions, conditioning the numerical model to each of the objective functions, and rejecting all parameter combinations that do not conform to the intersection area of all objective functions. The use of ML allows for sampling of millions of parameter combinations. The methodology leads to conditioning the prior parameter distributions to various observed data, ensuring that the posterior parameter distribution complies with a Bayesian framework. This methodology can be applied to other projects such as field (re-)developments as well.

BIBLIOGRAPHY

Steineder, Dominik, Vanegas, Gisela, Clemens, Torsten, and Zechner, Markus. "Deriving Alkali Polymer Parameter Distributions from Core Flooding by Applying Machine Learning in a Bayesian Framework to Simulate Incremental Oil Recovery." *Paper Presented at the SPE Europec*, Virtual, December 2020. https://doi.org/10.2118/200578-MS

6.4.3.7 Application-28: Field Development Optimization Using Machine Learning Methods to Identify the Optimal Water Flooding Regime

6.4.3.7.1 Background of the Application

This paper presents a novel approach to optimizing waterflood management in brownfields using a hybrid proxy model based on ML methods and geological–physical–mathematical reservoir modeling. The model considers the mutual influence of wells and predicts oil production based on reservoir pressure maintained by injection wells. The paper evaluates the effectiveness of various ML methods for production prediction and identifies the optimal prediction horizon of 6–12 months. The paper also presents an algorithm for optimizing injection volume distribution, using the simplicial homology global optimization (SHGO) algorithm for global optimization and the COBYLA algorithm for local optimization. The approach was tested on a pilot section of a field in Western Siberia, resulting in a 7% increase in daily oil production relative to the baseline scenario. The paper concludes that the developed approach can be used to optimize waterflooding at brownfields.

6.4.3.7.2 Description of the Workflow and Machine Learning Model

This paper explores the application of ML methods to optimize field development by identifying the optimal water flooding regime. The study develops a hybrid geological–physical–mathematical proxy model that integrates ML algorithms, such as temporal convolutional networks (TCN) and gradient boosting, to predict oil production under different water injection regimes. The model is trained on historical production and injection data from the terrigenous formations of the West Siberian field. The ML models are coupled with optimization algorithms, including SHGO, to allocate optimal injection volumes to wells, considering infrastructural constraints and maximizing oil recovery.

6.4.3.7.3 Advantages and Limitations of the Synchronized Algorithm

The ML approach offers a faster and computationally efficient alternative to traditional hydrodynamic simulations for optimizing water flooding operations. The use of TCN and gradient boosting models allows for high prediction accuracy while reducing the computational burden. However, the success of the models depends on the quality of the input data, and the selection of appropriate boundary conditions is critical for effective optimization. The SHGO algorithm provides robust optimization solutions but may require significant computational resources when dealing with complex reservoir models. Despite these challenges, the integrated approach demonstrates effectiveness in enhancing oil recovery in brownfield operations.

6.4.3.7.4 Results and Discussion

This paper presents a novel approach to optimizing waterflood management in brownfields using a hybrid proxy model based on ML methods and geological–physical–mathematical reservoir modeling. The paper evaluates the effectiveness of various ML methods for production prediction, identifying the XGBoost, LSTM, and TCN NNs as the most effective models. The paper also presents an algorithm for optimizing injection volume distribution, using the SHGO algorithm for global optimization and the COBYLA algorithm for local optimization. The approach was tested on a pilot section of a field in Western Siberia, resulting in a 7% increase in daily oil production relative to the baseline scenario. The paper concludes that the developed approach can be used to optimize waterflooding at brownfields. The results of this study have practical implications for the oil and gas industry, as the approach can be applied to improve the efficiency of waterflood management in fields with a long history of development.

BIBLIOGRAPHY

Timonov, Alexey Vasilievich, Shabonas, Arturas Rimo, and Schmidt, Sergey Alexandrovich. "Field Development Optimization Using Machine Learning Methods to Identify the Optimal Water Flooding Regime." *Paper Presented at the SPE Russian Petroleum Technology Conference*, Virtual, October 2021. https://doi.org/10.2118/206533-MS

6.4.4 Unconventional and Other Applications

6.4.4.1 Application-29: Using Machine Learning to Predict Production at a Peace River Thermal EOR Site

6.4.4.1.1 Background of the Application

This paper presents an ML approach for predicting production rates in oil fields using daily measurements of various field parameters. The authors propose a two-step prediction strategy that first utilizes various ML methods to predict future values of temperature, pressure, and other measurements that affect production rates. Then, the predicted values are used to predict future production rates using a Lasso regression method. The authors also explore various techniques for dimensionality reduction and feature selection, including PCA and agglomerative hierarchical clustering. The results show that incorporating daily seismic data using PCA greatly improves the accuracy of the predictions. Overall, this approach provides a promising tool for automating data interpretation in modern, highly instrumented oil fields.

6.4.4.1.2 Description of the Workflow and Machine Learning Model

This method employs PCA, hierarchical clustering, and multiple regression models to predict oil production rates based on daily field measurements. PCA reduces the dimensionality of seismic data, capturing key variations with fewer components. Hierarchical clustering groups similar measurements, improving data organization. Lasso regression predicts intermediate measurements such as pressures and temperatures from controlled well settings. These intermediate predictions are then refined using RFR, which combines predictions from multiple DTs to estimate production rates. This method with two steps uses multiple layers of data and ML techniques to improve prediction accuracy.

6.4.4.1.3 Advantages and Limitations of the Synchronized Algorithm

The integration of PCA, hierarchical clustering, and regression models offers robust handling of high-dimensional data, automating feature extraction and prediction processes. PCA effectively reduces data complexity while keeping critical information, and hierarchical clustering organizes data into more meaningful groups. Lasso regression identifies and uses the most predictive variables, and RFR provides a flexible and accurate estimation of production rates. However, this method needs significant computational resources and relies on the quality of the input data a lot. Inconsistent or noisy data can change model's performance negatively, thus highlighting the importance of clean data preprocessing and validation to make sure for reliable predictions.

6.4.4.1.4 Results and Discussion

In conclusion, this paper presents an ML approach for predicting oil field production rates using daily measurements of various field parameters. The proposed two-step prediction strategy, which utilizes various ML methods, shows promise in automating data interpretation. Incorporating daily seismic data using PCA greatly improves the accuracy of the predictions. However, the paper also highlights the limitations of the current approach, including the need for more data and the difficulty in incorporating past information into the model. Future work includes exploring new ML techniques for time-series analysis and incorporating physics-based models into the prediction process. Overall, this approach provides a valuable tool for predicting production rates in modern, highly instrumented oil fields.

BIBLIOGRAPHY

Martin, Eileen, Wills, Peter, Hohl, Detlef, and Lopez, Jorge L. "Using Machine Learning to Predict Production at a Peace River Thermal EOR Site." *Paper Presented at the SPE Reservoir Simulation Conference*, Montgomery, TX, February 2017. https://doi.org/10.2118/182696-MS

6.4.4.2 Application-30: Saturation Forecast Using Machine Learning: Enabling Smarter Decision-Making Capabilities

6.4.4.2.1 Background of the Application

This paper presents a novel workflow for forecasting saturation in time-lapse saturation monitoring using a combination of ML algorithms. The workflow consists of a class-based machine learning (CBML) algorithm for classification of flow units and the Prophet model for time series forecasting. The workflow is designed to optimize cost, time, and risk in the oil industry by using historical data to forecast reservoir properties. The paper discusses the proof of concept for utilizing ML for this novel application and presents the results of applying the workflow to a field under production for 10 years.

6.4.4.2.2 Description of the Workflow and Machine Learning Model

This method combines class-based machine learning (CBML) with the Prophet model to predict oil saturation levels. CBML classifies petrophysical data into distinct flow units. This classification helps tailor the predictions to the unique properties of each flow unit. Following this, the Prophet model predicts future saturation levels based on historical data trends. This approach uses both classification and time series forecasting to provide information for optimizing reservoir management.

6.4.4.2.3 Advantages and Limitations of the Synchronized Algorithm

This combined CBML and Prophet method has several advantages. Capturing the unique characteristics of different flow units leads to more accurate forecasts. The Prophet model is particularly good at handling irregular time series data, which is common in oil field data collection. This method allows for optimized production and resource allocation through detailed and accurate predictions. However, the accuracy of the initial clustering is important, as errors can affect the final predictions. Integrating CBML and Prophet requires careful calibration, validation, and is computationally expensive. Despite these challenges, this approach offers improved predictive accuracy and operational efficiency in the oil industry.

6.4.4.2.4 Results and Discussion

The use of a saturation forecast can aid in making informed decisions about planning the logging across different wells. The workflow presented in this paper shows the application of two robust ML methods combined to provide a solution for a new avenue of interpretation in the oil industry. The ready availability of the algorithms makes it easier to implement ML for more and more of the processes, which rely on something that holds the key to future data. ML capabilities have today changed the way a user perceives and interprets large volumes of data and enable more informed and smart decision-making capabilities, which become a critical factor in today's business scenario where time and money are of the essence.

BIBLIOGRAPHY

Saumya, Sachit, Naqeeb, Ibrahim, Vij, Jitesh, Khambra, Isha, and Kumar, Apoorva. "Saturation Forecast Using Machine Learning: Enabling Smarter Decision-Making Capabilities." *Paper Presented at the Abu Dhabi International Petroleum Exhibition & Conference*, Abu Dhabi, UAE, November 2019. https://doi.org/10.2118/197419-MS

6.4.4.3 Application-31: Using Artificial Intelligence in Estimating Oil Recovery Factor

6.4.4.3.1 Background of the Application

This paper presents an application of ANNs in estimating the RF from basic information. The authors compare the results of two ANNs: the simple ANN and the sophisticated ANN. The simple ANN uses 10 input parameters, while the sophisticated ANN uses those 10 parameters as well as additional qualitative parameters related to field development actions and technology usage. The paper discusses the process of constructing and training the ANNs, including selecting the tool, constructing the NN, inserting I/O data and teaching, and testing the ANN. The results show that the ANNs provide a quick and easy estimate for the RF, with the sophisticated ANN giving more accurate results due to its consideration of additional parameters.

6.4.4.3.2 Description of the Workflow and Machine Learning Model

This method employs two types of ANNs—a simple ANN and a sophisticated ANN—to estimate RF from basic reservoir information. The simple ANN uses 10 fundamental parameters to quickly estimate the RF. The sophisticated ANN, on the other hand, incorporates additional parameters such as Reservoir Technology and Well Technology, providing a more nuanced prediction by considering the complexities of field development and technology use. These ANNs produce efficient and accurate predictions, letting the user make better decisions in reservoir management and development planning.

6.4.4.3.3 Advantages and Limitations of the Synchronized Algorithm

Using both the simple and sophisticated ANNs has the advantage of balancing the prediction of the RF estimation. The simple ANN, with its reliance on easily obtainable parameters, is particularly useful in the early stages of field appraisal, offering quick information with reasonable accuracy. The sophisticated ANN, by incorporating more detailed and advanced parameters, improves prediction accuracy and accounts for a larger range of influencing factors. However, these methods have their own set of challenges. The simple ANN might miss some important factors due to its limited input parameters, whereas the sophisticated ANN requires deep data and is computationally more expensive. Despite these limitations, the combined use of both models ensures a robust and adaptable framework for predicting RFs in diverse scenarios.

6.4.4.3.4 Results and Discussion

The authors have successfully demonstrated the use of ANNs in estimating RF from basic information. The results show that the ANNs provide a quick and easy estimate for RF, with the sophisticated ANN giving more accurate results due to its consideration of additional parameters related to field development actions and technology usage. The paper also provides recommendations for anyone looking to use ANNs in similar applications, such as optimizing the structure of the hidden layer and trying different configurations for each algorithm. Overall, this paper contributes to the growing body of literature on the use of AI methods in the upstream oil and gas industry.

BIBLIOGRAPHY

Noureldien, Darhim M., and El-Banbi, Ahmed H. "Using Artificial Intelligence in Estimating Oil Recovery Factor." *Paper Presented at the SPE North Africa Technical Conference and Exhibition*, Cairo, Egypt, September 2015. https://doi.org/10.2118/175867-MS

6.4.4.4 Application-32: A Developed Smart Technique to Predict Minimum Miscible Pressure—EOR Implications

6.4.4.4.1 Background of the Application

This document presents a tabular compilation of MMP data for CO_2-oil systems. The data are compiled from various sources and include some systems that have not been previously reported. The tabular presentation includes the CO_2 composition, oil composition, and MMP for each system. The input parameters were not explicitly mentioned in this document. However, various research papers were cited in which different input parameters were used for their studies.

6.4.4.4.2 Description of the Workflow and Machine Learning Model

In this method, a feed-forward artificial neural network (FF-ANN) combined with PSO to predict the minimum miscible pressure (MMP) was used in EOR processes.

The model is designed to take reservoir temperature, molecular weight of C5+, and compositions of gas, H_2S, C1, C2–C4, and N_2 as inputs. By using PSO for weight optimization in the NN, the model achieves high accuracy in forecasting MMP values, providing a powerful tool for optimizing gas injection strategies in reservoir management. This hybrid approach enables the handling of complex, nonlinear relationships within the data, ensuring precise and reliable predictions.

6.4.4.4.3 *Advantages and Limitations of the Synchronized Algorithm*

The PSO-optimized NN methodology is very advantageous in capturing the intricate patterns and dependencies in reservoir data, delivering highly accurate MMP predictions. Its ability to automatically tune model parameters through PSO reduces manual intervention and enhances model performance. This method's flexibility allows it to adapt to varying datasets and conditions, making it a robust solution for EOR optimization. However, the success of this approach is dependent upon the quality and coverage of the input data. Inaccurate or incomplete data can lead to less reliable predictions, lessening the importance of meticulous data preprocessing and validation to achieve desirable results.

6.4.4.4.4 *Results and Discussion*

In conclusion, this document provides a comprehensive compilation of MMP data for CO_2-oil systems. The tabular presentation includes the CO_2 composition, oil composition, and MMP for each system. The data are compiled from various sources and include some systems that have not been previously reported. The compilation can be used to evaluate the performance of various correlations and models for predicting MMP. Further research is needed to develop more accurate correlations and models for predicting MMP in CO_2-oil systems.

BIBLIOGRAPHY

Zendehboudi, S., Ahmadi, M.A., Bahadori, A., Shafiei, A., Babadagli, T. A Developed Smart Technique To Predict Minimum Miscible Pressure—EOR Implications. *Can. J. Chem. Eng*. 9999:1–13, 2013 https://doi.org/10.1002/cjce.21802

6.4.4.5 Application-33: Simulation of Kinetic Behavior of Natural Surfactants Adsorption Using a New Robust Approach

6.4.4.5.1 *Background of the Application*

This study presents a new model for predicting the kinetic adsorption of natural surfactants on sandstone and carbonate rocks. The model is based on the LS-SVM with coupled simulated annealing (CSA) algorithm used for optimization. The model was developed using a comprehensive database that covers a wide range of operating conditions. The proposed LS-SVM-CSA model has two parameters, μ and $\sigma 2$, which were optimized during model development. The performance of the model was compared with other well-known kinetic models found in the literature. The proposed model demonstrated the best performance and highest efficiency with the highest value of coefficient of determination, R^2, and the lowest average absolute relative deviation (AARD) and RMSE.

6.4.4.5.2 Description of the Workflow and Machine Learning Model

This method utilizes an ML pipeline to predict the kinetic adsorption density of natural surfactants on various rock types. The process begins with data preparation, where input variables such as surfactant density, concentration, temperature, time, and rock type are standardized. SVR with an RBF kernel, known for effectively handling nonlinear relationships, is used. Model performance is optimized using the tree-structured Parzen estimator (TPE) from the Hyperopt library, which fine-tunes hyperparameters through an iterative process. This combined approach with data preprocessing, SVR modeling, and hyperparameter optimization ensures accurate and reliable predictions.

6.4.4.5.3 Advantages and Limitations of the Synchronized Algorithm

This methodology is very advantageous at managing high-dimensional data, automating feature extraction, and refining predictions through multiple layers of ML techniques. Standardization ensures that all input variables are scaled appropriately, and the SVR with RBF kernel captures complex, nonlinear dependencies in the data. The Hyperopt TPE algorithm further improves the model by optimizing its hyperparameters for maximum performance. However, this method requires significant computational resources and is very sensitive to the input data quality. Inconsistent or noisy data can negatively impact the model's accuracy, negatively affecting the importance of thorough data preprocessing and validation to achieve reliable results.

6.4.4.5.4 Results and Discussion

In summary, the suggested LS-SVM-CSA model demonstrated excellent precision and dependability in forecasting the kinetic adsorption density of naturally occurring surfactants on carbonate and sandstone rocks. When compared to other well-known models, the model performed the best. The suggested model's appropriate efficiency was demonstrated by an investigation into its dependability throughout a range of operating situations. The issues in simulating the adsorption behavior of surfactants using earlier kinetic models can be resolved by the proposed LS-SVM-CSA model.

BIBLIOGRAPHY

Daryasafar, N., Omid Borazjani, O., Daryasafar, A. Simulation of kinetic behavior of natural surfactants adsorption using a new robust approach. *Journal of Chemometrics*. 2018; 32:e3031. https://doi.org/10.1002/cem.3031

6.4.4.6 Application-34: Time Dependent Neural Network-Based Proxy Modeling of SAGD Process

6.4.4.6.1 Background of the Application

This paper presents a time-dependent RBF NN algorithm for reducing the computational cost of full reservoir simulations in the context of a SAGD process. The algorithm provides reliable predictions for the entire field cumulative oil production rate, entire field water injection rate, entire field oil production rate, and entire field cumulative steam-to-oil ratio (SOR) for different combinations of operational parameters. The proposed proxy model assumes that the network's response increases as a power function of the distance between two M-dimensional points in the parameter space.

The RBFs are used to develop a time-dependent proxy model that provides good prediction accuracy if sufficient initial training data are provided. The method can significantly reduce the computational cost of full reservoir simulations.

6.4.4.6.2 Description of the Workflow and Machine Learning Model

In this method, an ML pipeline is used to predict production performance metrics of a SAGD process. The RBF NN is used, which is good at approximating nonlinear relationships and capturing the dynamic nature of the SAGD process. To ensure robustness and reliability, the data are split into training and testing sets, then standardized to maintain consistency across different input variable scales. This pipeline significantly reduces computational costs while at the same time maintaining high prediction accuracy for various combinations of operational parameters.

6.4.4.6.3 Advantages and Limitations of the Synchronized Algorithm

The main advantage of this methodology is that it excels in handling complex, multidimensional data, automating feature extraction, and refining predictions through a sophisticated NN model. Standardizing the input variables ensures uniform scaling, while the RBF NN effectively captures intricate, nonlinear dependencies within the data. However, this method needs considerable computational power and is highly sensitive to input data quality. Any inconsistencies or noise in the data can significantly affect the model's accuracy, highlighting the need for thorough data preprocessing and validating settings to achieve dependable results.

6.4.4.6.4 Results and Discussion

In this paper, a new time-dependent RBF NN algorithm was presented, which significantly reduces computational cost while maintaining good prediction accuracy for various combinations of operational parameters. The proposed algorithm can provide reliable results for the entire field cumulative oil production rate, entire field cumulative SOR, and entire field water injection rate. The proposed proxy model can provide reliable results if sufficient initial training data are provided; however, there is a limit for decreasing the initial LHD size, after which the proxy becomes much less useful. The algorithm can significantly reduce the computational cost of full reservoir simulations and provide a good forecast of the SAGD process.

BIBLIOGRAPHY

Fedutenko, E., Yang, C., Card, C., and Nghiem, L. X. "Time-Dependent Neural Network Based Proxy Modeling of SAGD Process." *Paper Presented at the SPE Heavy Oil Conference-Canada*, Calgary, AB, Canada, June 2014. https://doi.org/10.2118/170085-MS

6.4.4.7 Application-35: Steam Generators Optimization Using a Modified Quantum-Behaved Particle Swarm Optimization (QPSO) Algorithm

6.4.4.7.1 Background of the Application

This paper presents an optimization algorithm for steam generation in a heavy oil field with multiple steam generators. The algorithm is designed to maximize the sum of the efficiencies of all generators while delivering the required amount of steam for the field. The solution is data-driven and flexible, incorporating efficiency curves for each steam generator, lower and upper bounds on the amount of steam that each generator can produce, and the total amount of steam required for the field. Two optimization algorithms, PSO and quantum-behaved particle swarm optimization (QPSO), are employed to solve the steam generation optimization problem.

6.4.4.7.2 Description of the Workflow and Machine Learning Model

In this method, a QPSO algorithm is used to optimize the steam distribution across various generators to achieve maximum efficiency. The process begins by initializing a population of particles, where each particle represents a potential solution for steam distribution. Each generator's steam output is randomly set within predefined lower and upper bounds. The fitness function evaluates these particles based on their ability to meet the total steam requirement while maximizing efficiency. During the iterations, particles are updated by considering their personal best positions and the global best position, using a quantum behavior model to enhance exploration and convergence. This iterative process continues until the optimal steam distribution is identified, achieving a balance between required steam output and efficiency.

6.4.4.7.3 Advantages and Limitations of the Synchronized Algorithm

This QPSO-based method has many advantages, including efficient exploration of the solution space and robust convergence properties due to the quantum behavior model. The approach ensures that all potential solutions are thoroughly evaluated, using the mean best position of particles to guide the search toward optimal solutions. This method effectively handles complex optimization problems with multiple constraints, ensuring a high degree of accuracy in identifying the optimal steam distribution. However, the algorithm's performance is very dependent on the choice of parameters such as the population size, maximum iterations, and the alpha coefficient. Additionally, the computational expense can be significant, especially for large-scale problems, requiring careful consideration of computational resources and potential changes to parameter settings to balance accuracy and efficiency.

6.4.4.7.4 Results and Discussion

The optimization algorithm presented in this paper represents a step change in the way to approach steam generation optimization in heavy oil fields. By maintaining all units running and maximizing the total efficiency while delivering the required steam target, the algorithm produces an optimal solution that is superior to the current field practices of shutting down entire units. The QPSO algorithm provides a realistic and data-driven solution that can lead to a significant business impact. The flexibility and simplicity of the solution make it easily adaptable to various field constraints and knowledge.

BIBLIOGRAPHY

Hao, Minshen, and Popa, Andrei. "Steam Generators Optimization Using A Modified Quantum-behaved Particle Swarm Optimization (QPSO) Algorithm." *Paper Presented at the SPE Digital Energy Conference and Exhibition*, The Woodlands, TX, March 2015. https://doi.org/10.2118/173391-MS

6.4.4.8 Application-36: Artificial Neural Network Modeling of Cyclic Steam Injection Process in Naturally Fractured Reservoirs

6.4.4.8.1 Background of the Application

This paper presents the development of three expert systems using ANNs to study cyclic steam injection in naturally fractured heavy oil reservoirs. The forward model focuses on forecasting the cumulative production profile and viscosity contours, while the inverse model 1 predicts the injection design parameters. Inverse model 2 is developed to determine reservoir parameters such as reservoir porosity, permeability, and natural fracture spacing through a history-matching protocol. The ANN models have been shown to generate results with a good level of accuracy within seconds.

6.4.4.8.2 Description of the Workflow and Machine Learning Model

In this method, an ANN is used to forecast the cumulative oil production and viscosity contours in naturally fractured heavy oil reservoirs. For model training, a sequential NN with multiple hidden layers is employed to capture the intricate relationships between the input parameters and the target outputs. By optimizing the network's architecture and parameters, the model provides accurate predictions of reservoir performance. This robust data-driven method ensures swift and reliable forecasting, essential for effective reservoir management.

6.4.4.8.3 Advantages and Limitations of the Synchronized Algorithm

This ANN-based methodology is very successful in modeling complex, nonlinear interactions between various reservoir and design parameters. Standardization of input data guarantees consistency and improves model convergence, while the NN's architecture adeptly captures detailed patterns in the data. The approach is advantageous for its speed and precision in predicting oil production and viscosity changes. However, the success of this method is highly dependent on the quality and quantity of training data. Inaccurate or insufficient data can lead to negative model performance, thus highlighting the need for careful data preparation and validation for good information from the model.

6.4.4.8.4 Results and Discussion

In conclusion, the ANN models developed in this study have shown high accuracy in generating results for cyclic steam injection in naturally fractured heavy oil reservoirs. The forward model accurately forecasts the cumulative production profile and viscosity contours. Inverse model 1 predicts the injection design parameters with acceptable error percentages. Inverse model 2 successfully determines reservoir parameters such as reservoir porosity, permeability, and natural fracture spacing. The developed graphical user interface (GUI) is user-friendly and can be used without any knowledge of programming languages. This study demonstrates the potential of ANN models in reservoir engineering and their ability to provide efficient solutions.

BIBLIOGRAPHY

Ersahin, Ahmet, Gul, Sercan, Ertekin, Turgay, and Temizel, Cenk. "Artificial Neural Network Modeling of Cyclic Steam Injection Process in Naturally Fractured Reservoirs." *Paper Presented at the SPE Western Regional Meeting*, San Jose, CA, April 2019. https://doi.org/10.2118/195307-MS

6.4.4.9 Application-37: Applications of Artificial Intelligence in Oil and Gas Development

6.4.4.9.1 Background of the Application

This paper explores the application of AI in the oil and gas industry, with a focus on the use of AI to improve various aspects of oilfield development. The paper discusses the importance of understanding user needs and the collection and processing of data as key factors in realizing the application of AI in oil and gas fields. The authors also explore various AI algorithms and their limitations, highlighting the importance of selecting and applying the appropriate algorithm for a given problem. The paper concludes that AI has great potential to improve oilfield development, but more efforts are needed to integrate intelligent systems at various stages.

6.4.4.9.2 Description of the Workflow and Machine Learning Model

In this method, a backpropagation (BP) NN is used to predict wax appearance temperature (WAT) from input parameters such as active energy, kinematic viscosity, density, minimum pour point, and molecular weight. The data is standardized before training the NN with two hidden layers. The trained model predicts WAT on the test data, and its performance is measured using mean absolute error (MAE), providing an indication of prediction accuracy.

6.4.4.9.3 Advantages and Limitations of the Synchronized Algorithm

The BP NN can capture complex, nonlinear relationships between inputs and outputs, making it effective for predicting WAT. Its flexibility allows for fine-tuning to improve accuracy. However, training such networks can be computationally expensive and time-consuming. The model's success depends a lot on data quality, and it may struggle with real-world scenarios if the training data isn't comprehensive enough. Additionally, careful tuning is needed to avoid overfitting.

6.4.4.9.4 Results and Discussion

In conclusion, this paper highlights the potential benefits of applying AI in the oil and gas industry, particularly in the area of oilfield development. The authors emphasize the importance of understanding user needs and collecting and processing data, as well as selecting and applying the appropriate AI algorithm for a given problem. They also caution against blindly trusting the results provided by AI and instead advocate for analysis and interpretation to improve the accuracy and reliability of AI solutions. Overall, the paper underscores the need for more efforts to integrate intelligent systems at various stages to fully realize the potential of AI in the oil and gas industry.

BIBLIOGRAPHY

Li, H., Yu, H., Cao, N. et al. Applications of Artificial Intelligence in Oil and Gas Development. *Arch Computat Methods Eng* 28, 937–949 (2021). https://doi.org/10.1007/s11831-020-09402-8

6.4.4.10 Application-38: Applying Reservoir Simulation and AI Algorithms to Optimize Fracture Characterization and CO_2 EOR in Unconventional Reservoirs

6.4.4.10.1 Background of the Application

This paper presents an optimization approach for hydraulic fracturing design using an ANN and a Bayesian optimization algorithm. The objective is to maximize the production of oil and gas from shale formations by optimizing various fracturing parameters, such as fracture efficiency, fracture half-length, fracture aperture, fracture height, fracture conductivity, fracture water saturation, fracture compaction coefficient, and matrix compaction coefficient. The results demonstrate that the proposed approach can effectively optimize the hydraulic fracturing design and improve the production of oil and gas.

6.4.4.10.2 Description of the Workflow and Machine Learning Model

In this method, an ANN combined with Bayesian optimization is used to optimize hydraulic fracturing design parameters. The ANN models the complex relationships between parameters and production outcomes, learning from historical data to predict performance. Bayesian optimization is used in the method to fine-tune the input parameters by iteratively searching for the optimal configuration that maximizes production efficiency. The workflow encompasses loading the data from a CSV file, training the ANN, defining the optimization function, and running the Bayesian algorithm to derive the optimal fracturing design.

6.4.4.10.3 Advantages and Limitations of the Synchronized Algorithm

The combination of ANNs and Bayesian optimization has major advantages for optimizing hydraulic fracturing designs. The ANN excels at capturing nonlinear relationships within high-dimensional data, providing accurate predictions of production outcomes. Bayesian optimization complements this by exploring the parameter space, using probabilistic models to balance the method. However, this methodology is computationally expensive, particularly during the training phase of the ANN and the iterative process of Bayesian optimization. The quality of the optimization is dependent on the data quality. Moreover, the process requires careful tuning of hyperparameters and a substantial amount of computational resources to achieve the desired results.

6.4.4.10.4 Results and Discussion

In this paper, an optimization approach for hydraulic fracturing design using an ANN and a Bayesian optimization algorithm was presented. The approach was applied to optimize various fracturing parameters, and the results showed that the proposed approach can effectively improve the production of oil and gas from shale formations. The use of an ANN allows for the study of the complex nonlinear relationships between the input and output variables, while the Bayesian optimization algorithm enables the efficient search of the solution space. The proposed approach provides a useful tool for optimizing hydraulic fracturing design and maximizing the production of oil and gas.

BIBLIOGRAPHY

Wan, X.; Jin, L.; Azzolina, N.A.; Butler, S.K.; Yu, X.; Zhao, J. Applying Reservoir Simulation and Artificial Intelligence Algorithms to Optimize Fracture Characterization and CO2 Enhanced Oil Recovery in Unconventional Reservoirs: A Case Study in the Wolfcamp Formation. *Energies* 2022, 15, 8266. https://doi.org/10.3390/en15218266

6.4.4.11 Application-39: Neural Network-Based Process Coupling and Parameter Upscaling in Reactive Transport Simulations

6.4.4.11.1 Background of the Application

This paper explores the use of NNs to enhance reactive transport modeling in geochemical applications. Surrogate models, such as NNs, are used to express the complex correlations between input and output data, without the need for a description of underlying physical mechanisms. The article suggests that NNs can offer significant computational efficiency improvements over traditional geochemical algorithms. Unlike geochemical solvers, the computational cost of NNs is constant, making them an excellent option for parallel computing. In order to enhance the initial guesses for the chemical speciation calculations and ultimately replace the chemical solver at the advanced simulation stage, the article also suggests dynamically training NNs throughout the reactive transport simulation. Large-scale 3D reactive transport simulations of intricate natural systems and the advancement of digital twins, which are essential for experimental high-resolution reactive transport investigations, are made possible by the availability of precise and numerically effective NNs.

6.4.4.11.1.1 Computational Efficiency and Dynamic Training of NNs The anticipated boost in computing efficiency is the primary driver behind the use of surrogate models. The intricate relationships between input and output data can be expressed by a surrogate model without a description of the underlying physical systems. The computational cost of NNs is constant, making them an excellent option for parallel computing. The paper also proposes the dynamic training of NNs during the reactive transport simulation, which can improve the initial guesses for the chemical speciation calculations and eventually replace the chemical solver at the advanced simulation stage.

6.4.4.11.1.2 Significance of Coupled Reactive Transport in Geochemical Applications and Development of Digital Twins When simulating transient occurrences in complex multicomponent natural systems, this section emphasizes how crucial it is to consider the realistic portrayal of heterogeneities in 3D. The substantial computational expense of the complete speciation chemical solvers is one of the main barriers to taking actual geometries into consideration. According to the article, in natural and geo-engineered systems with strong geochemical gradients and nonlinear coupling between chemical reactivity, transport, and heat transfer phenomena, NNs can offer an effective and useful method for accurately estimating the critical time scales and the reaction pathways for geochemical phenomena. The article concludes by stating that access to accurate and numerically efficient NNs opens horizons for large-scale 3D reactive transport simulations of complex natural systems and further development of digital twins indispensable for experimental high-resolution reactive transport studies.

6.4.4.11.2 Description of the Workflow and Machine Learning Model

In this method, an NN is used to enhance reactive transport modeling by taking three input components (C1, C2, and C3) and predicting eleven outputs (Y1–Y11). The workflow involves loading data from a CSV file, training the NN on historical simulation data to learn the complex relationships between inputs and outputs, and using the trained model to predict new outputs. The NN's architecture efficiently captures the intricacies of reactive transport processes, providing a faster and more accurate alternative to traditional methods.

6.4.4.11.3 Advantages and Limitations of the Synchronized Algorithm

The NN method reduces computational time, enabling large-scale 3D simulations that were previously inaccessible. It ensures accurate predictions by effectively learning from training data. However, training the NN is computationally expensive and requires high-quality data and careful hyperparameter tuning. However, even for these challenges, the major improvements in speed and scalability make NNs a good tool for reactive transport modeling.

6.4.4.11.4 Results and Discussion

This paper explores the use of NNs to enhance reactive transport modeling in geochemical applications. The article suggests that NNs can offer significant computational efficiency improvements over traditional geochemical algorithms. The paper also proposes the dynamic training of NNs during the reactive transport simulation, which can improve the initial guesses for the chemical speciation calculations and eventually replace the chemical solver at the advanced simulation stage. The article highlights the importance of considering the realistic description of heterogeneities in 3D when simulating transient phenomena in complex multicomponent natural systems. Access to accurate and numerically efficient NNs opens horizons for large-scale 3D reactive transport simulations of complex natural systems, and further development of digital twins indispensable for experimental high-resolution reactive transport studies.

BIBLIOGRAPHY

Prasianakis, N.I., Haller, R., Mahrous, M., Poonoosamy, J., Pfingsten, W., Churakov, S.V., Neural network-based process coupling and parameter upscaling in reactive transport simulations, *Geochimica et Cosmochimica Acta* (2020), https://doi.org/10.1016/j.gca.2020.07.019

6.4.4.12 Application-40: Optimal Carbon Storage Reservoir Management through Deep Reinforcement Learning

6.4.4.12.1 Background of the Application

The best injection rate schedule that optimizes rewards while taking risk and cost restrictions into account is presented in this research as part of a DL-based RL system for managing carbon storage reservoirs. The framework develops function approximators of the Q function using a deep Q-learning method, distributed process-level models, and surrogate models of the multistate using a deep multitask learning technique. The framework's ability to discover the best rules for both passive and active management scenarios is demonstrated in the article. The results show that a variety of sequential optimization problems using high-dimensional, process-level models can be effectively solved with the help of the suggested framework.

6.4.4.12.2 Description of the Workflow and Machine Learning Model

This paper discusses the integration of AI and ML to improve production and operational efficiencies in deepwater subsea tiebacks. The primary focus is on utilizing AI to enhance decision-making and optimize critical operations, such as EOR, predictive maintenance, and operational efficiency. By leveraging real-time data from various sources like sensors, autonomous underwater vehicles (AUVs), and remotely operated vehicles (ROVs), AI algorithms, including model predictive control (MPC) and hybrid AI-Physics models, are applied to monitor and optimize production rates, predict equipment failures, and ensure optimal flow conditions. The AI system processes massive amounts of data from subsea monitoring systems to provide targeted recommendations that improve performance, reduce downtime, and minimize environmental impact.

6.4.4.12.3 Advantages and Limitations of the Synchronized Algorithm

The application of AI and ML offers significant advantages in optimizing subsea tiebacks. The use of model predictive control allows for continuous monitoring and real-time optimization, leading to improvements in production rates and reduced operational costs. Hybrid AI-Physics models further enhance decision-making by incorporating physics-based simulations, making them particularly valuable in complex systems with limited historical data. The primary advantage of these methodologies lies in their ability to process vast amounts of data and provide actionable insights that surpass human capabilities.

However, there are limitations to the approach. The implementation of AI systems requires substantial computational resources, and the success of these systems depends on the availability of high-quality, structured data. Additionally, the adoption of AI may face resistance within organizations due to the need for changes in work processes and reliance on pilot projects to demonstrate their value. Furthermore, the accuracy of predictions and optimizations depends on the precision of the underlying models, making continuous validation and updates necessary to maintain effectiveness.

6.4.4.12.4 Results and Discussion

This paper presents a DL-based RL framework for optimizing carbon storage reservoir management that can identify the optimal injection rate schedule while considering risk and cost constraints. The framework is shown to be effective in both passive and active management scenarios, and results indicate that it can serve as a practical tool for solving a wide range of sequential optimization problems involving high-dimensional, process-level models. Future work may involve exploring more advanced variants of the deep Q-learning algorithm and extending the framework to include multiple types of actions for injection and production rates.

BIBLIOGRAPHY

Sun, Alexander Y. Optimal carbon storage reservoir management through deep reinforcement learning, *Applied Energy*, 278, 2020, 115660, ISSN 0306-2619, https://doi.org/10.1016/j.apenergy.2020.115660

6.4.4.13 Application-41: Deep Neural Networks Predicting Oil Movement in a Development Unit

6.4.4.13.1 Background of the Application

This paper presents a novel approach for predicting reservoir performance using reduced order modeling (ROM) and data-driven forecasting (DDF) methods. The proposed method consists of two distinct parts: the dynamics model and the production rate model. The dynamics model includes a dimensionality reduction model and a latent dynamics model, which are used to forecast the 3D fields of pressure and fluid saturations over the computational cells of a conventional hydrodynamic model of a reservoir. The production rate model calculates the production rates at the wellheads, considering the influence of gas liberated from the oil. The trained metamodel requires no additional training to be applied to a new development unit and is much faster than a conventional reservoir modelling tool. The results show that the proposed method is accurate and computationally efficient, making it a useful tool for applied reservoir engineering.

6.4.4.13.2 Description of the Workflow and Machine Learning Model

This paper presents a novel data-driven modeling workflow that integrates DL and data analytics techniques for analyzing production profiles from horizontal well pairs and temperature profiles from vertical observation wells in SAGD operations. The study utilizes CNNs to capture complex, nonlinear relationships between the input variables, such as production and temperature data, and the output variables that describe the characteristics of shale barriers in the reservoir. Wavelet analysis and other data processing techniques are used to extract input features, and a parameterization scheme is proposed to effectively model the shale barrier distribution.

6.4.4.13.3 Advantages and Limitations of the Synchronized Algorithm

The integrated workflow demonstrates the potential for efficiently analyzing SAGD reservoir heterogeneities by incorporating both production and temperature data. The use of CNN enables automatic feature extraction and reduces the computational burden compared to traditional modeling methods. However, the performance of the CNN model depends on the quality of the training data, and it may struggle to resolve fine details of reservoir heterogeneities when multiple shale barriers are present. Additionally, the complexity of the input data requires careful preprocessing to ensure accurate predictions.

6.4.4.13.4 *Results and Discussion*

In this paper, a novel approach for predicting reservoir performance was presented using reduced-order modeling and data-driven forecasting methods. It was shown that the proposed method is accurate and computationally efficient, making it a useful tool for applied reservoir engineering. Further research is needed to expand the metamodel's operational envelope to more complex fluid systems and low-permeable reservoirs. The metamodels are also planned to be adapted for full-scale reservoirs and oilfields containing many development units, leading to more practical tools for field development optimization and history matching.

BIBLIOGRAPHY

Temirchev, P., Simonov, M., Kostoev, R., Burnaev, E., Oseledets, I., Akhmetov, A., Margarit, A., Sitnikov, A., Koroteev, D. Deep neural networks predicting oil movement in a development unit. *Journal of Petroleum Science and Engineering*, 184, 2020, 106513, ISSN 0920-4105, https://doi.org/10.1016/j.petrol.2019.106513

6.4.4.14 Application-42: Design of Warm Solvent Injection Processes for Heavy Oil Reservoirs: A Hybrid Workflow of Multi-objective Optimization and Proxy Models

6.4.4.14.1 *Background of the application*

This document discusses various methods and techniques used in reservoir management and optimization. The authors cover topics such as data-driven modeling techniques, approximate ensemble gradients, and AI. They also present case studies on water flooding optimization and fracture stimulation design. The document highlights the importance of understanding reservoir heterogeneity and the potential benefits of using ML for reservoir management.

6.4.4.14.2 *Description of the Workflow and Machine Learning Model*

This paper presents a data-driven analysis of natural gas EOR in unconventional shale oils. The study uses various data-driven techniques, such as PCA, ANNs, SVMs, and RFR, to analyze patterns in production data. These methods help to identify key parameters in the recovery process and optimize operational strategies. The data-driven models are compared with physics-based evaluation methods to assess their effectiveness in understanding the recovery mechanisms and improving EOR processes in unconventional shale reservoirs.

6.4.4.14.3 *Advantages and Limitations of the Synchronized Algorithm*

The data-driven methods provide practical solutions for optimizing EOR in unconventional shale reservoirs. They offer a fast and robust analysis of production data, allowing for better decision-making in field operations. However, the success of these methods depends on the quality and quantity of the available data. The study highlights that while data-driven approaches are effective in mature fields with sufficient data, they may struggle in fields where data is sparse. Additionally, these methods may oversimplify complex reservoir interactions that are better captured by physics-based models, making a combined approach ideal.

6.4.4.14.4 Results and Discussion

In conclusion, this document emphasizes the importance of data-driven modeling and AI in reservoir management and optimization. The authors demonstrate the potential benefits of these techniques in optimizing water flooding and fracture stimulation design. Additionally, they stress the importance of understanding reservoir heterogeneity and the potential for ML to provide insights into this complex phenomenon. Overall, this document provides valuable insights into the current state of the art in reservoir management and optimization.

BIBLIOGRAPHY

Ma, Zhiwei, Leung, Juliana Y. Design of warm solvent injection processes for heterogeneous heavy oil reservoirs: A hybrid workflow of multi-objective optimization and proxy models. *Journal of Petroleum Science and Engineering*, 191, 2020, 107186, ISSN 0920-4105, https://doi.org/10.1016/j.petrol.2020.107186

6.4.4.15 Application-43: Predictive Data Analytics Application for EOR in a Mature Field

6.4.4.15.1 Background of the Application

In this study, a top-down modeling (TDM) approach was used for the first time on a carbonate oil reservoir in the Middle East. The TDM has five interconnected data-driven NN models, each modeling a key dynamic parameter, and the output of one model is the input for the next model. The TDM is calibrated using available well data and is trained and tested for each well in the reservoir. The TDM provides a fast and reliable means for optimizing WAG parameters and can also optimize the location of infill wells. The TDM is useful in reservoir management and production forecasting.

6.4.4.15.2 Description of the Workflow and Machine Learning Model

This paper discusses the development and application of ANN models to assist with chemical enhanced oil recovery (cEOR) field implementations. The study constructed five ANN models using 1,100 training cases generated by the CMG-STARS reservoir simulator. The ANN models include both forward and inverse models. Model #1 is a forward model that predicts reservoir response, such as oil rate and cumulative oil production, based on known reservoir characteristics and design parameters. Models #2–#5 are inverse models designed for various purposes, including predicting reservoir characteristics, project design parameters, and economic evaluation metrics. The inverse models help optimize field implementation by providing critical insights into reservoir behavior under different chemical injection scenarios.

6.4.4.15.3 Advantages and Limitations of the Synchronized Algorithm

The ANN models developed in this study offer a significant reduction in computational time compared to traditional numerical simulation methods, with a four-order magnitude decrease. Model #1 shows a low prediction error of 5%, while the inverse models, although effective, display higher error rates (20%–40%) due to the non-unique nature of inverse solutions. The back-check methodology, which compares predictions from the inverse models with forward model outputs, helps validate the accuracy of the predictions. Despite their effectiveness, the ANN models' performance heavily depends on the quality of training data and careful model tuning, and the inverse models are prone to producing multiple valid solutions.

6.4.4.15.4 Results and Discussion

The TDM created in this work offers a quick and dependable way to optimize WAG parameters and can also be used to optimize infill well placement. The TDM can accurately forecast reservoir performance three months in advance and has been validated in both time and space. According to the TDM's predicted results, higher injection volumes of gas and water won't always affect oil production in a positive way; instead, a lower injection volume during the injection period will increase oil production. For various injection schemes, the scheme with a 3-month injection period has a higher oil production rate than the one with a 6-month injection period. The TDM can be used to forecast reservoir performance and to manage well and reservoir performance. When new data is obtained, it can be compared with the predicted data from the TDM to ensure accuracy of the model and used to update the model.

BIBLIOGRAPHY

Yousef, Alklih Mohamad, Kavousi, Ghahfarokhi Payam, Alnuaimi, Marwan, Alatrach, Yara. Predictive data analytics application for enhanced oil recovery in a mature field in the Middle East. *Petroleum Exploration and Development*, 47 (2), 2020, 393–399, ISSN 1876-3804, https://doi.org/10.1016/S1876-3804(20)60056-8

6.4.4.16 Application-44: Saturation Forecast Using Machine Learning: Enabling Smarter Decision-Making Capabilities

6.4.4.16.1 Background of the Application

This paper presents a novel workflow for saturation forecast in time-lapse monitoring of reservoirs, which utilizes ML algorithms to analyze historical data. The workflow combines a class-based ML algorithm and time series modeling using the Prophet forecasting method to generate saturation forecasts for different flow units in a well. The CBML algorithm uses unsupervised clustering algorithms to classify petrophysical properties into different flow units, which are then used for time series modeling. The Prophet method is used to generate saturation forecasts based on an additive regression model that incorporates trends, seasonality, and holidays. The workflow is demonstrated on a well from a field that has been under production for ten years, and the results show a decent validation with the acquired saturation measurement. The workflow can be extended to generate saturation forecasts for multiple wells in the same field and can aid in field development planning.

The workflow comprises two different already available algorithms—a class-based machine learning algorithm (CBML) followed by time series modeling of the classes obtained through CBML. The CBML approach acquires knowledge from the training data and propagates, if and where applicable, to the new data. It eliminates the need for large training data and a high number of measurements and removes the subjectivity and inconsistency. In order to achieve the classification of flow units to drive time series modeling, a CBML algorithm has been used. The CBML methodology consists of an unsupervised clustering algorithm, which groups similar measurement vectors into a cluster. Advanced methods like cross-entropy clustering (CEC), Gaussian mixture model (GMM), and hidden Markov model (HMM) were used in the algorithm to answer the limitations of the single unsupervised clustering method. Once the classes are defined, time series modeling is run for each class/unit for the saturation forecast. An important step is to normalize the time domain as the saturation data historically has not been acquired in a defined time frame.

The workflow is demonstrated on a well from a field that has been under production for 10 years, and the results show a decent validation with the acquired saturation measurement. The workflow can be extended to generate saturation forecasts for multiple wells in the same field, and can aid in field development planning.

6.4.4.16.2 Description of the Workflow and Machine Learning Model

This paper explores the application of surrogate models in optimizing decision variables for cEOR, specifically surfactant-polymer (SP) flooding. The study compares the performance of various ML models, including multiple linear regression (MLR), Ridge and Lasso regression, SVR, ANNs, and tree-based algorithms like DTs, RF, gradient boosting, and XGBoost. The models were developed using a dataset of 202 points with seven input variables, such as polymer concentration, surfactant slug size, and salinity. The results demonstrated that tree-based ensemble models, along with SVR and ANN, provided the highest accuracy in predicting RF.

6.4.4.16.3 Advantages and Limitations of the Synchronized Algorithm

The surrogate models significantly reduce the computational cost and time associated with full reservoir simulations, making them ideal for real-time decision-making in cEOR processes. Tree-based ensemble techniques, such as RF and XGBoost, offer robust performance in handling nonlinear relationships and provide superior accuracy compared to traditional linear models. However, the effectiveness of these models depends on the quality and quantity of the input data, and they may struggle with extrapolating beyond the training dataset. Additionally, surrogate models are limited to the specific conditions of the reservoirs from which they are developed and cannot replace full-physics simulations entirely.

6.4.4.16.4 Results and Discussion

The workflow presented in this paper shows the application of two robust ML methods combined together to provide a solution for a new avenue of interpretation in the oil industry. The workflow can aid in making informed decisions about planning the logging across different wells and can optimize the resources, time, and cost. The more advanced applications lie in applying the workflow simultaneously to multiple wells in the same field and generating a saturation forecast for a flow unit/class across the wells. The workflow presented in this paper can be similarly extended to reservoir pressure forecasting in a producing well.

BIBLIOGRAPHY

Saumya, Sachit, Naqeeb, Ibrahim, Vij, Jitesh, Khambra, Isha, and Kumar, Apoorva. "Saturation Forecast Using Machine Learning: Enabling Smarter Decision-Making Capabilities." *Paper Presented at the Abu Dhabi International Petroleum Exhibition & Conference*, Abu Dhabi, UAE, November 2019. https://doi.org/10.2118/197419-MS

6.4.4.17 Application-45: Integration of Deep Learning and Data Analytics for SAGD Temperature and Production Analysis

6.4.4.17.1 Background of the Application

This paper proposes a workflow for characterizing shale heterogeneities in SAGD reservoirs using CNNs. The workflow integrates production and temperature data from observation wells into the input features of the CNN to predict the location and extent of shale barriers within the reservoir. The CNN is trained on a dataset of 2D heterogeneous SAGD reservoir models with varying numbers and sizes of shale barriers. Parametric bootstrapping is applied to sample multiple realizations of shale barrier parameters from the CNN estimations, which are then used to generate an ensemble of plausible shale barrier configurations. The proposed workflow is validated using three heterogeneous cases with different numbers of shale barriers. The results demonstrate that the workflow can characterize shale heterogeneities that are consistent with the production and temperature data.

6.4.4.17.2 Description of the Workflow and Machine Learning Model

This paper compares various ML algorithms to develop a forecasting tool for cyclic gas injection in hydraulically fractured wells. The study evaluates three ML models: RF, NNs, and MLR, each trained on a dataset generated from numerical simulation runs. The models use five operational parameters as input: injection rate, injection time, soaking time, cycle rate, and gas composition (CO_2, N_2, and CH_4). The ML models are designed to predict the economic efficiency of cyclic gas injection over several years of production. Data visualization techniques and feature importance analysis were used to optimize the models and improve prediction accuracy.

6.4.4.17.3 Advantages and Limitations of the Synchronized Algorithm

Among the three ML models, the RF algorithm demonstrated the best performance, with a coefficient of determination (R^2) of 0.93, outperforming NNs and MLR. The RF model offers significant advantages in terms of training speed and reduced tuning complexity. However, the study also notes that each model's accuracy depends on the quality of the simulation data and the careful selection of input features. While RFs provided the best overall performance, NNs showed potential for handling more complex, nonlinear relationships, albeit with a higher computational cost. MLR, though straightforward, exhibited the lowest accuracy and was limited by its inability to capture complex interactions between variables.

6.4.4.17.4 *Results and Discussion*

This paper presents a novel workflow for inferring uncertain shale barriers in SAGD reservoirs using DL. The workflow integrates production and temperature data from observation wells into the input features of the CNN to predict the location and extent of shale barriers within the reservoir. The CNN is trained on a dataset of 2D heterogeneous SAGD reservoir models with varying numbers and sizes of shale barriers. An ensemble of believable shale barrier realizations is established by sampling many shale parameter realizations from the CNN estimations using parametric bootstrapping. A few test scenarios are used to illustrate the viability of the suggested methodology, and the model predictions match the real target histories. The group of characterized models may be thought of as a collection of preliminary realizations that will be subjected to more thorough history matching. The workflow provides an alternate method for characterizing shale heterogeneities that are compatible with the production and temperature data.

BIBLIOGRAPHY

Ma, Zhiwei, and Leung, Juliana Y. "Integration of Deep Learning and Data Analytics for SAGD Temperature and Production Analysis." *Paper Presented at the SPE Reservoir Simulation Conference*, Galveston, TX, April 2019. https://doi.org/10.2118/193829-MS

6.4.4.18 Application-46: Data-Driven Analysis of Natural Gas EOR in Unconventional Shale Oils

6.4.4.18.1 *Background of the Application*

This paper evaluates the performance of three different ML techniques for predicting oil recovery: SVR, RFR, and ANN. The input parameters for the study are not explicitly mentioned in the given text, but the models are evaluated based on MSE, median average error (MAE), and coefficient of determination (R^2). The best results are obtained using SVR, while the performance of ANN and RFR models is average.

6.4.4.18.2 *Description of the Workflow and Machine Learning Model*

This paper introduces the non-Arrhenius machine learning approach (NAMLA) for predicting heavy oil combustion kinetics. The study uses a locally weighted linear regression model to predict the rate of oxygen (O_2) consumption during in situ combustion (ISC) directly from ramped temperature oxidation (RTO) experimental data. NAMLA is designed to overcome the limitations of traditional Arrhenius-based reaction models by using ML to model the combustion kinetics, thus eliminating the need for calibration of reaction parameters. The approach was applied to both simulated and experimental data from heavy oil samples, demonstrating its ability to predict O_2 consumption curves with high accuracy.

6.4.4.18.3 *Advantages and Limitations of the Synchronized Algorithm*

The NAMLA model offers significant advantages, particularly in handling sparse experimental data and providing uncertainty quantification for O_2 consumption predictions. Compared to interpolation-based methods, NAMLA performs better when predicting kinetics for data points outside the range of experimental measurements.

However, the model's reliance on ML requires significant computational resources, and the accuracy of predictions depends heavily on the quality of the training data. While NAMLA shows strong potential for improving ISC reaction modeling, it may struggle with very sparse data and requires careful selection of hyperparameters to ensure stability and accuracy.

6.4.4.18.4 Results and Discussion

In conclusion, this study has demonstrated the effectiveness of SVR for predicting oil recovery. The authors also compare the results of this model to those obtained using RFR and ANN techniques. The performance of each model is evaluated based on MSE, MAE, and R^2 criteria. The results show that SVR outperforms both ANN and RFR models for this specific dataset.

BIBLIOGRAPHY

Temizel, Cenk, Balaji, Karthik, Canbaz, Celal Hakan, Palabiyik, Yildiray, Moreno, Raul, Rabiei, Minou, Zhou, Zifu, and Ranjith, Rahul. "Data-Driven Analysis of Natural Gas EOR in Unconventional Shale Oils." *Paper Presented at the SPE Oklahoma City Oil and Gas Symposium*, Oklahoma City, OK, April 2019. https://doi.org/10.2118/195194-MS

6.4.4.19 Application-47: Development and Application of an Artificial Neural Network Tool for Chemical EOR Field Implementations

6.4.4.19.1 Background of the Application

This paper presents a project design model for evaluating and predicting project design parameters when the reservoir response is not available. By using reservoir characteristics, project duration, and targeted cumulative oil volume, the user can predict project design parameters. Five ANN models were developed, with Model #1 predicting reservoir response with an average overall error of 5.9%, Model #2 predicting reservoir characteristics with an overall prediction error of 18%, and Model #3 predicting design parameters with an overall prediction error of 28%. Models #4 and #5 predict project design parameters for economic feasibility studies prior to field execution. Model #4 is tailored for situations where the cumulative oil volume at the end of the project and the project duration are targeted as inputs, while Model #5 is used in situations where the cumulative oil volume at the end of the project is targeted, along with the project duration and design parameters.

6.4.4.19.2 Description of the Workflow and Machine Learning Model

This paper presents a deep RL framework for optimal carbon storage reservoir management. The framework employs a deep Q-learning network (DQN) to optimize the multiperiod planning of CO_2 injection strategies in carbon sequestration reservoirs. The DQN agent is trained to maximize rewards by learning from high-dimensional inputs and leveraging past experiences. Additionally, deep multitask learning (DeepMTL) is used to approximate high-dimensional, multistate transition functions, enabling efficient decision-making under complex, dynamic conditions. The framework is applied to two management strategies: monitoring-only and active brine extraction, designed to mitigate risks related to pressure buildup in the reservoir.

6.4.4.19.3 Advantages and Limitations of the Synchronized Algorithm

The deep RL framework provides a powerful tool for managing carbon storage reservoirs, allowing for efficient exploration of high-dimensional decision spaces. The DQN agent is capable of learning optimal policies while reducing the need for extensive manual intervention. However, the performance of the model depends on the quality of the training data and the complexity of the underlying system. Additionally, deep RL can be computationally expensive, particularly when dealing with large-scale, real-time applications. The effectiveness of the DeepMTL approach in approximating state transitions also requires careful tuning to ensure accuracy.

6.4.4.19.4 Results and Discussion

In conclusion, the developed ANN models showed a significant reduction in computational time and a useful tool for predicting project design parameters for economic feasibility studies. However, the predicted design parameters need to be validated through the back-check loop, and modeling cEOR is a nontrivial process that needs to be tailored to the project objective. The proposed back-check methodology for inverse models is crucial to validate and overcome the misleading interpretation of the predicted non-unique solutions of the inverse models, and conducting careful and meaningful data processing is imperative when feeding numerical results to train the ANN models.

BIBLIOGRAPHY

Abdullah, Mohammad, Emami-Meybodi, Hamid, and Ertekin, Turgay. "Development and Application of an Artificial Neural Network Tool for Chemical EOR Field Implementations." *Paper Presented at the SPE Europec featured at 81st EAGE Conference and Exhibition*, London, England, UK, June 2019. https://doi.org/10.2118/195492-MS

6.4.4.20 Application-48: Surrogate-Based Analysis of Chemical Enhanced Oil Recovery—A Comparative Analysis of ML Model Performance

6.4.4.20.1 Background of the Application

This study explores the use of data-driven ML models to predict residual oil saturation (RF) for surfactant-polymer enhanced oil recovery (SP EOR). Eleven different ML models are developed and compared using seven input parameters, including surfactant concentration, slug size, polymer concentration in surfactant slug, polymer concentration in polymer drive, polymer drive size, salinity of the polymer drive, and the Kv/Kh ratio. The performance of the models is evaluated using statistical methods and graphical methods, including parity plots and bar charts. The study concludes that ensemble methods, such as SVR, extremely randomized trees (ERT), gradient boosting regression (GBR), XGBoost regression, and ANNs, outperform all other models in terms of prediction accuracy and generalization ability. The study also shows that feature ranking using CART techniques can provide insight into the most influential variables affecting RF for SP EOR.

6.4.4.20.2 *Description of the Workflow and Machine Learning Model*

This paper introduces a novel approach for predicting the dynamics of multiphase fluid flow in oil reservoirs using deep NNs. The study presents a workflow called "Metamodel" that leverages a combination of variational autoencoder (VAE) and RNN models to approximate the complex 3D reservoir simulation results. The metamodel projects the dynamics of the system into a latent variable space, where the RNN predicts the flow dynamics. This approach significantly reduces computational time compared to traditional finite-difference hydrodynamical simulations (FDHS) while maintaining accuracy in predicting fluid flow rates, pressure distribution, and fluid saturations within the reservoir.

6.4.4.20.3 *Advantages and Limitations of the Synchronized Algorithm*

The metamodel provides a faster and computationally efficient alternative to conventional reservoir modeling techniques, making it ideal for scenario screening and optimization in oilfield development. The use of VAE for dimensionality reduction and RNN for latent dynamics forecasting ensures that the model captures complex relationships in the reservoir dynamics. However, the approach assumes certain symmetries in the reservoir geometry, which may limit its application to more heterogeneous reservoirs. Additionally, the model's performance depends on the quality of the training data and the accuracy of the underlying FDHS solutions used for training.

6.4.4.20.4 *Results and Discussion*

In conclusion, this study demonstrates that data-driven ML models can be used to accurately predict the residual oil saturation for surfactant–polymer enhanced oil recovery. The study shows that ensemble methods such as SVR, ERT, GBR, XGBoost regression, and ANN outperform other models in terms of prediction accuracy and generalization ability. Feature ranking using CART techniques provides insight into the most influential variables affecting RF for SP EOR. The study also highlights the importance of model interpretability and feature ranking for the SP EOR process, achieved with SHAP plots and CART, as opposed to statistical design of experiments with ANOVA. Overall, the study shows that data-driven models can provide fast and accurate approximations of full-field simulation models, making them suitable for real-time decision-making and optimization in reservoir management workflows.

BIBLIOGRAPHY

Idogun, Akpevwe Kelvin, Ujah, Ruth Oyanu, and James, Lesley Anne. "Surrogate-Based Analysis of Chemical Enhanced Oil Recovery – A Comparative Analysis of Machine Learning Model Performance." *Paper Presented at the SPE Nigeria Annual International Conference and Exhibition*, Lagos, Nigeria, August 2021. https://doi.org/10.2118/208452-MS

6.4.4.21 Application-49: Comparison of ML Algorithms for the Development of a Forecasting Tool for Cyclic Gas Injection in Hydraulically-Fractured Wells

6.4.4.21.1 *Background of the Application*

This paper explores the use of ML algorithms to predict efficiency values for cyclic gas injection in EOR applications. The study uses MLR, RFs, and NNs to develop accurate forecasting tools for cyclic gas injection. The models are based on a large set of numerical simulation runs and input parameters such as gas injection concentrations, cycle rate limit, duration of injection, and injection rate. The results show that RFs and NNs have comparable accuracy, while MLR resulted in the lowest accuracy. RFs have several advantages such as training speed and less number of tuning parameters involved.

6.4.4.21.2 *Description of the Workflow and Machine Learning Model*

This paper introduces a hybrid workflow for optimizing warm solvent injection (WSI) processes in heterogeneous heavy oil reservoirs. The workflow employs an MOO framework combined with proxy models to optimize operational parameters. The non-dominated sorting genetic algorithm II (NSGA-II) is used for the MOO, focusing on maximizing oil RF and minimizing the cumulative solvent-to-oil ratio (cSOR). The proxy models are built using response surface methodology (RSM) and ANNs to reduce computational costs while maintaining the accuracy of the optimization process.

6.4.4.21.3 *Advantages and Limitations of the Synchronized Algorithm*

The hybrid workflow provides a systematic approach for optimizing WSI processes in complex reservoir settings, significantly reducing the need for expensive simulations. The integration of NSGA-II allows for efficient handling of multiple conflicting objectives, while the proxy models ensure that the optimization process remains computationally feasible. However, the performance of the workflow depends on the quality of the initial training data and the accuracy of the proxy models. Additionally, the complexity of the workflow may require careful tuning of the optimization algorithm to avoid local optima and ensure robust solutions.

6.4.4.21.4 *Results and Discussion*

The study concludes that ML algorithms can be used to develop accurate forecasting tools for cyclic gas injection in EOR applications. RFs and NNs have comparable accuracy, while MLR resulted in the lowest accuracy. RFs have several advantages such as training speed and less number of tuning parameters involved. The accuracy of the models can be visualized by plotting the predicted vs real values for efficiency at different intervals. However, the decrease in accuracy with longer period of efficiency calculations can be attributed to the error propagation in the numerical simulations. The study suggests that the use of ML algorithms can improve the efficiency and accuracy of EOR applications.

BIBLIOGRAPHY

Seilkhanov, Anuar; Otynshin, Daniyar, Novopokrovskiy, Georgiy, and Artun, Emre. "Comparison of Machine Learning Algorithms for the Development of a Forecasting Tool for Cyclic Gas Injection in Hydraulically-Fractured Wells." *Paper Presented at the SPE Annual Caspian Technical Conference*, Virtual, October 2020. https://doi.org/10.2118/202507-MS

6.4.4.22 Application-50: Predicting Heavy Oil Combustion Kinetics with Machine Learning

6.4.4.22.1 Background of the Application

This paper presents a new approach for predicting heavy oil combustion kinetics using ML techniques. The proposed method, called NAMLA, allows for the quantification of uncertainty in the estimated conversion rate and has comparable performance to the established interpolation-based method. Through experiments with the NAMLA model, its ability to accurately synthesize O_2 consumption data directly from experimental inputs is demonstrated and applied to kinetics parameter optimization workflows. The NAMLA model also offers improved coverage of the conversion/temperature surface, reducing uncertainty and improving the accuracy of estimated conversion curves. Future work could include adaptive adjustment of the weight kernel width parameter based on data density or combining the interpolation and ML methods to optimize experimental design and calibration of reaction models.

6.4.4.22.2 Description of the Workflow and Machine Learning Model

This paper presents the application of predictive data analytics for EOR in a mature carbonate field located in the Middle East. The study utilizes a TDM approach, incorporating ML techniques such as ANNs to develop interconnected data-driven models. These models are built using more than 8 years of development, production, and injection data, as well as well test and log data from over 37 wells. The TDM approach involves four main steps: data gathering and preparation, model building, model training and validation, and model prediction. The models were validated using blind testing, demonstrating high accuracy in predicting reservoir performance under different operational scenarios.

6.4.4.22.3 Advantages and Limitations of the Synchronized Algorithm

The TDM approach offers several advantages, including reduced computation time and resource consumption compared to traditional numerical simulation methods. The use of interconnected ANN models enables the TDM to handle complex reservoir dynamics, making it a valuable tool for reservoir management and decision-making. However, the accuracy of the models depends on the quality of the input data, and the approach may struggle with highly heterogeneous reservoirs. Additionally, the reliance on historical data for training means that the models may need continuous updates to remain relevant as new data becomes available.

6.4.4.22.4 Results and Discussion

In conclusion, the NAMLA model offers a promising method for predicting heavy oil combustion kinetics and quantifying uncertainty in experimental data. While the interpolation-based method may have better performance in some cases, the NAMLA model's ability to quantify uncertainty makes it a valuable tool for experimental design and analysis. The extrapolation capabilities of the NAMLA model also make it suitable for predicting conversion rates outside the range of experimentally observed heating rates. Future work could focus on improving the speed of the NAMLA model and exploring its potential for optimizing experimental design and calibration of reaction models.

BIBLIOGRAPHY

Anderson, Timothy I., Li, Yunan, and Kovscek, Anthony R. "Predicting Heavy Oil Combustion Kinetics with Machine Learning." *Paper Presented at the SPE Western Regional Meeting*, Virtual, April 2021. https://doi.org/10.2118/200853-MS

7 AI Applications in Facilities, Pipelines, and Metering

Ekrem Alagoz, Ferit Ozdaban, Serkan Gozel, Mouad Al Krmagi, Emre Can Dündar, Mehmet Melih Oskay, Celal Hakan Canbaz, Luigi Saputelli, Mufrettin Murat Sari, Salih Tutun, and Cenk Temizel

7.1 INTRODUCTION

Artificial intelligence (AI) and machine learning (ML) have emerged as powerful tools with the potential to revolutionize various aspects of the oil and gas industry. In the realm of facilities, pipelines, and metering, AI/ML can be applied to enhance operational efficiency, improve safety, and optimize asset management. This chapter delves into the integration of AI/ML methodologies within these critical areas of the oil and gas sector. It provides a comprehensive overview of the different workflows that can be enhanced through AI/ML applications, including facilities management, pipeline integrity, reservoir monitoring, subsea operations, and scaling/corrosion prevention. By understanding the potential benefits of AI/ML in facilities, pipelines, and metering, industry professionals can harness these technologies to drive innovation, optimize processes, and improve overall performance.

7.2 CATEGORIZATION OF WORKFLOWS USED IN FACILITIES, PIPELINES, AND METERING

7.2.1 Facilities

AI/ML can be applied to facilities management to improve efficiency, reduce costs, and enhance safety. One of the most promising applications is predictive maintenance, which involves using AI/ML algorithms to analyze sensor data and predict equipment failures before they occur. By identifying potential failures in advance, maintenance teams can schedule repairs proactively, reducing downtime and preventing costly emergency repairs.

Another valuable application of AI/ML in facilities management is anomaly detection. By analyzing sensor data for abnormal patterns, AI/ML algorithms can

DOI: 10.1201/9781003617327-7

detect potential equipment malfunctions or security breaches. This enables early intervention and prevents costly failures or security risks. Additionally, AI/ML can be used to optimize energy consumption, inventory management, and asset management, further improving the overall efficiency and cost-effectiveness of facilities operations.

7.2.2 Pipeline

AI/ML can be applied to pipeline management to improve safety, efficiency, and reduce costs. One of the most critical applications is leak detection, which involves using AI/ML algorithms to analyze sensor data and identify potential leaks in pipelines. By detecting leaks early, operators can prevent environmental damage, minimize product loss, and avoid costly repairs.

Additionally, AI/ML can be used for corrosion monitoring and integrity assessment of pipelines. By analyzing inspection data and sensor readings, AI/ML algorithms can identify areas of corrosion and assess the overall integrity of the pipeline. This information can be used to prioritize maintenance activities and prevent pipeline failures. Furthermore, AI/ML can optimize pipeline operations by analyzing historical data and real-time conditions to determine the optimal flow rates and pressures, reducing energy consumption and improving efficiency.

7.2.3 Reservoir-Related Applications

AI/ML can be applied to reservoir management to improve understanding of reservoir behavior, optimize production, and enhance overall efficiency. One of the key applications is reservoir simulation, which involves using AI/ML algorithms to analyze reservoir data and create more accurate models of reservoir behavior. These models can be used to predict future production, optimize well placement, and evaluate different EOR strategies.

Another valuable application of AI/ML in reservoir management is production optimization. By analyzing reservoir data and production history, AI/ML algorithms can identify opportunities to improve production rates and reduce costs. This can involve optimizing well spacing, adjusting production rates, or implementing advanced EOR techniques. Furthermore, AI/ML can be used to enhance reservoir characterization by analyzing seismic data, well logs, and production data to create more detailed and accurate reservoir models.

7.2.4 Subsea Applications

AI/ML can be applied to subsea operations to improve efficiency, safety, and reduce costs. One of the key applications is subsea equipment monitoring, which involves using AI/ML algorithms to analyze sensor data and monitor the health and performance of subsea equipment, such as manifolds and control systems. By identifying potential issues early, operators can schedule maintenance proactively and prevent costly failures.

Another valuable application of AI/ML in subsea operations is subsea inspection, which involves using AI/ML algorithms to analyze inspection data and identify defects or structural damage in subsea structures. This can help ensure the safety and integrity of subsea infrastructure and prevent costly repairs or replacements. Furthermore, AI/ML can be used to optimize the operation of autonomous underwater vehicles (AUVs) by enabling them to navigate autonomously and perform tasks such as inspection and data collection.

7.2.5 Scaling/Corrosion

AI/ML can be applied to address scaling and corrosion issues in oil and gas facilities, pipelines, and equipment. One of the key applications is scaling prediction, which involves using AI/ML algorithms to analyze water chemistry data, operating conditions, and other relevant factors to predict the likelihood of scaling formation. By predicting scaling formation, operators can implement preventive measures to mitigate its impact and avoid costly downtime.

Another valuable application of AI/ML is corrosion monitoring and prevention. By analyzing corrosion data, sensor readings, and operating conditions, AI/ML algorithms can identify areas at risk of corrosion and optimize corrosion prevention strategies. This can involve adjusting water chemistry, implementing corrosion inhibitors, or modifying operating conditions to minimize corrosion rates and extend the lifespan of assets.

7.3 OVERVIEW OF AI METHODOLOGIES

AI, ML, and DL are rapidly evolving technologies with the potential to revolutionize various industries, including oil and gas. AI refers to the development of intelligent systems capable of reasoning, learning, and problem-solving. ML is a subset of AI that focuses on training algorithms on data to make predictions or decisions. DL, a further specialization of ML, involves training artificial neural networks (ANNs) with multiple layers to learn complex patterns from data.

AI encompasses a wide range of techniques, such as supervised and unsupervised learning, reinforcement learning (RL), convolutional neural networks (CNNs), recurrent neural networks (RNNs), and generative adversarial networks (GANs). These methodologies offer powerful tools for analyzing large datasets, identifying patterns and trends, and making data-driven decisions.

7.4 APPLICATIONS OF AI

7.4.1 Facilities

7.4.1.1 Application-1: Research on Calculation Model of Bottom of the Well Pressure Based on Machine Learning

7.4.1.1.1 Background of the Application

In this research, a series of ML-based techniques for monitoring the bottom of the well pressure of managed pressure drilling (MPD) are proposed. Through the examination of the hydraulic model, the viability of integrating an AI algorithm into the bottom of the well pressure calculation and monitoring method is investigated. The supported vector regression machine (SA-SVR), which is used to optimize the monitoring method of the bottom of the well pressure of MPD, is improved by combining the simulated annealing (SA) algorithm with the support vector regression (SVR) machine. This model is used to calculate and analyze the bottom of the well pressure, providing theoretical support for bottom of the well pressure monitoring of drilling operations, and guiding safe drilling at the construction site.

7.4.1.1.2 Description of the Workflow and Machine Learning Model

The authors used the SA technique to improve its hyperparameters and the SVR algorithm to perform regression analysis on the bottom well pressure. In oil and gas exploration and development, drilling in complex formations with a narrow density window has become a technical challenge. However, PWD instruments will distort or even make mistakes in the measurement due to the high temperature and high pressure at the well bottom. Therefore, in order to develop a precise model to forecast the bottom well pressure using 1) hydrostatic column pressure, ML techniques are required, 2) a decrease of annulus pressure, and 3) back pressure on the surface.

SVR is one of the supervised ML algorithms based on the statistical learning theory that can address this challenge when combined with SA. The SVR algorithm is suitable for regression analysis. With a radial basis function (RBF) kernel function, SVR is able to map the input dataset into a high-dimensional space for nonlinear regression. The algorithm aims to match different samples during model training by determining an optimal hyperplane that has the minimum deviation of separation between the samples.

The Monte Carlo iterative solution strategy serves as the foundation for the SA algorithm, a random optimization technique. It provides an efficient approximate solution approach for problems with non-deterministic polynomial complexity by overcoming optimization processes that tend to fall into local minimum faults and the dependence on beginning values. The combination of the SA algorithm and SVR regression is good at seeking the global optimal solution.

7.4.1.1.3 Advantages and Limitations of the Synchronized Algorithm

SVR is a learning machine algorithm specially designed for the case of limited samples. It can convert the actual problem into a high-dimensional feature space by nonlinear transformation, which solves the dimensionality problem and ensures a good generalization ability. Moreover, the complexity of the SVR algorithm is independent of the sample dimension. The limitation of SVR is that it is sensitive to the choice of parameters and kernel functions, so the SA algorithm is needed to do the optimization. The SA algorithm has the probability to jump out of the trap of local optimum and find the global optimal solution, so it can easily optimize the parameters of SVR. However, the SA algorithm also has disadvantages, such as slow convergence, long execution time, parameter sensitivity, as well as performance highly dependent on initial values.

7.4.1.1.4 Results and Discussion

This study has effectively developed an ML-based bottom of the well pressure monitoring model that can reliably track the well pressure without the usage of PWD instruments and forecast the well pressure following PWD equipment failure. This model offers a set of useful theoretical underpinnings for the MPD monitoring method's bottom of the well pressure, which has improved the safety of PWD drilling instruments and compensated for the shortcomings of the bottom of the well pressure monitoring technology. The safe density window issue is resolved, and drilling activities are appropriately guided by precise measurement of the well's bottom pressure characteristics, preventing complicated mishaps.

BIBLIOGRAPHY

Liang, Haibo, Liu, Gang, Zou, Jialing, Bai, Jing, and Jiang, Yingjun. Research on calculation model of bottom of the well pressure based on machine learning. *Future Generation Computer Systems* 124, 2021, 80–90, ISSN 0167-739X, https://doi.org/10.1016/j.future.2021.05.011

7.4.1.2 Application-2: Prediction of Pressure in Different Two-Phase Flow Conditions: Machine Learning Applications

7.4.1.2.1 Background of the Application

This article focuses on the accurate prediction of pressure in two-phase flow systems, specifically in the petroleum industry. Due to the complex nature of two-phase gas–liquid flow, computational determination of bottom-hole pressure is difficult and time-consuming. The article proposes two novel ML approaches, gradient tree boosting (GTB) and extreme learning machine (ELM), to address this challenge. The comparison of the proposed algorithms to actual pressure values demonstrates that ML approaches have excellent performance in oil field calculations. Additionally, sensitivity analysis is performed on input variables to determine the most effective parameter in determining bottom-hole pressure.

7.4.1.2.2 Description of the Workflow and Machine Learning Model

The authors employed GTB and ELM to perform the prediction of bottom-hole pressure using the well-head pressure data. To ensure the quality, they collected the data of 458 real bottom-hole pressure values, as well as eight inputs: wellhead pressure, oil flow rate, water flow rate, gas flow rate, average deviation, average angle, measured depth, and true vertical depth. They developed two novel, highly accurate models that utilize ML methods (GTB and ELM) to predict bottom-hole pressure using easily-to-get data from the wellhead.

Boosting trees are boosting methods that use an additive model (i.e., a linear combination of basis functions) with a forward stepwise algorithm. It uses a decision tree as the basis function. Since the linear combination of trees can fit the training data well, even if the relationship between inputs and outputs is complex, the gradient boosting tree is a highly functional learning algorithm.

ELM is a class of ML systems or methods built on a feedforward neural network (FNN), which is applicable to supervised and unsupervised learning problems. ELM is considered as a special class of FNN, which is characterized by the fact that the weights of the nodes in the hidden layer are randomly or artificially given and do not need to be updated, and the learning process only calculates the output weights. In terms of learning rate and generalization ability, the ELM algorithm is considered to have advantages when compared to other shallow learning systems, such as single layer perceptron or support vector machine (SVM).

7.4.1.2.3 Advantages and Limitations of the Synchronized Algorithm

In general, complex and accurate calculations are required to obtain bottom-hole pressure from easily collected data on the wellhead, so replacing manual calculation with the ML method becomes a challenge. Fortunately, both GTB and ELM exhibit extremely high accuracy (R-square values of 1 and 0.999) with a very small error rate (mean relative errors lower than 4%). However, these two methods are not 100% perfect. For example, GBT cannot support parallel computing due to dependencies between weak learning machines, and single ELM's learning performance is unstable. Both methods require the initial parameters (connection weight bias value, number of nodes, etc.) to be tuned carefully to get the best results.

7.4.1.2.4 Results and Discussion

This article focuses on the accurate prediction of pressure in two-phase flow systems, specifically in the petroleum industry. This study has successfully created an ML-based bottom of the well pressure monitoring model that can forecast the well pressure after PWD equipment failure and track the well pressure reliably without the need for PWD sensors. In order to increase the safety of PWD drilling tools and make up for the shortcomings of the bottom of the well pressure monitoring technology, this model provides a set of helpful theoretical foundations for the bottom of the well pressure monitoring method of MPD. By precisely measuring the well's bottom pressure characteristics, the safe density window problem is fixed, and drilling operations are suitably directed, averting complex accidents.

BIBLIOGRAPHY

Khamehchi, Ehsan, and Bemani, Amin. Prediction of pressure in different two-phase flow conditions: Machine learning applications. *Measurement* 173, 2021, 108665, ISSN 0263-2241. https://doi.org/10.1016/j.measurement.2020.108665

7.4.1.3 Application-3: Machine Learning Approach to Improve Calculated Bottom-Hole Pressure

7.4.1.3.1 Background of the Application

This paper discusses a machine-learning approach to improve calculated bottom-hole flowing pressure (BHFP) in oil and gas production wells. Due to the high installation cost, many wells lack BHFP gauges, and operators must rely on empirical correlations to estimate BHFP based on surface pressure readings. However, these estimates often over- or underestimate the actual BHFP by significant margins. In this study, a machine-learning model that corrects the pressure inferred by empirical correlations is trained using data from 11 producing horizontal wells in the Uinta basin that have downhole real-time pressure sensors. An input array comprising surface production rates, production ratios, pressure readings, and, if relevant, the gas-lift-injection rate is used to seed the model. The remaining 10 wells are used to train the model, and each well is used to test the predictions. The random forest (RF) model performed the best, reducing the root mean squared error (RMSE) in the pressure gradient by an average of 29.7% across all test wells. The findings indicate that ML techniques have a bright future in helping reservoir engineers and boosting the accuracy of BHFP forecasts.

7.4.1.3.2 Description of the Workflow and Machine Learning Model

To increase the accuracy of the BHFP estimated by the empirical or mechanistic approaches on the feasible level, the authors used three ML techniques. Data-driven approaches for the multiphase wellbore flow have drawn more attention in recent years. Many oil and gas production wells lack BHFP gauges due to their comparatively high installation costs. Often by hundreds of psi, the predicted BHFP constantly overestimates or underestimates the actual flowing pressure at the sandface.

In order to translate the flowing pressure measurements at the surface into BHFP estimations, operators usually depend on empirical correlations from the literature. Surface production rates (oil, water, and gas), production ratios (gas/oil, water/oil, and gas/liquid), pressure data (surface flowing pressure and the empirical BHFP estimate), and gas-lift-injection rate (if applicable) are all included in the input array. While being trained on the remaining ten wells, the ML model's predictions are tested on one well at a time. Using RF, SVR, and linear regression, a cross-validation method was created to test various ML architecture configurations.

One independent variable and one dependent variable make up a linear function. X and Y are the independent and dependent variables, respectively. In ML, support vector machines, or SVMs, are among the most well-liked and often applied algorithms for handling classification problems. Finding a hyperplane in an n-dimensional space that clearly classifies the data points is the aim of an SVM algorithm. Support vectors are the data points that are closest to the hyperplane on each side of it. These affect the hyperplane's orientation and position, which aids in the construction of the SVM. Finding a hyperplane in an n-dimensional space that clearly classifies the data points is the aim of an SVM algorithm. The data points on either side of the hyperplane that are closest to the hyperplane are called support vectors. RF is a commonly used ML algorithm, which combines the output of multiple decision trees to reach a single result. Its ease of use and flexibility have fueled its adoption, as it handles both classification and regression problems.

7.4.1.3.3 Advantages and Limitations of the Synchronized Algorithm

For the bulk of the test datapoints, the SVR-ensemble and RF approaches generated extremely accurate uncertainty models and significantly improved the projected flowing pressure. As additional data points are used in the training process, the approaches' accuracy can increase. To reduce the errors caused by inaccurate data, better outlier identification and correction methods are preferred. One might expand the method's use in the presence of larger datasets by adding additional features like fluid properties, flow path descriptions, and pipe diameters.

7.4.1.3.4 Results and Discussion

This study demonstrates the potential of ML methods to improve calculated bottom-hole pressure estimates in oil and gas production wells. By training a machine-learning model on data from wells with downhole real-time pressure readings, the model can correct the pressure estimated by empirical correlations, reducing the RMSE in the pressure gradient by 29.7% on average over all the test wells. The best performance was achieved by the RF model. This study's results have profound implications for reservoir-engineering studies that rely on BHFP information. While the ML models demonstrated excellent performance, caution should be exercised as they have the potential to exhibit poor generality outside of the training data range.

BIBLIOGRAPHY

Eltahan, Esmail, Ganjdanesh, Reza, Yu, Wei, Sepehrnoori, Kamy, Williams, Ryan, and Nohavitsa, Jack. "Machine Learning Approach to Improve Calculated Bottom-hole Pressure." Paper presented at the *SPE/AAPG/SEG Unconventional Resources Technology Conference*, Houston, Texas, USA, July 2021. doi: https://doi.org/10.15530/urtec-2021-5645

7.4.1.4 Application-4: Increasing Production Efficiency via Compressor Failure Predictive Analytics Using Machine Learning

7.4.1.4.1 Background of the Application

The paper presents an advanced ML approach for predicting compressor failure and its root cause with sufficient warning to allow for intervention. To categorize aberrant operation and identify the precise factors causing instability, the method relies on rate of change-based outlier identification and one-class SVMs. R is used to train and test the algorithms, and the Alteryx platform is used to plan a near-real-time online implementation that generates new forecasts every 10 minutes. After the results are displayed on a Spotfire dashboard, the model automatically sends the end user an email alerting them to any unusual activity. Production delays brought on by down-time from rotating equipment breakdowns could be greatly reduced with the use of predictive maintenance.

7.4.1.4.2 Description of the Workflow and Machine Learning Model

In this method, one-class SVM is used for anomaly detection, leveraging extensive statistical feature engineering. The one-class SVM model, trained on selected features, classifies data points as normal or anomalous based on the learned data boundary. The model outputs an anomaly score for each data point, allowing for the identification of outliers. By averaging the contributions of each feature to the anomalies, the method ranks the features according to their influence on the output, helping to pinpoint the most important contributors to the detected anomalies.

7.4.1.4.3 Advantages and Limitations of the Synchronized Algorithm

Combining one-class SVM with statistical feature engineering provides a good tool for detecting anomalies without needing labeled data. The diverse features improve the model's ability to differentiate normal from anomalous patterns. However, the method is computationally expensive and relies heavily on the quality of feature engineering. Poor feature selection can lower performance, and the assumption that most data points are normal may not always hold true, making the model's accuracy a lot dependent on them.

7.4.1.4.4 Results and Discussion

The model's median detection time of 7 hours indicates an improvement in identification efficiency. Investigation and corrective action are feasible with an upper detection time of a few days. The industry's current periodic compressor downtime might be reduced, and maintenance techniques could be revolutionized by applying ML to compressor failure. The combination of big data-driven approach with systems-level modeling is quite promising, and it is being explored in this work.

BIBLIOGRAPHY

Pandya, D., Srivastava, A., Doherty, A., Sundareshwar, S., Needham, C., Chaudry, A., and KrishnaIyer, S. "Increasing Production Efficiency via Compressor Failure Predictive Analytics Using Machine Learning." Paper presented at the *Offshore Technology Conference*, Houston, Texas, USA, April 2018. doi: https://doi.org/10.4043/28990-MS

7.4.1.5 Application-5: Enhancing Well, Reservoir and Facilities Management WRFM Opportunity Identification with Data Driven Techniques

7.4.1.5.1 Background of the Application

The paper presents a suite of workflows that are collectively referred to as WRFM Opportunity Finder (WOF), which generates a ranked list of opportunities across the well, reservoir, and facilities management (WRFM) opportunity spectrum. The WOF was developed for a mature waterflooded asset with over 500 active wells and over 30 years of production history. It uses data-driven techniques to identify and rank opportunities, offering transparent logic that is scalable and replicable across a wide range of subsurface settings. The WOF algorithms are largely comprised of rule-based workflows with occasional use of ML in intermediate steps. The WOF accelerates identification of low-capex/no-capex opportunities using existing data, promoting maximization of returns on already made investments and hence lending resilience to business in the low oil price environment.

7.4.1.5.2 Description of the Workflow and Machine Learning Model

In this method, a *RandomForestClassifier* is used to predict potential candidates for various well operations. It involves preprocessing steps like feature creation and splitting the dataset into training and testing sets. Models are trained and evaluated for different well types, with performance assessed using accuracy scores and classification reports. The final output includes identifying and plotting candidates for operations such as bean-up, stimulation, water shut-off, Add-Perf, and surveillance.

7.4.1.5.3 Advantages and Limitations of the Synchronized Algorithm

This method effectively uses *RandomForestClassifier* for accurate prediction across different well operations. It benefits from tailored feature engineering, ensuring accurate model training and easy result interpretation through visualizations. However, the usage is limited by its dependency on a single classifier, and the success of the models is highly dependent on the quality of the entered features and labels, which require careful design and expertise in the area.

7.4.1.5.4 Results and Discussion

The WOF has been successfully deployed in a large mature asset with over 500 active wells and over 30 years of production history. Oil output could increase by 1–2% as a result of the 90% reduction in data preparation and integration time and the ability to prioritize the best prospects. Using a number of independent tests, the integrated visualization dashboard enables rapid and transparent validation of the opportunities that have been found and their rating basis. A variety of business delivery components, including the workover and data collection plan, exception-based surveillance, and the facilities debottlenecking plan, will be influenced by the findings of WOF. The WOF can provide a visible, scalable, and replicable method of opportunity identification across the WRFM opportunity spectrum and can be tailored for various subsurface situations with very slight adjustments.

BIBLIOGRAPHY

Ujjwal, Manu, Modi, Gaurav, and Simha, Srungeer. "Enhancing Well, Reservoir and Facilities Management WRFM Opportunity Identification with Data Driven Techniques." Paper presented at the *SPE/IATMI Asia Pacific Oil & Gas Conference and Exhibition*, Virtual, October 2021. doi: https://doi.org/10.2118/205596-MS

7.4.1.6 Application-6: Automated ESP Failure Root Cause Identification and Analyses Using ML and NLP Technologies

7.4.1.6.1 Background of the Application

This paper presents a novel approach to automate the root-cause identification and analysis of electrical submersible pump (ESP) failures. The proposed approach uses an ensemble of ML algorithms to process structured and unstructured data from equipment, production, operations, and failure reports. The ML algorithms used in the workflow include linear and nonlinear algorithms such as XGBoost, AdaBoost, RF, and multinomial logistic classifier, along with data-processing techniques like stratified shuffle split and SMOTE. The objective is to predict the failure component of the ESP and determine the root-cause sequence of a failure, thereby reducing turnaround time for analysis, improving ESP design, and determining equipment that is fit for purpose for an enhanced design of ESP to improve equipment reliability. The workflow outputs the ESP failure component on unseen test data by choosing the best model from the collection, which gives the highest prediction accuracy. The workflow can predict multiple types/classes of failure components, ranging from two to eight classes, and the algorithm has been tested and is capable of performing classification if the failures experienced by the ESP are of only two types or more, up to eight different types of failure classes.

7.4.1.6.2 Description of the Workflow and Machine Learning Model

A couple of ML algorithms are used to predict failure components in a dataset containing both numerical and categorical features. The preprocessing steps include imputation, scaling, and one-hot encoding to handle missing values and categorical data. Stratified shuffle split is employed to maintain class distribution across training and test sets, while SMOTE is used to address class imbalance. Various classifiers such as RandomForest, AdaBoost, XGBoost, and logistic regression are evaluated to determine the most accurate model.

7.4.1.6.3 Advantages and Limitations of the Synchronized Algorithm

This method effectively uses both ensemble learning and comprehensive data preprocessing to improve prediction accuracy. The use of SMOTE ensures that class imbalance is mitigated, resulting in more balanced models. Evaluating multiple classifiers allows for selecting the best-performing model suited to the dataset. However, the process is computationally expensive, especially with SMOTE. Additionally, model performance is highly dependent on the quality of preprocessing and the choice of hyperparameters, therefore making a need for careful tuning and validation.

7.4.1.6.4 Results and Discussion

In this paper, a novel AI workflow is presented that utilizes an ensemble of ML algorithms to automate root-cause identification and analysis of ESP failures. The workflow processes structured and unstructured data from various stages of the ESP life cycle, including equipment, production, operations, and failure reports. The ML algorithms used in the workflow include linear and nonlinear algorithms such as XGBoost, AdaBoost, RF, and multinomial logistic classifier, along with data-processing techniques like stratified shuffle split and SMOTE. The workflow outputs the ESP failure component on unseen test data by choosing the best model from the collection, which gives the highest prediction accuracy. The objective is to predict the failure component of the ESP and determine the root-cause sequence of a failure, thereby reducing turnaround time for analysis, improving ESP design, and determining equipment that is fit for purpose for an enhanced design of ESP to improve equipment reliability. Through this approach, the possibility of automating the DIFA process by leveraging information from the past and aiming to mitigate failure events in the future was able to be justified. The result achieved by this ML workflow can help improve the existing DIFA process to be more effective and efficient. A more efficient DIFA process will give both the ESP provider and well operator an opportunity to focus on operational improvement that can allow a fit-for-purpose ESP to optimally run and the field in general to produce in a more sustainable manner.

BIBLIOGRAPHY

Karnik, Saniya, Yenuganti, Navya, Jusri, Bonang Firmansyah, Gupta, Supriya, Nirgudkar, Prasanna, Mohajer, Mahyar, and Malik, Asim. "Automated ESP Failure Root Cause

Identification and Analyses Using Machine Learning and Natural Language Processing Technologies." Paper presented at the *SPE Gulf Coast Section Electric Submersible Pumps Symposium*, Virtual and The Woodlands, Texas, USA, October 2021. doi: https://doi.org/10.2118/204519-MS

7.4.1.7 Application-7: AI Applied in Sucker Rod Pumping Wells: Intelligent Dynamometer Card Generation, Diagnosis, and Failure Detection

7.4.1.7.1 Background of the Application

This paper proposes a novel AI method for monitoring and diagnosing sucker-rod pumping wells using DL algorithms. Traditional methods for measuring the dynamometer card using load and displacement sensors are costly and difficult to maintain. In order to solve this problem, the authors use cutting-edge DL techniques to create a general model that creates the dynamometer card from electrical information. Additionally, using hundreds of dynamometer cards labeled with various classifications, they create a DL model for automated diagnosis. Hundreds of sucker rod-pumped wells in various PetroChina oilfields were used to test the recently created AI models. The results of the field tests revealed that the models and the actual dynamometer were more than 90% identical. These artificially intelligent models will play a significant role in "Intelligent Fields" and can be integrated into IIoT edge computing devices for automated control and diagnosis.

7.4.1.7.2 Description of the Workflow and Machine Learning Model

In this method, a series of DL models is used to analyze and interpret dynamometer and electrical data related to well operations. First, a convolutional autoencoder is trained on dynamometer images to learn a compressed representation, which is then used to reconstruct the images. The autoencoder's encoder is used together with a dense neural network (NN) to create a "card generator" that can predict dynamometer card images from electrical data inputs. And, a CNN is trained to classify well conditions based on these generated dynamometer cards, producing automated diagnostics of well conditions.

7.4.1.7.3 Advantages and Limitations of the Synchronized Algorithm

This method has advantages like the power of convolutional networks to capture dependencies in dynamometer card images, allowing for accurate reconstruction and classification. The use of an autoencoder enables us to efficiently extract features, while the card generator model integrates data sources to provide information. However, the training process is computationally expensive, requiring significant resources and time. And the model's performance is sensitive to the quality of input data and the setting of network hyperparameters, thus careful preprocessing and confirmation of the features are needed for the method to perform well.

7.4.1.7.4 Results and Discussion

The authors present an application of AI technology to generate the dynamometer card data directly from electrical power curves and to automatically diagnose the well working conditions in real time. The DL algorithms described in this paper can help reduce the investment in card acquisition sensors and automate diagnosis, improving the operational efficiency and budget. The proposed models have been tested in hundreds of wells in different oilfields in PetroChina with satisfactory results, demonstrating the potential for this technology to play an important role in the "Intelligent Fields" of the future.

BIBLIOGRAPHY

Peng, Yi "Artificial Intelligence Applied in Sucker Rod Pumping Wells: Intelligent Dynamometer Card Generation, Diagnosis, and Failure Detection Using Deep Neural Networks." Paper presented at the *SPE Annual Technical Conference and Exhibition, Calgary*, Alberta, Canada, September 2019. doi: https://doi.org/10.2118/196159-MS

7.4.1.8 Application-8: Increasing Facility Uptime Using Machine Learning and Physics-Based Hybrid Analytics in a Dynamic Digital Twin

7.4.1.8.1 Background of the Application

This paper discusses the integration of physics-based models and ML models with field sensors to enable real-time monitoring, predictive analytics, and automation of offshore production facilities. The use of hybrid analytics allows the generation of synthetic data from high-fidelity physics-based models to train ML models, which can then be used to estimate measurements in locations where real sensors are not available or are expensive to install. The paper presents case studies on the use of hybrid analytics for virtual sensors, real-time production insights, and production optimization using RL. The paper concludes that a dynamic digital twin that integrates physics-based models and ML models with field sensors can provide an end-to-end solution for increasing facility uptime, enabling autonomous operations, and overcoming challenges involved in maintaining an updated physics-based model.

7.4.1.8.2 Description of the Workflow and Machine Learning Model

The author employed a long short-term memory (LSTM) model to increase facility uptime in offshore oil and gas production. The LSTM model takes inputs as real time and historical data from field sensors such as hydrocarbon quantity, quality, pressure, and temperature. Then the trained LSTM model predicts the future states of the system, enabling condition-based maintenance (CBM) and production optimization.

The model starts with loading the training and validation data files synthesized by the dynamic digital twin, which replicates the physical offshore facility data. Then the input data is reshaped to fit the LSTM input format. Using both the synthesized and real data, the LSTM network is trained using the Adam optimizer and the use (mean squared error) loss function. Once trained, the LSTM model is used to predict the future states of the system, and the model's summary is plotted and displayed at the end.

7.4.1.8.3 Advantages and Limitations of the Synchronized Algorithm

Implementation of dynamic digital twin for LSTM and the model to increase facility uptime in offshore oil and gas production will enable predictive maintenance so that no unplanned shutdowns take place. Therefore, implementing a dynamic digital twin in oil and gas to maximize the facility's uptime through predictive maintenance is necessary. The LSTM allows the system to capture temporal dependencies in the data and increase the effectiveness of CBM. However, training and maintaining the dynamic digital twin for real-time data acquisition from the facility and processing requires powerful hardware and specialized expertise. The continuous calibration and updating of the physics-based model to synthesize real-time data for the LSTM model requires high computational requirements.

7.4.1.8.4 Results and Discussion

In conclusion, this paper presents the use of hybrid analytics, which integrates high-fidelity physics-based models with ML models, as a means of significantly improving the performance of offshore production facilities. The paper demonstrates that the use of hybrid analytics enables the development of virtual sensors, real-time production insights, and production optimization using RL. The paper also highlights that the use of a dynamic digital twin that integrates physics-based models and ML models with field sensors can provide an end-to-end solution for increasing facility uptime, enabling autonomous operations, and overcoming challenges involved in maintaining an updated physics-based model. Overall, this paper emphasizes the importance of integrating physics-based models and ML models with field sensors in offshore production facilities to achieve improved performance and safe operations.

BIBLIOGRAPHY

Srikonda, Rohit, Rastogi, Ankur, and Oestensen, Haavard. "Increasing Facility Uptime Using Machine Learning and Physics-Based Hybrid Analytics in a Dynamic Digital Twin." Paper presented at the *Offshore Technology Conference*, Houston, Texas, USA, May 2020. doi: https://doi.org/10.4043/30723-MS

7.4.1.9 Application-9: Physics Based Machine Learning for Annular Blowout Preventer Health Monitoring

7.4.1.9.1 Background of the Application

This paper presents a physics-based machine learning (PBML) approach for condition and performance monitoring of the blowout preventer (BOP) annular preventer. The BOP annular preventer plays a crucial role in well pressure control during drilling operations, but its reliability accounts for a significant portion of BOP system failures. The proposed PBML model efficiently characterizes the system and describes the annular preventer behavior. The model is created using a physics-based approach to describe the functional relationship between annular closing pressure and cylinder displacement. The model parameters are identified in real time, and the resulting coefficients are clustered to track system changes and detect functional degradation. The developed PBML model can monitor the health of the BOP annular in real time, enabling accurate assessment and forecasting of annular health.

7.4.1.9.2 Description of the Workflow and Machine Learning Model

PBML is a hybrid approach that combines principles from physics with ML techniques to create more accurate and interpretable models. By combining physical laws and constraints into ML frameworks, PBML leverages both theory-driven and data-driven insights. This principle is used to identify real-time parameters and their relationships, such as the one between annular closing pressure and cylinder displacement. Then, ML methods cluster the outputs together to predict the remaining life of BOP annular.

7.4.1.9.3 Advantages and Limitations of the Synchronized Algorithm

PBML models are interpretable and robust, especially in areas where physical principles are well-known. Areas such as fluid dynamics, material science, and climate modeling can be given as examples. PBML also tends to be more data-efficient, as the already known physical laws reduce the need for deep datasets. The integration of physical laws into ML models can be complex and computationally expensive, requiring deep expertise in both areas. This method can also limit the flexibility of models, making it less suitable for areas where physical laws are not well understood.

7.4.1.9.4 Results and Discussion

This work presents a method for condition and performance monitoring of the BOP annular preventer using a PBML approach. The developed PBML model combines physics-based methods with ML modeling methods to accurately assess and forecast annular health. The model parameters, which have physical meanings, can be mapped into BOP annular health regions based on original equipment manufacturer (OEM) BOP design parameters. The resulting coefficients are clustered to track system changes and detect functional degradation. The proposed methodology represents an innovative method in health monitoring as it efficiently characterizes the system and puts all the physical characteristics of the system into visually understandable parameters. The developed model can be used to reduce downtime related to BOPs, improving uptime in drilling and completion operations.

BIBLIOGRAPHY

Mallek, Moadh, Wassar, Taoufik, and Franchek, Matthew A. "Physics Based Machine Learning for Annular Blowout Preventer Health Monitoring." Paper presented at the Offshore Technology Conference, Houston, Texas, USA, May 2020. doi: https://doi.org/10.4043/30684-MS

7.4.1.10 Application-10: Smart Autopilot Drone System for Surface Surveillance and Anomaly Detection via Customizable Deep Neural Network

7.4.1.10.1 Background of the Application

In this paper, the authors present a new system for surface surveillance and anomaly detection using customized drones equipped with high-definition cameras, thermal imaging, and gas sensing devices. The proposed system is cloud-enabled and uses autopiloted drones that are connected to onboard computers. The authors have developed a series of advanced algorithms for processing and diagnosing the image, thermal, and gas sensing data collected by the drones in real time or near real time. These algorithms include accurate 2D geospatial aerial mapping, anomaly detection, and classification for events such as oil leaks, gas leaks, facility failures, and human activities. The authors have also customized and parallelized object detection DL models for low-profile multi-core single board computers.

7.4.1.10.2 Description of the Workflow and Machine Learning Model

The author employed a residual convolutional neural network to develop a smart autopilot drone system for surface surveillance and anomaly detection. The system leverages CNNs for anomaly detection and classification. Taking high-definition camera images and thermal images, which are the data collected by the autopilot drone during the flight as inputs, the residual convolutional neural network extracts any feature classes present in the inputs, matches them with potential anomaly classes, and then outputs the predicted anomaly class with the probability of the class.

First, the feature classes are loaded into the system, and the images that are captured by the drone's onboard cameras and sensors are reprocessed by normalizing pixel values and splitting the dataset into training, validation, and test sets to match the dimensions of the shape the model expects. After the residual convolutional neural network model is built using the convolutional layers for feature extraction and dense layers for classification, the model is compiled using the Adam optimizer and categorical cross-entropy loss function, and to prevent overfitting, the system defines callbacks for early stopping. Upon training, the model's performance is evaluated, and the best model is selected based on the performance metrics. The results are visualized by overlaying the predicted anomalies on the original images using the classify_anomaly function.

7.4.1.10.3 Advantages and Limitations of the Synchronized Algorithm

The smart autopilot drone system provides full automated mass surveillance, data acquisition, and anomaly detection, significantly reducing human need, which reduces the operational costs of facilities. Compared to the SVMs, the system outperforms them by its reported 99% ROC in multi-task binary classification of anomalies. With high-definition camera images and thermal images, the smart autopilot drone system can detect a wide range of anomalies, including oil and gas leaks, facility failures, and unauthorized human activities. However, maintaining the smart autopilot drone system requires resources such as strong internet connectivity to process real-time data transmission and cloud-based processing, and high-quality cameras and thermal sensors to provide quality data for the model, since poor-quality data can lead to suboptimal model performance.

7.4.1.10.4 Results and Discussion

For field surveillance and anomaly detection, the suggested solution is a complete, cloud-enabled, human-free autopilot drone system. It accomplishes complete automation of diagnostics, asset situational awareness, data collection and storage, and mass surveillance. Other popular drone platforms can easily adopt and use the system architecture, particularly the onboard and cloud compute engines. The authors have demonstrated the effectiveness of the system in a case study for a remote field in Latin America. They have achieved a 99% ROC in the multi-task binary classification of anomalies, which was significantly more promising than the SVM baseline with an ROC of 92%. The authors believe that with carefully designed software systems, the visual anomaly detection tasks, ranging from data acquisition to final report, can be largely automated with minimal human intervention, using AI-powered algorithms.

BIBLIOGRAPHY

Zhai, Xiang, Liu, Kui, Nash, William, and Castineira, David. "Smart Autopilot Drone System for Surface Surveillance and Anomaly Detection via Customizable Deep Neural Network." Paper presented at the *International Petroleum Technology Conference*, Dhahran, Kingdom of Saudi Arabia, January 2020. doi: https://doi.org/10.2523/IPTC-20111-MS

7.4.1.11 Application-11: Preventing Slugging by Tuning Choke through Machine Learning

7.4.1.11.1 Background of the Application

This paper presents an ML-based approach to predict and prevent slugging in gas-lift wells. The paper shows that slugs can be successfully forecasted up to 5 hours in advance using historical data and an LSTM network. The proposed method involves forecasting the well state from the current time to a future time, based on 12 input parameters. The LSTM network is employed to model the evolution of the well state over time and to recognize slugs in the evolution of the wellhead pressure. The results of this work show that the proposed method can accurately predict slugs, which can help operators take proactive actions to mitigate or avoid them.

7.4.1.11.2 Description of the Workflow and Machine Learning Model

This method uses an LSTM network to predict slugging in gas-lift wells. LSTM networks are ideal for modeling time series data and handling sequences. The model utilizes 12 input parameters to predict future well states and detect potential slugging events. The model can predict slugs up to 5 hours in advance, giving time to take important measures to intercept and optimize production.

7.4.1.11.3 Advantages and Limitations of the Synchronized Algorithm

LSTM networks have advantages in handling sequential data and long-term dependencies, making them effective for predicting slugging. They improve with more data and consider a comprehensive set of input parameters, enhancing their robustness. However, they require significant computational resources and expertise to train. Their accuracy depends on the quality of historical data, and they may struggle with sudden changes, because they are more specialized in predicting in the long term.

7.4.1.11.4 Results and Discussion

This paper presents a novel approach to predict and prevent slugging in gas-lift wells, using ML techniques. The proposed method employs an LSTM network to model the evolution of the well state over time and to forecast the occurrence of slugs up to 5 hours in advance. The results show that the proposed method can accurately recognize slugs in the evolution of the wellhead pressure and can help operators take proactive actions to mitigate or avoid them. The findings of this work have important implications for the oil and gas industry, as they can lead to improved production efficiency and reduced downtime.

BIBLIOGRAPHY

Bangert, Patrick. "Preventing Slugging by Tuning Choke through Machine Learning." Paper presented at the *International Petroleum Technology Conference*, Dhahran, Kingdom of Saudi Arabia, January 2020. doi: https://doi.org/10.2523/IPTC-19931-Abstract

7.4.1.12 Application-12: Smart Oilfield Safety Net: An Integrated System for Prediction of Asset Integrity Opportunities

7.4.1.12.1 Background of the Application

This paper presents an integrated system that can predict potential damage to assets by identifying telltale signatures of loss of containment (LOC) events through analytics. The system collects historical and current data from a variety of sources and stores them in a semantically enhanced system of record. ML techniques and rule-based systems are used to analyze the data, and advanced visualization environments are employed to minimize labor-intensive and error-prone data comprehension and decision-making. The system not only determines an asset's health but also automates LOC prevention decision-making.

7.4.1.12.2 Description of the Workflow and Machine Learning Model

This paper presents a comprehensive system designed to predict asset integrity opportunities in the oilfield using an integrated approach that combines asset management, predictive analytics, and contextual visualization. The system leverages ML techniques to analyze vast amounts of data from multiple sources, including time series, sensor data, and imagery, to predict potential LOC events and corrosion conditions. By employing a predictive analytics engine, the system identifies patterns and features that contribute to asset degradation and uses these insights to make proactive maintenance recommendations. The integration of semantic inferencing enhances the ability to detect complex correlations between data points, while contextual visualization facilitates human decision-making by displaying relevant information in a comprehensible format.

7.4.1.12.3 Advantages and Limitations of the Synchronized Algorithm

The integrated system offers significant advantages in improving asset integrity management. The use of ML enables the system to predict potential failures and anomalies with high accuracy, allowing for early interventions that can prevent catastrophic events. The semantic inferencing component enhances the system's ability to process and correlate heterogeneous data sources, making it highly adaptable to new environments. Additionally, the contextual visualization tool aids operators in understanding complex data and making informed decisions.

However, the system's effectiveness depends on the quality and completeness of the input data. Incomplete or inconsistent data could lead to inaccurate predictions. The computational complexity of integrating and processing vast amounts of heterogeneous data is another challenge, as it requires significant computational resources and robust algorithms to handle the complexity efficiently.

7.4.1.12.4 Results and Discussion

The Smart Oilfield Safety Net (SOSNet) presented in this paper is a step change in real-time situation awareness. It provides means to capture actionable information, detect train of event signatures in big data, and communicate and act in ways that best suit stakeholders. The system enables rapid identification of non-conformances, risk ranking, and recommendations, greatly enhancing decision support. SOSNet is already advancing the state-of-the-art in asset integrity management and can be an augmentation to all the safety systems and safeguards in place today to prevent an incident from starting or getting too large to cause major damage to humans, environment, and property. The benefits of automatically predicting caution and action alerts from measurement data at very high accuracy levels are enormous and highly valuable and can save human lives, minimize economic losses, and achieve superior safety records.

BIBLIOGRAPHY

Killen, N., House, R., Sankur, V., Thigpen, B., Chelmis, C., Prasanna, V. K., Neumann, U., Raghavendra, R., Yao, K., You, S., and Ershaghi, I. "Smart Oilfield Safety Net: An Integrated System for Prediction of Asset Integrity Opportunities." Paper presented at the *SPE Intelligent Energy International Conference and Exhibition*, Aberdeen, Scotland, UK, September 2016. doi: https://doi.org/10.2118/181048-MS

7.4.1.13 Application-13: Using Deep Learning and Computer Vision Techniques to Improve Facility Corrosion Risk Management Systems

7.4.1.13.1 Background of the Application

This paper presents a novel approach to detect and classify various types of corrosion defects in oil and gas facilities using state-of-the-art computer vision and DL techniques. The proposed method uses object detection and classification architectures, namely Fast Region-proposed Convolutional Neural Network (R-CNN) with transfer-learned Resnet and SSD with transfer-learned MobileNets, to accurately detect the location and type of corrosion defects in RGB images. The approach was evaluated using a limited dataset of 36 images and achieved an overall accuracy of 83.3%. The proposed method has the potential to lead to more cost-effective and quicker recommendations of preventive or corrective measures for corrosion defects.

7.4.1.13.2 Description of the Workflow and Machine Learning Model

The author employed an ML algorithm using an R-CNN with a transfer learning model to detect and reduce the corrosion risk in oil and gas facilities. The R-CNN takes the RGB images of corrosion defects as input data, and after training, the model learns to classify images of oil and gas facilities as having corrosion risk.

The algorithm starts with data acquisition and preprocessing, where the preprocessed RGB training and validation data images are loaded from the file directories and saved as proper datasets suitable for training. Then, the dataset is divided into training and validation sets. Then, the ResNet-50 backbone model is defined since the model is the core of the R-CNN algorithm. During training, the parameters are optimized using the stochastic gradient descent (SGD) with a specific learning rate and momentum. The training loop iterates over the dataset, computing losses, and updating the model weights. To improve training efficiency, the learning rate scheduler adjusts the learning rate based on the model's performance. At the end, the model is used in the validation phase to predict corrosion defects in new images, and the results are displayed.

7.4.1.13.3 Advantages and Limitations of the Synchronized Algorithm

The use of an R-CNN with a ResNet-50 backbone and transfer learning for corrosion detection on oil and gas facilities allows the model to utilize pre-trained weights and reduce the need for large amounts of training data, which leads to efficient training and improved accuracy. The R-CNN is particularly effective in extracting common patterns and anomalies in images; thus, it is a reliable approach in this case. However, the accuracy of the model is heavily dependent on the quality and quantity of the dataset. The dataset should be diverse and should include augmented data such as flipped, rotated, and in different lighting conditions to increase the accuracy of the model.

7.4.1.13.4 Results and Discussion

In conclusion, this paper demonstrates the effectiveness of using computer vision and DL techniques for detecting and classifying corrosion defects in oil and gas facilities. The proposed approach shows promising results and has the potential to improve the accuracy and efficiency of corrosion detection, leading to cost-effective and quicker recommendations for preventive or corrective measures. Future work will involve expanding the dataset to encompass more corrosion defect types and improving the accuracy of the model.

BIBLIOGRAPHY

Ejimuda, C. C., and Ejimuda, C. M. "Using Deep Learning and Computer Vision Techniques to Improve Facility Corrosion Risk Management Systems." Paper presented at the *SPE Western Regional Meeting, Garden Grove*, California, USA, April 2018. doi: https://doi.org/10.2118/190036-MS

7.4.1.14 Application-14: A Digital Oilfield Comprehensive Study: Automated Intelligent Production Network Optimization

7.4.1.14.1 Background of the Application

This paper proposes a workflow to simplify the model deployment for automated intelligent production network optimization. The workflow uses technologies from a flow assurance simulator, Python API toolkit, open-source ML packages, and a commercial visualization dashboard. The workflow consists of three steps: creation of a digital twin of the well and network model, an automated calibration workflow, and creation of a customizable optimization workflow based on field KPIs.

7.4.1.14.2 Description of the Workflow and Machine Learning Model

This paper addresses the challenges associated with production optimization at the network level, which is often a complex and time-consuming process. Leveraging advanced technologies such as a flow assurance simulator, Python Application Programming Interface (API) toolkit, and open-source ML libraries, the study proposes a workflow that simplifies model deployment and automates production network optimization. The approach integrates digital twin models, automated calibration, and a customizable optimization framework to enhance field-level decision-making. The results demonstrate significant improvements in time efficiency and data accuracy, allowing for faster and more reliable engineering decisions.

7.4.1.14.3 Advantages and Limitations of the Synchronized Algorithm

The study successfully developed automated workflows that significantly reduce the risk of data quality issues and the time required for production network optimization. The automation reduced manual input time from 15 minutes per well to just 3 seconds per well and decreased network model merging time from 30 minutes to 2 minutes. Additionally, ML auto-calibration workflows achieved high accuracy in predicting key parameters, such as liquid PI and reservoir pressure, but struggled with friction factor predictions due to limited relevance in the test cases. The deployment of these automated systems allows engineers to make more informed decisions with greater efficiency, ultimately enhancing production performance while reducing operational risks.

7.4.1.14.4 Results and Discussion

The proposed workflow successfully reduced the time consumed for well modeling and network model merging. The workflow also eliminated data input errors and inconsistencies, increasing data integrity and consistency. The model was successfully updated with the latest production data, and the well Inflow Performance Relationship (IPR)s' liquid Productivity Index (PI), reservoir pressure, and holdup factor are predicted from ML with more than 90% accuracy. The workflow also saves more than 90% of manual individual runs to a digital comprehensive optimization. By creating a personalized network surveillance dashboard, engineers can make rapid decisions based on the results derived from a daily optimization run.

BIBLIOGRAPHY

Naufal, Aulia Ahmad, and Metra, Sabrina. "A Digital Oilfield Comprehensive Study: Automated Intelligent Production Network Optimization." Paper presented at the *SPE/IATMI Asia Pacific Oil & Gas Conference and Exhibition*, Virtual, October 2021. doi: https://doi.org/10.2118/205735-MS

7.4.2 Pipeline

7.4.2.1 Application-15: Prediction of Oil Flow Rate through Orifice Flow Meters: Optimized Machine-Learning Techniques

7.4.2.1.1 Background of the Application

The measurement and control of flow rate through transportation infrastructure is crucial for maintaining quality, continuity, and detecting anomalous conditions in industries such as energy production, chemical, and oil and gas. While accurate and advanced measuring equipment is widely available, industries still seek ways to improve the accuracy and precision of older meters fitted to their systems. In order to offer correct flow rate forecasts over a large and intricate pipeline system with several orifice meters, a novel and extremely accurate hybrid model integrating ML and optimizer techniques is developed and utilized in this work. Pressure, temperature, kinematic viscosity, square root of differential pressure, oil specific gravity, beta ratio, base sediment, and water are the input variables that the models assess. The maximum prediction accuracy was attained by the combined artificial-bee colony (ABC) and firefly (FF) swarm-type optimizers using multi-layer perceptron (MLP) algorithms and distance-weighted K-nearest-neighbor algorithms. Enhancing hydrocarbon accounting and lowering errors and possible losses in the amounts of oil passing through the pipeline and related tank storage system are two benefits of producing more accurate and dependable flow measurements.

7.4.2.1.2 Description of the Workflow and Machine Learning Model

The authors employed the distance-weighted K-nearest neighbor (DWKNN) algorithm to perform the prediction of oil flow rate through orifice flow meters. They used ABC and FF swarm-type optimizers to optimize their DWKNN models. To ensure quality, continuity, and rapidly detect anomalous conditions, flow measurement of throughput is critical for managing and controlling processes associated with industrial flow transmission systems. They developed a novel, highly accurate hybrid model that combines ML and optimization algorithms to predict flow rates in complex piping systems using multiple orifice flow meters.

The DWKNN algorithm is one of the supervised ML algorithms based on the statistical learning theory that can address this challenge when combined with ABC, and it has been successfully applied to a wide range of complex data classification tasks. This algorithm finds a specified number (K) of data records in a dataset that are nearest in terms of their input variable values to each data record in the dataset to be predicted. Moreover, the Euclidean distance is used to weight each record's contribution, thus finding a more accurate model. The behavior of foraging bees in their quest to find the best nectar sources in the vicinity of their colonies is the basis for the ABC optimizer. A predetermined initial population is dispersed across a predetermined search region. It is divided into bees with distinct roles, such as scout-type bees who go farther in search of new, advantageous sites and onlooker-type bees that change their positions at the conclusion of each cycle in response to messages from other swarm members. The ABC optimizer's implementation sequence is summarized in Figure 7.1.

7.4.2.1.3 Advantages and Limitations of the Synchronized Algorithm

In general, complex and accurate calculations are required to obtain oil flow data from orifice flow meter readings, so replacing manual calculation with the ML method becomes a challenge. Therefore, the combination of KNN and ABC algorithm can offer an effective solution to predict the oil flow rate. KNN is strongly preferred for this task because KNN is both insensitive to outliers and friendly to irregular boundaries. ABC is chosen because it is an excellent global optimization algorithm with simple principles and few parameters (based on bee colony intelligence behavior).

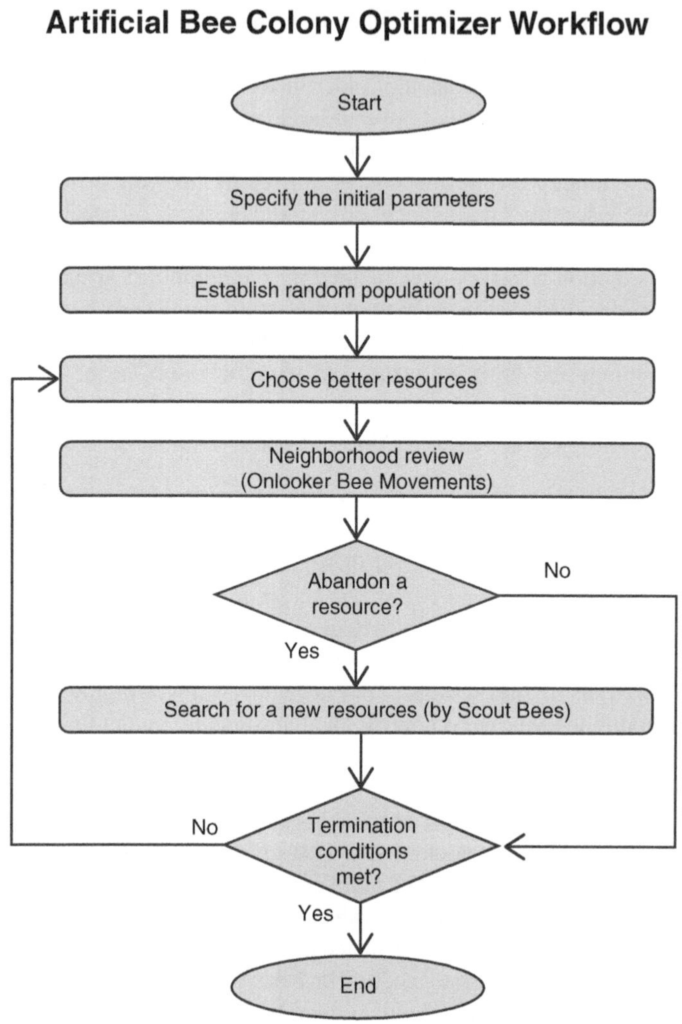

FIGURE 7.1 Artificial-bee colony workflow.

However, the artificial bee colony algorithm still has limitations, such as easy to premature convergence, low optimization speed in the later period of the optimization.

7.4.2.1.4 Results and Discussion

In order to forecast flow rate via a large and intricate pipeline system with several orifice meters, this study proposes a novel and incredibly accurate hybrid model that combines ML and optimizer techniques. Because of the accuracy attained in forecasting flow rate, complicated and unreliable empirical formulas can be avoided. The methodology can be extended to other systems and industry domains with similar requirements for flow rate measurement, and it can be applied to the optimization of flow rate in the oil and gas industry. The results demonstrate that the coupled ABC and FF swarm-type optimizers with the distance-weighted K-nearest-neighbor and MLP algorithms provide a reliable and accurate method for predicting flow rate, which is of great importance for hydrocarbon accounting and reducing losses in oil and gas transportation.

BIBLIOGRAPHY

Farsi, Mohammad, Barjouei, Hossein Shojaei, Wood, David A., Ghorbani, Hamzeh, Mohamadian, Nima, Davoodi, Shadfar, Nasriani, Hamid Reza, and Alvar, Mehdi Ahmadi. Prediction of oil flow rate through orifice flow meters: Optimized machine-learning techniques. *Measurement* 174, 2021, 108943, ISSN 0263-2241. https://doi.org/10.1016/j.measurement.2020.108943

7.4.2.2 Application-16: Digital Twin for Oil Pipeline Risk Estimation Using Prognostic and Machine Learning Techniques

7.4.2.2.1 Background of the Application

This paper proposes a digital twin based on ML and prognostics algorithms model to analyze and predict the risk probability rate of oil pipeline systems. The proposed work creates a virtual intelligent integrated automated control system to predict the risk rate in the oil industry by integrating entire transmission lines through enhanced wireless information networks in remote locations. The calculation of risk conditions in relation to pressure data for the assessment of remaining useful life (RUL) is the major focus of prognostics. Dirichlet process clustering and canopy clustering are used to separate the abnormal pressure drop and rise when the abnormality of the pressure attribute is considered for risk probability estimation in prognostic analysis. In order to provide on-time control action on the entire oil pipeline system through effective wireless data communication between the server and the oil substations, the features are extracted using a variety of learning techniques, and the best feature probability rates are assessed using a kernel-based SVM algorithm. This is done using multiple oil substation data integration platforms.

7.4.2.2.2 Description of the Workflow and Machine Learning Model

The authors employed kernel-based SVM algorithm to control the entire oil pipeline system between the server and the oil substations with associated learning algorithms that analyze data for classification analysis. The problem of vulnerability of pipelines control system would add the risk probability rate, which therefore makes system maintenance more difficult. Attention is attracted to the earlier prediction of accident risks by undertaking the entire pipeline. As a result, ML methods are needed to establish an accurate model for evaluating the remaining useful life (RUL), using 1) abnormal oil pressure, 2) sudden variations by external factors, and 3) normal pressure.

SVM is one of the supervised ML algorithms based on the statistical learning theory that can address this challenge when combined with clustering. The SVM algorithm is suitable for classification analysis. There are two common kernels in SVMs – linear and polynomial. In this study, the oil pipeline pressure condition is measured using the same set of data using both the linear and polynomial kernels. Following training of the two distinct kernel-based SVM types, two scores—accuracy, precision, recall, and F1-Measure—were produced in order to measure the classifier's performance.

SVMs frequently use the polynomial kernel, a kernel function that allows nonlinear model learning by representing the similarity of vectors (training samples) in a feature space over polynomials of the original variables. Characteristics that are derived from pre-existing characteristics in the dataset are called polynomial features. The polynomial features will be x, x^2, and x^3, for instance, If a dataset has a single feature, x, and polynomial features up to degree 3 need to be identified, each feature would be modified similarly if more were present.

7.4.2.2.3 Advantages and Limitations of the Synchronized Algorithm

Because SVM effectively increases the classifier's resilience by focusing on interval maximization. Back propagation NNs, in contrast, have the drawbacks of being slow, failing because of the local minimum, and overfitting because of the recently added data. It provides an efficient approximate solution approach for problems with non-deterministic polynomial complexity by overcoming optimization processes that tend to fall into local minimum faults and the dependence on beginning values. However, when compared to other methods used to handle text data, SVM does not provide the best performance when handling text structures. Performance suffers as a result of the loss of sequential information. Furthermore, it does not provide much explanation as the confidence of prediction is important in several applications.

7.4.2.2.4 Results and Discussion

The proposed digital twin based on ML and prognostics algorithms model provides a reliable and efficient method for predicting the risk rate in the oil industry. The integration of entire transmission lines through enhanced wireless information networks in remote locations along with the manifold learning methods and kernel-based SVM algorithms provides on-time control action on the entire oil pipeline system. The proposed work creates a virtual intelligent integrated automated control system to predict the risk rate in the oil industry, which can help in reducing catastrophic failures and costly downtime. The future work can focus on improving the accuracy of the risk probability estimation and exploring the potential of the proposed method in other industries as well.

BIBLIOGRAPHY

Priyanka, E. B., Thangavel, S., Gao, Xiao-Zhi, Sivakumar, N. S. Digital twin for oil pipeline risk estimation using prognostic and machine learning techniques. *Journal of Industrial Information Integration* 26, 2022, 100272, ISSN 2452-414X. https://doi.org/10.1016/j.jii.2021.100272

7.4.2.3 Application-17: Deep Autoencoder-Derived Features Applied in Virtual Flow Metering for Sucker-Rod Pumping Wells

7.4.2.3.1 Background of the Application

This paper presents an innovative method of virtual flow metering (VFM) for sucker-rod pumping wells using state-of-the-art AI algorithms. Conventional and unconventional wells become stripper wells, which are defined as producing no more than 35 barrels of liquid equivalent per day, as reservoirs are depleted. In comparison to the existing low production, the typical metering method is not cost-effective due to its significant investment in facilities, equipment, sensors, and continuous maintenance. The suggested approach predicts the real-time production rate with a very low cost and dependable accuracy by utilizing a wealth of real-time field and well data obtained from IIoT and digital field facilities.

7.4.2.3.2 Description of the Workflow and Machine Learning Model

This method employs an autoencoder and a multi-layer perceptron neural network (MLPNN) to predict pump efficiency. The autoencoder compresses input data into a lower-dimensional space, and the resulting features are fed into the MLPNN to predict pump efficiency. The autoencoder captures key patterns in the data, and the MLPNN leverages these patterns for accurate predictions, evaluated using mean absolute error (MAE) on test data.

7.4.2.3.3 Advantages and Limitations of the Synchronized Algorithm

Combining an autoencoder with an MLPNN enhances predictive accuracy by focusing on the most relevant features. The autoencoder reduces data dimensionality, while the MLPNN captures complex relationships within the compressed data. However, this approach needs significant computational resources and relies on high-quality, representative training data. Divergences between training data and real-world conditions can affect accuracy, thus needing continuous updates.

7.4.2.3.4 Results and Discussion

With R^2 greater than 0.92, MAE less than 0.5 of wells producing less than 5 m^3/day, RMSE less than 1.6 of wells producing 5–10 m^3/day, and RMSE less than 2.2 of wells producing more than 10 m^3/day, the suggested VFM method has been tested and validated in multiple fields in one operator's fields in China. It has also shown very high accuracy. Additionally, the model has been tested on hundreds of recently produced wells, and when compared to high-resolution real-time metering equipment, it has an inaccuracy of 10%. As AI technology advances and training datasets grow, this approach may serve as a fallback or substitute to assist the operator in reducing the cost of investment and upkeep for both conventional and unconventional fields during the period of low oil prices.

BIBLIOGRAPHY

Peng, Yi, Zhang, Jianjun, Xiong, Chunming, Shi, Junfeng, Zhao, Ruidong, Zhang, Xishun, Xu, Dongping, Li, Qiming, Deng, Feng, Chen, Shiwen, Liu, Meng, Chen, Guanhong, and Wang, Cai. "Deep Autoencoder-Derived Features Applied in Virtual Flow Metering for Sucker-Rod Pumping Wells." Paper presented at the *SPE/IATMI Asia Pacific Oil & Gas Conference and Exhibition*, Bali, Indonesia, October 2019. doi: https://doi.org/10.2118/196288-MS

7.4.2.4 Application-18: Machine-Learning Distributed-Temperature-Sensing-Based Pipeline Leak Detection

7.4.2.4.1 Background of the Application

This paper presents an ML method for pipeline leak detection using distributed temperature sensing (DTS) technology. The proposed ML method, called machine learning transient alarming (MLTA), uses a training dataset to extract typical patterns and detect temperature deviations. The method was tested on a non-insulated, wet oil 6" pipeline buried 1 m below the ground surface in northern Germany. The results showed that the MLTA method is capable of detecting pipeline leaks with high accuracy and low waste. The proposed method provides a strong tool to achieve high probability of detection and low nuisance alarm rate by extracting the usual thermal behavior of the pipeline under normal operating conditions and comparing the measured temperatures to the known behavior.

7.4.2.4.2 Description of the Workflow and Machine Learning Model

This method analyzes temperature data along various distances to detect anomalies, potentially indicating leaks. It prepares the data for model training, and K-nearest neighbors (KNN) regression models are trained for each distance. These models predict temperatures for the test data. The predictions are then evaluated using mean squared error (MSE) for each distance. Anomalies are identified by setting a threshold based on the MSE values, flagging distances with high MSEs as potential leak points.

7.4.2.4.3 Advantages and Limitations of the Synchronized Algorithm

This method has advantages like the ability to customize models for each distance and capture nonlinear patterns in the data. However, it has downsides as well, such as the computational cost of training multiple models and sensitivity to the choice of the number of neighbors. The method also relies on the assumption that the training and test data distributions are similar, which may not always be true in real-world scenarios.

7.4.2.4.4 Results and Discussion

The proposed MLTA method has been successfully tested for pipeline leak detection using DTS technology. The method has proven to be effective in detecting pipeline leaks with high accuracy and low waste. The MLTA method extracts typical patterns from a training dataset and detects temperature deviations from the current situation using the pre-trained model. The MLTA method provides a strong tool to achieve a high probability of detection and a low nuisance alarm rate by extracting the usual thermal behavior of the pipeline under normal operating conditions and comparing the measured temperatures to the known behavior. The proposed method can contribute to environmental protection, health, safety, and economic efficiency.

BIBLIOGRAPHY

Sfar Zaoui, Wissem, Lauber, Thomas, Pohl, Clemens, Kerk, Michael, Glaeser, Thomas, and Jelinek, Wolfgang. "Machine-Learning Distributed-Temperature-Sensing-Based

Pipeline Leak Detection." Paper presented at the *Abu Dhabi International Petroleum Exhibition & Conference*, Abu Dhabi, UAE, November 2020. doi: https://doi.org/10.2118/202905-MS

7.4.2.5 Application-19: Pipeline Performance Data Analysis through Machine Learning with Hydraulics Model Assisted Training

7.4.2.5.1 Background of the Application

This paper discusses the use of ML in pipeline data analysis. The authors constructed three ML models using ANN methods, which use nominal flow rate and head loss as input and pipeline roughness change, internal diameter change, and potential leaks (location and leakage rate) as output. The models were trained using half of the data generated from the hydraulics model and validated using the remaining data. This approach has the advantage of using ML to disclose flow conditions like ID, roughness, and leakage directly from pressure and flow rate measurement data, eliminating the requirement for comparison with hydraulic models. The authors show how pipeline data analysis and operation analysis can benefit from the application of AI and ML.

7.4.2.5.2 Description of the Workflow and Machine Learning Model

This method uses a NN that models pipeline characteristics like roughness change, diameter change, leak location, leak rate, and friction factor. It uses input features such as flow rate, head loss, Reynolds number, and other related metrics. The model is trained on scaled data, with a structure comprising two hidden layers and an output layer to learn the relationships between the inputs and outputs.

7.4.2.5.3 Advantages and Limitations of the Synchronized Algorithm

The model has advantages such as handling complex, nonlinear relationships and incorporating early stopping to prevent overfitting. However, it requires large datasets and significant computational resources. Additionally, the model's predictions can be difficult to interpret and make meaning of, which can be a limitation in some applications where understanding the decision-making process is important.

7.4.2.5.4 Results and Discussion

The authors constructed and trained three ANN, which reached an acceptable error level. The results demonstrate that the detection of pipeline roughness change, internal diameter change, and leakage can be efficiently and accurately detected using the constructed and trained NNs. Complicated inverse problems were successfully solved using intelligent ML methods. The method is promising in solving more complicated problems, such as the combined effects of multiple influences, such as roughness, diameter change, and leakage. The method will be further refined to improve efficiency and accuracy. Uncertainty propagation will also be studied to develop the technology in using real measurement data instead of hydraulic equations.

BIBLIOGRAPHY

Cheng, David, and Zeosky, Daniel S. "Pipeline Performance Data Analysis through Machine Learning with Hydraulics Model Assisted Training." Paper presented at the *PSIG Annual Meeting*, London, UK, May 2019.

7.4.2.6 Application-20: Data Driven Smart Monitoring for Pipeline Integrity Assessment

7.4.2.6.1 Background of the Application

The oil and gas industry strives for efficiency and safety in transportation systems, but the complexity of assets makes it difficult to derive a theoretical framework for managing control parameters. To address this, the use of digital transformation and big data is becoming more prevalent for real-time monitoring. "Data-driven" relations are inferred from large datasets recorded throughout an asset's lifecycle to predict the evolution of asset integrity. This paper summarizes a research project for the design, implementation, and testing of a "machine learning" approach to vibroacoustic data, recorded continuously along a pipeline, for monitoring pipeline integrity. The approach involves observing descriptors over time intervals, learning their safe range of variation, and using classification procedures to automatically obtain a discrete set of operational statuses. The classification criteria are then applied to new data to highlight system anomalies.

7.4.2.6.2 Description of the Workflow and Machine Learning Model

The method used is the Gaussian Mixture Model (GMM), an algorithm applied for clustering, density estimation, and anomaly detection. It models the data as a mixture of multiple Gaussian distributions, assigning probabilities to each data point for belonging to each cluster. The expectation-maximization (EM) algorithm is used to iteratively update the parameters to help improve the probabilities of correct pipeline integrity outcomes.

7.4.2.6.3 Advantages and Limitations of the Synchronized Algorithm

GMM has advantages like its flexibility in modeling complex, overlapping clusters, and providing probabilistic clustering assignments. However, it requires specifying the number of components in advance, is sensitive to initial conditions, and assumes normal distribution within clusters. Which may not fit all datasets. It can also be computationally expensive for large datasets.

7.4.2.6.4 Results and Discussion

By identifying false alarms and redefining the range of operating scenarios for the upstream system, operators' feedback can update the suggested system model for pipeline integrity monitoring, which is not static. Predictive maintenance, automated anomaly detection, and operational procedure optimization can all benefit from the medium- to long-term development and updating of data-driven models. Additionally, new technologies, professional figures, and operational and business models can be introduced by integrating the monitoring experience of various assets.

BIBLIOGRAPHY

Giunta, Giuseppe, Nielsen, Keld Lund, Bernasconi, Giancarlo, Bondi, Luca, and Korubo, Barry. "Data Driven Smart Monitoring for Pipeline Integrity Assessment." Paper presented at the *Abu Dhabi International Petroleum Exhibition & Conference*, Abu Dhabi, UAE, November 2019. doi: https://doi.org/10.2118/197327-MS

7.4.2.7 Application-21: Prediction of Remaining Life in Pipes Using Machine Learning from Thickness Measurements

7.4.2.7.1 Background of the Application

This paper presents a data-driven approach to predicting non-conformance in aging offshore oilfield platforms. The approach is based on thickness measurements taken by inspectors at various thickness measurement locations (TMLs) over time and employs ML techniques. The paper outlines the problem of predicting the need for action based on thickness measurements, as well as the methodologies employed, including classification and regression models. The results of the study demonstrate the viability of using ML approaches to predict non-conformance and can help in prioritizing inspection schedules.

7.4.2.7.2 Description of the Workflow and Machine Learning Model

In this method, two different ML algorithms are used to approach to same problem. Linear regression from classification models and RF regression from regression models are used. These algorithms are trained and tested on the time gap and periodic thickness of pipes to predict when an intervention is needed. Linear regression is used to predict the value of a variable based on the value of another variable, and RF regression is a method that uses multiple decision trees to predict a continuous target variable.

7.4.2.7.3 Advantages and Limitations of the Synchronized Algorithm

Linear regression is a simple and interpretable, which is efficient in predicting output with linear relations to inputs. Thus, it is suitable for explaining linear relations between thickness and time. However, it assumes linearity, is sensitive to outliers, and may be overfit with high dimensionality, limiting it for complex patterns. RF regression excels in handling nonlinear relationships and interactions and is robust to overfitting and outliers. Despite these advantages, it is computationally expensive, less interpretable than linear regression, and requires careful tuning to reach the desired performance.

7.4.2.7.4 Results and Discussion

This study demonstrates the effectiveness of using a data-driven approach to predicting non-conformance in aging offshore oilfield platforms. By applying ML techniques, action alerts are accurately predicted based on thickness measurements at TMLs. This approach can aid in prioritizing inspection schedules and help to minimize or eliminate non-conformance. The results of the study show that using ML is an effective way to capture potential non-conformances in TMLs and can lead to a more efficient and effective inspection program.

BIBLIOGRAPHY

Sharifi, Mona, Yao, Ke-Thia, Raghavendra, S., Ershaghi, Iraj, House, Robert, and Blouin, Jacques. "Prediction of Remaining Life in Pipes using Machine Learning from Thickness Measurements." Paper presented at the *SPE Western Regional Meeting, Garden Grove*, California, USA, April 2015. doi: https://doi.org/10.2118/173995-MS

7.4.2.8 Application-22: Remote Sensing for Leak Detection and Quantification

7.4.2.8.1 Background of the Application

Leak detection and repair is a challenging activity, even in the onshore asset environment, with fugitive emission detection reliability continuing to be difficult despite the introduction of innovative technology. This paper presents a full-service end-to-end technology platform that can improve leak detection and measurement activity through the deployment of hardware and software environments. The platform combines machine-based emissions detection and algorithmic quantification with unmanned aerial vehicles and in situ monitoring to offer an approach that enables the operator to increase survey frequency, get more accurate results at a reduced cost, and carry out the work in a way that is consistent with low-risk and safe operations. Significant decreases in false positives and negatives have been achieved as a result of the findings, and datasets that compare volumetric emissions before and after leak repair operations have been completed enable precise indications of greenhouse gas reduction.

7.4.2.8.2 Description of the Workflow and Machine Learning Model

Key features are used to train an RF classifier for gas detection, with the data split into training and testing sets. Visual features like shape, texture, and color intensity train another RF classifier to distinguish gas emissions from artifacts. Both models' performances are assessed with accuracy scores and classification reports, confirming improved survey frequency, reliability, and cost-effectiveness.

7.4.2.8.3 Advantages and Limitations of the Synchronized Algorithm

The method's advantages are its difficulty against overfitting and improved detection using RF classifiers. This enables more frequent, reliable surveys and better decision-making with reduced false positives. However, there are some limitations like the need for high-quality training data, computational intensity of the models, and the cost of obtaining labeled data. Despite these setbacks, this method enhances detection accuracy and operational efficiency in emission monitoring.

7.4.2.8.4 Results and Discussion

This paper demonstrates that it is possible to automate fugitive emissions surveying in a way that is positively disruptive to current state effectiveness, productivity, and cost arrangements. The technology provides a solution that allows the operator to increase survey frequency, obtain more reliable results at lower cost, and perform the work in a manner consistent with safe and low-risk operations. Remote surveying is possible under circumstances of both in situ remote monitoring and UAV operations, which are selected depending on the scope of the survey and required outcomes. Migrating from an academic calculation of volumetric emissions to a scientific and accurate inductive calculation based on field data collected adds a fact base to greenhouse gas reduction reporting that is also equally transformative. Testing has shown that the platform can be deployed under a variety of atmospheric and background conditions and is effective across a range of hydrocarbon gases, including VOCs and

CO_2. The results have been significant in improving safety, cost, data reliability, and insight.

BIBLIOGRAPHY

Amin, Laith "Remote Sensing for Leak Detection and Quantification." Paper presented at the *SPE Offshore Europe Conference and Exhibition*, Aberdeen, UK, September 2019. doi: https://doi.org/10.2118/195794-MS

7.4.2.9 Application-23: The Development of Leak Detection Model in Subsea Gas Pipelines Using ML

7.4.2.9.1 Background of the Application

The transportation of crude and refined petroleum, such as natural gas, is mainly done through pipelines worldwide. However, monitoring the health condition of pipelines at offshore locations is challenging. Therefore, detecting leaks in real-time is essential. This study proposes a flow simulation-based ML algorithm to detect the location and size of leaks in gas pipelines. The proposed method analyzes pressure, mass flow, and temperature changes due to leaks in the pipeline using a dynamic model that simulates the behavior of pipelines. The model is trained using a deep NN algorithm to identify the optimal learning model. The proposed procedure can be applied to various pipelines and support more efficient operation by detecting leaks in real-time.

7.4.2.9.2 Description of the Workflow and Machine Learning Model

This method predicts the location and size of a leak in a pipeline using a NN model. Initially, the dataset is loaded and prepared, with features such as pressure and flow rates used as inputs. The data is split into training and testing sets and standardized to optimize model performance. A NN is then made with an input layer, three hidden layers, and an output layer for the two target variables. The model is compiled with the Adam optimizer and MSE loss function.

7.4.2.9.3 Advantages and Limitations of the Synchronized Algorithm

This method offers advantages like the ability to model complex relationships and prevent overfitting with early stopping. Standardizing features enhances training efficiency, and the Adam optimizer ensures robust learning. However, limitations include the need for large datasets, the black-box nature of NNs, and the dependency on hyperparameter tuning. Early stopping could also lead to underfitting if not carefully monitored.

7.4.2.9.4 Results and Discussion

The proposed modeling and simulation-based ML method showed excellent accuracy in detecting the location and size of leaks in gas pipelines. The method can be applied to various pipelines and supports more efficient operation by detecting leaks in real time. The proposed flowchart for leak detection in the gas pipeline can help pipeline operators to identify the location and size of leaks promptly. Further research can be done to improve the accuracy of the model for leak sizes of 0.5 cm.

BIBLIOGRAPHY

Kim, Juhyun, Chae, Minju, Han, Jinju, Park, Simon, and Lee, Youngsoo. The development of leak detection model in subsea gas pipeline using machine learning. *Journal of Natural Gas Science and Engineering*, 94, 2021, 104134, ISSN 1875-5100. https://doi.org/10.1016/j.jngse.2021.104134

7.4.3 Scaling/Corrosion

7.4.3.1 Application-24: Automated Corrosion Mapping AI & Machine Learning

7.4.3.1.1 Background of the Application

By analyzing air corrosion in full-facility imaging using AI and ML algorithms, this study aims to lower risk and increase the effectiveness of fabric maintenance. Before analysis, the ML system is trained using manual ground truth inputs. By examining every pixel in the building, the algorithm finds, categorizes, and reports any corrosion that is found, assigning problems to particular pipes or pieces of equipment. In just a few weeks after data collection, the facility's health can be thoroughly and impartially assessed with the use of this instrument.

7.4.3.1.2 Description of the Workflow and Machine Learning Model

The author applied the use of a U-Net convolutional NN for full-facility imagery to detect and classify atmospheric corrosion, in particular, under harsh marine environments like that of the Gulf of Mexico. This workflow, through ML and AI, was applied to fabric maintenance to minimize risk and enhance efficiency, to automatically identify light, medium, and heavy corrosion using the pre-trained CNN model.

The workflow starts by extracting high-resolution test and training file images and preprocessing them to normalize pixel values between zero and one. The images are then divided into three sets: training, validation, and testing. The design of the model would follow an encoder–decoder, whereby the convolutional layers will serve the purpose of feature extraction and the upsampling layers for the reconstruction of the segmented output. Model optimization and loss function for this model are Adam and a sparse categorical cross-entropy function, respectively. The performance metric is set at the mean intersection over the union. With sufficient data, the model is then trained and tested from the validation set, and hyperparameters are tuned. Finally, the results with the test data are visually demonstrated using the function "plot_sample_with_color" by overlaying the predicted image over the top of the test image.

7.4.3.1.3 Advantages and Limitations of the Synchronized Algorithm

The U-Net convolutional NN captures minute details on the image with the model's encoder-decoder architecture, which enables precise segmentation of corrosion areas. The U-Net convolutional NN is an effective method for extracting patterns from imagery and identifying them with high accuracy, making it a reliable tool for this case. However, the model's effectiveness highly depends on the quality and quantity of the training data that is given. For the best results, the training data should include the sample corrosion image from various angles and lighting conditions to ensure that the model has a more general prediction capacity and covers potential real-world scenarios.

7.4.3.1.4 Results and Discussion

The innovative digital environment created with this project provides the ability to break down a facility by regions or blocks, quantify the corrosion severity in each location, and report surface areas of each level of corrosion severity. Implementing this AI-centric solution allows for a consistent standard of measure across various facilities and assets, unlike the individual interpretation of conventional inspector walk-through efforts that range in experience and capabilities. The ability to digitally capture, assess, and quantify areas of degradation in coating, insulation damage, or other fabric maintenance deficiencies in a repeatable and precise manner over an entire asset allows for detailed planning on an asset-wide basis.

BIBLIOGRAPHY

Majors, Marc, Harrington, Travis, Ferguson, Eric, Dunne, Toby, Potiris, Steve, Vlaskine, Vsevolod, Mohammed, Jaffar, Bargoti, Suchet, Naqshbandi, Masood, and Ahsan, Nasir. "Automated Corrosion Mapping AI & Machine Learning." Paper presented at the *Abu Dhabi International Petroleum Exhibition & Conference*, Abu Dhabi, UAE, November 2020. doi: https://doi.org/10.2118/202801-MS

7.4.3.2 Application-25: A Physic-Based Scaling Curve of Gas Production under Compaction: A Leading Way to Better ML Prediction

7.4.3.2.1 Background of the Application

In order to identify gas production under compaction, this research suggests a straightforward analytical method. The solution collapses the production profiles of various wells into a single generic curve after scaling them. Later on, this curve may be used as the "learning" function for ML predictions based on physics.

The growing abundance of big data in the oil and gas sector is highlighted in the article as a significant potential to gain a deeper understanding of hydrocarbon resources. A vast amount of data may be transformed into field economics and production forecasts using ML. However, the chosen learning function—which frequently ignores the physical components of the problem—determines the quality of ML predictions. In order to predict future gas output under extreme compaction, the research proposes an improved ML method using a physics-based function.

In this study, the coupled Darcy–Biot equation for vertical gas wells under reservoir compaction is solved in order to develop a physics-based master curve. The authors assume that the flow is radial, and the porosity is transiently changing with the reduction in pore pressure due to gas production. Finally, the authors reduce the complexity of the coupled nonlinear equation to two scaling optimization parameters: a mass scaling factor to scale the recovery factor and a time scaling factor to scale the diffusion time.

The authors use a field example from Indonesia's KLX field to validate their concept. Subsidence is a byproduct of the massive volume of gas produced by this process. The authors capture the universal scaling factors that reflect the behavior of gas flow under reservoir compaction by combining the production profiles of all wells that are currently in operation into a single master curve. Additionally, the authors can use the generated master curve as the machine-learning model's learning function to diagnose and predict other domains that experience similar events in the future.

7.4.3.2.2 Description of the Workflow and Machine Learning Model

In this method, an MLP neural network is used to predict the production rate based on porosity, permeability, flow rate, and compressibility data. The MLP is tuned using a genetic algorithm (GA) to optimize the network's hyperparameters, specifically the number and size of the hidden layers. This use of algorithms leverages their capability to learn complex nonlinear relationships and their strength in efficiently searching high-dimensional spaces for optimal solutions.

7.4.3.2.3 Advantages and Limitations of the Synchronized Algorithm

The MLP NN is very good in capturing intricate patterns and interactions among features, making it appropriate for predicting production rates. By using the GA for hyperparameter tuning, the model's performance is significantly enhanced, leading to more accurate predictions. However, the training process can be computationally expensive, especially with larger datasets and numerous generations in the GA. And, the model's performance is sensitive to the choice of initial parameters and the working of the GA, which would need extra attention to their setup.

7.4.3.2.4 Results and Discussion

In order to simplify the prediction process, an ML technique can be developed using the general scaling that the study gives to explain gas generation under compaction. The suggested physics-based approach can predict future gas production under extreme compaction and considers physical features of the issue. The authors solved the coupled Darcy–Biot equation for vertical gas wells under reservoir compaction in order to create a physics-based master curve. They used a field instance from Indonesia's KLX field, which generates a lot of gas with subsidence as a byproduct, to validate the model. In order to forecast and diagnose additional domains that experience the similar events in the future, the suggested method can use the generated master curve as the learning function for the ML model.

BIBLIOGRAPHY

Sajjad, Farasdaq, Chandra, Steven, Naja, Savinatun, and Suganda, Wingky. "A Physic-Based Scaling Curve of Gas Production Under Compaction: A Leading Way to Better Machine-Learning Prediction." Paper presented at the *SPE/IATMI Asia Pacific Oil & Gas Conference and Exhibition*, Bali, Indonesia, October 2019. doi: https://doi.org/10.2118/196254-MS

7.4.3.3 Application-26: Machine Learning Based Predictive Models for CO_2 Corrosion in Pipelines with Various Bending Angles

7.4.3.3.1 Background of the Application

This paper presents two ML models, light gradient boosting machine (LightGBM) and MLPNN, for predicting CO_2 corrosion in pipelines with different bending angles. The models were trained on a dataset of 77,745 data points generated by a computational fluid dynamics (CFD) model that took into account various environmental conditions and geometries of the pipeline. The models were found to be more efficient and faster than CFD models, while retaining similar accuracy. The authors also developed a web application based on these models for pipeline design and monitoring applications.

7.4.3.3.2 Description of the Workflow and Machine Learning Model

This method uses two ML models, LightGBM and MLPNN, to predict the maximum CO_2 corrosion rate in pipelines. After splitting the data, LightGBM is trained using gradient boosting, while the MLPNN uses dense layers and backpropagation. Model performance is evaluated using the R2 score. Both models are made for real-time predictions based on new data, ensuring reliable pipeline integrity assessments.

7.4.3.3.3 Advantages and Limitations of the Synchronized Algorithm

The combined use of these algorithms offers comprehensive corrosion rate predictions. LightGBM efficiently handles large datasets and diverse features, while MLPNN captures complex nonlinear relationships. This combination ensures correct and robust predictions. However, MLPNN training is computationally expensive, and both models rely on high-quality, representative training data. Divergence between training data and real-world conditions can impact accuracy, therefore a need for continuous model updates.

7.4.3.3.4 Results and Discussion

This paper provides an innovative solution for predicting CO_2 corrosion in pipelines with different bending angles using ML models. The LightGBM and MLPNN models developed in this study offer higher efficiency and faster prediction times compared to traditional CFD models. This can result in significant time and cost savings for the oil and gas industry. The authors also developed a web application based on these models as a tool for pipeline design and monitoring applications.

BIBLIOGRAPHY

Yang, Huihui, Lu, Ligang, and Tsai, Kuochen. "Machine Learning Based Predictive Models for CO2 Corrosion in Pipelines With Various Bending Angles." Paper presented at the *SPE Annual Technical Conference and Exhibition*, Virtual, October 2020. doi: https://doi.org/10.2118/201275-MS

7.4.3.4 Application-27: Estimating Corrosion Growth Rate for Underground Pipeline: A Machine Learning Based Approach

7.4.3.4.1 Background of the Application

This paper presents a model for estimating corrosion growth rates for underground assets in North America. The model uses a combination of environmental variables, including atmospheric conditions, AC interference, proximity to roads, railways, and bodies of water, soil parameters, and magnetic anomalies. Three ML approaches were used to develop the model: generalized linear models, eXtreme boosted trees, and NNs. The model was trained on 340,947 GirthWeldAddresses and can be used to estimate corrosion growth rates for any latitude and longitude pair in North America. The paper also highlights areas for further improvement, including incorporating data related to cathodic protection, cross-validation of models, and incorporating improvements in AC interference.

7.4.3.4.2 Description of the Workflow and Machine Learning Model

In this method, a pipeline with several models is used to make predictions on the corrosion growth rate from a dataset comprising both numerical and categorical features. In this case, the preprocessing steps include scaling the numerical data and one-hot encoding the categorical ones so that the features are well-formatted for the input models. Three models are involved: a generalized linear model, an XGBoost regressor, and an NN. The RMSE is calculated for each model to determine predictive accuracy on the holdout test set.

7.4.3.4.3 Advantages and Limitations of the Synchronized Algorithm

This method has advantages like comprehensive preprocessing and the evaluation of diverse models, allowing for comparison and selection of the best-performing model. Utilizing pipelines ensures the integration of preprocessing steps, reducing the risk of errors and enhancing reproducibility. However, the NN model's training process is computationally expensive and requires careful tuning of hyperparameters. The performance of all models heavily relies on the quality of handling of preprocessed data. Despite these limitations, the use of different model architectures provides valuable information about the dataset and ensures more reliable predictions.

7.4.3.4.4 Results and Discussion

This paper presents a model for estimating corrosion growth rates for underground assets in North America. The model uses a combination of environmental variables and was developed using three ML approaches. The model was trained on 340,947 GirthWeldAddresses and can be used to estimate corrosion growth rates for any latitude and longitude pair in North America. Incorporating additional data related to cathodic protection, cross-validation of models, and improvements in AC interference could further improve the model's accuracy. The authors suggest that future work should focus on developing a comprehensive NN approach that scales to any corrosion modeling situation. Overall, this work represents an important step forward in the field of corrosion prediction and provides a valuable tool for asset managers seeking to better understand and mitigate corrosion risk.

BIBLIOGRAPHY

Mazzella, Joseph, Hayden, Thomas, Krissa, Len, and Tsaprailis, Haralampos. "Estimating Corrosion Growth Rate for Underground Pipeline: A Machine Learning Based Approach." Paper presented at the *CORROSION 2019*, Nashville, Tennessee, USA, March 2019.

7.4.3.5 Application-28: Artificial Intelligence to Enhance Corrosion under Insulation Inspection

7.4.3.5.1 Background of the Application

This paper presents a solution for the challenge of detecting corrosion under insulation (CUI) in the oil and gas industry. This well-known industrial problem is made worse by the present deficiency of suitable inspection technologies. The strategy employed in this study was to use a specially designed AI based on ML to improve the output of existing established and field-approved thermography technologies. In order to predict corrosion and the factors that lead to this degradation, thermal anomaly features are extracted from thermal images that have been captured over time. These features are then correlated with corrosion and irregularities in the structural integrity of assets that were visually verified during the initial learning phase of the ML algorithm. Additionally, the method offers improved safety through non-contact online examination, is economical, and is safe through remote inspection.

7.4.3.5.2 Description of the Workflow and Machine Learning Model

The author employed an autoencoder NN, including an adversarial autoencoder (AAE) to improve the detection of CUI in the oil and gas industry. The system takes the thermal images of facility pipelines captured by a thermographic camera as input data, and using the adversarial autoencoder, the system detects any anomalies indicative of CUI.

The workflow starts with processing the thermal images of facility pipelines by resizing images to be 150 × 150 to match the model's input format and normalizing the pixel values of the train and validation images between 0 and 1. The autoencoder NN architecture is modeled to include an encoder part that compresses the input image into a lower-dimensional latent space and a decoder part that reconstructs the original image. The unsupervised trained autoencoder NN model can learn background patterns present in thermal images. During the test, the autoencoder deals with the newly obtained thermal images, and the reconstructed images are subtracted from the original inputs to emphasize any kind of anomaly indicative of CUI.

7.4.3.5.3 Advantages and Limitations of the Synchronized Algorithm

The autoencoder NN-based approach for CUI detection in industrial facilities adds a lot of value in detecting CUI since it is a non-destructive and fast technique covering a large inspection area. Also, since the model is trained in an unsupervised manner, the model does not require extensive labeled datasets, which can be challenging to obtain when working with industrial applications. This approach also reduces the need for manual inspections and associated operational disruptions because of the continuous monitoring of the facility. However, the effectiveness of the model is highly dependent on the quality and quantity of the training data. A large and diverse amount of data is required in terms of covering different locations, different processes such as hot or cold insulation, and different seasons/timing. It is crucial to have data that is ground-trothed for training purposes.

7.4.3.5.4 *Results and Discussion*

By applying AI to enable and improve the output of proven and field-approved technologies, such as thermography, the work reported in this paper addresses the current deficiency of technologies to monitor the existence of CUI. Additional advantages of this strategy include cost savings by lowering the complexity of asset preparation (scaffolding) and downtime, as well as safety improvement through non-contact online inspection. The predictive accuracy of the AI technique coupled with a commercial field approved thermographic camera returned results in the range of 85–90%. These results are a significant step in addressing the challenges of CUI detection and monitoring. Future work will focus on further improving the accuracy and efficiency of the detection system and its implementation in real-world scenarios.

BIBLIOGRAPHY

Amer, Ayman, Al-Shehri, Ali, Cunningham, Vincent, Saiari, Hamad, Alshamrany, Abdulaziz, and Meshaikhis, Ali. "Artificial Intelligence to Enhance Corrosion Under Insulation Inspection." Paper presented at the *Abu Dhabi International Petroleum Exhibition & Conference*, Abu Dhabi, UAE, November 2020. doi: https://doi.org/10.2118/202783-MS

7.4.4 Subsea Applications

7.4.4.1 Application-29: Improving Deepwater Facility Uptime Using Machine Learning Approach

7.4.4.1.1 *Background of the Application*

Unplanned deepwater facility outages have resulted in large production losses for the oil and gas sector. The current surveillance technologies aren't always sufficient to spot and anticipate anomalous activity that can result in an unscheduled facility closure. This research focuses on a deepwater facility that has multiple gas compressors in series and multiple oil export pipeline pumps in parallel. This work proposes an intelligent prognostic solution that uses an ML framework to identify key causal process variables and automatically predict upcoming facility outages. For the kind of data available for the deepwater facility, an anomaly detection method based on principal component analysis (PCA) and autoencoder (AE) algorithms was found to perform better. The adaptive ML models predicted unplanned events in 6 months of period, providing adequate early warning to take proactive actions. Case studies shown in the paper present diagnostic charts and identified early indicators that were found in agreement with pre-alarms generated by the existing alarm system, thus validating the ML solution.

7.4.4.1.2 *Description of the Workflow and Machine Learning Model*

The authors employed two ML methods to lower the risk of unplanned downtime. This work proposes an intelligent prognostic solution that uses an ML framework to identify key causal process variables and automatically predict upcoming facility outages. To find, clean, and process real-time data for model training and prediction, a methodical workflow was created.

In order to suggest appropriate mitigation measures, the ML framework additionally facilitated the study of the underlying causes of downtime. Each sub-system boundary and sensor list on which the ML model was trained was chosen using knowledge derived from a physical understanding of the process. The accuracy of the trained models was then evaluated by cross-validating these models. Lastly, the alarm database was utilized to verify the ML models' accuracy and pinpoint the underlying reasons for unscheduled shutdowns.

For the kind of data available for the deep-water facility, anomaly detection based on PCA and AE methods was found to perform better. An ANN type called an autoencoder is used to efficiently learn how to code unlabeled data (unsupervised learning). By trying to regenerate the input from the encoding, the encoding is verified and improved. The autoencoder trains the network to disregard irrelevant data, or "noise," in order to learn a representation (encoding) for a set of data, usually for dimensionality reduction.

7.4.4.1.3 Advantages and Limitations of the Synchronized Algorithm

Data examples are used to automatically train autoencoders. It indicates that it is simple to train certain instances of the algorithm that will function effectively with a particular kind of input and that no new engineering is needed—just the right training data. Autoencoders, however, will perform poorly when it comes to picture compression. The autoencoder will produce decent compression results on data that is similar to the training set because it has been trained on a specific collection of data, but it will not perform well as a general-purpose image compressor. Compression methods such as JPEG will perform significantly better.

7.4.4.1.4 Results and Discussion

The suggested adaptive ML approach has the potential to completely transform topside process monitoring. The value obtained when testing the suggested remedy is recorded. By using a cloud-based model deployment platform to train, predict, and initiate automatic model updates, the ML framework can be scaled for a sustainable solution that spans many process systems and facilities, makes predictions every minute, and maintains the model evergreen. This paper describes an automated workflow to identify and predict unplanned shutdowns before they happen, diagnose, and provide actionable insights. The conclusion outlines future research prospects. The existing model can continue to predict a variety of events, and as enough occurrences are gathered over time, more sophisticated supervised ML techniques can be created and used.

BIBLIOGRAPHY

Singh, Ajay, Sankaran, Sathish, Ambre, Sachin, Srikonda, Rohit, and Houston, Zach. "Improving Deepwater Facility Uptime Using Machine Learning Approach." Paper presented at the *SPE Annual Technical Conference and Exhibition*, Calgary, Alberta, Canada, September 2019. doi: https://doi.org/10.2118/195875-MS

7.4.4.2 Application-30: Offshore Water Treatment KPIs Using Machine Learning Techniques

7.4.4.2.1 Background of the Application

This paper explores the use of ML techniques to monitor and predict key performance indicators (KPIs) for offshore water treatment facilities. Sulfate removal units (SRUs) are commonly used to mitigate reservoir souring and scaling, but monitoring water quality can be challenging. Current practices rely on pressure drop and a constant cleaning interval frequency to perform maintenance, which can result in reduced membrane life or severe fouling. By using ML techniques, a prediction model can be developed based on both simulation and real-time field data. This model tracks and monitors KPIs such as pressure, membrane fouling factor (FF), and permeate sulfate concentration, providing estimates on when maintenance is required and optimizing membrane performance.

7.4.4.2.2 Description of the Workflow and Machine Learning Model

In this method, an MLP regressor is used to predict the FF in offshore water treatment based on input features. The MLP is a type of NN that can capture complex nonlinear relationships between input features and the target variable, which makes it well for this specific job. Additionally, the local outlier factor (LOF) is used for detecting anomalies in membrane performance by identifying unusual data points through comparison of their local density with that of their neighbors.

7.4.4.2.3 Advantages and Limitations of the Synchronized Algorithm

The MLP regressor has advantages such as its ability to model complex nonlinear relationships and its flexibility in handling various types of data. However, it can be prone to overfitting if not properly regularized and typically requires more data for effective training. Additionally, MLPs can be computationally expensive and require careful tuning of their parameters. LOF effectively detects anomalies without assuming specifications but is sensitive to parameter settings. And also, it may have problems with high-dimensional data.

7.4.4.2.4 Results and Discussion

The use of ML techniques for predicting and monitoring KPIs in offshore water treatment facilities has proven to be a valuable tool. The developed prediction model allows for enhanced visibility on membrane performance, aiding in decision-making for frequency of clean in place (CIP), troubleshooting, and optimizing membrane performance. The model also serves as a backup measurement for sulfate analyzers, reducing the risk of reservoir souring. Anomaly detection has been developed to understand deviations in membrane performance and screen out data outliers. By integrating ML models into the operator training simulator (OTS), operators can make informed decisions in the field based on membrane parameters, resulting in cost savings and optimized membrane performance.

BIBLIOGRAPHY

Flores, Lauren, Morles, Martin, and Chen, Cheng. "Offshore Water Treatment KPIs Using Machine Learning Techniques." Paper presented at the *SPE Annual Technical Conference and Exhibition*, Dubai, UAE, September 2021. doi: https://doi.org/10.2118/206173-MS

7.4.4.3 Application-31: Digital Twin of Subsea Pipelines: Conceptual Design Integrating IoT, Machine Learning and Data Analytics

7.4.4.3.1 Background of the Application

The conceptual architecture of a digital twin of subsea pipeline systems is presented in this work, combining ML-based predictive maintenance, field sensor data analytics, and computational modeling. The digital twin is an integrated system that uses ML-based predictive maintenance techniques, sophisticated data analytics to extract valuable insights, and inexpensive Internet of Things sensors to collect system data. The field sensor data installed on the physical system is used to calibrate the computational model, which is initially produced in the finite element (FE) model. Any change in pipe behavior brought on by abrupt changes in loading from high pressure, slugging, or leaks is predicted by the computational model. The proposed digital twin model assists oil and gas field operators in minimizing the OPEX with a predictive maintenance schedule when it is needed to avoid failure in the pipeline system.

7.4.4.3.2 Description of the Workflow and Machine Learning Model

This method uses LSTM networks to predict stress in subsea pipelines from sensor data. The data is normalized and used to train the LSTM model, which captures temporal dependencies. Predicted stress values are compared with measured values, and deviations are analyzed to schedule predictive maintenance based on significant stress deviations.

7.4.4.3.3 Advantages and Limitations of the Synchronized Algorithm

The LSTM-based method models complex temporal patterns in sensor data, necessary for accurate stress prediction of pipelines. It combines multiple sensor inputs for a comprehensive pipeline assessment. However, training LSTM models is computationally expensive and requires large, labeled datasets. The model's performance depends on data quality and consistency between training and real-world conditions.

7.4.4.3.4 Results and Discussion

This paper has demonstrated the effectiveness of the digital twin model for subsea pipeline systems in predicting the maintenance strategy based on the deviation between the model output data and sensor measurement data using ML algorithms. The predictive maintenance strategy based on ML algorithms can significantly reduce the OPEX by optimizing the maintenance schedule. The proposed digital twin model can be used to monitor and maintain the subsea pipeline system based on actual sensor data from the physical field using ML. The authors conclude that with the low-cost sensor development and the advancement of AI, digital twin models can accurately predict the remaining fatigue life of the asset, as well as predict the maintenance requirement based on the data-driven digital model.

BIBLIOGRAPHY

Bhowmik, Subrata. "Digital Twin of Subsea Pipelines: Conceptual Design Integrating IoT, Machine Learning and Data Analytics." Paper presented at the *Offshore Technology Conference*, Houston, Texas, May 2019. doi: https://doi.org/10.4043/29455-MS

7.4.4.4 Application-32: Machine Learning-Based Optimization for Subsea Pipeline Route Design

7.4.4.4.1 Background of the Application

The selection of a subsea pipeline route is a vital task that can significantly affect the overall project cost. The traditional method of manually selecting a pipeline route does not necessarily lead to an optimized engineering cost. This paper proposes a novel hybrid optimization method based on an RL algorithm for designing an optimal pipeline route. The proposed pipeline route selection method uses an RL algorithm, a particular type of ML method, to train a pipeline system that would optimize the route selection of subsea pipelines. The proposed optimal route design is compared with one of the popular multi-objective optimization methods named GA. The proposed method saves up to 20% of pipeline length for a complex problem with 15 crossings and 31 free spans. The proposed RL-based optimization method provides the optimal routes, considering different aspects of the design and the costs associated with the various factors to stabilize a pipeline.

7.4.4.4.2 Description of the Workflow and Machine Learning Model

The author employed an RL algorithm to optimize the route selection of subsea pipelines. This method integrates an environment using the Gym library, which has the input parameters seabed bathymetry (m, meters), static obstacles such as spans and crossings (m, meters), geotechnical data, and environmental loadings such as wave (m/s, meters per second) and current forces. The RL algorithm utilizes a Q-Learning model-free method to determine the discretization scheme's cross fragments and mutation points during the optimization process to minimize the associated costs.

After the development of the custom environment using the Gym library, the RL agent can simulate and interact with the subsea conditions. The custom subsea environment provides a realistic and dynamic simulation for the RL agent using the mentioned parameters. The RL agent is trained through episodes, where the episode number is dependent on the number of states, actions, base episodes, and the scaling factor. Through each episode, the RL agent explores various routes, where it receives feedback via rewards depending on the cost-effectiveness of the chosen route. After each feedback, the RL agent updates the Q-table, which stores the expected utility of actions taken in specific states. As the RL agent keeps exploring the environment, the Q-table is continuously updated, and eventually the agent finds the most optimal subsea route that minimizes operational costs and maximizes stability and safety.

7.4.4.4.3 Advantages and Limitations of the Synchronized Algorithm

In general, if the custom Gym environment provides a realistic and accurate representation of the actual subsea bathymetry, geotechnical data, and geotechnical hazards, Environmental loading such as wave and current, the RL agent provides an optimal route and can reduce pipeline installation and operational costs significantly. Also, this RL framework can be applied to various pipeline projects with different complexities and environmental conditions, making it a flexible tool for reducing the cost and finding the optimal subsea route. However, the RL algorithm heavily depends on the quality and accuracy of the custom Gym subsea environment. Using incomplete, less detailed, or inaccurate data might lead to suboptimal routing decisions.

7.4.4.4.4 *Results and Discussion*

The proposed RL-based optimization method provides a faster and cost-efficient solution for subsea pipeline route optimization. In comparison to the traditional route selection approach, it drastically reduces the pipeline's installation and operating expenses by up to 20%. In addition to other limitations like the minimum pipeline route length, static barriers, pipeline crossings, and free-span section length, the optimization approach based on RL considers the on-bottom stability criteria. The optimization method's cost function manages the reduction of mitigating processes' duration and expense at the same time. The suggested strategy greatly increases convergence speed and accuracy and is faster, more economical, and mixed-type. The suggested approach can shorten the pipeline's design period and lower its installation and operating expenses.

BIBLIOGRAPHY

Bhowmik, Subrata "Machine Learning-Based Optimization for Subsea Pipeline Route Design." Paper presented at the *Offshore Technology Conference*, Virtual and Houston, Texas, August 2021. doi: https://doi.org/10.4043/31031-MS

7.4.4.5 Application-33: Artificial Intelligence Transforms Offshore Analog Fields into Digital Fields

7.4.4.5.1 *Background of the Application*

This paper discusses the use of AI and computer vision to transform offshore analog fields into digital fields. The objective is to capture analog meter readings in a digital format efficiently, reliably, and accurately. The authors detail a process that uses AI to capture analog field-gauge data resulting in a dramatic reduction in cost and an increase in reliability. The solution was implemented in the Cheleken oil field in the Caspian Sea offshore Turkmenistan. During the field trial, operators took pictures of the gauges at given intervals, and the acquired images were processed using AI and deep-learning computer techniques.

7.4.4.5.2 *Description of the Workflow and Machine Learning Model*

In this method, image processing and ML techniques are used to analyze industrial data and enhance production predictions. Initially, gauge readings are extracted from images using OpenCV to isolate the relevant part of the image, while QR code data is decoded to retrieve additional information. These image processing techniques are used together with a ML model that uses pressure and temperature data to predict production rates. By training a linear regression model on this data, production outcomes can be forecasted with a certain degree of accuracy, providing valuable insights for operational decisions.

7.4.4.5.3 *Advantages and Limitations of the Synchronized Algorithm*

This method offers several advantages, like the straightforward implementation of image processing techniques with OpenCV and the simplicity of linear regression for predictive modeling. The combination of visual data extraction and statistical modeling helps in making decisions based on data quickly. However, the approach also has its limitations. The accuracy of predictions is heavily reliant on the quality and relevance of the input data, and the model's performance might be less with noisy or inconsistent data. Additionally, the linear regression model might not capture complex nonlinear relationships in the data, limiting its prediction power in more complex scenarios.

7.4.4.5.4 Results and Discussion

In conclusion, the use of AI and computer vision to capture analog field-gauge data in digital format proved to be an effective solution for offshore oil fields. The solution resulted in a more reliable prediction of production rates from the virtual flowmeter. It also reduced the time and cost of acquisition when acquiring pressure and temperature readings from analog gauges. The authors recommend prioritizing and defining objectives early in the project to achieve the best results. They also recommend training operators in basic photographic techniques to improve the quality of photos, thus improving the accuracy of the results. Overall, this solution provides a more efficient and reliable method for capturing and analyzing data in offshore oil fields.

BIBLIOGRAPHY

Carpenter, Chris. Artificial Intelligence Transforms Offshore Analog Fields Into Digital Fields. *Journal of Petroleum Technology* 72 2020, 71–72. doi: https://doi.org/10.2118/0120-0071-JPT

7.4.5 Reservoir-Related Applications

7.4.5.1 Application-34: Hybrid SVM-CIPSO Methods for Optimal Operation of Reservoir Considering Unknown Future Condition

7.4.5.1.1 Background of the Application

In order to solve the reservoir operation optimization problem under unknown water inflow conditions, this research suggests innovative hybrid approaches. The SVM model is combined with the improved particle swarm optimization (IPSO) algorithm. By explicitly meeting the problem requirements, the constrained improved PSO (CIPSO) algorithm reduces the search space and processing cost of the IPSO algorithm while increasing its efficiency. For the hybrid methods, two strategies are taken into consideration. The first method, called SVM–CIPSO1, uses the SVM model to predict water inflows into the dam reservoir. The CIPSO algorithm is then utilized to address the reservoir operation optimization problem. However, in the second approach, named SVM–CIPSO2, the CIPSO algorithm is applied to solve the reservoir operation optimization problem using historical data, and the optimal water release values are used as input and output data to create an SVM model for predicting optimal water release from the reservoir for future conditions. The results indicate the capability of the proposed methods to predict optimal water release values for future conditions with acceptable accuracy.

7.4.5.1.2 Description of the Workflow and Machine Learning Model

In this method, an SVM with an RBF kernel is used to predict optimal water release from a reservoir based on inflow and rainfall data. SVM is effective for capturing complex patterns between the inputs and the target variable (output). A CIPSO algorithm is also used to fine-tune the SVM parameters. CIPSO enhances the optimization by efficiently narrowing down the search space, making the model more accurate and less computationally expensive.

7.4.5.1.3 Advantages and Limitations of the Synchronized Algorithm

The SVM with RBF kernel can model complex relationships and is robust against overfitting, making it ideal for predicting reservoir operations. However, it requires careful tuning of parameters and can be slow with large datasets. CIPSO improves the SVM by optimizing these parameters efficiently, leading to better performance. Despite its benefits, CIPSO's performance depends a lot on its settings, and it might need a couple of iterations to find the best solution. Thus, it is a little computationally expensive.

7.4.5.1.4 Results and Discussion

The proposed hybrid methods show good capability to predict optimal water release values for future conditions. The SVM–CIPSO2 method performs better than the SVM–CIPSO1 method, with the best reliability, resiliency, vulnerability, and sustainability index values obtained from the SVM–CIPSO2 method. The IPSO algorithm with the CIPSO modification is effective in solving the reservoir operation optimization problem. The SVM and ANN models are useful for predicting water inflow or release values for future conditions. The proposed hybrid methods can help in the optimal management and allocation of water resources.

BIBLIOGRAPHY

Moeini, Ramtin and Babaei, Mohammad. Hybrid SVM-CIPSO methods for optimal operation of reservoir considering unknown future condition. *Applied Soft Computing* 95, 2020, 106572, ISSN 1568-4946. https://doi.org/10.1016/j.asoc.2020.106572

7.4.5.2 Application-35: Machine Learning Models to Predict Gas Hydrate Plugging Risks Using Flow Loop and Field Data

7.4.5.2.1 Background of the Application

This paper presents ML models to predict gas hydrate plugging risks using flow loop and field data. The models were developed to help flow assurance engineers simulate gas hydrate formation and transportability in subsea oil and gas pipelines. Several ML techniques, including regression, classification, feature learning with SVM and NNs, were applied to analyze hydrate tests conducted at pilot-scale flow loop facilities and transient operation field data. The hydrate fraction and likelihood of hydrate plugging in the pipeline are produced by the classification/regression model based on flow loop test data using a number of independent input factors. To ascertain the degree of hydrate resistance to flow during restart or dead oil displacement following production shutdown, a semi-supervised learning model was implemented using field data as input.

7.4.5.2.2 Description of the Workflow and Machine Learning Model

In this method, a NN is used to predict hydrate plugging risk and hydrate volume fraction based on input features. NNs capture complex patterns in data through multiple layers of interconnected neurons, improving accuracy via backpropagation, where errors are continuously minimized. This method uses the DL capability of NNs to model nonlinear relationships, using layers to prevent overfitting.

7.4.5.2.3 Advantages and Limitations of the Synchronized Algorithm

NNs offer high accuracy by modeling complex interactions and preventing overfitting through dropout layers. They are versatile, handling both classification and regression tasks effectively. However, they are significantly computationally expensive and need expertise in DL techniques for training and tuning. While they are often seen as "black box" models, limiting understanding of its workings, feature importance analysis techniques can show us some information behind the model's decisions.

7.4.5.2.4 Results and Discussion

In order to assess hydrate risks in submerged oil and gas pipelines, this study shows how ML models can be used in hydrate management. This ML method can help create sophisticated hydrate management plans in addition to more intricate transient multiphase flow simulations. The flow loop-based ML model was utilized to evaluate the hydrate risks in a real-world field and showed strong prediction accuracy in test and validation procedures. It was shown that the ML model based on field data could create field danger maps. The cost burden of hydrate control measures in subsea oil and gas pipelines could be greatly decreased by the ML technology.

BIBLIOGRAPHY

Qin, Hao, Srivastava, Vishal, Wang, Hua, Zerpa, Luis E., and Koh, Carolyn A. "Machine Learning Models to Predict Gas Hydrate Plugging Risks Using Flowloop and Field Data." Paper presented at the *Offshore Technology Conference*, Houston, Texas, May 2019. doi: https://doi.org/10.4043/29411-MS

7.4.5.3 Application-36: Decision Tree Regressions for Estimating Liquid Holdup in Two-Phase Gas-Liquid Flows

7.4.5.3.1 Background of the Application

Two-phase gas and liquid movement is typical in oil and gas wells. The first step in looking into production well problems is estimating liquid holdup. The ratio of a pipe element's liquid volume to its volume is known as liquid holdup. This makes it difficult to calculate the physical parameters of the mixture and the pressure decrease in two-phase flow. When building oil and gas facilities, predicting liquid holdup is crucial. Numerous investigations in the past have produced numerous empirical relationships with varying levels of accuracy. In contrast to current empirical correlations, ML techniques have lately been investigated to increase accuracy. These methods, however, disregarded the input parameters' heuristic feature significance in relation to the anticipated liquid holdup values. This work suggests using boosted decision tree regression (BDTR), an ML predictive model, to forecast liquid holdup in two-phase gas-liquid flows in gas and oil wells. Experimental data gathered from the literature has been used to train, test, and assess the BDTR model. According to the results, the fuzzy logic model and the best empirical correlations for estimating liquid holdup in gas–liquid multiphase flows are both outperformed by the suggested BDTR model. The superficial gas velocity (Vsg) is the most crucial input element for the constructed model's liquid holdup estimation. The empirical correlations' flow condition constraints do not affect this ML model.

7.4.5.3.2 Description of the Workflow and Machine Learning Model

In this method, writers use a gradient boosting regressor (GBR) [or boosted decision tree regression (BDTR)]to predict the target variable "HL" based on features. BDTR improves its accuracy through iterative learning, correcting errors from previous iterations. The method leverages feature importance analysis to determine the contribution of each feature to the model's predictions, providing valuable information about the data's underlying workings and relationships.

7.4.5.3.3 Advantages and Limitations of the Synchronized Algorithm

BDTR has advantages like robust performance, effectively handling overfitting, and modeling complex feature interactions. It delivers high accuracy and strong performance metrics like mean squared error and R-squared values. However, it is computationally expensive and requires meticulous hyperparameter tuning to avoid overfitting. Additionally, while understanding these models can be difficult, feature importance metrics help identify the most influential variables, partially helping to address this limitation.

7.4.5.3.4 Results and Discussion

The proposed BDTR model outperforms the best empirical correlations and the fuzzy logic model used in estimating liquid holdup in gas–liquid multiphase flows. The most important input feature in estimating liquid holdup for the built model is the superficial gas velocity (Vsg). The proposed model can be extended to predict other properties or to solve other regression problems in upstream oil and gas processes. The proposed predictive model can easily be integrated into simulators. The BDTR model provides an interpretable representation of the feature importance of the input features used in building the model. This helps in having a better understanding of the issues associated with the liquid holdup studies, such as the liquid loading phenomenon.

BIBLIOGRAPHY

Almashan, Meshal, Narusue, Yoshiaki, and Morikawa, Hiroyuki. "Decision Tree Regressions for Estimating Liquid Holdup in Two-Phase Gas-Liquid Flows." Paper presented at the *Abu Dhabi International Petroleum Exhibition & Conference*, Abu Dhabi, UAE, November 2020. doi: https://doi.org/10.2118/203448-MS

7.4.5.4 Application-37: Well GOR Prediction from Surface Gas Composition in Shale Reservoirs

7.4.5.4.1 Background of the Application

This paper presents an ML approach to predict gas to oil ratio (GOR) in shale reservoirs using early production surface gas data. The study involves building a large unconventional reservoir fluid database with more than 2400 samples from shale reservoirs in Canada, Argentina, and the USA. The surface gas compositional data and well location are input features to the ML model. The GOR predictions obtained from both the Canada-focused ML model and the generic ML model are good, with a mean percentage error (MAPE) of 4.31 and 4.86, respectively. The study confirms that surface gas data analysis is a practical approach for GOR prediction in shale reservoirs.

7.4.5.4.2 Description of the Workflow and Machine Learning Model

An RF is a method that constructs multiple decision trees during training and combines their outputs to make more accurate predictions. Each tree is built from a sample of the data, and random subsets of features are considered for each split. This randomness results in diversity among the trees and helps reduce overfitting. This approach allows for more accurate gas-to-oil prediction in shale reservoirs with limited fluid data.

7.4.5.4.3 Advantages and Limitations of the Synchronized Algorithm

RFs have advantages like high accuracy and robustness against overfitting. They handle both classification and regression tasks well and can process complex data effectively. However, they can be computationally expensive, requiring more memory and processing power than other models. They lack the interpretability of single decision trees. Moreover, they may have a hard time with imbalanced datasets, often sliding toward the majority class unless specific techniques are applied.

7.4.5.4.4 Results and Discussion

This study confirms that early production surface gas data can be used to predict well GOR in shale reservoirs, providing an economical alternative for the sampling challenges during early field development. The ML models developed in this study ensure that an adequate GOR prediction for shale reservoirs with limited fluid data is possible. The generic ML model demonstrates good performance without shale dependency on the GOR prediction, provided the up-to-date database used in the study. Therefore, the developed ML model can be applied to shale reservoirs with limited fluid data on different shale plays.

BIBLIOGRAPHY

Cely, Alexandra, Zaostrovski, Andrei, Yang, Tao, Uleberg, Knut, and Kopal, Margarete. "Well GOR Prediction from Surface Gas Composition in Shale Reservoirs." Paper presented at the *SPE Annual Technical Conference and Exhibition*, Dubai, UAE, September 2021. doi: https://doi.org/10.2118/205842-MS

7.4.5.5 Application-38: A Novel Well Test Data Analyzer and Process Optimizer Using Artificial Intelligence and Machine Learning Techniques

7.4.5.5.1 Background of the Application

This paper proposes a data-driven solution using ML and AI techniques to improve the efficiency and accuracy of well testing in the oil and gas industry. The proposed methodology involves a supervised well test validation model and an unsupervised well anomaly model that can predict well test validation flags and identify the reasons for well test rejections in real time during well test execution. The methodology has been evaluated on a mature oil-producing field consisting of over 500 producing strings and produced from multiple reservoirs. The results show that the proposed methodology has the potential to reduce well test rejection percentage through quick validations and re-testing, and feed right validated well tests to the back allocation process on time. Furthermore, the models have the potential to reduce human bias in the validation process and increase confidence over time before it is deployed into real-time mode.

7.4.5.5.2 Description of the Workflow and Machine Learning Model

The author employed an unsupervised ML model using a K-means clustering algorithm to enhance the validation and analysis of well test data in the oil and gas industry. The K-means clustering algorithm takes the historical well test data as inputs and then identifies common issues such as GOR anomalies, liquid rate anomalies, and MPFM calibration issues.

The algorithm starts with reprocessing the dataset by scaling and standardizing the features imported from the historical well test data file using StandardScaler. The optimal number of clusters is then decided using the Elbow Method, facilitated by Yellowbrick's KElbowVisualizer algorithm. Then, the unsupervised K-means clustering algorithm is applied to the scaled feature dataset to identify patterns and anomalies in the well test results. Finally, the resulting clusters are used to analyze and identify common issues such as GOR anomalies, liquid rate anomalies, and MPFM calibration issues.

7.4.5.5.3 Advantages and Limitations of the Synchronized Algorithm

Using an unsupervised K-means clustering algorithm to detect common issues such as GOR anomalies, liquid rate anomalies, and MPFM calibration issues from the historical well test data significantly reduces the time required for well-test validation and optimizes operational efficiency. Also, using an automated unsupervised K-means clustering algorithm for detecting common anomalies in the oil and gas industry reduces human bias and error and provides a reliable validation process. When it comes to limitations, the model is heavily dependent on the quality and quantity of the data; incomplete and insufficient datasets can lead to inaccurate and suboptimal predictions.

7.4.5.5.4 Results and Discussion

This paper presents a novel approach to applying ML and AI techniques to well testing in the oil and gas industry. The proposed methodology has been demonstrated to be effective in reducing well test rejection percentage, optimizing resources required, improving the back-allocation process, and complying with well testing requirements in an asset where resources are limited. The proposed methodology has the potential to increase efficiency, reduce turnaround time, and improve performance in well testing operations. The authors suggest that further testing and validation are required before the methodology can be deployed in real-time mode, and they encourage future research in this area.

BIBLIOGRAPHY

Reddicharla, Nagaraju, Rachapudi, Subba Ramarao, Utama, Indra, Khan, Furqan Ahmed, Vanam, Prabhker Reddy, Al Nuimi, Mubarak, Saber, and Ali, Mayada Ali Sultan. "A Novel Well Test Data Analyzer and Process Optimizer Using Artificial Intelligence and Machine Learning Techniques." Paper presented at the *SPE Annual Technical Conference and Exhibition*, Dubai, UAE, September 2021. doi: https://doi.org/10.2118/206137-MS

7.4.5.6 Application-39: A Hybrid Artificial Intelligence Approach in Well Test Interpretation

7.4.5.6.1 Background of the Application

This paper presents an integrated approach to reservoir characterization based on the integration of seismic interpretation, well log analysis, and geological modeling. The approach is applied to a heterogeneous sandstone reservoir in an offshore environment, and the results demonstrate the effectiveness of the integrated approach in improving reservoir characterization accuracy. The paper discusses the challenges associated with traditional reservoir characterization techniques and highlights the benefits of the integrated approach.

7.4.5.6.2 Description of the Workflow and Machine Learning Model

In this method, an NN is used to predict target values based on well log data. The data undergoes scaling and splitting into training and testing sets. An NN model is trained, and its performance is evaluated using the MAE metric. Porosity levels are interpreted by incorporating fuzzy logic, adding a qualitative layer to the quantitative predictions. Additionally, model parameters are optimized with differential evolution to enhance prediction accuracy.

7.4.5.6.3 Advantages and Limitations of the Synchronized Algorithm

Combining NNs with fuzzy logic results in predictions that are both accurate and easy to understand. The NN captures complex patterns in the data, while fuzzy logic offers clear, quality information. Differential evolution helps in fine-tuning model parameters for improved performance. However, this approach can be computationally expensive and relies on the quality of the data. The fuzzy logic component, although it is helpful, can introduce some different results in each variation, which may affect the consistency of the results.

7.4.5.6.4 Results and Discussion

In conclusion, the integrated approach presented in this paper provides an effective means of reservoir characterization in complex reservoirs. The approach combines seismic interpretation, well log analysis, and geological modeling to provide accurate and comprehensive reservoir characterization. The results of the case study demonstrate the effectiveness of the integrated approach in improving reservoir characterization accuracy and reducing uncertainty. The approach has the potential to be applied to a wide range of reservoir types and environments and has the potential to significantly improve reservoir management and production optimization.

BIBLIOGRAPHY

Houze, O. P., and Allain, O. F. "A Hybrid Artificial Intelligence Approach in Well Test Interpretation." Paper presented at the *SPE Annual Technical Conference and Exhibition*, Washington, D.C., October 1992. doi: https://doi.org/10.2118/24733-MS

7.4.5.7 Application-40: A Novel Correlation to Predict Gas Flow Rates Utilizing Artificial Intelligence: An Industrial 4.0 Approach

7.4.5.7.1 Background of the Application

This paper presents a study on the prediction of gas flow rates through wellhead chokes in gas-condensate reservoirs. The study focuses on developing a correlation model using AI techniques, including ANN, functional network (FN), and adaptive neuro fuzzy inference system (ANFIS). The model takes flowing well-head pressure, upstream temperature, choke size, flow rate of condensate, and flow rate of water as inputs and estimates gas flow rates with accuracy greater than 90%. The model's performance was compared with several other models, and it was found to be more accurate and reliable.

7.4.5.7.2 Description of the Workflow and Machine Learning Model

ANNs are interconnected neurons arranged in layers, with each connection weighted and adjusted during training. The model learns from data by minimizing errors between its predictions and actual outcomes. In the case of gas lift wells, ANNs can be utilized to predict the desired output, such as gas production rates, by learning from historical data. This predictive capability is essential for optimizing the performance and efficiency of these operations.

7.4.5.7.3 Advantages and Limitations of the Synchronized Algorithm

ANNs have several advantages, including their ability to handle nonlinear and complex relationships, learn from data, and perform well across areas. However, ANNs also have certain limitations. They require large amounts of data and significant computational power for training. Moreover, ANNs can be challenging to interpret due to their complex structure, making it difficult to understand the behind the process.

7.4.5.7.4 Results and Discussion

In conclusion, this study presents a correlation model for predicting gas flow rates through wellhead chokes that outperforms other existing models. The model was developed using AI techniques, including ANN, FN, and ANFIS, with inputs of flowing well-head pressure, upstream temperature, choke size, flow rate of condensate, and flow rate of water. The results show that the model accurately estimates gas flow rates with an accuracy greater than 90%. The study highlights the importance of using AI techniques in the oil and gas industry to improve the accuracy and reliability of production forecasts. Further research can be done to incorporate more complex reservoir and production parameters into the model to improve its accuracy and applicability in other reservoirs.

BIBLIOGRAPHY

Kalam, Shams, Khan, Mohammad Rasheed, Tariq, Zeeshan, Siddique, Faisal Anwar, Abdulraheem, Abdulazeez, and Khan, Rizwan Ahmed. "A Novel Correlation to Predict Gas Flow Rates Utilizing Artificial Intelligence: An Industrial 4.0 Approach." Paper presented at the *SPE/PAPG Pakistan Section Annual Technical Symposium and Exhibition*, Islamabad, Pakistan, November 2019. doi: https://doi.org/10.2118/201170-MS

7.4.5.8 Application-41: Machine Learning Application for Oil Rate Prediction in Artificial Gas Lift Wells

7.4.5.8.1 Background of the Application

This study aims to develop a correlation that can accurately predict oil rate in artificial gas lift wells. The developed correlation is based on ML algorithms, namely ANFIS, SVMs, and ANN, providing a usable equation that can be applied to any field. The developed correlation outperforms all the current empirical correlations and also performs multiple times better than previously developed AI models.

7.4.5.8.2 Description of the Workflow and Machine Learning Model

ANN consists of interconnected neurons arranged in layers, with each connection weighted and adjusted during training. The model learns from data by minimizing errors between its predictions and actual outcomes. Thus, it can be used to predict the desired output from the previous data of gas lift wells.

7.4.5.8.3 Advantages and Limitations of the Synchronized Algorithm

ANNs have advantages that include their ability to handle nonlinear and complex relationships, learn from data, and perform well in various fields. However, they require large amounts of data and computational power, can be challenging to interpret, and may overfit if not properly regularized.

7.4.5.8.4 Results and Discussion

The results of this study show that ML algorithms can be effectively used to predict the flow rate of artificially lifted wells. The recently created correlation eliminates any doubts or misunderstandings regarding AI knowledge and software by being more accurate than existing empirical correlations and offering a functional equation that everyone may use to their field data. In addition to validating the well tests, this endeavor offers an industrial perspective on the function of data-driven computational models for the production reconnaissance scheme as a useful instrument to lessen production provision concerns.

BIBLIOGRAPHY

Khan, Mohammad Rasheed, Alnuaim, Sami, Tariq, Zeeshan, and Abdulraheem, Abdulazeez. "Machine Learning Application for Oil Rate Prediction in Artificial Gas Lift Wells." Paper presented at the *SPE Middle East Oil and Gas Show and Conference*, Manama, Bahrain, March 2019. doi: https://doi.org/10.2118/194713-MS

7.4.5.9 Application-42: Machine Learning Applications for a Qualitative Evaluation of the Fracture Network

7.4.5.9.1 Background of the Application

This paper presents a data-driven approach for obtaining detailed insights into the fracture network and dynamic behavior of reservoirs using a combination of tracer, completion, and stimulation data. The authors employed advanced ML techniques, specifically an MLP NN model, to process the data and predict the observed tracer value. Eleven horizontal wells were drilled in the Wolfcamp Formation at the Hydraulic Fracturing Test Site-1 in the Midland Basin, Texas, where the data was collected. In order to support decisions that maximize hydrocarbon recovery, the authors created a four-dimensional numerical and visual model of the fracture network and its flow behavior.

7.4.5.9.2 Description of the Workflow and Machine Learning Model

An MLP network is an ANN with a feedforward architecture consisting of one input, multiple hidden, and one output layers. Neurons apply activation functions to enable the NN to handle nonlinearity. MLPs are used for classification, regression, and pattern recognition tasks. In this case, it is being used to classify fracture networks and their flow behavior.

7.4.5.9.3 Advantages and Limitations of the Synchronized Algorithm

MLPs have many advantages; they are versatile and can model complex, nonlinear relationships, making them suitable for inspecting the complex fracture network. They can be scaled to handle large datasets and are universal approximators. However, training MLPs can be computationally expensive and time-consuming and also hard to interpret. They are easy to overfit and require careful tuning of hyperparameters.

7.4.5.9.4 Results and Discussion

This study demonstrates that a data-driven approach using advanced ML techniques can provide detailed insights into the fracture network and dynamic behavior of reservoirs. The results of this study show that the MLP model can capture the patterns inherent in the data and provide predictions with a high degree of accuracy. To give the most realistic picture of the subsurface fracture network, a thorough grasp of the limitations and uncertainties in these models is required, as the data employed in this study is limited. This study lays the groundwork for future investigations that will enable the fracture network to be examined from a variety of perspectives. Additionally, the data-centric and model-centric AI methodologies will be balanced to prevent an excessive focus on either data or model improvement. To present the most accurate, comprehensive picture of the fracture network, various ML algorithms and visualization strategies will be investigated. In the end, decisions that maximize hydrocarbon recovery and lessen environmental effects will be supported by these data-driven insights.

BIBLIOGRAPHY

Kumar, Abhash, Shih, Chung Yan, Liu, Guoxiang, Holcomb, Paul, Zhao, Song, Hammack, Richard, Ilconich, Jeffery, and Bromhal, Grant. "Machine Learning Applications for a Qualitative Evaluation of the Fracture Network in the Wolfcamp Shale Using Tracer and Completion Data." Paper presented at the *SPE/AAPG/SEG Unconventional Resources Technology Conference*, Houston, Texas, USA, July 2021. doi: https://doi.org/10.15530/urtec-2021-5159

7.4.5.10 Application-43: The Use of Novel Technology of Inflow Chemical Tracers in Continuous Production Surveillance of Horizontal Wells

7.4.5.10.1 Background of the Application

The paper describes the implementation of a new inflow chemical tracer technology at the Upper-Salym oilfield as part of the Smart Field project. The technology uses nanoparticle quantum dots to provide high accuracy in the quantitative analysis of fluid inflow. Markers are inserted into the polymer coating of proppant, and the technology involves the injection of marked polymer-coated proppant during multi-stage hydraulic fracturing. Once the well is producing, fluid samples are taken at the surface and tested in a laboratory using ML software. A flow profile of oil and water can be generated for each frac stage, providing new opportunities for more accurate reservoir characterization and better hydrocarbon recovery. The advantages of this technology include continuous production surveillance, quantitative analysis of oil and water phase from each fracturing stage, and no complex and risky well interventions or production shutdowns, making it substantially more cost-effective than existing conventional methods.

7.4.5.10.2 Description of the Workflow and Machine Learning Model

An RF classifier is used to predict the flow profiles of oil and water from well fluid data. The data includes parameters such as proppant size, drawdown pressure, fluid rate, bottom-hole pressure, GOR, water cut, permeability, porosity, and laser signal measurements. The RF algorithm, an ensemble of decision trees, uses predictions from multiple trees to enhance accuracy and robustness, providing reliable predictions of fluid flows.

7.4.5.10.3 Advantages and Limitations of the Synchronized Algorithm

The RF method offers high accuracy, difficulty in overfitting, and effectively handles large datasets. Its approach produces stable predictions by averaging multiple trees' outputs. However, it can be computationally expensive and less interpretable compared to simpler models, making the decision-making process more difficult. Despite these limitations, RFs are a good tool for predicting well fluid flow profiles.

7.4.5.10.4 Results and Discussion

The new generation of inflow markers allows for continuous production surveillance and quantitative analysis of the oil and water phases from each fracturing stage. This enables better decision-making to optimize production and make better decisions for water conformance interventions. The implementation of the Smart Fields technology systems in the Salym group of oilfields has led to optimization of oil production, remote monitoring for risk minimization, and reduction of operating costs.

BIBLIOGRAPHY

Chaplygin, Dmitry, Azamatov, Marat, Khamadaliev, Damir, Yashnev, Viktor, Novikov, Igor, Drobot, Albina, Buyanov, Anton, Ovchinnikov, Kirill, and Husein, Nadir. "The Use of Novel Technology of Inflow Chemical Tracers in Continuous Production Surveillance of Horizontal Wells." Paper presented at the *Abu Dhabi International Petroleum Exhibition & Conference*, Abu Dhabi, UAE, November 2020. doi: https://doi.org/10.2118/203345-MS

7.4.5.11 Application-44: Sensor Fusion and Machine Learning Techniques to Improve Water Cut Measurements Accuracy in Multiphase Application

7.4.5.11.1 Background of the Application

A multi-phase flow meter (MPFM) estimates the water liquid ratio (WLR) using a dedicated sensor based on a specific physical principle, such as electrical impedance. The accuracy of the WLR sensor can be influenced by the flow properties. To improve the reliability and accuracy of WLR estimations in multi-phase applications, an ML approach combined with multi-sensor data fusion has been implemented, utilizing the onboard sensor measurements of an MPFM.

7.4.5.11.2 Description of the Workflow and Machine Learning Model

RBF Kernel Ridge is an ML method that merges Kernel Ridge Regression with the RBF to manage nonlinear relationships in data. Ridge Regression adds a regularization term to the loss function to prevent overfitting, while the RBF kernel transforms data into a higher-dimensional space. This combination enables linear regression techniques to be used in nonlinear relationships like in the data for water cut measurements.

7.4.5.11.3 Advantages and Limitations of the Synchronized Algorithm

The advantages of RBF kernel ridge are its ability to model complex nonlinearities and its robustness to overfitting. However, it has limitations, such as high computational complexity for large datasets because of the operations on large kernel matrices.

7.4.5.11.4 Results and Discussion

The proposed approach using ML and multi-sensor data fusion has been shown to improve the accuracy of water cut measurements in multiphase applications. The algorithms have been trained using data collected from flow loop test facilities with different flow conditions. The best model was selected based on predictive performance, computational time, and robustness to outliers. The resulting model provides enhanced estimations equipped with confidence intervals that can be used for prediction quality assessment and associated risk management.

BIBLIOGRAPHY

Barbariol, Tommaso, Feltresi, Enrico, Susto, Gian Antonio, Tescaro, Diego, and Galvanin, Silvia. "Sensor Fusion and Machine Learning Techniques to Improve Water Cut Measurements Accuracy In Multiphase Application." Paper presented at the *SPE Annual Technical Conference and Exhibition*, Virtual, October 2020. doi: https://doi.org/10.2118/201295-MS

7.4.5.12 Application-45: Machine Learning Application for Wellbore Damage Removal in the Wilmington Field

7.4.5.12.1 Background of the Application

In order to help with the screening and selection of potential applicants for acid maintenance positions at the Wilmington Oil Field in southern California, this paper offers a case study that uses ML techniques. More than 500 producers and 3900 injector acid maintenance operations completed in the field served as the basis for training the ML algorithms, which also included predictors from a variety of categories. To find the predictor variables with the highest explanatory power, more than 100 were gathered and filtered using supervised feature selection. The economic viability of a producer acid maintenance job was then effectively predicted using a number of ML algorithms.

7.4.5.12.2 Description of the Workflow and Machine Learning Model

This method uses logistic regression, which is a supervised ML algorithm that accomplishes binary classification. It provides a binary outcome limited to two possible outcomes: yes/no, 0/1, or true/false from the input. The resulting model predicts the economic success of acid maintenance jobs in the Wilmington Oil Field.

7.4.5.12.3 Advantages and Limitations of the Synchronized Algorithm

Logistic regression is simple, interpretable, and efficient, providing the probability of outputs for binary and multiclass classification. It handles regularization well but assumes a linear relationship between features and outcomes. Logistic regression was chosen for its ease of deployment in screening future acid maintenance jobs. It can struggle with high-dimensional data, outliers, and complex interactions, making it less effective than advanced models for capturing intricate patterns.

7.4.5.12.4 Results and Discussion

This study shows that, even for smaller datasets, ML techniques can be useful analytical tools for the oil and gas sector. The 700+ producers and 400+ injectors in the Wilmington Field have all been actively screened for acid maintenance job applicants using the established ML model, which has improved the program's effectiveness and saved a substantial amount of money. The study's findings provide a range of potential uses that surveillance teams can implement in addition to more conventional methods.

BIBLIOGRAPHY

Kellogg, Ryan P., Chessum, William, and Kwong, Ryan. "Machine Learning Application for Wellbore Damage Removal in the Wilmington Field." Paper presented at the *SPE Western Regional Meeting, Garden Grove*, California, USA, April 2018. doi: https://doi.org/10.2118/190037-MS

7.4.5.13 Application-46: From Data to Actionable Insight: Monitoring Fugitive Methane Emissions at Oil and Gas Facilities Using Satellites

7.4.5.13.1 Background of the Application

Due to the low spatial resolution of research satellites that monitor gases, satellite-based greenhouse gas monitoring has historically been restricted to a regional or global scale. Consequently, it had been impossible to link emissions to particular plants. But in June 2016, GHGSat launched its first satellite, GHGSat-D, which was created especially to track emissions straight from industrial sites with a spatial resolution of less than 50 meters. Oil and gas businesses may now affordably monitor their installations spread across large areas thanks to this method.

7.4.5.13.2 Description of the Workflow and Machine Learning Model

The author employed a CNN) to detect gas leaks in industrial facilities based on the thermal and standard images captured by a satellite. The system takes the high-quality thermal and standard images from the satellite as inputs, then the CNN model predicts whether the facility is experiencing a gas leak or not.

The workflow begins with processing the satellite image data by resizing images to be 150 × 150 to match the model's input format and normalizing the pixel values of the train and validation images between 0 and 1. The CNN architecture consists of several convolutional layers followed by max-pooling layers, a flattening layer, dense layers, and a final dropout layer to prevent overfitting. After defining the model, the model is then compiled using the Adam optimizer and the binary cross-entropy loss function to classify the images as a leak (1) or as a non-leak (0) image. The model is trained using the accuracy metrics, and the model summary is printed at the end.

7.4.5.13.3 Advantages and Limitations of the Synchronized Algorithm

The CNN based approach for gas leak detection in industrial facilities includes automated and accurate detection of gas leaks, which can significantly reduce the risks of accidents in industrial areas. If the model is trained with accurate and sufficient thermal and standard image data, the model can reach high accuracy when it comes to predicting whether an industrial facility is experiencing a gas leak or not based on satellite data. In contrast, if the model is trained by poor-quality data, the model's accuracy would be negatively affected thus making the system less viable. To ensure the maximum accuracy of the model, variations in image quality, environmental factors, and different types of leaks must be well-represented in the training dataset.

7.4.5.13.4 Results and Discussion

The ability of GHGSat's technology to work together with other sources of data in an effective tiered monitoring system has been demonstrated to be more effective and lower cost than the typical three OGI campaigns per year mandated by several regulators. GHGSat's innovative products and applications, developed using analytics, AI and ML, provide customers with actionable insights to optimize the operation of the system. The technology's ability to monitor sources from space and the air with the same sensor is a capability unique to GHGSat.

BIBLIOGRAPHY

Gauthier, Jean-Francois, and Germain, Stephane. "From Data to Actionable Insight: Monitoring Fugitive Methane Emissions at Oil and Gas Facilities Using Satellites." Paper presented at the *Abu Dhabi International Petroleum Exhibition & Conference*, Abu Dhabi, UAE, November 2019. doi: https://doi.org/10.2118/197783-MS

7.4.5.14 Application-47: Slug Flow Root Cause Analysis: A Data-Driven Approach

7.4.5.14.1 Background of the Application

This paper presents a data-driven approach to root cause analysis of slug flow in a subsea production system. The approach is based on a slug severity measure, which is quantified by pressure spikes, and ML techniques with feature ranking properties. The method is applied to historical production data from sensors measuring various signals such as pressures, temperatures, flow rates, pipeline pressures, and temperatures. To better identify the fundamental core cause of slugging, the analysis's findings are combined with simulator studies and technical expertise. This method is new because it models and analyzes historical production data using ML techniques to identify the factors that contribute to phenomena like slug flow.

7.4.5.14.2 Description of the Workflow and Machine Learning Model

ML methods are used to analyze historical production data for identifying the root causes of slug flow in subsea fields. Sensor data like pressures and flow rates are standardized and fed into to method as training and test sets. An RF regressor predicts slug severity, and recursive feature elimination (RFE) identifies the most influential features. Feature importance is visualized to understand key drivers of slug flow.

7.4.5.14.3 Advantages and Limitations of the Synchronized Algorithm

The method uses the interpretability of RFs and the effective feature selection of RFE, simplifying the model of data. However, its performance depends on data quality and may miss some complex temporal patterns. Combining ML with domain expertise and simulator studies provides a good base for understanding slug flow despite previously stated limitations.

7.4.5.14.4 Results and Discussion

In conclusion, a data-driven approach was successfully applied to illuminate the root causes of slugging in a production system on the NCS. For additional root cause investigation, feature priority ranking might be employed to identify probable slugging causes by training models on real production data. Low, moderate, and high slugging were successfully distinguished by the slug severity scale. A well branch that produced a lot of water was found and closed after the liquid rate was shown to be the primary cause. The operator attested to the fact that slugging was lessened by closing this branch. This could bring the overall production of oil and gas back to what it was before the start of slugging and the subsequent choking of riser production. The data-driven methodology significantly speeds up the examination of high-dimensional, large-scale production data, offering insightful information for crucial slug mitigation and avoidance decisions.

BIBLIOGRAPHY

Sandnes, Anders T., Uglane, Vidar, and Grimstad, Bjarne. "Slug Flow Root Cause Analysis: A Data-Driven Approach." Paper presented at the *Offshore Technology Conference Brasil*, Rio de Janeiro, Brazil, October 2019. doi: https://doi.org/10.4043/29925-MS

7.4.5.15 Application-48: Investigating the Use of ML Models for Prediction of Pressure Grad and Flow Regimes in Multiphase Flow in Horizontal Pipes

7.4.5.15.1 Background of the Application

This paper explores the use of supervised ML models to predict pressure gradient and flow regimes in multiphase flow in horizontal pipes. The models take input parameters such as relative phase volume, bulk fluid flow rates, individual phase flow rates, conduit diameters, inclination, phase densities, and temperature. The dataset used for the study was culled from a multiphase flow data repository, with a filter applied for data corresponding to horizontal flow. The models used were logistic regression, SVMs, and PCA. The trained model exhibits potential for usage in the industry since it enables the automation of flow control systems and provides a more thorough method of addressing flow issues in pipeline and oilfield systems.

7.4.5.15.2 Description of the Workflow and Machine Learning Model

PCA reduces the feature dimensions. Logistic regression and SVM models are trained on the transformed data. The models' performance is then evaluated using classification reports and confusion matrices.

7.4.5.15.3 Advantages and Limitations of the Synchronized Algorithm

Logistic regression is more simplistic and easier to interpret, while SVM is powerful for high-dimensional data. PCA improves computational efficiency and reduces overfitting. However, logistic regression assumes linear relationships and may not handle complex patterns well. SVMs can be computationally intensive and require careful tuning. PCA might lead to some information loss.

7.4.5.15.4 Results and Discussion

The study explored the use of supervised ML models to predict pressure gradient and flow regimes in multiphase flow in horizontal pipes. The trained model shows promise for application in the industry as it allows for automation of systems used to control flow and affords a more comprehensive approach to mitigating flow problems in pipeline systems and flow systems in oilfields. The use of ML models can lead to more efficient and effective solutions for managing flow problems in multiphase systems.

BIBLIOGRAPHY

Isemin, Isemin, and Nkundu, King-Akanimo. "Investigating the Use of Machine Learning Models for the Prediction of Pressure Gradient and Flow Regimes in Multiphase Flow in Horizontal Pipes." Paper presented at the *SPE Nigeria Annual International Conference and Exhibition*, Lagos, Nigeria, August 2021. doi: https://doi.org/10.2118/208410-MS

7.4.5.16 Application-49: Real-Time Steam Allocation Workflow Using Machine Learning for Digital Heavy Oil Reservoirs

7.4.5.16.1 Background of the Application

The authors suggest a proactive steam allocation approach that learns how steam injection patterns affect heavy oil recovery through ML. KPIs are forecasted in the short term using data-analytic prediction models. For continual optimal steam allocation, the process is done every day. The results clearly demonstrate the usefulness of the suggested approach, as net present value (NPV) is raised by around 25% to 50% when compared to the base scenario, which employs a constant steam injection scheme.

7.4.5.16.2 Description of the Workflow and Machine Learning Model

The method is to create an NN model to predict oil production rates based on steam injection rates and steam-oil ratios. A GA from the DEAP library is used to optimize these parameters for maximizing NPV. The algorithm evaluates, selects, mutates, and breeds populations of potential solutions to find the optimal configuration iteratively.

7.4.5.16.3 Advantages and Limitations of the Synchronized Algorithm

This method has some advantages like handling complex, nonlinear relationships and optimizing multiple parameters simultaneously. However, it requires significant computational resources and time, particularly for training NNs and running evolutionary(genetic) algorithms. The results depend on the quality of input data and the design of the NN and GA parameters.

7.4.5.16.4 Results and Discussion

The proposed workflow resulted in an improvement of NPV by 42.4% and a decrease in the cumulative steam-oil ratio by 24% when tested on a Middle Eastern reservoir. The suggested workflow respects the flexible and adaptable steam supply, guaranteeing optimal oil recovery process efficiency. The workflow for intelligent steam allocation will have a significant impact on enhancing the digital heavy oil assets' operational efficiency, which will raise revenues and lessen carbon emissions.

BIBLIOGRAPHY

Sibaweihi, N., Patel, R. G., Guevara, J. L., Gates, I. D., and Trivedi, J. J. "Real-Time Steam Allocation Workflow Using Machine Learning for Digital Heavy Oil Reservoirs." Paper presented at the *SPE Western Regional Meeting*, San Jose, California, USA, April 2019. doi: https://doi.org/10.2118/195312-MS

7.4.5.17 Application-50: Fault Identification in a Subsea ESP Power Distribution System Using Electrical Waveform Monitoring

7.4.5.17.1 Background of the Application

This paper discusses the use of electrical waveform monitoring (EWM) technology to detect electrical or mechanical degradation of subsea electric submersible pumps (ESPs) in real time. The EWM technology continuously samples line voltage and phase current waveforms that feed electric motors and analyzes them using proprietary digital signal processing and ML algorithms. The system produces performance measurements such as motor load, efficiency, RPM, and output torque, along with machinery health measurements relating to the condition of the ESP, both for mechanical and electrical issues. The EWM technology provides a reliable source of evidence for troubleshooting and diagnosing failures and other trip events, reducing work-over costs and limiting production losses.

7.4.5.17.2 Description of the Workflow and Machine Learning Model

In this method, a combination of PCA and RF classifier is used to predict motor parameters. PCA reduces the dimensionality of features (voltage, current waveform, high-frequency data, and time-domain data) into five principal components. These components are then used to train an RF classifier for each target variable.

7.4.5.17.3 Advantages and Limitations of the Synchronized Algorithm

Using PCA and RF classifier offers efficient computation and robust predictions. PCA reduces dimensionality, minimizing overfitting risks, while RFs capture complex feature interactions accurately. However, PCA can lead to some information loss, and choosing the right number of components is very important. RFs are usually efficient, but they can also be computationally expensive with very large datasets.

7.4.5.17.4 Results and Discussion

The EWM technology has demonstrated the ability to diagnose the sources of ESP trips and to attribute them to top-side or subsea origin through the review of archived high-frequency electrical waveforms. The data-driven analytics generated by the EWM could potentially enable operators to alter the loading of ESPs, as needed, and to proactively schedule replacement or repair of these assets. The experience with the EWM technology has potential wider implications for the subsea industry. If the EWM technology provides the same advantages in other subsea assets as it has shown in the Perdido spar environment, there are numerous subsea assets that could benefit from this development. The EWM technology could reduce the operating costs and improve production planning for subsea oil and gas operators, with a particular benefit to deep sea assets.

BIBLIOGRAPHY

Obst, Larry, Merlino, Andrew, Parlos, Alex, and Dario Rubio. "Fault Identification in a Subsea ESP Power Distribution System Using Electrical Waveform Monitoring." Paper presented at the *SPE Gulf Coast Section Electric Submersible Pumps Symposium, Virtual and The Woodlands*, Texas, USA, October 2021. doi: https://doi.org/10.2118/204506-MS

8 AI Applications in Unconventionals

Ekrem Alagoz, Emre Can Dündar, Srikanth Tangirala, Mehmet Melih Oskay, Celal Hakan Canbaz, Luigi Saputelli, Mufrettin Murat Sari, Salih Tutun, and Cenk Temizel

8.1 INTRODUCTION

Chapter 8 opens with an introduction to the evolving landscape of unconventional reservoirs (UCRs), emphasizing the challenges posed by their complex geological and reservoir characteristics. Unconventional resources, such as shale gas, tight oil, and coalbed methane, have become increasingly important in the global energy mix. However, their extraction presents unique difficulties, including intricate rock formations, low permeability, and variable fluid properties. Traditional methods of reservoir characterization and management often fall short in addressing these complexities. As a result, the industry has turned to advanced technologies, including artificial intelligence (AI), machine learning (ML), and data analytics (DA), to enhance the understanding and development of these resources. These technologies offer the potential to improve decision-making, optimize production, and reduce operational risks.

The introduction also highlights the significant role that AI methodologies play in transforming conventional workflows within the industry. By leveraging large datasets and sophisticated algorithms, these tools can uncover patterns and insights that were previously inaccessible. This chapter aims to explore the categorization of workflows in UCRs, focusing on geological, reservoir, HF, and production forecasting aspects. Each of these areas presents its own set of challenges and opportunities, and the application of AI methodologies can lead to more efficient and effective resource management. The chapter will also provide an overview of the various AI techniques currently employed in the industry, setting the stage for a deeper understanding of their impact on unconventional reservoir development.

8.2 CATEGORIZATION OF WORKFLOWS USED IN UNCONVENTIONAL RESERVOIRS

In Section 8.2, the focus shifts to categorizing the various workflows used in UCRs, emphasizing the tailored approaches required to handle the unique characteristics of

 DOI: 10.1201/9781003617327-8

these resources. The geological aspects of UCRs are particularly complex, involving intricate rock formations with varying degrees of porosity and permeability. This section delves into how AI methodologies can be employed to enhance geological modeling and characterization, offering a more precise understanding of the subsurface environment. By analyzing vast amounts of geological data, ML algorithms can identify patterns and correlations that might be missed by traditional methods, leading to more accurate reservoir models and better-informed drilling decisions.

The section also explores workflows related to reservoir aspects, HF, and production forecasting. In reservoir management, AI tools are used to predict fluid behavior and optimize reservoir performance, considering the unconventional nature of the resources. HF, a critical process in unconventional resource extraction, benefits from data-driven approaches that optimize fracture design and placement. Finally, production forecasting, traditionally a challenging task due to the unpredictability of unconventional wells, is increasingly relying on AI techniques to improve accuracy and reliability. These workflows illustrate the profound impact of advanced technologies on the development and management of UCRs, offering new avenues for efficiency and productivity in the energy sector.

8.2.1 Geological Aspects

Geological aspects of UCRs, the complexity and heterogeneity of the subsurface, play a critical role in resource extraction. Unlike conventional reservoirs, which often have relatively straightforward geological structures, UCRs are characterized by intricate formations, including tight sands, shales, and coal seams. These formations exhibit low permeability and porosity, making the extraction of hydrocarbons more challenging. Understanding the geological framework is essential for successful reservoir development, and this is where AI and ML come into play. By integrating vast amounts of geological data—ranging from core samples to seismic surveys—AI tools can help create more accurate geological models, revealing the intricacies of subsurface formations. These models are instrumental in guiding drilling operations, as they enable more precise targeting of sweet spots within the reservoir, thereby maximizing recovery while minimizing operational risks.

Furthermore, AI techniques offer the ability to analyze and interpret complex geological data at a scale and speed that would be impossible with traditional methods. This section highlights how these technologies are used to identify patterns in rock properties, stratigraphy, and faulting, which are critical for understanding reservoir behavior. For instance, ML algorithms can be trained to recognize subtle geological features that may indicate the presence of natural fractures or variations in mineral content, both of which can significantly impact hydrocarbon flow. By enhancing geological characterization, AI not only improves the accuracy of reservoir models but also supports the optimization of well placement and the design of HF treatments. This integration of advanced technologies into geological workflows represents a significant advancement in the development of unconventional resources, providing a more nuanced understanding of the subsurface and contributing to more efficient resource extraction.

8.2.2 Reservoir Aspects

Reservoir aspects in unconventional resources present distinct challenges due to the unique characteristics of these formations, including low permeability, complex fluid behaviors, and high degrees of heterogeneity. Unlike conventional reservoirs, where fluid flow can be relatively predictable, UCRs often exhibit unpredictable and nonlinear production patterns. This complexity necessitates advanced approaches to reservoir management, where AI and ML play an increasingly vital role. By leveraging vast amounts of production data, pressure measurements, and fluid properties, AI tools can develop predictive models that simulate reservoir behavior with greater accuracy. These models help in understanding how hydrocarbons move through tight formations, allowing operators to optimize production strategies, enhance recovery rates, and minimize the economic risks associated with unconventional resource development.

AI methodologies also excel in optimizing reservoir performance by continuously adapting to new data and evolving reservoir conditions. Traditional reservoir management techniques often rely on static models that may not fully capture the dynamic nature of unconventional formations. In contrast, AI-driven models can be updated in real time as new data is collected, providing a more responsive approach to reservoir management. This adaptability is particularly valuable in UCRs, where production can decline rapidly and maintaining optimal recovery requires constant adjustments. Additionally, ML algorithms can identify trends and anomalies in reservoir data that may indicate potential issues, such as early water breakthrough or unexpected pressure drops, allowing for proactive interventions. By integrating AI into reservoir management workflows, operators can achieve a more nuanced understanding of reservoir dynamics, leading to more effective and efficient resource extraction in unconventional settings.

8.2.3 Hydraulic Fracturing

HF is a crucial process in the development of UCRs, enabling the extraction of hydrocarbons from formations with low permeability, such as shale and tight sands. The success of HF hinges on the ability to create and maintain fractures in the rock that can effectively facilitate the flow of O&G. Traditionally, designing these fractures involved a combination of empirical methods and trial-and-error approaches, which often led to suboptimal results. However, with the advent of AI and ML, the process of HF has become more precise and data-driven. By analyzing large datasets from previous fracturing operations, AI models can predict the optimal fracture design for a given geological formation, including the placement, orientation, and size of fractures. This data-driven approach helps maximize hydrocarbon recovery while minimizing the environmental impact and operational costs associated with fracturing operations.

Moreover, AI tools are transforming how real-time data is used during HF operations. Sensors placed in wells and on the surface collect vast amounts of data on pressure, temperature, and other critical parameters during the fracturing process. ML algorithms can process this data in real time, providing operators with immediate feedback on the effectiveness of the fracturing treatment. This capability allows for on-the-fly adjustments to the fracturing process, such as altering the injection rate or modifying the fracturing fluid composition to improve fracture propagation and connectivity. The integration of AI into HF not only enhances the precision and efficiency of the process but also contributes to safer and more sustainable operations by reducing the risk of unintended consequences, such as induced seismicity or excessive water usage.

8.2.4 Production Forecasting

Production forecasting in UCRs is a complex task due to the highly variable nature of these formations and the non-linear production profiles they exhibit. Unlike conventional reservoirs, where production rates tend to follow more predictable decline curves, unconventional wells often experience sharp declines in output, followed by long periods of low but sustained production. Accurately predicting these trends is crucial for effective reservoir management and economic planning. AI and ML have revolutionized production forecasting by enabling the analysis of vast datasets from well performance, geological characteristics, and operational conditions. These advanced models can identify patterns and relationships within the data that are not apparent through traditional methods, leading to more accurate and reliable production forecasts. This data-driven approach helps operators optimize production schedules, manage reserves more effectively, and make informed decisions about future drilling and completion activities.

In addition to improving the accuracy of production forecasts, AI tools offer the ability to continuously refine predictions as new data becomes available. Traditional forecasting methods often rely on static models that can quickly become outdated as reservoir conditions change. In contrast, ML algorithms can update predictions in real time, incorporating new production data and adjusting forecasts accordingly. This dynamic approach allows operators to respond more rapidly to unexpected changes in well performance, such as sudden drops in pressure or water production, which can significantly impact overall recovery. Furthermore, AI-driven forecasts can be used to simulate various operational scenarios, helping operators evaluate the potential outcomes of different strategies and select the most effective approach for maximizing production. By integrating AI into production forecasting workflows, the industry can achieve greater precision in managing UCRs, ultimately leading to improved efficiency and profitability in resource extraction.

8.3 OVERVIEW OF AI METHODOLOGIES

The integration of AI, ML, and DA methodologies into the O&G industry has significantly enhanced the ability to manage and optimize complex processes in UCRs. These advanced technologies allow for the analysis of vast and varied datasets, providing insights that were previously unattainable through traditional methods. AI techniques, such as neural networks (NNs), decision trees (DTs), and support vector machines (SVMs), have been particularly effective in predicting reservoir behavior, optimizing drilling operations, and improving production forecasting. DA, on the other hand, enables the extraction of meaningful patterns from massive datasets, facilitating better decision-making in real-time. Together, these methodologies contribute to a more nuanced understanding of reservoir dynamics, enabling more efficient resource extraction and reduced operational risks.

The application of AI methodologies extends beyond individual tasks to transform entire workflows within the industry. These technologies are increasingly being used to automate routine processes, enhance the accuracy of geological and reservoir models, and optimize HF designs. By integrating AI into daily operations, companies can achieve higher levels of efficiency, safety, and sustainability. Moreover, these methodologies enable continuous learning and improvement, as models are constantly updated with new data, ensuring that predictions and strategies remain relevant in an ever-changing reservoir environment. As the industry continues to embrace these technologies, the potential for AI to drive innovation and improve outcomes in unconventional resource development is immense, promising a future where operations are more data-driven, precise, and adaptive.

8.4 APPLICATIONS OF AI IN UNCONVENTIONAL RESERVOIRS

8.4.1 Geological Aspects

8.4.1.1 Application-1: Incorporated Artificial Intelligence and Digital Imaging System for Unconventional Reservoirs Characterization

8.4.1.1.1 Background of the Application

The paper emphasizes the significance of UCRs, such as tight gas sands, oil shales, and coalbed methane, which have become increasingly important in global O&G production. However, characterizing these reservoirs poses challenges due to their complex pore structures and low permeability–viscosity ratios. Traditional techniques are insufficient for these reservoirs, necessitating advanced methods such as digital rock analysis (DRA). DRA allows for detailed imaging of rock pore structures and the simulation of rock properties like porosity, permeability, and electrical properties. In recent years, the incorporation of AI and ML has shown great potential for enhancing DRA, particularly in tasks like image segmentation, rock type prediction, and reservoir property estimation. The integration of AI with DRA represents a significant advancement in reservoir characterization, offering improved efficiency and accuracy in analyzing UCRs.

8.4.1.1.2 Description of the Workflow and Machine Learning Model

The paper integrates AI techniques, particularly ML, with DRA to enhance reservoir characterization. The ML methods are primarily employed for image segmentation, feature detection, rock image classification, and rock property prediction. Commonly used ML algorithms in the study include K-means clustering, fuzzy c-means (FCM), artificial neural networks (ANNs), and SVMs. These algorithms help automate the process of analyzing digital rock images, reducing manual labor, and increasing the accuracy of reservoir characterization. For instance, the K-means clustering algorithm is used for segmenting pores and grains in rock images, while SVM is utilized for classifying fractures and vugs in carbonate reservoirs. Additionally, convolutional neural networks (CNNs) are mentioned as an advanced ML technique that can extract features automatically from digital images, further aiding in reservoir characterization.

8.4.1.1.3 Advantages and Limitations of the Synchronized Algorithm

The AI methodologies employed in this paper offer several advantages. First, they significantly reduce the manual effort required for DRA, streamlining processes such as image segmentation and rock classification. The automation provided by ML algorithms improves the consistency and accuracy of results. Additionally, ML techniques allow for the analysis of large datasets, which would be time-consuming and error-prone if done manually. For example, the use of K-means clustering and SVM enhances the ability to classify rock features like fractures and vugs with greater precision.

However, there are limitations to these methodologies. The accuracy of ML algorithms is highly dependent on the quality of the input data, meaning that noisy or low-resolution images can lead to incorrect classifications. Moreover, the choice of algorithm and its parameters can significantly affect the results, requiring careful tuning and expertise. While CNNs can automatically extract features, they require large amounts of labeled data for training, which may not always be available in reservoir studies. Despite these limitations, the incorporation of AI in DRA provides a promising approach to overcoming the challenges of unconventional reservoir characterization.

8.4.1.1.4 Results and Discussion

The paper concludes that DRA is a powerful tool for characterizing UCRs, as it provides detailed insights into pore structures and enables comprehensive reservoir studies through image observation and analysis. The case study presented in the paper validates the effectiveness of DRA, demonstrating that it can provide accurate simulations of rock properties such as porosity and permeability. Furthermore, the integration of AI techniques, specifically ML, with DRA shows great potential to further enhance the characterization of UCRs by automating and refining the analysis process. The paper highlights that this combination can significantly reduce the burden of manual work, allowing for more efficient and accurate reservoir studies. The authors anticipate that this approach will continue to evolve and find broader applications in the field of O&G exploration and production.

BIBLIOGRAPHY

Sun, Huafeng, and Belhaj, Hadi. "Incorporated Artificial Intelligence and Digital Imaging System for Unconventional Reservoirs Characterization." Paper presented at the *SPE Annual Technical Conference and Exhibition*, Calgary, Alberta, Canada, September 2019. doi: https://doi.org/10.2118/195834-MS

8.4.1.2 Application-2: Unconventional Reservoir Microstructural Analysis Using SEM and Machine Learning

8.4.1.2.1 Background of the Application

The paper highlights the importance of understanding the microstructure of UCRs, as it directly affects hydrocarbon storage, transport, and mechanical properties. SEM has been a valuable tool for visualizing these complex microstructures, yet analyzing SEM images quantitatively remains a challenge. Traditional methods for image analysis are labor-intensive, time-consuming, and prone to errors. To address these limitations, the authors explore the application of ML to analyze SEM images from UCRs. Specifically, they employ DL techniques, such as CNNs, to automate pixel labeling and feature identification, ultimately improving the accuracy and efficiency of microstructural analysis.

8.4.1.2.2 Description of the Workflow and Machine Learning Model

Using DL techniques, the paper utilizes CNNs to perform image classification and segmentation on SEM images of UCRs. The research uses a modified U-Net CNN architecture, which is well-suited for pixel labeling and feature identification. The dataset consists of over 28,000 SEM images from 18 unconventional regions, with image augmentation used to enhance the training process. The CNN model, specifically AlexNet for image classification and a modified U-Net for image segmentation, was trained using supervised learning with TL applied from pre-trained models on the ImageNet dataset. The network accurately identifies various features such as pores, organics, and minerals in SEM images, enabling more efficient and accurate microstructural analysis.

8.4.1.2.3 Advantages and Limitations of Synchronized Algorithm

The AI methodologies used in this paper provide several advantages. The implementation of CNNs for image segmentation and classification significantly reduces manual effort and accelerates the process of analyzing large datasets of SEM images. The use of TL enhances the network's performance by leveraging pre-trained models, which reduces the training time and the need for extensive labeled datasets. However, the methodology is limited by the resolution of the input images and the size of the labeled dataset, which can impact the model's accuracy. Additionally, overfitting remains a concern, particularly with deeper networks like Inception-V3. Despite these limitations, the results demonstrate that further investment in dataset development and model refinement could yield even better performance, making the approach highly promising for future applications in unconventional reservoir characterization.

8.4.1.2.4 Results and Discussion

The paper concludes that using ML, particularly DL, significantly enhances the analysis of SEM images in UCR studies. By employing CNNs, the research achieved promising results in identifying and segmenting various components of rock microstructures, such as pores, minerals, and organic matter, with an accuracy of up to 87%. The authors suggest that further development of the training dataset and the incorporation of higher-resolution images could improve the performance of their models. Overall, the study demonstrates the potential of ML to accelerate and enhance microstructural analysis in UCRs, paving the way for more robust and efficient reservoir characterization.

BIBLIOGRAPHY

Knaup, Amanda, Jernigen, Jeremy, Curtis, Mark, Sholeen, John, Borer, John, Sondergeld, Carl, and Rai, Chandra. "Unconventional Reservoir Microstructural Analysis Using SEM and Machine Learning." Paper presented at the *SPE/AAPG/SEG Unconventional Resources Technology Conference*, Denver, Colorado, USA, July 2019. doi: https://doi.org/10.15530/urtec-2019-638

8.4.1.3 Application-3: Application of Artificial Intelligence on Black Shale Lithofacies Prediction

8.4.1.3.1 Background of the Application

The paper addresses the prediction of shale lithofacies in the Marcellus Shale of the Appalachian Basin, focusing on the application of AI techniques. Shale lithofacies are critical in identifying productive shale gas zones and in guiding decisions for horizontal drilling and HF. Traditional methods for lithofacies prediction using well logs are limited due to the complex and nonlinear relationships between lithofacies and log data. The authors explore AI classifiers, including ANNs and SVMs, to handle these complexities. Additionally, they introduce a hierarchical decomposition method to improve the performance of AI models in predicting lithofacies from well log data.

8.4.1.3.2 Description of the Workflow and Machine Learning Model

The paper employs AI techniques, specifically ANNs and SVMs, for shale lithofacies prediction in the Marcellus Shale. ANNs and SVMs are used as classifiers to handle the complex, nonlinear relationships between lithofacies and well log data. The authors also introduce optimization algorithms such as genetic algorithms (GAs) and particle swarm optimization (PSO) to train the classifiers. A hierarchical decomposition method is proposed to address the challenges of multi-class classification, breaking down the problem into simpler sub-problems, which improves the performance of the AI models. This comprehensive AI workflow is designed to predict shale lithofacies from well log data with greater accuracy.

8.4.1.3.3 Advantages and Limitations of the Synchronized Algorithm

The AI methodologies used in this paper offer several advantages. ANNs and SVMs are powerful classifiers that can handle complex, nonlinear relationships, which are common in lithofacies prediction. The hierarchical decomposition method further enhances the accuracy of AI models by breaking down the multi-class classification problem into simpler sub-problems. This approach reduces the risk of misclassification and improves the overall performance of the AI models. However, the methodology is computationally intensive and requires careful tuning of parameters such as the penalty value in SVMs and the architecture of ANNs. Additionally, the hierarchical decomposition method introduces complexity, which may increase the risk of error propagation through the classification hierarchy. Despite these challenges, the AI methodologies presented in this paper provide a promising solution for shale lithofacies prediction.

8.4.1.3.4 Results and Discussion

The paper concludes that AI techniques, particularly ANN and SVM, are effective in predicting shale lithofacies from well log data in the Marcellus Shale. The hierarchical decomposition method presented in the paper enhances the performance of AI classifiers, especially in multi-class classification problems like lithofacies prediction. This approach successfully captures the nonlinear relationships between lithofacies and well logs, leading to more accurate predictions and aiding in the identification of productive shale gas zones. The authors suggest that their methodology could be applied to other shale plays, making it a valuable tool in UCR characterization.

BIBLIOGRAPHY

Wang, Guochang, Ju, Yiwen, Carr, Timothy R., Li, Chaofeng, and Cheng, Guojian. "Application of Artificial Intelligence on Black Shale Lithofacies Prediction in Marcellus Shale, Appalachian Basin." Paper presented at the *SPE/AAPG/SEG Unconventional Resources Technology Conference*, Denver, Colorado, USA, August 2014. doi: https://doi.org/10.15530/URTEC-2014-1935021

8.4.1.4 Application-4: A Machine Learning Workflow to Support the Identification of Subsurface Resource Analogs

8.4.1.4.1 Background of the Application

Because of the complexity, noise, and high dimensionality of subsurface datasets, traditional methods for selecting these analogs—typically carried out by domain experts—face significant challenges. This article aims to simplify the process by introducing an objective geostatistics-based ML workflow for analog selection. Finding subsurface resource analogs from mature subsurface datasets is essential for developing new prospects, especially when initial information is limited or absent.

Novel methodology, which includes pairwise similarity scores, group consistency scoring, and a new dissimilarity metric, offers a methodical and objective approach. These elements measure similarities inside and between groups in reduced dimensional areas, efficiently accounting for spatial and multivariate data linkages. In order to represent data in a reduced dimensional space based on the dissimilarity metric, the procedure starts with multidimensional scaling using inferential ML. Analog clusters and spatial analogs are then found inside this condensed space using density-based spatial clustering of applications with noise (DBSCAN). After that, scoring metrics provide useful diagnostics for resource research by quantifying and identifying similar data samples. Researchers demonstrate the efficacy of this workflow using wells from the Duvernay Formation and a test scenario with various well types common in UCRs, including infill, outlier, sparse, and centered wells. The successful identification and grouping of analog clusters based on geological properties and cumulative gas production underscore the potential of this workflow for practical application in the field.

8.4.1.4.2 Description of the Workflow and Machine Learning Model

This study introduces a geostatistics-based ML workflow aimed at identifying subsurface resource analogs. The workflow integrates both spatial and multivariate information, using a novel dissimilarity metric to capture these relationships. The process involves dimensionality reduction via metric multidimensional scaling (MDS), followed by clustering analysis using DBSCAN. The primary decision variables include spatial and multivariate features, which are optimized to identify resource analogs effectively. This approach also incorporates kriging to model spatial variability and to predict unsampled values in the reduced dimensional space.

8.4.1.4.3 Advantages and Limitations of the Synchronized Algorithm

The combination of geostatistics and ML in this workflow offers a robust framework for subsurface analog identification, addressing the challenges of high-dimensional data and spatial relationships. The multi-stage process, from MDS to DBSCAN clustering, allows for effective dimensionality reduction and the identification of meaningful clusters without the need for pre-specified cluster numbers. This makes the method well-suited for complex subsurface datasets. However, the computational demands can be significant, particularly in large-scale applications with numerous features and data points. Additionally, the accuracy of the clustering and kriging results is heavily dependent on the quality of the variogram models and the chosen parameters for the clustering algorithm.

8.4.1.4.4 Results and Discussion

A new workflow designed for selecting spatial and multivariate analogs in subsurface resource analysis, leveraging both geostatistical methods and ML. This approach stands as a robust alternative to traditional expert-led and statistical techniques, effectively tackling issues such as dimensionality challenges, objectivity, repeatability, and the need to incorporate spatial context. The workflow introduces a hybrid dissimilarity metric that integrates spatial and multivariate data considerations, alongside inferential ML to reduce dimensionality. This strategy allows for a more effective clustering of analog groups, addressing the complexities of high-dimensional datasets. The resulting framework is both robust and adaptable, enhancing the process of identifying subsurface resource analogs and supporting more accurate inferential analyses and decision-making.

This workflow offers an unbiased and consistent alternative to traditional expert-driven approaches, setting it apart from existing methods like close-ology and standard multivariate analysis. Although the workflow requires sufficient data for reliable variogram and density-based clustering, as well as adherence to assumptions like stationarity of variance and clustering scale, it provides a solid foundation for future research. This approach not only has the potential to be optimized for various datasets but also significantly contributes to the advancement of analog selection tools. Ultimately, this methodology improves the precision and consistency of analog identification, fostering better decision-making in subsurface resource exploration.

BIBLIOGRAPHY

Mabadeje, A. O., Salazar, J. J., Ochoa, J., Garland, L., and Pyrcz, M. J. A Machine Learning Workflow to Support the Identification of Subsurface Resource Analogs. *Energy Explor Exploit* 2024;42(2):603–625. doi:10.1177/01445987231210966

8.4.1.5 Application-5: Application of Various Machine Learning Techniques in Predicting Total Organic Carbon from Well Logs

8.4.1.5.1 Background of the Application

Unconventional resources have recently garnered significant attention, leading to an increased focus on predicting TOC, a key quality indicator. Traditionally, TOC is measured experimentally, but continuous data acquisition is often hampered by sampling limitations. Consequently, various empirical correlations have been proposed to estimate TOC, though concerns persist regarding their generalization and accuracy.

Using various ML techniques, such as formation resistivity (FR), spontaneous potential (SP), sonic travel time (Δt), bulk density (RHOB), neutron porosity (CNP), gamma ray (GR), and spectral logs of thorium (Th), uranium (Ur), and potassium (K), this paper investigates the development of models for predicting TOC based on well logs. Models were trained on data from one well and verified on datasets from two additional wells using more than 1,250 data points from the Devonian Duvernay shale. Three empirical correlations were used to compare the predictions of SVMs, random forests (RF), and DTs in this work.

8.4.1.5.2 Description of the Workflow and Machine Learning Model

The paper focuses on predicting TOC from well logs using various ML techniques. The ML models applied include SVM, DT, and RF. The data used consists of well log measurements such as FR, SP, Δt, RHOB, CNP, GR, and Th, Ur, and K. The optimization of these models is conducted through tuning hyperparameters such as kernel functions for SVM, tree depth for DT, and the number of trees for RF. The models are evaluated using correlation coefficients and average absolute percentage error (AAPE), with the RF model demonstrating the best performance.

8.4.1.5.3 Advantages and Limitations of the Synchronized Algorithm

The use of ML techniques for predicting TOC from well logs offers significant advantages. These models are capable of producing high-accuracy predictions without the need for costly core analysis or well interventions. The RF model, in particular, excels at handling complex interactions between features and provides more accurate results compared to existing empirical correlations. However, the computational cost of training these models, especially with large datasets, can be considerable. Additionally, the accuracy of the predictions depends on the quality of the input data and the choice of hyperparameters. While the ML models outperform traditional methods, careful preprocessing and validation are required to ensure the reliability of the results.

8.4.1.5.4 Results and Discussion

This study developed three models for predicting TOC in Devonian shale using conventional well logs and spectral GR logs, applying different ML techniques on approximately 1,250 data points from three wells. The ML methods utilized were SVM, DT, and RF. The findings indicate that all three AI algorithms provided accurate predictions; however, the RF model exhibited the highest accuracy. Specifically, the RF model achieved R values of 0.99 and 0.93 for the training and testing datasets, respectively, and AAPE values of 5.3% and 13.8% in the same order.

Furthermore, data from an additional well, which was not used during the training phase, was employed for model validation. For this validation dataset, the RF model demonstrated a correlation coefficient of 0.94 and an AAPE of 14%. When comparing the predictions of the AI models to three empirical correlations, the AI models showed superior accuracy, with AAPE values significantly lower than the empirical models, which had AAPEs exceeding 20%.

BIBLIOGRAPHY

Siddig, O., Ibrahim, A. F., and Elkatatny, S. Application of Various Machine Learning Techniques in Predicting Total Organic Carbon from Well Logs. *Comput Intell Neurosci* 2021 Aug 30;2021:7390055. doi: 10.1155/2021/7390055. PMID: 34545282; PMCID: PMC8448996

8.4.1.6 Application-6: Geological Facies Prediction Using Computed Tomography in a Machine Learning and Deep Learning Environment

8.4.1.6.1 Background of the Application

ML and DL have emerged as powerful tools for predictive modeling in various fields, including geological facies classification in UCRs. Traditional techniques often struggle to adapt from conventional to UCR analysis. This study leverages ML and DL to predict geological facies from computed tomography (CT) scans. The approach combines two methods: the first uses molecular weight, density, and porosity data from CT scans to learn from existing geological facies classifications, while the second employs DL on the CT scan images of core samples to predict geological facies.

The results demonstrate that these models significantly outperform traditional methods, achieving misclassification rates of less than 30%, which is notably lower than the error rates achieved by human classifiers. Additionally, unsupervised ML techniques provided the robust clustering of geological facies, showcasing the effectiveness of CT scans combined with advanced modeling techniques. By automating the classification process, the proposed methodology reduces human error and speeds up the workflow, allowing for faster and more accurate geological facies predictions. This advancement holds promise for improving business decisions and optimizing the understanding of unconventional plays.

8.4.1.6.2 Description of the Workflow and Machine Learning Model

This paper presents a methodology for predicting geological facies using a combination of ML and DL) techniques, applied to data from the CT scans of core samples. The ML approach leverages supervised learning through RF algorithms, which are trained using CT-derived features such as density, porosity, and photoelectric factor. The model is built using pre-existing facies classifications, aiming to automate the prediction of geological facies in future core samples.

In addition to traditional ML, the paper also explores unsupervised learning techniques, particularly hierarchical clustering, to group similar rock types based on CT scan data. This method provides insights into the variability of rock types early in the core analysis workflow.

For image-based classification, the paper applies DL techniques, specifically CNNs, to classify geological facies directly from CT scan images. The CNN model is trained on thousands of labeled images, allowing it to learn relevant features automatically and make predictions about facies from new CT images.

8.4.1.6.3 Advantages and Limitations of the Synchronized Algorithm

The combination of ML and DL offers a robust approach to geological facies prediction. The RF algorithm, used in supervised learning, is advantageous for its ability to handle large datasets with multiple features and deliver accurate predictions, reducing the time and effort required for core analysis. However, the accuracy of the predictions depends heavily on the quality and quantity of training data.

Unsupervised learning, through hierarchical clustering, provides a fast and effective way to group similar rock types without the need for pre-existing labels. This can be particularly useful in exploratory settings where little prior knowledge is available. The primary limitation of this method is its reliance on statistical similarity, which may not always correspond to meaningful geological differences.

DL, particularly with CNNs, is highly effective in recognizing complex patterns in CT images, making it a powerful tool for image-based facies prediction. The automatic feature extraction by the CNN significantly reduces the need for manual input from experts. However, the computational requirements for training CNN models are high, and the method requires large amounts of labeled data for effective training. Additionally, while CNNs can perform well on image recognition tasks, their interpretability is often limited, which can make it challenging to understand the decision-making process of the model.

8.4.1.6.4 Results and Discussion

ML provides a promising approach for the rapid characterization of rock facies, leveraging both supervised and unsupervised techniques. Supervised ML utilizes existing geological classifications to classify new rocks from the same formation, making it a useful tool for core description and facies analysis workflows. While it does not replace conventional methods, it can enhance core sample selection and expedite descriptive workflows by identifying rocks with similar compositions.

Unsupervised ML, on the other hand, helps determine the number of statistically significant rock types based on the CT data. DL, with its ability to classify and predict facies from CT scan images, shows potential to further streamline and improve these processes. It offers a powerful alternative to traditional methods, potentially replacing existing workflows with more efficient and automated solutions.

BIBLIOGRAPHY

Odi, Uchenna, and Nguyen, Thomas. "Geological Facies Prediction Using Computed Tomography in a Machine Learning and Deep Learning Environment." Paper presented at the SPE/AAPG/SEG Unconventional Resources Technology Conference, Houston, Texas, USA, July 2018. doi: https://doi.org/10.15530/URTEC-2018-2901881

8.4.1.7 Application-7: Improving Prediction of Total Organic Carbon by Employing Machine Learning

8.4.1.7.1 Background of the Application

Because TOC is directly correlated with total porosity and gas content, it is an essential indicator of UCR potential. Although TOC is often determined from core samples at particular depths, a thorough grasp of the vertical and lateral distribution of organic matter throughout the reservoir is necessary to estimate the resource potential. Conventional wireline logs are usually used to collect this data, which includes resistivity, transit time, density, and GRs. However, standard techniques, which depend on variations in resistivity, density, and sonic velocity between organic and inorganic materials, may not always work due to the complexity of UCRs.

In this study, ML methods such as ANNs and SVMs were employed to improve TOC predictions using GR, density, porosity, resistivity, and Δt log responses. Data from seven wells drilled in the McArthur Basin and Georgina Basin in Australia were analyzed. The comparison between traditional methods, multiple linear regression (MLR), and ML approaches revealed that ANN and SVM-based nonlinear predictors provided the most accurate TOC estimates, outperforming both MLR and traditional models. This suggests that ML techniques are more adept at handling the geological complexities affecting log responses and TOC relationships in these UCRs.

8.4.1.7.2 Description of the Workflow and Machine Learning Model

This work addresses the challenges of predicting TOC in unconventional gas reservoirs by employing ML techniques. The study explores the application of various methods, including MLR, ANNs, and SVMs, to improve TOC prediction from well log data. The ML models leverage data from several wireline logs, such as GR, density, resistivity, porosity, and Δt, to develop predictive models. The dataset used in the study includes data from seven wells in the McArthur and Georgina Basins in Australia.

The MLR approach models a linear relationship between the well log data and TOC, while the ANN and SVM methods capture the nonlinear relationships between the input features and TOC. The ANN employed in this study is a feedforward Multilayer Perceptron (MLP), which consists of multiple layers of neurons that process the input data and produce TOC predictions. The SVM model uses a kernel-based approach to map the input data into a higher-dimensional space where nonlinear relationships can be more easily captured and modeled. These ML techniques are trained using core TOC measurements for calibration and then evaluated based on their ability to generalize across wells with similar geological structures.

8.4.1.7.3 Advantages and Limitations of the Synchronized Algorithm

The application of ML methods in this study offers significant advantages for TOC prediction in complex geological settings. The nonlinear models, such as ANN and SVM, outperform traditional methods like MLR by capturing the intricate relationships between well log data and TOC that are not well represented by linear models.

This enables the creation of more accurate and generalizable TOC models, which are crucial for cross-well predictions in basins with similar geological characteristics.

One of the key advantages of the ANN and SVM approaches is their ability to handle the inherent nonlinearity in the well log data, providing better predictions even in geologically complex regions. The SVM, in particular, is effective at mapping the data into higher-dimensional spaces, which allows for more accurate predictions in regions where traditional methods fail.

However, these advanced ML techniques also come with limitations. Both ANN and SVM models require careful tuning of hyperparameters and sufficient training data to achieve optimal performance. The computational cost of training these models can be high, particularly for large datasets and complex architectures. Additionally, the success of these models is heavily dependent on the quality of the input data. Inconsistent or noisy well log data can reduce the accuracy of predictions, and the models may struggle to generalize to new wells that exhibit different geological conditions from the training data.

8.4.1.7.4 Results and Discussion

In this study, wireline logs from seven wells in the McArthur Basin and Georgina Basin were utilized to predict TOC. The study compared the prediction quality of traditional methods with ML techniques. Nonlinear predictors, specifically ANNs and SVMs, provided the most accurate TOC estimates, surpassing both MLR and traditional models. These nonlinear methods demonstrated an excellent ability to generalize basin-specific relationships, making them suitable for cross-well TOC prediction in wells with similar geological structures. Notably, in the McArthur Basin, the TOC model trained on one well was effectively applied to an adjacent well, resulting in substantially higher prediction accuracy.

The study also suggests that further improvement in TOC modeling can be achieved by preprocessing input data to identify groups with similar geological and geophysical characteristics. By splitting the dataset of core TOC and log data accordingly, group-specific TOC prediction models can be developed. This approach would enhance the reliability and generalization capacity of TOC models for unknown data with similar characteristics. Implementing clustering algorithms, such as self-organizing maps, can facilitate this data preprocessing and improve the accuracy of TOC predictions.

BIBLIOGRAPHY

Emelyanova, I., Pervukhina, M., Clennell, M. B., and Dewhurst, D. N. Improving prediction of Total Organic Carbon in prospective Australian basins by employing machine learning. *ASEG Extended Abstracts* 2016;2016(1):1–5. https://doi.org/10.1071/ASEG2016ab164

8.4.1.8 Application-8: Incorporated Artificial Intelligence and Digital Imaging System for Unconventional Reservoirs Characterization

8.4.1.8.1 Background of the Application

Despite being essential to the world's O&G supplies, UCRs are difficult to characterize because of their intricate pore structures and low permeability–viscosity ratios. When it comes to precisely identifying pore structures and calculating reservoir parameters, traditional methods frequently fall short. A viable substitute is DRA, which simulates several rock characteristics such as porosity, permeability, electrical properties, and elastic properties while also recording comprehensive pore structure data. In order to overcome these obstacles, the application of AI methods—specifically, ML—has become more popular. In geoscience and reservoir characterization tasks including feature extraction, rock type prediction, and reservoir property assessment, ML, a subset of AI, has shown promise.

The DRA methodology for reservoir characterization is introduced in this research, along with a review of popular ML methods used in DRA studies and a case study on the usage of DRA and ML to characterize a tight carbonate reservoir. The investigation shows how ML may improve the DRA process in a number of areas, such as result analysis, numerical simulation, image segmentation, feature recognition, and rock image classification. ML-based techniques greatly reduce manual processes in comparison to conventional DRA methods, enhancing outcomes. In addition to highlighting the future potential of AI in tackling oilfield difficulties, the combination of ML and DRA provides fresh insights into the characterization of UCRs.

8.4.1.8.2 Description of the Workflow and Machine Learning Model

The paper explores the characterization of UCRs using a combination of AI and DRA. The study focuses on integrating ML techniques with DRA to enhance reservoir characterization by analyzing complex pore structures and rock properties. The process involves multiple stages, starting from digital image acquisition, followed by image processing, numerical simulations, and result analysis.

For image processing, techniques such as K-means clustering and FCM clustering are employed for image segmentation. These methods help in separating pores, grains, and other phases in digital rock images. SVMs are used for feature detection, specifically to classify and distinguish between fractures and vugs in the rock samples. Additionally, the paper uses CNNs for automatic feature extraction from digital rock images.

The combination of these ML algorithms with DRA allows for a more automated and accurate assessment of rock properties, such as porosity, permeability, and other mechanical properties. The workflow described in the paper provides a detailed methodological framework that uses AI to overcome the limitations of traditional rock analysis methods.

8.4.1.8.3 Advantages and Limitations of the Synchronized Algorithm

The incorporation of AI techniques, particularly ML algorithms, into DRA offers significant advantages for unconventional reservoir characterization. The use of K-means and FCM clustering enables more precise image segmentation, while SVMs effectively distinguish between different rock features, such as fractures and vugs. CNNs further enhance the process by automating feature extraction, reducing the need for manual intervention.

One of the primary advantages of this approach is the ability to process and analyze large datasets of digital rock images with high accuracy. By automating feature detection and classification, the methodology reduces human error and increases the efficiency of reservoir characterization. Additionally, the combination of multiple ML algorithms provides a comprehensive understanding of the reservoir properties.

However, the limitations of this approach include the computational complexity associated with training ML models, especially with large datasets. The accuracy of the results is highly dependent on the quality of the input data and the models used for simulations. Furthermore, the implementation of these techniques requires significant computational resources, which may limit their application in real-time reservoir analysis. Despite these challenges, the integration of AI in DRA represents a promising advancement in the field of reservoir characterization.

8.4.1.8.4 Results and Discussion

DRA proves to be a highly effective tool for characterizing UCRs due to its ability to provide detailed information about pore structures and support reservoir studies through image observation and comprehensive analysis. A case study highlighted in this paper demonstrates the validity of DRA and outlines the procedure for its application. The technology shows significant potential as a primary solution for reservoir characterization, particularly when complemented by other methods. AI techniques, in general, can be integrated into DRA and reservoir characterization processes, alleviating unnecessary mental and physical burdens.

The synergy between DRA and AI offers new insights into UCR characterization, enhancing the ability to analyze and understand complex reservoirs. This combination is expected to rapidly advance and expand its applications within the O&G exploration and production fields, promising more efficient and detailed characterizations of unconventional resources.

BIBLIOGRAPHY

Sun, Huafeng, and Belhaj, Hadi. "Incorporated Artificial Intelligence and Digital Imaging System for Unconventional Reservoirs Characterization." Paper presented at the *SPE Annual Technical Conference and Exhibition*, Calgary, Alberta, Canada, September 2019. doi: https://doi.org/10.2118/195834-MS

8.4.1.9 Application-9: Machine Learning Prediction of Methane, Ethane, and Propane Solubility in Pure Water and Electrolyte Solutions

8.4.1.9.1 Background of the Application

HF is a crucial technology for extracting hydrocarbons from unconventional shale and tight gas reservoirs. However, one of the potential risks associated with HF is the upward migration of stray gas from deep subsurface layers to shallow aquifers. This stray gas can dissolve in groundwater, leading to chemical and biological reactions that might adversely impact groundwater quality and contribute to atmospheric emissions. Understanding the solubility of light hydrocarbons like methane, ethane, and propane in aqueous environments is vital for an accurate numerical modeling of flow and transport in the subsurface.

In this study, a comprehensive database of 2,129 experimental data points on the solubility of these gases in pure water and various electrolyte solutions, across a wide range of temperatures and pressures, was compiled. ML algorithms, specifically regression tree (RT) and boosted regression tree (BRT) optimized with Bayesian optimization (BO), were employed to predict gas solubility. The performance of these predictions was compared with experimental data and four established thermodynamic models. The results indicate that the BRT-BO approach is highly accurate, with a coefficient of determination (R^2) of 0.99 and a mean squared error (MSE) of 9.97×10^{-8}. The statistical validation further confirmed the reliability of the model.

8.4.1.9.2 Description of the Workflow and Machine Learning Model

This paper investigates the prediction of methane, ethane, and propane solubility in pure water and electrolyte solutions using ML methods. The study focuses on applying RT and BRT algorithms tuned with BO to predict the solubility of light hydrocarbons under various conditions. The data used for model development consists of 2129 experimental data points, including solubility values for methane, ethane, and propane in different aqueous environments. The ML models are designed to handle nonlinear relationships between input parameters, such as pressure, temperature, and salt concentrations, and the solubility outputs. The optimization of hyperparameters using BO ensures that the models perform efficiently and provide accurate predictions across a wide range of conditions.

8.4.1.9.3 Advantages and Limitations of the Synchronized Algorithm

The combination of RT and BRT algorithms, optimized using Bayesian methods, presents several advantages for predicting the solubility of light hydrocarbons in aquatic systems. These ML models excel at capturing complex nonlinear relationships between input variables and provide high prediction accuracy, as evidenced by the coefficient of determination (R^2) value of 0.99 and a low MSE of 9.97×10^{-8}. Additionally, the BRT–BO model's ability to handle various conditions, such as different pressures, temperatures, and salt concentrations, makes it highly versatile for geochemical applications. However, the model's performance depends heavily on the quality and completeness of the training data. The computational complexity of the BO process may also pose challenges, particularly when dealing with large datasets or when attempting to fine-tune multiple hyperparameters simultaneously. Despite these challenges, the proposed ML approach offers a robust alternative to traditional thermodynamic models, particularly in scenarios requiring fast and reliable solubility predictions.

8.4.1.9.4 Results and Discussion

Using RT and BRT methods, a model was created that uses Bayesian Optimization (BO) to determine the solubility of methane, ethane, and propane in aquatic systems. Light hydrocarbon solubility can be estimated using the RT–BO and BRT–BO models over a wide range of temperatures, pressures, and salt concentrations. With an MSE of 9.97×10^{-8} and a coefficient of determination (R^2) of 0.99, the BRT–BO model outperformed the others in terms of accuracy. The experimental dataset, which consists of 2,129 data points on hydrocarbon solubility in different electrolyte solutions and pure water, closely matches the predictions made by the BRT–BO model.

When compared to four well-known thermodynamic models, the BRT–BO model's performance demonstrated a good predictive capability. Furthermore, with the bulk of data points (5.26% outliers) falling into the valid domain, the leverage strategy validated the validity of the model. This indicates that for light hydrocarbon solubility estimates, the BRT–BO model is a solid and trustworthy substitute for conventional techniques. Applications like the numerical modeling of subsurface stray gas migration, creating environmental risk management plans, improving gas extraction and processing, and creating plans to reduce methane and other light hydrocarbon emissions into the atmosphere are among its most beneficial uses.

BIBLIOGRAPHY

Kooti, G., Taherdangkoo, R., Chen, C. et al. Machine learning prediction of methane, ethane, and propane solubility in pure water and electrolyte solutions: Implications for stray gas migration modeling. *Acta Geochim* 2024;2024. https://doi.org/10.1007/s11631-024-00680-8

8.4.1.10 Application-10: Modeling Interfacial Tension of N_2/CO_2 Mixture + n-Alkanes with Machine Learning Methods

8.4.1.10.1 Background of the Application

A major contributor to greenhouse gas emissions is the burning of fossil fuels in power plants and oil refineries, as well as the flaring or venting of generated gases in oil fields. In both conventional and UCRs, effective use of flue and greenhouse gases can improve O&G recovery and provide opportunities for CO_2 sequestration. For enhanced oil recovery (EOR) procedures to be implemented successfully, the interfacial tension (IFT) between injected gases and crude oils must be accurately estimated. Using ML techniques, this work examines the IFT between n-alkanes and a CO_2/N_2 mixture at different pressures and temperatures. A dataset of 268 IFT values was compiled from existing literature, with pressure, temperature, the carbon number of n-alkanes, and the mole fraction of N_2 as input parameters. Six ML methods—radial basis function (RBF), adaptive neuro-fuzzy inference system (ANFIS), least squares support vector machine (LSSVM), RF, MLP, and extremely randomized trees (Extra-Trees)—were employed alongside four optimization techniques—colliding bodies optimization (CBO), PSO, Levenberg–Marquardt (LM) algorithm, and coupled simulated annealing (CSA)—to model the IFT.

Among these, the RBF model demonstrated exceptional precision in predicting IFT values with an average absolute relative error of 0.77%, outperforming all other models. Sensitivity analysis revealed that pressure and the carbon number of n-alkanes have the greatest impact on the IFT of the CO_2/N_2 mixture with n-alkanes. The study also evaluated the applicability of the RBF model and the utilized IFT database using the leverage method, confirming the robustness and reliability of the RBF model in predicting IFT values for various conditions.

8.4.1.10.2 Description of the Workflow and Machine Learning Model

This study focuses on the accurate modeling of IFT between CO_2/N_2 mixtures and n-alkanes at various pressures and temperatures, which is crucial for EOR operations. The study employs several ML techniques, including RBF neural network, LSSVM,

MLP, ANFIS, RF, and Extra-Trees. The ML models are optimized using advanced algorithms like PSO, CSA, and the LM algorithm. These models predict the IFT with high accuracy, using inputs such as pressure, temperature, carbon number of n-alkanes, and mole fraction of N2. The optimized models enable the estimation of IFT without the need for expensive and time-consuming experimental methods, making them valuable for practical applications in the petroleum industry.

8.4.1.10.3 Advantages and Limitations of the Synchronized Algorithm

The use of various ML models, particularly the RBF neural network, allows for accurate predictions of IFT with minimal error, outperforming traditional models. These ML techniques are capable of capturing the complex nonlinear relationships between the input variables and the target IFT, which leads to improved predictive performance. The optimization algorithms, such as PSO and CSA, further enhance the models by fine-tuning their parameters, resulting in more reliable predictions. However, the complexity and variety of models require careful selection and tuning to avoid overfitting and ensure generalization. The computational cost associated with running these ML models, especially in the case of large datasets, can also be significant. Additionally, the success of these models heavily depends on the quality and quantity of the input data, and their accuracy may be compromised if the data range is limited or contains outliers.

8.4.1.10.4 Results and Discussion

In this paper, ML methods were employed to model the IFT of a CO_2/N_2 mixture with n-alkanes across varying pressures and temperatures. A dataset of 268 IFT values was compiled from literature, with pressure, temperature, carbon number, and the mole fraction of N_2 as input parameters. Six well-known ML methods—RBF, LSSVM, ANFIS, MLP, RF, and Extra-trees—were utilized along with four optimization techniques—CBO, LM algorithm, PSO, enhanced colliding bodies optimization (ECBO) and CSA. The RBF model was found to be the most accurate, with an average absolute percentage relative error (AAPRE) of 0.77%, surpassing all other models. It effectively captured the trend of decreasing IFT with increasing pressure and temperature.

The results also indicated that LSSVM–CSA, MLP–LM, Extra-trees, ANFIS–PSO, ANFIS–ECBO, and RF models achieved high accuracy, following the RBF model. Sensitivity analysis revealed that pressure had the most significant impact on IFT, followed by the carbon number of n-alkanes, temperature, and mole fraction of N_2. Pressure and temperature were found to decrease the IFT of the CO_2/N_2 mixture with n-alkanes. Additionally, the leverage approach identified only six data points outside the model's applicability range, underscoring the high reliability of the RBF model for predicting IFT.

BIBLIOGRAPHY

Salehi, E., Mohammadi, M.-R., Hemmati-Sarapardeh, A., Mahdavi, V. R., Gentzis, T., Liu, B., and Ostadhassan, M. Modeling Interfacial Tension of N2/CO2 Mixture + n-Alkanes with Machine Learning Methods: Application to EOR in Conventional and Unconventional Reservoirs by Flue Gas Injection. *Minerals* 2022;12:252. https://doi.org/10.3390/min12020252

8.4.1.11 Application-11: Understanding the Controlling Factors for CO_2 Sequestration in Depleted Shale Reservoirs Using Data Analytics and Machine Learning

8.4.1.11.1 Background of the Application

Carbon capture and sequestration (CCS) is poised to develop into an industry that rivals, if not surpasses, the current O&G sector. CCS involves capturing carbon dioxide from sources such as refineries, industrial facilities, and power plants, and then storing it in underground formations. UCRs, including depleted shale formations, are potential candidates for this storage. Although CCS is crucial for the O&G industry, the understanding of carbon dioxide sequestration in UCRs remains limited. This paper aims to identify the key factors influencing CO_2 sequestration in depleted shale reservoirs through DA and ML.

A large dataset from shale reservoirs, including multiple numerical simulation scenarios, was used in the study. A DA study was carried out to examine the connections between the primary response variable, cumulative CO_2 injected, and reservoir and operational parameters, which are regarded as predictor variables, following a completeness check of the input data. The cumulative CO_2 injected was predicted using a variety of ML models, such as MLR, RTs, bagging, RFs, and boosting. To determine which factors were most important for the CO_2 sequestration performance in shale reservoirs, the relative value of several variables was evaluated.

The analysis demonstrated significant relationships between reservoir and operational parameters and between the predictor variables and the cumulative CO_2 injected. Operational parameters showed a consistent, monotonic relationship with CO_2 injection volumes. Among the ML methods, RFs delivered the most accurate predictions, aligning with its theoretical strengths, while RTs performed poorly due to overfitting. The screening results indicated that stimulated reservoir volume fracture permeability was the most influential factor for CO_2 sequestration efficiency. These findings will help exploration and production companies understand the factors affecting the effectiveness of CO_2 sequestration in shale reservoirs.

8.4.1.11.2 Description of the Workflow and Machine Learning Model

This paper employs various ML models to predict the performance of CO_2 sequestration in depleted shale reservoirs. The models used include MLR, RTs, bagging, RFs, and boosting. These models were built to analyze the relationship between reservoir parameters and operational parameters as predictors and cumulative CO_2 injected as the response variable. The dataset, consisting of numerical simulation scenarios, was processed and analyzed using these ML techniques. The primary goal was to identify the key parameters that influence the CO_2 sequestration performance and develop predictive models that can estimate the amount of CO_2 sequestered based on these parameters.

The methodology begins with exploratory data analysis (EDA) to visualize the relationships between various reservoir and operational parameters and the cumulative CO_2 injected. Following this, multiple ML models were trained on the dataset, which included both linear and tree-based methods. The models were evaluated using performance metrics such as Root Mean Square Error (RMSE) and R-squared (R^2), and the most effective model was selected for further analysis.

8.4.1.11.3 Advantages and Limitations of the Synchronized Algorithm

The use of ML models in this study offers several advantages. RFs, in particular, demonstrated high predictive accuracy, making them effective for analyzing complex relationships between variables without requiring linearity assumptions. The ensemble methods, such as bagging and boosting, further improved model performance by reducing variance and enhancing prediction accuracy. These models are capable of identifying the most influential parameters, which is critical for optimizing CO_2 sequestration strategies.

However, the application of ML models also comes with limitations. The complexity of tree-based methods, especially when combined with bagging or boosting, can lead to high computational costs and require careful tuning of hyperparameters to avoid overfitting. Additionally, the quality of predictions depends on the comprehensiveness of the dataset and the accuracy of the numerical simulations used to generate it. Although RFs performed well, simpler models like RTs struggled with overfitting, highlighting the need for robust model validation techniques.

8.4.1.11.4 Results and Discussion

This paper explores the use of DA and ML techniques on a comprehensive shale reservoir dataset to uncover patterns and develop predictive models. The study focuses on understanding the relationships between reservoir and operational parameters and their impact on the volume of CO_2 sequestered. By employing regression and tree-based ML methods, the research aims to identify key variables influencing the CO_2 sequestration performance. The analysis revealed that operational parameters exhibit a higher skewness compared to reservoir parameters, suggesting a more significant impact on CO_2 sequestration. Moreover, operational parameters showed a greater linear association with the response variable, with cumulative CO_2 injected having a modest positive correlation with various operational parameters.

From a ML perspective, the study found that regression trees effectively rank the importance of different parameters in CO_2 sequestration. Among the ML techniques, RF demonstrated the highest predictive accuracy, making it the most reliable method for this application. Both RF and boosting models proved useful for identifying critical variables affecting CO_2 sequestration. The analysis highlighted that stimulated reservoir volume fracture permeability is the most influential factor, emphasizing the importance of operational parameters in optimizing CO_2 sequestration performance.

BIBLIOGRAPHY

Baabbad, Hassan, Khaled, Hassan, Artun, Emre, and Kulga, Burak. "Understanding the Controlling Factors for CO2 Sequestration in Depleted Shale Reservoirs Using Data Analytics and Machine Learning." Paper presented at the *SPE EuropEC - Europe Energy Conference featured at the 83rd EAGE Annual Conference & Exhibition*, Madrid, Spain, June 2022. doi: https://doi.org/10.2118/209717-MS

8.4.2 Reservoir Aspects

8.4.2.1 Application-12: A Novel Type of Neural Networks for Feature Engineering of Geological Data

8.4.2.1.1 Background of the Application

The paper introduces a new type of neural network designed to address the specific challenges encountered in geological DA, particularly in the context of rock engineering. Traditional NNs often struggle with geological datasets due to issues such as limited sample sizes, repetitive values, and the sensitivity of models to initial weights and biases. These challenges hinder the ability of conventional models to generalize effectively, which is crucial for tasks like predicting rock strength. The authors propose a novel neural network structure that re-engineers the basic elements of the CNN model, aiming to extract more information from each individual data point without being influenced by other data points. The research includes both simulation-based verification and case studies focused on predicting tensile strength in coal and shear strength in gas hydrate-bearing sediments, demonstrating the network's superior performance compared to CNNs.

8.4.2.1.2 Description of the Workflow and Machine Learning Model

The paper introduces a new type of neural network structure specifically designed for geological data, addressing the limitations of CNNs when dealing with small or repetitive datasets. This new network differs from traditional approaches by independently processing each data point, extracting maximum information without being influenced by other data points. The model is implemented in MATLAB and Python, and it is tested through simulations and case studies. The network uses a parallel structure where each data point has its own set of weights and biases, which prevents the model from overfitting and ensures better generalization. The study applies this new network to predict the tensile strength of coal and the shear strength of gas hydrate-bearing sediments, demonstrating its effectiveness in both simulation and real-world data scenarios.

8.4.2.1.3 Advantages and Limitations of the Synchronized Algorithm

The advantages of the new neural network structure are evident in its ability to handle small, repetitive datasets, which are common in geological studies. Unlike CNNs, which can struggle with such data, the new model processes each data point independently, reducing the impact of noisy or redundant information. This approach also reduces the risk of overfitting, as the model does not rely on a single set of weights and biases for all data points. However, the methodology is computationally intensive, as it requires separate processing for each data point. Additionally, the model's performance heavily depends on the quality of the input data, and further refinement may be necessary for applications to larger datasets. Despite these limitations, the new neural network structure presents a significant improvement over traditional models in the context of geological DA, particularly in tasks involving limited or complex datasets.

8.4.2.1.4 Results and Discussion

The study concludes that the newly developed neural network structure significantly outperforms CNNs in the context of geological DA. The network's ability to independently process each data point allows it to better capture the underlying information, making it more effective for tasks with limited datasets or repetitive values. The case studies on coal and gas hydrate-bearing sediments confirm the model's accuracy in predicting tensile and shear strength, with the new network structure delivering higher accuracy and lower error rates than traditional models. The authors highlight the potential of their approach to be applied in various geoscience and rock engineering problems, especially where data is scarce or difficult to obtain. The paper suggests that future research could explore the application of this method to larger datasets and different geological properties.

BIBLIOGRAPHY

Jiang, Lishuai, Yang, Zhao, Golsanami, Naser, Chen, Lianjun, and Yan, Weichao. A Novel Type of Neural Networks for Feature Engineering of Geological Data: Case Studies of Coal and Gas Hydrate-Bearing Sediments. *Geosci Front* 2020;11(5):1511–1531, ISSN 1674-9871. https://doi.org/10.1016/j.gsf.2020.04.016

8.4.2.2 Application-13: Permeability Prediction and Diagenesis in Tight Carbonates Using Machine Learning Techniques

8.4.2.2.1 Background of the Application

This paper focuses on the prediction of permeability in carbonate reservoirs, a critical factor in understanding and optimizing hydrocarbon production. Carbonate reservoirs present unique challenges due to their complex pore structures, which are significantly influenced by diagenetic processes such as dissolution, cementation, and dolomitization. Traditional methods of permeability prediction, which rely on well logs and core sample data, are often inadequate due to the heterogeneous nature of carbonate formations. The authors propose using ML techniques to improve permeability prediction by integrating various geological and petrophysical data, including diagenetic factors, into the models. By doing so, the study aims to provide more accurate and reliable permeability estimates, which are essential for reservoir characterization and production planning.

8.4.2.2.2 Description of the Workflow and Machine Learning Model

The paper applies ML techniques to predict permeability in carbonate reservoirs by incorporating geological, petrophysical, and diagenetic data. The ML models used include ANNs, SVMs, and RFs. These models are trained on datasets that combine well logs, core data, and diagenetic information, allowing them to capture the complex relationships between permeability and the various factors influencing it. The inclusion of diagenetic processes, such as dissolution and cementation, is a key feature of the study, as these processes have a significant impact on the pore structures in carbonate reservoirs. The models are evaluated using cross-validation and performance metrics such as MSE and R^2 to ensure their accuracy and reliability in predicting permeability.

8.4.2.2.3 Advantages and Limitations of Synchronized Algorithm

The ML methodologies used in this study offer several advantages. First, they can handle large and complex datasets, integrating multiple types of data to improve the accuracy of permeability predictions. The inclusion of diagenetic factors allows the models to account for the effects of geological processes on reservoir properties, providing a more comprehensive understanding of permeability variations. However, the methodologies also have limitations. The performance of ML models depends heavily on the quality and quantity of the input data, and poor data quality can lead to inaccurate predictions. Additionally, ML models can be computationally intensive, requiring significant resources for training and validation. Despite these challenges, the study demonstrates that ML techniques offer a promising approach to improving permeability prediction in carbonate reservoirs.

8.4.2.2.4 Results and Discussion

The paper concludes that ML techniques provide a significant improvement in predicting permeability in carbonate reservoirs compared to traditional methods. The incorporation of diagenetic factors into the ML models enhances their predictive accuracy, enabling better characterization of the complex pore structures found in carbonate formations. The results of the study demonstrate that ML models can effectively integrate multiple types of geological and petrophysical data to produce more accurate permeability predictions. The authors suggest that their approach can be further developed and applied to other types of reservoirs, potentially improving hydrocarbon recovery and reservoir management across a wide range of geological settings.

BIBLIOGRAPHY

Al Khalifah, H., Glover, P. W. J., and Lorinczi, P. Permeability Prediction and Diagenesis in Tight Carbonates Using Machine Learning Techniques. *Mar Pet Geol* 2020;112:104096, ISSN 0264-8172. https://doi.org/10.1016/j.marpetgeo.2019.104096

8.4.2.3 Application-14: A New Artificial Intelligence Based Empirical Correlation to Predict Sonic Travel Time

8.4.2.3.1 Background of the Application

This paper presents a novel approach to predicting sonic travel time using AI, specifically ANNs. Sonic logs are crucial for estimating rock elastic parameters, which are used to determine in-situ stresses within reservoirs. However, acquiring sonic logs can be time-consuming and expensive. Therefore, the paper aims to develop an accurate model to predict sonic travel time using commonly available well logs, such as GR, bulk density, and CNP. The authors compare the performance of their ANN-based model with existing empirical correlations and demonstrate the superior accuracy of their approach. The study also derives an empirical correlation from the ANN model to make the prediction process more accessible for field applications.

8.4.2.3.2 Description of the Workflow and Machine Learning Model

The study employs ANNs to predict sonic travel time from commonly available well logs. The workflow involves training the ANN model on a dataset that includes gamma ray, bulk density, CNP logs. The ANN is designed to learn complex relationships between these input features and the sonic travel time, which is the target variable. Feature selection is performed to ensure that the most relevant data are used in training the model, and the ANN is optimized for accuracy. The performance of the ANN model is evaluated against existing empirical correlations through metrics such as correlation coefficient and accuracy. Additionally, the study derives an empirical correlation from the ANN model to simplify its use in practical applications where direct sonic log measurements are not available.

8.4.2.3.3 Advantages and Limitations of the Synchronized Algorithm

The ANN model offers several advantages over traditional empirical correlations. It can capture complex, non-linear relationships between well log data and sonic travel time, leading to higher prediction accuracy. The model's ability to generalize from a broad set of well log data ensures that it can be applied across different reservoirs. However, the approach also has limitations. The performance of the ANN is dependent on the quality and representativeness of the training data—insufficient or biased data can affect its accuracy. Additionally, while the ANN model provides high accuracy, it requires significant computational resources for training and validation. The derived empirical correlation simplifies the prediction process but may not capture all nuances of the ANN model's performance. Future work should focus on refining the ANN model and addressing its limitations to enhance its applicability in various field conditions.

8.4.2.3.4 Results and Discussion

The study concludes that the proposed ANN model effectively predicts sonic travel time with a high degree of accuracy. The model, tested on real field data, outperformed traditional empirical correlations, achieving a correlation coefficient of 0.96 and an accuracy of 96%. The derived empirical correlation also provided results that were consistent with the ANN model, making it a valuable tool for geo-mechanical engineers in the absence of direct sonic log data. The authors suggest that their AI-based approach can significantly enhance the prediction of sonic travel times, ultimately improving the accuracy of reservoir characterizations and operational decisions.

BIBLIOGRAPHY

Tariq, Zeeshan, Elkatatny, Salaheldin, Mahmoud, Mohamed, and Abdulraheem, Abdulazeez. "A New Artificial Intelligence Based Empirical Correlation to Predict Sonic Travel Time." Paper presented at the *International Petroleum Technology Conference*, Bangkok, Thailand, November 2016. doi: https://doi.org/10.2523/IPTC-19005-MS

8.4.2.4 Application-15: Exploratory Analysis of Machine Learning Methods in Predicting Subsurface Temperature and Geothermal Gradient

8.4.2.4.1 Background of the Application

The paper explores the application of ML methods to predict subsurface temperature and geothermal gradients in the northeastern United States. Subsurface temperature is a critical factor in geothermal energy exploration and O&G industry applications. Traditionally, physics-based models, particularly thermal conductivity models, have been used to estimate subsurface temperatures. However, these models rely on assumptions that can introduce uncertainties, making them less reliable in complex geological settings. The authors evaluate various ML models, including Extreme Gradient Boosting (XGBoost), RF, Ridge regression, and DNNs, for their ability to predict subsurface temperature using bottom-hole temperature (BHT) data from over 20,750 O&G wells. The study compares the performance of these ML models to traditional physics-based models, aiming to improve prediction accuracy and identify geothermally active regions.

8.4.2.4.2 Description of the Workflow and Machine Learning Model

The paper utilizes ML models, including XGBoost, RF, Ridge regression, and DNNs, to predict subsurface temperature and geothermal gradients. The models are trained on a dataset of BHT measurements from over 20,750 wells in the northeastern United States. XGBoost and RF, both ensemble-based methods, are highlighted for their superior performance in handling the target imbalance in the dataset, which consists mostly of shallower wells. The study also employs Gaussian kernel-weighted K-nearest neighbor (KNN) regression to interpolate the features across the region, enabling the creation of continuous temperature-at-depth maps. Hyper-parameter tuning for each ML model is performed using cross-validation to optimize their performance.

8.4.2.4.3 Advantages and Limitations of the Synchronized Algorithm

The ML methodologies applied in this paper offer several advantages over traditional physics-based models. XGBoost and RF, in particular, are capable of handling complex, non-linear relationships in the data, leading to more accurate predictions of subsurface temperature and geothermal gradients. The ability of these models to manage imbalanced datasets and integrate diverse data sources makes them well-suited for geothermal exploration. However, the performance of ML models depends on the availability and quality of training data, and they require extensive computational resources for training and tuning. Despite these challenges, the study demonstrates that ML models can outperform traditional methods, offering a promising approach to improving geothermal energy exploration and subsurface temperature prediction.

8.4.2.4.4 Results and Discussion

The study concludes that ML models, particularly XGBoost and RF, outperform traditional physics-based models in predicting subsurface temperature and geothermal gradients. The XGBoost model provided the highest accuracy, with a mean absolute error of 3.21°C, demonstrating that ML can offer a reliable alternative to physics-based methods. The authors also generate geothermal gradient maps for the northeastern United States, identifying regions with high geothermal potential, such as parts of West Virginia and New York. The paper suggests that ML models can significantly enhance geothermal energy exploration by providing more accurate subsurface temperature predictions, reducing uncertainty, and guiding future geothermal development efforts.

BIBLIOGRAPHY

Shahdi, A., Lee, S., and Karpatne, A. et al. Exploratory Analysis of Machine Learning Methods in Predicting Subsurface Temperature and Geothermal Gradient of Northeastern United States. *Geotherm Energy* 2021;9:18. https://doi.org/10.1186/s40517-021-00200-4

8.4.2.5 Application-16: Machine Learning Based Shale Gas Proration Synchronized Computing Method

8.4.2.5.1 Background of the Application

The paper discusses the challenges of shale gas production optimization, particularly in determining appropriate proration methods for shale gas wells in the Fuling shale gas field in China. Traditional methods, such as numerical simulations, are data-intensive, costly, and time-consuming. These methods require well testing and engineering data, which are often incomplete or estimated based on assumptions. In response, the authors propose a synchronized ML-based method for shale gas proration computation. This method combines ML algorithms such as XGBoost and KNN with a parameter optimization algorithm to improve accuracy while reducing data requirements. The study uses data from 165 wells in the Fuling shale gas field, aiming to provide a more efficient and accurate approach for determining proration in shale gas wells.

8.4.2.5.2 Description of the Workflow and Machine Learning Model

The paper applies a synchronized ML method to compute shale gas proration, combining XGBoost and KNN algorithms. XGBoost, a gradient boosting DT algorithm, is used to calculate preliminary proration based on geological, engineering, and well testing data from 165 wells in the Fuling shale gas field. The preliminary results generated by XGBoost are then incorporated into a new dataset and used as an independent variable for the KNN algorithm. KNN calculates the ultimate proration by identifying similar wells based on their Euclidean distance in the feature space. The model's performance is optimized through grid search for hyperparameter tuning, ensuring the best combination of parameters for each algorithm. The synchronized method is designed to provide more accurate proration predictions while requiring fewer input parameters than traditional methods.

8.4.2.5.3 Advantages and Limitations of the Synchronized Algorithm

The synchronized ML approach offers several advantages over traditional proration methods. By combining XGBoost and KNN, the model improves the accuracy of proration predictions while reducing the amount of input data required. The use of ML allows for efficient handling of large datasets and reduces the reliance on costly and time-consuming well testing. Additionally, the grid search method for hyper-parameter tuning ensures that the model is optimized for performance. However, the methodology also has limitations. The accuracy of the KNN algorithm can be affected by the quality of the input data, and the model's performance may decrease if the dataset is too small or contains significant noise. Despite these challenges, the synchronized ML method demonstrates a promising approach to shale gas proration prediction, offering a balance between accuracy and efficiency.

8.4.2.5.4 Results and Discussion

The study concludes that the proposed synchronized ML method offers a significant improvement in the accuracy of shale gas proration prediction. By combining XGBoost and KNN algorithms, the model achieved an accuracy of 91.8%, which is competitive with traditional numerical simulation methods. The model requires fewer input parameters, streamlining the computation process and reducing the need for extensive well testing data. The results demonstrate that the synchronized approach provides a viable alternative for proration determination in shale gas fields, offering both efficiency and accuracy. The authors suggest that this methodology could be applied to other shale gas fields and further refined with additional data.

BIBLIOGRAPHY

Chen, Zan, Zhu, Xueqian, Fang, Yexin, and Li, Jitong. "Machine Learning Based Shale Gas Proration Synchronized Computing Method." Paper presented at the *SPE/AAPG/SEG Unconventional Resources Technology Conference*, Virtual, July 2020. doi: https://doi.org/10.15530/urtec-2020-2552

8.4.2.6 Application-17: Application of Artificial Intelligence Methods to Underground Gas Storage Control

8.4.2.6.1 Background of the Application

The paper addresses the challenge of optimizing underground gas storage (UGS) control, specifically focusing on improving the techno-economic efficiency of UGS operations. Traditional methods of optimizing UGS control, which rely heavily on reservoir simulators, often result in high computational costs and are considered black-box problems. To overcome these limitations, the paper proposes a novel algorithm that combines AI with control theory and reservoir simulation. This approach leverages a parameterized decision tree integrated with ML-based optimization methods to automate and improve UGS control. By applying AI techniques, the goal is to maximize the energy extracted from stored gas during production cycles without additional financial investment, thereby optimizing the overall operation of UGS systems.

8.4.2.6.2 Description of the Workflow and Machine Learning Model

The paper employs AI methods, particularly a combination of parameterized DTs and ML-based optimization, to optimize the control of UGS. The DT models the control of UGS by representing reservoir and production properties, with branches corresponding to limit values of specific attributes. Instead of setting these values arbitrarily, the model optimizes them using ML algorithms. The integration of the DT with the ECLIPSE reservoir simulator enables real-time updates of well flow rates during production, which are optimized through iterative simulations. Python is used to link the reservoir simulator with the ML optimization tool, creating a fully automated process for determining the optimal control of UGS operations.

8.4.2.6.3 Advantages and Limitations of the Synchronized Algorithm

The AI methodology offers several advantages in optimizing UGS control. The parameterized DT allows for a clear and interpretable control scheme that can be adapted to specific cases. The use of ML for parameter optimization significantly reduces the time and computational resources required, making it possible to handle complex reservoir simulations efficiently. However, the reliance on reservoir simulators still imposes a high computational cost, especially for large-scale problems. Additionally, the DT-based model may require significant domain knowledge to set up effectively. Despite these challenges, the methodology demonstrates a promising approach to improving UGS operations, offering both increased energy production and improved gas quality without additional investment.

8.4.2.6.4 Results and Discussion

The study concludes that the proposed AI-based optimization method for UGS control significantly enhances the efficiency of gas storage operations. The developed algorithm successfully automates the determination of optimal control parameters, leading to an increase in energy production by 2.4% compared to traditional methods. The intelligent control also reduces the nitrogen content in the produced gas, thereby improving gas quality. The results suggest that AI methods, such as the parameterized DT combined with reservoir simulations, provide a reliable and efficient solution for UGS control optimization. The authors highlight the potential for further application of this method to other production processes, demonstrating the adaptability and effectiveness of AI in complex engineering challenges.

BIBLIOGRAPHY

Kuk, Edyta "Application of Artificial Intelligence Methods to Underground Gas Storage Control." Paper presented at the *SPE Annual Technical Conference and Exhibition, Calgary*, Alberta, Canada, September 2019. doi: https://doi.org/10.2118/200305-STU

8.4.2.7 Application-18: Multilateral Wells Evaluation Utilizing Artificial Intelligence

8.4.2.7.1 Background of the Application

The paper explores the evaluation of multilateral wells using AI techniques, particularly focusing on predicting oil flow rates. Multilateral wells, known for their complex structures and diverse configurations, are a technological advancement that enhances reservoir drainage and productivity. However, evaluating their performance accurately remains challenging due to the limitations of traditional analytical models and correlations. The authors propose the use of ANNs as a modern approach to address these challenges, leveraging field data and reservoir parameters to predict the average oil flow rate of multilateral wells. The study also compares the results of the ANN model with those obtained from Borisov's correlation, a traditional method used for estimating multilateral well productivity.

8.4.2.7.2 Description of the Workflow and Machine Learning Model

The paper utilizes ANNs to predict the oil flow rates of multilateral wells, using data collected from 174 wells in the Middle East. The ANN model is trained on various surface and reservoir parameters, including effective lateral length, open-hole size, choke size, reservoir pressure, flowing wellhead pressure, and average permeability. The model is trained using the Gradient Descent with Adaptive Learning Rate Backpropagation algorithm, which iteratively optimizes the weights of the network to minimize the error between predicted and actual values. The study uses 70% of the data for training and 30% for testing, with a focus on achieving high prediction accuracy. The ANN model's performance is evaluated through statistical error analysis, comparing the predicted oil flow rates with actual field data.

8.4.2.7.3 Advantages and Limitations of the Synchronized Algorithm

The use of ANN provides several advantages over traditional methods. ANN models are capable of handling complex, non-linear relationships in the data, making them well-suited for predicting oil flow rates in multilateral wells with diverse configurations. The flexibility of ANN allows for the integration of various parameters, improving the accuracy of predictions. However, the methodology also has limitations, including the need for a large dataset to train the model effectively and the computational resources required for training. Additionally, the performance of the ANN model depends on the quality of the input data, which may affect the accuracy of predictions in cases where data is noisy or incomplete. Despite these challenges, the ANN approach demonstrates superior predictive accuracy compared to traditional methods, making it a promising tool for evaluating multilateral well performance.

8.4.2.7.4 Results and Discussion

The study concludes that ANN-based models provide a significant improvement in predicting oil flow rates for multilateral wells compared to traditional methods such as Borisov's correlation. The ANN model achieved a correlation coefficient of 0.97 and an overall error of 7.85%, demonstrating strong agreement with actual field data. In contrast, the Borisov's correlation resulted in an error of more than 50%, highlighting the limitations of traditional approaches in handling the complexity of multilateral wells. The authors emphasize that AI techniques, particularly ANN, offer a reliable and accurate solution for evaluating multilateral well performance, making them a valuable tool in optimizing oil production.

BIBLIOGRAPHY

Buhulaigah, Ahmed, Al-Mashhad, Ali S., Al-Arifi, Sulaiman A., Al-Kadem, Mohammed S., and Al-Dabbous, Mohammed S. "Multilateral Wells Evaluation Utilizing Artificial Intelligence." Paper presented at the *SPE Middle East Oil & Gas Show and Conference*, Manama, Kingdom of Bahrain, March 2017. doi: https://doi.org/10.2118/183688-MS

8.4.2.8 Application-19: Effect of Wettability on Waterflood Recovery for Crude-Oil/Brine/Rock Systems

8.4.2.8.1 Background of the Application

This paper investigates the impact of wettability on oil recovery during waterflooding in crude-oil/brine/rock (COBR) systems, focusing on Berea sandstone. Wettability, which influences how water and oil distribute within a porous medium, is a key factor affecting the efficiency of waterflood recovery. The study highlights the challenge of controlling and characterizing wettability in laboratory experiments. By altering wettability through the interaction of crude oil and brine with rock surfaces, the authors conducted over 50 waterflood tests under varying conditions. The results show that maximum oil recovery was achieved at weakly water-wet conditions, suggesting that wettability plays a crucial role in determining recovery efficiency.

8.4.2.8.2 Description of the Workflow and Machine Learning Model

The paper uses experimental methods to examine the effect of wettability on oil recovery in COBR systems. The study employs crude oil as a wettability-altering agent, with the degree of wettability controlled by varying brine composition, aging temperature, and initial water saturation (Swi). Berea sandstone cores were used for the waterflood experiments, and the wettability of the samples was measured using a modified Amott method, which involves spontaneous imbibition and forced displacement of oil and water. The core samples were aged under different conditions to achieve a spectrum of wettability states, ranging from strongly water-wet to oil-wet. The experiments aimed to determine the relationship between wettability and oil recovery during waterflooding.

8.4.2.8.3 Advantages and Limitations of the Synchronized Algorithm

The experimental methodology provides a reproducible approach to studying the effect of wettability on oil recovery in COBR systems. By using crude oil as the wettability-altering agent, the study offers a more realistic representation of reservoir conditions compared to chemical treatments. However, the methodology also has limitations, particularly in the challenge of achieving consistent wettability control. The variability in crude oil composition and the sensitivity of wettability to factors such as aging temperature and brine composition can introduce uncertainties in the results. Despite these challenges, the study demonstrates the critical role of wettability in determining waterflood recovery efficiency and provides valuable insights into optimizing waterflooding operations.

8.4.2.8.4 Results and Discussion

The study concludes that wettability is a dominant factor in waterflood recovery, with weakly water-wet conditions leading to the highest oil recovery rates. The authors found that the interaction between crude oil, brine composition, and rock surfaces can significantly alter wettability, affecting both oil recovery and the coproduction of oil and water. The findings emphasize the importance of carefully controlling wettability in waterflooding operations to optimize recovery efficiency. The study suggests that further research is needed to develop standardized procedures for generating consistent wettability conditions in laboratory experiments.

BIBLIOGRAPHY

Jadhunandan, P. P., and Morrow, N. R. Effect of Wettability on Waterflood Recovery for Crude-Oil/Brine/Rock Systems. *SPE Res Eng* 1995;10:40–46. doi: https://doi.org/10.2118/22597-PA

8.4.2.9 Application-20: An Artificial Intelligence Decision Support System for Unconventional Field Development Design

8.4.2.9.1 Background of the Application

This paper introduces an AI Decision Support System (IDSS) specifically designed for optimizing field development in unconventional O&G reservoirs. The authors emphasize that field development in UCRs involves complex, high-dimensional decision-making under significant geological and economic uncertainties. The proposed IDSS integrates ML and optimization techniques across three stacked AI layers: Predictive, Prescriptive, and Cognitive. The Predictive unit leverages advanced ML and DL methods to forecast the outcomes of various field designs. The Prescriptive unit identifies optimal solutions by searching through these predicted outcomes, while the Cognitive unit refines these solutions by aligning them with the decision-maker's utility function. The paper focuses on the Predictive unit's role in generating reliable and efficient development designs, aiming to reduce costs and improve production performance.

8.4.2.9.2 Description of the Workflow and Machine Learning Model

The paper utilizes a multi-layered IDSS to optimize field development designs for UCRs. The IDSS comprises three AI layers: Predictive, Prescriptive, and Cognitive. The Predictive unit uses ML models, such as XGBoost, to predict well performance and estimate key production metrics like EUR and short-term production. The Prescriptive unit employs optimization techniques to identify the most suitable field development plans based on predefined project goals and risks. The Cognitive unit uses reinforcement learning to refine the solutions by learning from expert decision-makers' preferences, thus personalizing the optimization process. The paper demonstrates the system's ability to handle high-dimensional data, integrating geological, reservoir, and economic factors to generate actionable recommendations.

8.4.2.9.3 Advantages and Limitations of the Synchronized Algorithm

The IDSS methodology provides several advantages in field development design. ML models, particularly XGBoost, offer fast and accurate predictions of production outcomes, allowing the system to evaluate thousands of design scenarios within hours. The multi-layered approach ensures that the optimization process is both data-driven and expert-informed, reducing the risk of suboptimal decisions. However, the methodology also faces challenges, including the need for clean and comprehensive data to train the ML models effectively. Additionally, while the Cognitive unit adds value by incorporating expert preferences, it requires consistent and high-quality input from decision-makers to function optimally. Despite these challenges, IDSS offers a robust framework for optimizing field development in UCRs, enhancing both efficiency and production performance.

8.4.2.9.4 Results and Discussion

The study concludes that the application of the IDSS in unconventional field development design can significantly enhance decision-making efficiency. The Predictive unit's ability to simulate thousands of field design scenarios within a short time frame reduces the average cost per barrel of production by 15%. The results demonstrate that the system can provide more accurate and actionable insights compared to traditional methods, which rely heavily on numerical reservoir simulations. The authors highlight that the introduction of a Cognitive unit further enhances decision-making by adapting the optimal designs to specific expert preferences and project goals. The paper suggests that this technology paves the way for broader applications of AI-driven decision support systems in the O&G industry.

BIBLIOGRAPHY

Tamimi, Naser, Samani, Shirin, Minaei, Mohsen, and Harirchi, Farshad. "An Artificial Intelligence Decision Support System for Unconventional Field Development Design." Paper presented at the *SPE/AAPG/SEG Unconventional Resources Technology Conference*, Denver, Colorado, USA, July 2019. doi: https://doi.org/10.15530/urtec-2019-249

8.4.2.10 Application-21: An Integrated Machine Learning Framework for Optimizing Unconventional Resources Development

8.4.2.10.1 Background of the Application

This paper introduces an integrated ML framework designed to optimize the development of unconventional O&G resources. The authors emphasize that while UCRs, particularly shale formations, hold significant potential, their development is challenging due to complex geology, high heterogeneity, and the need for precise well spacing and completion designs. Traditional methods often fall short of addressing these challenges effectively, leading to suboptimal development strategies. The proposed framework leverages big data technologies and advanced ML algorithms to analyze heterogeneous datasets from geology, completion, development, and production domains. The framework is designed to be scalable and flexible, capable of processing large volumes of data to provide insights into optimal well placement, completion design, and production strategies. The system has been applied extensively in the Permian Basin, yielding promising results that demonstrate its utility in real-world scenarios.

8.4.2.10.2 Description of the Workflow and Machine Learning Model

The paper applies an integrated ML framework that combines big data technologies with advanced ML algorithms to optimize unconventional resource development. The framework is built on a data warehouse that handles large volumes of heterogeneous data, including geological, completion, development, and production information. Data transformation techniques are employed to standardize and regularize the data, making it suitable for ML applications. The ML models used in the framework include ensemble learning algorithms, such as RFs and gradient boosting (GB), which are particularly effective in handling complex, non-linear relationships in the data. The framework also incorporates visualization tools to facilitate the interactive analysis and interpretation of the results, helping engineers make informed decisions on well placement and completion designs.

8.4.2.10.3 Advantages and Limitations of the Synchronized Algorithm

The integrated ML framework offers several advantages over traditional methods. Its ability to handle large and diverse datasets allows for more accurate and reliable predictions, while the use of ensemble learning techniques enhances model performance by combining multiple algorithms to improve accuracy. The integration of visualization tools further aids in understanding the results and identifying key drivers of well performance. However, the framework also faces challenges, particularly in the need for high-quality data and significant computational resources to process and analyze the data effectively. Despite these limitations, the framework demonstrates significant potential for optimizing unconventional resource development, providing a more data-driven and efficient approach to field development planning.

8.4.2.10.4 Results and Discussion

The study concludes that the integrated ML framework provides significant advantages in optimizing the development of UCRs. By leveraging a comprehensive big data infrastructure and state-of-the-art ML techniques, the framework enables more accurate predictions and better decision-making compared to traditional methods. The results from the Permian Basin case studies show that the framework can effectively optimize well spacing and completion designs, leading to improved production performance and reduced costs. The authors suggest that the flexibility and scalability of the framework make it applicable to other unconventional plays, offering a powerful tool for enhancing resource development efficiency across the industry.

BIBLIOGRAPHY

Zhou, Hui, and Lascaud, Benjamin. "An Integrated Machine Learning Framework for Optimizing Unconventional Resources Development." Paper presented at the *SPE/AAPG/SEG Unconventional Resources Technology Conference*, Denver, Colorado, USA, July 2019. doi: https://doi.org/10.15530/urtec-2019-319

8.4.2.11 Application-22: Are Unconventional Well Performance Gains Exhausted?

8.4.2.11.1 Background of the Application

The paper investigates the factors driving year-over-year production improvements in major US unconventional plays, such as the Bakken, Eagle Ford, Midland Wolfberry, and Delaware Wolfbone. Over the past decade, unconventional well performance has dramatically improved due to longer lateral lengths, more intense completions, and drilling in higher-quality areas. The study applies ML models to analyze well performance, combining interpreted subsurface data, completion parameters, and well spacing metrics. By focusing on the drivers of these production improvements, the research aims to determine whether the recent gains can continue or if the key innovations that drove past performance have reached their limits. The study also explores how ML, combined with explainability tools like Shapley Additive Explanations (SHAP) values, can help quantify the impact of operator design choices on production outcomes.

8.4.2.11.2 Description of the Workflow and Machine Learning Model

The paper uses DT-based ML models to analyze well performance across four major US unconventional plays. The models incorporate a wide range of data, including subsurface geology, well spacing, lateral length, proppant tonnage, and completion fluid volumes. The ML approach allows for the identification of key production drivers and their contribution to well performance. After initial model training, surrogate models are built using SHAP values to explain the impact of each feature on production forecasts. By breaking down the contributions of completions, lateral length, and geology, the models provide insights into how these factors influence cumulative oil production, helping operators optimize future development strategies.

8.4.2.11.3 Advantages and Limitations of the Synchronized Algorithm

The ML methodology offers several advantages, including the ability to handle complex, non-linear relationships between variables and to generate explanations for model predictions through SHAP values. This approach allows for a more detailed analysis of the factors driving well performance, providing operators with actionable insights. However, the methodology also faces limitations, such as the need for high-quality data and the challenge of interpreting the results in the context of real-world operational constraints. Additionally, the diminishing returns of longer laterals and larger completions highlight the need for continued innovation to maintain productivity gains. Despite these challenges, the study demonstrates the power of ML in optimizing unconventional well performance and guiding future development decisions.

8.4.2.11.4 Results and Discussion

The study concludes that future performance gains in unconventional oil plays are likely to be smaller compared to the previous decade. The key innovations of longer laterals and larger completions, which drove significant production increases in the past, have plateaued since 2017 across the major plays. The research shows that while longer laterals still contribute to production improvements, their impact has diminished, and the upsizing of completions has reached its productive limit. The findings suggest that unconventional well productivity will likely face challenges as inventory depletion reduces rock quality, although targeted development planning and operational efficiencies could still improve returns. The authors emphasize the importance of ML tools in providing actionable insights for future field development.

BIBLIOGRAPHY

Cross, T., Chaplin, J., Sathaye, K., and Cui, A. "Are Unconventional Well Performance Gains Exhausted? Investigating the Drivers of Year-over-Year Production Improvements Across the Major US Unconventional Plays Using Machine Learning." Paper presented at the *SPE/AAPG/SEG Unconventional Resources Technology Conference*, Houston, Texas, USA, July 2021. doi: https://doi.org/10.15530/urtec-2021-5263

8.4.2.12 Application-23: Production Optimization of the CO_2 Huff-N-Puff Process in an Unconventional Reservoir Using a Machine Learning Based Proxy

8.4.2.12.1 Background of the Application

This paper investigates the optimization of the CO_2 HnP process in UCRs using ML-based proxy models. The main objective is to maximize the net present value (NPV) for life-cycle production by optimizing design variables such as CO_2 injection rate, production bottom hole pressure (BHP), and the duration of injection and production periods. The study utilizes ML techniques, specifically least-squares support vector regression (LS-SVR) and Gaussian process regression (GPR), to build proxy models that approximate the NPV from reservoir simulations. These proxy models are then used in an optimization framework to iteratively improve the design variables, leading to a more efficient and cost-effective approach to EOR in hydraulically fractured unconventional oil reservoirs. The research demonstrates that the ML proxies are significantly faster and computationally more efficient than direct simulation methods, with similar accuracy.

8.4.2.12.2 Description of the Workflow and Machine Learning Model

The paper applies ML methods, specifically LS-SVR and GPR, to optimize the CO_2 HnP process in UCRs. The proxy models are trained on simulation data to approximate the NPV based on input design variables, such as CO_2 injection rate, BHP, and the duration of injection and production periods. These proxy models are then used in an iterative optimization process, employing sequential quadratic programming (SQP) to maximize NPV. The ML models significantly reduce computational costs by approximating the forward simulations, allowing for faster convergence and more efficient optimization of the CO_2 HnP process.

8.4.2.12.3 Advantages and Limitations of the Synchronized Algorithm

The ML methodology offers several advantages, including significant reductions in computational time and the ability to handle complex optimization problems with multiple design variables. The LS-SVR and GPR models provide accurate approximations of NPV, enabling efficient optimization without the need for time-consuming direct simulations. However, the methodology is limited by the accuracy of the proxy models, which depend on the quality of the training data and the choice of hyperparameters. Additionally, the study focuses on a single-well scenario, and further research is needed to extend the methodology to multi-well and full-field optimizations. Despite these challenges, the ML approach demonstrates substantial potential in enhancing the efficiency and effectiveness of the CO_2 HnP process in UCRs.

8.4.2.12.4 Results and Discussion

The study concludes that the application of ML-based proxy models, particularly LS-SVR and GPR, significantly enhances the optimization process for the CO_2 HnP process in UCRs. The results show that both proxy models efficiently approximate NPV and provide accurate optimization of the design variables, reducing the computational time by at least five times compared to traditional simulation-based optimization. The research highlights the importance of proxy-based optimization in improving the economic viability of EOR methods in UCRs. Future work is recommended to extend the methodology to multi-well scenarios and consider the inclusion of geomechanical effects to further refine the optimization process.

BIBLIOGRAPHY

Almasov, Azad, Onur, Mustafa, and Reynolds, Albert C. "Production Optimization of the CO2 Huff-n-Puff Process in an Unconventional Reservoir Using a Machine Learning Based Proxy." Paper presented at the *SPE Improved Oil Recovery Conference*, Virtual, August 2020. doi: https://doi.org/10.2118/200360-MS

8.4.2.13 Application-24: A Framework for Predicting the Gas-Bearing Distribution of Unconventional Reservoirs by Deep Learning

8.4.2.13.1 Background of the Application

Predicting UCRs using multicomponent seismic data is possible, but it is a difficult undertaking. Despite the potential benefits of ML, especially DL, the accuracy of ML in reservoir prediction is highly reliant on the volume of training data and the intricacy of the model architecture. In order to overcome these obstacles, this study makes use of a compact convolutional neural network with a self-attention mechanism (SACNN) and TL. Through data sample expansion and model performance optimization, a framework for UCR prediction was created. The efficacy of the synthetic data used as the source data was first evaluated using the SACNN model for both O&G-reservoirs. A real dataset was then produced by maximizing real multicomponent seismic characteristics. Data from synthetic gas reservoirs was then transferred to the real dataset to create the TL dataset. Lastly, the distribution of gas in tight sandstone gas reserves was predicted using the SACNN model. The results showed that the suggested approach outperformed traditional ML models, predicting unusual reservoir distributions with less error. In addition, the model outperformed models trained just on synthetic (R2 = 0.9389) or actual (R2 = 0.9627) data in terms of prediction performance (R2 = 0.9731) on the testing dataset. Furthermore, the robustness and efficiency of the suggested model were validated using uncertainty analysis. The foundation for building data-driven models for energy conversion and usage is laid by this framework.

8.4.2.13.2 Description of the Workflow and Machine Learning Model

This study presents a DL framework for predicting the gas-bearing distribution in UCRs using multicomponent seismic data. The main AI technique employed is a SACNN model. This model is enhanced by a self-attention mechanism, which improves feature extraction from seismic attributes, leading to more accurate predictions of reservoir characteristics. The framework incorporates TL to leverage synthetic data for training, which is then transferred to real seismic data. The data optimization process includes the use of Principal Component Analysis (PCA) and Independent Component Analysis (ICA) to reduce redundancy and enhance the quality of input data. The methodology is applied to predict gas-bearing distribution in a tight sandstone gas reservoir, optimizing the accuracy of the predictions through a combination of DL and unsupervised learning techniques.

8.4.2.13.3 Advantages and Limitations of the Synchronized Algorithm

The SACNN provides significant advantages in reservoir prediction by combining the strengths of convolutional networks with self-attention mechanisms. This integration enhances the model's ability to focus on critical features, leading to more accurate predictions. The use of TL allows the model to benefit from synthetic data, addressing the challenge of limited real data samples. The PCA and ICA optimization steps further improve the model's performance by focusing on the most relevant seismic attributes. However, the complexity of the SACNN model increases computational demands, particularly when dealing with large datasets. Moreover, the effectiveness of TL depends on the similarity between synthetic and real datasets, which may introduce variability in prediction accuracy if the datasets are not closely aligned.

8.4.2.13.4 Results and Discussion

This study introduces an innovative prediction framework for UCR distribution, addressing challenges posed by limited sample data size and model performance in data-driven models. By integrating TL and DL techniques, this approach increases sample data volume and enhances model accuracy. Applied to a tight sandstone gas reservoir, the framework produced promising results. The streamlined SACNN model, developed for predicting O&G-bearing distribution using multicomponent seismic attribute data (MSAD), outperformed conventional ML methods, achieving reduced error and superior prediction capabilities. Notably, the SACNN model demonstrated commendable performance with an R^2 of 0.9627 when tested on real datasets. Optimization techniques such as PCA and ICA were employed to refine the original seismic attributes, reducing dimensionality and emphasizing reservoir-sensitive characteristics, which significantly boosted the prediction accuracy.

To address the limited sample data, TL was leveraged to transfer gas reservoir information from synthetic datasets, resulting in a model that outperformed those trained solely on either synthetic or real datasets, achieving an R^2 of 0.9731 on the testing dataset. The TL–SACNN model provided a clearer delineation of the gas reservoir distribution boundary, validating the effectiveness of TL in enhancing prediction accuracy. Furthermore, the robustness of the constructed model was confirmed through uncertainty analysis, underscoring the viability of applying TL to UCR prediction. Collectively, these results demonstrate the potential of this framework to overcome data limitations and improve predictive performance in UCR studies.

BIBLIOGRAPHY

Yang, J., Lin, N., Zhang, K. et al. A Framework for Predicting the Gas-Bearing Distribution of Unconventional Reservoirs by Deep Learning. *Nat Resour Res* 2024;33:1625–1655. https://doi.org/10.1007/s11053-024-10345-1

8.4.2.14 Application-25: Predicting Reservoir Fluid Properties from Drill Cuttings: Application of GPC-UV Method for Reservoir Fluid Analysis

8.4.2.14.1 Background of the Application

Every well that is dug yields drill cuttings, which are a useful source of data for geochemical research that estimates important oil characteristics including viscosity and API density. Drilling with oil-based mud makes it more difficult to identify reservoir fluids because of contamination, although the majority of study has focused on cuttings from wells using water-based mud. Because of this, there is now a substantial business requirement for trustworthy techniques to forecast reservoir fluid characteristics from drill cuttings in these situations. A major breakthrough was made in 2015 with the introduction of Gel Permeation Chromatography (GPC) in conjunction with ultraviolet (UV) absorbance detection.

By using a multiple-wavelength technique instead of a fixed-wavelength detection, this work improves the GPC–UV approach and strengthens the relationship between GPC–UV data and reservoir fluid characteristics. The samples studied are polluted with oil-based mud since the drill cuttings came from wells that used it. Oil peaks are successfully identified by the GPC–UV technique, but the oil-based mud response manifests as a baseline artifact. These findings can be used to predict API gravity from new samples and subjectively identify fluid kinds by comparing them with GPC–UV data from stock tank oil and known reservoir fluid attributes. A digital approach combining ML and an extensive GPC–UV database is being developed to significantly improve forecast accuracy.

8.4.2.14.2 Description of the Workflow and Machine Learning Model

This paper presents a novel approach to predicting reservoir fluid properties from drill cuttings using a combination of GPC coupled with UV absorbance detection. The core of the methodology is the extraction of reservoir fluids from contaminated drill cuttings and their analysis through GPC–UV. This method focuses on separating the molecular components of the fluids and using UV absorbance to detect the presence of reservoir hydrocarbons, particularly aromatic compounds.

To enhance the predictive capabilities of this method, the study proposes the integration of ML techniques. These techniques are employed to establish correlations between GPC–UV signals (such as signal strength and retention time) and key reservoir fluid properties, including API gravity and viscosity. The ML models are trained on a dataset of GPC–UV measurements from both stock tank oils and drill cuttings, enabling the automation of reservoir fluid property predictions based on the extracted fluid data.

The ML approach leverages isoabsorbance analysis, a method that visualizes UV absorbance across multiple wavelengths and retention times, providing a three-dimensional perspective of the fluid's chemical composition. This digital solution not only enhances the accuracy of reservoir fluid predictions but also allows for the identification of fluid types (oil, gas, or water) from drill cuttings that have been contaminated by oil-based mud.

8.4.2.14.3 Advantages and Limitations of the Synchronized Algorithm

The incorporation of ML into the GPC–UV method offers significant advantages in predicting reservoir fluid properties. One of the key benefits is the ability to automate the interpretation of complex GPC–UV data, enabling more accurate and efficient predictions of properties such as API gravity and viscosity. This is particularly valuable in cases where traditional methods struggle due to contamination from oil-based mud, as the GPC–UV method can effectively isolate relevant hydrocarbon signals.

The use of isoabsorbance analysis in combination with ML techniques provides a powerful tool for qualitative and quantitative reservoir fluid analysis. It allows for the identification of fluid types and the prediction of key properties from a diverse range of reservoir conditions and fluid compositions. Additionally, the ML models can be continuously improved with more data, leading to increasingly accurate predictions over time.

However, the methodology has its limitations. The quality of the predictions is highly dependent on the accuracy of the underlying GPC–UV measurements and the robustness of the ML models. Poor-quality drill cuttings can result in weak signals, which may compromise the predictive accuracy of the ML models. Furthermore, the computational demands of training and fine-tuning ML algorithms can be significant, particularly as the dataset of GPC–UV measurements expands. Finally, the generalizability of the ML models across different reservoirs and fluid types remains a challenge, necessitating ongoing refinement and validation.

8.4.2.14.4 Results and Discussion

A new GPC coupled with UV absorbance detection equipment has been developed, building upon previous research and extensive in-house testing and verification. The single-wavelength GPC–UV measurements for fluid samples from the Norwegian Continental Shelf demonstrate a strong correlation between API gravity and signal strength. However, when applying multiple-wavelength GPC–UV measurements to fluid samples from various global assets, the method shows good consistency between API gravity and signal strength for oil samples but exhibits weaker consistency for gas condensate samples.

The multiple-wavelength GPC–UV isoabsorbance analysis proves to be a robust tool for qualitatively identifying fluid types and properties. While the GPC–UV method is effective for analyzing extracts from drill cuttings, it does experience reduced signal strength, and poor-quality cutting samples can complicate the prediction of fluid properties.

BIBLIOGRAPHY

Yang, Tao, Cely, Alexandra, Moore, Julian, and Michael, Eric. "Predicting Reservoir Fluid Properties from Drill Cuttings: Application of GPC-UV Method for Reservoir Fluid Analysis." Paper presented at the *International Petroleum Technology Conference*, Dhahran, Saudi Arabia, February 2024. doi: https://doi.org/10.2523/IPTC-23357-MS

8.4.3 Hydraulic Fracturing

8.4.3.1 Application-26: Well Spacing Optimization for Permian Basin Based on Integrated Hydraulic Fracturing, Reservoir Simulation and Machine Learning Study

8.4.3.1.1 Background of the Application

This paper presents a comprehensive study on well spacing optimization in the Permian Basin, combining HF simulation, reservoir simulation, geomechanics, and ML. UCRs like the Permian Basin pose significant challenges due to their low permeability and complex geological structures, making well spacing optimization crucial for maximizing oil production. Traditional physics-based models often fall short in handling the complexity of these reservoirs, leading to uncertainties in well placement and production forecasts. To address this, the study introduces a workflow that integrates high-fidelity physics simulations with ML techniques to optimize well spacing. This approach enables more accurate predictions of production outcomes and well interference, helping operators make informed decisions in unconventional field development.

8.4.3.1.2 Description of the Workflow and Machine Learning Model

The paper utilizes a combination of physics-based simulations and ML to optimize well spacing in the Permian Basin. HF simulation, reservoir simulation, and geomechanics simulation are performed using Schlumberger's INTERSECT, KINETIX Shale, and Visage software, respectively. These simulations are used to model the behavior of parent and child wells under different well spacing scenarios, ranging from 200 feet to 1000 feet. The resulting data, including completion, geology, reservoir, and production information, are then fed into ML models. Ensemble regression methods such as GB, RF, and Extra-Tree regression are employed to predict cumulative oil production and identify optimal well spacing for different geological conditions.

8.4.3.1.3 Advantages and Limitations of the Synchronized Algorithm

The integrated methodology offers several advantages. By combining high-fidelity physics simulations with ML, the workflow captures complex interactions between wells, geological formations, and reservoir properties, leading to more accurate well spacing optimization. ML enables the analysis of large datasets, identifying key features that influence production outcomes. However, the methodology also faces limitations, including the high computational cost of the physics simulations and the challenge of feature selection in ML. The accuracy of the models depends on the quality of the input data, and there is a need for domain expertise to interpret the results effectively. Despite these challenges, the integrated approach provides a powerful tool for optimizing well spacing in UCRs, improving production efficiency while mitigating risks.

8.4.3.1.4 Results and Discussion

The study concludes that the integration of HF simulation, reservoir simulation, geomechanics, and ML offers a robust solution for optimizing well spacing in the

Permian Basin. The ML models achieved good predictive accuracy, with the workflow demonstrating its effectiveness in capturing the complexities of UCRs. The results show that the optimized well spacing leads to improved production performance, reducing the risk of well interference and maximizing resource recovery. The authors emphasize that this workflow can be adapted to other unconventional fields, providing a valuable tool for optimizing well spacing and enhancing production efficiency in various geological settings.

BIBLIOGRAPHY

Wang, Leizheng, and Sun, Alexander Y. "Well Spacing Optimization for Permian Basin Based on Integrated Hydraulic Fracturing, Reservoir Simulation and Machine Learning Study." Paper presented at the *SPE/AAPG/SEG Unconventional Resources Technology Conference*, Virtual, July 2020. doi: https://doi.org/10.15530/urtec-2020-3104

8.4.3.2 Application-27: Artificial Intelligence Assisted Hydraulic Fracturing Design in Shale Gas Reservoir

8.4.3.2.1 Background of the Application

This paper investigates the use of AI to enhance HF design in shale gas reservoirs, specifically focusing on the Marcellus Shale. Traditional HF design relies on numerical simulation to optimize key parameters like stage numbers, injection pressure, and proppant amount. However, the complex nature of shale gas reservoirs, including storage and transport mechanisms such as adsorption/desorption and diffusion, complicates these simulations. To address these challenges, the authors propose an AI-assisted model that utilizes NNs to analyze field data, perform history matching, and predict gas production under different fracturing design scenarios. This approach aims to improve the efficiency and accuracy of HF design in shale gas reservoirs.

8.4.3.2.2 Description of the Workflow and Machine Learning Model

The paper applies AI, specifically ANNs, to assist in HF design for shale gas reservoirs. The AI model is built on a spatial-temporal database that integrates static and dynamic field data, including reservoir characteristics, completion data, stimulation data, and production information from over 150 Marcellus shale wells. Key parameters influencing gas production are identified through KPI (Key Parameter Indicator) analysis, and the ANN model is trained using a feed-forward backpropagation algorithm. The training process involves 80% of the data for training, 10% for validation, and 10% for verification. The ANN model simulates gas production for different fracturing design scenarios, allowing engineers to compare the outcomes and optimize HF parameters.

8.4.3.2.3 Advantages and Limitations of the Synchronized Algorithm

The AI methodology provides several advantages over traditional HF design approaches. The ANN model can handle complex relationships between multiple variables, leading to more accurate predictions of gas production compared to conventional simulation methods. Additionally, the AI model allows for rapid scenario testing, with simulation times of under one second, significantly reducing the computational cost associated with numerical simulations. However, the methodology is highly dependent on the quality of the input data, and the model's performance can be affected by noise or inaccuracies in the data. Furthermore, the complexity of the ANN model requires careful tuning and validation to ensure reliable predictions. Despite these challenges, the AI-assisted approach offers a promising solution for optimizing HF design in shale gas reservoirs, improving both efficiency and accuracy.

8.4.3.2.4 Results and Discussion

The study concludes that AI-assisted HF design, particularly using NNs, offers significant advantages over traditional numerical simulation methods. The neural network model demonstrated strong predictive accuracy, with a correlation coefficient of 0.93699, and was able to simulate 200 different fracturing design scenarios in under a second. The results show that by optimizing key parameters like stage number, cluster number, and injected volumes, the AI model can predict cumulative gas production with high accuracy, providing a powerful tool for optimizing HF designs. The authors suggest that AI techniques have great potential to revolutionize shale gas reservoir development by enabling more efficient and cost-effective design processes.

BIBLIOGRAPHY

He, Qin, Zhong, Zhi, Alabboodi, Maher, and Wang, Guochang. "Artificial Intelligence Assisted Hydraulic Fracturing Design in Shale Gas Reservoir." Paper presented at the *SPE Eastern Regional Meeting*, Charleston, West Virginia, USA, October 2019. doi: https://doi.org/10.2118/196608-MS

8.4.3.3 Application-28: Application of Machine Learning for Closure Pressure Determination

8.4.3.3.1 Background of the Application

This paper explores the application of ML, specifically ANNs, to determine closure pressure in HF treatments. Closure pressure, which is closely related to the minimum horizontal stress, plays a critical role in the design and success of HF operations. Traditional methods for closure pressure determination, such as the square-root of time plot and G-function plot, often involve subjective interpretation based on the experience of the engineer analyzing the data. The authors propose using ML to minimize subjectivity and improve the accuracy of closure pressure predictions. The ANN model is trained on data from five wells using high-frequency bottom-hole pressure readings collected during mini-frac tests, aiming to automate the process and enhance the reliability of closure pressure determination.

8.4.3.3.2 Description of the Workflow and Machine Learning Model

The paper utilizes an ANN to predict closure pressure, using high-frequency bottom-hole pressure data from five wells. The ANN architecture consists of an input layer, a hidden layer, and an output layer, where each value of bottom-hole pressure is treated as a separate feature. The model is trained using data from three wells, with data from two additional wells used for testing and validation. The ANN is optimized using the feed-forward algorithm for calculating the error (J(θ)) and the back-propagation algorithm for minimizing the error through gradient descent. The training set error and test set error are evaluated for various values of the regularization parameter (λ) to identify the optimal configuration for predicting closure pressure.

8.4.3.3.3 Advantages and Limitations of the Synchronized Algorithm

The ML approach provides several advantages over traditional methods for closure pressure determination. The ANN model reduces subjectivity by automating the interpretation process, which can lead to more consistent and reliable predictions. The use of ML also enables the handling of complex data patterns that may not be easily captured by conventional analytical methods. However, the model's accuracy is limited by the size of the training dataset, and small datasets can lead to issues like overfitting or getting stuck in local minima. Despite these challenges, the study demonstrates that with larger datasets and further optimization, ML has the potential to significantly improve closure pressure determination, providing a valuable tool for HF design.

8.4.3.3.4 Results and Discussion

The study concludes that the application of ML, particularly ANNs, provides a more objective approach to closure pressure determination compared to traditional analytical methods. Although the model's accuracy was limited by the small dataset, ANNs successfully predicted closure pressure with reasonable accuracy, demonstrating the potential of ML in reducing subjectivity in this critical aspect of HF design. The authors acknowledge that the model's performance could be significantly improved with a larger dataset, and they recommend further research to refine the methodology. Overall, the study highlights the potential of ML to complement and enhance existing techniques for closure pressure determination in HF operations.

BIBLIOGRAPHY

Nande, Soumitra "Application of Machine Learning for Closure Pressure Determination." Paper presented at the *SPE Annual Technical Conference and Exhibition*, Dallas, Texas, USA, September 2018. doi: https://doi.org/10.2118/194042-STU

8.4.3.4 Application-29: Data-Driven Model for Hydraulic Fracturing Design Optimization: Focus on Building Digital Database and Production Forecast

8.4.3.4.1 Background of the Application

The paper explores the development of a data-driven model aimed at optimizing HF design through the use of ML (ML) techniques. With the increasing number of multistage fractured wells, particularly in UCRs, there is a vast amount of data available for developing predictive models. Traditional HF design methods often struggle with non-uniform production across different stages, resulting in inefficiencies. The authors propose a workflow that integrates field data from over 5000 wells in Western Siberia, Russia, into a comprehensive digital database. This database contains detailed information on the reservoir, well, and fracturing design parameters.

The study focuses on addressing data preprocessing challenges, such as missing values and OD, and leverages ML algorithms, particularly the CatBoost algorithm, to predict cumulative oil production, achieving an R^2 value of 0.815.

8.4.3.4.2 Description of the Workflow and Machine Learning Model

The paper applies ML algorithms to optimize HF design by leveraging a digital database containing data from over 5000 wells in Western Siberia. The CatBoost algorithm, a GBDT model, is employed for predicting cumulative oil production based on 92 input variables, which include reservoir, well, and fracturing design parameters. Data preprocessing techniques, such as collaborative filtering for handling missing values and clustering for OD, are utilized to enhance the quality of the dataset. The model's performance is evaluated through cross-validation, achieving an R^2 value of 0.815, demonstrating its effectiveness in handling complex datasets and improving production forecasts.

8.4.3.4.3 Advantages and Limitations of the Synchronized Algorithm

The ML methodology offers several advantages in optimizing HF design. DT-based algorithms like CatBoost are particularly effective in handling the heterogeneity of the input parameters, leading to accurate predictions of cumulative oil production. The use of ML also allows for the efficient processing of large datasets, incorporating various types of data to enhance prediction accuracy. However, the methodology faces challenges related to overfitting, particularly with small or noisy datasets. Additionally, the performance of the model is dependent on the quality of the input data, and significant preprocessing is required to handle missing values and outliers. Despite these limitations, the study demonstrates that ML techniques, when applied to a well-structured dataset, can significantly improve the optimization of HF design.

8.4.3.4.4 Results and Discussion

The study concludes that ML models, particularly DTs-based algorithms like CatBoost, outperform other methods in predicting cumulative oil production based on fracturing design parameters. The data-driven approach significantly enhances the accuracy of production forecasts, with the CatBoost model achieving an R^2 value of 0.815. The study emphasizes the importance of high-quality, comprehensive field data for building robust predictive models. The authors highlight the potential for further optimization in fracturing design by applying ML to a larger dataset and suggest that the next phase of their research will focus on solving the inverse problem: determining the optimal fracturing design parameters to maximize production.

BIBLIOGRAPHY

Morozov, Anton D., Popkov, Dmitry O., Duplyakov, Victor M., Mutalova, Renata F., Osiptsov, Andrei A., Vainshtein, Albert L., Burnaev, Evgeny V., Shel, Egor V., and Paderin, Grigory V. Data-Driven Model for Hydraulic Fracturing Design Optimization: Focus on Building Digital Database and Production Forecast. *J Petrol Sci Eng* 2020;194:107504, ISSN 0920-4105, https://doi.org/10.1016/j.petrol.2020.107504.

8.4.3.5 Application-30: The Use of Flowback Data for Estimating Dynamic Fracture Volume and Its Correlation with Completion-Design Parameters

8.4.3.5.1 Background of the Application

The paper focuses on analyzing flowback data to estimate dynamic fracture volume and its correlation with completion-design parameters (CDPs) in the Eagle Ford formation. Hydraulic fracturing, combined with horizontal drilling, has unlocked vast UCRs, but understanding the relationship between CDPs and fracturing efficiency remains challenging. The authors propose a methodology to analyze flowback data from 22 wells to evaluate the change in effective fracture volume (Vef) and investigate correlations between Vef and various CDPs, such as the gross perforated interval (GPI) and proppant concentration. By developing a new iterative approach that accounts for dynamic fracture porosity and compressibility during flowback, the study aims to improve the accuracy of Vef estimates and provide insights into how CDPs affect fracture closure.

8.4.3.5.2 Description of the Workflow and Machine Learning Model

The study correlates completion-design parameters (CDPs) with effective fracture volume (Vef) and fracture closure rate (FCR) using MLR and an iterative methodology. Initial effective fracture volume (Vefi) is estimated using the harmonic decline (HD) model and flowback data from 22 wells in the Eagle Ford formation. To take into consideration dynamic variations in fracture porosity and compressibility during flowback, an iterative technique is created. MLR is used to link Vefi and FCR with a number of CDPs, such as the gross perforated interval (GPI), proppant concentration, total injected water volume, and shut-in time. The paper investigates how CDPs affect fracture closure and presents new dimensionless type curves for assessing FCR.

8.4.3.5.3 Advantages and Limitations of the Synchronized Algorithm

The ML and iterative methodologies used in this study provide several advantages. The iterative approach captures dynamic changes in fracture properties during flowback, leading to more accurate estimates of fracture volume loss. MLR allows for the quantification of the relative impact of different CDPs on fracture closure and effective fracture volume, enabling better decision-making in fracturing operations. However, the methodology is limited by the relatively small dataset of 22 wells, which may impact the generalizability of the results. Additionally, the iterative approach is computationally intensive, requiring significant resources for accurate modeling. Despite these limitations, the study demonstrates the potential of combining flowback data analysis with advanced statistical techniques to optimize HF design.

8.4.3.5.4 Results and Discussion

The study concludes that the developed correlations between CDPs and Vef offer valuable insights into optimizing HF operations in the Eagle Ford formation. The GPI is identified as having the most significant effect on Vef, while proppant concentration is shown to play a crucial role in fracture closure during flowback. The proposed iterative approach for estimating fracture volume provides a more accurate representation of the dynamic changes occurring during flowback, enhancing the ability to predict well performance and optimize future fracturing operations. The authors recommend applying the developed correlations to other formations and expanding the dataset to further refine the methodology.

BIBLIOGRAPHY

Moussa, Tamer, Dehghanpour, Hassan, Yingkun Fu, and Ezulike, Obinna. The Use of Flowback Data for Estimating Dynamic Fracture Volume and Its Correlation with Completion-Design Parameters: Eagle Ford Cases. *J Petrol Sci Eng* 2020;195:107584, ISSN 0920-4105. https://doi.org/10.1016/j.petrol.2020.107584.

8.4.3.6 Application-31: Quantifying the Impact of Fracturing Chemicals on Production Performance Using Machine Learning

8.4.3.6.1 Background of the Application

This paper addresses the impact of fracturing chemicals on production performance in shale reservoirs, with a specific focus on the Powder River Basin in Wyoming. The completion design of HF operations involves the selection of a wide array of chemicals, including acids, scale dissolvers, and clay control agents. Although chemicals make up only a small percentage of the fracturing fluid by volume, they are crucial for optimizing production performance. The paper proposes using ML to analyze the contribution of different chemicals to production outcomes over short, mid, and long-term periods. By leveraging feature selection methods, such as SelectKBest, tree-based regression, and mutual information regression, the study aims to create a data-driven framework that can help completion engineers optimize the selection and use of chemicals in HF designs.

8.4.3.6.2 Description of the Workflow and Machine Learning Model

The paper applies various ML methods, including SelectKBest, tree-based regression, mutual information regression, and recursive feature elimination (RFE), to analyze the impact of fracturing chemicals on production performance. The dataset comprises production and completion data from 238 wells in the Powder River Basin, with variables such as cumulative O&G volumes, chemical additive amounts, proppant used, and petrophysical properties. The analysis focuses on predicting short-term (6 months), mid-term (12 months), and long-term (24 months) production outcomes. The ML models identify the most significant features and rank them based on their influence on production. The study utilizes hierarchical clustering to group wells by rock quality, allowing for a more granular understanding of how chemical additives affect production performance in different geological contexts.

8.4.3.6.3 Advantages and Limitations of the Synchronized Algorithm

The ML methodologies employed in this study offer several advantages. Feature selection methods, such as mutual information regression and tree-based regression, help to identify the most important chemicals in the fracturing process, allowing engineers to optimize treatment designs. Additionally, hierarchical clustering enables the identification of trends in chemical usage across different rock qualities, providing valuable insights into how these additives interact with geological formations. However, the study also faces limitations, particularly in the relatively small dataset of 238 wells, which may affect the generalizability of the results. The accuracy of the models is also dependent on the quality of the input data, and further research is recommended to refine the approach. Despite these challenges, the ML methods demonstrate their potential to improve HF design and optimize chemical usage for enhanced production performance.

8.4.3.6.4 Results and Discussion

The study concludes that fracturing chemicals play a significant role in driving well productivity, with certain chemicals showing varying levels of importance depending on the time frame of the production performance. The ML models revealed that chemicals consistently ranked among the top variables influencing production, even when compared with traditional completion and petrophysical parameters. The analysis also showed that the importance of specific chemicals, such as acids, varied depending on the quality of the rock and the production timeline. The authors suggest that the results of this study can be used to optimize fracturing designs and recommend further research to refine the data-driven models by incorporating larger datasets and exploring other basins.

BIBLIOGRAPHY

Rastogi, Ayush, and Sharma, Akash. "Quantifying the Impact of Fracturing Chemicals on Production Performance Using Machine Learning." Paper presented at the *SPE Liquids-Rich Basins Conference - North America*, Odessa, Texas, USA, November 2019. doi: https://doi.org/10.2118/197095-MS

8.4.3.7 Application-32: Selection of Candidate Wells for Re-fracturing in Tight Gas Sand Reservoirs Using Fuzzy Inference

8.4.3.7.1 Background of the Application

This paper presents a fuzzy-logic-based decision-making protocol designed to identify candidate wells for re-fracturing in tight gas sand reservoirs. Tight gas sands represent a major source of UCRs, but initial HF treatments often fail to achieve optimal recovery. Re-fracturing has emerged as a cost-effective option to improve recovery, but selecting the best candidates for re-fracturing can be challenging due to the complexity and variability of reservoir conditions. The authors propose using fuzzy logic to deal with the inherent imprecision and subjectivity in this decision-making process. By applying fuzzy set theory, fuzzy if-then rules, and fuzzy reasoning, the study integrates five indexes—fracture quality, reservoir characteristics, operational parameters, initial conditions, and production—to evaluate the re-fracturing potential of wells. The fuzzy inference system is designed to mimic human reasoning, allowing for more flexible and realistic decision-making in the selection of re-fracturing candidates.

8.4.3.7.2 *Description of the Workflow and Machine Learning Model*

The paper utilizes fuzzy logic to build a decision-making protocol for selecting re-fracturing candidates in tight gas sand reservoirs. The fuzzy inference system integrates five indexes—fracture quality, reservoir characteristics, operational parameters, initial conditions, and production—each of which is represented by three parameters. These parameters are grouped into categories (low, medium, and high) and are evaluated using trapezoidal membership functions. The fuzzy inference system processes these inputs through fuzzy if-then rules to output a re-fracturing potential index for each well. The system is designed to handle the imprecision inherent in geological data and to provide a flexible decision-making framework that mimics human reasoning. The study also uses the Mamdani inference method, which aggregates the outcomes of fuzzy rules and converts them into crisp values, providing a quantitative measure of re-fracturing potential.

8.4.3.7.3 *Advantages and Limitations of the Synchronized Algorithm*

The fuzzy logic methodology offers several advantages over traditional re-fracturing candidate selection methods. It allows for the incorporation of multiple, imprecise parameters, enabling a more flexible and comprehensive evaluation of well potential. The fuzzy inference system mimics human reasoning, making it particularly effective for handling subjective and uncertain data. However, the methodology also has limitations, particularly in the need for expert knowledge to define the fuzzy rules and membership functions. The accuracy of the results is dependent on the quality of the input data and the proper calibration of the fuzzy system. Despite these challenges, the study demonstrates that fuzzy logic provides a powerful tool for optimizing re-fracturing decisions in tight gas sand reservoirs, offering a practical and adaptable solution for complex reservoir conditions.

8.4.3.7.4 *Results and Discussion*

The study concludes that the fuzzy-logic-based decision-making protocol successfully identifies candidate wells for re-fracturing, offering a practical alternative to more traditional methods that rely on production data alone. By incorporating multiple parameters related to fracture quality, reservoir characteristics, and operational conditions, the fuzzy inference system provides a more comprehensive evaluation of re-fracturing potential. The case studies presented in the paper validate the effectiveness of this approach, demonstrating that wells with high re-fracturing potential can be accurately identified even in the absence of detailed hydraulic fracture information. The authors suggest that the methodology could be extended to other types of reservoirs by adjusting the input parameters and fuzzy rules to fit different geological settings.

BIBLIOGRAPHY

Artun, Emre, and Kulga, Burak. Selection of Candidate Wells for Re-Fracturing in Tight Gas sand Reservoirs Using Fuzzy Inference. *Pet Explor Dev* 2020;47(2):413–420, ISSN 1876-3804. https://doi.org/10.1016/S1876-3804(20)60058-1

8.4.3.8 Application-33: Experiential and Social Learning in Firms: The Case of Hydraulic Fracturing in the Bakken Shale

8.4.3.8.1 Background of the Application

This paper examines how firms learn to use new technologies, specifically HF, in the Bakken Shale in North Dakota between 2005 and 2012. The study utilizes data on input choices, profits, and information sets to analyze how oil companies have learned to make more profitable input choices over time. The focus is on understanding why firms capture only a fraction of the potential profits and how they balance learning from their own experience with information available from competitors. The results indicate that while firms learn from both their own experiences (learning-by-doing) and from others (social learning), they are slow to adopt new information and tend to prioritize lower uncertainty in profit expectations over experimentation.

8.4.3.8.2 Description of the Workflow and Machine Learning Model

The study employs an analytical approach combined with ML techniques to model and analyze the decision-making processes of firms in the Bakken Shale. The workflow begins by collecting and structuring data on input choices, profit margins, and the information sets available to the firms during the study period. This data is then used to develop a ML model that simulates the decision-making process, incorporating variables that represent both learning-by-doing and social learning. The model is designed to assess how firms adapt to new information and adjust their input choices to optimize profitability. Key aspects of the model include the analysis of the trade-offs firms make between immediate profits and the potential benefits of experimentation with new technologies. The output of the model is validated against historical data to ensure its accuracy in reflecting the firms' learning behavior over time.

8.4.3.8.3 Advantages and Limitations of the Synchronized Algorithm

The ML approach used in this study offers several advantages in analyzing the learning behaviors of firms. It allows for the integration of large datasets, capturing the complex interactions between firm decisions, market conditions, and technological advancements. This holistic view provides deeper insights into the learning dynamics that drive technology adoption. However, the approach also has limitations. The accuracy of the model is heavily dependent on the quality of the input data, and any gaps or biases in the historical data could affect the validity of the findings. Additionally, while the model provides valuable insights into the decision-making processes, it may oversimplify certain aspects of human behavior, such as risk aversion and the influence of organizational culture on learning. Future research could address these limitations by incorporating more qualitative data and exploring the role of external factors, such as regulatory changes or market shocks, in shaping learning behaviors in firms.

8.4.3.8.4 Results and Discussion

The study concludes that firms are slow to respond to new information and exhibit reluctance to experiment, often preferring input choices with lower profit uncertainty. This behavior limits their ability to maximize profits from HF in the Bakken Shale.

By 2012, firms captured only 60% of the potential profits, primarily due to underutilizing data from competitors and focusing on familiar input choices. The findings highlight significant barriers to learning in firms, suggesting that better incentives or improved access to competitor data could enhance the efficiency of technology adoption in HF and other industries. The study has implications for both policy-makers and firms aiming to improve learning behaviors in technology adoption.

BIBLIOGRAPHY

Covert, Thomas. Experiential and Social Learning in Firms: The Case of Hydraulic Fracturing in the Bakken Shale (February 22, 2015). Available at SSRN: https://ssrn.com/abstract=2481321 or http://doi.org/10.2139/ssrn.2481321

8.4.3.9 Application-34: Evaluation and Optimization of Completion Design using Machine Learning in an Unconventional Light Oil Play

8.4.3.9.1 Background of the Application

This paper focuses on the evaluation and optimization of HF performance in multi-fractured horizontal wells (MFHW) in the Viking Formation, Canada, using ML techniques. The study aims to address the challenges of optimizing well productivity in UCRs, where traditional methods often fall short due to the complex interplay of reservoir and completion parameters. By applying a Random Forest Model (RFM), the authors analyze a dataset of 875 MFHW across ten fields, using first-year oil production (IP365) as a well performance indicator. The model integrates geological, reservoir, and completion data to predict oil production and identify key production drivers, ultimately guiding engineers in optimizing completion designs for enhanced well productivity.

8.4.3.9.2 Description of the Workflow and Machine Learning Model

The paper employs a RFM to predict IP365 and optimize completion designs in the Viking Formation. The model integrates nine features, categorized into well geographic location, reservoir, and completion parameters, such as latitude, longitude, true vertical depth (TVD), net pay, average gas–oil ratio (GOR), completion length, proppant intensity, and proppant concentration. A dataset of 875 multi-fractured horizontal wells (MFHW) is used to train and validate the model, with 60% of the data allocated for training and 40% for testing. Feature importance, partial dependence plots (PDPs), and SHAP are applied to interpret the model's predictions and identify key drivers of well productivity.

8.4.3.9.3 Advantages and Limitations of the Synchronized Algorithm

The ML methodology offers several advantages, including the ability to handle large and complex datasets, capture non-linear relationships between variables, and provide interpretable results through model-agnostic methods like PDP and SHAP. The use of RF allows for robust predictions and the identification of key production drivers, such as completion length and proppant intensity, which can be optimized to enhance well productivity. However, the methodology also has limitations, such as the relatively low R^2 value (0.63), indicating that certain factors affecting production were not fully captured by the model. Additionally, the lack of detailed geological data in the dataset may have impacted the model's accuracy. Despite these challenges, the study demonstrates the potential of ML in optimizing completion designs and improving well performance in UCRs.

8.4.3.9.4 Results and Discussion

The study concludes that the application of ML, particularly the RFM, provides valuable insights into the optimization of well completion designs in the Viking Formation. The model identified completion length as the most significant production driver, followed by well geographic location and net pay. Wells with lateral lengths greater than 750 meters generally exhibited above-average production, while higher proppant intensities and concentrations showed diminishing returns beyond certain thresholds. The authors suggest that data-driven methodologies, such as ML, can significantly enhance the efficiency and accuracy of completion design optimization, leading to improved well performance and cost efficiencies in unconventional light oil plays.

BIBLIOGRAPHY

Porras, Luisa, Hawkes, Christopher, and Islam, Arshad. "Evaluation and Optimization of Completion Design using Machine Learning in an Unconventional Light Oil Play." Paper presented at the *SPE/AAPG/SEG Unconventional Resources Technology Conference*, Virtual, July 2020. doi: https://doi.org/10.15530/urtec-2020-2938

8.4.3.10 Application-35: Application of Deep Neural Network (DNN) in Reservoir Simulation for Hydraulic Fracturing and Production of Unconventional Wells

8.4.3.10.1 Background of the Application

A reservoir simulator based on DNNs is presented in this work with the goal of improving the precision and efficiency of production forecasts and HF. The goal is to build a scalable and automated method to successfully manage the complicated and nonlinear behavior of reservoirs. By combining several data sources, the DNN simulator offers a more thorough comprehension of reservoir dynamics, facilitating operational optimization and real-time decision-making.

Developing and improving production plans requires the use of reservoir simulation, a critical approach for forecasting fluid behavior in porous media, such as O&G reservoirs. With a focus on unconventional wells—horizontal wells that undergo HF to improve O&G flow—this study tackles problems like fluid dynamics, pressure prediction, and fracture geometry characterization during field development. Simulating these wells is made more difficult by the volume and unpredictability of data from sources such as engineering, geology, geophysics, and laboratory testing. A technique is offered that enhances conventional, laborious mathematical procedures by utilizing DNNs' capacity to interpret intricate data relationships. The simulations of three-dimensional conceptual and unconventional field models confirm that DL techniques offer both accuracy and speed, making them a valuable supplement to conventional reservoir simulation methods.

8.4.3.10.2 Description of the Workflow and Machine Learning Model

This paper presents the application of a DNN to improve the accuracy and efficiency of reservoir simulations, particularly for HF and production in unconventional wells. The DNN-based simulator is designed to handle the high-dimensional, nonlinear, and time-dependent complexities associated with these processes. By integrating data from multiple sources, such as geological, geophysical, petrophysical, and production data, the model learns the intricate relationships within the reservoir. The DNN architecture leverages convolutional layers for feature extraction, max-pooling layers for dimensionality reduction, and fully connected layers for final predictions. The system's performance is optimized using supervised learning, making it possible to predict reservoir behavior more accurately than conventional simulation methods.

8.4.3.10.3 Advantages and Limitations of the Synchronized Algorithm

The adoption of a DNN for reservoir simulation offers several advantages, particularly in terms of its ability to model complex nonlinear relationships that are difficult to capture using traditional finite difference-based simulators. The DNN's architecture allows for the automatic extraction of features from high-dimensional data, significantly reducing the need for manual feature engineering. Additionally, the model can generalize well to new data, making it a versatile tool for predicting reservoir performance under varying conditions. However, the computational requirements for training such models can be substantial, particularly as the complexity and size of the dataset increase. The quality of the results is also highly dependent on the availability of high-quality data for training, and the process of hyperparameter tuning can be time-consuming. Despite these challenges, the DNN-based approach shows promise for enhancing the accuracy and speed of reservoir simulations.

8.4.3.10.4 Results and Discussion

In this study, the development of a DNN-based reservoir simulator was undertaken with the goal of achieving precise predictions of reservoir performance during HF and production. By carefully orchestrating the processes of data preparation, feature selection, model training, and validation, diverse range of data was integrated—including geological, geophysical, and production information—into the simulator's framework. The architecture of the simulator, featuring convolutional layers, fully connected layers, and max-pooling layers, is designed to enhance the processing and abstraction of visual data, with each component contributing to the model's overall capability.

The approach involved a systematic workflow, beginning with a clearly defined objective and progressing through a detailed, multi-faceted data collection and interpretation strategy. While the preliminary results indicate a promising improvement in both the accuracy and speed of reservoir performance predictions, highlighting the DNN model's computational efficiency and ability to handle the complexities of HF and production more effectively than traditional simulators, there remains a need for continuous refinement. The success of the model depends significantly on the quality and comprehensiveness of the data, and further exploration and enhancement are essential for optimizing performance and achieving more detailed insights.

BIBLIOGRAPHY

Li, R., Fu, J., Gao, T., and Zhang, P. "Application of Deep Neural Network (DNN) in Reservoir Simulation for Hydraulic Fracturing and Production of Unconventional Wells." Paper presented at the *International Petroleum Technology Conference*, Dhahran, Saudi Arabia, February 2024. doi: https://doi.org/10.2523/IPTC-23759-MS

8.4.3.11 Application-36: Finding a Trend Out of Chaos, A Machine Learning Approach for Well Spacing Optimization

8.4.3.11.1 Background of the Application

ML methods are increasingly utilized to optimize field development in UCRs, but identifying meaningful trends amidst numerous influencing factors remains a challenge. Augmented AI addresses this by integrating expert knowledge with advanced AI frameworks to optimize well spacing and fracking operations. The approach begins with comprehensive data sculpting, including retrieval, cleaning, and quality control. Feature engineering generates relevant parameters for the ML models, which are then tuned and ensemble to improve robustness and accuracy. BO automates this iterative process, while an adaptive ensemble method combines predictions from various models to reduce variance and enhance performance.

The effectiveness of Augmented AI is demonstrated through a case study in the Midland Basin, where domain-specific feature engineering and initial feature selection identified key attributes for predicting well performance. The approach involved building an explainable ML model with uncertainty quantification, which was optimized and validated through sensitivity analyses. This model provided robust production forecasts and optimized well spacing recommendations, showcasing Augmented AI's ability to leverage big data and domain expertise for effective field development. The model's adaptability allows it to be applied and refined across different regions, even with limited well data.

8.4.3.11.2 Description of the Workflow and Machine Learning Model

This paper addresses well spacing optimization in UCRs using an advanced Augmented AI approach. The Augmented AI workflow integrates ML techniques with domain-specific knowledge to enhance well performance predictions. This approach employs an adaptive ensemble ML algorithm, which combines multiple regression models to improve prediction accuracy and robustness. The process begins with extensive data sculpting, which includes information retrieval, data cleaning, and feature engineering. Over 50 features related to geological, completion, and operational parameters are generated for the ML models.

The models are trained to predict initial production (IP) and EUR for wells in the Wolfcamp Formation. To optimize the model's performance, a BO technique is used for hyperparameter tuning. The workflow also includes uncertainty quantification and model explanation using SHAP values, which help deconvolute the effects of various well attributes, particularly well spacing, on production performance.

8.4.3.11.3 Advantages and Limitations of the Synchronized Algorithm

The Augmented AI methodology provides a powerful approach to well spacing optimization by combining ML with expert-based domain knowledge. The adaptive ensemble approach ensures that the strengths of multiple models are leveraged, leading to more accurate and robust predictions. This method allows for the effective handling of the complex interactions between well attributes and production outcomes. Additionally, the integration of SHAP values for model explanation enhances the transparency and interpretability of the ML models, making it easier to understand the impact of specific features on well performance.

However, the methodology also has its limitations. The computational complexity of the ensemble model and the iterative nature of BO can lead to longer processing times, particularly for large datasets with numerous features. Moreover, the accuracy of the predictions heavily depends on the quality of the input data, including geological, completion, and operational parameters. In cases where data is incomplete or noisy, the performance of the model may be affected. Finally, while SHAP values provide valuable insights into feature importance, interpreting these explanations in the context of complex reservoir dynamics can still be challenging.

8.4.3.11.4 Results and Discussion

This paper introduces an innovative model utilizing Augmented AI to optimize well spacing by integrating big data sculpting, domain and physics-induced feature engineering, and ML. The case study in the Midland Basin highlights the process of domain-induced feature engineering to identify critical features for well performance prediction. Initial feature selection through correlation analysis was followed by the development of a trusted, explainable ML model enhanced with uncertainty quantification. After conducting sensitivity analyses, Augmented AI refined the model's attributes, generating a comprehensive report and visualizing the results.

Data sculpting, powered by computer vision and natural language processing, facilitated the efficient extraction of datasets from unstructured sources, significantly reducing processing time from months to days. Various groups of feature engineering, including both basic attributes and those derived from domain knowledge, were used to build robust ML models. The feature space was smartly reduced before applying supervised learning algorithms. The model's learning was transferred from basin-wide models to specific regions, proving the adaptability of Augmented AI for optimizing well spacing and frac parameters. With existing models for major basins such as Midland and Delaware, the approach can be applied to different regions, even with limited well data, and refined with more accurate private data. Sensitivity analysis further examined the impact of well spacing on specific geological and fracking parameters, supported by both long-term and short-term development plans.

BIBLIOGRAPHY

Ma, Zheren, Davani, Ehsan, Ma, Xiaodan, Lee, Hanna, Arslan, Izzet, Zhai, Xiang, Darabi, Hamed, and Castineira, David. "Finding a Trend Out of Chaos, A Machine

Learning Approach for Well Spacing Optimization." Paper presented at the *SPE Annual Technical Conference and Exhibition*, Virtual, October 2020. doi: https://doi.org/10.2118/201698-MS

8.4.3.12 Application-37: Optimization of Abnormal Hydraulic Fracturing Conditions of Unconventional Natural Gas Reservoirs Based on a Surrogate Model

8.4.3.12.1 Background of the Application

Abnormal conditions significantly impact the efficiency of HF in UCRs, leading to issues such as pressure channeling, casing deformation, and proppant plugging. To address these challenges, this paper introduces a novel ML-based method for optimizing fracturing operations under abnormal conditions. Initially, key factors influencing these conditions are identified through hybrid controlling analysis. A surrogate model is then developed to predict the probability of abnormal conditions occurring, rather than simply their presence or absence. Following this, a ML-based optimization algorithm is employed to reduce the likelihood of these conditions, acknowledging their potential inevitability during the fracturing process.

The proposed method demonstrates superior performance compared to traditional approaches, showing significant improvements in minimizing the probability of abnormal conditions. The results indicate that this novel methodology is better suited to meet the practical needs of operations in uncertain environments, providing a more effective and responsive solution for optimizing HF under challenging conditions.

8.4.3.12.2 Description of the Workflow and Machine Learning Model

This paper presents a ML-based optimization approach to address abnormal conditions during HF in unconventional natural gas reservoirs. The methodology employs a hybrid ML framework consisting of various algorithms, including SVMs, DTs, RFs, Extra-Trees, GB, and AdaBoost, to predict the occurrence of abnormal working conditions such as pressure channeling, casing deformation, and proppant plugging. The model is trained on historical fracturing data, including parameters such as average cluster spacing, pump stop pressure gradient, and construction displacement. By leveraging a surrogate model, the study predicts the probability of abnormal conditions occurring rather than providing a binary classification. The optimization process then minimizes the occurrence probability using a combination of GAs, PSO, and differential evolution (DE).

8.4.3.12.3 Advantages and Limitations of the Synchronized Algorithm

The ML-based surrogate model provides an innovative approach to optimizing HF operations by accurately predicting the likelihood of abnormal conditions and allowing for the probabilistic optimization of fracturing parameters. This method offers significant advantages, such as improving operational safety and efficiency, by adjusting parameters to minimize the probability of adverse conditions. Additionally, ensemble learning techniques, such as RFs and AdaBoost, enhance model performance by integrating weak learners to create more robust predictions. However, the complexity of the optimization process, particularly when using evolutionary algorithms like GA, PSO, and DE, can lead to increased computational overhead.

Moreover, the success of the model depends on the quality of the input data and the accuracy of the surrogate model, making data preprocessing and feature selection crucial to achieving reliable optimization results.

8.4.3.12.4 Results and Discussion

This paper presents a ML-based approach to address abnormal conditions encountered during the HF of unconventional natural gas reservoirs. Initially, the study establishes a ML model to analyze the primary factors influencing abnormal conditions such as pressure channeling, casing deformation, and proppant plugging. The research then employs various ML methods to predict these abnormal conditions and utilizes three optimization algorithms—GO, PSO, and DE—to refine fracturing parameters and reduce the likelihood of these conditions. The analysis identifies key factors affecting each type of abnormal condition: for pressure channeling, they include fracturing fluid amount and sand ratio; for casing deformation, they involve liquid strength and mechanical properties; and for proppant plugging, they relate to sand dosage and construction displacement. The study concludes that the RF, additional tree, and AdaBoost algorithms offer the highest accuracy in predicting abnormal conditions.

In evaluating the optimization models, the research finds that different evolutionary algorithms yield varying results in minimizing abnormal conditions. It emphasizes that appropriate parameter adjustments in HF can significantly reduce the probability of pressure channeling, casing deformation, and proppant plugging, with specific strategies such as increasing average segment length proving effective. The study also notes that factors like in-situ stress distribution and well completion might influence abnormal conditions, suggesting the need for future research to integrate spatiotemporal data into predictive models. This will aim to enhance the accuracy and effectiveness of optimization strategies for managing abnormal conditions in HF operations.

BIBLIOGRAPHY

Yang, S., Han, J., Liu, L., Wang, X., Yin, L., and Ci, J. Optimization of Abnormal Hydraulic Fracturing Conditions of Unconventional Natural Gas Reservoirs Based on a Surrogate Model. *Processes* 2024;12:918. https://doi.org/10.3390/pr12050918

8.4.4 Forecasting Production

8.4.4.1 Application-38: Prediction of Single-Well Production Rate after Hydraulic Fracturing in Unconventional Gas Reservoirs Based on Ensemble Learning Model

8.4.4.1.1 Background of the Application

To overcome the challenges associated with predicting single-well production in tight gas and shale gas reservoirs following HF, AI methods were employed. Various ML algorithms, including RF, Extra-Trees), lightweight gradient boosting machines (LightGBM), gradient boosting regression (GBR), and linear regression (LR), were integrated with reservoir geology, engineering parameters, and production data to develop several foundational models for forecasting unconventional gas well production. These models were evaluated for accuracy, and improvements in their predictive performance and generalizability were achieved through the ensemble approach of combining multiple ML models.

The paper further demonstrated the application of the ensemble model by selecting two representative tight and shale gas reservoirs for well production forecasting. A comparative analysis with actual production data showed that the blending model performed notably well: for tight gas reservoir A, it achieved a mean absolute error (MAE) of 0.8419, an MSE of 1.0930, and an R^2 score of 0.8812. For shale gas reservoir B, the model attained an MAE of 1.4841, an MSE of 3.1629, and an R^2 score of 0.9524. These results indicate that the ensemble model approach provides superior predictive accuracy and reliability compared to individual ML algorithms, effectively handling high-dimensional, large-scale, and imbalanced data, and offering robust scientific validation and technical support for assessing well productivity in tight and shale gas wells.

8.4.4.1.2 Description of the Workflow and Machine Learning Model

This paper focuses on predicting the single-well production rate after hydraulic fracturing in unconventional gas reservoirs using an ensemble learning approach. The study integrates several ML algorithms, including RF, Extra-Trees, LightGBM, GBR, and Linear Regression (LR). These models are trained using geological, reservoir, and engineering data to forecast the production of unconventional gas wells. The ensemble model combines the outputs of these individual models through a blending technique to improve prediction accuracy and generalization. The key variables considered in the modeling process include reservoir type, effective thickness, porosity, gas saturation, fluid volume, and proppant volume.

8.4.4.1.3 Advantages and Limitations of the Synchronized Algorithm

The ensemble learning approach effectively improves the prediction accuracy by leveraging the strengths of multiple ML algorithms. The blending method, in particular, enhances the generalization ability of the model, resulting in higher predictive accuracy compared to individual models. This approach is highly suitable for handling complex and high-dimensional data, which are common in unconventional gas reservoir production forecasting. However, the computational demands of training multiple models and performing ensemble predictions can be high. Additionally, the quality of predictions is dependent on the accuracy and completeness of the input data, as well as the chosen hyperparameters for each model. Despite these challenges, the ensemble learning methodology provides a robust and reliable framework for predicting production rates in unconventional gas reservoirs.

8.4.4.1.4 Results and Discussion

This article introduces a refined production prediction model and methodology for forecasting the single-well gas rates of tight gas and shale gas reservoirs using AI. The proposed approach integrates predictions from various base models, including RF, Extra-Trees, LightGBM, and GBR. By leveraging ensemble learning, the model enhances both the accuracy and generalizability of the predictions.

The ensemble model, which combines the strengths of multiple basic models, demonstrates improved performance in predicting UCR gas rates. For tight gas, the ensemble model achieved an accuracy rate of 0.8812, while for shale gas, it reached an accuracy rate of 0.9524. Given the complexity of influencing factors and the limited number of effective samples for gas production predictions, future research will focus on integrating data with knowledge graphs. This approach will use graph embedding techniques to represent the temporal and spatial relationships of oil well production and fuse these features with DL-derived features through embedding layers. The goal is to develop a comprehensive prediction model that combines data and knowledge for more accurate forecasting of gas well production rates.

BIBLIOGRAPHY

Ye, F., Li, X., Zhang, N., and Xu, F. Prediction of Single-Well Production Rate after Hydraulic Fracturing in Unconventional Gas Reservoirs Based on Ensemble Learning Model. *Processes* 2024;12:1194. https://doi.org/10.3390/pr12061194

8.4.4.2 Application-39: Controlling Factors of Shale Gas Production: What Can Artificial Intelligence Tell Us?

8.4.4.2.1 Background of the Application

This paper investigates the use of AI and data mining techniques to study the controlling factors of shale gas production in the Marcellus Shale. Traditional reservoir simulation models often rely on assumptions based on coalbed methane production, which may not accurately reflect shale gas behavior. The authors propose using AI techniques, specifically NNs, to analyze field data and dynamically identify the key performance-impacting factors in shale gas production. The study integrates various data sources, including well location, reservoir characteristics, completion, and production data, to build a comprehensive database. This AI-based approach aims to enhance the understanding of shale gas production mechanisms and assist in optimizing reservoir management.

8.4.4.2.2 Description of the Workflow and Machine Learning Model

The paper utilizes AI, specifically NNs, to analyze shale gas production data from over 300 horizontal wells in the Marcellus Shale. The neural network models are trained to predict cumulative gas production at 6, 12, 18, and 24 months, with input data including well location, reservoir characteristics, completion details, and geomechanical properties. KPI analysis is performed to identify the most influential parameters for each production period. The study employs a feed-forward backpropagation neural network architecture with ReLU activation functions in the hidden layers and a logistic activation function in the output layer. The NNs are trained using 80% of the dataset, with the remaining 20% used for testing and validation.

8.4.4.2.3 Advantages and Limitations of the Synchronized Algorithm

The AI methodology offers several advantages over traditional sensitivity analysis methods. NNs can capture the complex, non-linear relationships between various parameters, allowing for a more comprehensive understanding of the factors that influence shale gas production. The KPI analysis provides valuable insights into the dynamic nature of these factors over time, highlighting the shift in importance from reservoir characteristics to completion and stimulation parameters as production progresses. However, the methodology is limited by the availability and quality of the input data, and the neural network models require careful tuning to avoid overfitting. Despite these challenges, the study demonstrates that AI techniques can significantly enhance the analysis of shale gas production and improve decision-making in reservoir management.

8.4.4.2.4 Results and Discussion

The study concludes that AI-based models, particularly NNs, provide valuable insights into the controlling factors of shale gas production. The KPI analysis revealed that reservoir quality is a critical factor in the early stages of production, while completion and stimulation parameters become more significant over time. The results demonstrate that AI techniques can outperform traditional methods in identifying key parameters that influence production, enabling more informed decision-making for shale reservoir development. The authors suggest that AI models can be further refined and applied to other shale plays, offering a powerful tool for optimizing production in UCRs.

BIBLIOGRAPHY

He, Qin, Bruno, Jonathan, and Zhong, Zhi. "Controlling Factors of Shale Gas Production: What can Artificial Intelligence Tell Us?." Paper presented at the *SPE Eastern Regional Meeting*, Charleston, West Virginia, USA, October 2019. doi: https://doi.org/10.2118/196577-MS

8.4.4.3 Application-40: Application of Machine Learning for Production Forecasting for Unconventional Resources

8.4.4.3.1 Background of the Application

This paper explores the application of ML, specifically LSTM networks, to improve production forecasting in unconventional resources. Traditional decline curve analysis (DCA), often used for forecasting production rates, faces challenges in unconventional fields due to operational changes, complex flow behavior, and limited historical data. The authors propose a data-driven ML approach that leverages time series data from initial production periods to predict future oil rates and cumulative production. The study demonstrates how LSTM, with its ability to handle sequential data and capture long-term dependencies, offers a more dynamic and accurate alternative to conventional forecasting methods.

8.4.4.3.2 Description of the Workflow and Machine Learning Model

The paper applies LSTM networks to predict oil production rates in unconventional resources, using time series data from the initial three months of production. The LSTM model is trained on tubing pressure and latency information, along with data from similar wells selected dynamically based on correlation metrics. To improve prediction accuracy, the study employs ensemble learning, combining two separate models: one that predicts the oil rate directly and another that forecasts cumulative production. The models are integrated through a weighted average, with the weights optimized using historical data from over 300 wells. The LSTM architecture includes three stacked layers to increase the network's representational power, and the model is validated through hindcasting on a 21-month production period.

8.4.4.3.3 Advantages and Limitations of the Synchronized Algorithm

The LSTM-based methodology offers several advantages over traditional DCA. LSTM networks can capture long-term dependencies and handle complex, non-linear relationships in the data, leading to more accurate production forecasts. The use of ensemble learning further enhances prediction reliability by compensating for over-prediction and under-prediction tendencies in the individual models. However, the methodology also has limitations, including the need for significant computational resources and the reliance on dynamic data without incorporating static geological or completion parameters. Despite these challenges, the LSTM approach demonstrates a promising improvement in production forecasting for unconventional resources, with the potential for further refinement and broader application.

8.4.4.3.4 Results and Discussion

The study concludes that ML, particularly LSTM, significantly enhances production forecasting accuracy in unconventional resources. By incorporating ensemble learning and feature engineering, the LSTM model effectively addresses the limitations of traditional DCA, such as over-prediction and sensitivity to operational changes. The hindcasting results from over 300 wells show that the LSTM model outperforms traditional methods, achieving a mean prediction error of less than 0.2% over the first two years of production. The authors suggest that further improvements could be made by integrating static geological data and refining the ensemble approach to reduce variance in the predictions.

BIBLIOGRAPHY

Zhan, Cheng, Sankaran, Sathish, LeMoine, Vincent, Graybill, Jeremy, and Mey, Didi-Ooi Sher. "Application of Machine Learning for Production Forecasting for Unconventional Resources." Paper presented at the *SPE/AAPG/SEG Unconventional Resources Technology Conference*, Denver, Colorado, USA, July 2019. doi: https://doi.org/10.15530/urtec-2019-47

8.4.4.4 Application-41: Hybrid Modeling in Unconventional Reservoirs to Forecast Estimated Ultimate Recovery

8.4.4.4.1 Background of the Application

This paper explores the use of hybrid modeling approaches to forecast the EUR in unconventional shale reservoirs. It emphasizes that while purely data-driven ML models have shown promise in conventional, data-rich reservoirs, they struggle with the high uncertainty and variability present in UCRs. To address this, the study integrates physics-based reservoir simulation with ML techniques. The hybrid model leverages synthetic data generated through full-physics simulations to train a ML model, which then forecasts production in sections of the reservoir with limited data. This approach aims to improve the accuracy and scalability of EUR forecasts in shale reservoirs, providing a more reliable framework for field development planning.

8.4.4.4.2 Description of the Workflow and Machine Learning Model

The paper utilizes a hybrid modeling approach that combines physics-based reservoir simulation with ML techniques to forecast EUR in shale reservoirs. The workflow involves generating synthetic data using a full-physics reservoir simulator, which is then used to train a ML model. This trained model is applied to other sections of the reservoir with different properties, where data is scarce. The ML component primarily focuses on production forecasting, while the physics-based model helps reduce uncertainty and improve the reliability of predictions. The hybrid model balances the advantages of both approaches, leveraging the precision of physics-based simulations and the scalability of ML.

8.4.4.4.3 Advantages and Limitations of the Synchronized Algorithm

The hybrid modeling methodology offers several advantages, including enhanced accuracy in EUR forecasts and the ability to handle uncertainty in UCRs. By integrating physics-based models, the hybrid approach compensates for the limitations of purely data-driven models, which often struggle with high variability and limited data. However, the methodology also has limitations, such as the computational intensity of physics-based simulations and the reliance on synthetic data for training. The success of the hybrid model depends on the quality of the simulation data and the complexity of the reservoir being modeled. Despite these challenges, the study demonstrates that hybrid modeling is a promising solution for improving production forecasts in UCRs.

8.4.4.4.4 Results and Discussion

The study concludes that hybrid modeling significantly improves the accuracy of EUR forecasts in UCRs. By integrating physics-based simulations with ML, the hybrid approach reduces the uncertainty inherent in purely data-driven models. The results demonstrate that hybrid models can consistently forecast EUR with acceptable accuracy, even in data-starved sections of the reservoir. This methodology offers a scalable and repeatable solution for UCRs, where limited data and high variability have traditionally hindered reliable production forecasting. The authors suggest that further research could refine this approach by incorporating more complex geological features and real-field data.

BIBLIOGRAPHY

Temizel, Cenk, Canbaz, Celal Hakan, Balaji, Karthik, Ozesen, Ahsen, Yanidis, Kirill, Alsaheib, Hasanain, AlSulaiman, Nouf, Basri, Mustafa, and Jama, Nayif. "Hybrid Modeling In Unconventional Reservoirs To Forecast Estimated Ultimate Recovery." Paper presented at the *Offshore Technology Conference*, Virtual and Houston, Texas, August 2021. doi: https://doi.org/10.4043/31010-MS

8.4.4.5 Application-42: Comprehensive Review of Data-Driven Approaches for Forecasting Production from Unconventional Reservoirs

8.4.4.5.1 Background of the Application

Despite the abundance of data, predicting well production from UCRs is a difficult task because the underlying physics is not well understood. Data Analytics Techniques (DATs) have recently become a viable method for predicting output in UCRs. Some of these strategies use data-driven techniques to resolve uncertainties and heterogeneities while combining DAT with physics-based models to capture the fundamental physical principles of fluid flow in porous media. By contrasting predictive models across different algorithms, validation standards, input data, well numbers, and formation types, this overview looks at the uses and effectiveness of DAT in production forecasting.Each model's advantages and disadvantages are listed, as well as the difficulties and prospects for more study in this area.

According to the investigation, models based on ML can anticipate output from UCRs with reasonable accuracy. Two dimensionless metrics, the coefficient of determination (R2) and mean absolute percentage error (MAPE), were used to assess performance. The majority of models exhibit a low error rate, and there is a significant correlation between the anticipated and actual production data. In particular, over 80% of the models have an R2 greater than 0.6, and about 65% of the models have a MAPE of less than 20%. These results imply that DAT can improve the robustness and dependability of production forecasting for non-traditional resources. There is still opportunity for development, though, including the creation of new ML algorithms, the integration of DAT with models based on physics, and the establishment of multi-perspective methods for assessing model performance.

8.4.4.5.2 Description of the Workflow and Machine Learning Model

This paper provides an extensive review of data-driven approaches, specifically focusing on the application of AI, ML, and DA techniques for forecasting production from UCRs. The study highlights the challenges associated with predicting well production due to the complex nature of UCRs and the significant variability in reservoir properties.

The paper categorizes the AI techniques into supervised and unsupervised learning methods. Supervised learning techniques, including regression models such as ANNs, Support Vector Regression (SVR), RF, and XGBoost, are primarily employed to model the relationship between historical production data and various reservoir and operational parameters. Unsupervised learning methods are used for clustering and identifying patterns in the data without predefined labels.

Additionally, the study discusses the integration of physics-based models with data-driven approaches, which combines the physical understanding of fluid flow with the predictive power of ML algorithms. This hybrid approach is particularly useful for capturing the uncertainties and heterogeneities inherent in UCRs.

8.4.4.5.3 Advantages and Limitations of the Synchronized Algorithm

The data-driven approaches reviewed in this paper demonstrate significant potential in improving the accuracy and reliability of production forecasting for UCRs.

The use of supervised learning models, such as ANNs, RF, and XGBoost, allows for the effective handling of large datasets and the modeling of complex, nonlinear relationships between input variables and production outcomes. These models are particularly valuable for their ability to incorporate various types of data, including geological, geophysical, and operational information.

One of the main advantages of using AI techniques is their capacity to learn from historical data and make predictions that account for the inherent uncertainties in UCR behavior. The integration of physics-based models with ML methods enhances the robustness of the predictions by grounding them in established physical principles.

However, the application of these methodologies also comes with challenges. The computational complexity of training and optimizing ML models can be substantial, especially when dealing with large datasets and numerous decision variables. Additionally, the performance of these models is highly dependent on the quality and representativeness of the input data. Inadequate or biased data can lead to inaccurate predictions, limiting the utility of the models in real-world scenarios.

8.4.4.5.4 Results and Discussion

Engineering systems can be categorized into two types: clear box systems, where the physics is well understood, and dark box systems, where data is available, but the underlying physics is not well comprehended. Predicting well production from UCRs, which is crucial for developing these assets, falls into the dark box category. A practical approach to addressing this challenge is through data-driven techniques that can manage the complex and poorly understood components of the predictive model. DATs, particularly those utilizing ML algorithms, offer the ability to learn from data and make predictions without needing explicit rules or equations.

This work provides ML algorithms, for production forecasting in UCRs. It summarizes the key features of predictive models developed using ML algorithms, based on 95 research studies, including research papers, review papers, and theses published between 2012 and 2023. These studies were categorized by the ML algorithm used, formation type, and performance metrics. The most commonly used ML algorithms for production forecasting were NNs, RFs & logistic regression (LogR), and gradient boosting machines (GBMs) & SVMs, which accounted for 40%, 28%, and 20% of the studies, respectively. The Eagle Ford Formation was the most studied, followed by Marcellus, Bakken, and Wolfcamp formations. The analysis also revealed that there was no standard approach for validating predictive models, with R^2 and MAPE being the most common benchmarks. Despite these challenges, ML models showed acceptable performance, with around 65% of the models having a MAPE of less than 20% and 80% achieving an R^2 greater than 0.6, particularly in the Bakken Formation.

While DAT with ML algorithms have the potential to enhance the reliability and robustness of production forecasting from UCRs, several limitations must be considered. The quality and quantity of data are critical, as ML algorithms require extensive and reliable datasets for effective training and validation. In UCRs, data is often limited, noisy, or incomplete, which can compromise the accuracy of predictions. Moreover, the interpretability of ML models is a significant challenge, as they are often seen as "black boxes," limiting the understanding of how predictions are made and what factors influence outcomes. This lack of transparency can hinder the adoption of these models in the industry. Additionally, the generalization and transferability of ML models are concerns, as these models are typically trained on specific datasets that may not capture the full range of conditions found in UCRs. To improve model performance, there is a need for comprehensive datasets that reflect the variability of reservoir characteristics over time. Future opportunities in this field include developing automated workflows for model development across different reservoirs, creating new ML algorithms, combining physics-based models with DAT, and establishing multi-perspective approaches for model comparison.

BIBLIOGRAPHY

Rahmanifard, H., and Gates, I. A Comprehensive Review of Data-Driven Approaches for Forecasting Production from Unconventional Reservoirs: Best Practices and Future Directions. *Artif Intell Rev* 2024;57:213. https://doi.org/10.1007/s10462-024-10865-5

8.4.4.6 Application-43: AI Assisted Shale Gas Production Performance Evaluation

8.4.4.6.1 Background of the Application

Shale gas reservoirs play a crucial role in global hydrocarbon production, particularly in the United States. With the rapid expansion of these reservoirs, understanding effective methods for modeling production and evaluating performance has become essential. In recent times, AI and ML have emerged as prominent techniques for analyzing production performance. The exploration and production of hydrocarbons involve a constant influx of data from both subsurface and surface operations. The growing volume of this data enhances the computational power and accuracy of AI and ML applications through a data-driven approach. ML techniques, including supervised and unsupervised methods, ANNs, RFs, SVMs, boosting techniques, clustering methods, and network-based architectures, can all be leveraged for performance evaluation. This paper provides a comprehensive literature review on the application of AI and ML for evaluating and modeling shale gas production performance.

8.4.4.6.2 Description of the Workflow and Machine Learning Model

This paper addresses the evaluation of shale gas production performance using various AI techniques. The focus is on applying ML models to optimize production forecasting, well completion, and reservoir characterization. The primary methodologies employed include RF, Gradient Boosting Trees (GBTs), SVMs, and ANNs. These techniques are used to model the complex interactions between reservoir properties, hydraulic fracturing, and production outcomes.

The RF approach is utilized for its ability to manage high-dimensional data, handling multiple decision trees to improve the accuracy of production forecasts. GBTs are used for their robustness in optimizing model performance by combining weak predictive models into a strong one. SVM is applied to classify well production outcomes based on reservoir characteristics, and ANN models are used for TSA of shale gas production rates.

The methodologies rely on data-driven approaches to analyze and model various aspects of shale gas production, including production rates, well spacing, and fracture permeability. The ultimate goal is to improve production efficiency and optimize well performance by leveraging these ML techniques.

8.4.4.6.3 Advantages and Limitations of the Synchronized Algorithm

The advantages of the ML methodologies used in this study include their ability to handle large, complex datasets and model nonlinear relationships between various factors affecting shale gas production. RF and GBTs offer robustness against overfitting and provide high accuracy in predictions, making them suitable for forecasting production rates. SVMs are effective in classifying production outcomes and identifying key features that influence well performance, while ANNs excel at capturing temporal patterns in production data.

However, these methodologies also come with limitations. The computational complexity of ensemble methods like RF and GBTs can be high, especially when dealing with large-scale datasets. SVMs may struggle with high-dimensional data unless appropriate kernel functions are selected. ANNs, particularly DL models, require large amounts of data to avoid overfitting and may suffer from issues such as vanishing gradients, which can affect the accuracy of predictions.

The effectiveness of these ML models is heavily dependent on the quality and quantity of the data used, as well as the proper tuning of hyperparameters to optimize model performance. Despite these challenges, the application of AI techniques in shale gas production shows promise in enhancing production efficiency and improving decision-making processes.

8.4.4.6.4 Results and Discussion

This paper provides a comprehensive review of ML techniques applied to the evaluation and modeling of shale gas production performance. The significance of this topic lies in its potential to improve the assessment of various methods used in shale gas extraction. Despite the increasing interest in this field, research on the application of ML specifically for shale gas production remains relatively sparse. Various ML approaches have been successfully utilized for tasks such as production forecasting, fracture design optimization, uncertainty assessment, history matching, sweet spot identification, gas characterization, and TOC impact analysis. Among these methods, ANNs have shown particular promise in providing accurate predictions by leveraging large datasets. However, challenges arise with smaller datasets, as ML algorithms generally require more extensive data to perform effectively.

To address these challenges, integrating ML algorithms with physics-based approaches may enhance predictive accuracy even when data is limited. Future research should focus on implementing ensemble learning techniques that combine multiple ML methods to further improve shale gas production performance. Additionally, more analysis is needed to identify the most effective ML approach to serve as a baseline for modeling shale gas productivity. These steps will help overcome current limitations and advance the application of ML in shale gas development and production studies.

BIBLIOGRAPHY

Syed, F. I., Muther, T., and Dahaghi, A. K. et al. AI Assisted Shale Gas Production Performance Evaluation. *J Petrol Explor Prod Technol* 2021;11:3509–3519. https://doi.org/10.1007/s13202-021-01253-8

8.4.4.7 Application-44: Application of a Deep Learning Network for Joint Prediction of Associated Fluid Production in Unconventional Hydrocarbon Development

8.4.4.7.1 Background of the Application

ML techniques have become increasingly popular in various O&G applications. Among these, time series-based predictive forecasting using DL–ML strategies has gained considerable attention for its ability to generalize temporal or sequence-based patterns within data. As interest in hydrocarbon production grows, there are new opportunities to apply DL–ML to other areas of O&G development where dynamic and temporal dependencies are relevant and impact production forecasting. This study introduces a framework that combines both supervised and unsupervised ML approaches to jointly predict produced water and natural gas volumes associated with oil production from UCRs, using a time series approach. The focus is on the Spraberry and Wolfcamp Formations in the Midland Basin of the U.S., utilizing a DNN model with LSTM layers.

The model can accurately reproduce and forecast produced water and natural gas volumes for wells on a monthly basis, achieving 91 percent joint prediction accuracy with minimal performance differences between training and test datasets. The model effectively replicates production profiles, even in cases with irregular trends. By integrating this model with an Arps decline model, cumulative production forecasts for oil, gas, and water at both the well and basin levels are generated. These forecasts consider well completion, decline curves, and spatial and reservoir characteristics. Such model-derived outlooks can assist operators in developing management and remedial strategies for fluid volumes anticipated in unconventional O&G development.

8.4.4.7.2 Description of the Workflow and Machine Learning Model

This work focuses on the application of ML techniques to predict associated fluid production in unconventional O&G reservoirs. Specifically, the study employs a DL approach using a LSTM network to model the time-series nature of oil, gas, and water production. This model aims to predict the volumes of produced water and natural gas alongside oil production at a monthly resolution. The input data includes well performance attributes, completion details, and spatial and reservoir attributes, which are processed and normalized to create a robust predictive model. The study also incorporates unsupervised learning techniques, such as K-means clustering, to group wells with similar characteristics, which further improves the accuracy of predictions. This hybrid approach leverages both supervised and unsupervised ML techniques to enhance the predictive capabilities of the model.

8.4.4.7.3 Advantages and Limitations of the Synchronized Algorithm

The use of a DL model, particularly an LSTM network, offers significant advantages in capturing the temporal dependencies inherent in time-series data. This allows the model to accurately forecast the production profiles of oil, water, and natural gas over time. The integration of clustering techniques enhances the model's ability to generalize across wells with varying characteristics, making it adaptable to different reservoir conditions. The joint prediction capability of the model provides a more holistic view of production trends, which is valuable for operational decision-making in unconventional O&G development. However, the computational complexity of training DL models, especially with large datasets, can be a challenge. Additionally, the performance of the model heavily depends on the quality and completeness of the input data, as missing or inaccurate data can significantly impact the accuracy of the predictions. The clustering approach, while useful, may also introduce some uncertainty if the clusters do not perfectly capture the underlying reservoir heterogeneity.

8.4.4.7.4 Results and Discussion

In this paper, a novel data-driven modeling framework that integrates both supervised and unsupervised ML techniques is presented. The findings indicate that this hybrid approach is effective in creating a time series predictive model capable of either reproducing or forecasting cumulative volumes of natural gas and water produced alongside oil at the well level. The supervised learning component focuses on developing a DL model that provides reliable estimates of produced water and natural gas based on well completion and placement. Meanwhile, the unsupervised learning component classifies wells into groups, facilitating an easy method to derive Arps Decline parameters, well completion specifics, and reservoir characteristics for each cluster. This combination enables accurate forecasts of oil, water, and natural gas production at the well level, which can be scaled up to pad, field, or basin levels. The framework has been demonstrated using data from the "Wolfberry" area within the Midland Basin, but it is adaptable to other mature unconventional oil and gas regions due to its reliance on widely available datasets.

The study also addresses significant environmental concerns related to shale oil and gas development, such as water usage, induced seismicity from wastewater disposal, and the flaring or venting of natural gas. The presented framework can aid in developing management and remediation strategies based on expected fluid volumes. The results highlight the variability in water and gas volumes produced due to wellbore design and placement, underscoring the complexity of forecasting. The data provided in tables offer valuable insights for estimating fluid production and building future production outlooks. Future research could address limitations and explore opportunities for improvement, such as incorporating geologic data, integrating operational time series data, and developing separate models for predicting water and gas production. Enhancing the model with these elements could further refine predictions and support more effective decision-making in unconventional oil and gas development.

BIBLIOGRAPHY

Vikara, D., and Khanna, V. Application of a Deep Learning Network for Joint Prediction of Associated Fluid Production in Unconventional Hydrocarbon Development. *Processes* 2022;10:740. https://doi.org/10.3390/pr10040740

8.4.4.8 Application-45: Application of Machine Learning for Productivity Prediction in Tight Gas Reservoirs

8.4.4.8.1 Background of the Application

Accurate prediction of well productivity is crucial for developing effective reservoir management plans. Traditional methods often fall short in the context of tight gas reservoirs, leading this paper to quantitatively assess the relationship between absolute open flow and key parameters for Linxing tight gas reservoirs using statistical analysis. By integrating reservoir engineering theories, the study identifies the primary controlling factors and introduces a novel ML-based method for predicting well productivity in tight gas reservoirs. The effectiveness of this predictive model is evaluated through ML techniques and field data analysis, alongside traditional DCA, to forecast the EUR of individual wells within an acceptable range.

The study reveals ten critical parameters, such as gas saturation, as dominant factors influencing well productivity, with geological elements proving more significant than fracturing parameters in this context. The ML models assessed—SVR, Back Propagation (BP), and RF—demonstrate varying levels of accuracy, with R2 values of 0.72, 0.87, and 0.91, respectively. The results indicate that the RF model offers the most precise predictions, particularly for medium and high-production wells, while also highlighting that water production significantly impacts low-producing wells. This study validates the reliability and practical value of the RF model for predicting recoverable reserves, confirming its efficacy for real-world applications.

8.4.4.8.2 Description of the Workflow and Machine Learning Model

This paper focuses on leveraging ML techniques to improve the accuracy of well productivity prediction in tight gas reservoirs. The study evaluates the relationships between key reservoir parameters and well productivity, using three ML models: SVR, BPNNs, and RF. These models are employed to predict absolute open flow rates based on factors such as gas saturation, permeability, and porosity. The research optimizes the model parameters using a grid search method, enhancing prediction accuracy, and validating the models with a 10-fold cross-validation approach.

8.4.4.8.3 Advantages and Limitations of the Synchronized Algorithm

The use of ML models, particularly the RFM, provides a robust approach to predicting well productivity in tight gas reservoirs. The models allow for handling complex, nonlinear relationships among multiple parameters, improving the accuracy of predictions compared to traditional methods. The RFM shows the highest predictive performance with better generalization and resistance to overfitting. However, the SVR model demonstrated limitations in stability and adaptability, particularly for low-producing wells. Moreover, the computational cost increases with the complexity of the models, and the accuracy of the predictions heavily depends on the quality of the input data and the selection of appropriate model parameters.

8.4.4.8.4 Results and Discussion

The statistical analysis method efficiently identifies the primary factors influencing well productivity and provides a quantitative assessment of these factors. This study evaluates ten parameters, including gas saturation and reservoir thickness, to understand their impact on gas well productivity within the block. The findings suggest that geological factors are more influential than fracturing factors in predicting well productivity.

The RF model proves to be highly effective in predicting the absolute open flow of gas wells and distinguishing between different well types. With an RMSE of 3.98 and an R2 value of 0.91, the RF model demonstrates superior prediction accuracy compared to other models, including BP and SVR, which exhibits the largest prediction error. Practically, the RF model successfully identifies medium- and high-production wells with up to 100% accuracy, making it the recommended choice for productivity prediction in Linxing gas reservoirs. Additionally, the study highlights that deviations in initial production can lead to inaccuracies in predictions using typical decline curves. The RF model's 12-month cumulative production forecasts for four wells are more precise than those generated by the decline curve model, underscoring the effectiveness of ML techniques in enhancing production predictions.

BIBLIOGRAPHY

Fang, M., Shi, H., Li, H., and Liu, T. Application of Machine Learning for Productivity Prediction in Tight Gas Reservoirs. *Energies* 1916;2024:17. https://doi.org/10.3390/en17081916

8.4.4.9 Application-46: Applied Transfer Learning for Production Forecasting in Shale Reservoirs

8.4.4.9.1 Background of the Application

ML has been successfully applied to various sectors within the O&G industry, although these successes have not always translated to effective production forecasting for UCRs due to the limited availability of early-stage production data. This lack of data presents a significant challenge, but TL offers a promising solution. TL allows for the transfer of knowledge from well-established regions with abundant data to less explored areas with limited information, thereby addressing the data sparsity issue.

This study discusses the theoretical background and practical application of TL to enhance production forecasting in UCRs. By employing similarity metrics to identify relevant candidates for TL based on key drivers such as reservoir mechanisms and subsurface structures, the approach enables the reuse of learned parameters from data-rich fields. Training models on these related fields allows the foundational features to be applied to areas with sparse data, thus enriching the models. The study details the process of selecting related sites for TL, highlights the challenges, and provides a step-by-step approach for improving forecasting accuracy. Although there is limited research on TL specifically for O&G reservoirs, this study demonstrates its potential to significantly enhance forecasting models by leveraging knowledge from well-studied areas, particularly in unconventional shale reservoirs.

8.4.4.9.2 Description of the Workflow and Machine Learning Model

This paper investigates the application of TL for production forecasting in shale reservoirs. The study utilizes ML and DL techniques to address challenges posed by limited production data in UCRs. Transfer learning is employed to leverage knowledge from well-established reservoirs with abundant data and apply it to regions with sparse data. Two primary tasks are explored: history matching of half fracture length and gas production forecasting. The ML models used in the study include GB and DNNs. These models are first trained on a source domain (e.g., HF simulation data) and then applied to target domains (e.g., actual field data from the Eagle Ford shale) to predict critical reservoir parameters such as half fracture length and gas production rates.

8.4.4.9.3 Advantages and Limitations of the Synchronized Algorithm

The TL approach provides a powerful framework for production forecasting in shale reservoirs with sparse data. By utilizing knowledge from related reservoirs with rich datasets, TL significantly enhances model accuracy and reduces uncertainty. The methodology also allows for the efficient reuse of trained models, saving computational time and resources. The use of GB and DNNs in this context enables the handling of complex relationships between geological, reservoir, and production variables. However, the approach has its limitations. The effectiveness of TL heavily depends on the similarity between the source and target domains; if the characteristics of the target domain differ significantly from the source domain, the model's performance may decline. Additionally, the computational demands of training complex ML/DL models can be substantial, especially when large datasets or numerous variables are involved.

8.4.4.9.4 Results and Discussion

TL was employed in the Eagle Ford production analysis in two key areas: predicting half fracture length and estimating gas production. Both applications illustrate the effectiveness of TL in approximating gas production and cumulative volume.

In the context of half fracture length prediction, TL facilitated the history matching of gas production by applying insights gained from probabilistic HF simulations to the Eagle Ford production data. This approach enabled the adjustment of historical production models to better reflect observed outcomes. For gas production prediction, TL was used to forecast gas output in locations that had not yet been drilled. This was achieved by leveraging ML and DL models developed from shale data in La Salle County and applying these models to adjacent counties to estimate cumulative gas volumes.

BIBLIOGRAPHY

Odi, Uchenna, Ayeni, Kola, Alsulaiman, Nouf, Reddy, Karri, Ball, Kathy, Basri, Mustafa, and Temizel, Cenk. "Applied Transfer Learning for Production Forecasting in Shale Reservoirs." Paper presented at the *SPE Middle East Oil & Gas Show and Conference, event canceled*, November 2021. doi: https://doi.org/10.2118/204784-MS

8.4.4.10 Application-47: Deep Learning-Based Production Forecasting and Data Assimilation in Unconventional Reservoir

8.4.4.10.1 Background of the Application

UCRs, such as those producing shale oil, play a crucial role in meeting global energy demands. However, forecasting oil production from these reservoirs remains particularly challenging due to the complexity of fracture networks, low matrix porosity, and permeability. Traditional forecasting methods, including the Arps decline model and reservoir simulation techniques, struggle with the highly nonlinear and heterogeneous nature of these rocks and fluids, leading to significant deviations and limited applicability for UCRs. To enhance forecasting accuracy, data-driven approaches are essential. In this study, data from two major shale oil-producing regions—the Eagle Ford and the Bakken—were utilized. The Bakken data was used for training and testing the models, while the Eagle Ford data served for validation. A random search method optimized the model parameters, and a sliding window technique was employed to determine an appropriate window size for predicting future values.

The study developed and compared six hybrid deep learning models, including gated recurrent units (GRUs), LSTM, and temporal convolutional networks (TCNs). These models were designed to capture both spatial and temporal patterns in oil production data. Among the models tested, the TCN–GRU combination demonstrated superior performance both statistically and computationally. This robust model not only enhances decision-making capabilities but also reduces the overall forecasting costs, providing a more effective tool for managing production in unconventional shale oil reservoirs.

8.4.4.10.2 Description of the Workflow and Machine Learning Model

This paper explores DL-based forecasting and data assimilation in unconventional shale oil production systems. The research integrates various DL methods, including GRUs, LSTM, and TCNs, to model and predict oil production. These models are implemented individually and in hybrid forms, such as GRU–TCN and LSTM–TCN, to capture both short-term and long-term dependencies in time-series data. The study focuses on improving prediction accuracy by leveraging historical production data from shale oil fields. The optimization process involves hyperparameter tuning using the random search method, and model performance is evaluated based on metrics such as the determination coefficient (R^2) and normalized root mean squared error (NRMSE).

8.4.4.10.3 Advantages and Limitations of the Synchronized Algorithm

The hybrid models combining TCNs with GRU or LSTM provide a powerful approach to forecasting shale oil production by effectively capturing both temporal and spatial patterns. These models outperform traditional single models, offering better accuracy in predicting production trends. The advantage of using TCNs lies in its ability to handle long sequences without loss of information, while GRU and LSTM excel in managing nonlinearity and learning from sequential dependencies. However, the computational cost of training these complex hybrid models is significant, requiring considerable resources, especially for large-scale time-series data.

Additionally, the quality of the predictions heavily depends on the accuracy and completeness of the historical production data used for training.

8.4.4.10.4 Results and Discussion

The performance of individual and hybrid DL models was thoroughly examined using production data from ten wells, five of which were set aside for blind validation and the other five for training and testing Single-Entity Production Data (SEPD) and DL models. The TCN–LSTM and TCN–GRU models both showed excellent accuracy and low error. Each model's efficiency and resilience, as well as its computational costs, were evaluated. TCN–GRU outperformed the other models in terms of computational efficiency, evaluation metrics (NMAE, WAPE, and NRMSE), and determination coefficient (R^2). The robustness and dependability of the DL models were evaluated using K-fold cross-validation, which verified that the TCN–GRU model works well with a range of production data types, including those with short or extended production histories.

For shale oil production forecasting, where precise forecasts are essential for making well-informed decisions with substantial financial and environmental ramifications, the TCN–GRU model's efficacy is especially advantageous. The study showed that spatial and temporal data may be efficiently captured by both individual and hybrid DL models, with high accuracy at minimal computing costs. The successful use of automated history matching to predict shale oil output may lower total operational expenses. Future research will concentrate on multivariate production estimates that take into account field factors including annular pressure, gas/oil ratio, choke size, and bottomhole flowing temperature and pressure.

BIBLIOGRAPHY

Tripathi, Bineet Kumar, Kumar, Indrajeet, Kumar, Sumit, and Singh, Anugrah. Deep Learning–Based Production Forecasting and Data Assimilation in Unconventional Reservoir. *SPE J* 2024. https://doi.org/10.2118/223074-PA

8.4.4.11 Application-48: Production Forecasting in Tight Gas Reservoirs Using Long Short-Term Memory Methods (LSTM)

8.4.4.11.1 Background of the Application

Forecasting the EUR for extremely tight gas reservoirs, which exhibit long-term transient behaviors, poses significant challenges. Traditional methods used for such predictions have limitations, prompting researchers to explore newer techniques like LSTM–DL methods. This study evaluates the performance of LSTM estimations against those provided by a physics-based reservoir simulation model.

To achieve reliable EUR forecasts, the study generates unconventional tight gas reservoir data through simulation and applies LSTM–DL techniques designed for sequential data analysis. Concurrently, a reservoir simulation model is created from the same data for comparative purposes. The LSTM model, with its ability to incorporate operational interventions without disrupting the ML framework, offers predictions that closely match those from the physics-based model. The results indicate that the AI-driven LSTM model provides forecasts with comparable accuracy to the physics-based approach but with the added advantage of extended long-term forecasting capabilities. These data-driven LSTM models hold significant potential for analyzing tight gas reservoirs with time series well information, thereby enhancing recovery estimates and guiding better decision-making in reservoir management. This study highlights the benefits of DL methods and suggests further exploration of various ANNs could offer substantial advantages in the O&G industry.

8.4.4.11.2 Description of the Workflow and Machine Learning Model

The work focuses on forecasting gas production in tight gas reservoirs using advanced ML techniques. Specifically, the study employs LSTM networks, a type of recurrent neural network (RNN), to predict gas production rates over time. LSTM networks are particularly well-suited for time-series forecasting because they can capture long-term dependencies in sequential data, such as historical production data from tight gas reservoirs. The study utilizes an extensive dataset of gas production from over 13,000 wells, incorporating various features such as gas–water ratio (GWR), gas–oil ratio (GOR), water–oil ratio (WOR), lateral length, fracture stages, fracture fluid, proppant, and API gravity. The data is preprocessed using feature engineering techniques, including time-series transformations, to enhance the predictive power of the LSTM model.

The methodology involves training the LSTM model on historical production data, with the goal of predicting future gas production. The model's performance is compared with other ML algorithms, such as GB variants, to evaluate its accuracy and robustness. The study also explores different feature lists and preprocessing strategies to optimize the LSTM model's performance. The final model architecture consists of multiple LSTM layers with dropout regularization to prevent overfitting, followed by dense layers for output prediction.

8.4.4.11.3 Advantages and Limitations of the Synchronized Algorithm

The LSTM-based methodology provides a significant advantage in capturing the complex temporal dependencies in gas production data, making it well-suited for long-term production forecasting in tight gas reservoirs. LSTM networks are capable of modeling both short-term and long-term patterns in the data, which can improve the accuracy of predictions compared to traditional methods such as DCA and other ML algorithms. Additionally, the inclusion of time-series transformations and feature engineering enhances the model's ability to capture relevant trends in the data, further improving the quality of the forecasts.

However, the effectiveness of the LSTM model is highly dependent on the quantity and quality of the available data. For example, the study found that LSTM models require a minimum of nine years of production data to achieve acceptable predictive performance, and twelve years for more reliable forecasts. The computational complexity of training LSTM models can also be a limitation, particularly for large datasets with many features. Moreover, while LSTM models offer superior performance in certain scenarios, they may not always outperform other ML techniques, such as GB, depending on the characteristics of the dataset and the specific task at hand. Therefore, a hybrid approach that combines LSTM with other ML models could offer a more robust solution for production forecasting in tight gas reservoirs.

8.4.4.11.4 *Results and Discussion*

In this study, gas production data from a tight gas resource was utilized to train an LSTM– DL model. The workflow included a thorough quality check of over 20,000 wells, eliminating those with fewer than five years of production data. Several parameters were evaluated for inclusion in the model, focusing on those with the strongest correlation to gas production data. Key parameters included GWR, WOR, GOR, lateral length, fracture stages, frac fluid pumped, proppant, and API gravity. Time-series transformations and custom features were developed to align with the production data's characteristics.

Initial results showed that traditional ML models, like various GB methods, outperformed the LSTM model in terms of accuracy for predicting tight gas production. However, the LSTM model still performed competitively against GB variants, such as the gradient boost model champion. Other models, such as EUREQA (symbolic regression software), demonstrated around 80% forecasting accuracy. The study suggests that combining LSTM with prominent ML algorithms, like GB, could enhance predictive accuracy. The effectiveness of the LSTM model depends significantly on the quality and length of the production data. Despite its limitations, the hybrid approach of combining LSTM with traditional methods improved both the efficiency and speed of gas production forecasting compared to conventional techniques.

BIBLIOGRAPHY

Qoqandi, Afrah, Alfaleh, Omar, Ramadan, Moemen, and Odi, Uchenna. "Production Forecasting in Tight Gas Reservoirs Using Long Short-Term Memory Methods (LSTM)." Paper presented at the *Middle East Oil, Gas and Geosciences Show*, Manama, Bahrain, February 2023. doi: https://doi.org/10.2118/213343-MS

8.4.4.12 Application-49: Suitability of Different Machine Learning Outlier Detection Algorithms to Improve Shale Gas Production Data for Effective Decline Curve Analysis

8.4.4.12.1 *Background of the Application*

Despite their large reserves, shale gas reservoirs are difficult to evaluate economically because of their complicated driving processes, demanding drilling and completion setups, and challenging production circumstances. Because it just uses production history, DCA has long been thought of as a simple technique for forecasting output in UCRs. The selection of a suitable DCA model to fit the production behavior of shale gas wells is difficult, and this uncertainty is exacerbated by noisy production data from multiple shut-ins for water removal and varying choke widths. Addressing this data noise is necessary for DCA prediction to be effective.

In order to identify the best ML outlier identification technique for improving the quality of production data, this study looked at twelve different algorithms. Five methods were deemed to be insufficient because they eliminated large segments of the data instead of addressing isolated outlier spots. Assuming that 20% of the production data were outliers, the remaining seven algorithms were examined. The techniques that worked best for enhancing the quality of the data for DCA were the angular-based outlier factor, KNN, and clustered-based outlier factor algorithms.

On the other hand, algorithms for subspace OD and stochastic outlier selection performed worse.

8.4.4.12.2 Description of the Workflow and Machine Learning Model

This study investigates the suitability of various ML OD algorithms to enhance shale gas production data for effective DCA. The analysis is centered on improving the quality of production data by identifying and removing outliers, which can significantly affect the prediction accuracy of DCA models. Twelve ML algorithms, including Isolation Forest, Local Outlier Factor (LOF), and PCA, are evaluated for their effectiveness in detecting outliers. The cleaned data is then subjected to eight DCA models, such as the Hyperbolic Arps and Power Law Exponential models, to assess the impact of outlier removal on production forecasting. The ML methods and DCA models are integrated into a workflow that optimizes the forecasting accuracy of shale gas production data.

8.4.4.12.3 Advantages and Limitations of the Synchronized Algorithm

The application of ML OD algorithms in shale gas production data improves the reliability of DCA by removing noise and erroneous data points. The multi-algorithm approach allows for a comprehensive evaluation of different OD techniques, ensuring that the most effective algorithms are identified for enhancing data quality. This leads to more accurate predictions of production rates and ultimate recovery. However, the complexity of the ML algorithms, particularly in terms of computational resources and parameter tuning, can be a limitation. Additionally, the effectiveness of the DCA models depends on the accuracy of the initial production data and the choice of OD algorithm, which may vary across different datasets.

8.4.4.12.4 Results and Discussion

This study evaluated the effectiveness of 12 ML algorithms for OD in improving shale gas production data for more accurate DCA. The methodology presupposed that 20% of the production data consisted of outliers, with each algorithm designed to identify these outliers based on its specific approach. Eight DCA models were then applied to the actual production data from three shale gas wells, both before and after removing the identified noise, to determine the most effective combination of OD algorithms and DCA models for reducing uncertainties in production forecasting and reserve estimation. The study led to several conclusions.

It was found that while most OD algorithms are generally applicable, some are less effective for enhancing production data prior to DCA, particularly Linear-base algorithms, IF, and Histogram-Based Outlier Detection (HBOD), which tend to remove entire segments of data and complicate DCA application. In contrast, CBOD, KNN, and Angle-Based Outlier Detection (ABOD) emerged as the most effective algorithms, smoothing the production profile by identifying scattered outliers without disrupting data trends. LOF proved useful for production profiles with isolated outliers but could distort trends with high threshold values. Conversely, Subspace Outlier Score (SOS) and Subspace Outlier Detection (SOD) were less effective as they did not detect all scattered data points, leading to similar fitting results post-DCA as those before noise removal. Among the DCA models, the Arps, Duong, and

Wang models exhibited less sensitivity to noise removal, while SEPD, Power Law Exponential (PLE), and Logistic Growth Model (LGM) models showed considerable variation in forecasting accuracy depending on the OD algorithm used. The study also highlighted the importance of optimizing the OD algorithm's threshold based on the noise level in the production data and recommended using multiple DCA models to evaluate shale gas reserves due to their differing assumptions and structures.

BIBLIOGRAPHY

Yehia, T., Wahba, A., Mostafa, S., and Mahmoud, O. Suitability of Different Machine Learning Outlier Detection Algorithms to Improve Shale Gas Production Data for Effective Decline Curve Analysis. *Energies* 2022;15:8835. https://doi.org/10.3390/en15238835

8.4.4.13 Application-50: Well Performance from Numerical Methods to Machine Learning Approach: Applications in Multiple Fractured Shale Reservoirs

8.4.4.13.1 Background of the Application

Horizontal well fracturing technology is extensively utilized in UCRs like tight or shale oil and gas fields. To further enhance oil recovery in these reservoirs, methods such as HnP miscible gas injection are employed. However, the complexities of hydraulic fracture properties and multiphase flow0 present significant challenges in accurately predicting well performance, especially with traditional numerical methods. To address this, this paper develops two ML-based methods for forecasting well production performance in UCRs, with a particular focus on EOR pilot projects. The first method involves using ANNs to analyze extensive data from UCRs, including HF parameters, well completion details, and production data. Feature selection is conducted to identify key factors, and ANNs are then used to model the relationship between these factors and well production performance.

The second method applies TSA, which is well-suited for understanding the time-dependent properties of UCRs, such as fluid properties and reservoir pressure. Data from over 10,000 wells across various fractured shale reservoirs, including Bakken, Eagle Ford, and Barnett, are utilized for training and testing. Results show a strong correlation between predicted and actual well performance, with overall R values exceeding 0.8 for both methods, indicating their reliability in forecasting well performance. Notably, for EOR field cases like HnP miscible gas injection, TSA provides more accurate results compared to ANNs. This paper offers a comprehensive evaluation of ML techniques for multiple fractured shale reservoirs, presenting a more efficient and robust alternative to conventional numerical methods for well performance evaluation.

8.4.4.13.2 Description of the Workflow and Machine Learning Model

This paper focuses on the prediction of well performance in multiple fractured shale reservoirs, particularly in the context of EOR using ML techniques. The study employs two distinct AI methodologies: ANNs and TSA. The ANN approach is used to uncover hidden patterns and relationships in large datasets that include well completion data, HF details, and production history. Key features such as cumulative production, initial production rates, and well perforation data are used as inputs to predict future well production.

The ANN model incorporates four hidden layers, each with 80 neurons, and utilizes ReLU as the activation function. The model is trained using a large dataset collected from thousands of wells across several shale formations, including Bakken, Eagle Ford, and Barnett. The output of the ANN model is the future production performance over a one-year period.

On the other hand, the TSA method focuses on the production data as a function of time. It uses AutoRegressive Integrated Moving Average (ARIMA) models to account for temporal dependencies in the data, which is particularly beneficial for scenarios where EOR methods have been applied. Hyperparameters for the ARIMA model are determined using partial autocorrelation and autocorrelation methods, followed by a grid search to refine the model parameters.

8.4.4.13.3 Advantages and Limitations of the Synchronized Algorithm

The combination of ANN and TSA provides a robust approach to predicting well performance in fractured shale reservoirs. The ANN model excels at handling large datasets and can uncover complex relationships between input features and production outcomes. This makes it highly effective in forecasting production performance across various shale formations. However, the ANN model requires significant computational resources, and its accuracy is sensitive to the quality and preprocessing of input data. Additionally, the risk of overfitting is present, especially in complex NNs with multiple layers.

TSA offers an advantage in dealing with temporal data, especially when production trends change due to EOR techniques. The ability of the TSA model to update its training data over time allows it to adapt to new data, making it more dynamic than traditional ML models. However, the challenge lies in selecting appropriate hyperparameters, which requires careful analysis and tuning. The ARIMA model also has limitations when dealing with non-stationary data, and its performance can be affected by seasonal variations in production trends.

8.4.4.13.4 Results and Discussion

In this study, two ML methods were employed to predict well production performance, offering a more robust alternative to conventional numerical models. The first method uses ANNs incorporating various features such as well completion, hydraulic fracturing, and production data. A modified DNN was developed with feature selection to enhance prediction accuracy by filtering out noisy features. This approach demonstrates that appropriate feature selection can significantly improve model performance.

The second method involves TSA, focusing solely on well production performance as the outcome influenced by factors like well completion and hydraulic fracturing. TSA is particularly effective for handling production data, especially in EOR scenarios, due to its ability to update training data as a function of time. This results in accurate predictions, surpassing traditional numerical models like multiphase flow models and DCA. Despite their advantages, both ML methods have limitations. For example, in TSA, selecting optimal hyperparameters (p, q, d) remains challenging.

Although methods such as autocorrelation functions and the "MODEL_ARIMA.FIT" package help, they often yield similar predictive accuracies. Future research should focus on developing additional criteria to identify more suitable parameters for TSA.

BIBLIOGRAPHY

Liu, Kailei, Xu, Boyue, Kim, Changjea, and Fu, Jing. Well Performance from Numerical Methods to Machine Learning Approach: Applications in Multiple Fractured Shale Reservoirs, *Geofluids* 2021;3169456:13. https://doi.org/10.1155/2021/3169456

9 AI Applications in the Fourth Industry Revolution

Ekrem Alagoz, Ruoyu "Tagore" Cheng, Mouad Al Krmagi, Ferit Ozdaban, Serkan Gozel, Emre Can Dündar, Mehmet Melih Oskay, Celal Hakan Canbaz, Luigi Saputelli, Mufrettin Murat Sari, Salih Tutun, and Cenk Temizel

9.1 INTRODUCTION

Artificial intelligence (AI) is rapidly emerging as powerful tool with the potential to revolutionize the oil and gas industry. The integration of AI technologies, along with other digital innovations, is driving the Fourth Industrial Revolution (IR 4.0) and enabling the creation of smart fields.

This chapter delves into the application of AI within the context of IR 4.0 and smart fields. It provides a comprehensive overview of various workflows that can be enhanced through AI, including industrial automation, field development and optimization, well management, and other relevant areas. By understanding the potential benefits of AI in IR 4.0, industry professionals can harness these technologies to drive innovation, improve efficiency, and optimize operations.

9.2 CATEGORIZATION OF WORKFLOWS USED IN IR 4.0 AND SMART FIELDS

9.2.1 Industrial Revolution 4.0

AI can be applied to various aspects of IR 4.0 in the oil and gas industry. By analyzing sensor data and identifying patterns, AI algorithms can predict equipment failures before they occur, allowing for proactive maintenance and reducing downtime. Additionally, AI can be used to optimize production processes by identifying bottlenecks, improving energy efficiency, and reducing waste. This can lead to significant cost savings and increased productivity.

Furthermore, AI can automate quality control (QC) inspections, ensuring product consistency and identifying defects early. This can help improve product quality and reduce customer complaints. AI can also optimize supply chain operations by

 DOI: 10.1201/9781003617327-9

improving demand forecasting, inventory management, and transportation planning. This can lead to reduced costs and improved customer satisfaction. Finally, AI can be used to identify safety hazards, assess risks, and improve workplace safety. This can help to prevent accidents and injuries, and reduce liability.

9.2.2 Oil Field Development, Planning, and Optimization

AI can be applied to various aspects of oil field development, planning, and optimization. By analyzing reservoir data, production history, and other relevant factors, AI algorithms can optimize well placement, improve reservoir characterization, and enhance production efficiency. Additionally, AI can be used to predict future production, optimize field development plans, and identify opportunities for enhanced oil recovery (EOR). This can help to maximize oil production and reduce costs.

Furthermore, AI can be used to optimize drilling operations by predicting drilling challenges, improving drilling efficiency, and reducing non-productive time. AI can also be applied to optimize production operations by analyzing production data and identifying opportunities to improve well performance, reduce downtime, and optimize field operations. This can help to increase production rates and reduce operating costs.

9.2.3 Well-Related Applications

AI can be applied to various aspects of well management to improve efficiency, reduce costs, and optimize production. By analyzing well data, such as pressure, flow rate, and downhole conditions, AI algorithms can predict well performance, identify potential issues, and optimize well operations. Additionally, AI can be used to optimize drilling operations by predicting drilling challenges, improving drilling efficiency, and reducing non-productive time.

Furthermore, AI can be applied to optimize production operations by analyzing production data and identifying opportunities to improve well performance, reduce downtime, and optimize field operations. This can involve optimizing well spacing, adjusting production rates, or implementing advanced EOR techniques. AI can also be used to monitor well integrity and identify potential safety hazards, helping to prevent equipment failures and ensure safe operations.

9.2.4 Other Applications

AI and ML offer transformative applications across various facets of the oil and gas industry, enabling enhanced efficiency, cost reduction, and sustainability. In safety and risk management, these technologies play a vital role in identifying safety hazards, assessing risks, and improving workplace safety measures, thereby helping to prevent accidents, injuries, and potential liabilities. AI is equally impactful in environmental monitoring, where it tracks parameters such as air quality, water quality, and soil contamination, ensuring compliance with environmental regulations and fostering environmental protection.

Additionally, in supply chain management, AI optimizes operations by improving demand forecasting, inventory management, and transportation planning, which reduces costs and enhances operational efficiency. In the financial domain, these technologies enable the analysis of financial data to identify trends and uncover opportunities, aiding in improved financial performance and informed investment decisions. By harnessing the capabilities of AI, oil and gas companies can address key challenges, streamline operations, and align their strategies with sustainable and forward-thinking practices.

9.3 OVERVIEW OF AI METHODOLOGIES

Artificial intelligence (AI), machine learning (ML), and deep learning (DL) represent the technological backbone of the Fourth Industrial Revolution (4IR). These methodologies are enabling the automation and augmentation of complex decision-making processes across various industries, including the energy sector. AI serves as the overarching concept that encompasses the ability of machines to mimic human intelligence, while ML refers to the subset of AI that focuses on the development of algorithms capable of learning from data without being explicitly programmed. DL, as a more advanced subfield of ML, involves neural networks with multiple layers that allow for the extraction of hierarchical features from large and unstructured datasets.

These technologies are applied through supervised, unsupervised, and reinforcement learning paradigms, each serving different purposes based on the nature of the problem and the availability of labeled data. Supervised learning is widely used in predictive modeling and classification tasks, while unsupervised learning helps uncover hidden patterns and clusters within datasets. Reinforcement learning, on the other hand, is particularly useful in scenarios involving dynamic decision-making under uncertainty, such as production optimization or reservoir management. Recent advancements have also seen the integration of hybrid models that combine physics-based simulators with data-driven algorithms, further enhancing the capabilities of digital oilfields and smart production systems.

9.4 APPLICATIONS OF AI

The Fourth Industrial Revolution (4IR) has accelerated the digital transformation of the oil and gas sector, enabling a shift from reactive to predictive and autonomous operations. At the core of this transformation lies the integration of AI, ML, and DL into virtually every aspect of the upstream value chain. From surface operations to subsurface analytics, AI-powered applications are helping companies enhance efficiency, minimize environmental impact, and extract deeper insights from operational data.

In the industrial domain, several pioneering solutions have been deployed to support intelligent sensor networks, operational reliability, and data integrity. For example, recurrent autoencoders have been employed to validate historical sensor data and detect anomalies in real time, ensuring consistent measurement quality even with aging equipment. Open-source tools and cross-industry methodologies are also playing a critical role in accelerating digital adoption, as seen in solutions that apply industrial automation principles from manufacturing to upstream oilfield operations. Additionally, intelligent models have been developed to predict gas flow rates and preserve crude quality specifications using AI-driven correlations, while digital production optimization platforms leverage real-time data and feedback control to maintain performance targets under fluctuating operational conditions.

Field development, planning, and optimization have also benefited significantly from AI integration. Machine learning models are being used to support early-phase planning by evaluating development scenarios, forecasting production potential, and minimizing economic risk. Neural networks assist in the ranking of development plans by simulating long-term behavior based on geological, geophysical, and operational variables. Horizontal shale wells, particularly in complex plays like the Permian Basin, are analyzed using hybrid approaches that combine pattern recognition, multivariate statistics, and geologic data to forecast EUR more accurately. Ensemble methods, fuzzy logic, genetic algorithms, and quantum-based optimizers are further applied to improve smart well performance, enhance conformance control, optimize injection and production strategies, and predict dynamic market indicators such as oil prices.

AI's application in well-related technologies is equally profound. Predictive models have been developed to estimate hydraulic properties in complex carbonate formations and to calculate water saturation in challenging lithologies, thereby reducing the need for expensive laboratory analysis. Deep learning algorithms enable detailed characterizations of remaining oil distribution during ultra-high water cut stages, and BP neural networks have been used to predict residual oil parameters across entire reservoir regions. Operational challenges like wax formation, flow control, leak detection, and cavitation are also addressed using machine learning, where models help diagnose early warning signs based on historical and real-time input. Advanced diagnostics for sucker rod pumping systems and gas kick detection are driven by real-time monitoring and data-driven anomaly recognition, improving safety and reducing unplanned downtime. Furthermore, intelligent control systems, including fuzzy controllers and optimization algorithms, regulate wellhead pressure and improve production stability.

In addition to the core areas of field operations and well planning, AI is increasingly employed in supporting domains that contribute to strategic decision-making. For instance, artificial neural networks have been used to model gas production in unconventional reservoirs, identify multiphase flow regimes, and predict liquid holdup in horizontal wells. Fiber-optic data integrated with machine learning techniques has enabled multilayer reservoir modeling, while convolutional and recurrent neural networks have been adopted to simulate reservoir behavior under complex production scenarios. Comparative studies between decline curve analysis and recursive neural networks illustrate how deep learning can outperform traditional methods in production forecasting. AI has also proven useful in lithology identification using semi-supervised learning, porous media characterization from 2D imaging, and early warning systems for production anomalies. Moreover, digital twins, Bayesian networks, and probabilistic AI models are facilitating better pipeline risk assessment and spatial prediction of oil and gas accumulations, providing a strategic advantage in asset management.

The convergence of these diverse applications clearly demonstrates that AI, ML, and DL are not merely supplemental tools but foundational enablers of the Fourth Industrial Revolution in the oil and gas industry. As data continues to grow in volume and complexity, the ability to harness it through intelligent models will remain essential for driving innovation, resilience, and sustainability across upstream operations.

9.4.1 Industrial Revolution 4.0

9.4.1.1 Application-1: Recurrent Auto Encoders for Automatic Sensor Validation; Tomorrows Data with Yesterday's Sensors

9.4.1.1.1 Background of the Application

This paper presents a novel approach for validating time-series data collected from multiple sensors at the wellsite using a DL-based method. The proposed method uses a type of Recurrent Neural Network (RNN) called Long Short-Term Memory (LSTM) organized into an Auto-Encoder (AE) architecture. The AE model is trained on unlabeled data, which is preprocessed to reduce noise and improve data quality. The model then reconstructs non-defective data when defective data are accepted as input. The anomalies can be identified by analyzing the input imperfect series with reconstructed perfect series. The process enables the calculation of reconstruction error, which helps identify anomalies and assist in system diagnostics, all of which can then be used for sensor validation.

9.4.1.1.2 Description of the Workflow and Machine Learning Model

In this method, an AE with LSTM networks is used to validate drilling rig sensor data. The LSTM networks capture temporal dependencies in the time-series data. The AE encodes input sequences into a compact form and then reconstructs them. The reconstruction error, the difference between the original and reconstructed data, is used to identify anomalies in the sensor readings.

9.4.1.1.3 Advantages and Limitations of the Synchronized Algorithm

The LSTM-based AE method manages sequential data and long-term dependencies, crucial for accurate anomaly detection in sensor data. It can handle variable-length sequences, providing flexibility. However, training these models require a lot of computational power and expertise in this area. The method's accuracy relies on the input data being high quality, and it may struggle with sudden changes or noise. Injecting artificial errors for testing is essential but challenging.

9.4.1.1.4 Results and Discussion

This paper proposes a DL-based approach for validating sensor data collected from drilling rigs, which typically suffer from missing data, noisy data, and collinearity. The proposed method uses a type of RNN called LSTM organized into an AE architecture to reconstruct non-defective data when defective data are accepted as input. The anomalies can be identified by analyzing the input imperfect series with reconstructed perfect series. The authors demonstrate that the proposed AE model can be generalized to any multivariate sensor system for diagnostic anomaly detection. Additionally, analyzing the context vector by clustering methods can help identify different states of the system, which can be marked as healthy or faulty based on previous knowledge. The authors conclude that this approach opens up possibilities for other areas of the industry to adopt this method for anomaly detection with unlabeled sensor observations.

BIBLIOGRAPHY

Gupta, Soumya, Chatar, Crispin, and Celaya, José R. "Recurrent Auto Encoders for Automatic Sensor Validation; Tomorrows Data with Yesterday's Sensors." Paper presented at the *IADC/SPE International Drilling Conference and Exhibition*, Galveston, Texas, USA, March 2020. doi: https://doi.org/10.2118/199615-MS

9.4.1.2 Application-2: Acceleration of 4IR Driven Digital Transformation through Pen Source: Methods and Parallel Industries Knowledge Reapplication in the Field

9.4.1.2.1 Background of the Application

The paper explores the potential of open-source software in accelerating the digital transformation of the oil and gas industry. The authors argue that companies relying solely on proprietary platforms for developing AI solutions risk being left behind, as the rapid development and adoption of AI in various sectors is largely driven by the global community of academicians and data scientists working on open-source platforms. The paper highlights the importance of embracing open-source platforms in the development of flexible and agile software that can leverage better methods that may emerge in the future. It also showcases examples of how the reapplication of AI methods from other industries can be useful in the oil and gas sector.

9.4.1.2.2 Description of the Workflow and Machine Learning Model

The author used DL models to analyze images and time-series data. The Convolutional Neural Network (CNN) extracts feature from images for classification. LSTM networks capture temporal dependencies in sequential data, ideal for sequence prediction. The A Sequence-to-Sequence (Seq2Seq) model, with LSTM layers, handles variable-length input and output sequences. Lastly, the Generative Adversarial Network (GAN) generates realistic images from latent space, aiding in synthetic data creation.

9.4.1.2.3 Advantages and Limitations of the Synchronized Algorithm

DL models like CNNs, LSTMs, Seq2Seq, and GANs handle complex data effectively. CNNs are excellent for image classification, LSTMs for time-series prediction, Seq2Seq for sequence mapping, and GANs for generating high-quality synthetic data. However, these models require significant computational power and large labeled datasets. They can also be complex and less interpretable. Despite these challenges, their high performance makes them essential in modern AI applications.

9.4.1.2.4 *Results and Discussion*

DL models like CNNs, LSTMs, Seq2Seq, and GANs handle complex data effectively. CNNs are excellent for image classification, LSTMs for time-series prediction, Seq2Seq for sequence mapping, and GANs for generating high-quality synthetic data. However, these models require significant computational power and large labeled datasets. They can also be complex and less interpretable. Despite these challenges, their high performance makes them essential in modern AI applications.

BIBLIOGRAPHY

Haroon, Sammy Sarblund, Viswanathan, Aruna, Alyamkin, Sergey, and Shenoy, Ramachandra. "Acceleration of 4IR Driven Digital Transformation Through Open Source: Methods and Parallel Industries Knowledge Reapplication in the Field." Paper presented at the *Offshore Technology Conference*, Houston, Texas, USA, May 2020. doi: https://doi.org/10.4043/30760-MS

9.4.1.3 Application-3: A Novel Correlation to Predict Gas Flow Rates Utilizing Artificial Intelligence: An Industrial 4.0 Approach

9.4.1.3.1 *Background of the Application*

This paper presents a study on predicting gas flow rates in a gas field containing ten wells using AI and ML techniques. The input parameters include flowing wellhead pressure, flowing wellhead temperature, choke size, condensate flow rate, and water flow rate. The study employs artificial neural networks (ANNs), functional networks (FNs), and ANFIS methods to develop a correlation that accurately predicts gas flow rates with an accuracy in excess of 90%. The results demonstrate that the proposed approach can provide an effective and accurate solution for gas flow rate prediction in gas fields.

9.4.1.3.2 *Description of the Workflow and Machine Learning Model*

ANN, FN, and Adaptive Neuro-Fuzzy Inference System (ANFIS) methods were used to predict gas flow rates. The ANN captures complex patterns between input parameters and gas flow rates through the multiple layers of neurons and activation functions. The FN leverages both domain knowledge and data to construct a network that approximately detects the functional relationships, providing robust predictions. The ANFIS combines NNs with fuzzy logic principles, forming an inference system that can adapt and enhances prediction accuracy by accommodating the data's inherent uncertainties and imprecisions.

9.4.1.3.3 *Advantages and Limitations of the Synchronized Algorithm*

The ANN method excels at capturing intricate nonlinear relationships within the dataset, offering high flexibility and strong performance across various scenarios. However, it is significantly computational expensive and can be prone to overfitting without proper regularization. The FN method benefits from integrating domain knowledge with data, allowing it to build reliable models with potentially less data, but it may not perform as needed with highly complex patterns without sufficient domain expertise.

9.4.1.3.4 Results and Discussion

In this study, AI- and ML-based approach was proposed to predict gas flow rates in a gas field containing ten wells. The proposed approach uses ANN, FN, and ANFIS methods to develop a correlation that accurately predicts gas flow rates. The accuracy of the proposed correlation is in excess of 90%, which demonstrates the effectiveness and accuracy of the proposed approach for gas flow rate prediction in gas fields. The proposed approach can provide an efficient and reliable solution for gas flow rate prediction, which can be used to optimize production and enhance the economic viability of gas fields.

BIBLIOGRAPHY

Kalam, Shams, Khan, Mohammad Rasheed, Tariq, Zeeshan, Siddique, Faisal Anwar, Abdulraheem, Abdulazeez, and Khan, Rizwan Ahmed. "A Novel Correlation to Predict Gas Flow Rates Utilizing Artificial Intelligence: An Industrial 4.0 Approach." Paper presented at the *SPE/PAPG Pakistan Section Annual Technical Symposium and Exhibition*, Islamabad, Pakistan, November 2019. doi: https://doi.org/10.2118/201170-MS

9.4.1.4 Application-4: Sustaining Crude Quality Specifications by Leveraging IR 4.0 Solution

9.4.1.4.1 Background of the Application

In order to maintain crude quality standards for the Saudi Aramco supply chain, which includes more than 50 gas–oil separation plants (GOSPs), pipelines, and terminals, this study proposes an integrated system that makes use of IR 4.0. Throughout the supply chain, big data analysis is necessary to maintain requirements like the water and salt contents for Arabian crude. The Crude Quality Monitoring Solution (CQMS), a customized solution created by Saudi Aramco to address this issue, uses over 800 vital data streams per minute, categorizes the data based on the degree of risk affecting crude quality standards, and sends out automated alerts for proactive corrective action. The paper also describes several examples where the solution notifications have proactively remediated process disturbances by up to 20% at upstream and downstream facilities while ensuring asset integrity. The solution deployment has also substantially improved the operational efficiency across the network by benchmarking critical data streams.

9.4.1.4.2 Description of the Workflow and Machine Learning Model

The author used a Random Forest classifier to monitor and predict crude oil quality by analyzing data from GOSPs. The classifier evaluates various parameters such as pressure, flow, and temperature to detect potential risks affecting crude quality. Historical data is used to train the model, enabling it to identify patterns and forecast future issues. The primary aim is to enable timely interventions and maintain high crude quality standards by continuously assessing incoming data streams.

9.4.1.4.3 Advantages and Limitations of the Synchronized Algorithm

The Random Forest classifier provides notable advantages for monitoring crude oil quality. It effectively captures complex patterns and relationships within the data, crucial for accurate predictions. Additionally, the model is robust against data variations like missing values or outliers. However, this method also presents challenges, including the need for significant computational resources and the time required for training and execution, particularly with large datasets. Furthermore, the model's complexity can make interpreting results difficult, posing challenges in explaining the decision-making process to stakeholders. Despite these limitations, the predictive capabilities of the classifier are essential for proactive crude quality management.

9.4.1.4.4 Results and Discussion

The CQMS has demonstrated its effectiveness in maintaining crude quality specifications across the Saudi Aramco supply chain. By analyzing big data in real-time and utilizing ML and AI, the solution has enabled the company to proactively remediate process disturbances, improve operational efficiency, and ensure asset integrity. With the continuous enhancement of the solution capabilities, Saudi Aramco is maximizing the potential of its crude network and optimizing operations across the entire value chain.

BIBLIOGRAPHY

Ahmed, Syed, Al-Zubail, Ahmad, Al-Jeshi, Majed, Yousef, Khaled, Musabbeh, Alya, Mousa, Saad, Bukhari, Adeeb, Seraihi, Emad, and Alamri, Sultan H. "Sustaining Crude Quality Specifications by Leveraging IR 4.0 Solution." Paper presented at the *Abu Dhabi International Petroleum Exhibition & Conference*, Abu Dhabi, UAE, November 2019. doi: https://doi.org/10.2118/197577-MS

9.4.1.5 Application-5: Digital Production Optimization Program

9.4.1.5.1 Background of the Application

The application of the Digital Production Optimization program in the oil and gas sector is covered in this article. Through forecasts, operations, and hydrocarbon accounting, the program seeks to attain end-to-end production excellence. To meet production goals and lower the volume of deferments, the program makes use of real-time tracking, remote monitoring, and predictive analytics. Additionally, the program records events, production data, and well data to facilitate improved operational and financial performance analysis. Through improved accuracy, analysis, and collaboration, the program reduces costs. The application uses strict business procedures, powerful analytics, and high-fidelity input data to support decision-making. The program has several value drivers and benefit ranges, including production increase up to 3%, deferment reduction of 2–6%, Operations and Maintenance (O&M) cost reduction of 1–2%, and IT cost reduction of 10–20%.

9.4.1.5.2 Description of the Workflow and Machine Learning Model

This paper discusses the implementation of a comprehensive digital production optimization program aimed at enhancing operational excellence in unconventional resource production. The system integrates real-time tracking, remote monitoring, and predictive analytics to optimize production processes, reduce deferment, and improve cost-efficiency. The optimization process involves several interconnected components, including hydrocarbon allocation, production forecasting, and deferment management. Advanced analytics and ML models are employed to forecast production volumes and identify potential deferment risks. These models are trained on historical production data and sensor readings from wells to accurately predict future production outcomes. The system also supports automated field data capture and enterprise-wide reporting, ensuring accurate hydrocarbon accounting and revenue allocation.

9.4.1.5.3 Advantages and Limitations of the Synchronized Algorithm

The digital production optimization program provides a robust framework for improving production efficiency and reducing operational risks. The use of predictive analytics allows for more accurate production forecasting, enabling proactive decision-making and timely intervention in case of potential deferment. Real-time data integration enhances the system's ability to track production performance and identify inefficiencies across the network. However, the system's effectiveness is heavily reliant on the quality and availability of real-time data from sensors and wells. Inconsistent or missing data could lead to inaccurate predictions and suboptimal decision-making. Additionally, the complexity of integrating various data sources and ensuring the system's scalability across multiple assets poses challenges that require continuous improvement and monitoring.

9.4.1.5.4 Results and Discussion

In conclusion, the implementation of the Digital Production Optimization program has enabled the oil and gas industry to achieve production targets, reduce deferment volume, and drive cost efficiencies. The program has enabled better collaboration, analysis, and accuracy by capturing well data, production data, and events. The program has several value drivers and benefit ranges, including production increase up to 3%, deferment reduction of 2–6%, O&M cost reduction of 1–2%, and IT cost reduction of 10–20%. The program's success demonstrates that the use of real-time tracking, remote monitoring, and predictive analytics can lead to significant improvements in operational and financial performance.

BIBLIOGRAPHY

Hamad, Mohammed, Bawardi, Manal, Noaimi, Reem, Lahoti, Vishal, and Anderson, Michel. "Digital Production Optimization Program." Paper presented at the *SPE Middle East Oil and Gas Show and Conference*, Manama, Bahrain, March 2019. doi: https://doi.org/10.2118/194857-MS

9.4.2 Oil Field Development, Planning, and Optimization

9.4.2.1 Application-6: A Machine Learning Application for Field Planning

9.4.2.1.1 Background of the Application

The topic of well site planning is addressed in the paper by presenting a field planning framework based on ML. To create a prediction tool for well productivity, unsupervised learning is utilized to comprehend the data's features. A regression model is then developed and trained on the data that is available in a marginal oilfield. Using an adaption of modern portfolio theory (MPT), this data-based prediction tool is employed in a workflow to optimize well placements under uncertainty and assess suggested well sites. Both processes can be used in plays with limited well drainage zones. The framework's ability to manage big datasets makes it appealing to shales.

9.4.2.1.2 Description of the Workflow and Machine Learning Model

The Radial Base Function (RBF) network was used to create the regression model. This is a means of unsupervised learning. The most important part of the problem is the decision for the portfolio. The decisive factor in this is the comparison and trade-off between the advantages and disadvantages of well selection zones and the benefits. MPT is applied to this analysis; however, unlike financial analysis, the drilling exploration problem only makes a distinction between whether or not a well is drilled and no other consideration, i.e., it simply outputs a Boolean value. RBFs are used as activation functions in RBF networks, which are ANNs used in ML. A linear combination of the RBFs of the inputs and neuron parameters is the network's output. There are numerous applications for RBF networks, such as system control, classification, time-series prediction, and function approximation.

9.4.2.1.3 Advantages and Limitations of the Synchronized Algorithm

The advantage lies in the full use of the original and constructed variables to obtain more objective experimental results. In this paper, several constructed variables were used to help push the logic progress. There are constructed variables like Interquartile Range (IQR), Reservoir Capacity, and Estimated Ultimate Recovery (EUR). Those variables not only play roles in the early stage such as data sorting and collecting, but also provides basic logic direction to deeper analysis and the construction of framework. It can also be observed that different methods were used, including Principal Component Analysis (PCA) to reduce certain variables. That is another advantage of the RBF.

The limitation is that the RBF is comparatively complicated when building the code, especially when trying to input certain variables into universal templates. There might be deviation and high variance if the RBF is wrongfully used without proceeding test and preparation.

9.4.2.1.4 Results and Discussion

The workflows presented in this paper provide a systematic process to optimize well locations and evaluate locations in a given plan. The workflows explicitly consider uncertainty resulting from well productivity and use data typically acquired in the field, but not used currently for field planning. The regression model developed in this paper provides a valuable integration of data from multiple sources, which is a valuable activity not typically undertaken in organizationally siloed operators. This integration activity often accords an opportunity to provide QC and validation of data. By using ML-based models, the paper provides a faster and less expensive alternative to traditional static and dynamic models for field planning.

BIBLIOGRAPHY

Kumar, Amit. "A Machine Learning Application for Field Planning." Paper presented at the *Offshore Technology Conference*, Houston, Texas, May 2019. doi: https://doi.org/10.4043/29224-MS

9.4.2.2 Application-7: Application of Artificial Intelligence in Oilfield Operation and Intervention

9.4.2.2.1 Background of the Application

The authors of this research suggest an AI method for keeping an eye on and assisting with oilfield activities. The performance of operational operations is deciphered using real-time data obtained from remote and difficult oilfields. To obtain the necessary intelligence, the authors combine the robustness of fuzzy logic reasoning, the learning capacity of NNs, and the meta-heuristic optimization searching speed of an improved method. This enables proactive and anticipatory intervention in the management and operations of oilfields.

9.4.2.2.2 Description of the Workflow and Machine Learning Model

The hybrid application of metaheuristic optimization algorithms and AI is the basic background and the main technical tool of this paper. The most important feature of this ENFS-related paper lies in the application of fuzzy systems to NNs. In the meantime, unsupervised learning is the main tone of this model application. The front end of the framework maintains the basic structure of the NN, and the fuzzy system accesses after the NN for information filtering and optimization.

A collaborative body has been formed by combining fuzzy logic (FL), neural network (NN) theory, GAs, random reasoning (PR), confidence networks, chaos theory, and partial learning theory. the cooperation of different approaches. The approaches are complimentary rather than competitive in this regard. All of the different approaches have different functions in a collaborative body. This partnership leads to the development of hybrid intelligent systems. The two soft computing techniques—fuzzy logic and NNs—are both significant intelligent control techniques. By merging them to enhance each other's advantages and disadvantages, a collaborative body known as a fuzzy NN is created.

9.4.2.2.3 Advantages and Limitations of the Synchronized Algorithm

The advantage of the fussy method is that it combines the strengths of NN system and fuzzy system, it has great superiority in dealing with nonlinearity and fuzziness, and has great potential in intelligent information processing; which makes more and more experts and scholars devote themselves to this field and make fruitful research results.

The limitation is that although the overall interpretability of the model is somewhat enhanced compared to the use of NNs alone, it is still more difficult to understand the process of the model and the nature of the causal links between variables.

9.4.2.2.4 Results and Discussion

The upstream oil and gas industry have invested heavily in new technologies to address the challenges associated with oilfield operation, intervention, and production optimization. However, there is a need to understand the dynamic nature of the industry and adopt the multitude of emerging technologies capable of meeting these demands. Using real-time data from difficult and remote oilfields, the authors suggest an AI method for tracking and assisting in oilfield activities. To obtain the necessary intelligence, the method combines the robustness of fuzzy logic reasoning, the learning capacity of NNs, and the meta-heuristic optimization searching speed of an improved algorithm. This enables proactive and anticipatory intervention in the management and operations of oilfields.

BIBLIOGRAPHY

Akanji, L. T., and Ofi, I. "Application of Artificial Intelligence in Oilfield Operation and Intervention." Paper presented at the *SPE Intelligent Energy International Conference and Exhibition*, Aberdeen, Scotland, UK, September 2016. doi: https://doi.org/10.2118/181116-MS

9.4.2.3 Application-8: Neural Network Method for Comprehensive Evaluation of Oil Field Development Planning Scheme

9.4.2.3.1 Background of the Application

The choice of oil field development plan is related to the proper economic production and effectiveness of technology implementation. However, the opposing and agreeing indicators and variables under the influence of multiple factors make it difficult to rely on individual variables for comprehensive selection and evaluation. With the application and background of the proposed fuzzy decision-making method, fuzzy integral evaluation method and gray multi-objective method as well as fuzzy hierarchical weighted analysis method, it is necessary to propose an evaluation model that is free from excessive randomness and subjectivity. The NN theory has the inherent nature of quantitative and qualitative reference indicators imitating human brain functions and thinking patterns. In this paper, a formal attempt is made to construct such a model to achieve a professional evaluation and provide a comprehensive and effective analysis, aiming to obtain objective results as much as possible.

9.4.2.3.2 Description of the Workflow and Machine Learning Model

Firstly, the overall framework of this paper is a three-layer back propagation (BP) NN with multiple input units, multiple hidden units and single output units, the output of which is used for comprehensive evaluation of oil field development. The data are first normalized and divided into cost-based indicators, benefit-to-benefit and moderate indicators. And for the indicators that are not predictable and quantifiable and can only be evaluated qualitatively, the evaluation scale is adopted to transpose the matrix for comprehensive evaluation. Then enter the construction of the model, determine the set of indicators and vector theoretical domain and then input the desired output vector of the attribute matrix X. The Gnu indicator type transformation variables are then used to initiate learning.

9.4.2.3.3 *Advantages and Limitations of the Synchronized Algorithm*

The output of the NN is almost identical to that obtained by the fuzzy hierarchical evaluation method, which shows that the learned NN has well acquired and stored the knowledge, experience and judgment of the evaluation experts. By storing the ANNBP model in the knowledge base, the model can be used for comprehensive evaluation of other development solutions. The results can be obtained immediately by simply entering the matrix of values of the index attributes of the case to be evaluated. It can be seen that the application of the NN method for comprehensive evaluation of oilfield development planning is fully feasible.

The disadvantage of the model may be that the parameters are more complicated to adjust and require careful debugging. Also, the standardization and quantification of traditional evaluation metrics is important.

9.4.2.3.4 *Results and Discussion*

As a conclusion, the NN approach can emphasize the interdependence between data and has a strong adaptive nature. Also, the model is very modifiable and the NN approach is simpler and does not require large-scale adjustment of relevant parameters. The NN method can also better simulate the evaluation process of experts has a broad application prospect.

BIBLIOGRAPHY

Kang, X. J., Yuan, A. W., Gao, J. et al (2006). Neural network method for comprehensive evaluation of oilfield development planning scheme. *Spec Oil Gas Res* 13(2):48–50

9.4.2.4 Application-9: Horizontal Shale well EUR Determination Integrating Geology, Machine Learning, Pattern Recognition and Multi Variate Statistics Focused on the Permian Basin

9.4.2.4.1 *Background of the Application*

In order to ascertain an area's actual economic potential, this study offers a data-intensive methodology for identifying the type-curve and EUR/well for that location. Using a variety of metrics from adjacent horizontal shale wells as inputs, the methodology combines geology, ML, pattern recognition, and statistical analysis. The process makes use of local geological data, ML to pinpoint crucial well parameters that improve output, and pattern recognition as a quality assurance check. The existing research on the relationship between well productivity and other well characteristics is used to validate the conclusions. In order to pick wells with the best completions, type-curves were created for every Wolfcamp bench in the area of interest.

9.4.2.4.2 *Description of the Workflow and Machine Learning Model*

ANN is used in this article to obtain the research findings. In order to discover crucial well parameters that result in improved output, this work first applies ML to local geological data. After ensuring that the number of wells in each category is statistically significant, a pattern recognition step is carried out. Because the conclusion is not based only on ML results, this also acts as a QC measure. The available research on the relationship between well production and other well parameters is used to validate the conclusions. To create a type curve for the target zone or zones in the area of interest, wells with the best (controllable) parameters are chosen.

A hardware and/or software system that mimics how neurons function in the human brain is called an ANN. ANNs are a type of DL technology that is also referred to as AI. Many processors working in parallel and organized in tiers are often used in an ANN. Similar to optic nerves in human visual processing, the first tier gets the raw input data. Similar to how neurons farther away from the optic nerve receive signals from those closer to it, each succeeding layer receives the output from the tier before it instead of the raw input. The system's output is generated by the last tier.

9.4.2.4.3 Advantages and Limitations of the Synchronized Algorithm

The advantage of the ANN is that it has the self-learning function, which is of particular importance for prediction. It is expected that the future ANN computer will provide economic prediction, market prediction, and benefit prediction for human beings, and its application future is very ambitious.

The limitations of the ANN is that the interpretability of NNs is relatively poor, and a relatively reliable and predictive result may be obtained after comparison, but the correlation between the independent and dependent variables is very ambiguous. In addition, problems such as gradient explosion can easily arise if reasonable regularization is not used in the prediction process. There is a certain complexity and expertise in the tuning and application of parameters of NNs.

9.4.2.4.4 Results and Discussion

This study prescribes a novel scientific data-intensive methodology to identify the type-curve and EUR/well for an area, thereby determining the area's true economic potential. The methodology integrates geology, ML, pattern recognition, and statistical analysis using various parameters of nearby horizontal shale wells as inputs. The study emphasizes the importance of tying in the controllable well parameters to well production for the optimum evaluation of shale assets. Once this relationship is established, the type-curve determination and the EUR estimation can be done more accurately. The methodology presented in this paper can be used to make the type-curve/EUR determination process scientific, systematic, and seamless, and it can identify the area of interest's true economic potential with regards to horizontal shale well development.

BIBLIOGRAPHY

Gaurav, Abhishek. "Horizontal Shale Well EUR Determination Integrating Geology, Machine Learning, Pattern Recognition and MultiVariate Statistics Focused on the Permian Basin." Paper presented at the *SPE Liquids-Rich Basins Conference - North America*, Midland, Texas, USA, September 2017. doi: https://doi.org/10.2118/187494-MS

9.4.2.5 Application-10: Smart-Well Production Optimization Using an Ensemble-Based Method

9.4.2.5.1 Background of the Application

This paper discusses the application of ensemble-based optimization methods in two complex problems related to production optimization. In the first problem, the optimization objective is to minimize cumulative water production over a 10-year production period while maintaining a constant liquid-production rate. The second problem involves multilateral wells with intelligent completion valves (ICVs) installed at each lateral junction, and the optimization algorithm was successful in reducing water production. The authors demonstrate that a proper choice of control variables can be important to the success of the optimization.

9.4.2.5.2 Description of the Workflow and Machine Learning Model

In this method, an ensemble optimization is used to iteratively correct control variables. Starting with first controls for an ensemble of members, the algorithm changes these controls and evaluates an objective function for each changed set. The cross-covariance between the controls and their objective values is calculated to adjust the mean controls. This iterative process continues until convergence, aiming to optimize the objective function through systematic corrections.

9.4.2.5.3 Advantages and Limitations of the Synchronized Algorithm

Ensemble optimization has numerous advantages such as exploring a broad range of solutions and improving accuracy. They achieve this through continuous correction. It is effective in handling complex, multi-dimensional problems, using the power of ensembles to find optimal solutions. However, the method requires significant computational resources and depends heavily on well-defined initial controls and objective functions. The iterative nature of the process can be heavy on resources, and the success of the optimization is dependent on the accuracy of the initial setup and the definition of the objective function.

9.4.2.5.4 Results and Discussion

The authors have successfully demonstrated the application of ensemble-based optimization methods in two complex production optimization problems. The results show that this approach can be relatively efficient and effective in reducing water production and improving reservoir performance. The authors emphasize the importance of considering the interaction between wells in the total field when computing the optimal control-valve settings. The ensemble-based approach can be an attractive option, especially for large-scale problems where other methods may not be practical.

BIBLIOGRAPHY

Su, Ho-Jeen, and Oliver, Dean S. (2010). Smart-well production optimization using an ensemble-based method. *SPE Res Eval & Eng* 13, 884–892. doi: https://doi.org/10.2118/126072-PA

9.4.2.6 Application-11: Production-System Optimization of Gas Fields Using Hybrid Fuzzy/Genetic Approach

9.4.2.6.1 Background of the Application

Because of the nonlinear character of the problem and the intricate relationships between operational parameters, optimizing gas production systems is a challenging undertaking. The objective and constraint functions of conventional approaches, which are typically expressed in exact mathematical forms, are unable to account for uncertainties related to imprecise or ambiguous information. To account for these uncertainties, a fuzzy nonlinear programming technique is presented in this study and applied to various optimization procedures. In particular, the fuzzy h-formulation is used in conjunction with a hybrid coevolutionary GA to optimize the design of gas production systems. A viable fuzzy domain is used to formulate the many competing objective and constraint functions for gas field production systems.

9.4.2.6.2 Description of the Workflow and Machine Learning Model

In this method, a GA was used to optimize decision variables. The process begins by initializing a population with random values within specified limits. It employs exponential ranking selection to choose individuals based on their fitness scores. Selected individuals undergo a mixture, where parts of their decision variables are exchanged to create children. These children also include mutation, where random changes are introduced to continue the diversity. The algorithm iteratively improves the population by selecting, crossing, and mutating individuals to optimize the objective function.

9.4.2.6.3 Advantages and Limitations of the Synchronized Algorithm

GA is preferred in this method due to its robustness in searching large and complex spaces for optimal solutions. It effectively continues diversity in the population, reducing the risk of early convergence. However, the method can be computationally expensive, particularly for large populations or many generations. The success of the algorithm depends on correctly setting parameters like rates of mutation and crossover. Additionally, GAs may require significant fine-tuning and multiple runs to achieve the best results, which can need a long time.

9.4.2.6.4 Results and Discussion

In the planning and optimization of gas production processes, the method suggested in this research is a potent instrument for cost savings. The synthetic-optimization approach can provide a new option that is significantly better than the current conventional procedures while also identifying a global compromise solution. A synthetic real-world example that produces better outcomes validates the suggested approach's dependability.

BIBLIOGRAPHY

Park, Hui-June, Lim, Jong-Se, Roh, Jeongyong, Kang, Joo M., and Min, Bae-Hyun. (2010). Production-system optimization of gas fields using hybrid fuzzy/genetic approach. *SPE J* 15, 417–425. doi: https://doi.org/10.2118/100179-PA

9.4.2.7 Application-12: Predicting Oil Price Movements: A Dynamic Artificial Neural Network Approach

9.4.2.7.1 Background of the Application

This paper discusses the challenges in predicting oil prices due to the complex interactions between economic and other factors. To address this issue, the authors develop a dynamic ANN approach known as Nonlinear Auto-Regressive model with eXogenous input (NARX). The NARX model is optimized by identifying dummy variables, which help the inclusion of qualitative factors and time delays. The authors compare the results of the NARX model to those of time-series and static ANN models, and find that the NARX model is more accurate in predicting oil prices and shocks. The authors conclude that a better understanding of the likely future behavior of oil prices can reduce vulnerability of the economy from fluctuations and changing conditions in the oil market.

9.4.2.7.2 Description of the Workflow and Machine Learning Model

This paper presents a dynamic approach to predicting oil price movements using ANNs, with a particular focus on NARX. The NARX model is designed to capture the time dependencies in oil price data by considering both past values of the oil price and external factors, such as macroeconomic indicators. The study employs a multi-step forecasting strategy to predict future oil prices, leveraging historical data and exogenous variables. The model is trained on a large dataset of historical oil prices and relevant economic indicators, and its performance is compared with traditional time-series models such as Autoregressive Integrated Moving Average (ARIMA) and static ANNs. The ultimate goal is to improve the accuracy of oil price forecasts by incorporating dynamic modeling techniques that account for the complex relationships between oil prices and external influences.

9.4.2.7.3 Advantages and Limitations of the Synchronized Algorithm

The use of a NARX model offers several advantages in predicting oil price movements. By incorporating time dependencies and external factors, the model provides more accurate forecasts compared to traditional time-series models. The dynamic nature of the NARX model allows it to capture the non-linear relationships between oil prices and influencing factors, making it well-suited for complex economic systems. Additionally, the model's flexibility in incorporating various external variables enhances its predictive power.

However, the NARX model's performance is heavily dependent on the quality and quantity of input data. Insufficient or noisy data can lead to inaccurate predictions, and the model may struggle to generalize when faced with sudden market shifts or external shocks. Moreover, the training process can be computationally expensive, particularly when optimizing the model's hyperparameters. Despite these challenges, the NARX model demonstrates significant potential in forecasting oil prices, especially when used in conjunction with other predictive methods.

9.4.2.7.4 Results and Discussion

In order to increase the modeling and precision of oil price and shock forecasts, the authors of this research create a dynamic ANN technique called NARX. By defining dummy variables that facilitate the addition of qualitative aspects and time delays, the NARX model is optimized. The authors discover that the NARX model is more accurate in forecasting shocks and oil prices when compared to time-series and static ANN models. The authors concluded that their strategy can be applied to lessen the economy's susceptibility to changes and oscillations in the oil market.

BIBLIOGRAPHY

Godarzi, A. A., Amiri, R. M., Talaei, A., & Jamasb, T. (2014). Predicting oil price movements: A dynamic Artificial Neural Network approach. *Energy Policy*, 68, 371–382. ISSN 0301-4215, doi: 10.1016/j.enpol.2013.12.049

9.4.2.8 Application-13: Forecast Model of Oilfield Indexes Based on IPSO GNN

9.4.2.8.1 Background of the Application

The combined prediction model improved particle swarm optimization (IPSO) algorithm and gray neural network (GNN) (IPSO–GNN) is proposed for the oilfield development index prediction problem, and the a and u parameters of GNN are optimized by IPSO, which improves the shortcomings of GNN and effectively ensures the prediction accuracy. The water content of the oil field development index is used as an example. The simulation results show that the prediction accuracy of this model is higher than that of the gray prediction model, GNN and BPNN model, and it also shows the feasibility and effectiveness of this method. It also shows the feasibility and effectiveness of this method.

9.4.2.8.2 Description of the Workflow and Machine Learning Model s

9.4.2.8.2.1 GNN

Gray system theory is a prediction method proposed by Professor Deng Julong of the Huazhong University of Science and Technology for dealing with inaccurate data, partly known and partly unknown. In the process of oil field development and production, due to the diversity of its development indicators themselves, and the interconnection and constraints between each indicator to be measured and its influencing factors, in this many and in such a large number of complex factors and relationships, there will be partly known and partly unknown, so it can be regarded as a gray system. The most widely used model in gray systems is the Gray Model (GM) GNN, which is based on the idea of solving the gray model and optimizing its specific parameters through the unique nonlinear mapping capability of NNs. The different values of a and u in the gray GM model will lead to different prediction values. The NN has a good nonlinear reflection ability, and it continuously corrects the weights a and u through the input and output. The problem of inaccurate predictions due to different values of a and u can be solved effectively.

9.4.2.8.2.2 IPSO

The traditional particle swarm optimization (PSO) algorithm is a population intelligence optimization algorithm in addition to the fish swarm and ant colony algorithms, which was proposed by Kennedy and Eberhart in 1995. This algorithm originated from the predatory behavior of birds and was inspired by their predatory behavior characteristics to solve optimization problems. The PS initializes a population of particles in a solvable space, where each particle represents a potential optimal solution to the optimal problem. The particle is characterized by its fitness value, position, and velocity. The fitness of a particle is reflected by its strength or weakness, which is obtained by solving the fitness function. When a particle moves in the solution space, its position needs to be updated according to the individual extremum pbest and the population extremum gbest. When the position of a particle is updated, the fitness value is recalculated and the position of the particle is updated based on the individual and group extremes and the fitness value of the new particle, pbest and gbest. At the same time, in order to maintain the diversity of the population, the mutation operation is considered for introduction in the traditional particle swarm optimization algorithm: each time a particle is updated, it is not necessary to set the initialization of the particle, but to do it with the particles can jump out of the previously searched local optimal solution position and obtain a larger search space, thus enabling the iterative process can effectively expand the previously shrinking population space and increase the population diversity. This allows the iterative process to effectively expand the previously shrinking population space and increase the population diversity.

9.4.2.8.3 Advantages and Limitations of the Synchronized Algorithm

The lineage of this paper is purpose-oriented shaped, and the multifaceted exploration of the variability of the indicator factors in the oilfield development process is carried out to model the process. The combination of the numerical simulation technology and reservoir engineering technology for prediction breaks the conventional pipeline-fitting prediction process, and parallel simulation gives a new way of thinking about oilfield indicator prediction, but this paper hopes to explore further in generality. Gray theory and NN have been developed for many years and have corresponding mature theoretical systems. Gray theory has better small sample data processing ability and NN has excellent nonlinear mapping ability, which makes them both widely used in information processing, but due to the different characteristics of both, their respective actual data processing results are sometimes not ideal. However, due to their different characteristics, the actual data processing results of each are sometimes not very satisfactory. The effective combination of the two methods to build a combined GNN model can make up for the shortcomings of a single model, thus obtaining better data processing capability and higher prediction accuracy.

To this end, an improved PSO algorithm is proposed based on the traditional PSO algorithm, in which the inertia weights w is optimized and the variational operation in GA is introduced to establish the improved particle swarm optimization (IPSO) algorithm, and finally the parameters a and u in GNN are optimized with this algorithm, so that the IPSO–GNN combined prediction model is established. The experimental results show that the prediction accuracy of the combined IPSO–GNN-based model is higher than that of each individual model.

9.4.2.8.4 Results and Discussion

In order to effectively solve the index prediction in the process of oil field development, this paper integrates gray theory, NN and PSO algorithm. Firstly, the IPSO algorithm is proposed, which introduces the genetic variation operator to improve the inertia weight (w) of PSO. The inertia weight (w) of PSO is improved, and then the IPSO–GNN model is proposed. The IPSO–GNN model is proposed to optimize the a and u parameters in the GNN model by the IPSO algorithm. The water content of the oil field development index is taken as the prediction object. The experimental results show that the IPSO–GNN model has high prediction accuracy, and this method is effective and feasible for the field development index. The experimental results show that the IPSO–GNN model has high prediction accuracy and this method is effective and feasible, which can provide new insights and ideas for oilfield index prediction. The results show that the IPSO–GNN model has high prediction accuracy, and this method is effective and feasible, which can provide new insights and ideas for oilfield index prediction.

BIBLIOGRAPHY

Kennedy, J., and Eberhart, R. "Particle Swarm Optimization." *Proceedings of ICNN'95 - International Conference on Neural Networks*, Perth, WA, Australia, 1995, pp. 1942–1948 vol. 4. doi: 10.1109/ICNN.1995.488968

Yan, H. Y., Fu, J. Y., and Dong, J. H. et al (2014). Forecast model of oilfield indexes based on IPSO–GNN. *Sci Technol Eng* 14(15):197–202.

9.4.2.9 Application-14: Oilfield Development Indicators Prediction Based on Radial Basis Process Neural Network

9.4.2.9.1 Background of the Application

So far, the existing methods for predicting oilfield development indexes can hardly reflect the influence of the actual time accumulation effect on the index prediction. Therefore, in order to improve the accuracy of oilfield development index prediction, this paper proposes a radial-based process neuron network-based oilfield development dynamic index prediction model and applies it to the prediction of the actual oilfield development dynamic index. The results of the case study show that the proposed radial basis process neuron network is a practical method for predicting oilfield development indexes with high accuracy and high speed.

9.4.2.9.2 Description of the Workflow and Machine Learning Model

The network is a feedforward NN that achieves nonlinear mapping by varying the parameters of the nonlinear transformation function of neurons and improves the network by linearizing the connection weight adjustment. The network is a feedforward NN. In recent years, it has been widely used in pattern recognition, data mining, etc.

The radial basis process neuron network model is a 3-layer forward structure. The first layer is the input layer, which consists of the signal source nodes and completes the input of the time-varying function to the network; the middle layer is the hidden layer of the radial-based process neuron, and the transformation function is the radial-based kernel function, and the kernel center is also a time-dependent function; the third layer is the output layer, which responds to the input pattern of the process interval. Since the radial basis process neuron has to complete the aggregation operation on time, the computational process is more complicated. For this reason, a set of function orthogonal bases is introduced in the input space, and the input function is represented as the expansion form of this set of orthogonal bases, so that the orthogonality of the basic functions can be used to simplify the aggregation operation process.

9.4.2.9.3 Advantages and Limitations of the Synchronized Algorithm

The fuzzy equivalence clustering and correlation analysis method is used to select the monthly oil production and liquid production as the output development index. After screening by fuzzy clustering analysis, the monthly oil and liquid production are selected as the output development indexes, and the remaining geological reserves, the total number of recovered wells, the monthly injection and extraction ratio, the total kernel function, the number of injection wells started, the number of new wells started, and the number of effective wells of old wells are taken as the input factors. The historical values of oil and fluid production and influencing factors in the whole A3 oilfield during the period of exceptionally high-water content from 1990 to 2013 were selected as samples. The data from 1990–2011 were used as 22 training samples, and that from 2012–2013 were used as 2 test samples.

9.4.2.9.4 Results and Discussion

Since there are many factors affecting oil field production, and the relationship between them is usually nonlinear, uncertain and complex, and it is difficult for the traditional oil field development prediction methods to reflect the impact of the time accumulation effect on the actual existence of indicators, this paper proposes the prediction of oil field development indicators based on radial basis process neuron network. This paper proposes a radial-based process neuron network for predicting oilfield development indexes. The analysis shows that the model has strong approximation ability and greatly improves the prediction accuracy.

BIBLIOGRAPHY

Xu, S. H., Bi, C. C., Zhang, Y. et al (2015). Oilfield development indicators prediction based on radial basis process neural network. *Comput Technol Autom* 3, 52–54.

9.4.2.10 Application-15: Hierarchical Optimization Research Based on Fuzzy Clustering Analysis

9.4.2.10.1 Background of the Application

China has the largest number of oil fields developed with water injection in the world. Among the current main oil fields, the eastern oil fields are increasingly characterized by high water injection as water injection development continues, especially with increasing differences in the degree of mobilization between layers. The reorganization of the layer system can obviously avoid inter-layer interference and improve the degree of recovery of the field to reduce the water content. In this paper, the H oilfield is used as an example, having been developed for a long time and currently in the high water-bearing stage. The regrouping of the layer systems is done to reduce the interlayer conflicts and to improve the degree of water recovery. The layer system optimization methods include the fuzzy cluster analysis method, gray correlation analysis method, and NN algorithm. This paper uses the gray correlation analysis method to determine the evaluation index, applies the fuzzy cluster analysis method to optimize the layer system, evaluates multiple influencing factors of the reservoir, and finally gets a comprehensive evaluation index and classifies the reservoir accordingly.

9.4.2.10.2 Description of the Workflow and Machine Learning Model

Firstly, fuzzy cluster analysis is performed, which means that a fuzzy mathematical language is used to describe and classify things according to certain requirements. The fuzzy relationship between samples is determined quantitatively using fuzzy mathematical methods. After performing data preprocessing, a fuzzy similarity matrix is established and then cluster analysis is performed. After that, fuzzy cluster analysis is used to study the optimization of the high water-bearing stage layer system. In the optimization session using fuzzy cluster analysis, the correlation of each index was analyzed by using the gray correlation analysis method. Also, the degree of extraction reflects the degree of reservoir goodness to a large extent. The closer the correlation degree is to 1, the greater the influence of the indicator on the degree of recovery. Finally, the optimization scheme of the formation system is given based on the numerical simulation study. According to the characteristics of oil and gas water distribution, the well network encryption is adjusted.

9.4.2.10.3 Advantages and Limitations of the Synchronized Algorithm

The advantage of fuzzy analysis is that it can pre-analyze the correlation of indicators, which helps the subsequent model building. Fuzzy evaluation deals with fuzzy evaluation objects through precise numerical means, and can make a more scientific, reasonable, and close to the actual quantitative evaluation of the information that contains information presenting fuzziness. The evaluation result is a vector, not a point value, and contains richer information, which can not only portray the evaluated object more accurately but also be further processed to obtain reference information. The calculation's complexity and the indicator weight vector's extremely subjective determination are its limitations. The relative affiliation weight coefficient is frequently modest and the weight vector does not match the fuzzy matrix R when the indicator set U is higher, i.e., the number of indicator sets were larger, within the condition constraint that the weight vector and is 1. The outcome will be an extremely fuzzy phenomenon with low resolution, making it impossible to determine which affiliation is higher. This could potentially lead to evaluation failure, which the hierarchical fuzzy evaluation approach can help to improve.

9.4.2.10.4 Results and Discussion

According to the physical properties' information of each small layer of the reservoir and the remaining reserve abundance distribution, the fuzzy clustering analysis was applied to the layers with similar reservoir properties. The reservoir was initially divided into two sets of development formations for development. This optimization method proved to be effective. On the basis of layer system optimization, the remaining oil saturation and the remaining reserve abundance of each small layer were combined. and residual reserves, three sets of well network reconfiguration schemes were developed. and predicted the indexes of different schemes by numerical simulation. The results of the application show that the optimized system exhibits a high level of performance, and it is highly comprehensive, scientific, and has certain guiding significance for reservoir development.

BIBLIOGRAPHY

Feng, G. Q., Pan, L. Y., Kong, B. et al (2018) Hierarchical optimization research based on fuzzy clustering analysis. *Eval Dev Oil Gas Reserv* 3, 30–39.

9.4.2.11 Application-16: Measures Optimization for Oil and Water Well Based on Quantum Particle Swarm Optimization

9.4.2.11.1 Background of the Application

China has the largest number of oil fields developed with water injection in the world, and the eastern oil fields, which are the main oil fields, are an important base for water injection extraction. However, as this water injection development proceeds, many fields enter a period of high-water content with large differences in the degree of mobilization between layers. In contrast, layer reorganization can improve this non-homogeneity, avoid inter-layer interference and reduce the water yield. Additionally, this paper conducts a specific analysis for the H oilfield, where the longitudinal inhomogeneity is very severe, leading to a diversity of remaining oil distribution. In order to improve the effect, this paper uses gray correlation analysis to determine the evaluation index, applies fuzzy cluster analysis to optimize the formation system, evaluates multiple influencing factors of the reservoir, and finally obtains a comprehensive evaluation index and classifies the reservoir accordingly.

9.4.2.11.2 Description of the Workflow and Machine Learning Model

This paper specifically aims to provide a new way to solve the problem by establishing a measure scheme preference method based on quantum particle swarm optimization (QPSO). There are 12 commonly used oil and water well measures, and firstly, a table of data related to the commonly used measure types is presented. Then, a function is designed for the degree of adaptation, and the objective function is the current year output ratio and cumulative output ratio. After the basic adaptation function design is completed, the optimization process design is carried out. Firstly, the parameters are initialized, and the optimization space is reasonably planned in terms of population size and limited number of steps. After that, particle initialization and compilation strategy design are carried out in order to enhance the diversity of examples. Considering the high-dimensional complexity of the measure scheme preference problem, the variation strategy is adopted to improve it to target all dimensions on individuals, i.e., if the variation probability is satisfied, a particle is randomly selected and initialized to a new particle. This problem is a multivariate, multi-constrained, nonlinear, and strongly coupled continuous optimization problem in high-dimensional space with high complexity. The optimization is implemented by the above strategy.

9.4.2.11.3 Advantages and Limitations of the Synchronized Algorithm

The advantage is that the introduction of quantum mechanical methods broadens the basic pool of ideas and methods. The model in this paper uses a multivariate, multi-constrained, nonlinear, and strongly coupled relationship of the continuum problem itself complexity and difficulty accuracy are raised to the desired level.

Also, the disadvantage is also very obvious that the use of this method is too jumpy and bold. The use of a certain threshold, even in the information age, can avoid the complicated manual calculation, and its practical value is always in doubt.

9.4.2.11.4 Results and Discussion

The algorithm proposed in the paper establishes a search mechanism through quantum mechanics theory with certain global retrieval capability. Moreover, this scheme invokes specific examples for numerical exploration in addition to theoretical construction, which can provide a reasonable basis for the development and planning of oil fields, especially water–oil fields.

BIBLIOGRAPHY

Sun, H., and Li, P. C. (2016). Measures optimization for oil and water well based on quantum particle swarm optimization. *Comput Technol Dev* 26(9), 78–82.

9.4.2.12 Application-17: Optimization of Designing of the Oil Field Exploiting Based on Fuzzy Mathematics and BP Neural Network

9.4.2.12.1 Background of the Application

The importance of oil as one of the most important resources in the 21st century for social and political and economic stability cannot be overstated, and the importance of determining oil extraction options that are directly related to oil production is also very evident. This paper uses a combination of the fuzzy integrated evaluation method and NN ideas for the learning of historical scenarios to resume a more systematic scientific and rational decision model.

9.4.2.12.2 Description of the Workflow and Machine Learning Model

Firstly, the data selection is carried out, and the final recovery rate, the annual rate of decline, the maximum oil recovery progress, and other rates as technical indicators are used as the economic indicators for evaluation in combination with theoretical and practical experience. After that, single indicators are selected from the superior affiliation as the principle for comparison. This important fuzzy mathematical theory of affiliation is introduced as a standard to measure. The ranking of the solutions according to the degree of affiliation is used as the output value of the learning sample for the training of the NN. Furthermore, the strong nonlinear dynamic processing capability of the NN is utilized to take an organizational self-learning approach from knowledge acquisition to structure construction. Each NN is fully connected between adjacent layers, while no connection is made between each neuron in each layer.

9.4.2.12.3 Advantages and Limitations of the Synchronized Algorithm

This process continues until the error between the ideal output and the actual output of the network is less than a certain amount, or until a predetermined number of times is learned.

9.4.2.12.4 Results and Discussion

Objective and reasonable are the important features of the fuzzy comprehensive evaluation method used to measure oilfield development programs comprehensively. The expression presented in the form of correlation values has a strong quantitative metric character. The strong function approximation capability of NNs can reflect complex linear relationships between the output and input of an object, and even generalize to arbitrary forms of nonlinear relationships. The organic combination of the two approaches makes the two theoretical strengths and weaknesses complementary, allowing both an accurate understanding of the assessment approach based on fuzzy theory and the accomplishment of reasonable generalization and generalization.

BIBLIOGRAPHY

Xiao, D. R., and Pan, H. (2010) Optimization of designing of the oil field exploiting based on fuzzy mathematics and BP neural network. *Microcomput Inf* 26(6), 209–211.

9.4.2.13 Application-18: Study on Diagnosis Model of Oilfield Abnormal Well Based on Feedback Dynamic Neural Network

9.4.2.13.1 Background of the Application

In the oilfield production process, many factors are involved, such as surface equipment factors and downhole equipment factors. Only when each factor reaches a balanced state, the well can operate normally. However, due to the uncertainty of the well and the complexity of the functional structure of the equipment, it is difficult for traditional technical analysis to deal with abnormal wells in real time. To ensure oilfield safety, a model for diagnosing abnormal wells in oilfields needs to be established. Combined with PSO algorithm, the learning algorithm is optimized to improve the convergence speed and robustness. There are many kinds of feedback dynamic NN models, and this paper studies the network model with three-layer structure and proposes the oilfield abnormal well diagnosis model based on feedback dynamic NN, which can judge the abnormal wells and has the feasibility of practical operation.

9.4.2.13.2 Description of the Workflow and Machine Learning Model

The input layer has n nodes, which are used to complete the input of time-varying signals and the delayed feedback of signals from the hidden layer; the hidden layer has m dynamic neuron nodes, whose function is to perform spatially weighted aggregation and excitation operations on the input time-varying signals and receive the signals from the feedback layer; the feedback layer has the same number of nodes as the hidden layer, i.e., m nodes, and the last layer is the output layer. The function of the feedback layer is to receive the time-delayed output from the hidden layer as the input signal and send the signal back to the hidden layer; the output layer performs spatial weighted aggregation, time accumulation, and excitation operations on the time-varying output signal.

In this paper, considering the complexity, diversity, and learning efficiency of the objective function, a QPSO algorithm is used, and the movement of particle positions is realized by a quantum revolving gate. The change of the traditional particle movement speed is transformed into the update of the quantum revolving gate angle, and the change of the position is transformed into the update of the quantum bit probability amplitude. The global solution of the network parameters in the time-varying space is realized to improve the learning ability of the model. With the feedback dynamic NN topology determined, the QPSO algorithm is used to determine the network connection power function.

9.4.2.13.3 Advantages and Limitations of the Synchronized Algorithm

BPNN is a multilayer feedforward network trained by error back propagation algorithm without feedback link, which is one of the most widely used NN models at present. In this paper, BPNN and feedback dynamic NN models are analyzed separately for oilfield abnormal well injection pressure data. Firstly, the normal well injection pressure data are used to train the two models respectively, and then, the abnormal injection well pressure data are used as test samples. Under the same premise, the feedback dynamic NN model outperforms the BPNN model in terms of accuracy, improves the speed of diagnosis, and implements global optimization techniques.

9.4.2.13.4 Results and Discussion

In this paper, a feedback dynamic NN is combined with the PSO algorithm, and an optimization algorithm is proposed to apply the method in practice, demonstrating improved learning efficiency and network stability. However, the feedback dynamic neural network has to consider the combination of forward transmission and feedback in information processing. Therefore, the information processing mechanism and process of feedback dynamic NN is an important topic for the next research.

BIBLIOGRAPHY

Li, T. N. (2015). Study on diagnosis model of oilfield abnormal well based on feedback dynamic neural network. *Comput Technol Autom* 2, 114–116.

9.4.2.14 Application-19: A Working Condition Diagnosis Model of Sucker Rod Pumping Wells Based on Big Data Deep Learning

9.4.2.14.1 Background of the Application

In the current big data context, CNNs are successful in the field of classification and recognition of images, especially in the fields of handwriting recognition, face recognition and other graphic recognition with good recognition results. As an important branch in the field of NNs, CNNs are characterized by the fact that the features of each layer are obtained from the local regions of the previous layer through the excitation of convolutional kernels with shared weights. This feature makes CNNs more suitable for the learning and representation of image features than other neural network methods. Combined with the current mainstream CNN technology, by graphing oil well production parameters and combining surface and downhole workings with scatterplot shapes, a working diagnosis model is established to realize automatic diagnosis and abnormality alarm of single-well workings, which can identify oil well working problems more accurately and efficiently and take optimization measures in a more targeted manner.

9.4.2.14.2 Description of the Workflow and Machine Learning Model

The surface power diagram of a pumping unit well is a graph consisting of a suspended point load and displacement, which is mixed with interference signals such as the friction and vibration of the oil rod and wellbore, and cannot accurately describe the working state of the pumping unit. It is necessary to consider the working state and graphics of both surface and downhole pumping diagrams to fully characterize the various features. Since the chances of two or more equipment failures of the same intensity causing symmetric changes in the shape of the power diagram occurring simultaneously are small, the pumping pump power diagram can be described by differential curves using the finite difference numerical solution method. The finite difference numerical solution method is used to solve the fluctuation equation finely to realize the downhole pumping power diagram simulation.

In ML, a CNN is a deep feed-forward ANN. The fundamental structure of CNN typically consists of two layers. The first layer is called the feature extraction layer, and it is where the features of the local region are extracted. Each neuron's input is coupled to the previous layer's local receptive domain. The second layer is the feature mapping layer, where each computational layer of the network consists of multiple feature mappings, each of which is a plane with equal weights for all neurons on the plane. This is done after the local feature has been extracted and the position relationship between it and other features has been established. The Feature Mapping shift is invariant because the feature mapping structure employs a sigmoid function with a modest influence function kernel as the activation function of the CNN. In addition, since the neurons on a mapping plane share the weights, the number of self-referential parameters of the network is reduced. Because the neurons on a single mapping surface share weights, the network has fewer free parameters. A computational layer for quadratic extraction and local averaging comes after each convolutional layer in a CNN. The feature resolution is decreased by this special two-feature extraction structure.

9.4.2.14.3 Advantages and Limitations of the Synchronized Algorithm

The advantage of CNN is that it first shares convolutional kernel and is stress-free for high-dimensional data processing. Secondly, the features need to be manually selected and trained with good weights, i.e., the features need to be classified well. The disadvantages of CNNs are that they need to be tuned, they require a large sample size, they require a Graphics Processing Unit (GPU) for training, and their physical meaning is unclear.

9.4.2.14.4 Results and Discussion

The use of CNN pattern recognition technology to establish a diagnostic model of pumping machine well workings and develop a software system for automatic workings identification and diagnosis is an important element in the development of an intelligent oilfield information technology. Compared with the traditional diagnostic method, this model can improve the feature details by handing over the feature extraction to the mature convolution-pooling technology; at the same time, the graphical method is combined with the production parameters to extract as many working conditions as possible and improve the working condition recognition accuracy. With the iterative improvement of the model in the process of application, it will effectively improve the timely and accurate rate of pumping well early warning diagnosis and working condition management.

BIBLIOGRAPHY

Wang, X., He, Y., Li, F., Dou, X., Wang, Z., Xu, H., and Fu, L. (2019). "A Working Condition Diagnosis Model of Sucker Rod Pumping Wells Based on Big Data Deep Learning." In: *International Petroleum Technology Conference* 26–28 March, Beijing, China, pp 1–10.

9.4.2.15 Application-20: Machines Performance Algorithmic Modeling for Anticipating Machines Health Using Real Time Condition Monitoring Data

9.4.2.15.1 Background of the Application

This paper presents a study on the health condition prediction of a Gas Turbine using ML algorithms. The focus of this study is on a RB211 Gas Generator, a two-spool design machine derived from the Rolls Royce RB211 aircraft, which is widely used in the offshore applications due to its lightweight and quick and easy removal. The paper describes the important parameters that are measured from the core engine and supporting auxiliary systems such as GG compressor discharge pressure, lube oil quality, fuel gas composition/quality, intake air purity and temperature, and vibration. The authors employed Support Vector Machine (SVM) learning algorithm with RBF kernel to train the model, which was based on the historical data collected from the site Distributed Control System (DCS) for 18 months. The results show that the SVM model can accurately predict the health condition of the Gas Turbine using pressure, temperature, vibration, and speed as input features.

9.4.2.15.2 Description of the Workflow and Machine Learning Model

In this method, a SVM model with a Gaussian RBF kernel is used to predict target variables based on input features. The SVM is a supervised learning algorithm that can handle classification and regression tasks. They achieve this through finding the optimal hyperplane that maximizes the margin between different classes. The data undergoes preprocessing steps such as handling missing values, normalization, and encoding of categorical variables before being split into training and testing sets.

9.4.2.15.3 Advantages and Limitations of the Synchronized Algorithm

The SVM model with an RBF kernel has many advantages, including its effectiveness in high-dimensional spaces and its ability to handle non-linear relationships through the kernel trick. This allows the model to create complex decision limits. Thus, making it good for classification tasks with non-linear data distributions. However, the SVM model can be computationally expensive, particularly with large datasets, and requires careful setting of hyperparameters such as the regulating parameter and kernel parameters to achieve optimum performance.

9.4.2.15.4 Results and Discussion

This paper demonstrates the potential of using ML algorithms for the condition-based monitoring of the Gas Turbine health. The SVM learning algorithm with RBF kernel has shown to be suitable for predicting the health condition of the Gas Turbine using pressure, temperature, vibration, and speed as input features. The results indicate that the model can accurately classify the machine health status as normal, danger, or failure. This study can be further extended to consider more input features that are important for the machine health and performance. The use of ML algorithms for the Gas Turbine health prediction can lead to improved plant operation, production, and cost.

BIBLIOGRAPHY

Almadhoun, Wael, and Alashqar, Ameed. "Machines Performance Algorithmic Modeling for Anticipating Machines Health Using Real Time Condition Monitoring Data." Paper presented at the *Abu Dhabi International Petroleum Exhibition & Conference*, Abu Dhabi, UAE, November 2017. doi: https://doi.org/10.2118/188803-MS

9.4.2.16 Application-21: Injection and Production Optimization for Tengiz Platform Sour Gas Injection by Reactive and Proactive Conformance Completions Control

9.4.2.16.1 Background of the Application

The Tengiz Sour Gas Injection project has been successful in improving the oil recovery and disposal of produced gas. However, with increased gas injection, a high gas-to-oil ratio (GOR) has become a challenge. Shutting-in high GOR producers reduces the number of active producers and results in production loss. To overcome this, zonal production and injection control capability were introduced through conformance control completions. A reactive completion control strategy is used to shut-in the well when the gas breakthrough occurs. The objective of this study was to establish an efficient optimization workflow to improve vertical/areal sweep, thereby maximizing recovery under operation constraints. This will be enabled through the proactive and reactive control completions that have been installed in many production/injection wells.

9.4.2.16.2 Description of the Workflow and Machine Learning Model

In this method, a Random Forest Regressor and a Genetic Algorithm (GA) are used to predict the recovery factor in oil and gas production based on various input features. The Random Forest algorithm, an ensemble learning method, constructs multiple decision trees and combines their predictions to enhance accuracy and reduce overfitting. Additionally, a GA is used to optimize the features to the solution for maximizing the predicted recovery. The GA iteratively evolves a population of potential solutions using operations like selection, crossover, and mutation, aiming to identify the optimal set of parameters that give us the highest recovery.

9.4.2.16.3 Advantages and Limitations of the Synchronized Algorithm

The Random Forest Regressor offers robustness to overfitting by averaging multiple decision trees and can handle many input features without extensive pre-processing. It is also resilient to noisy data and provides valuable information into feature importance. However, this method can be computationally expensive, especially with large datasets and numerous trees. The GA enhances the optimization process by exploring a wide solution space and converging on the most effective set of input features for maximizing recovery. However, GAs can be computationally demanding and need a careful setting of parameters like population size, crossover rate, and mutation rate.

9.4.2.16.4 Results and Discussion

The findings indicated that rule-based field management (FM) optimization was the greatest option for a long-term objective function improvement, while a hybrid strategy that took schedule-based optimization into account was the best way to optimize short-term goals. Applying optimization results to field operational guidelines is made more feasible using the hybrid technique. The hybrid optimization of both schedule and operation rules, the application of model-based optimization results to field operations, and the lessons gained from the use of optimization techniques ranging from traditional GA to ML-supported techniques are all examples of this study's novelty. In order to improve recovery and sustain production under operational restrictions, the study offers a workable optimization strategy. There are three ways in which this study is novel: 1) Field operations have adopted the model-based optimization result from this work, and greater recovery has been observed. 2) Both schedule and operation rule hybrid optimization offered practicality in terms of optimization performance and field operation applicability. 3) It offers insights gleaned via the use of optimization methods ranging from traditional GAs to methods aided by ML.

BIBLIOGRAPHY

Tanaka, Shusei, Rousset, Matthieu, Ghomian, Yousef, Azhigaliyeva, Aizada, Bopiyev, Chingiz, Yechshanov, Ilyas, and Dehghani, Kaveh. "Injection and Production Optimization for Tengiz Platform Sour Gas Injection by Reactive and Proactive Conformance Completions Control." Paper presented at the *SPE Reservoir Simulation Conference*, On-Demand, October 2021. doi: https://doi.org/10.2118/203995-MS

9.4.2.17 Application-22: Improving Deepwater Facility Uptime Using Machine Learning Approach

9.4.2.17.1 Background of the Application

The paper discusses the problems faced by deep-water oil and gas facilities due to unplanned downtime, which can result in significant production losses and increased operational costs. The authors propose an intelligent prognostic solution using a ML framework for the automatic prediction of impending facility downtime and the identification of key causative process variables. They developed a systematic workflow to identify, cleanse, and process real-time data for both model training and prediction. Several ML methods were evaluated, and anomaly detection based on PCA and AE algorithms were found to perform better for the type of data available for the deep-water facility. The adaptive ML models provide early warning of impending events, with several predictions found as early as 30 minutes to 2 hours, providing adequate time to take proactive actions.

9.4.2.17.2 Description of the Workflow and Machine Learning Model

In this project, the author used advanced ML techniques to predict downtime in deep-water oil and gas facilities. By analyzing real-time sensor data with PCA and AE algorithms, the system can detect anomalies that indicate potential failures. The PCA method reduces data dimensionality to highlight significant patterns, while the AE method, a type of deep neural network, learns to encode and decode the data to spot deviations from normal behavior. Together, these approaches enable the proactive identification of issues, providing operators with early warnings and enough time to mitigate risks.

9.4.2.17.3 Advantages and Limitations of the Synchronized Algorithm

The combination of PCA and AE offers robust anomaly detection for managing complex sensor data in deep-water facilities. PCA's simplicity and efficiency in highlighting key data variations make it a valuable tool for initial anomaly detection. Meanwhile, the ability of AE to handle non-linear data relationships enhances the system's predictive power. However, these methods require extensive computational resources and expert knowledge to implement effectively. Despite these challenges, the increased prediction accuracy and early warning capabilities significantly improve facility uptime, making operations more reliable and safer.

9.4.2.17.4 Results and Discussion

The topside process surveillance for deep-water facilities could be completely transformed by the suggested adaptive ML approach. Using a cloud-based model deployment platform to train, predict, and initiate automatic model updates, the authors explain how the ML framework can be scaled for a sustainable solution that spans multiple process systems and facilities, makes predictions every minute, and maintains the model evergreen. The existing model can continue to predict other occurrences, and more sophisticated ML techniques based on supervised learning can be created and used over time after enough data have been gathered.

BIBLIOGRAPHY

Singh, Ajay, Sankaran, Sathish, Ambre, Sachin, Srikonda, Rohit, and Houston, Zach. "Improving Deepwater Facility Uptime Using Machine Learning Approach." Paper presented at the *SPE Annual Technical Conference and Exhibition*, Calgary, Alberta, Canada, September 2019. doi: https://doi.org/10.2118/195875-MS

9.4.2.18 Application-23: Applications of Smart Proxies for Subsurface Modeling

9.4.2.18.1 Background of the Application

This paper discusses the use of proxy modeling as an alternative to numerical simulation models in subsurface modeling and the earth science domain. Specifically, the paper demonstrates the application of smart proxy models for reservoir simulation models in the petroleum industry. Reservoir simulation models are commonly used in petroleum reservoir management workflows to analyze the fluid flow behavior in porous media and make operational decisions. However, these models are computationally expensive and time-consuming to run, making it difficult to study different operational scenarios and assess uncertainty. The paper argues that the development of smart proxies for reservoir simulation models would allow for a more realistic subsurface modeling study without sacrificing accuracy.

9.4.2.18.2 Description of the Workflow and Machine Learning Model

In this method, feed-forward ANNs are used to create smart proxy models for reservoir simulation. Input parameters are normalized and used to train the ANNs. The model predicts key outputs like bottom-hole pressure, gas production, and water production, offering a fast and efficient alternative to traditional simulation methods. The ANN's ability to learn complex, non-linear data relationships improve the accuracy of these predictions.

9.4.2.18.3 Advantages and Limitations of the Synchronized Algorithm

Using ANNs in this method greatly reduces computational time, saving up to 98.9% compared to traditional simulations. The NN captures complex data relationships, ensuring high prediction accuracy, which is good for rapid modeling and forecasting. However, the method depends on high-quality historical data for training and may require regular updates to remain accurate under changing conditions. Thus, an important limit is set because of the dependency on data quality and quantity, making it necessary for ongoing data collection and model retraining.

9.4.2.18.4 Results and Discussion

This paper presents two case studies demonstrating the effectiveness of smart proxy modeling for subsurface modeling. The first case study involves building a proxy model using ANNs for the purpose of estimating well pressure and production, while the second case study focuses on the prediction of the reservoir pressure and phase saturation distribution at the injection stage and post-injection stage. The results of these case studies demonstrate the advantage of using smart proxies for addressing the high computational cost and time associated with numerical simulation models. Furthermore, these proxy models predict the outputs of reservoir simulation models with high accuracy.

BIBLIOGRAPHY

Shahkarami, Alireza, and Mohaghegh, Shahab. (2020). Applications of smart proxies for subsurface modeling. *Petroleum Exploration Dev* 47(2), 400–412, ISSN 1876-3804. https://doi.org/10.1016/S1876-3804(20)60057-X

9.4.3 Well Related Applications

9.4.3.1 Application-24: Prediction of Hydraulic Properties in Carbonate Reservoirs Using Artificial Neural Network

9.4.3.1.1 Background of the Application

Using BPNNs and well log data from various areas across the globe, this study investigates the application of ANNs for accurately predicting porosity and permeability in carbonate reservoirs. ANN considers developing two scenarios. In the first case, all accessible logging data is used directly as an ANN's input. The diagenesis parameter, which is effectively determined from logging data, is added as an additional input to the ANN in the second scenario. Two models are created for each scenario: one predicts permeability in carbonate reservoirs, and the other predicts porosity. Data sets from stratigraphic carbonate reservoirs and different geographic locations are used to train and evaluate the models.

9.4.3.1.2 *Description of the Workflow and Machine Learning Model*

Designing a perfect and effective program for the development of an oil and gas field requires a precise calculation of porosity and permeability. Porosity and permeability can be found using a variety of techniques, such as log-derived models and laboratory observations. An effective method for resolving a number of geological and engineering problems is the ANN. The BPNN in the current study uses ANNs to help estimate porosity and permeability in carbonate reservoirs with excellent accuracy on well log data from various fields globally. BP is an algorithm that performs gradient descent on multi-layer ANNs, which means that the gradient of the loss function is calculated using the chain rule with the weights of each layer of the network as variables to update the weights to minimize the loss function.

9.4.3.1.3 *Advantages and Limitations of the Synchronized Algorithm*

The number of neurons reaches a certain number, and the number of weight parameters reaches a certain number. It is obviously unrealistic to set the parameters of the NN manually. Thus, the power of big data can be used to correct the parameters of the NN through statistical methods. The advantage of the statistical method is that it saves the trouble of setting the rules manually.

The corresponding disadvantage is that the accuracy can be flawed. Therefore, it is important to pay attention to the environment of use when using statistical methods.

9.4.3.1.4 *Results and Discussion*

The developed models in this study using ANN provide an accurate prediction for porosity and permeability in carbonate reservoirs. High-accuracy models were created utilizing BPNNs on well log data from various fields across the globe. For ANN to develop, two possibilities were taken into consideration. In the first case, all accessible logging data is used directly as an ANN's input. The diagenesis parameter, which is effectively determined from logging data, is added as an additional input to the ANN in the second scenario. Two models are created for each scenario: one predicts permeability in carbonate reservoirs, and the other predicts porosity. The models were trained and tested with data sets from various geographic locations and stratigraphic carbonate reservoirs. The results show that the models are successful in predicting the porosity and permeability for carbonate reservoirs, with good correlation coefficients obtained between the actual and predicted values.

BIBLIOGRAPHY

Salem, Khalaf G., Gad, Khalaf, Abdulaziz, Abdulaziz M., Aziz, Abdel, Sattar, Abdel Sattar Abdel, and Dahab, A. "Prediction of Hydraulic Properties in Carbonate Reservoirs

Using Artificial Neural Network." Paper presented at the *Abu Dhabi International Petroleum Exhibition & Conference*, Abu Dhabi, UAE, November 2018. doi: https://doi.org/10.2118/193007-MS

9.4.3.2 Application-25: Machine Learning Derived Correlation to Determine Water Saturation in Complex Lithologies

9.4.3.2.1 Background of the Application

This paper presents a new empirical correlation to determine water saturation in carbonate reservoirs using ML techniques. The importance of water saturation in petrophysical and reservoir engineering computations cannot be overstated, and accurately determining it is crucial in making vital field development decisions. While empirical correlations have been formulated in the past, they require log-derived data, which is not a precise representation, leading to a lack of consensus among log analysts about which model can be universally used. With less uncertainty in the parameters it uses than current models, this article suggests using computer-generated algorithms, fuzzy logic, and neural networking to create a correlation that is easy to use and practical. Several ML methods, including as the ANFIS and ANN, are used to assess water saturation using core dean-stark data as the output and traditional wireline log data as input parameters. When compared to experimental core readings, the established correlation can detect water saturation with a precision of roughly 94%. It also performs exceptionally well in the prediction mode when given log data that has been impacted by shale.

9.4.3.2.2 Description of the Workflow and Machine Learning Model

An ANN method was employed to model water saturation in a system. The neural network consists of an input layer, a hidden layer utilizing the sigmoid activation function, and an output layer for prediction. The input features are fed into the network, where weights and biases are applied to compute the output. The hidden layer's output is then used to calculate the normalized water saturation, which is then transformed to actual water saturation. Additionally, the ANFIS is integrated for rule-based prediction, combining NNs with fuzzy logic.

9.4.3.2.3 Advantages and Limitations of the Synchronized Algorithm

The main advantage of this method is it can model complex, nonlinear relationships between input variables and water saturation, providing high accuracy in predictions. The integration of ANFIS allows for incorporating human-like reasoning, improving interpretability and adaptability. However, the approach requires significant number of computational resources and large datasets for training. Additionally, the complexity of the model can make it difficult to understand how the decisions are made, creating challenges in scenarios where model transparency and interpretability are important.

9.4.3.2.4 Results and Discussion

In conclusion, this paper successfully demonstrates that ML techniques can predict water saturation with a high degree of accuracy in carbonate reservoirs. The new empirical correlation proposed in this paper can be applied to any case with input parameters within the developed model's range, without needing an intricate software. The results show that ANFIS provides slightly better results than ANN, but the estimation is not significantly affected if ANN is used. Furthermore, the NN has produced a practical working equation that can be used to determine water saturation in carbonate reservoirs. Future research can explore the possibility of gaining better results by using other techniques, such as SVMs and FN.

BIBLIOGRAPHY

Khan, Mohammad Rasheed, Tariq, Zeeshan, and Abdulraheem, Abdulazeez. "Machine Learning Derived Correlation to Determine Water Saturation in Complex Lithologies." Paper presented at the *SPE Kingdom of Saudi Arabia Annual Technical Symposium and Exhibition*, Dammam, Saudi Arabia, April 2018. doi: https://doi.org/10.2118/192307-MS

9.4.3.3 Application-26: Remaining Oil Distribution and its Description Technology during Ultra-High Water Cut Stage

9.4.3.3.1 Background of the Application

Chinese oil fields are mostly terrestrial deposits, with large variations in permeability between, within and on the surface of reservoirs. Oil recovery in the high water-bearing period of oil fields will be a long development process, and tapping the remaining oil becomes the main task for the development of old oil fields. After the oil field enters the late development of high-water content, the remaining oil distribution becomes more and more complex, which makes it increasingly difficult for the oil field to stabilize the production and adjust the tapping potential. The distribution of remaining oil is related to the sedimentary micro-phase, while the distribution of residual oil is related to many factors such as reservoir inhomogeneity, fluid inhomogeneity, faults, development factors (injection and extraction relationship, and well network deployment).

The content of residual oil research in the high water-bearing stage should not only clarify the exact location and quantity of residual oil distribution, but also its causes and characteristics of distribution, and adopt corresponding excavation techniques to enhance the development potential of the oilfield according to the residual oil distribution law.

9.4.3.3.2 Description of the Workflow and Machine Learning Model

BPNN is a concept proposed by scientists led by Rumelhart and McClelland in 1986, which is a multi-layer feed-forward NN trained according to the error BP algorithm and is one of the most widely used NN models.

The formation and distribution of residual oil in high water-bearing period is mainly controlled by two major factors, geology and development factors. The geological factors mainly refer to sedimentary micro-phase, reservoir micro-characteristics, macro inhomogeneity, reservoir microstructure, fluid properties, and oil formation. Development factors mainly refer to the injection and extraction system. All the factors are interrelated and mutually constrained, and together control the distribution of remaining oil.

9.4.3.3.3 Advantages and Limitations of the Synchronized Algorithm

The strength of the research approach used in this literature is the strong focus on basic research methods, using a variety of intra-disciplinary research methods for inquiry. A comparison between traditional Chinese literature and international perspectives is also conducted. The reader is impressed by the use of 3D seismic methods, numerical simulation of reservoirs, calculation methods of oil drive efficiency and wave coefficients, dynamic analysis methods, sediment micro-phase methods, and inspection and observation well methods. The exploration of micro studies is carried out, while macro studies are emphasized.

However, the article also has some flaws; the full text of the paper only carries out the construction of the framework and does not carry out specific and micro-analysis and testimony. Relevant industry experiences and reference directions are summarized, but examples and considerations of research are lacking.

9.4.3.3.4 Results and Discussion

1. The research methods of residual oil distribution in the high water-bearing period should form a complete system to make up for the shortcomings of a certain set of schemes. In practical application, various methods should be used to conduct comprehensive research according to specific situations.
2. There are many factors affecting and controlling the distribution of residual oil in the high water-bearing period, which should be considered from two main aspects, geology and development.
3. Strengthen the research on high water-bearing reservoirs, continuously deepen the understanding of remaining oil potential, and find the right direction of tapping in order to develop high-quality and high-efficiency plans in a targeted manner and improve the reservoir development effect.
4. The well network is already relatively dense after the oilfield enters the extraordinarily high water-bearing period. After the field enters the extraordinarily high water-bearing period, the well network is already relatively dense, so it is not very useful to densify the well network to improve the final recovery rate of the oil field.

BIBLIOGRAPHY

Du, Q. L., and Zhu, L. H. (2009). Remaining oil distribution and its description technology during ultra-high water cut stage in La-Sa-Xing oilfi1eld. *Daqing Pet Geol Dev* 5, 99–105

Rumelhart, D. E., and McClelland, J. L. (1986). *Parallel Distributed Processing, Explorations in the Microstructure of Cognition I and II.* Cambridge: MIT Press. https://doi.org/10.7551/mitpress/5236.001.0001

9.4.3.4 Application-27: Parameters Prediction of Remaining Oil of All Oil Region Based on BP Neural Networks

9.4.3.4.1 Background of the Application

One of the key aspects of residual oil research is whether the residual oil can be accurately identified, especially when the oil field enters the high water-bearing period, the accurate identification and evaluation of the residual. The accurate identification and evaluation of residual oil distribution is a difficult point, and is of great importance to the exploration and development of oil fields. NN theory is an emerging edge and cross-discipline; from the information theory point of view, it is another kind of information processing tools. ANN is a kind of model established from the microstructure and function of the human brain nervous system simulation, with the ability to simulate a part of the human image thinking, its characteristics are mainly non-linear characteristics, learning ability, and self-adaptability.

It is an important way to simulate human intelligence. The paper exploits the outstanding nonlinear approximation capability and self-learning function of ANNs, which can adequately approximate arbitrarily complex nonlinear relationships by learning from past historical data. The paper applies the BPNN method to the study of residual oil distribution in oil fields and obtains more satisfactory results.

9.4.3.4.2 Description of the Workflow and Machine Learning Model

BPNN, the learning process of error BP algorithm, consists of two processes: forward propagation of information and backward propagation of error. BP network can learn and store a large number of nonlinear input-output pattern mapping relationships without the need for a mathematical model of such mapping relationships to be known. The BPNN model topology includes the results of input layer, hidden layer, and output layer.

A residual oil distribution prediction model was developed using a BPNN: three-dimensional geographical coordinates (horizontal coordinates, vertical coordinates, depth) are used as the input parameters of the network, and the porosity, permeability, and saturation of the rock as the output parameters of the network. as the input parameters of the network, and the porosity, permeability and saturation of the rock as the output parameters of the network, and establish a non-linear network mapping relationship network to predict the distribution of remaining oil in the study area.

9.4.3.4.3 Advantages and Limitations of the Synchronized Algorithm

The advantage of the method is its ability to build the algorithm architecturally for implementation in different zones and in different time zones. Moreover, the controllability of the parameters further extends the range of applications. The limitation is that the pre-processing of data is relatively complicated, and the change of the tuning model is not intuitive and easy to ghost.

9.4.3.4.4 Results and Discussion

The research on the formation and distribution of residual oil in China and abroad shows that it is very important to study the formation and distribution of residual oil in China and abroad. The paper applies a BPNN algorithm to predict the remaining oil distribution of oil wells in a plant area of the Daqing oilfield. The method introduced in the paper uses the outstanding nonlinear fitting ability of NNs to directly establish a nonlinear mapping relationship between the three-dimensional geographic location of oil wells and the porosity, permeability, and saturation. The method is intuitive and does not require much information on known data, only the location of the well is required as input data, which is simple and easy to implement.

BIBLIOGRAPHY

Liu, J., Gao, H. Y., and Wang, T. T. et al (2013). Parameters prediction of remaining oil of aII oil region in Daqing based on BP neural networks. *Inner Mong Petrochem Ind* 23, 47–50.

9.4.3.5 Application-28: Predicting Wax Formation Using Artificial Neural Network

9.4.3.5.1 Background of the Application

This paper presents a study on the use of ANNs to predict the Wax Appearance Temperature (WAT) of various hydrocarbon mixtures. The input parameters used in this study are Molecular Weight (MW), Activation Energy (Ea), Density (D), Kinematic Viscosity (KV), and Minimum Pour Point (MPP). The ANN model was trained and tested using experimental data from different sources and was found to provide the best predictions compared to other Equation of State (EOS) and thermodynamic models. The results of this study show the potential of using ANN to predict WAT for hydrocarbon mixtures.

9.4.3.5.2 Description of the Workflow and Machine Learning Model

In this method, a NN model is used to make predictions based on input features. The model has a simple structure with a hidden layer containing four nodes and an output layer. The hidden layer uses the tanh activation function to handle nonlinear relationships in the data. Training is performed using the stochastic gradient descent (SGD) optimizer and the mean squared error (MSE) loss function. The model's parameters are updated iteratively to improve its accuracy, with performance validated on a separate dataset to ensure it generalizes well.

9.4.3.5.3 Advantages and Limitations of the Synchronized Algorithm

NNs have several benefits, such as the ability to capture complex patterns in the data and improve accuracy through iterative training. They are effective for diverse datasets and use validation sets to prevent overfitting, making sure the model works well on new data. However, they require significant computational power and time, and their success depends on the quality of the input data. Additionally, the model's performance is influenced by the careful selection and tuning of hyperparameters, such as the number of layers and nodes, activation functions, and optimization methods.

9.4.3.5.4 Results and Discussion

In conclusion, this study has demonstrated the effectiveness of using ANNs to predict the WAT of hydrocarbon mixtures. The combination of MW, D, and Activation Energy (Ea) as input parameters was found to provide the best prediction of WAT. The ANN model outperformed other EOS and thermodynamic models in terms of accuracy and reliability, with an Average Absolute Deviation (AAD%) of 0.4%. The results of this study indicate the potential of using ANN as a reliable and efficient tool for predicting WAT in flow assurance applications.

BIBLIOGRAPHY

Adeyemi, B. J., and Sulaimon, A. A. "Predicting Wax Formation Using Artificial Neural Network." Paper presented at the *Nigeria Annual International Conference and Exhibition*, Lagos, Nigeria, August 2012. doi: https://doi.org/10.2118/163026-MS

9.4.3.6 Application-29: Downhole Signal Compression and Surface Reconstruction Based on Dictionary Machine Learning

9.4.3.6.1 Background of the Application

The oil and gas industry is shifting from traditional heavy machinery to high-accuracy and high-efficiency technology, which involves installing miniaturized but powerful sensors into downhole tools. However, communication between downhole tools and surface units remains far behind. Mud-pulse telemetry is still the dominant communication method, with a physical bit rate of only 10–20 bits per second under ideal conditions. As the well depth and formation complexity increase, the performance of mud pulse deteriorates, resulting in an effective bit rate as low as 1 bit per second. Lossless data compression based on universal protocols applied to any input signals has been widely used by the industry to mitigate the communication bottleneck. However, such compression results in low compression ratios. In this paper, a smart, lossy data compression and reconstruction methodology is presented based on the theory and algorithms of Compressed Sensing. This methodology can sparsely represent signals with a set of over-complete basis functions called a dictionary. An improved K-SVD algorithm was applied to learn the dictionary from a calibration database of over 1,000 pump-out jobs. The proposed method can compress and reconstruct the downhole densitometry data using 30% fewer bits to achieve a reconstruction mean-square-error that is 18 times smaller than the conventional method of sampling, averaging, and surface interpolation.

9.4.3.6.2 Description of the Workflow and Machine Learning Model

In this method, dictionary learning for sparse coding is used, starting with a dictionary matrix initialized with KK atoms (the number of atoms in the dictionary matrix). The process begins with a random dictionary, which is refined through iterations. Sparse coding is performed using orthogonal matching pursuit (OMP) to represent each data point with a limited number of non-zero coefficients. Following this, the dictionary is updated by adjusting each atom to better fit the data, normalizing each atom after updates. This iterative process continues until the dictionary effectively represents the data, typically over a decided number of iterations.

9.4.3.6.3 Advantages and Limitations of the Synchronized Algorithm

Dictionary learning has many advantages, such as creating efficient, sparse representations of data, which improve storage and computational efficiency. It is particularly useful for high-dimensional data, providing compact representations and reducing noise. The iterative refinement ensures that the dictionary adapts to the data, improving accuracy with each iteration. However, this approach can be computationally expensive, especially with large datasets and dictionaries. The success depends on the quality of the initial dictionary and careful selection of parameters. The convergence process may require numerous iterations, making it heavy on resources.

9.4.3.6.4 Results and Discussion

This paper presents a smart, lossy data compression, and reconstruction methodology for mitigating the communication bottleneck in the Logging While Drilling (LWD), Measurement While Drilling (MWD). The proposed method can sparsely represent signals with a set of over-complete basis functions called a dictionary. The results show that the improved K-SVD algorithm can learn the dictionary from a calibration database of over 1,000 pump-out jobs. Compared to the conventional method of sampling, averaging, and surface interpolation, the proposed method can use 30% fewer bits to achieve a reconstruction mean-square-error that is 18 times smaller than with the conventional approach. Moreover, the proposed method can accurately transmit critical information in complex pump-outs, such as multi-phase flow, bubbling, or sensor malfunctions. The proposed method has the potential to improve the upstream accuracy and efficiency of the oil and gas industry by orders of magnitude.

BIBLIOGRAPHY

Li, Jian, Dai, Bin, Jones, Christopher M., Samson, Etienne M., and Gascooke, Darren. "Downhole Signal Compression and Surface Reconstruction Based on Dictionary Machine Learning." Paper presented at the *SPWLA 61st Annual Logging Symposium*, Virtual Online Webinar, June 2020. doi: https://doi.org/10.30632/SPWLA-5088

9.4.3.7 Application-30: Prediction of Multilateral Inflow Control Valve Flow Performance Using Machine Learning

9.4.3.7.1 Background of the Application

This paper presents a novel approach to improve the performance of multilateral wells by predicting the optimal ICV settings. The model uses ML techniques to predict the well flow profile based on a limited number of data points. The approach is tested on the real field data from a multilateral well in the Middle East, and the results show that the model can accurately predict the well flow profile at all possible ICV setting combinations. This enables operators to locate the optimal ICV setting combination for a newly drilled or underperforming multilateral well. The proposed approach can improve the performance of multilateral wells, increase production, and reduce operational costs.

9.4.3.7.2 Description of the Workflow and Machine Learning Model

The author used Fully connected neural networks (FCNNs) that are made up of layers of interconnected neurons. Each connection has a weight that gets adjusted as the model learns from data. For predicting well flow profiles, FCNNs can forecast outputs like the surface and downhole flow rates of oil and water by learning from historical and synthetic data. This helps in finding the best ICV settings to boost well performance and efficiency.

9.4.3.7.3 Advantages and Limitations of the Synchronized Algorithm

FCNNs are great at handling complex, non-linear relationships and learning directly from data, which makes them perfect for predicting well performance. However, they need a lot of data and computing power to train effectively. They can also be hard to understand exactly how they're making their predictions, which can be a downside when you need to explain the results.

9.4.3.7.4 Results and Discussion

In this paper, a novel approach is presented to improve the performance of multilateral wells by predicting the optimal ICV settings using ML techniques. The proposed approach was tested on the real field data from a multilateral well in the Middle East and was found to be effective in accurately predicting the well flow profile at all possible ICV setting combinations. This enables operators to locate the optimal ICV setting combination for a newly drilled or underperforming multilateral well, leading to improved well performance, increased production, and reduced operational costs. The approach presented in this paper can be applied to other wells and fields, and it can also be extended to include additional variables such as reservoir pressure, temperature, and fluid composition to further improve the accuracy of the predictions.

BIBLIOGRAPHY

Aljubran, Mohammad Jabs, and Horne, Roland. "Prediction of Multilateral Inflow Control Valve Flow Performance Using Machine Learning." Paper presented at the *SPE Annual Technical Conference and Exhibition*, Calgary, Alberta, Canada, September 2019. doi: https://doi.org/10.2118/196003-MS

9.4.3.8 Application-31: Utilizing Bayesian Optimization and Machine Learning to Find the Best Inflow Control Design for Horizontal Wells

9.4.3.8.1 Background of the Application

This paper presents a method for optimizing the design of inflow control devices (ICDs) in horizontal wells using Bayesian optimization and ML techniques. Uneven inflow issues, such as early water breakthrough and gas coning, can lead to reduced well productivity. ICDs are placed along the lateral section of a well to control zonal production and improve well productivity. However, choosing an effective ICD design is a time-consuming and laborious task. The proposed workflow automates the optimization process of the ICD design evaluation and reduces the time required by 50% compared to the manual trial and error process. The workflow utilizes a surrogate model based on Gaussian Process Regression (GPR) and an appropriate acquisition function to design the next sample to be evaluated. Two case studies demonstrate that the proposed approach optimizes the ICD design for a given reservoir model, resulting in an improved oil recovery and reduced water or gas production.

9.4.3.8.2 Description of the Workflow and Machine Learning Model

The method used is Bayesian Optimization with Gaussian Process Regression (GPR), a technique suited for optimization problems. Bayesian Optimization efficiently explores the parameter space using GPR to model the objective function. It iteratively selects the most promising points to evaluate, balancing exploration through an acquisition function. This approach is particularly useful for optimizing complex, expensive, and noisy functions, such as those found in reservoir simulations.

9.4.3.8.3 Advantages and Limitations of the Synchronized Algorithm

Bayesian Optimization with GPR have several advantages like efficiently finding optimal with fewer evaluations and handling noisy objective functions effectively. It is particularly well-suited for problems where evaluations are expensive, making it a powerful tool for optimizing ICD completion designs. However, this method requires a careful selection of the prior and kernel functions in GPR, and it can be computationally expensive for high-dimensional parameter spaces. Additionally, Bayesian Optimization assumes a smooth objective function, which may not be true in all cases.

9.4.3.8.4 Results and Discussion

The proposed approach for optimizing ICDs through Bayesian optimization and ML techniques provides a faster and more computationally efficient way to reach the ICD optimum design for a given reservoir model. The Intelligent Sequential Sampling algorithm, based on Bayesian Optimization and Gaussian Processes as surrogate models, automates the optimization process of ICD design evaluation workflows. The approach improves the quality of the optimal solution while minimizing the number of evaluations required, and obviates the need for a human expert to endure the effort of selecting all the design candidates to be evaluated. The approach has been demonstrated to be effective in two case studies, resulting in improved oil recovery and reduced water or gas production. The accuracy of the reservoir simulation model is critical, and detailed reservoir characterization is recommended to improve the predictability of reservoir behavior and enable better decisions leading to better well performance and enhanced recovery.

BIBLIOGRAPHY

Shahkarami, Alireza, Friedrichs, Meagan, Iyer, Naresh, Izadi, Ghazal, Klenner, Robert, Meyer, Elena, Becerra Moreno, Oscar, and Murrell, Glen. "Utilizing Bayesian Optimization and Machine Learning to Find the Best Inflow Control Design for Horizontal Wells." Paper presented at the *Offshore Technology Conference*, Houston, Texas, USA, May 2020. doi: https://doi.org/10.4043/30716-MS

9.4.3.9 Application-32: Development and Prediction of Cavitation in Centrifugal Pumps Using Industrial Internet of Things Technology

9.4.3.9.1 Background of the Application

This paper proposes a cheaper and more efficient system to address cavitation in centrifugal pumps in the oil and gas industry, which is a widespread problem that causes enormous downtime costs and safety hazards. The proposed system involves constructing a test rig of the centrifugal pump, integrating and programming sensors using C++ and Arduino microcontroller, and analyzing telemetry data and predicted the likelihood of cavitation or no cavitation. The paper also discusses the use of Python and ML techniques to further improve pump cavitation prediction accuracy.

9.4.3.9.2 Description of the Workflow and Machine Learning Model

The author utilized a method that uses Random Forest Classifiers to predict cavitation in centrifugal pumps based on sensor data. The data is standardized and used to train the Random Forest model, which handles various input parameters. Predicted cavitation events are compared to actual occurrences to enhance maintenance scheduling and pump efficiency.

9.4.3.9.3 Advantages and Limitations of the Synchronized Algorithm

The Random Forest method effectively manages diverse sensor inputs, crucial for accurate cavitation prediction. It provides a comprehensive view of pump conditions using ensemble learning. However, training can be computationally intensive, and requires high-quality and representative data. The model's performance depends on the alignment between training data and real-world conditions.

9.4.3.9.4 Results and Discussion

In conclusion, this paper has presented a unique method of prediction and control of cavitation in centrifugal pumps using Industrial Internet of Things technology. The proposed system is cheaper, more efficient, and capable of addressing cavitation in a timely manner, which is crucial in the oil and gas industry. The preliminary results from the current methodology are promising, and the use of Python and ML techniques is expected to further improve the accuracy of pump cavitation prediction.

BIBLIOGRAPHY

Ejimuda, Chinonyelum. "Development and Prediction of Cavitation in Centrifugal Pumps using Industrial Internet of Things Technology." Paper presented at the *SPE Nigeria Annual International Conference and Exhibition*, Virtual, August 2020. doi: https://doi.org/10.2118/203661-MS

9.4.3.10 Application-33: Sand Control Optimization for Rubiales field: Trade-Off between Sand Control, Flow Performance and Mechanical Integrity

9.4.3.10.1 Background of the Application

This paper presents a comprehensive study on the sand control design for cold primary heavy oil production in horizontal wells. The study involves a detailed analysis of sand particle size distribution (PSD) and composition of fines in the development area. Based on the PSD variation and existing design criteria, several seamed slotted-liner configurations are proposed and tested for flow and sand control performance. The study also considers hydro-thermo-mechanical phenomena such as plugging, scaling, and corrosion. The results of large-scale sand retention tests (SRTs) and finite element analysis indicate that the optimized aperture size and the application of seamed slots can mitigate plugging and improve flow performance.

9.4.3.10.2 Description of the Workflow and Machine Learning Model

The method used is K-Means Clustering, a widely adopted algorithm for partitioning data into groups based on feature similarity. It minimizes the variance within clusters and maximizes the variance between clusters by iteratively adjusting cluster centers. In this study, K-Means classifies the PSD data into four categories, aiding in the proposal of appropriate slotted-liner configurations for sand control. By grouping similar PSD profiles, K-Means helps target specific design requirements.

9.4.3.10.3 Advantages and Limitations of the Synchronized Algorithm

K-Means Clustering is good for this method due to its simplicity, speed, and efficiency with large datasets. It's easy to implement and understand. However, it requires the number of clusters to be specified in advance and is sensitive to the initial center placement, potentially leading to suboptimal clustering. It also assumes that clusters are spherical and of similar size, which might not suit all datasets. Despite these limitations, K-Means is effective for many practical clustering applications.

9.4.3.10.4 Results and Discussion

In conclusion, the study presents a comprehensive sand control design workflow for cold primary heavy oil production in horizontal wells. The proposed seamed slotted-liner configurations and optimized aperture size are effective in mitigating plugging and improving flow performance. The study also highlights the importance of considering hydro-thermo-mechanical phenomena in the sand control design. The results of this study can guide the design and selection of sand control screens for similar reservoirs.

BIBLIOGRAPHY

Mayorga Cespedes, E. A., Roostaei, M., Uzcátegui, A. A., Soroush, M., Izadi, H., Hosseini, S. A., Schroeder, B., Mahmoudi, M., Gomez, D. M., Mora, E. and Alpire, J. "Sand Control Optimization for Rubiales Field: Trade-Off Between Sand Control, Flow Performance and Mechanical Integrity." Paper presented at the *SPE Latin American and Caribbean Petroleum Engineering Conference*, Virtual, July 2020. doi: https://doi.org/10.2118/199062-MS

9.4.3.11 Application-34: Downhole Leak Detection: Introducing a New Wireline Array Noise Tool

9.4.3.11.1 Background of the Application

This paper presents a new generation wireline Array Noise Tool (ANT) that is designed to detect downhole acoustic/vibration activities originating from the fluid-structure friction flow. One of the main applications of ANT is to locate leak sources in well completions and tubulars, particularly in well integrity (WI) and plug & abandonment (P&A) operations. The novel sensor matrix and system configuration, together with three novel data processing methods, are developed to address the primary challenges associated with leak detection in noisy and complex downhole environments. The proposed methods include a multi-dimensional ML classification module, wave velocity-domain decomposition method, and Bayesian likelihood analysis. The tool has been validated through lab tests, field logs, and commercial jobs in multiple regions worldwide.

9.4.3.11.2 Description of the Workflow and Machine Learning Model

The objective of the ML model in this workflow is to analyze downhole acoustic signals to detect leaks using the wireline ANT. The author started the process with data conditioning, where a moving average filter is applied to smooth the raw acoustic signals, reducing noise and improving the signal quality. Next, a wave velocity-domain decomposition method is used to isolate frequency ranges associated with potential leaks. This enhances the signal-to-noise ratio (SNR) by filtering out irrelevant noise. Finally, a Bayesian likelihood analysis is conducted to estimate the probability of leaks at various depths. This analysis combines prior knowledge with the processed data to provide a clear visualization of leak probabilities along the well's depth.

9.4.3.11.3 Advantages and Limitations of the Synchronized Algorithm

The advantages of this methodology include improved data quality through noise reduction and enhanced leak detection by isolating relevant frequencies. The Bayesian analysis offers a robust way to estimate leak probabilities, increasing confidence in the results. However, the noise reduction process might remove some important signal details, and the frequency isolation assumes distinct leak frequencies, which might not always be accurate. Despite these limitations, this approach provides a reliable way to identify leaks and can be further refined with additional field data.

9.4.3.11.4 Results and Discussion

In conclusion, the new wireline ANT has been successfully developed and verified for downhole leak detection and monitoring. With the novel sensor matrix and data processing methods, the ANT has addressed the challenges of weak acoustic leak identification that is contaminated by strong road noise signals, non-stationary acoustic measurements, and downhole seismic noise. The tool has been successfully deployed and tested in various field cases, including multi-annulus leak cases, low flow rate leakage cases, and gas production well field examples, with a leak detection success rate of over 95%. The ANT provides a reliable and efficient solution for leak detection in well completions and tubulars in various downhole environments.

BIBLIOGRAPHY

Yang, Qinshan, Zhao, Jinsong, and Rourke, Marvin. "Downhole Leak Detection: Introducing A New Wireline Array Noise Tool." Paper presented at the *SPE/ICoTA Well Intervention Conference and Exhibition, The Woodlands*, Texas, USA, March 2019. doi: https://doi.org/10.2118/194264-MS

9.4.3.12 Application-35: Shape Recognition of Remained Oil Based on BP Neural Network

9.4.3.12.1 Background of the Application

In the field of oil field development, the method of oil field development determines the high or low recovery of oil and gas. Oilfield development methods are mainly divided into primary, secondary and tertiary oil recovery. Due to the structural characteristics of the reservoir pore throat, the physical properties of the rock and the viscosity of the crude oil, the underground residual oil that cannot be recovered after a certain extraction method is called residual oil. In order to further study the residual oil, petroleum geology researchers classify the residual oil into different forms and study the suitable replacement methods for different forms of residual oil, so as to fully exploit the potential of oil and gas reservoirs and improve the oil repelling efficiency of oil fields. Therefore, the research on the classification and identification of residual oil forms in microscopic replacement experiments can provide an important theoretical basis for the correct formulation of the development plan of secondary and tertiary oil recovery in oil fields, so as to improve the recovery rate of oil fields.

9.4.3.12.2 Description of the Workflow and Machine Learning Model

Based on the extracted morphological feature parameters, each remaining oil target in the repulsion model image is identified one by one, and the specific remaining oil distribution of the whole model can be obtained. However, in the experimental analysis stage, due to the large amount of data, it is therefore important to use a classifier that can learn intelligently and quickly and accurately for the remaining oil morphology classification and identification work. ANN has a strong ability of self-adaptation, learning, and fault tolerance. In a manner similar to human thinking, it obtains output results by analyzing potential patterns between a series of input and output data provided in advance and then extrapolating new input parameters based on these patterns. This provides a new idea to solve the nonlinear residual oil morphology identification problem. BPNN is a feed-forward NN with an input layer, an implicit layer, and an output layer. It uses the fast descent method in nonlinear programming to make the network weights and threshold parameters vary along the negative gradient direction of the error function in order to minimize the mean square difference value between the actual output value and the desired output value of the network to obtain the appropriate NN structure parameters, and finally, the new morphological feature parameters are computed and discriminated by the trained network parameters to obtain the recognition results.

9.4.3.12.3 Advantages and Limitations of the Synchronized Algorithm

The research on the classification and identification of residual oil morphology in microscopic replacement experiments can provide an important theoretical basis for the correct formulation of the development plan of secondary and tertiary oil recovery in oilfields, thus improving the recovery rate of oilfields. The limitation is the complex use of mathematical algorithm theory, which requires deep theoretical skills.

9.4.3.12.4 Results and Discussion

The remaining oil morphology distribution, with strong fuzziness and uncertainty, and BPNN has the ability of self-learning, self-adaptive, association, and fuzzy judgment recognition. Through the experiment, the classification recognition result of BPNN for the remaining oil is basically consistent with its actual morphology. This indicates that the BPNN has certain reliability in the classification and recognition of the remaining oil distribution. Therefore, using BPNN to quickly and correctly discriminate the morphology of remaining oil provides an important theoretical basis for the oil field to develop a reasonable way to continue exploitation and improve the oil drive efficiency of the oil field, which has great practical application significance.

BIBLIOGRAPHY

Mao, G. Q., Teng, Q. Z., Wu, Y. et al (2014). Shape recognition of remained oil based on BP neural network. *J Terahertz Sci Electron Inf Technol* 12(6):858–864.

9.4.3.13 Application-36: Predicting Wax Formation Using Artificial Neural Network

9.4.3.13.1 Background of the Application

In deepwater oilfield production operations, wax formation is an issue. Wax deposition may result in clogged flowlines, production delays, the need for more pumping power to ensure smooth flow, or even the total abandonment of the flowline system. In order to prevent and mitigate the ensuing wax deposits, it is crucial to accurately estimate the conditions under which wax formation will begin. An ANN method was employed in this study to forecast WAT for different hydrocarbon blends. The model was validated with experimental data after being trained using various combinations of twelve distinct hydrocarbon fluids. The ANN method outperforms the current algorithms in predicting WAT. Compared to the current models, the ANN's average absolute deviation (AAD%) was lower. The combination of molecular weight, density, and activation energy as input parameters provided the best prediction.

9.4.3.13.2 Description of the Workflow and Machine Learning Model

ANNs were originally designed to mimic the information processing functions of neuronal networks in the brain. A NN consists of a set of interconnected artificial neurons that use connectionist computational methods to process information. ANNs are adaptive systems that change their structure during the learning phase by internal or external information about the network. Modern NNs are a tool for modeling nonlinear statistics and do not contain preconceived ideas about the way the model is structured. They are used to model complex relationships between inputs and outputs, or to find patterns in data. ANNs learn to solve problems by developing a memory capable of associating a large number of input patterns with the resulting output or effect set. The ANN was trained to develop a solution system by the given example.

The most important aspect of ANNs is their ability to find ambiguous patterns in the data that are difficult to detect for standard observations and standard statistical methods. While traditional correlations are based on simple models that often must be extended by adding terms and constants to become flexible enough to fit experimental data, NNs are surprisingly adaptive.

9.4.3.13.3 Advantages and Limitations of the Synchronized Algorithm

Ongoing education NNs strive for constant learning and performance improvement. After training, the system can generate outputs without requiring full inputs. The application or program gets easier to use as it is used. Information Recovery Retrieval is the primary advantage of employing online data management or cloud services. The complete software backup is accessible online in the event that the hardware is damaged or fails. As a result, the system doesn't require constant training. One of the frequent benefits of NNs is their ability to multitask; the new, sophisticated algorithms are made to be able to multitask and deliver different results. Multiple tasks can be scrolled through simultaneously by users.

The derivations cannot be tracked and verified by human analysts, even if the results are true. Since most NNs are black-box systems that rely on experience rather than a predetermined process to produce outcomes, they are challenging to alter. Several theorems are employed to approximate the results, providing only one probable value. None of the theories are entirely appropriate to provide the desired outcome in every situation. One of the startling issues with NNs is this ambiguity.

9.4.3.13.3.1 Dependency On Data

The machine will behave in accordance with the data that is supplied to it. The accuracy of the output increases with the amount of data used in the training process. The dependence on data is one of the main drawbacks of NNs, as some people have to be on the maintenance side to see it. As there are errors in the data, the results will be wrong, which can pose a serious threat.

9.4.3.13.4 Results and Discussion

Table 9.1 shows the WAT predictions and error estimates when each thermodynamic property is used. The predictions for the network with density (D) as the input parameter are more accurate than when using the other properties. The error analysis shows that density provides a minimum percentage error of 6.68% and a minimum squared error (MSE) of 0.04. ANN methods have been used to predict the starting conditions for wax precipitation. Combining MW, D, and activation energy as input parameters allows for the best prediction of moisture content. The weight of the ANN model function can be saved to provide fast and reliable predictions for new reservoir fluids. The model can be modified as new experimental data become available. The newly

TABLE 9.1
WAT Prediction Using One Thermodynamic Property

	Expt.	AE	KV	MW	D	MPP
Sample No.			(°C)			
1	31	37.0	35.9	35.1	33.1	32.1
2	39	38.4	383	37.5	37.1	34.5
5	40	37.6	35.6	33.7	34.8	41.2
8	38	38.3	382	37.9	37.2	33.7
MSE		0.05	0.05	0.07	0.04	0.05
Error (%)		°6.90	°7.29	°8.25	°6.68	°7.32

predicted data should be within the training range of values. However, the availability of more experimental data will increase the efficiency and improve the prediction performance of the network.

BIBLIOGRAPHY

Adeyemi, B. J., and Sulaimon, A. A. "Predicting Wax Formation Using Artificial Neural Network." In *Nigeria Annual International Conference and Exhibition*, 6–8 August. Society of Petroleum Engineers, Lagos, pp 1–8. 2012.

9.4.3.14 Application-37: A Working Condition Diagnosis Model of Sucker Rod Pumping Wells Based on Big Data Deep Learning

9.4.3.14.1 Background of the Application

Timely and accurate monitoring of the working conditions of pumping rod pump wells is of great importance to oil production. With the development of smart oilfield, more and more sensors are installed on the wells and the monitoring data are continuously transmitted to the data center to form big data. This study aims to establish a new generation of intelligent diagnostic models to monitor the working conditions of the pumping rod pump wells by using the big data collected during the production of oil wells and DL technology.

9.4.3.14.2 Description of the Workflow and Machine Learning Model

CNN is a class of deep neural networks in DL. It is most commonly used to analyze data with grid patterns, such as images. In a CNN, each neuron is only connected to a small, localized patch of the previous layer—just as neurons in the animal visual cortex respond to stimuli within a limited "receptive field." CNNs do not require preprocessing of image data to obtain features such as texture and shape, but rather do not require manual effort and prior knowledge to design and extract features. It takes the pixel matrix of an image directly as input and learns how to extract features to classify the image. Based on these features, CNNs have been applied to image and video recognition, image classification, natural language processing, and many other fields. A typical CNN architecture consists of an input layer, an output layer, multiple convolutional layers, a pooling layer, and a fully connected layer. During the training or testing of a CNN model, each input image will be passed through a series of convolutional layers with filters (also called kernels), pooling layers, and fully connected layers, and an activation function is applied to calculate the probability value for each classification.

For CNN architectures, a convolutional layer is used to extract the intrinsic features from the image and generate a series of feature mappings. The pooling layer is used to reduce the dimensionality of the feature mappings, reduce computational complexity, and increase computational speed while maintaining accuracy. A flat layer is used to map features into a vector for further computation. The information in the feature map is integrated layer by layer using a fully connected layer, and finally the probabilities of the inputs belonging to each class are output to the neurons in the output layer. The number of convolutional layers, the number of pooling layers, the number of fully connected layers, and the relationship between each layer, as well as the parameter settings for each layer are all factors controlled by the CNN design. Systematically exploring the impact of each parameter on CNN performance is very complex. In this study, working condition diagnostic CNNs were designed by analyzing current popular and successful CNN architectures and comparing the differences in the problems addressed by each architecture.

9.4.3.14.3 Advantages and Limitations of the Synchronized Algorithm

There is a discrepancy between training validation accuracy (99%) and operational accuracy (90%). This situation stems mainly from the following reasons: first, the CNN training process is a repetitive learning process. Each dynamometer card in the dataset was repeatedly learned by CNN 10 times. This allows it to achieve 99% training and verification accuracy. In real applications, the dynamometer cards are newly generated and have not been learned by the CNN before, so the accuracy rate decreases. Second, although the dataset contains many representative dynamometer cards for each operating condition, it is far from covering all possible cards. To further close the gap in accuracy, the number of samples in the dataset needs to be continuously increased.

9.4.3.14.4 Results and Discussion

Based on the trained CNN model, a sucker rod pump well operating condition monitoring software was developed. The back-end of the software is run at a local workstation for 724 hours to diagnose the working conditions of the wells. The front end is used to display the results of the working condition diagnosis to the oilfield user. The software also has a feedback learning workflow that periodically updates the CNN model to improve its performance. The software has been applied to an oilfield and field trials have shown that the actual accuracy is greater than 90%. The application results show that the proposed CNN pumping rod pump well working condition diagnosis is a promising method.

BIBLIOGRAPHY

Wang, X., He, Y., Li, F., Dou, X., Wang Z., Xu, H., and Fu, L. "A Working Condition Diagnosis Model of Sucker Rod Pumping Wells Based on Big Data Deep Learning. In: *International Petroleum Technology Conference* 26–28 March, Beijing, China, pp 1–10. 2019.

9.4.3.15 Application-38: An Improved Genetic Algorithm Optimization Fuzzy Controller Applied to the Wellhead Back Pressure Control System

9.4.3.15.1 Background of the Application

The paper proposes an improved GA optimization fuzzy controller to overcome the limitations of the traditional PID controller for wellhead back pressure control. The wellhead back pressure control system has strong nonlinearity and time variability, which poses a challenge for precise closed-loop control. The authors establish a wellhead back pressure control model and calculate the transfer function. Then, they design an improved GA to optimize the highly nonlinear fuzzy control rules in the fuzzy PID controller. Finally, they compare the developed model with four traditional controllers to demonstrate the superiority of the proposed controller in terms of time response parameters, speed, stability, and robustness. The improved GA optimization fuzzy controller significantly improves the performance of the wellhead back pressure control system.

9.4.3.15.2 Description of the Workflow and Machine Learning Model

The author used a fuzzy logic controller, boosted by a GA, to manage the wellhead back pressure in drilling operations. Essentially, it takes real-time pressure data and tweaks the control settings on the fly to keep things running smoothly and safely. The GA helps by fine-tuning the fuzzy control rules, making sure the system can handle the tricky and ever-changing conditions of drilling. The ultimate goal is to keep the pressure right where it needs to be, enhancing the efficiency and safety of the drilling process.

9.4.3.15.3 Advantages and Limitations of the Synchronized Algorithm

Combining fuzzy logic with a GA brings some great benefits for controlling wellhead back pressure. This setup is really good at adapting to the complex, unpredictable nature of drilling, ensuring precise control even with fluctuating inputs. Plus, it boosts the system's accuracy and robustness. On the flip side, it does need a lot of computing power and can be pretty hard to understand and explain how it works. However, despite these challenges, this method significantly improves control performance and makes drilling operations safer and more efficient.

9.4.3.15.4 Results and Discussion

The authors propose an improved GA optimization fuzzy controller for the wellhead back pressure control system. They establish a wellhead back pressure control model and design an improved GA to optimize fuzzy control rules in the fuzzy PID controller. The proposed controller exhibits superior performance in terms of time response parameters, speed, stability, and robustness compared to traditional controllers. The results demonstrate the effectiveness of the proposed controller in improving the performance of the wellhead back pressure control system.

BIBLIOGRAPHY

Liang, Haibo, Zou, Jialing, Zuo, Kai, and Khan, Muhammad Junaid. 2020. An improved genetic algorithm optimization fuzzy controller applied to the wellhead back pressure control system. *Mech Syst Sig Process*, 142, 106708, ISSN 0888-3270. https://doi.org/10.1016/j.ymssp.2020.106708

9.4.3.16 Application-39: Advanced Real-Time Gas Kick Detection Using Machine Learning Technology

9.4.3.16.1 Background of the Application

This paper proposes a method for the early monitoring of gas kick using AI and ML methods. It relies entirely on the data generated during the drilling process, providing a pure data-driven approach for the early monitoring of gas kick. The proposed method involves dimensionality reduction of data features, screening out parameters related to gas kick monitoring and applying PCA to find the relationship between those parameters. An ANN is then used to identify gas kick through the analyzed data. The proposed method is verified using field data, and the results show that the method can realize early monitoring of gas kick, which has great significance for the safety of drilling operations.

9.4.3.16.2 Description of the Workflow and Machine Learning Model

In this method, an ANN combined with PCA is used for real-time gas kick detection in drilling operations. PCA is used for dimensionality reduction, transforming the input features into a set of principal components. The ANN is then trained on these principal components to learn patterns and make predictions about gas kicks, utilizing a structure that includes multiple layers to capture complex relationships in the data.

9.4.3.16.3 Advantages and Limitations of the Synchronized Algorithm

The ANN–PCA model has the advantage of efficiently handling high-dimensional data, reducing computational load through PCA while, in the meantime, maintaining the ability to detect complex patterns via the ANN. This makes it good for real-time applications where quick and accurate detection is crucial. However, the model requires significant amounts of data for training to ensure accuracy and robustness. The black-box nature of NNs makes it hard to interpret the model. Additionally, while PCA aids in reducing dimensionality, it may also have an effect which some information may be lost.

9.4.3.16.4 Results and Discussion

In conclusion, this paper has proposed an effective early-stage monitoring and warning system for deep-water gas kick to ensure the safety of drilling operations. The method relies entirely on the data generated during the drilling process, providing a pure data-driven approach for early monitoring of gas kick and avoiding the shortcomings of relying on personnel and short discovery time in the traditional method. Through the dimensionality reduction of data features, parameters related to gas kick monitoring were screened out to provide certain guidance for the early monitoring of gas kick in the actual drilling process. Through the verification of the cases, the method realizes the early monitoring of gas kick, which has great significance for the safety of drilling operations.

BIBLIOGRAPHY

Yang, Jin, Sun, Ting, Zhao, Ying, Borujeni, Ali Takbiri, Shi, Haidong, and Yang, Hao. "Advanced Real-Time Gas Kick Detection Using Machine Learning Technology." Paper presented at the *29th International Ocean and Polar Engineering Conference*, Honolulu, Hawaii, USA, June 2019.

9.4.3.17 Application-40: Semi-Supervised Learning for Lithology Identification Using Laplacian Support Vector Machine

9.4.3.17.1 Background of the Application

The accurate identification of lithology is essential in petroleum exploration and production. In this paper, a semi-supervised ML approach using Laplacian support vector machine (LapSVM) is proposed for lithology identification. The approach utilizes unlabeled well logging data and a small subset of labeled data to train the LapSVM model, which is then used to classify the remaining unlabeled data. K-means clustering is used to select the most informative samples for labeling. Feature similarity and depth similarity are also incorporated to improve classification accuracy. The proposed approach is tested on real well logging data from a carbonate reservoir in the Middle East. The results show that the proposed approach outperforms other commonly used ML methods, especially for minority lithology classes.

9.4.3.17.2 Description of the Workflow and Machine Learning Model

In this method, a semi-supervised learning approach is used for lithology identification using a LapSVM. The model incorporates both labeled and unlabeled data to improve classification performance. Key components include building similarity matrices in feature and geographic spaces to capture both data characteristics and spatial relationships. These matrices are used to form a Laplacian matrix, which helps the LapSVM optimize a decision function.

9.4.3.17.3 Advantages and Limitations of the Synchronized Algorithm

The LapSVM approach uses both labeled and unlabeled data that enhances the model's ability to generalize from limited labeled samples. This method is particularly useful in scenarios where acquiring labeled data is expensive or time-consuming. However, building similarity matrices and optimizing the LapSVM can be computationally expensive. Additionally, the choice of parameters for the similarity matrices and the kernel function significantly impacts performance, requiring careful tuning and checking of settings to ensure optimal results.

9.4.3.17.4 Results and Discussion

In this paper, a semi-supervised ML approach using LapSVM was proposed for lithology identification. The approach utilizes unlabeled well logging data and a small subset of labeled data to train the LapSVM model, which is then used to classify the remaining unlabeled data. K-means clustering is used to select the most informative samples for labeling. Feature similarity and depth similarity are also incorporated to improve classification accuracy. The proposed approach is tested on real well logging data from a carbonate reservoir in the Middle East. The results show that the proposed approach outperforms other commonly used ML methods, especially for minority lithology classes. The proposed approach has the potential to significantly improve the accuracy and efficiency of lithology identification, which can ultimately contribute to better decision making in petroleum exploration and production.

BIBLIOGRAPHY

Li, Z., Kang, Y., Feng, D., Wang, X. M., Lv, W., Chang, J., and Zheng, W. X. (2020). Semi-supervised learning for lithology identification using Laplacian support vector machine. *J Pet Sci Eng*, 195, 107510. ISSN 0920-4105. https://doi.org/10.1016/j.petrol.2020.107510

9.4.4 Other Applications

9.4.4.1 Application-41: Predicting Natural Gas Production Using Artificial Neural Network

9.4.4.1.1 Background of the Application

This study introduces a novel method for utilizing an ANN to forecast natural gas output in the United States. The process entails creating a NN model to predict the availability of natural gas in the United States through 2020. Drilling, production, economic, and exploratory data are all included in the model's input pool. The input data was preprocessed using functional transformation and normalization, and the NN was made simpler using the sensitivity analysis of the input variables and dimension reduction techniques. The quick-propagation learning algorithm was used to successfully train the model using annual data from 1950 to 1989. The natural gas production rate is the NN's intended output. Excellent agreement was found between the actual and expected output rates. The NN model has use as a short-term as well as a long-term predictive tool of natural gas supply.

9.4.4.1.2 Description of the Workflow and Machine Learning Model

The ANN is the main means used in this paper, including the calling of relevant packages, the cleaning and analysis of data, and the construction of each level of the neural network. Among them, the sensitivity analysis in the data processing part is very helpful for the selection of variables, and the more relevant ones are successfully applied. The use of this dimensionality reduction method successfully reduces the amount of operations and improves the accuracy of correlation. An econometric approach, called stochastic modeling or time-series analysis, was used to develop forecasting models for the NN input parameters. A hardware and/or software system that mimics how neurons function in the human brain is called an ANN. ANNs are a type of DL technology that is also referred to as AI. Many processors working in parallel and organized in tiers are often used in an ANN. Similar to optic nerves in human visual processing, the first tier gets the raw input data. Similar to how neurons farther away from the optic nerve receive signals from those closer to it, each succeeding layer receives the output from the tier before it instead of the raw input. The system's output is generated by the last tier.

9.4.4.1.3 Advantages and Limitations of the Synchronized Algorithm

The advantage of the ANN is that it has the self-learning function, which is of particular importance for prediction. It is expected that the future ANN computer will provide economic prediction, market prediction, and benefit prediction for human beings, and its application future is very ambitious.

The limitations of the ANN is that the interpretability of NNs is relatively poor, and a relatively reliable and predictive result may be obtained after comparison, but the correlation between the independent and dependent variables is very ambiguous. In addition, problems such as gradient explosion can easily arise if reasonable regularization is not used in the prediction process. There is a certain complexity and expertise in the tuning and application of parameters of NNs.

9.4.4.1.4 Results and Discussion

The ANN model developed in this study provides an accurate and reliable method for predicting natural gas production in the United States. The model can be used as a short-term as well as a long-term predictive tool of natural gas supply, and can also be used to examine quantitatively the effects of the various physical and economic factors on future gas production. The methodology presented in this paper can be extended to other areas of science, engineering, and finance, and can help in the development of more accurate and reliable predictive models in these fields.

BIBLIOGRAPHY

Al-Fattah, S. M., and Startzman, R. A. "Predicting Natural Gas Production Using Artificial Neural Network." Paper presented at the *SPE Hydrocarbon Economics and Evaluation Symposium*, Dallas, Texas, April 2001. doi: https://doi.org/10.2118/68593-MS

9.4.4.2 Application-42: Artificial Neural Network Models for Identifying Flow Regimes and Predicting Liquid Holdup in Horizontal Multiphase Flow

9.4.4.2.1 Background of the Application

This paper presents ANN models for identifying flow regimes and predicting liquid holdup in horizontal multiphase flow. The models were developed using 199 experimental data sets and three-layer BPNNs). Inputs to the models included superficial gas and liquid velocities, pressure, temperature, and fluid properties. The results showed that the developed models provided better predictions and higher accuracy than empirical correlations previously developed for these data groups.

9.4.4.2.2 Description of the Workflow and Machine Learning Model

This paper presents two ANN models to identify the flow regime and calculate the liquid holdup in horizontal multi-phase flow. These models are developed with 199 experimental data sets and with three-layer BPNNs).

BP is an algorithm that performs gradient descent on multi-layer ANNs, which means that the gradient of the loss function is calculated using the chain rule with the weights of each layer of the network as variables to update the weights to minimize the loss function.

9.4.4.2.3 Advantages and Limitations of the Synchronized Algorithm

The number of neurons reaches a certain number, and the number of weight parameters reaches a certain number. It is obviously unrealistic to set the parameters of the NN manually. Thus, the power of big data can be used to correct the parameters of the NN through statistical methods. The advantage of the statistical method is that it saves the trouble of setting the rules manually.

The corresponding disadvantage is that the accuracy can be flawed. Therefore, it is important to pay attention to the environment of use when using statistical methods.

9.4.4.2.4 Results and Discussion

The ANN models developed in this paper provide reliable predictions of flow regimes and liquid holdup in horizontal multiphase flow. The flow-regime model predicted correctly for more than 97% of the data points, while the liquid-holdup model outperformed published models with an average absolute percent error of 9.407, a standard deviation of 8.544, and a correlation coefficient of 0.9896. The use of ANN models in multiphase flow is becoming increasingly popular due to their ability to recognize complex patterns within available data and provide accurate predictions.

BIBLIOGRAPHY

Osman, El-Sayed A. (2004). "Artificial neural network models for identifying flow regimes and predicting liquid holdup in horizontal multiphase flow." *SPE Prod Facil* 19: 33–40. doi: https://doi.org/10.2118/86910-PA

9.4.4.3 Application-43: A Fiber-Optic Assisted Multilayer Perceptron Reservoir Production Modeling: A Machine Learning Approach in Prediction of Gas Production from Shales

9.4.4.3.1 Background of the Application

This study aims to predict daily gas production from the Marcellus Shale using a predictive data-driven model that utilizes distributed temperature sensing (DTS) and distributed acoustic sensing (DAS) fiber-optic systems. The study used 1320 DTS measurements along the lateral of the well MIP-3H, which were upscaled to a stage scale by an averaging method. A multi-layer perceptron neural network (MLPNN) was trained with stage-based daily DTS data and daily flowing time to predict gas production for the next day. The study also carried out a sensitivity analysis to identify the most influential stages in predicting gas production. Results show that DAS data is more correlated with sensitivity analysis results than production logging data.

9.4.4.3.2 Description of the Workflow and Machine Learning Model

The code reads a file, cleans the data by handling missing values and outliers, encodes the data, and then splits it into training and testing sets. It then uses a Multi-Layer Perceptron (MLP) classifier to perform DL on the dataset. Finally, it plots the Receiver Operating Characteristic (ROC) curve to evaluate the performance of the classifier.

9.4.4.3.3 Advantages and Limitations of the Synchronized Algorithm

The code uses MLP, a powerful DL technique that can handle complex non-linear relationships between features and target variables. The code performs data cleaning and outlier handling, which can improve the accuracy and reliability of the model.

The ROC curve provides an intuitive visualization of the model's performance. The code only uses one ML model, and it is not clear if this is the best model for this particular dataset. The code does not perform feature selection, which could lead to overfitting or reduce the accuracy of the model. The code does not provide a thorough explanation of the data, including the meaning and relevance of each feature, which could limit the interpretability of the results.

9.4.4.3.4 Results and Discussion

The study presents a predictive data-driven model that utilizes DTS and DAS fiber-optic systems to predict daily gas production from the Marcellus Shale. A multi-layer perceptron NN was trained with daily flowing time and stage-based DTS data to predict gas production for the next day. The sensitivity analysis carried out in the study identified the most influential stages in predicting gas production, and the results suggest that DAS data is more correlated with sensitivity analysis results than production logging data. The study shows that the use of fiber-optic systems can provide valuable insights and improve the accuracy of predicting gas production in unconventional reservoirs.

BIBLIOGRAPHY

Ghahfarokhi, Payam Kavousi, Carr, Timothy, Bhattacharya, Shuvajit, Elliott, Justin, Shahkarami, Alireza, and Martin, Keithan. "A Fiber-Optic Assisted Multilayer Perceptron Reservoir Production Modeling: A Machine Learning Approach in Prediction of Gas Production From the Marcellus Shale." Paper presented at the *SPE/AAPG/SEG Unconventional Resources Technology Conference*, Houston, Texas, USA, July 2018. doi: https://doi.org/10.15530/URTEC-2018-2902641

9.4.4.4 Application-44: Simulating the Behavior of Reservoirs with Convolutional and Recurrent Neural Networks

9.4.4.4.1 Background of the Application

The potential of ML algorithms to learn reservoir behavior from data was investigated using RNNs and CNNs in this paper. RNNs are able to retain information in the memory of previous patterns and information contained in the previous behavior of phenomena being modeled, while CNNs have shown promise in sequence modeling domains. RNN and CNN were used for activities that would typically be handled by a reservoir simulator. The link between the physical quantities of interest from subsurface reservoirs was formulated as a sequence mapping issue. In order to examine their capacities in a range of tasks relevant to reservoir engineers, the performance of a CNN layer in comparison to an RNN was carefully assessed. According to preliminary findings, CNNs can predict information sequences just as well as RNNs and are equally dependable when drawing conclusions about situations the algorithm hasn't encountered during training.

9.4.4.4.2 Description of the Workflow and Machine Learning Model

The workflow in this study involves loading the required data files, defining the RNN and CNN models using the Keras library, and training and evaluating the models.

The input parameters used in the models include the choke profile, bottom-hole pressure (BHP), drill stem test (DST) data, and pressure and choke settings of approximately 12000 minutes. The output of the models is the prediction of BHP. The RNN model used in this study consists of a single LSTM layer with 32 neurons followed by a dense output layer. The input shape of the model is (None, 1), which means that the model can handle input sequences of varying lengths. The RNN model is trained using the mean squared error loss function and the Adam optimizer. The CNN model used in this study consists of a single convolutional layer with 32 filters and a kernel size of 5, followed by a max pooling layer, a flatten layer, and a dense output layer. The input shape of the model is (None, 1), which means that the model can handle input sequences of varying lengths. The CNN model is also trained using the mean squared error loss function and the Adam optimizer.

9.4.4.4.3 Advantages and Limitations of the Synchronized Algorithm

The study demonstrates the potential of using DL models, specifically RNN and CNN, to simulate the behavior of reservoirs. Both models were found to be capable of modeling sequences of information and making inferences to cases that have not been seen by the algorithm during training. This has implications for the oil and gas industry, as effective use of digital technologies and automation is expected to reduce capital expenditure up to 20% and better position companies that implement these technologies for growth opportunities.

One of the advantages of using RNNs is their ability to retain information in the memory of previous patterns and information contained in the previous behavior of phenomena being modeled. On the other hand, CNNs are useful for sequence modeling domains, and they have the advantage of allowing for parallel computing. The CNN memory can be controlled and has more interpretive capacity than the RNN memory.

However, the study also has some limitations. The number of parameters in a network is a function of many factors, including the number of variables modeled, number of neurons/filters, and the number of layers. This makes the parameter count problem and architecture-specific. In addition, the authors note that the information contained in the choke profile, BHP, and DST data is limited, and more data may be needed for improved accuracy. Finally, the units of the input parameters may vary depending on the source of the data.

9.4.4.4.4 Results and Discussion

This paper investigated the potential of ML algorithms, specifically RNNs and CNNs, to learn reservoir behavior from data. The results suggest that CNNs are as capable as RNNs in modeling sequences of information, even when the reservoir modeler is faced with incomplete knowledge of the underlying physical system. By formulating the relationship between physical quantities of interest from subsurface reservoirs as a sequence mapping problem, the study found that RNN and CNN were able to perform tasks that traditionally would be modeled by a reservoir simulator. The performance of a CNN layer as compared to RNN was evaluated systematically to investigate their capabilities in a variety of tasks of interest to the reservoir engineer. These findings are promising for the oil and gas industry, as the effective use of digital technologies and automation is expected to reduce capital expenditure up to 20% and better position companies that implement these technologies for growth opportunities.

BIBLIOGRAPHY

Alakeely, Abdullah, and Horne, Roland N. (2020). "Simulating the Behavior of Reservoirs with Convolutional and Recurrent Neural Networks." *SPE Res Eval Eng* 23, 992–1005. doi: https://doi.org/10.2118/201193-PA

9.4.4.5 Application-45: Comparison of Decline Curve Analysis DCA with Recursive Neural Networks (RecNN) for the Production Forecast of Multiple Wells

9.4.4.5.1 Background of the Application

This research suggests a new method for employing recursive neural networks (RecNN) to simulate time-series related problems like production forecast. Conventional techniques, such as decline curve analysis (DCA) and numerical simulations, lack flexibility when modeling complex physics and necessitate a deep understanding of the subject. The study used RNN and DCA to forecast production from both single and multiple wells. For production forecasting, a number of LSTM-based sequence-to-sequence models, including many-to-one, many-to-many, and one-to-one, were effectively deployed. To create more features to aid in the training process, feature engineering was done. It was shown that the created RNN-based sequence-to-sequence models could be easily expanded to simulate additional time-series-related issues including failure prediction and condition-based maintenance.

9.4.4.5.2 Description of the Workflow and Machine Learning Model

In this method, an LSTM neural network is used for time-series prediction, specifically targeting the forecasting of production rates over time. The LSTM model is especially good at capturing temporal dependencies in data, making it well for predicting future values based on historical sequences. The LSTM structure consists of multiple hidden layers designed to learn the complex temporal patterns within the data, ultimately producing a sequence of predicted production rates.

9.4.4.5.3 Advantages and Limitations of the Synchronized Algorithm

The main advantage of using an LSTM model is its ability to handle sequential data with long-term dependencies, making it suited for time-series forecasting. The model can adapt to varying patterns over time, providing robust predictions even in the presence of some error in the data. However, LSTM models require deep training data and significant computational resource because of their complexity. Additionally, the interpretability of LSTM models can be challenging, as the internal workings of the network are not as understandable as traditional DCA methods.

9.4.4.5.4 Results and Discussion

This study proposes a data-driven approach for evaluating well production rates for single and aggregated well time-series models based on available LSTM algorithms in open-source libraries. The developed approach was validated and compared with actual data history and traditional DCA methods, showing much better trend forecasting and fewer errors. The approach considers complex operation scenarios that could reflect more unconventional reservoir characterization, which DCA methods cannot implement. Compared with numerical simulation physics-based models, this NN approach gives solutions faster and is a good alternative to provide quick and robust results beside of empirical and numerical simulations.

BIBLIOGRAPHY

Sun, J., Ma, X., and Kazi, M. "Comparison of Decline Curve Analysis DCA with Recursive Neural Networks RNN for Production Forecast of Multiple Wells." Paper presented at the *SPE Western Regional Meeting, Garden Grove*, California, USA, April 2018. doi: https://doi.org/10.2118/190104-MS

9.4.4.6 Application-46: Application of Artificial Intelligence to Forecast Hydrocarbon Production from Shales

9.4.4.6.1 Background of the Application

This paper discusses the use of surrogate models for predicting oil and gas production from hydraulically fractured low permeability reservoirs. The study compares three types of surrogate models, namely Least Square Support Vector Machine (LSSVM), ANNs, and Response Surface Model (RSM), in terms of their performance and accuracy. The study shows that RSM and LSSVM have very accurate oil recovery forecasting capabilities, while LSSVM shows the best performance for complex GOR behavior. The paper emphasizes the importance of surrogate models in reservoir engineering and highlights the potential of using AI techniques to analyze and forecast hydrocarbon production.

9.4.4.6.2 Description of the Workflow and Machine Learning Model

The goal of this ML model is to predict hydrocarbon production from shale reservoirs using eight key input parameters. The author started the process with data preparation, where the data is divided into training and testing sets and then standardized to ensure consistency. Next, three different surrogate models—LSSVM, ANNs, and RSM—are trained on the prepared data. The LSSVM and ANN models are optimized using PSO to fine-tune their parameters, while the RSM uses a polynomial equation to fit the data. The performance of these models is then evaluated using metrics like the coefficient of determination (R2) and the Normalized Root Mean Square Error (NRMSE) to ensure accurate predictions.

9.4.4.6.3 Advantages and Limitations of the Synchronized Algorithm

This methodology offers several benefits, including the ability to handle complex relationships between the input parameters and the output using advanced ML techniques. The optimization process with PSO helps in achieving the best possible model performance, ensuring more accurate hydrocarbon production forecasts. However, there are limitations as well. The training process for ANN and LSSVM can be computationally intensive, and the models might require fine-tuning to avoid overfitting. Additionally, the polynomial approach of RSM may not capture all the complexities of the data. Despite these challenges, the combined use of LSSVM, ANN, and RSM provides a robust framework for predicting hydrocarbon production, which can be improved with more data and advanced optimization techniques.

9.4.4.6.4 Results and Discussion

This study demonstrates the successful application of surrogate models for predicting oil and gas production from hydraulically fractured low permeability reservoirs. The study shows that LSSVM and RSM are the most accurate models for forecasting oil recovery and LSSVM is the most accurate model for predicting gas-oil ratio. The study shows that all three surrogate models can be used as reliable proxy reservoir models and can provide fast fluid recovery forecasts and sensitivity analyses. The paper highlights the potential of using AI techniques such as ANN and LSSVM to improve the accuracy of reservoir engineering models. Overall, the study provides valuable insights into the use of surrogate models and AI techniques for analyzing and forecasting hydrocarbon production.

BIBLIOGRAPHY

Panja, Palash, Velasco, Raul, Pathak, Manas, and Deo, Milind. (2018). Application of artificial intelligence to forecast hydrocarbon production from shales. *Petroleum*, 4(1), 75–89, ISSN 2405-6561. https://doi.org/10.1016/j.petlm.2017.11.003

9.4.4.7 Application-47: Machine Learning for the Predicting Properties of Porous Media from 2d X-ray Images

9.4.4.7.1 Background of the Application

This study suggests a technique for utilizing CNNs with micro-computed tomography (micro-CT) X-ray images as input data to estimate the physical characteristics of porous material. Porosity, specific surface area, and average pore size are among the physical characteristics considered in this work. Greyscale and binary images, along with the associated calculated attributes, are used to train the suggested CNN architecture independently. When the model is trained using binary images, the relative error in determining porosity, surface area, and average pore size is less than 6%; when greyscale photos are used, the error is less than 7%. With possible uses in geology, materials science, petroleum engineering, and other disciplines, this approach offers a quick and automated means of estimating the physical characteristics of porous media.

9.4.4.7.2 Description of the Workflow and Machine Learning Model

In this method, CNNs are used to estimate the physical properties of porous media from micro-CT X-ray images. The images are normalized and used to train the CNN, which identifies spatial features relevant to porosity, specific surface area, and average pore size. The model's predictions, derived from both binary and greyscale images, are compared to measured values, showing high accuracy with relative errors.

9.4.4.7.3 Advantages and Limitations of the Synchronized Algorithm

The CNN-based method detects spatial patterns in micro-CT images, important for the accurate estimation of porous media. It automates the extraction of features, handling large datasets effectively. However, training CNNs is computationally expensive and requires high-quality, representative data. Inconsistent or noisy data can affect model performance negatively, emphasizing the need for thorough data preprocessing.

9.4.4.7.4 Results and Discussion

In this work, the authors propose a CNN-based method for estimating the physical properties of porous media using micro-CT X-ray images. The results show that the CNN model can accurately estimate porosity, specific surface area, and average pore size with a low relative error. The proposed method provides a rapid and automated way to estimate the physical properties of porous media, which has potential applications in various fields. The authors suggest that future work could focus on improving the accuracy of the CNN model by incorporating more advanced image processing techniques and exploring the use of other ML algorithms. Overall, this work contributes to the development of more efficient and accurate methods for characterizing porous media.

BIBLIOGRAPHY

Alqahtani, N., Alzubaidi, F., Armstrong, R. T., Swietojanski, P., and Mostaghimi, P. 2020. Machine learning for predicting properties of porous media from 2D X-ray images. *J Pet Sci Eng*, 184, 106514, ISSN 0920-4105, https://doi.org/10.1016/j.petrol.2019.106514

9.4.4.8 Application-48: A Novel Early Warning System of Oil Production Based on Machine Learning

9.4.4.8.1 Background of the Application

This paper proposes an effective early warning system (EWS) based on ML to evaluate the production condition of different reservoirs and predict the possibility of anomalous decline in mature oilfields. The EWS is established by selecting factors that can affect the abnormal decline and comparing their influence using grey relational analysis. The ML model is optimized based on the variation of accuracy under different input parameter numbers, and artificial samples are added to the training process to improve the prediction accuracy. The EWS can effectively guide the oilfield production strategy and is a useful tool for oil production monitoring in mature oil fields.

9.4.4.8.2 Description of the Workflow and Machine Learning Model

In this method, grey relational analysis (GRA) and NNs are used to create an early warning system (EWS) for oil production anomalies. The input data is normalized and analyzed using GRA to identify key features, which are then used to train a NN model. This model predicts potential production declines, allowing for earlier measures to be taken based on significant deviations from expected performance.

9.4.4.8.3 Advantages and Limitations of the Synchronized Algorithm

The GRA and NN combination identifies and uses critical production factors, enhancing the accuracy of anomaly detection. GRA makes sure that the selection of features is relevant, while the NN captures complex relationships within the data. However, this method requires deep historical data and consistent data quality to maintain accurate predictions. Also, the model may need frequent updates to remain effective under changing conditions.

9.4.4.8.4 Results and Discussion

In conclusion, the proposed EWS based on ML is an effective method for predicting abnormal oil production decline and maintaining stable oil production in mature Chinese oil fields. The main factors that cause abnormal oil production decline are the production composition, and the appropriate data input form and number of input features are crucial for achieving high prediction accuracy. Furthermore, adding artificial samples can be an effective way to overcome the problem of unbalanced sample size and improve the prediction accuracy. The EWS can provide early warning of abnormal oil production decline, allowing petroleum engineers to take steps in advance to prevent financial loss.

BIBLIOGRAPHY

Ma, Kang, Jiang, Hanqiao, Li, Junjian, Zhang, Rongda, Zhang, Lufeng, Fang, Wenchao, Shen, Kangqi, and Dong, Rencheng. "A Novel Early Warning System of Oil Production Based on Machine Learning." Paper presented at the *Abu Dhabi International Petroleum Exhibition & Conference*, Abu Dhabi, UAE, November 2019. doi: https://doi.org/10.2118/197365-MS

9.4.4.9 Application-49: Digital Twin for Oil Pipeline Risk Estimation Using Prognostic and Machine Learning Techniques

9.4.4.9.1 Background of the Application

This paper presents a digital twin-based approach to analyze and predict the risk probability rate of oil pipeline systems. The proposed approach uses ML and prognostics algorithms to model the pipeline system's behavior and predict the likelihood of failure. The system uses pressure data from sensors in the oil pipeline system as input and employs several AI methods such as prognostic-manifold learning, Dirichlet Process Clustering, Canopy clustering, and Kernel-based SVM algorithm to predict the probability of failure. The results show that the proposed approach can accurately predict the risk probability rate of the oil pipeline system.

9.4.4.9.2 Description of the Workflow and Machine Learning Model

In this project, the author used advanced ML techniques to predict risk probabilities in oil pipelines. By analyzing real-time sensor data with PCA and SVM algorithms, the system can effectively identify potential issues. PCA reduces data dimensionality, emphasizing critical patterns, while SVM, a powerful classification method, separates normal from abnormal pressure conditions. This combined approach allows for accurate and timely risk assessments, providing operators with valuable insights to preemptively address any emerging problems.

9.4.4.9.3 Advantages and Limitations of the Synchronized Algorithm

The integration of PCA and SVM offers a robust framework for managing complex pressure data in oil pipelines. PCA's ability to simplify data by highlighting significant variations makes it an excellent tool for initial pattern recognition. SVM's strength in handling complex data relationships further refines the system's predictive capabilities. However, implementing these methods demands substantial computational power and specialized expertise. Despite these hurdles, the enhanced predictive accuracy and early warning system significantly boost pipeline reliability, ensuring safer and more efficient operations.

9.4.4.9.4 Results and Discussion

In this paper, a digital twin-based approach is proposed to predict the risk probability rate of oil pipeline systems using ML and prognostics algorithms. The proposed approach uses pressure data from sensors in the oil pipeline system as input and employs several AI methods such as prognostic-manifold learning, Dirichlet Process Clustering, Canopy clustering, and Kernel-based SVM algorithm to predict the probability of failure. The results showed that the proposed approach can accurately predict the risk probability rate of the oil pipeline system. The proposed approach can be useful for pipeline operators to prioritize maintenance activities, reduce downtime, and prevent accidents. Future work can focus on incorporating additional data sources and developing more sophisticated algorithms to improve the accuracy of the predictions.

BIBLIOGRAPHY

Priyanka, E. B., Thangavel, S., Gao, Xiao-Zhi, and Sivakumar, N. S. (2022). Digital twin for oil pipeline risk estimation using prognostic and machine learning techniques. *J Ind Inf Integr*, 26, 100272, ISSN 2452-414X, https://doi.org/10.1016/j.jii.2021.100272

9.4.4.10 Application-50: Spatial Prediction of Oil and Gas Distribution Using Tree Augmented Bayesian Network

9.4.4.10.1 Background of the Application

Exploration efficiency, exploration target optimization, and exploration risk reduction all depend on the spatial prediction of hydrocarbon distribution. The Tree Augmented Bayesian Network (TAN) approach is presented in this research as a means of forecasting the spatial distribution of oil and gas resources. Geological data, exploration and drilling findings, and geophysical data are among the seven categories of geoscientific information that are used to determine the TAN structure. The trained TAN model is used to determine the target layer's hydrocarbon-bearing probability, and a probability map of the predicted hydrocarbon spatial distribution is produced. The method is applied to the first member of Dongying Formation (Ed1) of Nanpu Depression in Bohai Bay Basin as an example, and the results show that it can serve to visualize the relationship between various geological factors, predict the spatial distribution of hydrocarbons, minimize exploration risk, optimize exploration targets, and provide a theoretical basis for drilling decision-making.

9.4.4.10.2 Description of the Workflow and Machine Learning Model

In this method, the TAN, a type of Bayesian network effective in handling uncertainty in spatial data, is used. TAN models are particularly effective for making probabilistic predictions and classifications. They go beyond traditional Bayesian networks by allowing each node to have an additional parent, forming a tree-like structure. This technique predicts the spatial distribution of hydrocarbons by integrating geological data, drilling results, and geophysical data, resulting in a probability map indicating hydrocarbon presence.

9.4.4.10.3 Advantages and Limitations of the Synchronized Algorithm

The TAN has advantages for spatial predictions as it effectively captures relationships between geological variables and provides interpretable results. It generates probabilistic outputs that help visualize geological risks and optimize exploration efforts. However, TAN assumes that data can be discretized into a few categories, which may not always reflect complex geological formations accurately. Also, while it reduces complexity more than fully connected networks, it would still have problems with very high-dimensional datasets and may require more detailed input for variable selection and data preparation.

9.4.4.10.4 Results and Discussion

This study suggests a TAN-based approach to forecast the spatial distribution of hydrocarbons. Seismic interpretation, basin simulation, and data from 222 exploratory wells are used to train the TAN structure. A probability map of the predicted hydrocarbon spatial distribution is produced by using the trained TAN model to compute the target layer's hydrocarbon-bearing posterior probability. There are two benefits to the suggested approach: (1) the network topological structure makes it possible to see and understand the link between geological variables; (2) the Bayesian Network has a strong theoretical basis. The application findings demonstrate that the suggested TAN technique can give a theoretical foundation for drilling decision-making, decrease exploratory risk, optimize exploration targets, and forecast the geographical distribution of hydrocarbons. However, there is still room for improvement in improving the accuracy of model prediction.

BIBLIOGRAPHY

Ren, Hong-Jia, Wang, Xian-Chang, Guo, Qiu-Lin, Guo, Xiao-Xin, and Zhang, Rui. (2020). Spatial prediction of oil and gas distribution using Tree Augmented Bayesian network, *Comput Geosci*, 142, 104518, ISSN 0098-3004. https://doi.org/10.1016/j.cageo.2020.104518

10 AI Applications in Renewable Resources

Cenk Temizel, Salih Tutun, Celal Hakan Canbaz, Ekrem Alagoz, Emre Can Dündar, Mufrettin Murat Sari, Luigi Saputelli, and Mehmet Melih Oskay

10.1 INTRODUCTION

The integration of artificial intelligence (AI), machine learning (ML), and deep learning (DL) in renewable resources is revolutionizing how clean energy is harnessed and managed. As the world increasingly shifts toward sustainable energy sources, these technologies play a crucial role in optimizing the performance, efficiency, and reliability of renewable energy (RE) systems. AI and ML are being applied across various RE domains—such as solar, wind, geothermal, and bioenergy—helping to predict energy production, enhance grid integration, and minimize operational costs. These technologies enable more precise forecasting, real-time monitoring, and intelligent decision-making, ultimately leading to more efficient energy generation and distribution.

In this chapter, the application of AI, ML, and DL methodologies across various renewable energy (RE) technologies will be explored. The focus will be on how these advanced tools are being used to improve solar technology, optimize wind energy production, enhance geothermal energy systems, and advance bioenergy applications. By examining these areas, the transformative impact of AI and ML on the renewable energy sector will be highlighted, demonstrating their potential to accelerate the global transition to sustainable energy solutions.

10.1.1 Solar Technology

The application of AI and ML in solar technology is transforming the way solar energy is harnessed, managed, and optimized. One of the key areas where AI and ML are making a significant impact is in solar energy forecasting. Accurate prediction of solar power generation is critical for effective grid management and energy planning. ML algorithms analyze historical weather data, solar radiation patterns, and other environmental factors to predict solar energy output with high accuracy. These predictions enable grid operators to better balance supply and demand, reducing the reliance on fossil fuels and enhancing the stability of the power grid.

 DOI: 10.1201/9781003617327-10

AI and ML are also being utilized to optimize the performance of solar photovoltaic (PV) systems. These technologies can analyze data from solar panels in real time to detect inefficiencies, identify potential faults, and optimize energy output. For instance, AI-driven algorithms can adjust the angle of solar panels or optimize energy storage based on real-time conditions, maximizing the efficiency of energy capture and usage. Furthermore, AI and ML are instrumental in advancing the design and manufacturing processes of solar cells, leading to the development of more efficient and cost-effective solar technologies. By integrating AI and ML into solar energy systems, the RE sector can significantly improve the efficiency, reliability, and scalability of solar power generation.

10.1.2 Wind Energy

AI and ML are playing a pivotal role in advancing wind energy technology by optimizing the operation, maintenance, and overall efficiency of wind farms. One of the most significant applications is in wind power forecasting, where ML algorithms analyze vast datasets, including historical wind speed, atmospheric conditions, and turbine performance, to predict future wind energy output. Accurate forecasting is essential for grid stability and efficient energy distribution, as it helps operators make informed decisions about energy storage, load balancing, and grid integration. AI-driven models can predict short-term fluctuations and long-term trends in wind power generation, enabling more reliable and efficient management of wind energy resources.

In addition to forecasting, AI and ML are being used to enhance the performance and longevity of wind turbines. Predictive maintenance is a critical area where these technologies are making a substantial impact. By analyzing real-time data from turbine sensors, AI algorithms can detect anomalies and predict potential failures before they occur, reducing downtime and maintenance costs. Furthermore, AI can optimize turbine operation by adjusting the blade angles and rotational speeds based on real-time wind conditions, maximizing energy capture and minimizing wear and tear. This not only improves the efficiency of individual turbines but also enhances the overall performance of wind farms, contributing to more sustainable and cost-effective wind energy production.

10.1.3 Geothermal and Heat Transfer

AI and ML are increasingly being integrated into geothermal energy systems and heat transfer processes to improve efficiency, reliability, and resource management. Geothermal energy, which involves harnessing the heat from beneath the Earth's surface, presents unique challenges due to the complex and variable nature of subsurface conditions. AI and ML technologies are helping to address these challenges by enabling more accurate modeling and prediction of geothermal reservoirs. ML algorithms analyze geological, geophysical, and thermal data to predict the location and behavior of geothermal resources. This predictive capability allows for more efficient exploration, reducing the risk and cost associated with drilling and resource extraction.

In heat transfer processes, which are crucial for both geothermal systems and broader energy applications, AI and ML are used to optimize the design and operation of heat exchangers and other thermal systems. These technologies can model complex heat transfer dynamics in real time, allowing for precise control and optimization of thermal processes. For instance, AI-driven models can adjust the flow rates, temperatures, and pressures in a geothermal plant to maximize energy extraction while minimizing operational inefficiencies. Additionally, AI can be used to predict and prevent scaling, corrosion, and other issues that can degrade the performance of heat transfer systems, extending their lifespan and improving overall system reliability. By integrating AI and ML into geothermal energy and heat transfer processes, the RE sector can achieve greater efficiency and sustainability in harnessing Earth's thermal resources.

10.1.4 Bioenergy and Other Applications

AI and ML technologies are making significant strides in the bioenergy sector, helping to optimize the production, conversion, and utilization of biological resources for energy. Bioenergy involves the use of organic materials, such as plant biomass, agricultural waste, and algae, to produce RE in the form of biofuels, biogas, and bioelectricity. One of the key applications of AI and ML in bioenergy is in optimizing the feedstock supply chain. ML algorithms analyze data from various sources, including crop yields, weather patterns, and market prices, to predict the availability and cost of biomass resources. This enables more efficient planning and management of the bioenergy supply chain, ensuring a consistent and cost-effective supply of feedstock for energy production.

In addition to supply chain optimization, AI and ML are being used to enhance the efficiency of bioenergy conversion processes. For example, AI algorithms can optimize the parameters of biochemical and thermochemical conversion processes, such as fermentation, gasification, and pyrolysis, to maximize energy output and minimize waste. These technologies can also monitor and control bioenergy production systems in real time, adjusting operational conditions to respond to changes in feedstock composition or process dynamics. Beyond bioenergy, AI and ML are being applied to other RE applications, such as energy storage, grid integration, and energy efficiency. By leveraging AI and ML, the bioenergy sector and other RE fields can achieve greater sustainability, reliability, and efficiency in the transition to a low-carbon energy future.

10.2 OVERVIEW OF AI METHODOLOGIES

The application of AI, ML, and DL methodologies in RE is driving significant advancements across various domains, including solar, wind, geothermal, and bioenergy. These methodologies provide powerful tools for analyzing complex datasets, identifying patterns, and making predictions that enhance the efficiency and reliability of RE systems. In the context of renewable resources, AI techniques are employed for tasks such as forecasting energy production, optimizing system performance, and managing resources effectively.

AI methodologies often involve rule-based systems and decision trees (DTs) that help in automating decision-making processes and optimizing operations based on pre-defined criteria. ML, particularly supervised and unsupervised learning, is extensively used to analyze historical and real-time data, enabling the prediction of energy outputs, equipment failures, and resource availability. For instance, regression models, clustering techniques, and neural networks (NNs) are commonly applied to improve the accuracy of energy production forecasts and optimize the operation of RE systems.

DL, a subset of ML, leverages multi-layered NNs to model complex relationships and patterns in large datasets. In RE, DL techniques are particularly valuable for tasks that involve image and signal processing, such as analyzing satellite imagery for solar radiation assessment or processing seismic data for geothermal exploration. The continuous advancement of AI methodologies is crucial for overcoming the challenges associated with RE deployment, improving the integration of renewable resources into the energy grid, and ultimately contributing to a more sustainable energy future.

10.3 APPLICATIONS OF AI

10.3.1 Solar Technology

10.3.1.1 Application-1: The Effective Solar Panel Tracker to Obtain Maximum Energy Optimization by Using Innovative Machine Learning

10.3.1.1.1 Background of the Application

This paper proposes an innovative machine learning (IML) method for optimizing solar trackers. The proposed method is compared with existing methods, including optical efficiency and performance optimization (OEPO), novel optical optimization model (NOOM), optimization based on reinforcement learning (OBRL), and adaptive neuro-fuzzy inference system (ANFIS). The comparison is based on the level of integrated tracking, distributed tracking, combined tracking, heat-based tracking, and high-power solar tracking. The proposed IML method achieved better results than the existing methods in all categories.

10.3.1.1.2 Description of the Workflow and Machine Learning Model

This paper presents an IML model designed to enhance the tracking of solar panels for maximum energy optimization. The proposed model utilizes ML techniques to monitor and adjust the positioning of solar panels to achieve optimal energy absorption. The tracking system focuses on several modes of operation, including integrated tracking, distributed tracking, combined tracking, heat-based tracking, and high-power solar tracking. The ML model continuously evaluates the solar panel's angle and adjusts it to maximize the amount of solar energy collected. The system leverages data from multiple inputs, including the solar panel's position, environmental conditions, and energy requirements, to optimize its operation.

10.3.1.1.3 Advantages and Limitations of the Synchronized Algorithm

The proposed IML model demonstrated superior performance in optimizing solar energy collection compared to existing methods such as OEPO, NOOM, and

ANFIS. The IML model achieved high tracking efficiency in various modes, including 87.20% for integrated tracking, 91.59% for distributed tracking, and 93.77% for high-power solar tracking. However, the system's performance is dependent on the accuracy of the input data and the complexity of the environmental conditions. While the IML model offers significant improvements in energy optimization, the need for continuous monitoring and adjustment may present challenges in large-scale implementations.

10.3.1.1.4 Results and Discussion

In conclusion, the proposed IML method for solar tracker optimization achieved better results than the existing methods. The results demonstrate the potential of ML in improving the efficiency of solar energy generation. The proposed method can be used to optimize different types of solar trackers and can contribute to the widespread adoption of solar energy. However, further research is needed to validate the proposed method in real-world scenarios and to explore the potential of other AI methods in solar energy optimization.

BIBLIOGRAPHY

Kaushal, R. K., V. Velmurugan, A. P. M. Mahendran, B. Singh, and J. A. Dhanraj. The Effective Solar Panel Tracker to Obtain Maximum Energy Optimization by Using Innovative Machine Learning. *2023 International Conference on Artificial Intelligence and Knowledge Discovery in Concurrent Engineering (ICECONF)*, Chennai, India, 2023, pp. 1–8. doi: 10.1109/ICECONF57129.2023.10084249

10.3.1.2 Application-2: Machine Learning in Renewable Energy Application: Intelligence System for Solar Panel Cleaning

10.3.1.2.1 Background of the Application

In order to address the effect of dust on solar panel efficiency and output power, this paper suggests an electrical system that includes a dust sensor, an Arduino Uno M-controller, two end races to indicate the panel edges, a circuit L298D that serves as a bridge to power the engine, and a DC motor for the ballet's translatory movement. The suggested method divides the required level of cleanliness into four groups using ML algorithms that classify the amount of cleanliness based on the sensor reading. The article also covers a number of ML techniques that have been used in the cleaning of solar panels, including DTs and NNs.

10.3.1.2.2 Description of the Workflow and Machine Learning Model

This paper explores the development of an automatic cleaning system for PV solar panels using ML algorithms. The proposed system involves two phases: sensor-based data collection and classification using supervised learning. The K-Nearest Neighbors (KNN) algorithm is employed to classify dust levels into four categories: no dust, little dust, dusty, and very dusty. The sensor readings are taken during both day and night to improve accuracy. The ML model uses the data gathered from the sensor to predict the appropriate cleaning action required, with an accuracy rate of 95% during nighttime readings.

10.3.1.2.3 Advantages and Limitations of the Synchronized Algorithm

The ML-based system offers an efficient and automated solution for maintaining solar panel cleanliness, which is crucial for optimizing energy output. The KNN algorithm performs well in classifying dust levels, especially when the sensor readings are stable during nighttime. However, the system's performance is affected by environmental conditions such as heat and light, which can cause fluctuations in daytime sensor readings. Additionally, the model requires a well-calibrated sensor system and sufficient training data to ensure accurate classification. While effective, the system may need further enhancements to handle diverse environmental challenges more robustly.

10.3.1.2.4 Results and Discussion

This research suggests an electrical system that includes a dust sensor, an Arduino Uno M-controller, two end races to indicate the panel edges, a circuit L298D that serves as a bridge to power the engine, and a DC motor for the ballet's translatory movement. The system employs ML algorithms to categorize the required level of cleanliness into four groups based on the sensor's reading. In terms of precisely identifying and categorizing the amount of dust on solar panels, the technology exhibits encouraging results. The study also explores how ML techniques, such as DTs and NNs, might increase the dependability and efficiency of solar panel cleaning systems.

BIBLIOGRAPHY

Al-Dahoud, Ahmad, Mohamed Fezari, and Ali Aldahoud. Machine Learning in Renewable Energy Application: Intelligence System for Solar Panel Cleaning. Transactions on Environment and Development, 2023, p. 472–478. World Scientific and Engineering Academy and Society (WSEAS) https://doi.org/10.37394/232015.2023.19.45

10.3.1.3 Application-3: Daily Solar Irradiation Estimation over a Mountainous Area Using Artificial Neural Networks

10.3.1.3.1 Background of the Application

In this paper, the authors develop an artificial neural network (ANN) model to estimate daily global irradiation on complex terrain. The model considers various input parameters such as longitude, latitude, altitude, slope, azimuth, day of year, daily extraterrestrial solar radiation, and daily clearness index for the reference station. The ARD methodology is employed to determine the most important geographic variable in the estimation of daily global radiation. The model performance is satisfactory, with an RMSE of 6.0% when comparing the estimated values with the measured ones over the whole validation dataset.

10.3.1.3.2 Description of the Workflow and Machine Learning Model

This paper presents a methodology for estimating daily global solar irradiation over mountainous terrain using ANNs. The study leverages a multi-layer perceptron (MLP) ANN model, trained on data from 12 radiometric stations located in the Sierra Nevada National Park in Spain. The model incorporates topographical variables such as altitude, slope, and azimuth, along with meteorological inputs, to predict daily solar radiation levels. The ANN approach is employed as an alternative to traditional interpolation/extrapolation techniques, which often struggle with accuracy in complex terrains.

10.3.1.3.3 Advantages and Limitations of the Synchronized Algorithm

The ANN model provides an efficient and accurate method for estimating solar radiation over complex topographies. The use of the automatic relevance determination (ARD) methodology allows the model to select the most significant input variables, improving its predictive capabilities. The model achieves a root mean square error (RMSE) of 6.0% and a mean bias error (MBE) of 0.2%, demonstrating its reliability. However, the performance of the model is dependent on the quality of the training data and the proper selection of input variables. Additionally, the complexity of the terrain introduces challenges in capturing local variations in solar radiation, which may require further refinement of the model.

10.3.1.3.4 Results and Discussion

In order to deliver appropriate daily global radiation levels on complex terrain, the authors have demonstrated the feasibility of an ANN-based technique. The site altitude was identified as the most significant geographic variable in the estimation of daily global radiation by the ARD methodology used. It is also demonstrated that the DOY and the daily clearness index at a reference station are pertinent input variables. When comparing the estimated and measured values over the entire validation dataset, the ANN model that was constructed performs satisfactorily, with an RMSE of 6.0%. Although the methods used in this work calibrated the model proposed for the location under study, other mountainous regions with complicated terrain can benefit from the same approach.

BIBLIOGRAPHY

Bosch, J. L., G. López, and F. J. Batlles. Daily solar irradiation estimation over a mountainous area using artificial neural networks. *Renewable Energy*, 33, 7, 2008, 1622–1628, ISSN 0960-1481. https://doi.org/10.1016/j.renene.2007.09.012

10.3.1.4 Application-4: Modelling a Solar-Assisted Air-Conditioning System Installed in CIESOL Building Using an Artificial Neural Network

10.3.1.4.1 Background of the Application

This paper presents an ANN model for predicting the performance of a solar-assisted air-conditioning system. The model aims to estimate the efficiency of the absorption chiller and the global system, including the coefficient of performance (COP), cooling capacity, global system's efficiency 1, and global system's efficiency 2. The ANN model was trained using real data from an operating solar-assisted air-conditioning system, and the accuracy of the model was evaluated based on the RMSE and MBE of the predicted values. The results demonstrate accurate predictions from the ANN model, yielding an RMSE less than 1.9% and practically null deviation, which can be considered very satisfactory.

10.3.1.4.2 Description of the Workflow and Machine Learning Model

This paper models a solar-assisted air-conditioning system installed in the CIESOL building using ANNs. The system consists of a single-effect LiBr–H_2O absorption chiller powered by solar energy. The study aims to predict the efficiency of the chiller and the overall system using an ANN with a 7-8-4 topology, where the input layer includes seven variables such as temperatures, flow rates, and radiation intensity, the hidden layer has eight neurons, and the output layer predicts four key performance indicators (KPIs), including the COP and global system efficiency. The ANN was trained using real-world data collected from the CIESOL system during the cooling period of 2008.

10.3.1.4.3 Advantages and Limitations of the Synchronized Algorithm

The ANN model demonstrated strong predictive capabilities, achieving an RMSE of less than 1.9% across all KPIs. This level of accuracy highlights the ANN's effectiveness in capturing the complex nonlinear relationships between the system's input variables and its performance. However, the accuracy of the model is contingent upon the quality and completeness of the training data. Additionally, while the ANN model provides an efficient way to model the system, it requires careful tuning of hyperparameters and adequate data preprocessing to avoid overfitting. The model may also need to be recalibrated periodically as the system's operational conditions change.

10.3.1.4.4 Results and Discussion

This study outlines a method for configuring an ANN model to forecast a solar-assisted air conditioner's efficiency. Real data was used to train the ANN model, which proved to be useful in this context. This study's primary objective was to identify the distinct ANN model with the fewest possible input variables. With an RMSE of less than 1.9% and almost null deviation, the results show that the ANN model produced accurate predictions. This is quite gratifying. The ANN model used in this work was created for a 70 kW rated WFC SC 20-hot water driven single-effect LiBr–H_2O absorption chiller, although it is readily adaptable to other absorption chillers. Future studies in this field should focus on the use of the ANN model to predict the efficiency of the chiller and global system when the system works at various operation modes, considering the energy provided from all heat sources.

BIBLIOGRAPHY

Rosiek, S., and F. J. Batlles. Modelling a solar-assisted air-conditioning system installed in CIESOL building using an artificial neural network. *Renewable Energy*, 35, 12, 2010, 2894–2901. ISSN 0960-1481, https://doi.org/10.1016/j.renene.2010.04.018

10.3.1.5 Application-5: Forecasting Solar Energy Production Using Machine Learning

10.3.1.5.1 Background of the Application

An integrated ML model and statistical method for forecasting future solar power generation from RE facilities are presented in this research. By combining statistical analysis with ML techniques, the suggested hybrid model increases accuracy.

Two case studies are examined in the paper; the former is simulated for bigger PV farms with 100,000 PV cells and smaller farms with 1000 PV cells.

The AI methods employed in this paper include feature selection, filter and wrapper techniques, ensemble learning, and ML algorithms such as support vector machines (SVMs) and ANNs. The input parameters include time series data on PV solar panels, environmental variables such as weather conditions, and the output of PV solar panels.

10.3.1.5.2 Description of the Workflow and Machine Learning Model

This paper presents a hybrid model that integrates ML and statistical approaches for forecasting solar energy production. The study uses an ensemble of ML models, including SVMs and DL algorithms, to predict solar electricity generation based on historical weather data. The model incorporates a multimodel ensemble feature selection method, which combines both filter and wrapper techniques to select the most relevant features for improving prediction accuracy. The hybrid model is trained on data from multiple PV farms, with the goal of optimizing solar energy forecasts across various weather conditions.

10.3.1.5.3 Advantages and Limitations of the Synchronized Algorithm

The ensemble model outperforms individual ML models in terms of prediction accuracy, particularly in scenarios involving varying weather patterns. The integration of both ML and statistical methods enhances the model's ability to generalize across different datasets and time intervals. However, the success of the model depends on the quality of the input data and the proper tuning of the ensemble components. Additionally, the computational complexity of the ensemble model can be a limitation, especially when scaling to larger datasets or real-time forecasting applications.

10.3.1.5.4 Results and Discussion

In conclusion, the proposed ML model outperforms traditional methods in predicting solar power generation from RE plants. The hybrid model integrates ML methods and the statistical method, resulting in improved accuracy. Ensemble learning and feature selection techniques were also employed to improve the performance of the model. Future work includes the exploration of DL mechanisms to further improve the model's accuracy and performance.

BIBLIOGRAPHY

Vennila, C., A. Titus, T. S. Sudha, U. Sreenivasulu, N. P. R. Reddy, K. Jamal, … and A. Belay. Forecasting solar energy production using machine learning. *International Journal of Photoenergy*, 2022(1), 2022, 7797488. https://doi.org/10.1155/2022/7797488

10.3.1.6 Application-6: A Machine Learning-Based Novel Energy Optimization Algorithm in a Photovoltaic Solar Power System

10.3.1.6.1 Background of the Application

This paper presents a study on the performance of linear Fresnel reflectors (LFRs) for solar power generation. The proposed model incorporates nonlinear kernels in the convolution layers of a convolutional neural network (CNN) to optimize the performance of the LFR. The model is compared with existing models such as stacked autoencoders (SAE), ANNs, and long short-term memory (LSTM) models. The proposed model outperforms these models in real-world applications due to its ability to process local spatial and temporal correlations of data through parallel processing. The paper also discusses the design of the LFR and its potential advantages in terms of increasing energy consumption, operating hours, and thermal efficiency while reducing land use and costs per unit of energy.

10.3.1.6.2 Description of the Workflow and Machine Learning Model

This study introduces an ML-based energy optimization algorithm aimed at improving the performance of PV solar power systems. The proposed method leverages an ML approach, specifically utilizing CNNs, to optimize solar energy capture and system performance. The LFR is designed to maximize solar energy absorption by tracking sunlight in the north–south direction, and the CNN model is employed to predict and optimize power output based on input variables such as solar radiation, temperature, and panel characteristics. The model operates on the Carnot cycle to predict energy conversion efficiency, minimizing the need for manual calculations. By integrating ML into the optimization process, the system can automate predictions and adjustments, improving the overall energy output of the PV system.

10.3.1.6.3 Advantages and Limitations of the Synchronized Algorithm

The integration of CNNs into the PV system optimization process provided significant advantages in terms of automation and accuracy. The ML model demonstrated strong performance in predicting solar energy output and optimizing system settings, leading to increased energy efficiency and reduced human intervention. Additionally, the model outperformed traditional methods by maximizing electrical output while minimizing land usage and component costs. However, the approach has limitations. The complexity of the CNN model requires substantial computational resources, and the system's performance depends heavily on the quality of the input data, such as accurate measurements of solar radiation and environmental conditions. Moreover, while the proposed model showed improvements in simulation results, real-world applications may present challenges related to the variability of environmental factors and the cost of implementing such advanced systems at scale.

10.3.1.6.4 Results and Discussion

In conclusion, the proposed ML-based LFR model is a promising solution for improving the performance of LFR for solar power generation. The model outperforms existing models such as SAE, ANNs, and LSTM models in real-world applications due to its ability to process local spatial and temporal correlations of data through parallel processing. The improved LFR design is expected to increase energy consumption, operating hours, and thermal efficiency while reducing land use and costs per unit of energy. While the LFR has certain potential financial drawbacks, it is still worth investigating, especially in scenarios with high land costs where effective land use becomes even more advantageous. Further research is needed to address the constraints of the LFR and enhance its cost-effectiveness.

BIBLIOGRAPHY

Prasad, Kalapala, J. Samson Isaac, P. Ponsudha, N. Nithya, Santaji Krishna Shinde, S. Raja Gopal, Atul Sarojwal, K. Karthikumar, and Kibrom Menasbo Hadish. A machine learning-based novel energy optimization algorithm in a photovoltaic solar power system. *International Journal of Photoenergy*, 2022, 2022, 2845755. https://doi.org/10.1155/2022/2845755

10.3.1.7 Application-7: Solar Panel Modelling Through Computational Intelligence Techniques

10.3.1.7.1 Background of the Application

This paper presents a study on estimating the maximum power point (MPP) of a solar panel using various computational intelligence paradigms. The input variables used in the study are the temperature, voltage, and current of the panel at different working conditions. The computational intelligence paradigms used in this study include fuzzy C-means (FCM), radial basis function (RBF), k-nearest neighbors (k-NN), and feedforward neural networks (FFNs). The goal of the study is to compare the performance of these paradigms in estimating the MPP. The FFN model was found to achieve the lowest error, which was 2.20 and 7.55 times smaller than the error of k-NN and FCM, respectively, but comparable with the RBF error.

10.3.1.7.2 Description of the Workflow and Machine Learning Model

This study focuses on predicting the MPP of PV panels using various computational intelligence techniques. The MPP represents the most efficient operating condition for a solar panel. To predict the MPP, the study utilizes several soft-computing paradigms, including FCM clustering, RBF networks, FFNs, and k-NN. These models use easily measurable input variables such as voltage, current, and temperature to estimate the MPP, eliminating the need for more complex and expensive measurements like solar irradiance. The study compares the performance of these techniques in predicting MPP under different operational conditions of the solar panel.

10.3.1.7.3 Advantages and Limitations of The Synchronized Algorithm

The computational intelligence models, particularly FFN and RBF, demonstrated strong performance in predicting the MPP with high accuracy. FFN achieved the lowest prediction error among the tested models, followed closely by RBF. The ability of these models to generalize from the training data without requiring explicit input–output mapping models is a significant advantage. However, there are limitations to the approach. The FCM model, while simple, produced higher errors compared to other models, and k-NN, despite better performance, had higher computational costs. The success of these models is also dependent on the quality and distribution of the training data, and large errors occurred in regions of low current, presenting challenges for accurate prediction in those scenarios. Additionally, the need for extensive computational resources, particularly for FFN and k-NN, limits their real-time applicability in certain cases.

10.3.1.7.4 Results and Discussion

In conclusion, the study shows that the estimation of the MPP of a solar panel can be effectively achieved using computational intelligence techniques. Among the paradigms tested, the FFN model achieved the lowest error, demonstrating its potential for practical applications. The study also reveals that large errors occur in low current regions of the input space. To address this issue, a more advanced learning strategy, such as a hierarchical RBF model, could be used to improve the accuracy while preserving the simplicity of the approach. Overall, this study provides valuable insights into the use of computational intelligence techniques for the estimation of the MPP of a solar panel.

BIBLIOGRAPHY

Ferrari, Stefano, Massimo Lazzaroni, Vincenzo Piuri, Ayşe Salman, Loredana Cristaldi, Marco Faifer, and Sergio Toscani. Solar panel modelling through computational intelligence techniques. *Measurement*, 93, 2016, 572–580, ISSN 0263-2241. https://doi.org/10.1016/j.measurement.2016.07.032

10.3.1.8 Application-8: Artificial Intelligence-Based Forecasting & Optimization of Solar Cell Model

10.3.1.8.1 Background of the Application

This paper explores the use of various soft computing techniques for predicting solar radiation, including MLP (NN), ANFIS, and SVM. The paper discusses the different input parameters used in the study, such as surface air pressure ratio, temperature ratio, sunshine duration ratio, and relative humidity ratio. The results of the study are presented in graphs and tables, demonstrating the accuracy of the different soft computing techniques used.

10.3.1.8.2 Description of the Workflow and Machine Learning Model

This study focuses on forecasting and optimizing the performance of solar cells using various AI and soft computing techniques. The methods employed include ANNs, ANFIS, and SVMs, which are used to predict and optimize solar radiation and the performance of PV systems. Key input parameters for the models include sunshine duration, temperature, and humidity, while the clearness index is used as an output parameter. The ANFIS model, in particular, uses a combination of fuzzy logic and NNs to optimize the solar cell model's performance by dynamically adjusting parameters based on the input data. Additionally, the study incorporates least square support vector machine (LSSVM) for future solar radiation prediction, further enhancing the accuracy of the model.

10.3.1.8.3 Advantages and Limitations of the Synchronized Algorithm

The AI-based forecasting and optimization methods demonstrated significant advantages in accurately predicting solar radiation and optimizing solar cell performance. The ANFIS model, with its hybrid learning rule combining backpropagation and least-squares algorithms, provided high accuracy, particularly when using Gaussian membership functions. Additionally, LSSVM showed strong predictive capabilities with minimal errors, making it a valuable tool for future solar radiation forecasting. However, limitations were noted, including the computational complexity of these AI models and the need for extensive training data. The performance of the models is highly dependent on the quality of input data, and errors may occur under certain conditions, such as low solar radiation levels. Moreover, the need for tuning multiple parameters in these AI models can be resource-intensive.

10.3.1.8.4 Results and Discussion

This study has shown that the proposed LSSVM algorithm can accurately predict solar radiation using input parameters and ML techniques. The Gaussian RBF optimization method was found to produce the most accurate results. The study also found that the magnitude of the active current associated with PV inverters can be determined by solar irradiation. While the reactive current associated with the PV inverter can be controlled within certain limitations, the units of the parameters used in this study were not explicitly specified.

BIBLIOGRAPHY

Mohanty, Sthitapragyan, Prashanta Kumar Patra, Asit Mohanty, Meera Viswavandya, and Prakash K. Ray. Artificial intelligence-based forecasting & optimization of solar cell model, *Optik*, 181, 2019, 842–852, ISSN 0030-4026. https://doi.org/10.1016/j.ijleo.2018.12.176

10.3.1.9 Application-9: An Enhanced Productivity Prediction Model of Active Solar Still Using Artificial Neural Network and Harris Hawks Optimizer

10.3.1.9.1 Background of the Application

This paper presents a study on predicting and optimizing the performance of solar stills using a modified ANN based on Harris hawks optimizer (HHO–ANN). The input parameters consist of solar irradiance, ambient temperature, time, wind speed, and vapor velocity. The paper compares the HHO–ANN model with traditional ANN and SVM models and evaluates their performance using seven statistical error criteria. The study concludes that the HHO–ANN model provides the most accurate predictions for the productivity of three desalination units: active solar still, condenser, and passive solar still.

10.3.1.9.2 Description of the Workflow and Machine Learning Model

This study presents a novel productivity prediction model for active solar stills using a combination of ANNs and the HHO. The HHO algorithm, inspired by the predatory behavior of Harris Hawks, is employed to optimize the parameters of the ANN, improving its prediction accuracy. The proposed model, referred to as HHO–ANN, was compared with two other models: traditional ANN and SVM. These models were applied to predict the yield of three different distillation systems: passive solar still, active solar still, and an active solar still integrated with a condenser. The study found that the HHO–ANN model outperformed the other two models, achieving the highest accuracy in predicting the solar still yield, with a coefficient of determination (R^2) of 0.97 for the active still and 0.98 for the passive still.

10.3.1.9.3 Advantages and Limitations of the Synchronized Algorithm

The combination of ANN with the HHO offers several advantages, notably in optimizing the prediction accuracy for solar still productivity. The HHO–ANN model significantly improved the performance compared to traditional ANN and SVM, as it effectively identified the optimal parameters for the NN. This resulted in more accurate predictions of solar still yield across different configurations. However, the HHO–ANN model also presents some limitations, including its computational complexity, as the optimization process requires a considerable amount of computational resources. Additionally, the model's performance is highly dependent on the quality of the input data, and achieving the optimal configuration may require extensive experimentation with different hyperparameters.

10.3.1.9.4 Results and Discussion

The study shows that the HHO–ANN model outperforms the traditional ANN and SVM models in predicting the productivity of solar stills. The HHO–ANN model provides the most accurate predictions for the three desalination units: active solar still, condenser, and passive solar still. The study recommends using the HHO–ANN model to predict the productivity of solar stills without involving more costly and time-consuming experiments. The results of the study demonstrate the effectiveness of the HHO–ANN model in predicting and optimizing the performance of solar stills.

BIBLIOGRAPHY

Essa, F. A., Mohamed Abd Elaziz, and Ammar H. Elsheikh, An enhanced productivity prediction model of active solar still using artificial neural network and Harris Hawks optimizer, *Applied Thermal Engineering*, 170, 2020, 115020, ISSN 1359-4311. https://doi.org/10.1016/j.applthermaleng.2020.115020

10.3.1.10 Application-10: Performance Analysis of Energy Production of Large-Scale Solar Plants Based on Artificial Intelligence (Machine Learning) Technique

10.3.1.10.1 Background of the Application

This paper presents an autoregressive integrated moving average (ARIMA) model developed through ML techniques to predict/forecast the efficiency results of a large-scale solar plant in Pakistan. The model was trained using real-time data collected from the Quaid-e-Azam Solar Park (QASP), including power production, plane of array (POA), and performance ratio (PR) over a 1-year period. The paper predicts an overall increase in production by approximately 380–400 MWh after 10 years, with an increase in POA from 6.88 W/m^2 to 10 W/m^2 over the same period. The PR is predicted to decrease from about 76.6% to 73.2% over the 10-year period. The results demonstrate the effectiveness of the ARIMA model in predicting/forecasting the efficiency of the solar plant.

10.3.1.10.2 Description of the Workflow and Machine Learning Model

This study utilizes the ARIMA model, an ML-based statistical prediction technique, to forecast the performance of large-scale solar plants. Specifically, the ARIMA model is developed using Python to predict the PR, production amount (MWh), and POA of the QASP in Bahawalpur, Pakistan. By training the model on 1 year of real-time data from the solar park, the study forecasts the plant's energy production and operational performance over the next 10 years. The ARIMA model's predictive capabilities allow for efficient and effective long-term planning for the solar plant, providing insights into future energy production and performance trends.

10.3.1.10.3 Advantages and Limitations of the Synchronized Algorithm

The ARIMA model demonstrates several advantages in predicting the long-term performance of large-scale solar plants. Its statistical foundation allows it to utilize past data trends for future predictions, making it a reliable tool for forecasting energy production and operational performance. Compared to traditional forecasting methods, the ARIMA model is efficient and effective, delivering accurate predictions that can be utilized for strategic planning. However, the model also has limitations, including its reliance on historical data, which may not fully account for unexpected changes in environmental conditions or operational disruptions. Additionally, the accuracy of the ARIMA model is dependent on the quality and quantity of the input data, and its performance may vary based on the chosen parameters for the time series forecasting.

10.3.1.10.4 Results and Discussion

In this paper, an ARIMA model was developed to predict and forecast the efficiency results of a large-scale solar plant in Pakistan. The results show that the model is effective in predicting/forecasting power production, POA, and PR over a 10-year period. An overall increase in production by approximately 380–400 MWh after 10 years is predicted, with POA rising from 6.88 W/m² to 10 W/m² over the same period. The PR is predicted to decrease from about 76.6% to 73.2% over the 10-year period. It is concluded that the ARIMA model serves as an effective tool for predicting the efficiency of solar plants, and future studies are encouraged to explore similar intelligent techniques to enhance existing predictive models.

BIBLIOGRAPHY

Abubakar, M., Y. Che, L. Ivascu, F. M. Almasoudi, and I. Jamil Performance analysis of energy production of large-scale solar plants based on artificial intelligence (machine learning) *technique. Processes*, 10, 2022, 1843. https://doi.org/10.3390/pr10091843

10.3.1.11 Application-11: Modelling, Design, and Experimental Validation of a Grid-Connected Farmhouse Comprising a Photovoltaic and a Pumped Hydro Storage System

10.3.1.11.1 Background of the Application

In this paper, the authors propose a pumped hydroelectric storage (PHS) management system for farmhouses to reduce electricity costs. The system controls the pump power and the turbine flow rate, considering PV power generation, energy demand, water demand, PHS losses, and grid tariff. The proposed system was experimentally validated and simulated using MATLAB for 1 year to calculate the cost-benefit of the proposed system.

10.3.1.11.2 Description of the Workflow and Machine Learning Model

This study focuses on the design and implementation of a grid-connected PV system combined with a PHS system for a farmhouse. The research proposes using existing irrigation infrastructure, including a water well and reservoir, to store surplus PV energy. During the day, PV energy is used to pump water from the well to a reservoir, storing energy as gravitational potential energy. When needed, water is released back through a hydro turbine to generate electricity. The energy management system (EMS) controls the pump and turbine operations to optimize energy usage, minimizing electricity costs while also supporting irrigation needs. The system's performance is validated through both experimental testing and simulation over a year using real-world data on energy demand, weather conditions, and water requirements.

10.3.1.11.3 Advantages and Limitations of the Synchronized Algorithm

The combination of a PV system with a pumped hydro storage system offers several advantages for rural and agricultural applications. The PHS system is eco-friendly and cost-effective, utilizing existing farm infrastructure for energy storage rather than relying on batteries. The EMS enhances efficiency by dynamically adjusting operations based on real-time data, such as energy demand, weather forecasts, and water requirements. However, there are limitations to this approach. The system's effectiveness depends on accurate predictions and careful management of water levels, which can be impacted by factors like hydraulic losses, evaporation, and rainfall. Additionally, the implementation of the PHS system requires significant upfront capital investment, and the accuracy of the EMS relies on the quality of the forecast data.

10.3.1.11.4 Results and Discussion

The authors draw the conclusion that the suggested PHS management system can effectively control turbine flow rate and pump power to lower electricity expenses. By including the suggested PHS system with the PV system, the system can drastically lower the farmhouse's yearly electricity expenses. By calculating the PHS's efficiency under various circumstances, the comprehensive PHS model has helped the management system choose the pump power and turbine flow rate as efficiently as possible. In addition to generation and demand, the management system's design takes the PHS's efficiency into account. Farmhouses with reservoirs and water wells are advised to employ this method.

BIBLIOGRAPHY

Mousavi, Navid, Ganesh Kothapalli, Daryoush Habibi, Choton K. Das, and Ali Baniasadi. Modelling, design, and experimental validation of a grid-connected farmhouse comprising a photovoltaic and a pumped hydro storage system. *Energy Conversion and Management*, 210, 2020, 112675, ISSN 0196-8904. https://doi.org/10.1016/j.enconman.2020.112675

10.3.1.12 Application-12: Prediction of Daily Global Solar Radiation Using Different Machine Learning Algorithms: Evaluation and Comparison

10.3.1.12.1 Background of the Application

This paper proposes a method for predicting daily global solar radiation using AI techniques, including support vector regression (SVR), ANNs, and wavelet-coupled SVM. The input parameters are daily meteorological data, such as temperature, humidity, pressure, and wind speed. The output of the method is the prediction of daily global solar radiation. The paper compares and evaluates the performance of different techniques.

10.3.1.12.2 Description of the Workflow and Machine Learning Model

This study aims to predict daily global solar radiation in four Turkish provinces using four ML algorithms: SVM, ANN, Kernel Nearest Neighbor (k-NN), and DL. The input variables used to train the models include daily minimum and maximum ambient temperatures, cloud cover, daily extraterrestrial solar radiation, and day length. The data were collected over 2 years from meteorological stations in Kırklareli, Tokat, Nevşehir, and Karaman. The models were evaluated using seven statistical metrics, including R^2, RMSE, rRMSE, MBE, MABE, t-stat, and MAPE. Among the models, ANN provided the best prediction performance, followed by DL, SVM, and k-NN.

10.3.1.12.3 Advantages and Limitations of the Synchronized Algorithm

The ANN and DL models offered superior performance in predicting daily global solar radiation compared to the SVM and k-NN models. ANN, in particular, demonstrated the highest accuracy across all four provinces, achieving R^2 values as high as 0.932. DL was a close second, though its performance was limited by higher MBE values, which affected its t-stat scores. The study also found that while k-NN provided the least accurate predictions, all models showed good predictive capabilities overall. The primary limitations of these approaches include the need for large datasets and significant computational resources, especially for DL models. Additionally, the accuracy of predictions is highly dependent on the quality of input data, and the results can vary significantly between regions due to local climatic conditions.

10.3.1.12.4 Results and Discussion

In this paper, a method for predicting daily global solar radiation using AI techniques has been proposed. The method employs SVR, ANNs, and wavelet-coupled SVM. The input parameters are daily meteorological data, such as temperature, humidity, pressure, and wind speed. The output of the method is the prediction of daily global solar radiation. The results show that the proposed method outperforms other techniques in terms of predicting accuracy. This method has potential application in the field of solar energy.

BIBLIOGRAPHY

Ağbulut, Ümit, Ali Etem Gürel, and Yunus Biçen, Prediction of daily global solar radiation using different machine learning algorithms: Evaluation and comparison, *Renewable and Sustainable Energy Reviews*, 135, 2021, 110114, ISSN 1364-0321. https://doi.org/10.1016/j.rser.2020.110114

10.3.1.13 Application-13: Coupling Fluid Flow and Geomechanical Deformation Using AI & FEM Approaches

10.3.1.13.1 Background of the Application

In geomechanics and reservoir engineering, combining fluid flow with geomechanical deformation is a difficult task. In order to integrate fluid flow and geomechanical deformation, this study will combine elastoplastic and finite element method (FEM) approaches with ML and AI techniques to create a reliable and accurate numerical model.

The study uses fluid flow equations and FEM techniques to create an elastoplastic model that simulates the deformation of geological materials under stress. The geomechanics simulator receives updated porosity and permeability values from the flow simulator, which sends pore pressures to it. The two simulators are sequentially coupled. The particular challenge determines the frequency of these coupling phases, which need additional optimization. Since solving the geomechanical model numerically is computationally demanding, ML and AI techniques are used to decrease the frequency of coupling steps, resulting in significant computational time savings. AI and ML are also utilized to improve model accuracy, optimize input parameters, and shorten runtime. The classic coupled model and the AI-based coupled model are compared for validation.

This study demonstrates how combining fluid flow with geomechanical deformation through the use of ML, AI, elastoplastic, and FEM techniques is a promising line of inquiry that has the potential to revolutionize the comprehension of intricate geological processes. The AI-based numerical model that was created provides a useful and precise instrument for forecasting how geological materials would behave under stress, which helps to create more efficient natural resource management plans.

An inventive and effective way of connecting fluid flow and geomechanical deformation is introduced by the combination of ML and AI with elastoplastic and FEM techniques. With potential uses in sectors including mining, geothermal energy, and oil and gas exploration, the AI-based numerical model created in this work is a major breakthrough in geomechanics.

10.3.1.13.2 Description of the Workflow and Machine Learning Model

This study presents an ML model based on the LSSVM for predicting ocean wave elevations in a random sea state. The model incorporates both time and frequency domain characteristics of historical wave data. Using data following a JONSWAP spectrum, the time series data is transformed into the frequency domain using the fast Fourier transform (FFT) and divided into sub-frequency bands. The LSSVM model is trained separately for each band and used to predict future wave elevations. This method extends the effective prediction time length, making it more suitable for short-term real-time wave prediction. The study highlights the effectiveness of LSSVM in handling nonlinear wave propagation, outperforming traditional methods in complex wave environments.

10.3.1.13.3 Advantages and Limitations of the Synchronized Algorithm

The LSSVM model provides several advantages for ocean wave prediction, particularly in extending the prediction time length and handling nonlinearities in wave data. By dividing the wave data into sub-frequency bands, the model improves prediction accuracy, especially in complex wave conditions. However, the approach also has limitations, including the computational complexity of the frequency decomposition process and the need for careful parameter tuning. The accuracy of the model depends on the quality of the input data and the selection of appropriate sub-frequency bands. While LSSVM outperforms traditional methods in certain scenarios, its performance may vary in different sea states and require further validation in diverse ocean environments.

10.3.1.13.4 Results and Discussion

In summary, applying 4D geomechanics in reservoir development offers substantial benefits for understanding and mitigating geomechanical risks associated with development operations. This approach provides a comprehensive insight into subsurface conditions and the dynamic changes in reservoir behavior over time. A review of 3D/4D geomechanical models using AI and ML was conducted, leveraging available technologies and petrophysical log data. Properties from the 3D geomechanical model were integrated into the GigaPOWERS simulator to assess changes in stress and strain as reservoir pressure decreased due to production. A predictive geomechanical model for the reservoir was developed by combining the geological model with the pressure simulator, using high-resolution reservoir simulation to analyze pressure scenarios.

The results of the 4D geomechanical model in the study area revealed stress redistribution and strain concentration within the reservoir and overburden formations. As the reservoir compacted, the overburden layers stretched, creating a challenging drilling environment. Permeability and porosity reductions were examined in sandstone formations under different depletion scenarios. The stress-dependent permeability models indicated a clear decrease in permeability with increasing confining pressure. The results showed that variations in porosity and permeability depended on factors such as initial values, rock type, pore structure and geometry, cementation, and rock mechanical properties.

Finally, the coupled 4D flow and geomechanics model was utilized as a high-fidelity data generator to produce the training data required for developing an AI-based model (ANN model) to replace the Finite Element Geomechanics solver. An initial implementation demonstrated promising results in predicting permeability changes due to geomechanical variations when running the coupled model. The next step is to refine the AI-based model by considering additional input parameters and optimizing the NN hyperparameters (such as the number of layers and hidden neurons per layer). The goal is to couple the optimized AI model with a reservoir flow simulator, thereby streamlining the process of running coupled 4D flow and geomechanics models without sacrificing accuracy, eliminating the need for a high-fidelity Finite Element-based Geomechanics model at every coupling step.

BIBLIOGRAPHY

Hamid, Osman, Tameem Almani, Sulaiman Alqannas, and Ghalia Alshanbari. Coupling Fluid Flow and Geomechanical Deformation Using AI & FEM Approaches. Paper presented at the *International Petroleum Technology Conference*, Dhahran, Saudi Arabia, February 2024. doi: https://doi.org/10.2523/IPTC-23358-MS

10.3.2 Wind Energy

10.3.2.1 Application-14: Knowledge-Enhanced Deep Learning for Simulation of Tropical Cyclone Boundary-Layer Winds

10.3.2.1.1 Background of the Application

A knowledge-enhanced DL network is proposed in this paper to model the boundary-layer winds of tropical cyclones. The network makes use of machine-readable prior knowledge regarding the semi-empirical formulas and physics-based equations that govern the wind field inside a tropical cyclone boundary layer. Using typical storm characteristics, the created network gives the 3D boundary-layer wind field for an arbitrary tropical cyclone with great accuracy and computational efficiency. Using historical data from four storms, the trained model is evaluated, and high agreement between simulations and observations is achieved. The created network may be successfully implemented as part of early system warning and used in conjunction with the tropical cyclone hazards risk analysis framework.

10.3.2.1.2 Description of the Workflow and Machine Learning Model

This study aims to improve the simulation of tropical cyclone boundary-layer winds using a knowledge-enhanced DL algorithm. The method integrates physics-based equations and semi-empirical formulas into a deep neural network (DNN), improving regularization and reducing the need for large datasets. The DNN leverages input parameters such as storm coordinates, size, intensity, and other relevant storm metrics, along with simplified Navier–Stokes equations and prior knowledge, to predict the spatial distribution of boundary-layer winds. The training datasets are derived from a recent hurricane wind model, which incorporates realistic wind dynamics. This hybrid approach enables the network to simulate wind fields for arbitrary tropical cyclone scenarios efficiently and accurately.

10.3.2.1.3 Advantages and Limitations of the Synchronized Algorithm

The knowledge-enhanced DL approach demonstrated strong performance in simulating complex wind dynamics in tropical cyclone boundary layers. By incorporating both empirical data and physics-based knowledge, the model effectively reduces the reliance on large datasets, which are often expensive and difficult to obtain. The model achieves high accuracy with relatively small datasets while ensuring computational efficiency. However, the accuracy of the model is still subject to the quality of the training data, and it may struggle with scenarios where key environmental factors like wind shear are not fully accounted for. The use of idealized pressure profiles may also introduce discrepancies between observed and simulated wind fields.

Additionally, while the training process can be resource-intensive, the trained model provides rapid predictions once developed.

10.3.2.1.4 Results and Discussion

This research proposes a knowledge-enhanced DL network to model the boundary-layer winds of tropical cyclones using typical storm parameters. The 3D boundary-layer wind field is provided by the created network with great accuracy and computational efficiency. Using historical data from four storms, the trained model is evaluated, and high agreement between simulations and observations is achieved. The suggested network can be successfully deployed as part of early system warning and used in conjunction with the tropical cyclone risks risk analysis methodology. The performance of the created knowledge-enhanced DL network for simulating tropical cyclone boundary-layer winds can be further boosted in future work by improving training data and network architecture, as well as by breakthroughs in scientific and technical knowledge.

BIBLIOGRAPHY

Snaiki, Reda, and Wu Teng. Knowledge-enhanced deep learning for simulation of tropical cyclone boundary-layer winds, *Journal of Wind Engineering and Industrial Aerodynamics*, 194, 2019, 103983, ISSN 0167-6105. https://doi.org/10.1016/j.jweia.2019.103983

10.3.2.2 Application-15: Ultrasound-Based Identification of Damage in Wind Turbine Blades Using Novelty Detection

10.3.2.2.1 Background of the Application

This paper proposes the application of novelty detection (ND) methods for ultrasonic testing of wind turbine blades. The methods employed include support vector machine data description (SVDD), k-means clustering, and distance-based methods. The performance of these methods is evaluated using quantitative indexes such as the efficiency product and the receiver operating characteristic (ROC) curve. The proposed ND methods achieve discrimination efficiencies greater than 97% for the best-performing algorithm among the considered methods. A practical decision support system is also proposed that requires only signatures from the no-defect condition for classifier development.

10.3.2.2.2 Description of the Workflow and Machine Learning Model

This study explores the use of ND techniques combined with nondestructive ultrasound testing for identifying structural damage in wind turbine blades. ND is used to address the one-class classification problem, where only normal condition signatures are required for system training, making it well-suited for scenarios where abnormal conditions are rare or unknown. The study applies a combination of wavelet denoising and principal component analysis (PCA) to preprocess ultrasonic signals acquired from wind turbine blade specimens. The ND algorithms utilized include k-means clustering, one-class SVM, and distance-based methods. These methods are applied to detect structural anomalies such as delamination and dry matrix in the blades, providing a cost-effective solution for predictive maintenance in wind energy infrastructure.

10.3.2.2.3 Advantages and Limitations of the Synchronized Algorithm

The ND approach offers significant advantages, particularly in scenarios where it is difficult to obtain comprehensive data for all potential fault conditions. The one-class classification paradigm is effective in detecting anomalies based solely on normal condition data, which reduces the need for extensive labeled datasets. The integration of wavelet denoising and PCA enhances the signal-to-noise ratio and reduces computational complexity, making the method suitable for real-time applications. However, the performance of the ND methods is highly dependent on the quality and distribution of the training data. The study also highlights that while k-means and distance-based methods performed well, the support vector data description (SVDD) method exhibited a higher false-positive rate, which could impact reliability in certain cases. Additionally, the reliance on ultrasonic data requires consistent and accurate signal acquisition, and the implementation of these algorithms in embedded systems may require further optimization for practical use.

10.3.2.2.4 Results and Discussion

To sum up, this study suggests using a number of ND techniques to find anomalous signals in wind turbine blade non-destructive testing (NDT) data. High discrimination efficiencies are achieved by the suggested techniques, which could be used in embedded electronics to supply information to aid in decision-making throughout the evaluation process. Additionally, a workable decision support system that solely needs signatures from the no-defect condition to develop classifiers is suggested. In subsequent research, By applying deep learning parameters to gather the signals, flaws in wind turbine blades can be detected, thereby enhancing the efficacy of non-destructive (ND) methods for nondestructive testing (NDT).

BIBLIOGRAPHY

Oliveira, Moisés A., Eduardo F. Simas Filho, Maria C. S. Albuquerque, Ygor T. B. Santos, Ivan C. da Silva, and Cláudia T. T. Farias. Ultrasound-based identification of damage in wind turbine blades using novelty detection, *Ultrasonics*, 108, 2020, 106166, ISSN 0041-624X. https://doi.org/10.1016/j.ultras.2020.106166

10.3.2.3 Application-16: Swarm Intelligence-Based Differential Protection Scheme for Wind Integrated Transmission System

10.3.2.3.1 Background of the Application

This paper presents the application of the particle swarm optimization (PSO) technique for developing a secure differential protection function of a transmission system integrated with a wind farm. The objective function is designed to vary the system's operating conditions and fault cases to achieve an optimized threshold value. Control parameters for wind turbines, such as DC bus voltage regulator gain, converter current regulator gain, speed regulator gain, and pitch controller gain, are also considered. The proposed method's performance is evaluated for numerous cases, and its performance is compared with the existing technique. The simulation results confirm that the proposed method has desirable performance and can provide a secure decision during dynamic operating conditions of the power network.

10.3.2.3.2 Description of the Workflow and Machine Learning Model

This study proposes an adaptive differential protection scheme for wind-integrated transmission systems using swarm intelligence. The methodology focuses on addressing the limitations of fixed-threshold protection schemes, which are unreliable in wind-integrated systems due to dynamic variations in wind turbine control parameters. The proposed method employs PSO to determine an optimal threshold for fault detection. The PSO algorithm optimizes the threshold by considering variables such as fault location, inception angle, and fault resistance. The scheme was tested using a standard wind-integrated transmission system model simulated in MATLAB, and the performance was evaluated across various fault and system operating conditions. The adaptive threshold setting enhances the reliability of the protection system by dynamically adjusting to the varying conditions of the wind farm.

10.3.2.3.3 Advantages and Limitations of the Synchronized Algorithm

The application of PSO for adaptive threshold setting in differential protection provides several advantages, including the ability to handle dynamic variations in wind farm operations, which are often missed by conventional fixed-threshold schemes. The swarm intelligence approach optimizes fault detection across different fault types and operating conditions, leading to more reliable protection. Additionally, PSO offers fast convergence, with fewer iterations required to reach optimal solutions, making it computationally efficient. However, the method also has limitations, such as the complexity of accurately modeling the impact of fault resistance, location, and inception time on the protection scheme. Furthermore, the implementation of PSO in real-time applications could pose challenges due to the need for continuous monitoring and dynamic adjustments based on real-time data.

10.3.2.3.4 Results and Discussion

In order to create a secure differential protection function for a transmission system that is integrated with a wind farm, this research proposes a PSO technique. The strategy is tested in several scenarios and contrasted with other approaches. The findings of the simulation verify that the suggested approach performs as intended and can offer a safe choice under the power network's dynamic operating settings. The PSO methodology is used to optimize the threshold value of the suggested method while taking the fault scenarios and operational conditions of the system into account. The performance of the suggested approach is little impacted by changes in wind turbine control parameters, such as DC bus voltage regulator gain, converter current regulator gain, speed regulator gain, and pitch controller gain.

BIBLIOGRAPHY

Prasad, C. D., and Monalisa Biswal. Swarm intelligence-based differential protection scheme for wind integrated transmission system, *Computers & Electrical Engineering*, 86, 2020, 106709, ISSN 0045-7906. https://doi.org/10.1016/j.compeleceng.2020.106709

10.3.2.4 Application-17: Forecasting of Wave Energy in Canary Islands Based on Artificial Intelligence

10.3.2.4.1 Background of the Application

This paper presents a study on predicting the amount and distribution of wave power in deep ocean waters near the coast of Canary Islands. The study employs two AI techniques, namely ANNs and fuzzy inference systems (FISs), to model the 2D-Weibull distribution coefficients based on wave height and period. The models were trained using data collected from two buoys near the Canary Islands and validated using data from 2015 and 2016. The results show that both models achieved good accuracy, with mean relative errors less than 25% for both buoys. The developed models are efficient tools to compute wave power quickly and accurately, which can contribute to the enhancement of electric power systems using renewable energies.

10.3.2.4.2 Description of the Workflow and Machine Learning Model

This study explores the forecasting of wave energy in the Macaronesian region, particularly the Canary Islands, using ANNs and FISs. These soft computing techniques were applied to predict wave energy based on datasets obtained from deep-water buoys near the Canary Islands. The research utilized 18 years of wave data to develop and validate the models. The ANN-based model employed an MLP architecture, while the FIS-based model used subtractive clustering for prediction. Both models aimed to provide quick and accurate wave power predictions, facilitating the evaluation and deployment of wave energy converters (WECs) in deep ocean waters surrounding the islands.

10.3.2.4.3 Advantages and Limitations of the Synchronized Algorithm

The use of ANN and FIS models demonstrated several advantages in wave energy forecasting. ANN, with its universal approximation capability, provided high accuracy in predicting wave energy with minimal errors, particularly for significant wave height and peak period parameters. The FIS-based model offered an alternative approach with comparable performance, especially in handling nonlinear relationships. However, both models have limitations. The ANN model's accuracy depends on the proper selection of neurons and training data, and the FIS model's performance is sensitive to clustering parameters. Additionally, the models require extensive computational resources and data preprocessing. Despite these challenges, the study concludes that both ANN and FIS are effective tools for wave energy forecasting in oceanic deep waters.

10.3.2.4.4 Results and Discussion

In conclusion, this study demonstrated the effectiveness of ANNs and FISs in predicting wave power in deep ocean waters near the coast of Canary Islands. The models achieved good accuracy in predicting the amount and distribution of wave power, with mean relative errors less than 25% for both buoys. The developed models are efficient and cost-effective tools to compute wave power quickly and accurately, which can contribute to the enhancement of electric power systems using renewable energies. Further research is needed to investigate and improve the models for weather variability and specific conditions near the coast of Canary Islands.

BIBLIOGRAPHY

Deivis Avila, G. Nicolás Marichal, Isidro Padrón, Ramón Quiza, and Ángela Hernández. Forecasting of wave energy in Canary Islands based on Artificial Intelligence, *Applied Ocean Research*, 101, 2020, 102189, ISSN 0141-1187. https://doi.org/10.1016/j.apor.2020.102189

10.3.2.5 Application-18: Improvement of Wind Power Prediction from Meteorological Characterization with Machine Learning Models

10.3.2.5.1 Background of the Application

This paper is a literature review of various studies related to wind energy, focusing on topics such as wind turbine performance, power production, and forecasting. The reviewed studies employ various AI techniques, such as ensemble-based classifiers, DTs, and GB machines, to analyze and predict wind energy-related data. While the paper does not provide a comprehensive list of units for all parameters discussed in the reviewed studies, some units mentioned include pounds (lb), meters per second (m/s), and watts (W).

10.3.2.5.2 Description of the Workflow and Machine Learning Model

This study applies an ML approach to improve wind power prediction by incorporating meteorological data, such as hub-height wind speed, rotor-equivalent wind speed (REWS), and lapse rate. A DT regression model is employed to evaluate the effectiveness of these variables in predicting wind power. The research focuses on using atmospheric data to train regression trees that correlate power outputs to wind profiles and meteorological conditions. The study emphasizes the need for multiple calculations of wind speed at various levels of the rotor layer, highlighting the importance of vertical wind profile (VWP) classifications. The results show that the combination of rotor-equivalent wind speed and lapse rate improves prediction accuracy by 22% compared to traditional power curve methods, demonstrating the benefit of incorporating atmospheric criteria into wind power prediction.

10.3.2.5.3 Advantages and Limitations of the Synchronized Algorithm

The DT model demonstrates several advantages, including improved prediction accuracy and the ability to account for complex atmospheric conditions affecting wind power output. By considering variables like REWS and lapse rate, the model provides a more accurate representation of wind speed variations throughout the rotor layer, leading to better power predictions. However, the model also has limitations, such as the potential for overfitting and the need for extensive data to train the DTs effectively. Additionally, while the inclusion of atmospheric data improves accuracy, the model's performance may vary depending on the quality of the input data and the specific wind profiles encountered.

10.3.2.5.4 Results and Discussion

In conclusion, the reviewed studies demonstrate the potential of AI techniques in analyzing and predicting wind energy-related data, such as wind turbine performance, power production, and forecasting. However, it should be noted that the accuracy of these predictions may be affected by various factors, such as the quality of the input data and the choice of AI methods. Further research is needed to fully explore the capabilities and limitations of AI in the field of wind energy.

BIBLIOGRAPHY

Sasser, Christiana, Meilin Yu, and Ruben Delgado. Improvement of wind power prediction from meteorological characterization with machine learning models. *Renewable Energy*, 183, 2022, 491–501, ISSN 0960-1481. https://doi.org/10.1016/j.renene.2021.10.034

10.3.2.6 Application-19: Using Machine Learning and Computer Vision to Estimate the Angular Velocity of Wind Turbines in Smart Grids Remotely

10.3.2.6.1 Background of the Application

The paper explores various applications of AI and ML in the field of wind energy. The input parameters for these applications include wind speed, wind direction, turbine rotor speed, and other sensor data. The paper discusses several AI methods used in wind energy applications, such as NNs, SVMs, DTs, random forests (RFs), and clustering algorithms. The output of these models varies depending on the specific application but can include fault diagnosis, power generation prediction, wind turbine reliability analysis, and wind turbine blade ice detection.

10.3.2.6.2 Description of the Workflow and Machine Learning Model

This study proposes an ML and computer vision-based method for estimating the angular velocity of wind turbine blades in smart grids remotely. The research utilizes a combination of vision sensors and signal processing techniques, leveraging computer vision algorithms such as FAST (Features from Accelerated Segment Test), SIFT (Scale-Invariant Feature Transform), and SURF (Speeded Up Robust Features) for detecting the turbine hub and blades in consecutive frames of a video stream. These feature extraction methods are followed by classification using SVMs and an Autoencoder to improve the accuracy of blade detection. The proposed method also integrates Brute-Force (BF) and FLANN (Fast Library for Approximate Nearest Neighbors) matchers for feature matching. By analyzing the periodic presence of the blade in the video stream, the method estimates the angular velocity with 95.36% accuracy.

10.3.2.6.3 Advantages and Limitations of the Synchronized Algorithm

The use of computer vision and ML in this study provides several advantages, including the ability to remotely monitor and estimate the angular velocity of wind turbine blades with high accuracy. The integration of robust feature extraction techniques, such as SIFT and SURF, combined with ML classifiers like SVM, enhances the reliability of the proposed method. However, the approach also has limitations, such as the high computational cost associated with real-time image processing and the potential challenges in dealing with noisy data caused by camera vibrations in high wind conditions. The use of Autoencoder improves performance, but it requires significant computational resources and may not be feasible for all smart grid applications without optimization.

10.3.2.6.4 Results and Discussion

In conclusion, the paper demonstrates the potential for AI and ML to improve the efficiency and reliability of wind energy systems. By analyzing sensor data and predicting power generation, these technologies can help optimize wind turbine performance and reduce maintenance costs. However, the effectiveness of these methods is highly dependent on the quantity and quality of the input data, as well as the accuracy of the models themselves. As such, further research is needed to fully realize the benefits of AI and ML in the wind energy industry.

BIBLIOGRAPHY

Bahaghighat, Mahdi, Fereshteh Abedini, Qin Xin, Morteza Mohammadi Zanjireh, and Seyedali Mirjalili. Using machine learning and computer vision to estimate the angular velocity of wind turbines in smart grids remotely. *Energy Reports*, 7, 2021, 8561–8576, ISSN 2352-4847. https://doi.org/10.1016/j.egyr.2021.07.077

10.3.3 Geothermal and Heat Transfer

10.3.3.1 Application-20: Thermodynamic and Thermoeconomic Optimization of a Cooling Tower-Assisted Ground Source Heat Pump

10.3.3.1.1 Background of the Application

A novel approach to ground source heat pump (GSHP) system optimization with cooling tower assistance is presented in this research. The suggested approach takes into account both economic and thermodynamic factors while designing a system and choosing its components. Twelve decision variables, along with the proper technical and feasibility constraints, are used to design the optimization problem. Three optimization scenarios—thermodynamic, thermoeconomic, and multi-objective optimizations—were carried out using a genetic algorithm (GA). The findings indicated that an increase in the interest rate raises the cost of all systems' products. WP fluctuations have no discernible effect on the overall cost of the product, and thermodynamic optimization is cost-effective when operating in cooling mode for extended periods of time and/or when energy prices are high. A decision-making process determines the multi-objective optimization's ultimate solution. However, its results are somewhere between the results corresponding to the thermodynamic and thermoeconomic single-objective optimizations.

10.3.3.1.2 Description of the Workflow and Machine Learning Model

This paper presents the thermodynamic and thermoeconomic optimization of a cooling tower-assisted GSHP using a multi-objective optimization process. The study develops a thermodynamic model based on energy and exergy analyses and combines it with an economic model that follows the total revenue requirement (TRR) method. The optimization process is carried out using a GA, which aims to optimize both thermodynamic (energy efficiency) and thermoeconomic (cost efficiency) objectives simultaneously. The proposed system includes 12 decision variables that are optimized under different scenarios, such as single-objective thermodynamic optimization, single-objective thermoeconomic optimization, and multi-objective optimization.

10.3.3.1.3 Advantages and Limitations of the Synchronized Algorithm

The multi-objective optimization using a GA enables a balanced approach that considers both energy efficiency and cost minimization. The thermodynamic optimization focuses on reducing exergy destruction, leading to a higher COP but requires a larger capital investment. Conversely, the thermoeconomic optimization prioritizes reducing costs but may result in lower energy efficiency. The multi-objective optimization strikes a compromise between these two extremes, providing an efficient and cost-effective solution. However, the optimization process is computationally intensive and depends heavily on accurate modeling of both thermodynamic and economic factors.

10.3.3.1.4 Results and Discussion

A novel approach to cooling tower-assisted GSHP system optimization was introduced in this research. The suggested approach establishes the irreversibility (exergy annihilation) of the HGSHP system and addresses both the thermodynamic and economic aspects of system design and component selection. The findings demonstrated that multi-objective optimization is a generalization of single-objective optimization that takes thermodynamic and thermoeconomic goals into account at the same time. Whereas the thermoeconomic optimization solely considers financial resources, the thermodynamic optimization concentrates on making provisions for the finite energy supply. The multi-objective optimization considers both financial and energy resources. The basic case and three improved HGSHP systems' cooling costs were examined for their sensitivity to changes in the following input parameters: interest rate, annual cooling mode working hours, energy price, and WP. The findings indicated that increasing the interest rate raises the cost of all systems' products. WP fluctuations have no discernible effect on the overall cost of the product, and thermodynamic optimization is cost-effective when operating in cooling mode for extended periods of time and/or when energy prices are high.

BIBLIOGRAPHY

Sayyadi, Hoseyn, and Mostafa Nejatolahi. Thermodynamic and thermoeconomic optimization of a cooling tower-assisted ground source heat pump. *Geothermics*, 40, 3, 2011, 221–232, ISSN 0375-6505. https://doi.org/10.1016/j.geothermics.2011.06.003

10.3.3.2 Application-21: Modeling and Prediction of Geothermal Reservoir Temperature Behavior Using Evolutionary Design of Neural Networks

10.3.3.2.1 Background of the Application

This paper discusses the use of group method of data handling (GMDH)-type NNs to model the behavior of geothermal reservoir temperatures. The input data used in the model includes northing and easting coordinates, major depth, angle, and azimuth of each inside point of the wells. The output is the temperature of each inside point of the wells. The paper also describes the genetic design of the network and the training and testing sets used to evaluate the model's predictive ability.

10.3.3.2.2 Description of the Workflow and Machine Learning Model

This study utilizes GMDH-type NNs to model and predict geothermal reservoir temperature behavior. The GMDH algorithm, which is a self-organizing approach, constructs NNs using evolutionary methods combined with singular value decomposition (SVD) for optimal design. The model is trained on input–output data from geothermal wells, including variables such as depth, angle, and azimuth, to predict temperature at various points within the reservoir. GAs are employed to optimize both the network architecture and the coefficients of the model's quadratic expressions.

10.3.3.2.3 Advantages and Limitations of the Synchronized Algorithm

GMDH-type NNs offer a flexible and robust approach to modeling complex systems like geothermal reservoirs. They allow for self-organizing network construction, which eliminates the need for a pre-defined structure. The use of GAs enhances the optimization of the network, ensuring an efficient search for the best model configuration. However, the GMDH method can be computationally intensive, particularly during the training phase, as the evolutionary algorithms require multiple iterations to converge on an optimal solution. Additionally, the model's performance is highly dependent on the quality of the input data and the correct selection of hyperparameters.

10.3.3.2.4 Results and Discussion

The use of GMDH-type NNs proved to be a successful method for modeling the behavior of geothermal reservoir temperatures. The polynomial model developed using this method accurately predicted temperature behavior based on the input data parameters. The model was able to omit two factors, northing of the top point of the well and azimuth of each inside point in relation to the top point of the well, without affecting the network's output. This study provides insight into the use of NNs to model geothermal reservoir behavior and has potential for future applications in the field.

BIBLIOGRAPHY

Porkhial, S., M. Salehpour, H. Ashraf, and A. Jamali. Modeling and prediction of geothermal reservoir temperature behavior using evolutionary design of neural networks. *Geothermics*. 53, 2015, 320–327, ISSN 0375-6505. https://doi.org/10.1016/j.geothermics.2014.07.003

10.3.3.3 Application-22: A Review of Unsupervised Statistical Learning and Visual Analytics Techniques Applied to Performance Analysis of Non-residential Buildings

10.3.3.3.1 Background of the Application

This paper reviews various publications that utilize unsupervised ML and visual analytics in the analysis of non-residential building performance. The paper categorizes the publications based on ML technique, application, and sub-application of utilization. The paper highlights challenges related to interdisciplinary collaboration, differing sets of objectives, and a lack of reproducibility. The paper also discusses opportunities for better dialogue between research communities, the sharing of datasets, and the use of benchmarking data to evaluate proposed algorithms.

10.3.3.3.2 Description of the Workflow and Machine Learning Model

This paper reviews the use of unsupervised statistical learning techniques and visual analytics applied to performance analysis in non-residential buildings. The primary focus is on analyzing large datasets collected from buildings, such as smart meter data and other temporal or spatial performance data. The review covers five main categories of techniques: clustering, ND, motif and discord detection, rule extraction, and visual analytics. These techniques are often used in combination to identify patterns, detect anomalies, and optimize building performance without the need for labeled training data. The paper highlights the growing importance of unsupervised learning in the context of energy management, operational optimization, and anomaly detection across large building portfolios.

10.3.3.3.3 Advantages and Limitations of the Synchronized Algorithm

Unsupervised learning techniques provide significant advantages in handling vast amounts of building performance data with minimal manual intervention. Clustering, ND, and motif detection are particularly effective in discovering hidden patterns and anomalies that would be difficult to identify using traditional methods. Visual analytics further enhances the interpretability of the results by allowing human analysts to interact with the data visually. However, the review identifies several limitations, including the challenge of cross-domain collaboration due to differing terminologies and objectives in various research fields. Additionally, the lack of standardized benchmarking datasets and reproducibility in many studies makes it difficult to compare the performance of different algorithms. Further work is needed to improve reproducibility and create open datasets that can be used to benchmark new techniques.

10.3.3.3.4 Results and Discussion

In conclusion, this review highlights the increasing use of unsupervised ML and visual analytics in non-residential building performance analysis. The review categorizes publications based on their ML technique and application. The paper highlights several challenges facing the use of unsupervised ML in building performance analysis, including a noticeable divide between research communities, differing sets of objectives, and the lack of reproducible studies. The paper suggests that open, benchmarking datasets could be used to evaluate proposed algorithms and that cross-disciplinary venues could facilitate better communication between research communities.

BIBLIOGRAPHY

Miller, Clayton, Zoltán Nagy, and Arno Schlueter. A review of unsupervised statistical learning and visual analytics techniques applied to performance analysis of non-residential buildings. *Renewable and Sustainable Energy Reviews*, 81, Part 1, 2018, 1365–1377, ISSN 1364-0321. https://doi.org/10.1016/j.rser.2017.05.124

10.3.3.4 Application-23: Prediction of Small Hydropower Plant Power Production in Himreen Lake Dam (HLD) Using ANN

10.3.3.4.1 Background of the Application

This paper presents a study on the use of ANNs to predict power production in a small hydropower plant (SHP) located in Diyala/Iraq. The study utilized historical data on net head, flow rate, and power production collected over a 10-year period from 2005 to 2015. The data was preprocessed and analyzed statistically before being used to train the ANN model using the Levenberg–Marquardt algorithm. The model was able to accurately predict power output based on the input parameters of net head, flow rate, efficiency, and ground acceleration. The results showed a high correlation coefficient and a low RMSE, indicating the accuracy and generalization capability of the model.

10.3.3.4.2 Description of the Workflow and Machine Learning Model

This study focuses on predicting power production in an SHP at the Himreen Lake Dam in Diyala, Iraq, using ANNs. The research utilizes a feedforward back-propagation (FF-BP) ANN model to simulate and predict power generation based on input variables such as net turbine head, flow rate, and power production, using data collected over a 10-year period (2005–2015). The ANN model is trained to capture the complex, nonlinear relationships between these variables. The primary objective is to optimize the prediction of power production, which is influenced by variable hydrological and operational conditions at the dam. The ANN model demonstrates the potential to simulate and predict the performance of the hydropower plant with a high correlation coefficient ($R > 0.96$).

10.3.3.4.3 Advantages and Limitations of the Synchronized Algorithm

The use of ANN models in this study offers significant advantages in handling the nonlinear and complex nature of hydropower production data. The FF-BP model is particularly effective in predicting power production with high accuracy, as evidenced by the correlation coefficient of R > 0.96. This demonstrates the model's ability to generalize well to new input data, making it a powerful tool for performance prediction. However, there are limitations, including the reliance on extensive historical data for training the model. The success of the ANN approach is dependent on the quality and quantity of the input data, and challenges may arise when the model is applied to new conditions that were not well represented in the training data. Additionally, the selection of optimal network architecture, such as the number of hidden layers and neurons, requires a trial-and-error approach, which can be time-consuming and resource-intensive.

10.3.3.4.4 Results and Discussion

The study demonstrated the successful use of ANNs in predicting power production in an SHP, with a high correlation coefficient and low RMSE indicating the accuracy and generalization capability of the model. The model was trained using historical data on net head, flow rate, and power production collected over a 10-year period and was able to accurately predict power output based on input parameters of net head, flow rate, efficiency, and ground acceleration. The model has potential applications in predicting the behavior of other SHPs.

BIBLIOGRAPHY

Hammid, Ali Thaeer, Mohd Herwan Bin Sulaiman, and Ahmed N. Abdalla. Prediction of small hydropower plant power production in Himreen Lake dam (HLD) using artificial neural network. *Alexandria Engineering Journal*, 57, 1, 2018, 211–221, ISSN 1110-0168. https://doi.org/10.1016/j.aej.2016.12.011

10.3.3.5 Application-24: Machine Learning for Heat Transfer Correlations

10.3.3.5.1 Background of the Application

In this work, the correlation for high-order nonlinear convection heat transfer using ML is investigated. In particular, a cooling channel with variable rib roughness is the subject of the investigation. Using numerically simulated data, the ML model is trained to forecast the cooling channel's local convective heat transfer coefficient. Using a variety of novel channel shapes, the RF regression algorithm's interpolation ability is also examined. According to the results, the RF algorithm is efficient at managing a large number of features and prioritizing them, while the ML model can forecast the convection heat transfer coefficients with accuracy. This concept can be applied to other complicated phenomena, including multivariable heat transport problems. T

10.3.3.5.2 Description of the Workflow and Machine Learning Model

This study investigates the application of ML, specifically the RF algorithm, to predict convection heat transfer coefficients for high-order nonlinear problems. The research focuses on a cooling channel integrated with variable rib roughness, where traditional empirical correlations (EC) struggle to accurately model complex convective heat transfer phenomena. An ML model is trained using numerically simulated data from 243 different rib array geometries, incorporating six input features related to channel geometry and flow conditions. The RF model predicts local convection heat transfer coefficients with high accuracy, achieving an R^2 value of over 0.966 for the testing dataset. This approach demonstrates the potential of ML to replace traditional mathematical correlations, especially for complex thermal systems.

10.3.3.5.3 Advantages and Limitations of the Synchronized Algorithm

The use of RF in this study offers several advantages, including the ability to handle nonlinear, high-dimensional data and the flexibility to expand the parameter domain with ease. The RF model successfully predicted convection heat transfer coefficients for various rib geometries with high accuracy, outperforming traditional mathematical correlations. Moreover, the model's interpolation capabilities were effective, showing strong performance even when applied to new input data. However, limitations exist, such as the need for a large and well-distributed dataset to ensure accuracy. The model's performance is sensitive to the quality and diversity of the training data, and challenges remain in capturing highly complex flow patterns in regions far from the channel entrance. Additionally, the training process can be computationally intensive, particularly when optimizing hyperparameters like the number of estimators.

10.3.3.5.4 Results and Discussion

This paper demonstrates the versatility and high accuracy of ML regression as a convection heat transfer correlation. The RF algorithm is capable of accurately predicting the convection heat transfer coefficients of cooling channels with variable surface roughness, which has not been accomplished by any single mathematical-form correlation. The demonstrated framework can be widely adopted in various fields with high reliability and can be extended to other heat transfer problems involving multiple variables and complex phenomena. Retraining the ML model with additional data can significantly improve its accuracy. This study proves that ML regression is a promising approach for developing heat transfer correlations.

BIBLIOGRAPHY

Kwon, Beomjin, Faizan Ejaz, and Leslie K. Hwang. Machine learning for heat transfer correlations. *International Communications in Heat and Mass Transfer*, 116, 2020, 104694, ISSN 0735-1933. https://doi.org/10.1016/j.icheatmasstransfer.2020.104694

10.3.3.6 Application-25: Predicting Geothermal Heat Flow in Antarctica with a Machine Learning Approach

10.3.3.6.1 Background of the Application

The paper presents an ML-based approach to predict geothermal heat flow (GHF) in Antarctica. The model incorporates various geological and geophysical observables, such as surface heat flow, Moho depth, lithospheric thickness, magnetic field, topography, and proximity to geological features like volcanoes, ridges, trenches, and transform faults. The new model overcomes the limitations of using laterally constant thermal values and single geophysical methods. The algorithm learns from other continents with higher coverage of geophysical data and available heat flow measurements. The model is trained and tested for Australia and Antarctica, and the results demonstrate the suitability of the approach. The final model for Antarctica is based on all available measurements and considers 12 geological and geophysical observables.

10.3.3.6.2 Description of the Workflow and Machine Learning Model

This study introduces an ML approach to predict GHF in Antarctica, utilizing a Gradient Boosted Regression Tree (GBRT) algorithm. The approach is based on combining multiple geophysical and geological datasets to optimize predictions. The GBRT model is designed to find the optimal relationship between GHF and the geological observables, which include crustal thickness, tectonic units, and proximity to volcanic regions. This model builds on global datasets, complemented by regional data for Antarctica and its Gondwana neighbors, such as Australia and South Africa. The study emphasizes the importance of integrating reliable data sources to improve the accuracy of predictions, and it compares results with known GHF values from regions with more extensive measurement databases, like Australia. The final model produces a new GHF map for Antarctica with predicted values ranging from 35 to 156 mW/m^2.

10.3.3.6.3 Advantages and Limitations of the Synchronized Algorithm

The GBRT model offers significant advantages for predicting GHF, especially in regions with sparse data like Antarctica. By leveraging multiple geophysical and geological datasets, the model effectively captures the complex relationships between GHF and various geological features. The use of regional data improves prediction accuracy, as demonstrated by comparisons with known measurements in Australia. However, the approach has limitations, including the high computational demand of the GBRT algorithm and the need for careful selection of input features to avoid overfitting. Additionally, the accuracy of predictions is highly dependent on the quality of the input data, and the model may struggle to predict extreme values in areas with limited measurements, as seen in the underprediction of high GHF values in West Antarctica.

10.3.3.6.4 Results and Discussion

A new GHF model for Antarctica is presented in the research, taking into account data from 12 geological and geophysical observables and all accessible observations. The limitations of using single geophysical methods and laterally constant thermal values are overcome by the ML approach. Other continents with greater geophysical data coverage and accessible heat flow measurements are used to train the algorithm. Australia and Antarctica are used to train and test the model, and the outcomes show how appropriate the strategy is. Although there are still uncertainties, the model yields minimum and maximum results that can be utilized as input for additional sensitivity analysis, such as for ice flow and ice sheet dynamics. Future efforts should aim to improve the statistics of the prediction model and deliberate about which and how many features are reasonable and trustworthy. The addition of regional data is preferred for global data with limited accuracy in polar regions.

BIBLIOGRAPHY

Lösing, M., and J. Ebbing. Predicting geothermal heat flow in Antarctica with a machine learning approach. *Journal of Geophysical Research: Solid Earth*, 126, 2021, e2020JB021499. https://doi.org/10.1029/2020JB021499

10.3.3.7 Application-26: Three-Dimensional Seismic Characterization and Imaging of the Soda Lake Geothermal Field

10.3.3.7.1 Background of the Application

In this work, a multiscale connection-fusion U-shaped convolutional neural network (MCFU) is used to detect faults in seismic migration images in a novel way. In order to preserve fault location information more precisely, the MCFU architecture includes skip connections between encoders and decoders at various spatial resolution scales. In order to produce a final high-resolution fault map, the MCFU also creates fault maps at various spatial scales and uses a number of fusion processes. Through a number of examples of both synthetic and real seismic images, it is shown that the MCFU approach improves fault interpretability by producing faults that are cleaner and more continuous than those generated by conventional methods.

10.3.3.7.2 Description of the Workflow and Machine Learning Model

This study introduces a novel ML-based fault detection approach for seismic interpretation at the Soda Lake Geothermal Field. The method employs an MCFU, which enhances the traditional U-Net architecture by incorporating additional skip connections between different resolution levels of the network. This architecture enables the MCFU to leverage spatial information across multiple scales, improving fault detection accuracy. The MCFU generates independent fault maps at different spatial resolutions, which are fused to create a high-resolution final fault map. The model was tested on seismic migration images from the Soda Lake geothermal field, showing superior performance compared to conventional methods such as Schlumberger's ant-tracking technique and standard U-Net models.

10.3.3.7.3 Advantages and Limitations of the Synchronized Algorithm

The MCFU model demonstrates significant advantages in seismic fault detection, particularly in producing cleaner and more interpretable fault maps. The multiscale connection-fusion architecture allows for better spatial information preservation and reduces artifacts, leading to improved fault detection accuracy over conventional methods. Compared to the U-Net, the MCFU generates more reliable fault maps with fewer misidentifications, particularly for complex geological structures. However, the increased complexity of the MCFU architecture requires more computational resources and longer training times. Additionally, while the model's performance is strong in 2D seismic images, further work is needed to extend its application to 3D fault detection.

10.3.3.7.4 Results and Discussion

In conclusion, a novel fault detection method has been presented using the MCFU architecture, which incorporates multiscale skip connections and fusion operations. The MCFU produces fault maps with notably cleaner and more continuous faults compared with conventional approaches, leading to improved fault interpretability. The method outperforms both the state-of-the-art U-Net architecture and the ant-tracking method widely used in industry. Future work will focus on training the MCFU for 3D fault detection.

BIBLIOGRAPHY

Gao, Kai, Lianjie Huang, and Trenton Cladouhos, Three-dimensional seismic characterization and imaging of the Soda Lake geothermal field, *Geothermics*, 90, 2021, 101996, ISSN 0375-6505. https://doi.org/10.1016/j.geothermics.2020.101996

10.3.3.8 Application-27: Physics-Guided Machine Learning Approach to Characterizing Small-Scale Fractures in Geothermal Fields

10.3.3.8.1 Background of the Application

This paper presents a new method for fracture characterization using ML. The method, called the double-beam neural network (DBNN), interprets complex-valued double-beam interference (DBI) patterns as a discrete fracture network (DFN). The DBNN is an image-to-image learning method that directly maps DBI to DFN. The authors demonstrate the capability of the DBNN using a synthetic seismic dataset based on the Soda Lake geological model and show that the method can provide correct information on fracture orientation and the number of fracture sets. The location errors in the DBNN fractures are within 5 m using a 60 Hz P-wave data (wavelength ~40–50m) for targets at about 300 m in depth.

10.3.3.8.2 Description of the Workflow and Machine Learning Model

This study introduces a physics-guided ML approach to characterize small-scale fractures in geothermal fields. The method builds on the double-beam (DB) seismic technique, which focuses seismic and receiver beams to extract interference patterns from fracture-generated waves. These patterns contain information about fracture parameters, including orientation, density, and compliance. An NN architecture, called DBNN, is proposed to interpret the complex-valued DBI patterns and convert them into DFN information. The DBNN is trained using synthetic data based on a geological model of the Soda Lake geothermal field, where small-scale fractures are represented as linear slip boundaries. This method leverages the physics of seismic wave propagation to guide the ML model, ensuring that it aligns with underlying physical principles.

10.3.3.8.3 Advantages and Limitations of the Synchronized Algorithm

The DBNN approach offers significant advantages in fracture characterization, particularly in converting DBI patterns into DFN information with high accuracy. By integrating physics-guided learning, the model effectively handles the complex nonlinear relationships between seismic data and fracture parameters, making it more reliable than conventional methods. The DBNN is capable of detecting multiple fracture sets, providing detailed spatial information about fractures in geothermal fields. However, the approach also has limitations, including the requirement for extensive training data and the computational complexity of both the NN and the synthetic data generation process. Additionally, the accuracy of the model depends on the quality of the seismic data and the assumptions made during the data simulation process.

10.3.3.8.4 Results and Discussion

This paper presents a new method for fracture characterization using ML. The DBNN is an image-to-image learning method that directly maps complex-valued DBI patterns to a DFN. The authors demonstrate the capability of the DBNN using a synthetic seismic dataset based on the Soda Lake geological model and show that the method can provide correct information on fracture orientation and the number of fracture sets. The location errors in the DBNN fractures are within 5 m using a

60 Hz P-wave data (wavelength ~40–50m) for targets at about 300 m in depth. Future improvements may be possible using multiple frequencies. Overall, the DBNN method provides a promising approach to fracture characterization that could have important applications in geothermal energy, hydrocarbon reservoirs, and other subsurface engineering problems.

BIBLIOGRAPHY

Zheng, Yingcai, Jiaxuan Li, Rongrong Lin, Hao Hu, Kai Gao, and Lianjie Huang. *Proceedings, 46th Workshop on Geothermal Reservoir Engineering*, Stanford University, Stanford, California, February 15-17, 2021. SGP-TR-218

10.3.3.9 Application-28: Characterizing Signatures of Geothermal Exploration Data with Machine Learning Techniques: An Application to the Nevada Play Fairway Analysis

10.3.3.9.1 Background of the Application

This paper explores the use of supervised and unsupervised ML techniques, including PCA and k-means clustering, to analyze various geothermal system features such as geodetic strain rate, earthquakes, heatflow, fault density, fault slip rate, horizontal gravity gradient, and quaternary fault recency. By identifying correlations between these features and their respective spatial patterns, the paper aims to improve the identification of new geothermal systems and enhance exploration efforts. The results demonstrate the potential of ML in estimating the importance of geothermal system features and identifying principal correlations between features and measurements.

10.3.3.9.2 Description of the Workflow and Machine Learning Model

This study introduces ML techniques to the geothermal play fairway analysis for generating geothermal potential maps in the Nevada Great Basin region. The methodology combines supervised and unsupervised learning approaches to evaluate the influence of geological and geophysical features on geothermal favorability. The study employs ANNs, permutation analysis for feature selection, PCA, and k-means clustering to assess and optimize model performance. The supervised learning approach focuses on optimizing the combination of features to predict geothermal favorability, while the unsupervised learning aspect identifies patterns and correlations within the data. The aim is to enhance the original Nevada Play Fairway Analysis by incorporating ML methods to better predict blind geothermal systems.

10.3.3.9.3 Advantages and Limitations of the Synchronized Algorithm

The application of ML techniques in this study provides several advantages, including improved accuracy in predicting geothermal favorability and the ability to handle complex, nonlinear relationships between geological features. The supervised learning approach, particularly ANN, enables the automation and optimization of feature selection, reducing biases from manual weighting methods. The unsupervised methods, like PCA and k-means clustering, offer valuable insights into the underlying data patterns and correlations. However, limitations include the small sample size of training data, which can lead to overfitting, and the complexity of balancing numerical and categorical data types. Additionally, the performance of the ML models depends heavily on the quality of the input data, and addressing these challenges requires careful tuning and data augmentation.

10.3.3.9.4 Results and Discussion

This study's application of ML techniques has shown how supervised feature selection techniques can offer an automated means of identifying potential representations of training sample features in a learning issue. Principal correlations between features and measurements can be more accurately identified using unsupervised dimensionality reduction techniques like PCA. Sets of favorable structural settings, with and without known systems, as well as groups of positive and negative training sites, can be more easily identified by ignoring goal variables. In order to lessen the impact of erroneous outlier samples and features, this information is useful not only for selecting which features to include in the model but also for performing decorrelation and identifying extreme values. Overall, the results of this study demonstrate the potential of ML in geothermal exploration efforts and indicate which features may act as the best proxies for permeability/heat.

BIBLIOGRAPHY

Smith, Connor M., James E. Faulds, Stephen Brown, Mark Coolbaugh, Cary R. Lindsey, Sven Treitel, Bridget Ayling, Michael Fehler, Chen Gu, and Eli Mlawsky. Characterizing Signatures of Geothermal Exploration Data with Machine Learning Techniques: An Application to the Nevada Play Fairway Analysis. Proceedings, 46th Workshop on Geothermal Reservoir Engineering, Stanford University, Stanford, California, February 15-17, 2021. SGP-TR-218

10.3.4 Bioenergy and Other Applications

10.3.4.1 Application-29: Modeling and Optimization of Biogas Production on Saw Dust and Other Co-substrates Using Artificial Neural Network and Genetic Algorithm

10.3.4.1.1 Background of the Application

This paper describes the optimization of substrate profiles to improve biogas production using AI strategies. The work used an ANN model to predict the yield of biogas production based on the concentrations of five co-substrates— banana stem, cow dung, paper waste, rice bran, and sawdust. The optimization of the substrate profiles was carried out using GA running on Pro-optimizer software. The optimized substrate profile had a co-substrate profile of 25% banana stem, 25% cow dung, 25% paper waste, 5% rice bran, and 20% sawdust, with a biogas projected performance of 10,144 ml. Experimental validation of the optimized substrate profile resulted in a biogas performance of 10,280 ml, which was an improvement of 8.64% from the non-optimized process.

10.3.4.1.2 Description of the Workflow and Machine Learning Model

This paper presents the modeling and optimization of biogas production using a combination of ANNs and GAs. The study focuses on biogas production from mixed substrates, including sawdust, cow dung, banana stem, rice bran, and paper waste. A MLP ANN model with a topology of 5-2-1 was trained on data from 25 mini-pilot biogas fermentations to predict biogas production performance. The ANN model served as the fitness function in the GA optimization process, which was used to identify the optimal substrate combination for maximizing biogas yield.

10.3.4.1.3 Advantages and Limitations of the Synchronized Algorithm

The ANN model successfully captured the nonlinear relationships between substrate concentrations and biogas production, offering a powerful tool for bioprocess modeling. When coupled with GA, the approach enabled the optimization of substrate combinations, leading to an 8.64% increase in biogas yield compared to non-optimized systems. However, the effectiveness of the model depends on the quality of the training data, and the optimization process can be computationally intensive. Additionally, the ANN-GA approach requires careful tuning of parameters to avoid convergence on suboptimal solutions.

10.3.4.1.4 Results and Discussion

In conclusion, the optimization of substrate profiles using AI strategies was successful in improving biogas production. The use of an ANN model and GA optimization process led to the identification of the optimal substrate profile for maximum biogas production. The optimized substrate profile had a co-substrate profile of 25% banana stem, 25% cow dung, 25% paper waste, 5% rice bran, and 20% sawdust, with a biogas performance of 10,280 ml. Further research studies, including analysis of energy and economic balance, should be conducted to support its wide application at industrial scales.

BIBLIOGRAPHY

Gueguim Kana, E. B., J. K. Oloke, A. Lateef, M. O. Adesiyan. Modeling and optimization of biogas production on saw dust and other co-substrates using artificial neural network and genetic algorithm, *Renewable Energy*, 46, 2012, 276–281, ISSN 0960-1481. https://doi.org/10.1016/j.renene.2012.03.027

10.3.4.2 Application-30: Predictive Single-Step Kinetic Model of Biomass Devolatilization for CFD Applications: A Comparison Study of Empirical Correlations (EC), Artificial Neural Networks (ANN) and Random Forest (RF)

10.3.4.2.1 Background of the Application

This paper presents the development and evaluation of AI techniques for predicting the kinetic parameters of biomass devolatilization in computational fluid dynamics (CFD) applications. The input parameters used are cellulose fraction, hemicellulose fraction, lignin fraction, and heating rate. The AI techniques employed in this paper are simplex-lattice mixture design, ANNs, and RFs. The study evaluates the performance of the proposed models and identifies the relative impact of each input parameter. The results show that the RF model performs the best, while the EC model shows the worst performance. The paper also introduces a method for coupling the developed models with CFD simulation.

10.3.4.2.2 Description of the Workflow and Machine Learning Model

This paper presents a comparison of three approaches—EC, ANNs, and RF—for predicting kinetic parameters in a single-step biomass devolatilization model used in CFD applications. The study develops predictive models for key kinetic parameters such as frequency factor, activation energy, and reaction order based on biomass chemical compositions and heating conditions. The dataset used for training and validation includes over 100 experimental samples. ANN and RF models were trained to capture the complex and nonlinear relationships between the kinetic parameters and input variables, including the fractions of cellulose, hemicellulose, and lignin, as well as the heating rate. The performance of these models was then compared to the traditional EC method.

10.3.4.2.3 Advantages and Limitations of the Synchronized Algorithm

The study found that the RF model provided the best overall performance, achieving a determination coefficient (R^2) greater than 0.92 and a mean absolute percentage error (MAPE) below 5% across all kinetic parameters. The ANN model also performed well but was slightly less accurate than the RF model, particularly in predicting the reaction order. The EC method, while computationally efficient, struggled to capture the nonlinear relationships inherent in the data, resulting in lower accuracy ($R^2 < 0.80$). Despite the effectiveness of the RF model, it requires significant computational resources and careful parameter tuning. The ANN model offers a balance between performance and computational complexity but may require extensive training data to achieve high accuracy.

10.3.4.2.4 Results and Discussion

This paper presents a study on predicting kinetic parameters of biomass devolatilization in CFD applications using AI techniques. Three models—the EC model, the ANN model, and the RF model—were developed and evaluated. Results showed the RF model outperformed the others, while the EC model had the worst performance. The relative impact of each input parameter on the predicted kinetic parameters was also identified. Finally, a method for coupling the developed models with CFD simulation was introduced. This study provides a promising approach for predicting biomass devolatilization kinetics in CFD using AI.

BIBLIOGRAPHY

Xing J., H. Wang, K. Luo, S. Wang, Y. Bai, and J. Fan. Predictive single-step kinetic model of biomass devolatilization for CFD applications: A comparison study of empirical correlations (EC), artificial neural networks (ANN) and random forest (RF), *Renewable Energy* 2019. doi: https://doi.org/10.1016/j.renene.2018.12.088

10.3.4.3 Application-31: Prediction of the Ground Temperature with ANN, LS-SVM and Fuzzy LS-SVM for GSHP Application

10.3.4.3.1 Background of the Application

This paper proposes the use of computational intelligence algorithms to predict ground temperature distribution in the whole buried depth of the ground heat exchanger. The traditional method of measuring ground temperature is time-consuming and expensive. Therefore, the aim of this study is to provide an efficient and cost-effective alternative. The paper employs various AI methods, including ANNs, co-active neuro-fuzzy inference system, multi-nonlinear regression, fuzzy weighted pre-processing-based ANFIS, SVM, GAs, wavelet NNs, gene expression programming, and self-adaptive evolutionary algorithm. The results of the study show that the ANN model performs well in the prediction of ground temperature, especially in the training process of large datasets. The fuzzy logic-based least squares support vector machines (LS-SVM) model is also superior to the ANN model in terms of computational speed by nearly three times.

10.3.4.3.2 Description of the Workflow and Machine Learning Model

This paper presents three intelligent computation algorithms—ANNs, LS-SVM, and fuzzy LS-SVM—to predict ground temperature for GSHP applications. The ANN model used in this study is a backpropagation NN, which captures complex nonlinear relationships between input variables such as depth and time and the target ground temperature. The LS-SVM model, an advanced form of SVM, is optimized for processing small sample sizes and reduces the computational complexity by converting inequality constraints into equality constraints. The fuzzy LS-SVM model integrates fuzzy logic to improve prediction accuracy and computational efficiency, especially for large datasets. These models are trained and validated using ground temperature data collected from two boreholes in Chongqing, China.

10.3.4.3.3 Advantages and Limitations of the Synchronized Algorithm

The ANN model demonstrated strong generalization capabilities, particularly in large datasets, but required significant computational resources and time for training. LS-SVM, while efficient for small sample sizes, struggled with larger datasets due to memory constraints and slower computation. The introduction of fuzzy logic in the fuzzy LS-SVM model improved both prediction accuracy and computational efficiency, outperforming the standard LS-SVM model. However, the performance of all models is heavily dependent on the quality and quantity of input data. The ANN model, while accurate, requires careful tuning of hyperparameters and can be prone to overfitting if not properly managed. The fuzzy LS-SVM model provides a balance between computational efficiency and accuracy but is more complex to implement.

10.3.4.3.4 Results and Discussion

This study concludes by demonstrating that ground temperature distribution over the entire buried depth of the ground heat exchanger can be effectively predicted using computational intelligence techniques. Ground temperature can be accurately predicted using the ANN model, particularly when training on big datasets.

Nevertheless, the model takes a lot of time, particularly when k-fold validation is calculated. However, the LS-SVM model performs poorly in terms of computing speed and prediction accuracy. It is apparent that LS-SVM performs better when fuzzy theory is incorporated. In terms of computational speed, the generated fuzzy LS-SVM model outperforms the ANN model by about three times.

BIBLIOGRAPHY

Zhou, Shiyu, Xin Chu, Shubo Cao, Xiaoping Liu, and Yucheng Zhou. Prediction of the ground temperature with ANN, LS-SVM and fuzzy LS-SVM for GSHP application. *Geothermics*, 84, 2020, 101757, ISSN 0375-6505. https://doi.org/10.1016/j.geothermics.2019.101757

10.3.4.4 Application-32: Global Solar Irradiation Modelling and Prediction Using Machine Learning Models for Their Potential Use in Renewable Energy Applications

10.3.4.4.1 Background of the Application

This study explores the use of ML models to predict MGSI using easily obtainable input variables such as latitude, longitude, altitude, month, insolation, hours of sunshine, average temperature, average relative humidity, and rainfall. Three types of ML models, including RF, SVM, and ANN, were developed and tested. The ANN models provided the best adjustments, with MAPE values ranging from 3.9% to 13.8% for the stations used to develop the models and MAPE values between 4.1% and 12.5% for the overall Q2-8 phase. The developed models could be used to predict MGSI values in locations where data are available and can be extrapolated to nearby locations.

10.3.4.4.2 Description of the Workflow and Machine Learning Model

This study focuses on modeling and predicting MGSI in the Rias Baixas area of Galicia, Spain. The goal is to enhance RE applications through accurate solar irradiation forecasts. Three ML models were developed and compared: RF, SVM, and ANN. These models use input variables such as latitude, longitude, altitude, month, temperature, humidity, rainfall, insolation, and hours of sunshine to model MGSI. Two types of models were created: Type-1, which used commonly measured variables like temperature and humidity, and Type-2, which replaced some of these variables with insolation and hours of sunshine. The study demonstrated that ANN models consistently provided the best results for modeling and predicting solar irradiation, achieving an MAPE between 3.9% and 13.8%.

10.3.4.4.3 Advantages and Limitations of the Synchronized Algorithm

The application of ML models, particularly ANN, proved effective in modeling and predicting solar irradiation. ANN models excelled in capturing the complex relationships between input variables, leading to accurate predictions and outperforming both RF and SVM models. However, the performance of the models varied depending on the input variables used. Models incorporating insolation and hours of sunshine (Type-2) generally provided better results. A key limitation of this approach is the availability of certain input variables like insolation and hours of sunshine, which are not widely measured at all stations. Additionally, predicting solar irradiation one month ahead was more challenging than modeling it for the same month, leading to a decrease in model accuracy. The computational complexity and data requirements for training these models also pose challenges, especially when expanding the scope to new locations.

10.3.4.4.4 Results and Discussion

In this study, ML models were successfully developed to predict MGSI using easily obtainable input variables. The ANN models provided the best adjustments for all phases, with MAPE values ranging from 3.9% to 13.8% for the stations used to develop the models and MAPE values between 4.1% and 12.5% for the overall Q2-8 phase. The created models might be used to determine where solar farms and systems should be placed, or they could even be utilized as input data to forecast the amount of solar energy that will be produced. Predictive power may be lost if the produced models are used outside the domains under study. This should be confirmed by getting comparable parameters and applying them in other contexts. Lastly, additional measurement stations, new or different input variables, various datasets for training, validation, and querying, and studies of various parameter ranges for each model can all be used to improve the established models.

BIBLIOGRAPHY

Puga-Gil, D., G. Astray, E. Barreiro, J. F. Gálvez, and J. C. Mejuto Global solar irradiation modelling and prediction using machine learning models for their potential use in renewable energy applications. *Mathematics* 2022, 10, 4746. https://doi.org/10.3390/math10244746

10.3.4.5 Application-33: Optimization of the Energy Consumption in Activated Sludge Process using Deep Learning Selective Modeling

10.3.4.5.1 Background of the Application

This paper presents a methodology for optimizing energy consumption in wastewater treatment plants (WWTP) using DNNs and KPIs. The methodology involves selecting data based on KPIs, training and testing a DNN model, and using the model to optimize energy consumption for non-optimal data. The study shows that the DNN model coupled with KPI-based selection can accurately predict energy consumption and identify three classes of energy consumption: underconsumption, overconsumption, and optimal consumption. The results of this research can help reduce overconsumption and allow electricity generated from biogas to satisfy the energy needed for the aeration process.

10.3.4.5.2 Description of the Workflow and Machine Learning Model

This study focuses on optimizing energy consumption in WWTP using the activated sludge process. The approach leverages DL models to predict and optimize energy usage. A DNN was trained using selected data based on KPIs that measure both energy consumption and environmental compliance. The dataset includes pollution parameters (such as BOD5, COD, SS, NH_4), influent temperature, flow, and recirculated sludge flow. The DNN model was structured with multiple hidden layers, and the selection of the optimal configuration was achieved through trial and error. The primary objective is to create a model that can forecast energy consumption and identify potential energy savings across different operating scenarios in WWTPs.

10.3.4.5.3 Advantages and Limitations of the Synchronized Algorithm

The DL model, particularly the DNN, effectively captured complex relationships between input variables, allowing for accurate predictions of energy consumption in WWTPs. The model showed strong performance during training and testing, with high coefficients of determination (R^2 ranging from 74% to 94%) across different KPIs. However, the approach has several limitations. The main limitation is the small dataset, which restricts the generalizability of the model to other WWTPs. Additionally, the methodology relies heavily on specific KPIs for both the selection of training data and model evaluation, which may not capture all relevant factors affecting energy consumption. The transferability of the model to other facilities requires further investigation, and future work could explore the use of transfer learning to address this issue.

10.3.4.5.4 Results and Discussion

This study demonstrates that a methodology using DNN coupled with KPI-based selection can effectively optimize energy consumption in WWTP. The method identifies three classes of energy consumption: underconsumption, overconsumption, and optimal consumption, with clear limits between these classes. The research shows that the DNN model coupled with KPI-based selection can accurately predict energy consumption and reduce overconsumption, allowing electricity generated from biogas to satisfy the energy needed for the aeration process. Further work is needed to improve the selection method and test different ML models. The results of this research can be applied to reach energy self-sufficiency of WWTPs.

BIBLIOGRAPHY

Oulebsir, Rafik, Abdelouahab Lefkir, Abdelhamid Safri, and Abdelmalek Bermad. Optimization of the energy consumption in activated sludge process using deep learning selective modeling. *Biomass and Bioenergy*, 132, 2020, 105420, ISSN 0961-9534. https://doi.org/10.1016/j.biombioe.2019.105420

10.3.4.6 Application-34: Optimizing Generative AI Chatbots for Net-Zero Emissions Energy Internet-of-Things Infrastructure

10.3.4.6.1 Background of the Application

The rapid evolution of energy networks with distributed consumption, RE generation, and battery storage has led to a steady increase in the integration of Internet-of-Things (IoT) technology into energy infrastructure. It is possible to collect, process, analyze, and synthesize the data streams produced by this energy IoT infrastructure in order to facilitate well-informed decision-making, which will improve grid operations, lower costs, and move the country closer to net-zero carbon emissions.

However, the sheer amount of data frequently yields an excessive number of analytical results, which can make it difficult for energy grid managers to make wise decisions. By using conversational agents, also known as AI chatbots, to help human operators proactively analyze data and pinpoint critical decision points in energy networks, this difficulty can be lessened. In order to create AI chatbots that are optimized for the complicated information requirements of energy IoT systems and the goal of net-zero emissions, the latest developments in generative AI are built upon in this study. These chatbots will have natural language understanding and generation skills. The intent classifier, knowledge extractor, database retriever, cached hierarchical vector storage, secure prompting, and a conversational interface with a language generator are the six main components of the suggested system. The effectiveness of this method and the optimized generative AI chatbot in a real-world energy IoT infrastructure at a sizable, multi-campus postsecondary educational institution is empirically evaluated. The findings show that generative AI chatbots greatly reduce the complexity of energy IoT infrastructure, which helps to achieve net-zero carbon emissions and optimize grid operations.

10.3.4.6.2 Description of the Workflow and Machine Learning Model

This study proposes a novel methodology for optimizing generative AI chatbots tailored to the complex needs of net-zero emissions energy IoT infrastructure. The generative AI chatbots integrate six core modules: intent classifier, knowledge extractor, database retriever, cached hierarchical vector storage, secure prompting, and conversational interface with language generator. These chatbots are designed to assist human operators in analyzing large-scale energy data streams, providing informed decision-making opportunities for optimized grid operations. The approach leverages advanced generative AI techniques, combining natural language understanding and real-time data processing to deliver high-quality conversational responses. The chatbots are empirically evaluated in the real-world energy IoT infrastructure of a multi-campus tertiary institution, aiming to contribute to net-zero carbon emissions by 2029.

10.3.4.6.3 Advantages and Limitations of the Synchronized Algorithm

The proposed generative AI chatbots offer several advantages for energy IoT infrastructure. They enhance decision-making by providing real-time, contextually relevant insights while masking the complexities of energy data. The integration of hierarchical vector storage improves response times and reduces computational costs, and the secure prompting module ensures robust defenses against cyberattacks. However, the system's reliance on large language models (LLMs) like GPT-4 for natural language processing can introduce challenges, such as potential latency during high-demand periods and the need for ongoing fine-tuning to adapt to evolving energy policies. Additionally, while the hierarchical vector search optimizes performance, finding the optimal clustering parameters for different datasets may require extensive experimentation.

10.3.4.6.4 Results and Discussion

This article introduces a novel methodology designed to tackle the challenges of data and information complexity within energy IoT infrastructure. Leveraging the computational power of generative AI and conversational AI, the methodology ensures that human operators have access to accurate, relevant, and reliable information for making informed decisions regarding grid operations and related energy IoT activities. The methodology is composed of six key modules: intent classifier, knowledge extractor, database retriever, cached hierarchical vector storage, secure prompting, and a conversational interface with a language generator. It was empirically tested in a real-world environment, specifically within an energy IoT infrastructure at a large, multi-campus tertiary education institution. The results confirm the effectiveness of generative AI chatbots in meeting the complex demands of energy IoT infrastructure, thereby facilitating optimized grid operations and supporting the goal of achieving net-zero carbon emissions.

BIBLIOGRAPHY

Matharaarachchi, A., W. Mendis, K. Randunu, D. De Silva, G. Gamage, H. Moraliyage, N. Mills, and A. Jennings. Optimizing generative AI chatbots for net-zero emissions energy internet-of-things infrastructure. *Energies* 2024, 17, 1935. https://doi.org/10.3390/en17081935

10.3.4.7 Application-35: Comparison of Multiple Linear Regression, Artificial Neural Network, Extreme Learning Machine, and Support Vector Machine in Deriving Operation Rule of Hydropower Reservoir

10.3.4.7.1 Background of the Application

This paper explores the use of AI and ML methods in deriving the operation rule of a hydropower reservoir. The study utilizes multiple linear regression (MLR), ANN, extreme learning machine (ELM), and SVM to develop the operation rule. The optimal operation rule for the hydropower reservoir is obtained using these methods. The study shows that the three AI-based methods (ANN, ELM, and SVM) can provide better simulation results than the traditional scheduling graph method (SGM) and MLR. The dynamic programming method is found to provide the best scheduling results.

10.3.4.7.2 Description of the Workflow and Machine Learning Model

This study evaluates multiple methodologies for deriving operation rules for hydropower reservoirs, focusing on four different techniques: MLR, ANN, ELM, and SVM. The goal is to optimize the operation rules to maximize the multi-year average electric energy production in the Hongjiadu reservoir located in southwest China. The study leverages data from 1952 to 2015, applying each method to simulate the reservoir's operation rules. ANN, ELM, and SVM are categorized as AI techniques and are compared against traditional methods like MLR and the SGM. Each AI method employs different strategies to model complex relationships between variables such as water inflow, water levels, and power output.

10.3.4.7.3 Advantages and Limitations of the Synchronized Algorithm

The AI-based methods—ANN, ELM, and SVM—demonstrate superior performance in deriving operation rules for the hydropower reservoir compared to traditional methods like MLR and SGM. These AI methods are better equipped to handle non-linear relationships between input variables and can generalize well to new data, leading to improved long-term energy production. Among them, ELM exhibited the best performance, outperforming other methods in terms of both average power generation (APG) and generation guarantee rate (GGR). However, the complexity of the AI models requires careful tuning of parameters and substantial computational resources, especially for SVM, which has higher computational complexity compared to ELM and ANN. While AI methods offer significant advantages in accuracy and adaptability, they also require extensive data and may face challenges with overfitting or local minima, particularly for ANN. The reliance on large datasets and computational intensity is limitations that need to be addressed for broader real-world applications.

10.3.4.7.4 Results and Discussion

In conclusion, this study demonstrates the promise of AI and ML methods in deriving the operation rule of a hydropower reservoir. The results indicate that the three AI-based methods (ANN, ELM, and SVM) provide better simulation performances than SGM and MLR. The paper also highlights the importance of updating the reservoir operation rules on a monthly basis to reflect the changing environment and non-stationary conditions. Future research can focus on developing effective tools to choose appropriate model parameters.

BIBLIOGRAPHY

Niu, W.-J., Z.-K. Feng, B.-F. Feng, Y.-W. Min, C.-T. Cheng, and J.-Z. Zhou. Comparison of multiple linear regression, artificial neural network, extreme learning machine, and support vector machine in deriving operation rule of hydropower reservoir. *Water* 2019, 11, 88. https://doi.org/10.3390/w11010088

10.3.4.8 Application-36: Experimental and Numerical Assessment of Deterministic Nonlinear Ocean Waves Prediction Algorithms Using Non-uniformly Sampled Wave Gauges

10.3.4.8.1 Background of the Application

This study aims to assess and compare the accuracy of three ocean wave prediction algorithms based on three wave models, with different nonlinear properties, in predicting ocean surface oscillations that significantly affect the slope calculation and prevent evaluation of the corresponding prediction misfit. The algorithms were tested against data acquired for long-crested wave fields using wave gauges, in wave tank experiments, and simulations of similar experiments with a high-order numerical wave tank. The study found that the prediction accuracy of the algorithms converged with the duration of the assimilated surface elevation time series and that smaller data acquisition time steps yielded higher convergence rates. The improved choppy wave model (ICWM) provided improved predictions within the accessible prediction zone compared to the linear wave theory (LWT) with a corrected dispersion relation (LWT-CDR) and LWT models, with a maximum prediction misfit reduction of approximately 35% for an intermediate wave steepness, based on numerical data.

10.3.4.8.2 Description of the Workflow and Machine Learning Model

This study focuses on the deterministic prediction of nonlinear ocean surface waves using a combination of linear and weakly nonlinear wave prediction algorithms. The research applies these algorithms to long-crested irregular waves using data collected from wave gauges in laboratory experiments and numerical simulations. The primary models used in this study include the LWT and the ICWM, with the latter incorporating higher-order corrections to better capture nonlinear wave dynamics. The models are initialized using data from non-uniformly distributed wave gauges, simulating the typical spatial sampling that an optical sensor, like LIDAR, would capture. The goal of the study is to assess the accuracy of these prediction algorithms by comparing their performance in modeling ocean wave fields under various conditions.

10.3.4.8.3 Advantages and Limitations of the Synchronized Algorithm

The application of the ICWM and LWT algorithms demonstrates the potential to accurately predict ocean surface waves, with the ICWM providing improvements in modeling nonlinear wave behavior. The incorporation of nonlinear wave phase velocity in the ICWM model enhances the accuracy of predictions, especially in cases with higher wave steepness. However, the study highlights some limitations, including the computational intensity of the ICWM model due to its higher-order corrections. Additionally, while ICWM improves surface elevation representation, its effectiveness is limited to the region where input data are available. The reliance on non-uniformly sampled data from wave gauges introduces challenges in model initialization, and the accuracy of predictions is sensitive to the quality and distribution of the input data.

10.3.4.8.4 Results and Discussion

This study has shown that the ICWM provides improved predictions within the accessible prediction zone compared to the LWT with a corrected dispersion relation (LWT-CDR) and LWT models. The maximum prediction misfit reduction achieved by ICWM was approximately 35% for an intermediate wave steepness based on numerical data. The study also found that the prediction accuracy of the algorithms converged with the duration of the assimilated surface elevation time series and that smaller data acquisition time steps yielded higher convergence rates. These findings will help in applying the prediction algorithms to data acquired at sea using optical sensor systems and will aid in future research on short-crested waves.

BIBLIOGRAPHY

Desmars, N., F. Bonnefoy, S. T. Grilli, G. Ducrozet, Y. Perignon, C.-A. Guérin, and P. Ferrant. Experimental and numerical assessment of deterministic nonlinear ocean waves prediction algorithms using non-uniformly sampled wave gauges. *Ocean Engineering*, 212, 2020, 107659, ISSN 0029-8018. https://doi.org/10.1016/j.oceaneng.2020.107659

10.3.4.9 Application-37: Least Square Support Vector Machine and Multivariate Adaptive Regression Splines for Streamflow Prediction in Mountainous Basin Using Hydrometeorological Data as Inputs

10.3.4.9.1 Background of the Application

This study evaluates the accuracy of four data-driven models, OP-ELM, LSSVM, MARS, and M5 model tree (M5Tree), in predicting monthly streamflow of the Swat River Basin using hydro-meteorological inputs. Various input combinations derived from correlation analysis were applied. The validation and test statistics of the models are reported in Table 2 for Kalam Station and Table 3 for Chakdara Station. The results showed that LSSVM, MARS, and M5Tree models provided the best accuracy for the fourth scenario (S4) in the test period except for OP-ELM, which had a slightly better accuracy in S3 compared to the S4 model. The M5Tree-based model provided the lowest accuracy for the prediction of monthly streamflows of Kalam and Chakdara Stations. The linear structure of M5Tree might be the reason for its less accuracy compared to other models. In the case of Chakdara Station, all the models provided the best accuracy with input scenario 4 (S4). In both stations, OP-ELM, LSSVM, and MARS-based models were found to be better than SARIMA in monthly streamflow prediction.

10.3.4.9.2 Description of the Workflow and Machine Learning Model

This study investigates the application of various ML techniques, including LSSVM and multivariate adaptive regression splines (MARS), for predicting monthly streamflow in the Swat River Basin, Pakistan. The research compares four different heuristic methods: LSSVM, MARS, optimally pruned extreme learning machine (OP-ELM), and M5Tree. These models were applied to predict streamflow using precipitation and temperature data from the Kalam and Chakdara stations. The models were trained and validated on historical hydrometeorological data spanning from 1964 to 2008. LSSVM and MARS models were found to outperform OP-ELM and M5Tree in terms of accuracy for monthly streamflow prediction. The evaluation metrics included RMSE, mean absolute error (MAE), and Nash–Sutcliffe efficiency (NSE).

10.3.4.9.3 Advantages and Limitations of the Synchronized Algorithm

The LSSVM and MARS models demonstrated significant advantages in streamflow prediction, with superior accuracy compared to OP-ELM and M5Tree. LSSVM, in particular, reduced RMSE by up to 35% for the Kalam station, indicating its robustness in handling complex, nonlinear relationships in hydrological data. MARS also provided strong predictive performance by efficiently capturing nonlinear interactions between input variables. However, the study identified some limitations, including the sensitivity of the models to the quality of input data and the potential for overfitting when the number of input variables increases. Additionally, while LSSVM and MARS excelled in scenarios with diverse input data, the performance of OP-ELM and M5Tree suffered, particularly in cases where precipitation data had high skewness. The computational complexity and need for fine-tuning were also noted as challenges in deploying these models for real-time applications.

10.3.4.9.4 Results and Discussion

In order to estimate the monthly streamflow of Kalam and Chakdara stations at the Swat River Basin, the newly used data-driven model OP-ELM was analyzed and contrasted with other soft computing models, such as LSSVM, MARS, M5Tree, and SARIMA. The study discovered that when it came to forecasting peak streamflows, the LSSVM and MARS-based models outperformed the statistical model. This study also looked into the estimation of streamflows at downstream stations using data from the upstream station. The findings indicated that adding upstream data to model inputs significantly increased their accuracy. It was also discovered that, in every instance, the OP-ELM and M5Tree significantly overestimate or underestimate low or high streamflow values during the test period, whereas the LSSVM and MARS estimations closely match the corresponding observed values. Thus, it can be concluded that the LSSVM and MARS are capable of providing more accurate streamflow predictions compared to OP-ELM and M5Tree-based models with and/or without local data.

BIBLIOGRAPHY

Adnan, Rana Muhammad, Zhongmin Liang, Salim Heddam, Mohammad Zounemat-Kermani, Ozgur Kisi, and Binquan Li. Least square support vector machine and multivariate adaptive regression splines for streamflow prediction in mountainous basin using hydro-meteorological data as inputs. *Journal of Hydrology*, 586, 2020, 124371, ISSN 0022-1694. https://doi.org/10.1016/j.jhydrol.2019.124371

10.3.4.10 Application-38: Evolutionary Computational Intelligence Algorithm Coupled with Self-Tuning Predictive Model for Water Quality Index Determination

10.3.4.10.1 Background of the Application

This study aims to predict the water quality index (WQI) in Kinta River Basin, Malaysia, using computational intelligence techniques. The input parameters for the WQI prediction include dissolved oxygen, biochemical oxygen demand, chemical oxygen demand, total suspended solids, temperature, pH, and electrical conductivity. Several ML methods, including genetic programming (GP), extreme learning machine, and extreme gradient boosting (XGB), were employed for input variable selection, feature importance analysis, and prediction modeling. The best approach for predicting the WQI was determined by comparing the performance of different ML models. The study found that the proposed computational intelligence techniques are effective and satisfactory and can serve as a reliable prediction approach.

10.3.4.10.2 Description of the Workflow and Machine Learning Model

This study investigates the application of evolutionary computational intelligence algorithms to predict the WQI for the Kinta River Basin, Malaysia. The research employs XGB and GP to determine feature importance, followed by the use of extreme learning machine (ELM) for WQI prediction. The study compares standalone models such as XGB, GP, and ELM with hybrid models like GPELM and XGBELM. These hybrid models combine the feature selection capabilities of XGB and GP with the predictive power of ELM. The models are evaluated using metrics such as the coefficient of determination (R^2) and RMSE. The results show that the hybrid models outperform the standalone ones, with GPELM demonstrating superior accuracy in predicting WQI.

10.3.4.10.3 Advantages and Limitations of the Synchronized Algorithm

The hybrid modeling approach, particularly GPELM, significantly improved the prediction accuracy of the WQI, reducing RMSE by up to 9% compared to standalone models. The combination of evolutionary algorithms (GP and XGB) for feature selection and ELM for prediction capitalized on the strengths of both methodologies. The hybrid models reduced the number of input variables while maintaining high accuracy, making them efficient tools for WQI prediction. However, the study highlights certain limitations, such as the complexity of tuning multiple models and the computational intensity of the hybrid approach. Additionally, the performance of the models is highly dependent on the quality and representativeness of the input data, which may limit their applicability in different environmental contexts.

10.3.4.10.4 Results and Discussion

In conclusion, the computational intelligence techniques used in this study proved to be effective and satisfactory for predicting the WQI in Kinta River Basin, Malaysia. The use of GP and XGB as selection tools for input variables improved the performance of the prediction dramatically. The study also found that the extreme learning machine model with the important features selected by GP and XGB improved the performance of the prediction dramatically. Overall, the study reveals the advantages of using these ML techniques for input variable selection, feature importance analysis, and prediction modeling in highly nonlinear complex processes. The results suggest that other data intelligence models should be adopted for further improvements in the prediction of WQI.

BIBLIOGRAPHY

Abba, S. I., S. J. Hadi, S. S. Sammen, S. Q. Salih, R. A. Abdulkadir, Q. B. Pham, and Z. M. Yaseen Evolutionary computational intelligence algorithm coupled with self-tuning predictive model for water quality index determination. *Journal of Hydrology*, 587, 2020, 124974, ISSN 0022-1694. https://doi.org/10.1016/j.jhydrol.2020.124974

10.3.4.11 Application-39: Intelligent Identification of Effective Reservoirs Based on the Random Forest Classification Model

10.3.4.11.1 Background of the Application

This paper proposes a random forest classifier (RFC) model for identifying effective reservoirs in the real-time flood control operation of multi-reservoirs. The model is compared to an expert rule-based inductive reasoning (ERIR) model and several other ML models. The stability and accuracy of the RFC model are evaluated by using different flood samples. The criteria that have the greatest impact on the accuracy of the RFC model are determined. A case study of a multi-reservoir system in the Huaihe

River basin is presented to demonstrate the superiority of the RFC model in effective reservoir identification. The results show that the RFC model has the highest stability and classification accuracy among the models compared. The criteria that have the greatest impact on the accuracy of the RFC model are the contribution ratio to flood volume in the flood peak stage and the contribution ratio to the flood peak.

10.3.4.11.2 Description of the Workflow and Machine Learning Model

This study introduces an intelligent method for identifying effective reservoirs in real-time flood control operations using the random forest classification (RFC) model. The research focuses on improving the identification accuracy of effective reservoirs, which are critical to optimizing multi-reservoir flood control systems. The RFC model, an ensemble learning method that builds multiple DTs, is applied to classify reservoirs based on seven key criteria such as forecasted runoff depth, contribution to flood peak, and reservoir flood control pressure. The methodology is designed to avoid the uncertainties caused by traditional inference rule-based models (ERIR) and to provide a more stable and accurate solution for effective reservoir identification. The model's performance is evaluated using real-world data from the Huaihe River basin in China, with a focus on model stability, classification accuracy, and sensitivity analysis of the criteria.

10.3.4.11.3 Advantages and Limitations of the Synchronized Algorithm

The RFC model offers several advantages in effective reservoir identification, particularly in handling limited training samples and avoiding overfitting. Its ensemble nature provides higher classification accuracy compared to single models like DTs and SVMs. The model's stability is demonstrated through its consistent performance across varying sample sizes and floods. However, some limitations include the complexity of the model and the need for sufficient training data to ensure accurate predictions. While the RFC model outperforms the ERIR model in both stability and accuracy, its computational intensity and sensitivity to data quality remain challenges. Additionally, the study highlights that the success of the RFC model is contingent upon accurate flood forecasting and data reliability, which are critical for real-time flood control operations.

10.3.4.11.4 Results and Discussion

In terms of efficient reservoir identification, the RFC model put out in this study outperforms both the ERIR model and other ML models. The model has the maximum stability and is less susceptible to flood samples. The two most significant factors influencing the RFC model's classification accuracy are the contribution ratio to flood volume during the flood peak stage and the contribution ratio to the flood peak. The foundation for accurately identifying efficient reservoirs is the precision of flood predictions in real-time operations. However, there aren't enough flood samples in this basin; therefore, more research is required to confirm the findings in other flood control systems. Furthermore, further research is required to fully understand the RFC model's intricacy and the elements that influence it.

BIBLIOGRAPHY

Li, Jieyu, Ping-An Zhong, Minzhi Yang, Feilin Zhu, Juan Chen, Weifeng Liu, and Sunyu Xu. Intelligent identification of effective reservoirs based on the random forest classification model. *Journal of Hydrology*, 591, 2020, 125324, ISSN 0022-1694. https://doi.org/10.1016/j.jhydrol.2020.125324

10.3.4.12 Application-40: Estimating Spatio-temporal Air Temperature in London (UK) Using Machine Learning and Earth Observation Satellite Data

10.3.4.12.1 Background of the Application

This paper presents a novel approach to predict maximum temperature (Tmax) for London's grid cells using ML and Earth observation (EO) data. The paper investigates the performance of 54 regression models assembled from six regression models, six satellite products, three geospatial features, and 29 meteorological stations to select the parsimonious model to retrieve daily Tmax for London. The selected model used a gradient boosting (GB) algorithm calibrated with EO data to retrieve Tmax at high spatio-temporal resolution in an urban area. The paper also analyzes the spatial distribution of Tmax and provides evidence of the cool and warm effects from land cover elements on the Tmax variability at the local level. The research findings provide benefits for public health policies regarding adaptation and mitigation responses to climate change.

10.3.4.12.2 Description of the Workflow and Machine Learning Model

This study aims to estimate maximum air temperature (Tmax) across London at a high spatio-temporal resolution using ML models and EO satellite data. The study employed various regression methods, including traditional linear regression (LR) and ML models such as RF, GB, DTs, SVMs, and NNs. The dataset comprised satellite-derived predictors, such as land surface temperature (LST), normalized difference vegetation index (NDVI), and digital elevation model (DEM), along with meteorological station data for training the models. The best-performing model was the GB algorithm, which used five key predictors—LST, Julian day, NDVI, DEM, and solar zenith angle (SZA)—and achieved the highest accuracy with $R^2 = 0.68$, MAE = 1.60°C, and RMSE = 2.03°C.

10.3.4.12.3 Advantages and Limitations of the Synchronized Algorithm

The GB model demonstrated superior performance compared to other methods in predicting Tmax, especially in capturing the nonlinear relationships between satellite-derived predictors and air temperature. The model's ability to efficiently handle large datasets and complex interactions contributed to its high accuracy. However, the study also highlighted some limitations, including the challenges associated with data availability and quality, particularly with satellite imagery, which may be affected by cloud cover. The computational cost of training ML models, particularly GB, was also significant, making it resource-intensive. Additionally, while ML models like GB outperformed traditional methods, their complexity and the need for careful tuning of hyperparameters present challenges for widespread application.

10.3.4.12.4 Results and Discussion

This study offers a novel method for utilizing ML and EO data to predict Tmax in an urban region. The chosen model retrieved Tmax in London with great spatiotemporal resolution using the GB technique calibrated with EO data. The parsimonious model's predicted errors (RMSE = 2.03 °C and MAE = 1.60 °C) are similar to those found in earlier research. The results of the study have implications for public health policies concerning climate change adaptation and mitigation measures. They will help health professionals and epidemiologists better estimate the death load at the intra-city level caused by high temperatures. In order to prioritize measures to reduce heat-related mortality among the most vulnerable, practitioners and policymakers can also use this dataset to better understand the regional distribution of Tmax and identify UHI anomalies. In order to offer accurate local climate data for epidemiological research on the relationships between temperature and mortality, future studies utilizing sophisticated statistical approaches may investigate the creation of a suitable Tmax dataset for census-specific levels.

BIBLIOGRAPHY

dos Santos, Rochelle Schneider. Estimating spatio-temporal air temperature in London (UK) using machine learning and earth observation satellite data. *International Journal of Applied Earth Observation and Geoinformation*, 88, 2020, 102066, ISSN 1569-8432. https://doi.org/10.1016/j.jag.2020.102066

10.3.4.13 Application-41: Short-Term Prediction of Building Energy Consumption Employing an Improved Extreme Gradient Boosting Model: A Case Study of an Intake Tower

10.3.4.13.1 Background of the Application

This paper proposes a novel hybrid model, CEEMDAN-XGBoost, for predicting building energy consumption. The model combines the CEEMDAN method for denoising raw data with the XGBoost algorithm for prediction. The study uses daily energy consumption data from the City of Bloomington (CBU) Intake Tower in Indiana, USA, obtained from https://data.gov/. The data covers the period from January 1, 2011 to December 31, 2018. The performance of the CEEMDAN-XGBoost model is evaluated and compared to other models such as CEEMDAN-RF, RBFNN, PSO-SVM, and LSSVM.

10.3.4.13.2 Description of the Workflow and Machine Learning Model

This study proposes a novel hybrid model for short-term building energy consumption prediction, combining complete ensemble empirical mode decomposition with adaptive noise (CEEMDAN) and extreme gradient boosting (XGBoost). The CEEMDAN method is utilized to decompose raw energy consumption data into multiple smooth datasets, effectively reducing the noise present in the original data. Following data decomposition, XGBoost is employed to predict building energy consumption. The model is applied to daily energy consumption data from the City of Bloomington Intake Tower, which operates as a water transportation facility. The hybrid CEEMDAN–XGBoost model is evaluated against five benchmark models, including RF, SVMs, and radial basis function neural networks (RBFNNs). The results demonstrate that the proposed model achieves the highest prediction accuracy, with an MAPE of 4.85%.

10.3.4.13.3 Advantages and Limitations of the Synchronized Algorithm

The hybrid CEEMDAN–XGBoost model provides significant advantages in predicting building energy consumption. By combining data decomposition with a powerful tree-based ML algorithm, the model effectively handles the nonlinear characteristics of energy consumption time series data. The CEEMDAN method reduces noise in the dataset, allowing XGBoost to make more accurate predictions. The study shows that CEEMDAN–XGBoost outperforms traditional models such as SVM, RF, and RBFNN in terms of accuracy. However, the model's complexity and computational requirements are notable limitations. The process of data decomposition and model training is resource-intensive, and the performance is highly dependent on the quality of the input data. Additionally, the choice of parameters for both CEEMDAN and XGBoost requires careful tuning, which can be time-consuming.

10.3.4.13.4 Results and Discussion

In this study, the CEEMDAN–XGBoost model is proposed for predicting building energy consumption, and its performance is evaluated using daily energy consumption data from the CBU Intake Tower. The results show that the CEEMDAN–XGBoost model outperforms other models in terms of prediction accuracy, with an average MAPE of 5.31%. The model also has better prediction performance than the original XGBoost model, with an average MAPE reduced by 37.57%. The CEEMDAN–XGBoost model has good prediction performance in different datasets, including weekday and weekend datasets and datasets for different seasons. The sliding window length is found to be crucial for prediction accuracy, with the best performance achieved when the sliding window length is either 6 or 8. Overall, the CEEMDAN–XGBoost model is a promising approach for building energy consumption prediction.

BIBLIOGRAPHY

Hongfang, Lu, Feifei Cheng, Xin Ma, and Gang Hu. Short-term prediction of building energy consumption employing an improved extreme gradient boosting model: A case study of an intake tower, *Energy*, 203, 2020, 117756, ISSN 0360-5442. https://doi.org/10.1016/j.energy.2020.117756

10.3.4.14 Application-42: Hybrid Decision Tree-Based Machine Learning Models for Short-Term Water Quality Prediction

10.3.4.14.1 Background of the Application

This paper proposes two hybrid DT-based models, CEEMDAN–XGBoost and CEEMDAN–RF, for water quality prediction. The CEEMDAN in the two models is used to decompose the raw data with large fluctuations, so that the prediction performance of XGBoost and RF can be better. The paper focuses on predicting six water quality indicators—temperature, dissolved oxygen, pH value, specific conductance, turbidity, and FDOM. The data is collected from the Gales Creek site of the Tualatin River in Oregon, USA from May 1st to July 20th, 2019. The paper evaluates the prediction performance based on six error metrics—MAE, RMSE, MAPE, RMSPE, U1, and U2—and compares their performance across the different models. The results show that CEEMDAN–XGBoost has the best predictive performance, followed by CEEMDAN–RF.

10.3.4.14.2 Description of the Workflow and Machine Learning Model

This study introduces two novel hybrid DT-based ML models, CEEMDAN–XGBoost and CEEMDAN–RF, designed for short-term water quality prediction. The models combine the DT algorithms XGBoost and RF with the advanced data denoising technique CEEMDAN. These hybrid models were applied to predict six key water quality indicators: temperature, dissolved oxygen, pH value, specific conductance, turbidity, and fluorescent dissolved organic matter (FDOM) at the Gales Creek site in the Tualatin River Basin, Oregon. By integrating the CEEMDAN method, the models effectively reduce noise in the data, leading to more accurate predictions. The models were evaluated using six error metrics, demonstrating that CEEMDAN–RF performs best for temperature, dissolved oxygen, and specific conductance, while CEEMDAN–XGBoost excels in predicting pH, turbidity, and FDOM.

10.3.4.14.3 Advantages and Limitations of the Synchronized Algorithm

The hybrid CEEMDAN–XGBoost and CEEMDAN–RF models offer several advantages for short-term water quality prediction. The CEEMDAN technique enhances prediction accuracy by decomposing raw data into more stable components, which allows the DT-based models to achieve superior performance compared to traditional methods. CEEMDAN–RF performed best for specific indicators, achieving MAPEs as low as 0.69%, while CEEMDAN–XGBoost was more effective for others, with MAPEs as low as 0.27%. However, the models have limitations, including the computational complexity associated with CEEMDAN, which can be resource-intensive. Additionally, the performance of these models is dependent on the quality and distribution of the input data, and challenges remain in ensuring accuracy across all water quality indicators, particularly for turbidity, where prediction errors were higher than for other indicators.

10.3.4.14.4 Results and Discussion

In conclusion, this paper proposes two novel hybrid DT-based models, CEEMDAN–XGBoost and CEEMDAN–RF, for water quality prediction. The models use CEEMDAN to decompose the raw data and improve the prediction performance of XGBoost and RF. The paper evaluates the prediction performance of the models based on six error metrics and compares their performance across the different models. The results show that CEEMDAN–XGBoost has the best predictive performance, followed by CEEMDAN–RF. Future research can consider other factors that affect water quality, such as parallel computing to improve computing time.

BIBLIOGRAPHY

Hongfang Lu, and Xin Ma. Hybrid decision tree-based machine learning models for short-term water quality prediction, *Chemosphere*, 249, 2020, 126169, ISSN 0045-6535. https://doi.org/10.1016/j.chemosphere.2020.126169.

10.3.4.15 Application-43: Renewable Energy: Present Research and Future Scope of Artificial Intelligence

10.3.4.15.1 Background of the Application

This paper presents research on the application of AI methods in RE systems. The paper covers the use of AI in single RE systems such as wind and solar energy, as well as hybrid RE systems. Various AI methods are discussed, including BPNN, MLR, PCR, PLS, RBFNN, SVM, KNN, PSO, and GP, as well as hybrid AI approaches such as ANN-ARIMA, Fuzzy logic and ANN, BPNN and GA, and ANFIS and BPNN. The research discussed in this paper shows that AI approaches have a great potential to improve the performance of RE systems, and their proper utilization in future research can lead to further advancements in the field.

10.3.4.15.2 Description of the Workflow and Machine Learning Model

This paper reviews the application of AI in RE systems, focusing on wind, solar, geothermal, hydro, ocean, bioenergy, and hybrid RE sources. The research highlights various AI techniques, including ANNs, SVMs, fuzzy logic, and evolutionary algorithms (EAs), applied in areas such as power prediction, system optimization, and resource management. The paper also explores hybrid AI approaches, like neuro-fuzzy and neuro-genetic models, which combine different AI methods to enhance performance in RE applications. AI plays a crucial role in improving the efficiency and reliability of RE systems by providing more accurate predictions and optimizing resource usage.

10.3.4.15.3 Advantages and Limitations of the Synchronized Algorithm

The application of AI in RE systems offers several advantages, including improved accuracy in power prediction, system optimization, and enhanced operational efficiency. AI techniques like ANN and SVM have demonstrated strong performance in modeling complex relationships within RE systems, leading to better decision-making and resource management. However, the effectiveness of AI models is often limited by the quality and quantity of available data. Additionally, the complexity of AI algorithms and the need for extensive data preprocessing and parameter tuning can be resource-intensive. Hybrid AI approaches, while offering enhanced accuracy, also introduce additional layers of complexity and computational demands.

10.3.4.15.4 Results and Discussion

In conclusion, this paper highlights the potential of AI approaches in improving the performance of RE systems. The research discussed in this paper shows that AI methods such as BPNN, MLR, PCR, PLS, RBFNN, SVM, KNN, PSO, and GP have been successfully applied in various RE systems. Hybrid AI approaches such as ANN-ARIMA, fuzzy logic and ANN, BPNN and GA, and ANFIS and BPNN have also been used to achieve better performance. While most of the research discussed in this paper focuses on wind- and solar energy-based systems, there is a need for further research on the application of AI in other and hybrid RE sources. The implementation of novel and hybrid AI approaches has the potential to improve the performance of RE sources for the betterment of the world.

BIBLIOGRAPHY

Jha, S. K., J. Bilalovic, A. Jha, N. Patel, and H. Zhang. Renewable energy: Present research and future scope of Artificial Intelligence. *Renewable and Sustainable Energy Reviews*, 77, 2017, 297–317, ISSN 1364-0321. https://doi.org/10.1016/j.rser.2017.04.018

10.3.4.16 Application-44: A Machine Learning Model for Wave Prediction Based on Support Vector Machine

10.3.4.16.1 Background of the Application

In order to forecast ocean wave elevations in a random sea condition, the authors of this work propose a LS-SVM model. Both the frequency and time domain features of historical wave data are incorporated into the model. Wave data from an indoor wave tank experiment that follows a JONSWAP spectrum is used in the study. The quick Fourier transform was used to convert the measured time series into the frequency domain, and a filtering technique was used to separate them into five bands. Future time series were predicted using the LS-SVM model, which was independently trained on the time series for each band. For short-term real-time wave prediction applications, the suggested model is more effective because it considerably increases the prediction time.

10.3.4.16.2 Description of the Workflow and Machine Learning Model

This study presents a novel approach that combines AI and ML techniques with traditional FEM to model fluid flow and geomechanical deformation in subsurface reservoirs. The research focuses on developing a coupled AI-based model that leverages NNs to predict permeability changes due to geomechanical variations. The AI model is trained using high-fidelity data from coupled flow and FEM geomechanical simulations, with the goal of optimizing the computational performance and prediction accuracy of reservoir simulations. This approach aims to reduce the frequency of computationally intensive coupling steps while maintaining accuracy. The constructed NN model consists of multiple hidden layers and is trained to output permeability multipliers for subsequent flow-geomechanics coupling iterations, improving the efficiency of the simulation process.

10.3.4.16.3 Advantages and Limitations of the Synchronized Algorithm

The integration of AI and ML with FEM-based geomechanical modeling offers significant advantages, particularly in reducing the computational time required for high-fidelity simulations. By using an NN to predict changes in permeability, the model can optimize the coupling between fluid flow and geomechanics, providing faster and more efficient simulations without sacrificing accuracy. However, there are limitations to this approach, including the need for extensive training data from high-fidelity simulations and the complexity of constructing and tuning the NN. Additionally, while AI models can enhance prediction accuracy for specific scenarios, they may struggle with generalization when applied to new or unseen conditions. Further refinement of the model, including optimizing input parameters and network hyperparameters, is necessary to improve performance and robustness.

10.3.4.16.4 Results and Discussion

To predict short-term real-time ocean wave signals, an ML model based on the LSSVM algorithm was developed. The effectiveness of the model was validated using wave elevation signals measured during a wave tank test. Model parameters were optimized using a grid search algorithm, resulting in a final re-sampling frequency of 8 Hz, an input–output ratio of 16:1, and a model training time of 90 seconds. The maximum effective prediction time for a random wave elevation signal was found to be 3 seconds. To improve this, a frequency division strategy using bandpass filtering was introduced, decomposing the original elevation spectrum into five-time signals with narrower spectrum bandwidths. Predictions made for each of these signals showed that the effective prediction time increased to 7 seconds. The overall accuracy of the reconstructed wave time series was enhanced by employing this sub-frequency prediction strategy.

BIBLIOGRAPHY

Liu, Q., X. Feng, and T. Tang. A machine learning model for wave prediction based on support vector machine. *International Journal of Offshore and Polar Engineering* (ISSN 1053-5381) 32, 4, December 2022, 394–401. https://doi.org/10.17736/ijope.2022.jc879

10.3.4.17 Application-45: Development of Advanced Artificial Intelligence Models for Daily Rainfall Prediction

10.3.4.17.1 Background of the Application

This study aims to predict daily rainfall in the Hoa Binh Province using three advanced AI models: PSOANFIS, ANN, and SVM. Maximum temperature, minimum temperature, wind speed, relative humidity, and solar radiation were used as input parameters, and daily rainfall as an output parameter. The performance of the models was evaluated using R and MAE, SS, POD, CSI, and FAR. The study found that the SVM model provided the best prediction performance, highest value of R mean, and lowest value of MAE mean over 1000 Monte Carlo simulations. Furthermore, the SVM model was the most robust and efficient prediction model using the Monte Carlo approach.

10.3.4.17.2 Description of the Workflow and Machine Learning Model

This study explores the development and comparison of advanced AI models for the prediction of daily rainfall in Hoa Binh province, Vietnam. The research investigates three AI models: ANFIS optimized with PSO, ANNs, and SVMs. These models use meteorological variables such as maximum and minimum temperature, wind speed, relative humidity, and solar radiation as input parameters to predict daily rainfall. The models were trained and validated using meteorological data, with performance evaluated through various criteria such as the correlation coefficient (R), MAE, skill score (SS), probability of detection (POD), critical success index (CSI), and false alarm ratio (FAR). The results indicated that all models provided reasonable predictions, with SVM showing the best performance and robustness in accounting for input variability using the Monte Carlo approach.

10.3.4.17.3 Advantages and Limitations of the Synchronized Algorithm

The AI models demonstrated significant advantages in predicting daily rainfall, with SVM emerging as the most robust and accurate model. SVM's ability to handle nonlinear relationships between meteorological variables contributed to its superior prediction accuracy. The PSO–ANFIS model also showed strong performance by combining the learning capabilities of ANFIS with the optimization power of PSO, while ANN provided a flexible framework for prediction. However, limitations include the need for large datasets to train these models effectively and the complexity of tuning parameters, particularly for hybrid models like PSO–ANFIS. Additionally, the models' performance is sensitive to the quality and distribution of input data, and real-world applications may face challenges related to data variability and computational demands.

10.3.4.17.4 Results and Discussion

This study demonstrated that the SVM model is the most accurate and efficient method for predicting daily rainfall in the Hoa Binh Province. The study also highlights the importance of selecting appropriate input parameters, especially those related to cloud characteristics, for training AI models for daily rainfall prediction. While the study focused on a limited time period, the results indicate strong predictive ability for the models tested. Future studies should aim to validate the models for rainfall prediction during non-rain to rainy day transitions and refine the models with an integrated approach that includes cloud-related input variables.

BIBLIOGRAPHY

Pham, B. T., L. M. Le, T. T. Le, K. T. T. Bui, V. M. Le, H. B. Ly, and I. Prakash Development of advanced artificial intelligence models for daily rainfall prediction. *Atmospheric Research*, 237, 2020, 104845, ISSN 0169-8095. https://doi.org/10.1016/j.atmosres.2020.104845

10.3.4.18 Application-46: A Two-Stage Numerical Simulation Framework for Pumped-Storage Energy System

10.3.4.18.1 Background of the Application

In order to accurately describe the hydraulic-mechanical coupling behavior of pumped storage power stations (PSES), a novel two-stage numerical simulation framework is presented in this research. In order to increase the model's degree of freedom, the framework incorporates a non-viscous potential (NVP)-modified orifice equation (MOC) model, which uses parameter sensitivity analysis to find nine of the twenty-one parameters. The enhanced scatter search algorithm (ESSA) is used for intelligent parameter identification, which is based on opposition-based learning and chaotic local search techniques. To show that the integrated objective function created especially for PSES modeling is superior to the widely used normalized root mean square error (NRMSE)-type and normalized integral time-weighted mean absolute error (NITMAE)-type objective functions, comparative tests were carried out. The global searching efficiency of the proposed ISCA optimizer compared with state-of-the-art metaheuristic optimization algorithms such as PSO, GA, and original SCA was also investigated. The proposed numerical simulation framework for PSES was shown to achieve higher modeling precision than the conventional MOC model, particularly in the hydraulic pressure fitting performance.

10.3.4.18.2 Description of the Workflow and Machine Learning Model

A two-stage numerical simulation framework for the PSES is presented in this paper, with an emphasis on the dynamics of the hydraulic-mechanical coupling. In the first step, a nonlinear variable-parameter model (NVP-MOC) is created, which increases the degree of freedom by determining the parameters that are sensitive to the dynamic behavior of the system. An enhanced sine cosine algorithm (ISCA), which optimizes the parameter identification procedure, is incorporated in the second stage. Opposition-based learning and chaotic local search strategies are used to improve the ISCA's global exploration capabilities and initialization variety. On-site data was used to validate the framework, and comparison tests revealed notable gains in modeling accuracy over conventional techniques, especially when it came to describing transient phenomena like hydraulic pressure and rotation speed.

10.3.4.18.3 Advantages and Limitations of the Synchronized Algorithm

The proposed two-stage framework offers several advantages, including improved modeling precision for PSES, particularly under dynamic conditions involving large-scale operational transitions. The NVP-MOC model, with its increased degree of freedom, effectively captures complex system behaviors, while the ISCA provides robust parameter optimization, ensuring accurate simulations. However, the method also has limitations, such as the computational intensity required for solving the nonlinear equations and the need for extensive parameter sensitivity analysis to reduce dimensionality. Additionally, while the ISCA enhances global search performance, the complexity of the system may still pose challenges in real-time applications due to the heavy computation required.

10.3.4.18.4 Results and Discussion

In order to precisely describe the hydraulic–mechanical coupling behavior of PSES, a novel two-stage numerical simulation framework was presented in this research. The suggested framework consists of an intelligent parameter identification algorithm based on opposition-based learning and chaotic local search approaches, as well as an NVP-MOC model. Comparative analyses demonstrated that, particularly in hydraulic pressure fitting performance, the suggested framework outperformed the traditional MOC model in terms of modeling precision. It was demonstrated that the suggested integrated objective function outperformed the widely utilized NRMSE-type and NITMAE-type goal functions. It was shown that the suggested

ISCA optimizer outperformed cutting-edge metaheuristic optimization algorithms like PSO, GA, and original SCA in terms of global searching efficiency.

BIBLIOGRAPHY

Zheng, Yang, Qijuan Chen, Donglin Yan, and Wanying Liu. A two-stage numerical simulation framework for pumped-storage energy system. *Energy Conversion and Management*, 210, 2020, 112676, ISSN 0196-8904. https://doi.org/10.1016/j.enconman.2020.112676

10.3.4.19 Application-47: Development of a Machine Learning Forecast Model for Global Horizontal Irradiation Adapted to Tibet Based on Visible All-Sky Imaging

10.3.4.19.1 Background of the Application

This paper proposes a short-term prediction model for global horizontal irradiance (GHI) using ML algorithms. The main objective of the study is to predict GHI over the subsequent 1–6 hours, with a focus on the Yangbajing area of the Qinghai-Tibet Plateau. The study finds that cloud cover is the main factor affecting solar radiation reaching the surface, and the prediction accuracy of the model can be greatly improved by adding cloud cover time series as an input variable. The study also shows that the number of input features of input variables is critical to the prediction accuracy of the model. By using high-time resolution data and an appropriate number of input features, the model error can be greatly reduced, providing a new method with high precision for solar power generation and GHI prediction.

10.3.4.19.2 Description of the Workflow and Machine Learning Model

This study introduces an ML forecast model for GHI adapted to the unique environmental conditions of Tibet. The research leverages ground-based visible all-sky imaging and ML algorithms to improve the accuracy of GHI predictions. The study uses two primary ML models: RF and LSTM networks. The models incorporate key input variables, including cloud cover data obtained from all-sky imaging, as well as ground-based solar shortwave radiation measurements. These models predict short-term GHI values, specifically focusing on forecast horizons ranging from 1 to 6 hours. The research demonstrates that adding cloud cover data as an input variable significantly improves the prediction accuracy, with both RF and LSTM models showing substantial reductions in RMSE when cloud cover is included.

10.3.4.19.3 Advantages and Limitations of the Synchronized Algorithm

The RF and LSTM models offer several advantages in GHI prediction, particularly when combined with cloud cover data from ground-based imaging. The inclusion of cloud cover enhances the models' ability to predict rapid changes in solar irradiance, leading to improved accuracy, especially for shorter forecast horizons (e.g., 1 hour). LSTM, with its recurrent architecture, is particularly effective in capturing temporal dependencies in the data, while RF provides robust performance with a lower risk of overfitting. However, the models also face limitations, such as increased computational complexity and the need for large datasets to achieve optimal performance. Additionally, while cloud cover improves accuracy for short-term forecasts, its impact diminishes for longer forecast horizons (greater than 4 hours), limiting the models' effectiveness in such scenarios.

10.3.4.19.4 Results and Discussion

Building a short-term GHI prediction model in the Qinghai-Tibet Plateau using RF and LSTM ML algorithms is confirmed to be possible in this work. According to the study, cloud cover is the primary factor influencing the amount of solar radiation that reaches the surface. By including cloud cover time series as an input variable, the model's forecast accuracy can be significantly increased. The model error can be significantly decreased by using high-time resolution data and a suitable amount of input features, offering a new, highly accurate technique for solar power generation and GHI prediction. The study concludes that the proposed model can be applied to other regions with similar features, and the results can be used to optimize solar power generation and promote the development of RE.

BIBLIOGRAPHY

Wu, L., T. Chen, N. Ciren, D. Wang, H. Meng, M. Li, W. Zhao, J. Luo, X. Hu, S. Jia, et al. Development of a machine learning forecast model for global horizontal irradiation adapted to tibet based on visible all-sky imaging. *Remote Sensing*, 2023, 15, 2340. https://doi.org/10.3390/rs15092340

10.3.4.20 Application-48: Multiscale Groundwater Level Forecasting: Coupling New Machine Learning Approaches with Wavelet Transforms

10.3.4.20.1 Background of the Application

This work investigates the potential for producing precise groundwater level (GWL) forecasts using more recent ML techniques, such as RFs and variations of XGB. To improve the models' capacity to handle multiscale changes, multiscale data that had been retrieved using wavelet transforms (WTs) was input into them. Model parameter estimation was done automatically using Bayesian hyperparameter optimization. The created models were evaluated for GWL forecasting 1, 2, and 3 months in advance. The wavelet-based models' performance was evaluated against benchmark techniques like SVR models as well as standalone models. In contrast to SVR, new methods, particularly the XGB with wavelet transform (WT-XGBL), demonstrated comparable accuracy, according to a comparative analysis of standalone ML techniques. The coupling of WT further improved the performance for all ML approaches, and the improvement was more prominent for 3-month ahead GWL forecasts than 1-month ahead lead forecasts.

10.3.4.20.2 Description of the Workflow and Machine Learning Model

This study investigates the use of ML techniques combined with WTs for multiscale GWL forecasting. The research employs state-of-the-art ML methods, including XGB, RFs, and SVMs, to improve the prediction of GWL in Kumamoto City, Japan. The maximal overlap discrete wavelet transform (MODWT) is applied to decompose the input time series data, capturing various temporal scales of groundwater dynamics.

The study introduces a wavelet data-driven forecasting framework (WDDFF) that integrates the MODWT with ML models, enhancing the ability to forecast GWL at 1-, 2-, and 3-month lead times. Bayesian optimization is employed to automatically select hyperparameters for the XGB, RF, and SVM models, ensuring optimal performance. The model's effectiveness is demonstrated through comparisons with standalone ML models and conventional approaches, highlighting significant improvements in forecasting accuracy.

10.3.4.20.3 Advantages and Limitations of the Synchronized Algorithm

The integration of MODWT with advanced ML models, such as XGB and RF, offers several advantages in multiscale GWL forecasting. The wavelet decomposition allows the models to capture key temporal patterns and trends at different scales, resulting in more accurate predictions. The study demonstrates that the hybrid models significantly outperform standalone ML models, particularly in capturing short-term fluctuations and long-term trends. However, the approach also has limitations. The computational complexity of the wavelet-based models is higher than that of standalone models, and the choice of wavelet filters and decomposition levels can significantly impact model performance. Additionally, while Bayesian optimization helps streamline hyperparameter selection, it also increases the overall computational demand. The study also notes that wavelet-based models do not always guarantee superior performance and may require careful tuning to avoid overfitting.

10.3.4.20.4 Results and Discussion

A promising framework for GWL forecasting is the combination of ML techniques, particularly XGB variants, with Bayesian hyperparameter optimization and WT. This combination may merit additional research in the fields of hydrology and water resources for both short-term and long-term forecasting. Even though GWL modeling with ML techniques has advanced significantly over the past 20 years, there are still a few crucial problems that need to be resolved to encourage congruence between ML models and their conceptual and physical-based equivalents. In the fields of hydrology and water resources, such factors are anticipated to increase the acceptability of ML models and offer additional physical insights.

BIBLIOGRAPHY

Sakiur Rahman, A. T. M., Takahiro Hosono, John M. Quilty, Jayanta Das, and Amiya Basak. Multiscale groundwater level forecasting: Coupling new machine learning approaches with wavelet transforms. *Advances in Water Resources*, 141, 2020, 103595, ISSN 0309-1708. https://doi.org/10.1016/j.advwatres.2020.103595

10.3.4.21 Application-49: Evaluation of Machine Learning Models for Smart Grid Parameters: Performance Analysis of ARIMA and Bi-LSTM

10.3.4.21.1 Background of the Application

This paper investigates the use of two different ML models, ARIMA and Bi-LSTM (bidirectional long short-term memory), for the prediction of three important parameters in solar power generation: grid-connected power generation (MW), daily power generation (kWh), and radiance (MJ/m^2). The study utilized 80% of 1 year's real-time data on these parameters for training and validation purposes, while the remaining 20% of the data was used for testing. The study found that the Bi-LSTM model outperformed the ARIMA model in terms of accuracy and performance metrics for all three parameters. The Bi-LSTM model's ability to capture the complex relationships and patterns in the data was found to be superior to the ARIMA model's reliance on linear relationships.

10.3.4.21.2 Description of the Workflow and Machine Learning Model

This study evaluates the performance of two ML models, ARIMA and Bi-LSTM, for predicting solar power production in a smart grid context. The ARIMA model is a popular statistical approach used for time series forecasting, focusing on linear relationships in stationary data. On the other hand, the Bi-LSTM model, a type of recurrent NN, is designed to capture complex temporal dependencies in sequential data. The models were trained using 1 year of real-time solar power production data collected from a large-scale solar power plant. The performance of the models was compared based on key metrics such as MAE and RMSE. The results demonstrated that the Bi-LSTM model outperformed the ARIMA model in terms of accuracy, successfully identifying intricate patterns in the time series data.

10.3.4.21.3 Advantages and Limitations of the Synchronized Algorithm

The Bi-LSTM model offers several advantages over ARIMA, particularly in capturing complex temporal patterns in sequential data. Its bidirectional architecture allows the model to consider both past and future information, leading to more accurate predictions. Additionally, Bi-LSTM models are effective at handling non-stationary data, which is common in real-world energy production scenarios. However, the Bi-LSTM model has limitations, including high computational costs and the need for large datasets to achieve optimal performance. On the other hand, the ARIMA model is simpler and less computationally intensive but struggles with nonlinear relationships and non-stationary data. The study highlights the need for careful model selection based on the specific characteristics of the data and the forecasting requirements in smart grid applications.

10.3.4.21.4 Results and Discussion

In conclusion, this study demonstrates that the Bi-LSTM model offers more accurate and reliable predictions for grid-connected power generation, daily power generation, and radiance in solar power facilities. The study found that the Bi-LSTM model outperformed the ARIMA model in terms of accuracy and performance metrics for all three parameters. The study suggests that future research should investigate the effectiveness of other deep-learning models, such as CNNs and transformer-based models, for time series forecasting in RE. Furthermore, incorporating more relevant features, such as weather data and grid load information, could improve the accuracy of the models' predictions.

BIBLIOGRAPHY

Chen, Y., M. S. Bhutta, M. Abubakar, D. Xiao, F. M. Almasoudi, H. Naeem, and M. Faheem. Evaluation of machine learning models for smart grid parameters: Performance analysis of ARIMA and Bi-LSTM. *Sustainability* 2023, 15, 8555. https://doi.org/10.3390/su15118555

10.3.4.22 Application-50: Machine Learning-Based Surrogate Model for Accelerating Simulation Driven Optimization of Hydropower Kaplan Turbine

10.3.4.22.1 Background of the Application

This paper provides a review of various artificial intelligence/machine learning (AI) methods used in shape optimization. The paper discusses a range of methods, including dimensionality reduction techniques, generative design techniques, and optimization algorithms. The studies and methods cited in the paper aim to optimize the shape of objects to improve their performance, such as marine propellers and yacht hulls.

10.3.4.22.2 Description of the Workflow and Machine Learning Model

This study introduces an ML-based surrogate model to accelerate the simulation-driven optimization process of a Kaplan hydropower turbine. The approach integrates unsupervised feature extraction techniques, specifically Karhunen–Loève expansion, with supervised ML methods like Gaussian process regression (GPR) to create a lower-dimensional representation of the design space. This surrogate model is used to predict the performance of the turbine's runner blade by learning the nonlinear relationship between geometric features and the turbine's physics, particularly the relative tangential velocity at the leading and trailing edges. The optimization process leverages the surrogate model to efficiently explore the design space, reducing the computational cost typically associated with high-fidelity simulations like CFD. The method improves turbine efficiency from 56.98% to 90.73% while significantly reducing the number of required simulations run.

10.3.4.22.3 Advantages and Limitations of the Synchronized Algorithm

The ML-based surrogate model offers substantial advantages in optimizing complex hydropower turbine designs. By reducing the dimensionality of the design space and using surrogate modeling, the approach drastically reduces computational cost, allowing for faster and more efficient optimization. The use of GPR enables accurate predictions with a smaller training dataset, effectively handling the nonlinearity of the design problem. However, the approach has limitations, including the need for careful selection of training data to avoid overfitting and the computational intensity of the initial simulations required to generate the training data. Additionally, the success of the optimization process heavily depends on the accuracy of the feature extraction and the surrogate model's ability to generalize across the design space.

10.3.4.22.4 Results and Discussion

In conclusion, the review of AI methods for shape optimization shows that such methods have great potential for improving the performance of objects with complex shapes, such as marine propellers and yacht hulls. The paper highlights the importance of selecting appropriate input parameters and using a variety of AI methods to achieve optimal results. With further research and development, AI methods for shape optimization could revolutionize the design of complex objects in a range of fields.

BIBLIOGRAPHY

Masood, Zahid, Shahroz Khan, and Li Qian. Machine learning-based surrogate model for accelerating simulation-driven optimisation of hydropower Kaplan turbine, *Renewable Energy* 173, 2021, 827–848, ISSN 0960-1481. https://doi.org/10.1016/j.renene.2021.04.005

11 Future Directions in AI Applications in the Energy Industry

Mehmet Melih Oskay, Mesut Yurukcu, Ekrem Alagoz, Emre Can Dündar, Celal Hakan Canbaz, Luigi Saputelli, Mufrettin Murat Sari, Cenk Temizel, and Salih Tutun

11.1 INTRODUCTION

Artificial Intelligence (AI) has been in the industry (Munim, Z. H. et al., 2020) for decades to create innovation and increase effectiveness in the energy sector. AI's impact on the energy sector should not be overlooked. It fundamentally changes conventional energy networks by enhancing the management of resources, driving the shift to renewable energy, and enabling data-driven decision-making. Through automation and real-time analysis, AI has resulted in better, safer, and environmentally friendly energy operations. Also, AI works in harmony with other emerging technologies like the IoT (Rane, N. 2023), digital twins, and blockchain to create the building blocks of a smart, autonomous, and self-healing energy grid.

Previous chapters have discussed in detail the development and implementation of AI technologies in energy applications, which include optimization techniques for energy efficiency, machine learning algorithms for fault detection and predictive analytics, computer vision for infrastructure monitoring, the use of AI in the integration of renewable energy sources, and the improvement of grid reliability and asset management.

In the last chapter of this paper, we highlighted and discussed the future opportunities of AI in the energy sector. After the analysis of the current tendencies, the most important issues to solve and the most significant and interesting areas of the research, we propose a systematic recommendation for the reasonable and optimal use of AI in the generation of a sustainable energy future. In light of these observations, this chapter will discuss how AI is going to continue changing the energy sector, visualizing future possibilities with the help of new technologies and knowledge integration.

11.2 EMERGING TRENDS IN AI FOR ENERGY

Using AI with real-time data processing technologies is changing the way of collecting, processing, and analyzing energy data. They are using AI-based Big Data

DOI: 10.1201/9781003617327-11

analytics to transform energy forecasting, optimization, and system monitoring by looking at a huge volume of data to find ways to act (Jagatheesaperumal, S. K. 2021). IoT-enabled intelligent AI applications are already being adopted to bring smarter, automated energy efficiency, predictive maintenance, and real-time grid monitoring (see Figure 11.1) through smart sensor networks (Benfradj, A., et al., 2024). These capabilities improve the operational reliability of energy infrastructures and minimize inefficiencies in the system [De Silva, D., et al., 2020].

Digital twins powered by AI enables real-time simulation and optimization of energy infrastructure and predicts equipment failure for sustainable grid operations (Tadi, V., n.d.). Stream processing and event driven architecture are highlighted as key enablers of real-time analytics for adaptive energy distribution and near real-time decision-making. These technologies enable energy companies to transition from reactive to proactive energy management, reduce system downtime, and enhance resilience.

Furthermore, the AI automation of energy data integration is consistent with the new energy trends such as dynamic load balancing in smart grids, AI demand response programs, and AI predictive analytics for asset management. However, data standardization, cybersecurity risks and scalability issues need to be solved (Wolniak, R. et al., 2024). Using AI addresses data-quality control, encryption-based security measures and scalable cloud-based architectures are addressing these challenges. In this way, integrating AI with real-time data processing will enable the energy industry to gain automated decision-making, improve operational efficiency, and enhance

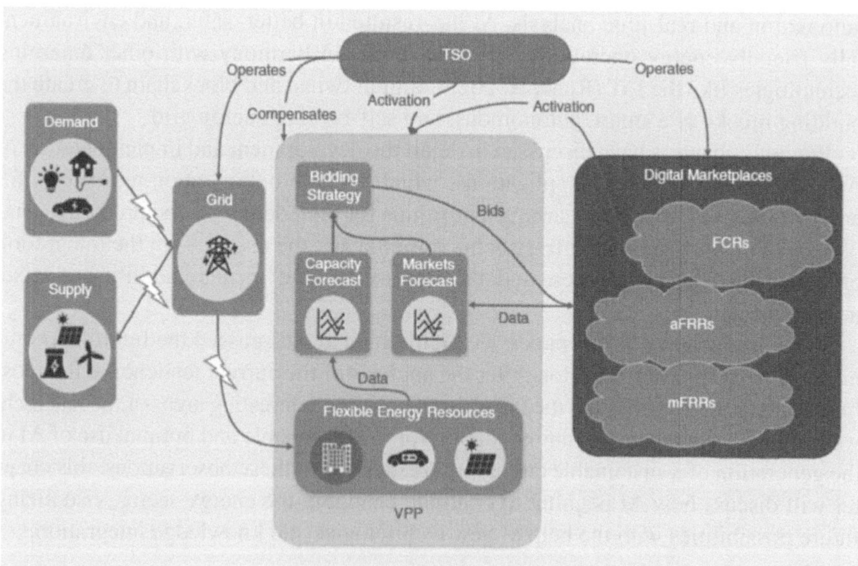

FIGURE 11.1 The VPP as a CPS. The arrows with the electricity symbol represent power flows in the physical world [De Silva, D., et al., 2020].

grid reliability (Karduri, R. K. R., et al., 2023). The results of the AI-based data integration study have shown that AI is a key enabler of the decentralized, adaptive, and scalable energy systems. AI-powered predictive modelling, edge computing, and distributed AI architecture advancements in the future will further revolutionize energy management and create a more resilient and sustainable future (Kallimani, R., et al., 2024).

11.3 CHALLENGES AND OPPORTUNITIES

However, while AI offers so much potential in the energy sector, there are still many challenges and obstacles that have to be overcome for the implementation to be successful and sustainable. This is perhaps one of the biggest challenges: to guarantee the further and uninterrupted availability of good, detailed, and reliable data. This is because AI models require good data to make accurate predictions, minimize costs, and effectively control energy assets. On the other hand, recent studies show that energy sector data is often dispersed, inadequate, or restricted due to proprietary interests, diverse methodologies, and the absence of coherent data sharing platforms. This fragmentation poses a major challenge to the scalability and effectiveness of AI applications in the sector (Farayola, O. A. et al., 2023).

In addition, the current research reveals the major problem of combining several data sets from different sources: weather forecasts, sensors, and market data (Benfradj, A., et al., 2024). Therefore, for the optimal integration of the data for a holistic analysis and improved operational intelligence of energy infrastructures, AI-driven data fusion techniques and real-time integration frameworks are vital. The use of digital twins, IoT, and blockchain-enabled data sharing can greatly improve these problems and increase the precision and flexibility of energy systems based on AI (Lezzi, M., et al., 2024). Real-time data analytics, distributed computing architectures, and scalable data governance are the enabling factors of efficient AI deployment, as recent studies by Park and Kim (2024) and Mitchell et al. (2022). As the application of AI increases, so do the concerns of ethics and regulation that are becoming increasingly important with the growing size of the AI implementation. However, while using AI in the energy sector, concerns such as algorithmic bias, data privacy, cybersecurity, and the transparency of decision-making frameworks cannot be ruled out. The lack of standardized global regulatory frameworks and governance policies for scalable and responsible AI deployment across the energy industry is a major issue. The governance frameworks are now urgently needed in decentralized energy markets, autonomous operations, and blockchain-enabled energy trading platforms (Park and Kim, 2024; Mitchell et al., 2022). In this regard, regulation, ethical AI standards, and cybersecurity guidelines must be properly developed to ensure that AI-powered energy systems are reliable (Li, C. 2021).

Additionally, the socio-economic effects of automation in the energy sector fueled by AI, including job losses and inequalities in technology availability, cannot be ignored. Research also points to the need for multidisciplinary cooperation of industry professionals, technologists, policymakers, and social scientists to tackle these multifaceted challenges (Dwivedi, Y. K. et al., 2023). For example, Rinku and Singh

(2023) solve the socio-economic impacts and workforce transitions with AI education, reskilling programs, and inclusive policymaking to avoid the negative impact while optimizing the positives of AI.

As for the second area, specific technical and operational challenges have been identified in specialized energy domains like offshore wind and renewable integration. AI systems are required to work independently and accurately in demanding conditions. In their paper, Mitchell et al. (2022) have highlighted the importance of well-defined governance frameworks and safety certification processes for autonomous robotic platforms in offshore energy infrastructures. These frameworks will enhance the safety, reliability, and the public acceptance of the AI-enabled autonomous operations. To address these diverse challenges, it needs proactive, collaborative, and multidisciplinary approaches. These approaches include stakeholders from academia, industry, and government developing strong data sharing protocols, ethical AI frameworks (see Table 11.1), cybersecurity measures, and regulatory oversight. However, to fully and responsibly leverage AI in the energy sector, such efforts are required; thus, through such a concerted effort, the sector can only be able to achieve sustainable, resilient, and socially equitable innovations (Dwivedi, Y. K. et al., 2023).

TABLE 11.1
Comparison of Prognostic Maintenance Techniques for Complex Systems [Afridi, Y. S., et al., 2022]

Methodology	Advantages	Disadvantages
Hidden Markov model (HMM)	• Performs effectively for fault diagnosis and system degradation analysis of random dynamical systems and nonstationary signals • Best suits the systems that are having multi-failure modes • Ability to distinguish between various faults of different bearing types, if trained properly • Since the algorithm can model various stages of degradation, hence it can take into account non-monotonous failure trends • Ability to model both temporal and spatial data • Ability to quickly handle the incomplete data sets • The RUI predictions by HMMs also provide the confidence intervals • Successful real-life implementations as the interpretation and understanding of the model are easy	• The model does not work for scenarios where the failure states are observable • Proportionate to the hidden states, extensive training data is required for accurate modeling • The algorithm is computationally expensive for situations where the number of hidden states is large • The efficiency of the prognostics depends upon the selection of the failure threshold • Cannot model unanticipated and novel faults • The HMM can only provide an estimation since the true state transitions cannot be observed

TABLE 11.1 (Continued)
Comparison of Prognostic Maintenance Techniques for Complex Systems [Afridi, Y. S., et al., 2022]

Methodology	Advantages	Disadvantages
SVMs	• High accuracy and robustness even with nonlinear and higher dimension input data • Dynamic SVMs have got the better ability to model nonstationary time series • Extended decision boundaries • Higher accuracy and efficiency • Performs exceptionally well for both large and small datasets • Ability to carry out a real-time analysis • Ability to be generalized without compromising the performance even against a limited number of learning patterns • Works extremely well for carrying out fault diagnosis in the electro-mechanical systems and machines • Proves to be an effective prognostic method even for limited samples • SVMs are more memory efficient	• There are no set parameters for the selection of the kernel function • The training requires to be carried out for both positive and negative examples • The performance degrades for cases where the number of features exceeds the number of training data samples • There is no accurate probabilistic explanation to an SVMs model • It does not directly provide the RUL • Requires fine-tuning of the parameters • The regression and classification outcome of SVMs is point estimates only • The problem of numerical stability may pop-ups for constrained quadratic programming
Artificial neural network (ANN)	• Ability to model nonlinear, multi-dimensional, complex systems • Robustness toward noise because of the ability to process nonlinear information • Ability to carry out unsupervised learning and model complex phenomena • Ability to efficiently tackle imprecise and complex data • The parallel model structure makes it computationally inexpensive and hence feasible for real-time operation • Ability to handle multivariate systems with speed and accuracy • Ability to sustain network damage due to the resilient and fault-tolerant behavior • Provision of confidence intervals	• Extensive pre-processing of input data is required for dimensionality reduction and complexity • A large quantity of training data is required encompassing the variability and ranges and must be a true reflection of the original data • Model perfection can be a time-consuming process • The process of decision-making in a trained network is left covert to the users • Failure diagnosis is based on the presumed threshold
Fuzzy logic	• Ability to carry out condition-based classification based on historic data • Ability to work effectively for complex systems and imprecise data • Performs exclusively in developing data models based on uncertainty • Interpretation of the model is simpler • Provision of the confidence interval	• Unable to provide exact time and probability of failure • Performance degrades in cases where the member functions arc complicated • Inability to self-learn and requires the help of domain experts for the development of fuzzy rules

11.4 FUTURE RESEARCH DIRECTIONS BY CONCEPTUAL THEMES

It is essential to specifically identify domains in which further investigation of the impact of artificial intelligence on the future can be most effective, to chart a sustainable path forward. The following conceptual themes are recognized as key areas for future enquiry in this realm, with ideas gleaned from recent literature and analyses.

11.4.1 Data Integration and Knowledge Fusion

Future work should also address the development of durable AI-based approaches for integration and analysis of multiple types of data from market conditions, weather forecasts (see Figure 11.2), IoT sensors, and historical energy assets performance (Benfradj, A., et al., 2024). The application of AI-based data fusion strategies, integrated with real-time analytics and security by blockchain, can solve the problem of data dispersion and, therefore, offer a holistic view of energy systems. The recent methodological innovations include combining the topic modelling techniques such as LDA and BERT as employed by Saheb and Dehghani (2022) which shows the need for contextualized data analysis for the enhancement of the accuracy and predictive capability of the AI models.

11.4.2 Autonomy and Decision Intelligence

In case of autonomous energy systems controlled by AI, research should be aimed at self-learning and adaptive algorithms that are able to make near instantaneous decision in conditions of real time (see Figure 11.3). This includes sophisticated

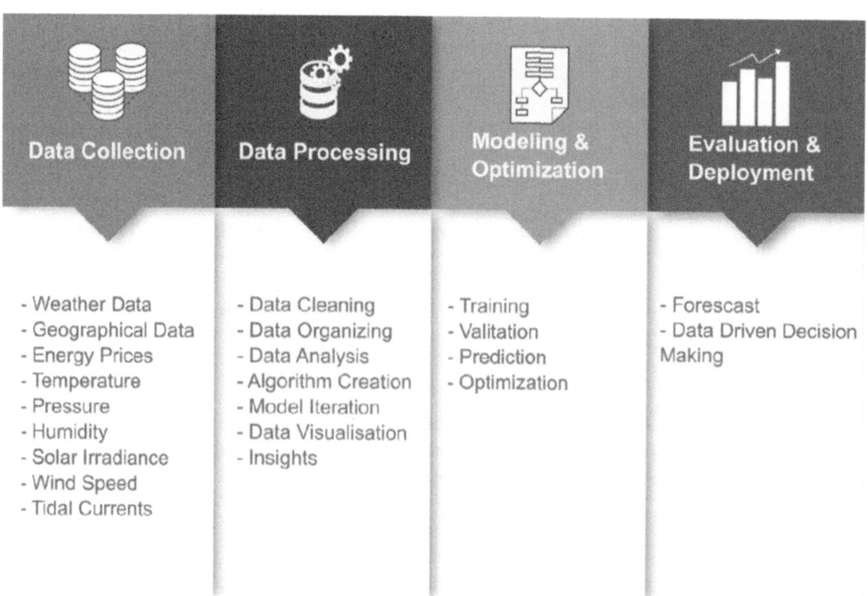

FIGURE 11.2 Steps in developing modeling and optimization tools for energy forecasting [Yousef, L. A., et al., 2023].

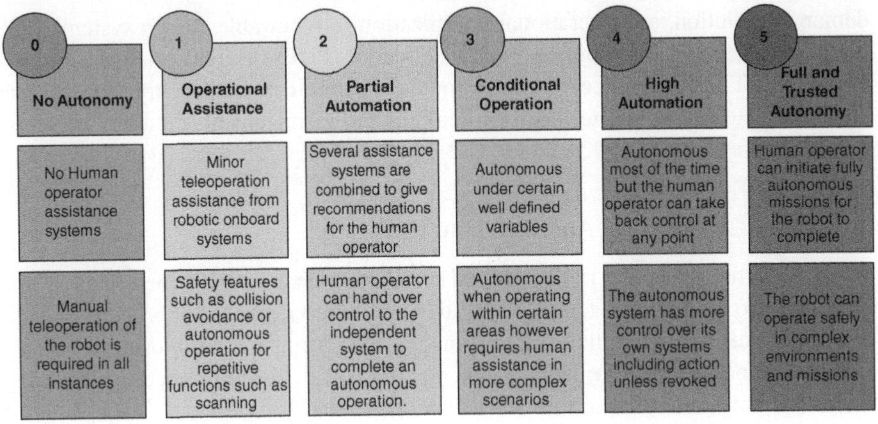

FIGURE 11.3 The spectrum of autonomy [Mitchell, D., et al., 2022].

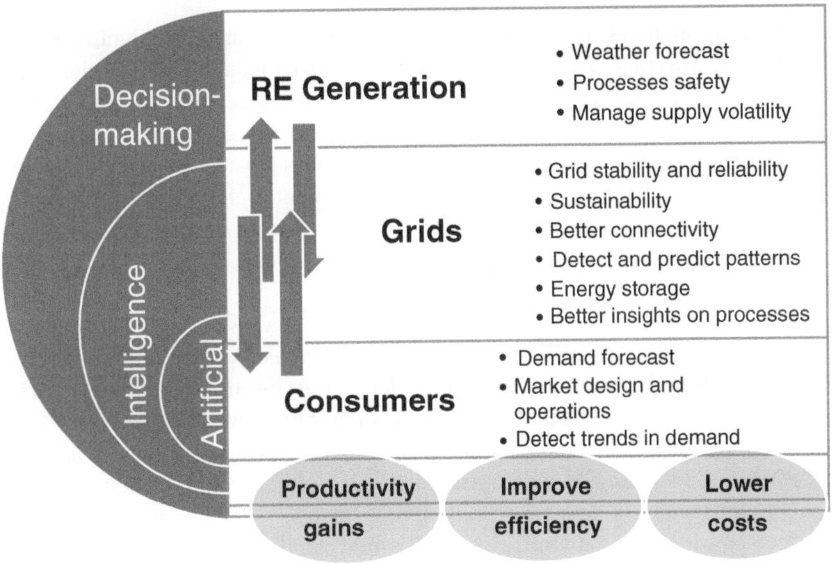

FIGURE 11.4 Energy system: interactions based on AI application [Şerban and Lytras 2020].

frameworks of reinforcement learning and digital twins for the autonomous control of energy resources, predictive maintenance, and stability operations. As per Mitchell et al. (2022), it is very important to embed these AI systems into regulatory and safety standards (Figure 11.4).

11.4.3 Advanced Predictive Modeling

A key future consists of the continuing innovation of the predictive AI models in integration with the physical knowledge-based models and data-driven models. These hybrid modeling approaches will increase the accuracy and efficiency of forecasting, demand prediction, and operational optimization of renewable energy systems. The integration of physics-informed neural networks (PINNs), contextualized embeddings (BERT), and advanced computational modeling can enhance these predictive capabilities to support better decision-making and management of energy systems' infrastructure (Farayola, O. A. et al., 2023).

11.4.4 Resilience and Adaptive Systems

In view of the increasing risks imposed by climate change and cyber-attacks, it is imperative to consider the feasibility of using AI-based resilience approaches. These AI models used for predictive maintenance, infrastructure monitoring, anomaly detection, and adaptive grid management are critical to the protection of energy infrastructure against disruptions (Daniel, C. et al., 2024).

11.4.5 Human–AI Collaboration

It is only through the integration of both human and artificial intelligence that AI-generated insights can be made practical, understandable, and implementable in the energy system operations (Figure 11.5). Current research is mainly on the development of decision support systems based on AI to assist human operators in decision-making concerning energy management, urban water management, climate risk management, and agricultural energy management (*Şerban and Lytras 2020*).

11.4.6 Energy Transition Pathways

Another important future direction of research is AI-assisted policy modeling for sustainable energy transitions. AI can assist policymakers and industry leaders in creating better and fair energy transition plans by trying to predict many different technological, economic, and social scenarios that might occur in the future. Future work should also focus on more collaborative approaches across different disciplines (i.e. economic modeling, technological innovation, and social dynamics) to address barriers and enable holistic energy transition pathways. These thematic research areas can be addressed with the help of AI, which can be the foundation for remitting the energy sector, driving sustainable and resilient innovations, and meeting technical, ethical, and socio-economic challenges. AI can be the foundation of reshaping the energy sector, driving sustainable and resilient innovations, and meeting technical, ethical, and socio-economic challenges across these thematic research areas (Dwivedi, Y. K. et al., 2023).

11.5 VISION FOR THE FUTURE OF AI IN ENERGY

AI will play an important role in future energy production, storage, consumption, and distribution which can be seen in Table 11.2. Because of this reason, collaboration between academia, industry stakeholders, and regulatory authorities is becoming crucial to realize the potential of AI fully. To build agility into the value network so that it can adapt easily to new breakthroughs as they emerge and align technology

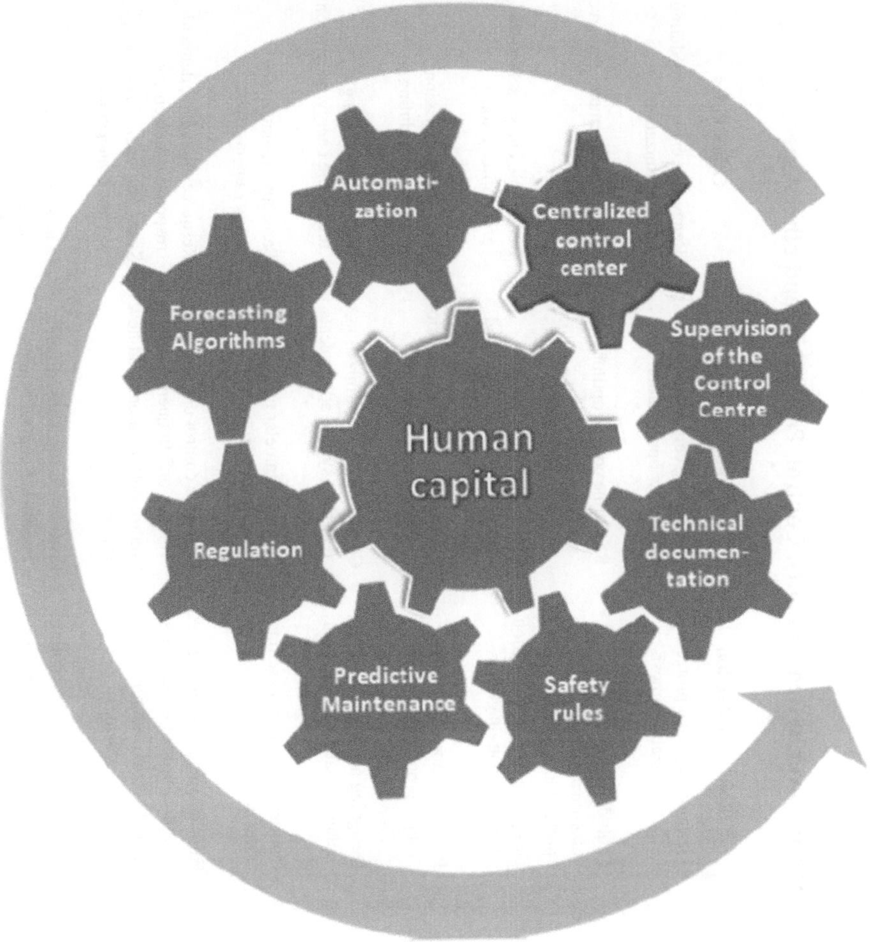

FIGURE 11.5 The importance of humans in AI processes [Şerban and Lytras 2020].

adoption with continually evolving market dynamics and regulatory environments (Chen, J., et al., 2023).

There are several key areas that will shape AI's trajectory in the energy sector. They will leverage the use of next-generation AI simulations and predictive analytics to enhance system resilience and operational efficiency. It is discussed by Chen et al. (2023) has its focus on integrating AI into the network itself to enable intelligent data perception, automated modelling, and distributed learning. The intrinsic AI strategy could enable self-learning, autonomous networks that optimize energy use dynamically and reduce energy consumption substantially, especially as the density of wireless communication networks such as 5G and future 6G technologies (Figure 11.6).

The establishment of AI-integrated policy frameworks is important. Governments and regulatory bodies must have agile adaptive policy frameworks that balance innovation with accountability and ethical considerations. The essential transparent AI

TABLE 11.2
Roles and Applications of Metaheuristic Algorithms in Power System Stability, Control, Protection, and other Aspects [Alhamrouni, I., et al., 2024]

Metaheuristic Algorithm	Role in Stability	Role in Protection	Role in Control	Other Roles
Particle Swarm Optimization (PSO)	Optimizing stability control strategies, tuning stability parameters	Designing optimal protection schemes, fault diagnosis	Optimizing control parameters, load flow optimization	Renewable energy scheduling, grid optimization
Artificial Bee Colony (ABC)	Enhancing stability through optimized control actions, tuning system parameters	Fault detection and classification, adaptive protection schemes	Real-time control optimization, voltage control	Energy management, resource allocation
Genetic Algorithms (GAs)	Optimization of stability control strategies, tuning of control parameters	Designing optimal protection settings, fault diagnosis	Solving complex control optimization problems, load flow optimization	Renewable energy scheduling, grid optimization
Ant Colony Optimization (ACO)	Enhancing system stability by finding optimal paths for power flow	Fault detection, designing robust protection schemes	Control parameter optimization, dynamic load balancing	Network optimization, resource scheduling
Differential Evolution (DF.)	Stability optimization through parameter tuning, dynamic stability analysis	Designing adaptive protection schemes, fault classification	Control strategy optimization, real-time system adjustments	Demand forecasting, renewable energy integration
Simulated Annealing (SA)	Stability enhancement by optimizing control actions, mitigating stability issues	Optimizing protection settings, fault detection	Control parameter tuning, load flow optimization	Preventive maintenance, energy management
Harmony Search (HS)	Stability control optimization, tuning system parameters	Fault diagnosis, designing adaptive protection schemes	Real-time control optimization, voltage, and frequency control	Resource allocation, demand response management
Firefly Algorithm (FA)	Stability enhancement through optimized control strategies, parameter tuning	Fault detection, adaptive protection schemes	Control optimization, load flow management	Smart grid operations, energy storage management
Bat Algorithm (BA)	Optimizing stability control parameters, enhancing dynamic stability	Fault classification, designing robust protection schemes	Real-time control adjustments, optimizing control actions	Renewable energy management, asset optimization

Cuckoo Search (CS)	Stability optimization, enhancing system resilience	Designing optimal protection settings, fault detection	Control parameter tuning, dynamic load balancing	Energy management, predictive maintenance
Grey Wolf Optimizer (GWO)	Stability control optimization, enhancing system stability	Fault detection and classification, adaptive protection settings	Control strategy optimization, voltage, and frequency regulation	Resource scheduling, demand forecasting
Whale Optimization Algorithm (WOA)	Stability enhancement through parameter optimization, dynamic stability analysis	Fault diagnosis, designing adaptive protection schemes	Real-time control optimization, load flow management	Smart grid operations, energy storage management
Dragonfly Algorithm (DA)	Enhancing stability by optimizing control parameters, dynamic stability enhancement	Fault detection, adaptive protection schemes	Control optimization, dynamic load balancing	Resource allocation, demand response management
Salp Swarm Algorithm (SSA)	Optimizing stability control actions, enhancing system resilience	Fault classification, designing robust protection schemes	Real-time control adjustments, optimizing control actions	Renewable energy management, asset optimization
Crow Search Algorithm (CSA)	Stability control optimization, enhancing system stability	Fault detection and classification, adaptive protection settings	Control strategy optimization, voltage, and frequency regulation	Resource scheduling, demand forecasting
Sine Cosine Algorithm (SCA)	Stability enhancement through parameter optimization, dynamic stability analysis	Fault diagnosis, designing adaptive protection schemes	Real-time control optimization, load flow management	Smart grid operations, energy storage management
Elephant Herding Optimization (EHO)	Optimizing stability control parameters, enhancing dynamic stability	Fault detection, adaptive protection schemes	Control optimization, dynamic load balancing	Resource allocation, demand response management
Moth-Flame Optimization (MFO)	Stability control optimization, enhancing system resilience	Fault detection and classification, adaptive protection settings	Control strategy optimization, voltage, and frequency regulation	Resource scheduling, demand forecasting

(Continued)

TABLE 11.2 (Continued)
Roles and Applications of Metaheuristic Algorithms in Power System Stability, Control, Protection, and other Aspects [Alhamrouni, I., et al., 2024]

Metaheuristic Algorithm	Role in Stability	Role in Protection	Role in Control	Other Roles
Grasshopper Optimization Algorithm (GOA)	Enhancing stability by optimizing control parameters, dynamic stability enhancement	Fault detection, adaptive protection schemes	Control optimization, dynamic load balancing	Resource allocation, demand response management
League Championship Algorithm (ICA)	Stability enhancement through parameter optimization, dynamic stability analysis	Fault diagnosis, designing adaptive protection schemes	Real-time control optimization, load flow management	Smart grid operations, energy storage management
Flower Pollination Algorithm (FPA)	Optimizing stability control actions, enhancing system resilience	Fault classification, designing robust protection schemes	Real-time control adjustments, optimizing control actions	Renewable energy management, asset optimization
Jaya Algorithm	Stability control optimization, enhancing system stability	Fault detection and classification, adaptive protection settings	Control strategy optimization, voltage, and frequency regulation	Resource scheduling, demand forecasting
Quantum PSO (QPSO)	Enhancing stability through quantum-based parameter optimization	Fault detection, adaptive protection schemes	Real-time control optimization, dynamic load balancing	Smart grid operations, energy storage management
Teaching-Learning-Based Optimization (TLBO)	Stability control optimization, enhancing dynamic stability	Fault detection and classification, adaptive protection settings	Control strategy optimization, voltage, and frequency regulation	Resource scheduling, demand forecasting
Shuffled Frog-Leaping Algorithm (SFLA)	Optimizing stability control parameters, enhancing system resilience	Fault detection, designing adaptive protection schemes	Control optimization, dynamic load balancing	Resource allocation, demand response management

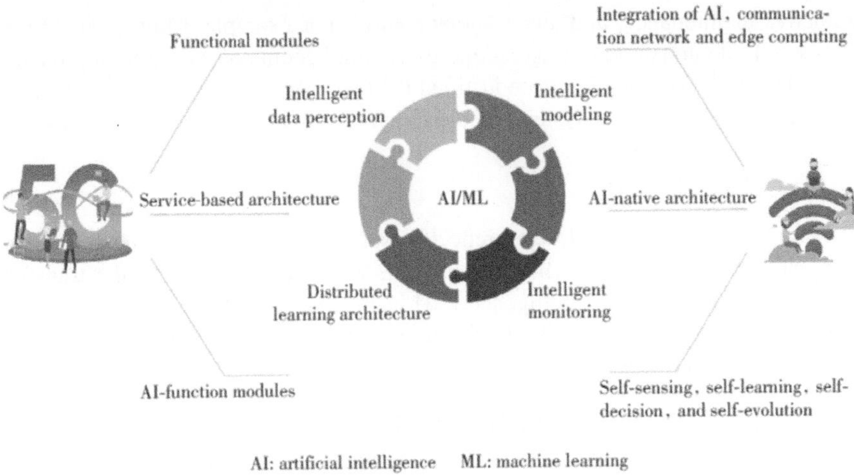

FIGURE 11.6 Transition of AI structure from 5G to future wireless network [Chen, J. et al., 2023].

FIGURE 11.7 Categorization of the Sustainable Development Goals (SDGs).

[Source: https://www.un.org].

governance mechanisms include XAI, clear data-sharing standards, privacy-preserving architectures, and strong cybersecurity frameworks. The collaborative efforts of regulatory bodies, energy providers, and AI developers should be on standardizing certification and compliance requirements for transparent and trustworthy AI adoption (Figure 11.7).

Another important future direction is therefore scalable AI deployment. They must be able to function collectively throughout immense and interconnected smart energy grids, which poses the problems of scalability, security, and adaptability (Yafei, X. et al., 2024). Recent studies have revealed that sophisticated AI-based energy management strategies can work effectively in real-time energy optimization, predictive maintenance, and autonomous control. For example, Chen et al. (2023) successfully demonstrated the great energy-saving benefits of AI-based approaches in wireless networks and therefore justified the feasibility of applying AI to the control of real-world complex systems. Therefore, AI in smart grids is indispensable.

11.6 CONCLUSION

In this final chapter, the following reflective summary of the key insights and lessons learned throughout this book is offered, with the emphasis on the transformative power of artificial intelligence (AI) in the energy industry. As can be seen Figure 11.8, Yousef et al., (2023) clearly summarized AI applications' gaps and challenges along with directions.

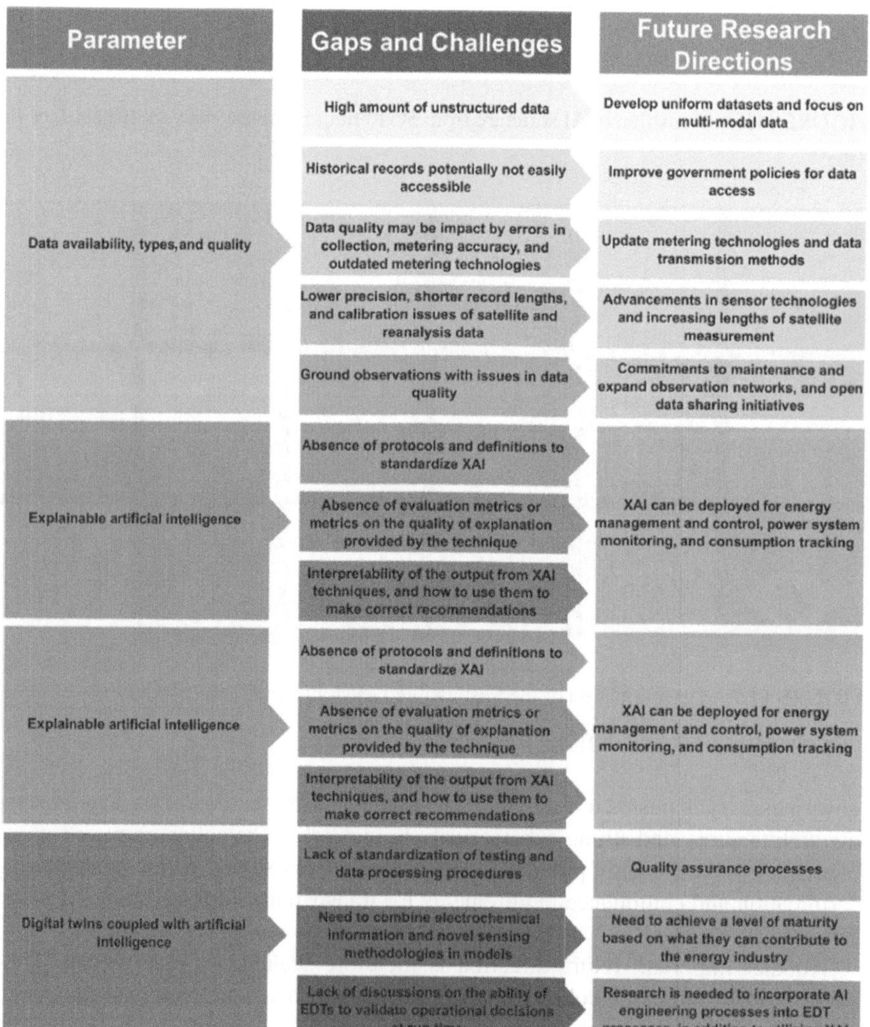

FIGURE 11.8 Summary of the main gaps, challenges, and directions [Yousef, L. A., et al., 2023].

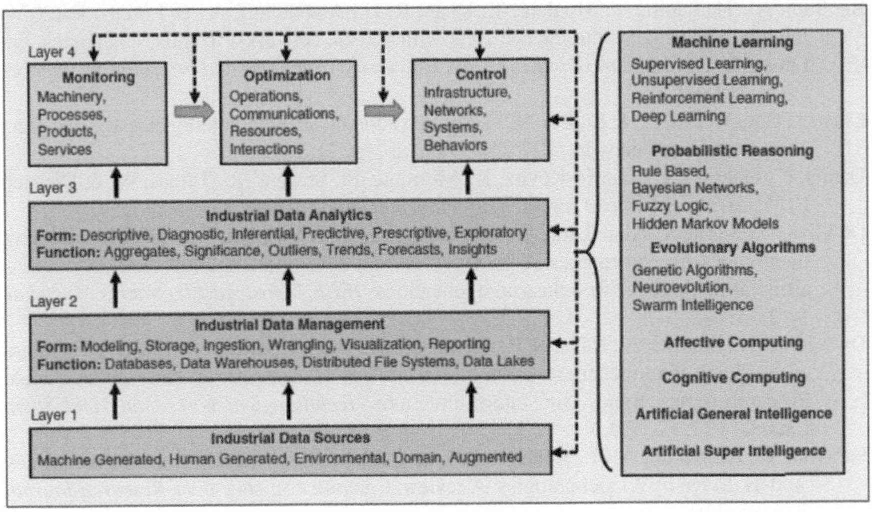

FIGURE 11.9 The reference framework for intelligent industrial informatics [De Silva, D., et al., 2020].

AI is not only a technological advancement but also a powerful force that can reshape how energy is produced, moved, stored, and used around the world (Naz, F. et al., 2022). The energy sector is on the cusp of an unprecedented era in which AI is enhancing operational efficiency and helping to address sustainability, resilience, and equity issues. By harnessing the potential of the emerging frontiers of AI, we are ideally placed to enhance energy optimization and predictive analytics, accelerate the transition to renewable and decentralized energy systems, improve grid security and stability by enhancing real-time decision-making, and improve energy policy development with the help of AI, to create a cleaner and more inclusive energy future as shown in Figure 11.9.

Energy leaders, researchers, policymakers, and industry stakeholders must ensure that there is continuous innovation, collaboration, and sustainable implementation of AI to realize its full potential to the fullest (Metaxiotis and Psarras, 2003). In its current form, policy frameworks, research agendas, and strategic initiatives of the energy industry must be re-proactively adapted to the changing environment of AI technologies so that AI can be effectively used for sustainability, security, and equitable progress. Thus, the energy sector can pave the way to a future of resilience, sustainability, and inclusive growth with the help of synergies between AI technologies and human ingenuity and redefine possible boundaries.

REFERENCES

Afridi, Y. S., Ahmad, K., & Hassan, L. (2022). Artificial intelligence based prognostic maintenance of renewable energy systems: A review of techniques, challenges, and future research directions. *International Journal of Energy Research*, 46(15), 21619–21642.

Alhamrouni, I., Abdul Kahar, N. H., Salem, M., Swadi, M., Zahroui, Y., Kadhim, D. J., & Alhuyi Nazari, M. (2024). A comprehensive review on the role of artificial intelligence in power system stability, control, and protection: Insights and future directions. *Applied Sciences*, 14(14), 6214.

Benfradj, A., Thaljaoui, A., Moulahi, T., Khan, R. U., Alabdulatif, A., & Lorenz, P. (2024). Integrating artificial intelligence (AI) with sensor networks: Trends, challenges, and future directions. *Journal of King Saud University-Computer and Information Sciences*, 36(1), 101892.

Chen, J., Gao, Y., Liu, Z., & Li, D. (2023). Future vision on artificial intelligence assisted green energy efficiency network. *ZTE Communications*, 21(2), 34.

Daniel, C., Gehin, J. C., Laurin-Kovitz, K., Morreale, B., Stevens, R., Tumas, W., & Dietrich, E. M. (2024). *Advanced research directions on AI for energy.*

De Silva, D., Sierla, S., Alahakoon, D., Osipov, E., Yu, X., & Vyatkin, V. (2020). Toward intelligent industrial informatics: A review of current developments and future directions of artificial intelligence in industrial applications. *IEEE Industrial Electronics Magazine*, 14(2), 57–72.

Dwivedi, Y. K., Sharma, A., Rana, N. P., Giannakis, M., Goel, P., & Dutot, V. (2023). Evolution of artificial intelligence research in technological forecasting and social change: Research topics, trends, and future directions. *Technological Forecasting and Social Change*, 192, 122579.

Farayola, O. A., Abdul, A. A., Irabor, B. O., & Okeleke, E. C. (2023). Innovative business models driven by AI technologies: A review. *Computer Science & IT Research Journal*, 4(2), 85–110.

Jagatheesaperumal, S. K. (2021). The Duo of Artificial Intelligence and Big Data for Industry 4.0: Review of Applications, Techniques, Challenges, and Future Research Directions.

Kallimani, R., Pai, K., Raghuwanshi, P., Iyer, S., & López, O. L. (2024). TinyML: Tools, applications, challenges, and future research directions. *Multimedia Tools and Applications*, 83(10), 29015–29045.

Karduri, R. K. R., & Ananth, C. (2023). Energy Efficiency in Industrial Sectors: Future Directions. Available at SSRN 4637816.

Lezzi, M., Del Vecchio, V., & Lazoi, M. (2024). Using blockchain technology for sustainability and secure data management in the energy industry: Implications and future research directions. *Sustainability*, 16(18), 7949.

Li, C. (2021). *AI-powered energy internet towards carbon neutrality: Challenges and opportunities.* Authorea Preprints.

Metaxiotis, K., & Psarras, J. (2003). Expert systems in business: Applications and future directions for the operations researcher. *Industrial Management & Data Systems*, 103(5), 361–368.

Mitchell, D., Blanche, J., Harper, S., Lim, T., Gupta, R., Zaki, O., & Flynn, D. (2022). A review: Challenges and opportunities for artificial intelligence and robotics in the offshore wind sector. *Energy and AI*, 8, 100146.

Munim, Z. H., Dushenko, M., Jimenez, V. J., Shakil, M. H., & Imset, M. (2020). Big data and artificial intelligence in the maritime industry: A bibliometric review and future research directions. *Maritime Policy & Management*, 47(5), 577–597.

Naz, F., Agrawal, R., Kumar, A., Gunasekaran, A., Majumdar, A., & Luthra, S. (2022). Reviewing the applications of artificial intelligence in sustainable supply chains: Exploring research propositions for future directions. *Business Strategy and the Environment*, 31(5), 2400–2423.

Park, C., & Kim, M. (2024). Utilization and challenges of artificial intelligence in the energy sector. *Energy & Environment*, 0958305X241258795. https://doi.org/10.1177/0958305X241258795

Rane, N. (2023). Integrating leading-edge artificial intelligence (AI), internet of things (IoT), and big data technologies for smart and sustainable architecture, engineering and construction (AEC) industry: Challenges and future directions. Engineering and Construction (AEC) Industry: Challenges and Future Directions (September 24, 2023).

Rinku, G. S. (2023). Artificial Intelligence in sustainable energy industry: Status quo, challenges, and opportunities. *EPRA International Journal of Multidisciplinary Research (IJMR)*, 9.

Saheb, T., Dehghani, M., & Saheb, T. (2022). Artificial intelligence for sustainable energy: A contextual topic modeling and content analysis. *Sustainable Computing: Informatics and Systems*, 35, 100699.

Şerban, A. C., & Lytras, M. D. (2020). Artificial intelligence for smart renewable energy sector in Europe—smart energy infrastructures for next generation smart cities. *IEEE Access*, 8, 77364–77377.

Tadi, V. (n.d.) *Revolutionizing data integration: The impact of AI and real-time technologies on modern data engineering efficiency and effectiveness.*

Wolniak, R., & Stecuła, K. (2024). Artificial intelligence in smart cities—Applications, barriers, and future directions: A review. *Smart Cities*, 7(3), 1346–1389.

Yafei, X., Wu, Y., Song, J., Gong, Y., & Lianga, P. (2024). Generative AI in industrial revolution: A comprehensive research on transformations, challenges, and future directions. *Journal of Knowledge Learning and Science Technology ISSN: 2959-6386 (online)*, 3(2), 11–20.

Yousef, L. A., Yousef, H., & Rocha-Meneses, L. (2023). Artificial intelligence for management of variable renewable energy systems: A review of current status and future directions. *Energies*, 16(24), 8057.

Index

Pages in *italics* refer to figures and pages in **bold** refer to tables.

H

I

K

L

M

N

O

T

U

V

W

For Product Safety Concerns and Information please contact our EU representative GPSR@taylorandfrancis.com Taylor & Francis Verlag GmbH, Kaufingerstraße 24, 80331 München, Germany